Stem Cells and Human Diseases

Rakesh K. Srivastava • Sharmila Shankar
Editors

Stem Cells and Human Diseases

Editors
Rakesh K. Srivastava
Department of Pharmacology, Toxicology and Therapeutics, and Medicine
The University of Kansas Medical Center
Rainbow Boulevard 3901
Kansas City, KS 66160
USA

Sharmila Shankar
Department of Pathology and Laboratory Medicine
The University of Kansas Medical Center
Rainbow Boulevard 3901
Kansas City, KS 66160
USA

ISBN 978-94-017-8471-9 ISBN 978-94-007-2801-1 (eBook)
DOI 10.1007/978-94-007-2801-1
Springer Dordrecht Heidelberg London New York

Printed on acid-free paper

Springer is part of Springer Science+Business Media (www.springer.com)

Preface

The main objective of this book is to provide a comprehensive review on stem cells and their role in tissue regeneration, homeostasis and therapy. In addition, the role of cancer stem cells in cancer initiation, progression and drug resistance are discussed. The cell signaling pathways and microRNA regulating stem cell self-renewal, tissue homeostasis and drug resistance are also mentioned. An increased understanding of stem cell behavior and biology along with rapid advancement of high throughput screening has led to the discovery and development of novel drugs that control stem cell self-renewal and differentiation. In near future, these molecules will be very useful for treating and preventing several human diseases.

The authors represent a diverse group of experts who have endeavored to provide a historical perspective on the generation of stem cells and their roles in tissue regeneration, therapy and disease initiation and progression, to allow clinicians to assimilate these facts into their treatment algorithms. For this purpose we have considered both normal and malignant stem cells. While progress on the clinical front has been slower than desired, the use of stem cells for tissue regeneration and disease management has great potential in human health. Overall, these reviews will provide a new understanding of the influence of stem cells in tissue regeneration, disease regulation, therapy and drug resistance in several human diseases.

We greatly appreciate the exceptional contributions of the authors, each of which reflects their commitment to the field of stem cells and human diseases.

The University of Kansas Medical Center
Kansas City, Kansas

Sharmila Shankar, Ph.D.
Rakesh K. Srivastava, Ph.D.

Contents

Chapter 1
Cancer Stem Cells: Biology, Perspectives and Therapeutic Implications

Brahma N. Singh, Sharmila Shankar, and Rakesh K. Srivastava

Contents

Abstract Cancer stem cells (CSCs) biology has come of age. The CSC theory is currently central to the field of cancer research, because it is not only a matter of academic interest but also crucial for the cancer therapy and prevention. Most cancers comprise of a heterogenous population of CSCs with marked differences in their proliferative potential as well as the ability to reconstitute the tumor upon transplantation. CSCs share a variety of biological properties with normal somatic stem cells in terms of self-renewal, the expression of specific cell surface markers and the utilization of common signaling pathways. Perhaps the most important and useful characteristic of CSCs is that of self-renewal. Through these properties,

B.N. Singh • R.K. Srivastava (✉)
Department of Pharmacology, Toxicology and Therapeutics, and Medicine,
The University of Kansas Cancer Center, The University of Kansas Medical Center,
3901 Rainbow Boulevard, Kansas City, KS 66160, MO, USA
e-mail: rsrivastava@kumc.edu

S. Shankar
Department of Pathology and Laboratory Medicine, The University of Kansas
Cancer Center, The University of Kansas Medical Center, 3901 Rainbow Boulevard,
Kansas City, KS 66160, MO, USA

R.K. Srivastava and S. Shankar (eds.), *Stem Cells and Human Diseases*,
DOI 10.1007/978-94-007-2801-1_1, © Springer Science+Business Media B.V. 2012

striking parallels can be found between CSCs and stem cells: tumors may often originate from the transformation of normal stem cells. This review will have significant ramifications for the biological basis and the therapeutic implications of the stem cell. In addition, dysregulation of CSC self-renewal is a likely requirement for the development of human cancers. Understanding the properties of, and exploring self-renewal, cell surface markers and signaling pathways specific to CSCs of different cancers, will lead to progress in therapy, intervention, and improvement of the prognosis of patients. In the near future, the evaluation of CSCs may be a routine part of practical diagnostic pathology.

Keywords Cancer stem cells • Stem cells • Notch • Sonic hedgehog • Wnt • Breasts cancer stem cells • Brain cancer stem cells • Prostate cancer stem cells • Pancreatic cancer stem cells • Progenitor cells

1.1 Introduction

Recent *in vitro* and *in vivo* research evidences have demonstrated that in hematologic and solid malignancies only a minority of cancer cells have the capacity to proliferate extensively and form new malignancies [3, 28, 86, 95]. These cancer stem cells (CSCs)/tumor-initiating cells (TICs) have been recognized and enriched on the basis of their expression of cell-surface markers (CSMs). Upon transplantation, TICs give rise to tumors comprising both new TICs as well as heterogeneous populations of non-tumorigenic cells reminiscent of the developmental hierarchy in the tissues from which the tumors arise. Most adult tissues are maintained over the lifetime of the host by somatic/or normal stem cells (SCs) that undergo expansion and differentiation to yield the functional elements of the organ [33]. Through self-renewal process, SCs are able to function over the lifespan of the host.

SCs are a class of undifferentiated cells that have the potential to perpetuate themselves through self-renewal and to generate mature cells of a specific tissue through differentiation. Commonly, SCs come from two main sources: embryos formed during the blastocyst phase of embryological development and adult tissue. Both types are generally characterized by their potency, or potential to differentiate into different cell types (such as skin, muscle, bone, etc.). In most tissues, SCs are rare. Although, it appears reasonable to propose that each tissue arises from a tissue-specific SC, the difficult identification, purification and isolation of these somatic SCs has been accomplished only in a few instances [12]. The genetic constraints on self-renewal restrict the expansion of SCs in normal tissues. Breakdowns in the guideline of self-renewal are likely a key event in the development of human malignancy as demonstrated by the fact that several cellular signaling pathways implicated in carcinogenesis. It also play a key role in normal SC self-renewal decisions. Thus, malignant tumors can be viewed as an abnormal organ in which a minority population of tumorigenic cancer cells have escaped the normal constraints on self-renewal giving rise to abnormally differentiated cancer cells that have lost the potential to form tumors. This new model for cancer has important implications for the study and treatment of tumors [12]. Not only is ruling the source of cancer cells essential for successful treatments, but if current treatments of cancer do not properly finish enough CSCs, the tumor will reappear. Including the possibility that the treatment of for instance, chemotherapy will consent only chemotherapy-resistant CSCs, then the ensuing tumor will most likely also be resistant to chemotherapy. If the cancer tumor is detected early enough, enough of the tumor can be killed off and marginalized with customary treatment. But as the tumor size increases, it becomes more and more difficult to remove the tumor without conferring resistance and leaving enough behind for the tumor to regenerate.

Some treatments with chemotherapeutic agent such as paclitaxel in ovarian cancer (a cancer usually discovered in late stages), may actually induce chemoresistance (55–75% relapse <2 years) [21]. It potentially does this by destroying only the cancer cells susceptible to the drug by targeting those that are $CD44^+$ positive, and allowing the cells which are unaffected by paclitaxel ($CD44^+$ negative) to regrow, even after a reduction in over a third of the total tumor size. There are studies, though, which show how paclitaxel can be used in combination with other ligands to affect the $CD44^+$ positive cells [7]. While paclitaxel alone, as of late, does not cure the cancer, it is effective at extending the survival time of the patients [21].

The recent discoveries that these are bone marrow [15, 65, 75], and purified CSCs such as hematopoietic stem cell (HSCs) [57, 58], can give rise to non-haematopoietic tissues suggests that these cells may have greater differentiation potential than was assumed previously. Conclusive experiments are required to determine whether the cells from the bone marrow that are capable of giving rise to different non-haematopoietic lineages are indeed HSCs or another population. If further studies support the idea of CSCs plasticity, this will undoubtedly open new frontiers for understanding the developmental potential of CSCs, as well as expand their therapeutic application.

1.2 Self-renewal and Cancer

Self-renewal is a cell division in which one or both of the resulting daughter cells remain undifferentiated and retain the ability to give rise to another SC with the same capability to proliferate as the parental cell [23, 26]. Self-renewal is crucial to SC function, because it is required by many types of SCs to persist for the lifetime of the animal. Moreover, whereas SCs isolated from different organs may differ in their progressive potential, all the SCs must have to self-renew and regulate the relative balance between self-renewal and differentiation [24]. Propagation (unlike self-renewal), does not require either daughter cell to be a SC nor to retain the ability to give rise to a differentiated progeny [25]. The dedicated progenitor cell is destined to stop multiplying as with each cell division its potential to proliferate decreases.

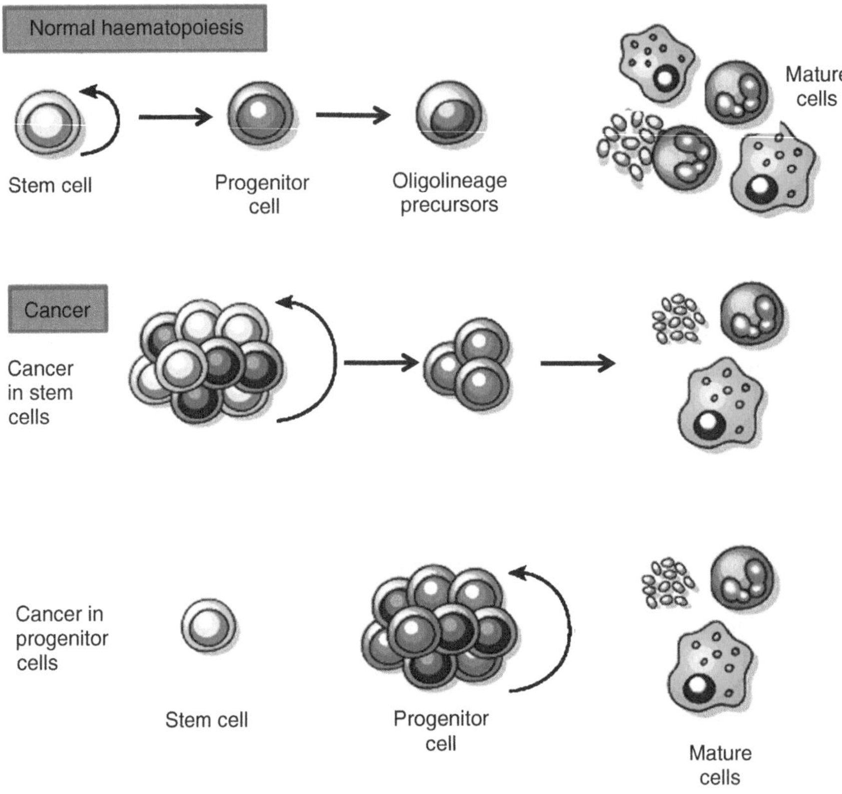

Fig. 1.1 Self-renewal during haematopoietic stem cell development and leukaemic transformation. Normal haematopoiesis, where signalling pathways that have been proposed to regulate selfrenewal are tightly regulated (*top*), during transformation of stem cells, the same mechanisms may be dysregulated to allow uncontrolled self-renewal (*middle*). Furthermore, if the transformation event occurs in progenitor cells, it must endow the progenitor cell with the self-renewal properties of a stem cell, because these progenitors would otherwise differentiate (*bottom*)

In the blood, both SCs and dedicated progenitor cells have an wide capacity to proliferate [19]. The normal SC self-renewal regulation is also fundamental to understanding the regulation of cancer cell proliferation, since cancer can be considered to be a disease of unregulated self-renewal. Although, up to 6–8 weeks devoted ancestor populations can maintain hematopoiesis [2, 8]. For example, a single HSC can restore the blood system for the life of the animal (Fig. 1.1). This incredible potential is a direct outcome of its capacity to self-renew.

Most tumors develop over a period of months to years and like normal tissues consist of heterogeneous populations of cells. The unregulated growth of tumors attributed to the serial acquisition of genetic events that resulted in the turning on genes promoting proliferation, silencing of genes involved in inhibiting proliferation, and avoiding of genes involved in programmed cell death. Another key event in tumorigenesis is the interruption of genes involved in the regulation of SC self-renewal.

Thus, some of the cancer cells within a tumor share with somatic SC the capacity to replicate without losing the potential to proliferate. Signaling pathways such as Bmi-1, Notch, Wnt and Sonic hedgehog (Shh) that have been identified to be involved in regulation of self-renewal in normal SC [6, 9, 94]. Recently, Reya et al. [81] demonstrated the requirement of normal HSC self-renewal decisions on Wnt-signaling through the canonical pathway. The capacity of purified Wnt3a to permit the *in vitro* expansion of transplantable HSCs has been observed [4, 49, 104].

The Polycomb and trithorax groups are parts of multimeric complexes that interact with chromatin leading to either a repressed or activated state of gene expression, respectively. Bmi-1, a member of the Polycomb group, targets the INK4a locus and overexpression of Bmi-1 consequences in inhibition of both p16 and p19Arf expression [47]. Post-natal mice deficient in the expression of Bmi-1 display failure of hematopoiesis and fetal liver and bone marrow SCs from Bmi-1 mice are able to contribute to recipient hematopoiesis only transiently indicating a primary defect in adult HSC self-renewal [61, 71, 99]. Bmi-1 also plays a key role in malignant hematopoiesis as HOXA9/MEIS 1 induced murine leukemia [61]. The importance of epigenetic events, such as modification of chromatin, in normal and malignant tissues is likely to remain a key focus of research on self-renewal property of SCs. Preliminary studies have examined the ability to reverse these epigenetic alterations through the transfer of nuclei from cells in a differentiated tissue into enucleated oocytes. Nuclei obtained from medulloblastoma tumor cells arising in Ptc1 heterozygous mice were transferred into enucleated oocytes [62]. Blastocysts derived from medulloblastoma were morphologically indistinguishable from those derived from control spleen cell nuclei without evidence of the uncontrolled proliferation. This study suggests that epigenetic reprogramming was responsible for the loss of the tumor cells' capacity to form tumors.

1.3 Cellular Origin of CSCs

The term 'CSC is an operational term defined as a cancer cell that has the potential to self-renew giving rise to another malignant SC as well as undergo differentiation to give rise to the phenotypically diverse non-tumorigenic. In foregoing years, the cell-of-origin for cancer SCs remains unclear: they may or may not be derived from their somatic cell counterpart. Recent evidence strongly favors a progenitor cell origin for many types of leukemic SCs in addition to the SC origin [64]. In solid tumors too, it is most likely that not only somatic SCs but also differentiating progenitor cells are capable of becoming CSCs [107].

The fact that multiple mutations are necessary for a cell to become cancerous [56] has implications for the cellular origin of cancer cells. As both progenitor cells and mature cells have a very restricted lifespan, it is unlikely that all of the mutations could occur during the life of these relatively short-lived cells. In addition, to maintain the disease, cancer cells must overcome the tight genomic constraints on both self-renewal as well as proliferation [66]. Because cancer SCs must possess the

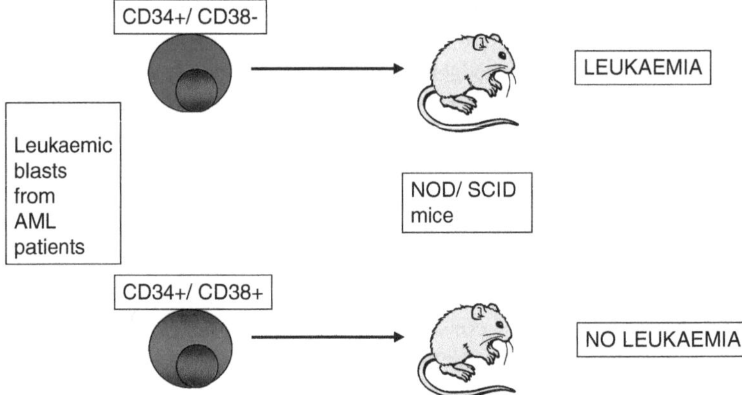

Fig. 1.2 Leukaemia stem cells exist in human acute myeloid leukaemia (AML). The cells capable of initiating human AML in NOD/SCID (non-obese diabetic/severe combined immunodeficiency) mice have a CD34$^+$CD38$^-$ phenotype in most AML subtypes, and thus have a phenotype similar to normal HSCs

ability to self-renew, it follows that they are derived either from self-renewing normal SCs—which could be transformed by altering only proliferative pathways—or from progenitor cells that have acquired the potential to self-renew as a result of oncogenic mutations. In case of most cancers, the target cell of transforming mutations is unknown; nevertheless, there is considerable evidence that certain types of leukaemia arise from mutations that accumulate in HSCs. The cells capable of initiating human acute myeloid leukaemia (AML) in NOD/SCID (non-obese diabetic/severe combined immunodeficiency) mice have a CD34$^+$CD38$^-$ phenotype in most AML subtypes, and thus have a phenotype similar to normal HSCs (Fig. 1.2) [14].

Feldman and Feldman [35] proposed a model of oncogene-induced plasticity for CSC origin by demonstrating reprogramming events triggered by a specific combination of oncogenes. [63] suggested that genomic instability is a driving force for transforming normal SCs to CSCs. In CSCs, a potential mechanism for cancer cell heterogeneity. A common phenotype for the LICs has been identified [10, 13, 14, 52]. Although the phenotype of the LIC is much related to that of the normal HSC, there are differences, including the differential expression of Thy1 and IL3 receptor a chain [13, 72]. These differences suggest that early mutations occurred in the HSCs and the final transforming events either alter the phenotype of the SCs or occur in early downstream progenitors.

A model of CML was reported recently in which the expression of the fusion product was targeted to myeloid/monocyitc progenitor cells using the hMRP-8 promoter. A subset of the hMRP8p210BCR/ABL mice develops a CML-like disease with elevated white cell counts and splenomegaly [48]. When crossed with hMRP-8bcl-2 mice, a proportion of the mutant mice developed a disease resembling AML. One explanation for this finding was that targeting the expression of the fusion protein to the committed progenitor instills in this population the capacity for self-renewal. Additional studies examining the ability of purified hMRP8p210BCR/ABL progenitors to reconstitute the disease upon transplantation into primary as well as secondary recipients may be helpful to distinguish between these two possibilities. As with AML, the phenotype of breast cancer TICs may be similar to that of normal breast epithelial stem or progenitor cells because early multipotent epithelial cells have been reported to exhibit a similar phenotype to that seen in the tumorigenic breast cancer cells [1, 38, 91].

1.4 Identifying Characteristic Cell Surface Markers

Although functions have yet to be determined for many of these early surface markers, their unique expression pattern and timing provide a useful tool for scientists to initially identify and isolate SCs from the source. These are CD34 in several kinds of leukemia, CD44 in pancreas, prostate, breast, colorectal, head/neck cancers and some bone sarcomas, to detect and isolate CSCs from among the uncountable cancer cells and stromal cells occupying the entire tumor tissue. Representative cell surface markers for human hematologic and solid cancers reported to date are listed in Table 1.1.

The cell surface sialomucin CD34 has been a focus of interest ever since it was found expressed on a small fraction of human bone marrow cells [22]. The $CD34^+$ -enriched cell population from marrow or mobilized peripheral blood appears responsible for most of the hematopoietic activity [22]. CD34 has therefore been considered to be the most critical marker for HSCs. CD34 expression on primitive cells is down-regulated as they differentiate into mature cells [92]. It is also found on clonogenic progenitors, however, and some lineage-committed cells. In contrast to the high endothelial venules for which CD34 serves as a ligand for l-selectin, CD34 is not the ligand for l-selectin in hematopoietic stem/progenitor cells and ligands for hematopoietic CD34 remain to be identified [59]. Few lung CSC markers have been identified to date. Two recent reports suggest that CD133, together with the pan-epithelial marker EpCAM, can be used to isolate human lung CSCs.

CD44, originally described as a leukocyte-homing receptor, includes a family of glycoproteins encoded by a single gene, which vary in size due to alternative splicing CD44 has been used as a CSC marker for leukemia and for a variety of solid cancers as described above. CD133, a 120 kDa, glycosylated protein containing five transmembrane domains, identified initially by the AC133 monoclonal Ab, which recognizes a $CD34^+$ subset of human HSCs [108]. CD133 is a specific marker of CSCs in a wide spectrum of malignant tumors including brain tumors, colorectal, pancreatic, breast, prostate, ovarian cancers [55], and some lung cancers. CD133 was first reported as a novel marker for human hematopoietic stem and progenitor cells [108]. Recent studies have offered evidence that CD133 expression is not limited to primitive blood cells, but defines unique cell populations in non-hematopoietic tissues as well. CD133+ progenitor cells from peripheral blood can be induced to differentiate into endothelial cells in vitro [93]. In addition, human neural SCs can

Table 1.1 Specific cell surface markers for human CSCs

S. no.	Type of cancer	Cell surface markers	Reference
1.	Pancreatic	CD133$^+$, CD44$^+$, CD24$^+$, Lgr5	[22, 55]
2.	Prostatic	CD44$^+$, integrin	[55, 59]
3.	Breast	CD44$^+$, CD24$^{-/low}$	[55, 106]
4.	Ovarian	CD44$^+$, MyD88$^+$	[55]
5.	Colon	CD133$^+$, CD44$^+$, CD166$^+$, E-CAM^{high}, Lgr5	[55, 101]
6.	AML	CD34$^+$, CD38$^-$	[43]
7.	Myeloproliferative disorder	CD117$^+$	[55]
8.	Glioblastoma	CD133$^+$, Nestin, CD15$^+$	[50]
9.	Medulloblastoma	CD133$^+$	[55]
10.	Hepatocellular cancer	CD133$^+$	[55]
11.	Head and neck squamous cell carcinoma	CD44$^+$	[102]
12.	Metastatic melanoma	CD20$^+$	[43]
13.	Bone sarcomas	Stro-1$^+$, CD105$^+$, CD44$^+$	[55]
14.	Lung	CD133$^+$	[108]

be directly isolated by using an anti-CD133 Ab [102]. The aldehyde dehydrogenase (ALDH) superfamily denotes a divergently related group of enzymes that metabolize a wide variety of endogenous and exogenous aldehydes, At least 17 functional genes and 3 pseudogenes have been identified in human genomes [90]. ALDH also contributes to the oxidation of retinol to retinoic acid, a modulator of cell proliferation, which may also modulate SC proliferation [44]. Murine and human hematopoietic SCs [51], murine neural SCs [27], normal and malignant human mammary SCs [36], and normal and malignant human colorectal SCs [32] display ALDH activity and express this enzyme, strongly suggesting that strong ALDH activity and/or antigen expression can be used as a marker for SCs in a variety of cancers.

Breast CSCs have been isolated from human breast tumors or breast cancer-derived pleural effusions using flow cytometry to find subpopulations of cells with a specific pattern of cell surface markers [CD44$^+$, CD24$^{-/low}$, ESA$^+$ (epithelial specific antigen)] but lacking expression of specific lineage markers (Lin$^-$) [43]. These cells expressed (EMT) markers and had higher tumorigenic potential than bulk tumor cells after transplantation in nonobese diabetic/severe combined immunodeficient (NOD/SCID) mice [54]. It has also been shown that single cell suspensions of CD44$^+$CD24$^{-/low}$Lin$^-$ cells from human breast cancers were able to proliferate extensively and form clonal nonadherent mammospheres in a low attachment *in vitro* culture system [77]. These mammospheres were more tumorigenic than established breast cancer-derived cell lines including MCF-7 and B3R [77]. PROCR, identified using gene expression profiling of primary breast cancers [87], is also a known marker of hematopoietic, neural, and embryonic stem cells. An additional marker, CD133, was identified for breast cancer stem cells isolated from cell lines generated from Brca1$^{-exon11}$/p53$^{+/-}$ mouse mammary tumors [106] and is a known marker of cancer stem cells in several organs including brain, blood, liver and prostate [45]. As in breast cancer, FACS analysis revealed heterogeneous surface marker expression for CD44, CD24, and ESA among pancreatic tumor cells.

Colon cancer cells such as $CD44^+CD166^+$ display a higher ability to form tumors in immunodeficient mice as compared to $CD44^+CD166^-$, $CD44^-CD166^-$ or $CD44^-CD166^-$ cell populations, making this an useful combination for the identification of colon CSCs [78]. Lgr5 is a Wnt target gene, exclusively expressed on colon CSCs and normal intestinal SCs and could thus also be a colon CSC marker [101]. On the basis of this hypothesis that the presence of tumor cells expressing SCs markers would affect the survival of glioblastoma patients, the expression of three SCs markers have been investigated: nestin, the prototypical marker for the identification of (NSCs); CD133 (cluster of differentiation 133), the most accredited marker for CSCs in various organs including the brain; and CD15, which is one of the most recently highlighted NSC markers and is also used to identify CSCs in central nervous system tumors [17]. Recent findings support the belief that cancer stem-like cells are responsible for tumor formation and ongoing growth. Differential expression was verified by Western blotting analysis of six interesting proteins, including the up-regulated Receptor-type tyrosine-protein phosphatase zeta, Tenascin-C, Chondroitin sulfate proteoglycan NG2, Podocalyxin-like protein 1 and CD90, and the down-regulated CD44. An improved understanding of these proteins may be important for earlier diagnosis and better therapeutic targeting of glioblastoma [40]. Neurons and glia cells also differentiate into abnormal cells with multiple differentiation markers and express many genes characteristic of NSCs and other stem cells, like the cell surface marker CD133, transcription factor SRY-related HMG-box gene 2 (Sox2) and neural RNA binding protein musashi 1 [41]. A population of cells in human brain tumors, medulloblastomas, astrocytomas, ependymomas and gangliogliomas that expresses the cell surface marker CD133 identified and elicit CSCs characteristics [88]; the $CD133^+$ isolated cells correspond to a small fraction of the entire brain tumor cell population, express the NSC marker nestin, exhibit increased self-renewal capacity, generate clonal tumor spheres in culture, and are capable of tumor initiation upon transplantation in to the brains of immune compromised mice. Sox2 is one of the key regulatory genes that maintain the pluripotency and self-renewal properties in embryonic SCs. Recently, Jia et al. [50] reported that Sox2 the potential to be a significant marker to evaluate the progression of prostate cancer and serve as a potentially useful target for prostate cancer therapy.

1.5 Pathways Regulating Stem Cell Self-renewal and Oncogenesis

It seems reasonable to propose that newly arising cancer cells appropriate the machinery for self-renewing cell division that is normally expressed in SCs. Evidence shows that many pathways that are naturally associated with cancer may also regulate normal SC development. For example, the prevention of apoptosis by

Table 1.2 CSC molecular signatures in different cancer types

Target type	Specific target	Cancer type	Use
Cell surface markers	CD34$^+$/CD38	AML	Identification has allowed for characterization of LSCs. Too broad to use as a target for chemotherapy but is very useful in identification for further characterization
	CD33$^+$	AML	Leukemia Gemtuzumab ozogamacin
	C-type lectin like molecule -1 (CLL-1)	AML	No clinical trials but efficacy seen *in vitro* and *in vivo* experimental studies
Signaling pathways	PI3K/Akt/mTOR	FDA approved therapy for renal cell carcinoma. Evidence that may be effective in other solid tumors	Temsirolimus, Everolimus FDA approved for renal cell Carcinoma
	Hedgehog	Evidence in basal cell carcinoma but has been identified as being up-regulated in many cancer types	Novel GDC-0449 and GANT-61
	HMG-CoA reductase	Increase ROS within cells leading to apoptosis, being investigated in many cancers including CML	Synergistic effect seen when imatinib and simvastatin in CML
	Wnt/β-catenin	AML, colon, pancreatic, breast, prostate, melanoma, glioblastoma etc	Celecoxib, Rofecoxib, Valdecoxib, AV-4126, ICG-001, Troglitazone and Rosiglitazone

enforced expression of the oncogene Bcl-2 results in increased numbers of HSCs *in vivo*, suggesting that cell death has a role in regulating the homeostasis of HSCs [29]. Other signaling pathways associated with oncogenesis, such as the Notch, AKT, Shh and Wnt signaling pathways, may also regulate SC self-renewal (Table 1.2) [94]. Notch activation in HSCs in SC culture using the ligand Jagged-1 have consistently increased the amount of primitive progenitor activity that can be observed *in vitro* and *in vivo* model systems, suggesting that Notch activation stimulates SC self-renewal, or at least the maintenance of multi-potentiality [53, 100]. Another pathway, Shh signaling has also been concerned in the regulation of self-renewal by the finding that populations highly enriched for human HSCs (CD34$^+$Lin–CD38$^-$) exhibit increased self-renewal in response to Shh stimulation *in vitro*, albeit in combination with other growth factors [9]. The involvement of both Notch and Shh in the self-renewal of HSCs is especially interesting in light of studies that implicate these pathways in the regulation of self-renewal of SCs from other tissues as well [80].

Our recent studies have been indicated that the aberrant reactivation of shh pathway is common event in pancreatic, prostate, brain, breast CSCs as downstream effectors of this pathway, Gli1, Gli2 and Gli3, induce genes that promote cellular proliferation and self-renewal. Therefore targeting Shh pathway could be a novel approach to prevent disease progression and metastatic spread. Small molecule inhibitors of Gli family proteins, such as GDC-0449, GANT-61, and other are used to block Shh signaling in human pancreatic, prostate, brain, breast CSCs that express Shh signaling components. Inhibition of the Shh signaling pathway by these inhibitory molecules induced significant cell death in CSCs. Gli1 and Gli2 expressions, promoter binding activity and Gli-luciferase reporter activity were also decreased. Increased level of Fas, DR4, DR5, caspase-3 and PARP cleavage were observed however expression of PDGFR-α and Bcl-2 was decreased following treatment of inhibitors. Silencing both Gli1 and Gli2 using shRNA abolished all the alterations produced by inhibitors. Collectively, these results highlight that these inhibitors induce apoptosis and cell death through inhibition of Gli family transcription factors (Gli1 and Gli2), and inhibition of SHH-signaling pathway, can be used for treatment of human pancreatic, prostate, brain, breast cancers.

One particularly interesting pathway that has also been shown to regulate both self-renewal and oncogenesis in different organs is the Wnt signalling pathway. Wnt proteins are intercellular signaling molecules [69] that regulate development in several organisms and contribute to cancer when dysregulated. The expression of Wnt proteins in the bone marrow suggests that they may influence SCs as well [16, 79]. Using highly purified mouse bone-marrow HSCs, overexpression of activated β-catenin (a downstream activator of the Wnt signalling pathway) in long-term cultures of HSCs expands the pool of transplantable HSCs determined by both phenotype and function (ability to reconstitute the haematopoietic system *in vivo*) was observed. Additionally, ectopic expression of Axin, an inhibitor of Wnt signaling, leads to inhibition of HSC proliferation, increased death *in vitro*, and reduced reconstitution *in vivo*. Soluble Wnt proteins from conditioned supernatants have also been shown to influence the proliferation of haematopoietic progenitors from mouse fetal liver and human bone marrow [6, 98]. Both *in vitro* and *in vivo* investigations into the PI3K/Akt/mTOR signaling pathways have also shown some potential for targeting CSCs. Integrin linked kinase (ILK) is also involved in phosphorylation of Akt and is upregulated in many malignancies including pancreatic cancer and AML blast cells [18, 67, 82, 85]. One of the hardest parts of targeting cancers is being able to target cells when they are quiescent. Interestingly, there is an over-expression of ILK during this phase which may play a part in the survival of cells or inhibition of apoptosis [67].

1.6 Stem Cells and Heterogeneity

Tumors are heterogeneous, but the mechanisms underlying this are unclear. Heterogeneity may result from mutations occurring early or late in a SC's maturation. For example, CML is believed to derive from an early progenitor SCs because

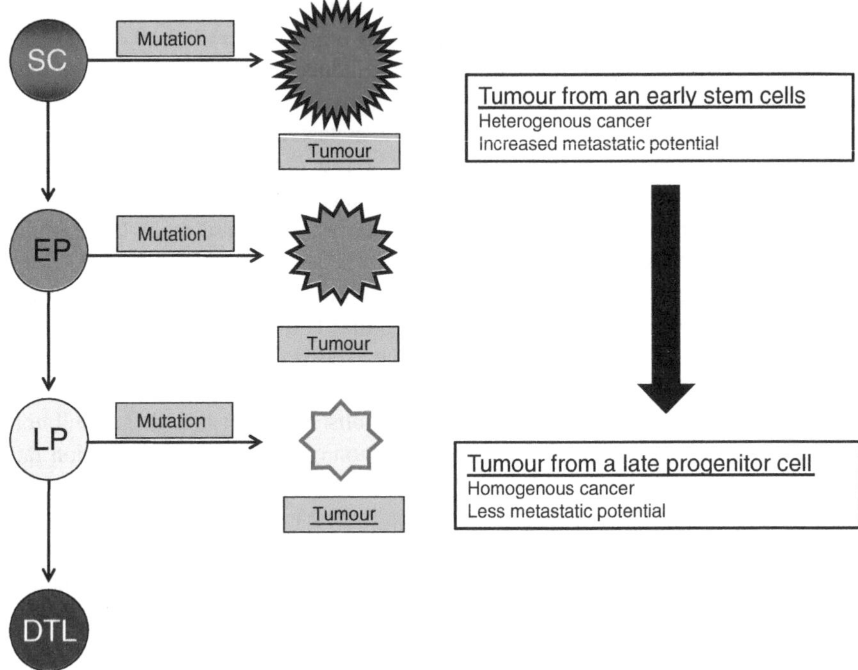

Fig. 1.3 In the stem cell model, only the stem cells or their progenitor cells have the ability to form tumours. Tumour characteristics vary depending on which cell undergoes the malignant transformation. *DTL* definitive tissue line, *EP* early progenitor, *LP* late progenitor, *SC* stem cell

its cytogenic marker (BCR-ABL) is present in several cell lineages, for example lymphoid, myeloid and platelet cells. Nevertheless, an abnormality in a late SC progenitor in the myeloid lineage at the promyelocytic stage results acute promyelocytic leukaemia [97]. Tumors derived from an early SC may develop a more heterogenous phenotype and have an increased metastatic potential. Mutations in late progenitor SCs may lead to tumors of a single cell type with reduced metastatic potential (Fig. 1.3).

As recently shown in an experiment, the mammary gland develops by differentiation from its mammary SC [84]. A diverse range of breast cancers may, hence, develop depending on where a mutation occurs in this pathway [30, 31]. Therefore, a SC model for estrogen receptor (ER) expression in breast cancer has been proposed, dividing breast cancer into three types, in an attempt to explain how ER-positive, ER-negative or heterogenous receptor status tumors can be created by mutations in the SC or progenitor cell populations. In early fetal life, SCs are ER negative, but presumably under the influence of oestrogen, progenitor cells that are both ER positive and ER negative can be identified at various times during growth, in particular during puberty and pregnancy [30].

Type 1 tumors develop from mutations in ER-negative stem/ progenitor cells, blocking differentiation and averting the development of ER-positive progenitors [30].

These tumors are poorly differentiated and seem to be more aggressive with a poorer prognosis. Less than 10% of these tumors are ER positive [30]. In the ER negative stem/progenitor cells Type 2 tumors are also derived from mutations. However, a variable percentage of the tumor will differentiate into ER-positive cells [31]. Antioestrogen therapy can produce a decrease in tumor size; however, the effect is short lived as ER-negative SCs are unaffected and tumor proliferation continues despite hormonal therapy [30]. This may explain why some ER-positive breast cancers continue to grow despite adjuvant hormonal therapy [30]. Type 3 tumors are well differentiated and result from mutations in ER-positive progenitor cells. Hormone replacement therapy use increases the risk of cancer formation. They respond best to antioestrogen therapy and have the best prognosis [30].

1.7 Therapeutic Implications

A practical importance of CSC heterogeneity is that strategies to induce apoptosis and cell death to treat cancer must address the unique survival mechanisms of the CSC within the cancer cell population. Most old-style cancer treatments have been developed and assayed based on their ability to kill most of the cancer cell population and result in tumor shrinkage. These treatments likely miss the CSCs, which have been shown in several cancer types to be quite resistant to standard chemotherapy and radiation. A prediction of the CSC model is that treatments that target the CSC will be required to result in an effective cure of cancer. As such, tumor shrinkage is not going to be a useful parameter to measure effectiveness of CSC therapies, and approaches to measure CSC burden will need to be devised.

The identification of CSCs has potential therapeutic implications [30]. As SCs are important for tissue growth and repair, they have developed highly efficient mechanisms for resistance to apoptosis [30]. Many have overexpression of anti-apoptotic genes such as Bcl-2 and may express drug-resistance transporter proteins such as MDR1 and ABC transporters [37, 70, 83]. These mechanisms permit normal SCs to become resistant to chemotherapy [3, 60, 88]. It has been proposed that CSCs also express these proteins at higher levels than the bulk population of tumor cells and may be more resistant to chemotherapeutic agents, allowing the reproliferation of tumors after chemotherapy [80]. Developing targeted therapies that are selectively toxic to CSCs while sparing normal SCs may lead to more effective treatment options for eradicating this crucial population of cells [70].

Although, there is clinical research based evidence that are indicating more clearly towards the origin of brain tumors from brain proliferative sections. The cell of origin for brain tumors is still unknown and these may vary from one tumor type to another or may be different in tumors occurring at different patient ages. One could claim that once the tumor exists, its cell of origin is not relevant, what is relevant is the CSCs and the directing therapy to this cell to effect a cure. The CSC hypothesis suggests that the CSC must be eliminated to cure cancer, but it is likely that different components of the tumor hierarchy will need to be targeted. This hypothesis suggests that current therapies spare CSCs leading to tumor regrowth and clinical recurrence. One key factor for treatment may be the cell cycle status of the SCs, as most currently available treatments target cells that are promptly cycling. Although, normal SCs, and leukaemic SCs, have been shown to be quiescent, the cell cycle status of solid CSCs has not yet been well characterized. If the brain tumor SC is relatively quiescent, these cells will probably require distinct therapy from tumor progenitors that are rapidly proliferating.

Some studies suggest that Shh, Bim-1, Notch, Wnt/β-catenin signaling leads to increased tolerance of DNA damage, thus conferring radiation resistance of CSCs [20, 105]. Wnt/β-catenin signaling activities the DNA damage response, and one transcriptional target of β-catenin signaling is survivin, which is known to encourage cellular survival in CSCs in response to apoptotic stimuli such as ionizing radiation [5]. The complex nature of CSC survival mechanisms extends beyond Wnt/β-catenin signaling. Shh and Notch signaling has also been implicated in prostrate, pancreatic, brain, and breast cancer's response to radiation injury and targeting this pathway has shown effective antitumor response in preclinical trials [34, 76]. Alternatively, some studies suggest that the level of compaction of chromatin dictates accessibility to genomic DNA and subsequent mediation of DNA damage responses and that a looser configuration of chromatin in SCs leads to accelerated DNA repair following injury. Such has been shown in embryonic SCs that have lower levels of the chromatin structural protein histone-1. Embryonic SCs with lower levels of histone-1, which results in less chromatin compaction, had enhanced recovery from DNA damage in comparison to differentiated cells [68]. Pancreatic and prostrate CSCs likely share some of the signaling cascades involved in cellular responses to DNA damage present in other SC systems; however, the specific responses and mechanisms involved in the chemotherapy and radiation resistance of prostate and pancreatic CSCs remain to be elucidated.

There has been much interest in modern technique microarray analysis of tumors, the development of prognostic and predictive markers, allowing tumor subtyping and the possibility of developing specific tumor treatments [42, 103, 109]. The identification of SCs in several cancers such as AML, pancreatic, breast cancers and CNS tumors raises the possibility that decision-making on the basis of microarray analysis of the bulk tumor population may not be entirely appropriate because the gene expression profile of the CSC may be different to the rest of the tumors [80]. By comparing gene expression profiles of CSCs, the bulk tumor cell population, normal SCs and normal tissue, it may be possible to identify therapeutic targets that preferentially attack CSCs [70].

Understanding CS cell biology may lead to insights into the causes and treatment of tumor metastasis. The metastatic ability of a tumor cell may be related to properties of the SC of origin [30]. For example, the cytokine receptor CXCR4 is expressed on haematopoietic SCs and interacts with cytokines CXCL12/SCDF that are secreted by bone marrow stromal cells. This attracts haematopoietic SCs to the bone marrow [31]. The CXCR4 cytokine is also overexpressed on metastatic breast cancer cells. This may direct them to the bone marrow and may be one of several potential explanations for the increased incidence of bone metastases in breast cancer.

Existing remedies have been developed largely against the bulk population of tumor cells because they are often identified by their ability to shrink tumors. Because most cells with a cancer have limited proliferative potential, an ability to shrink a tumor mainly reflects an ability to kill these cells. It seems that normal SCs from various tissues tend to be more resistant to chemotherapeutic agents than mature cell types from the same tissues [39]. The reasons for this are not clear, but may relate to high levels of expression of anti-apoptotic proteins or ABC transporters such as the multidrug resistance gene [74, 96, 110]. If the same were true of CSCs, then one would predict that these cells would be more resistant to chemotherapeutic agents than tumor cells with limited proliferative potential. Even therapies that cause complete regression of tumors might spare enough CSCs to allow regrowth of the tumors. Therapies that are more specifically directed against CSCs might result in much more durable responses and even cures of metastatic tumors.

The defining characteristics of a circulating cancer stem cell (CTSC) are its capacity for self-renewal and for initiation of distant metastases; some of these cells are also resistant to traditional chemotherapy. The concept that circulating tumor SCs can be identified and targeted is attractive and has major diagnostic, prognostic, and therapeutic implications for patients with metastatic cancer. The development of metastasis is a late event in the linear progression model of metastasis, as opposed to the parallel progression model that suggests dissemination of circulating tumor cells can be an early event [89]. Some patients with Dukes' stage A CRC had detectable mRNA of CEA/CK/CD133 but failed to show any differences in overall survival and disease-free survival regardless of their CEA/CK/CD133 status [46]. These findings suggest that dissemination of CTSCs may indeed be an early event but also indicate that the prognostic significance of these cells in association with early disease may be negligible. Additional studies are needed to shed more light on the prognostic significance of CTSCs that are detected in patients with early-stage disease.

Genomics may provide a powerful means for identifying drug targets in cancer cells. Although targeting genetic mutations does not require isolation of the SCs, there are likely to be differences in gene expression between CSCs and tumor cells with limited proliferative potential. The application of microarray analysis to malignant tumors has shown that patterns of gene expression can be used to group tumors into different categories, often reflecting different mutations [11, 73]. As a result, tumor types that cannot be renowned pathologically, but that can be renowned on the basis of differences in gene-expression profile, can be examined for differences in treatment sensitivity. However, gene-expression profiling is often conducted on tumor samples that contain a mixture of normal cells, highly proliferative cancer cells, and cancer cells with limited proliferation potential. These findings in a composite profile that may obscure differences between tumors, because the highly proliferative cells that drive tumorigenesis often represent a minority of cancer cells. Gene-expression profiling of CSCs would allow the profile to reflect the biology of the cells that are actually driving tumorigenesis. Micro dissection of morphologically homogeneous collections of cancer cells is one way of generating profiles that reflect more homogeneous collections of cells [60]. The next boundary will be to purify the CSCs from the whole tumor that keep unlimited proliferative potential and to perform gene-expression profiling on those cells. In addition to being a more efficient way of identifying new therapeutic and diagnostic targets, the profiling of CSCs might sharpen the differences in patterns observed between different tumors.

1.8 Conclusion

Self-renewal is the hallmark property of CSCs in normal and neoplastic tissues. Distinct signaling pathways control CSC self-renewal in different tissues. But perhaps within individual tissues, the same pathways are used consistently by both normal CSCs and cancer cells to regulate proliferation. For example, Wnt Shh, and Notch signaling pathways regulate the self-renewal of normal CSCs. Constitutive activation of these signaling pathways have been implicated in a number of epithelial cancers. The regulation and consequences of these pathways in normal and neoplastic cells need to be further elucidated. The discovery of CSCs in AML, pancreatic, prostrate, breast, brain cancers and some other tumors offers a new approach to understanding the biology of these conditions. Understanding the signaling pathways that are used by for normal SCs and neoplastic cells should facilitate the use of normal SCs for regenerative medicine and the identification of CSC targets for anticancer therapies. Further studies are needed to understand both normal and CSC development and whether CSCs are present in other tumor types. Ultimately, new prognostic and predictive markers, as well as targeted therapeutic strategies, may be developed to force tumors into permanent remission. There are many connections between SCs and cancer that are important to understand. Just as the signals that are known to control oncogenesis are providing clues about the control of self-renewal of normal SCs, studies of stem cell biology are lending insight into the origins of cancer and will ultimately yield new approaches to fight this disease.

Acknowledgements This work was supported in part by the grants from the National Institutes of Health (R01CA125262, RO1CA114469 and RO1CA125262-02S1), Susan G. Komen Breast Cancer Foundation, and Kansas Bioscience Authority.

References

1. Aigner S, Sthoeger ZM, Fogel M, Weber E, Zarn J, Ruppert M, Zeller Y, Vestweber D, Stahel R, Sammar M, Altevogt P (1997) CD24, a mucin-type glycoprotein, is a ligand for P-selectin on human tumor cells. Blood 89:3385–3395
2. Akashi K, Traver D, Miyamoto T, Weissman IL (2000) A clonogenic common myeloid progenitor that gives rise to all myeloid lineages. Nature 404:193–197
3. Al-Hajj M, Becker MW, Wicha M, Weissman I, Clarke MF (2004) Therapeutic implications of cancer stem cells. Curr Opin Genet Dev 14:43–47
4. Andl T, Reddy ST, Gaddapara T, Millar SE (2002) WNT signals are required for the initiation of hair follicle development. Dev Cell 2:643–653

5. Asanuma K, Moriai R, Yajima T, Yagihashi A, Yamada M, Kobayashi D, Watanabe N (2000) Survivin as a radioresistance factor in pancreatic cancer. Jpn J Cancer Res 91:1204–1209
6. Austin TW, Solar GP, Ziegler FC, Liem L, Matthews W (1997) A role for the Wnt gene family in hematopoiesis: expansion of multilineage progenitor cells. Blood 89:3624–3635
7. Auzenne E, Ghosh SC, Khodadadian M, Rivera B, Farquhar D, Price RE, Ravoori M, Kundra V, Freedman RS, Klostergaard J (2007) Hyaluronic acid-paclitaxel: antitumor efficacy against CD44(+) human ovarian carcinoma xenografts. Neoplasia 9:479–486
8. Baum CM, Weissman IL, Tsukamoto AS, Buckle AM, Peault B (1992) Isolation of a candidate human hematopoietic stem-cell population. Proc Natl Acad Sci USA 89:2804–2808
9. Bhardwaj G, Murdoch B, Wu D, Baker DP, Williams KP, Chadwick K, Ling LE, Karanu FN, Bhatia M (2001) Sonic hedgehog induces the proliferation of primitive human hematopoietic cells via BMP regulation. Nat Immunol 2:172–180
10. Bhatia M, Wang JC, Kapp U, Bonnet D, Dick JE (1997) Purification of primitive human hematopoietic cells capable of repopulating immune-deficient mice. Proc Natl Acad Sci USA 94:5320–5325
11. Bittner M, Meltzer P, Chen Y, Jiang Y, Seftor E, Hendrix M, Radmacher M, Simon R, Yakhini Z, Ben-Dor A, Sampas N, Dougherty E, Wang E, Marincola F, Gooden C, Lueders J, Glatfelter A, Pollock P, Carpten J, Gillanders E, Leja D, Dietrich K, Beaudry C, Berens M, Alberts D, Sondak V (2000) Molecular classification of cutaneous malignant melanoma by gene expression profiling. Nature 406:536–540
12. Bixby S, Kruger GM, Mosher JT, Joseph NM, Morrison SJ (2002) Cell-intrinsic differences between stem cells from different regions of the peripheral nervous system regulate the generation of neural diversity. Neuron 35:643–656
13. Blair A, Hogge DE, Ailles LE, Lansdorp PM, Sutherland HJ (1997) Lack of expression of Thy-1 (CD90) on acute myeloid leukemia cells with long-term proliferative ability in vitro and in vivo. Blood 89:3104–3112
14. Bonnet D, Dick JE (1997) Human acute myeloid leukemia is organized as a hierarchy that originates from a primitive hematopoietic cell. Nat Med 3:730–737
15. Brazelton TR, Rossi FM, Keshet GI, Blau HM (2000) From marrow to brain: expression of neuronal phenotypes in adult mice. Science 290:1775–1779
16. Cadigan KM, Nusse R (1997) Wnt signaling: a common theme in animal development. Genes Dev 11:3286–3305
17. Capela A, Temple S (2002) LeX/ssea-1 is expressed by adult mouse CNS stem cells, identifying them as nonependymal. Neuron 35:865–875
18. Chen X, Thakkar H, Tyan F, Gim S, Robinson H, Lee C, Pandey SK, Nwokorie C, Onwudiwe N, Srivastava RK (2001) Constitutively active Akt is an important regulator of TRAIL sensitivity in prostate cancer. Oncogene 20:6073–6083
19. Chen S, Do JT, Zhang Q, Yao S, Yan F, Peters EC, Scholer HR, Schultz PG, Ding S (2006) Self-renewal of embryonic stem cells by a small molecule. Proc Natl Acad Sci USA 103:17266–17271
20. Chen MS, Woodward WA, Behbod F, Peddibhotla S, Alfaro MP, Buchholz TA, Rosen JM (2007) Wnt/beta-catenin mediates radiation resistance of Sca1+ progenitors in an immortalized mammary gland cell line. J Cell Sci 120:468–477
21. Cheng L, Bao S, Rich JN (2010) Potential therapeutic implications of cancer stem cells in glioblastoma. Biochem Pharmacol 80:654–665
22. Civin CI, Strauss LC, Brovall C, Fackler MJ, Schwartz JF, Shaper JH (1984) Antigenic analysis of hematopoiesis. III. A hematopoietic progenitor cell surface antigen defined by a monoclonal antibody raised against KG-1a cells. J Immunol 133:157–165
23. Clarke MF (2005a) Self-renewal and solid-tumor stem cells. Biol Blood Marrow Transplant 11:14–16
24. Clarke MF (2005b) A self-renewal assay for cancer stem cells. Cancer Chemother Pharmacol 56(Suppl 1):64–68
25. Collins CA, Olsen I, Zammit PS, Heslop L, Petrie A, Partridge TA, Morgan JE (2005) Stem cell function, self-renewal, and behavioral heterogeneity of cells from the adult muscle satellite cell niche. Cell 122:289–301

26. Conway AE, Lindgren A, Galic Z, Pyle AD, Wu H, Zack JA, Pelligrini M, Teitell MA, Clark AT (2009) A self-renewal program controls the expansion of genetically unstable cancer stem cells in pluripotent stem cell-derived tumors. Stem Cells 27:18–28
27. Corti S, Locatelli F, Papadimitriou D, Donadoni C, Salani S, Del Bo R, Strazzer S, Bresolin N, Comi GP (2006) Identification of a primitive brain-derived neural stem cell population based on aldehyde dehydrogenase activity. Stem Cells 24:975–985
28. de Sousa EMF, Guessous I, Vermeulen L, Medema JP (2011) Cancer stem cells and future therapeutic implications. Rev Med Suisse 7:774–777
29. Domen J, Gandy KL, Weissman IL (1998) Systemic overexpression of BCL-2 in the hematopoietic system protects transgenic mice from the consequences of lethal irradiation. Blood 91:2272–2282
30. Dontu G, Al-Hajj M, Abdallah WM, Clarke MF, Wicha MS (2003) Stem cells in normal breast development and breast cancer. Cell Prolif 36(Suppl 1):59–72
31. Dontu G, El-Ashry D, Wicha MS (2004) Breast cancer, stem/progenitor cells and the estrogen receptor. Trends Endocrinol Metab 15:193–197
32. Dylla SJ, Beviglia L, Park IK, Chartier C, Raval J, Ngan L, Pickell K, Aguilar J, Lazetic S, Smith-Berdan S, Clarke MF, Hoey T, Lewicki J, Gurney AL (2008) Colorectal cancer stem cells are enriched in xenogeneic tumors following chemotherapy. PLoS One 3:e2428
33. Fabrizi E, di Martino S, Pelacchi F, Ricci-Vitiani L (2010) Therapeutic implications of colon cancer stem cells. World J Gastroenterol 16:3871–3877
34. Fan X, Matsui W, Khaki L, Stearns D, Chun J, Li YM, Eberhart CG (2006) Notch pathway inhibition depletes stem-like cells and blocks engraftment in embryonal brain tumors. Cancer Res 66:7445–7452
35. Feldman BJ, Feldman D (2001) The development of androgen-independent prostate cancer. Nat Rev Cancer 1:34–45
36. Ginestier C, Hur MH, Charafe-Jauffret E, Monville F, Dutcher J, Brown M, Jacquemier J, Viens P, Kleer CG, Liu S, Schott A, Hayes D, Birnbaum D, Wicha MS, Dontu G (2007) ALDH1 is a marker of normal and malignant human mammary stem cells and a predictor of poor clinical outcome. Cell Stem Cell 1:555–567
37. Golub TR (2001) Genome-wide views of cancer. N Engl J Med 344:601–602
38. Gudjonsson T, Villadsen R, Nielsen HL, Ronnov-Jessen L, Bissell MJ, Petersen OW (2002) Isolation, immortalization, and characterization of a human breast epithelial cell line with stem cell properties. Genes Dev 16:693–706
39. Harrison DE, Lerner CP (1991) Most primitive hematopoietic stem cells are stimulated to cycle rapidly after treatment with 5-fluorouracil. Blood 78:1237–1240
40. He J, Liu Y, Xie X, Zhu T, Soules M, DiMeco F, Vescovi AL, Fan X, Lubman DM (2010) Identification of cell surface glycoprotein markers for glioblastoma-derived stem-like cells using a lectin microarray and LC-MS/MS approach. J Proteome Res 9:2565–2572
41. Hemmati HD, Nakano I, Lazareff JA, Masterman-Smith M, Geschwind DH, Bronner-Fraser M, Kornblum HI (2003) Cancerous stem cells can arise from pediatric brain tumors. Proc Natl Acad Sci USA 100:15178–15183
42. Henrique D, Hirsinger E, Adam J, Le Roux I, Pourquie O, Ish-Horowicz D, Lewis J (1997) Maintenance of neuroepithelial progenitor cells by Delta-Notch signalling in the embryonic chick retina. Curr Biol 7:661–670
43. Herrlich P, Sleeman J, Wainwright D, Konig H, Sherman L, Hilberg F, Ponta H (1998) How tumor cells make use of CD44. Cell Adhes Commun 6:141–147
44. Huang EH, Hynes MJ, Zhang T, Ginestier C, Dontu G, Appelman H, Fields JZ, Wicha MS, Boman BM (2009) Aldehyde dehydrogenase 1 is a marker for normal and malignant human colonic stem cells (SC) and tracks SC overpopulation during colon tumorigenesis. Cancer Res 69:3382–3389
45. Hwang-Verslues WW, Kuo WH, Chang PH, Pan CC, Wang HH, Tsai ST, Jeng YM, Shew JY, Kung JT, Chen CH, Lee EY, Chang KJ, Lee WH (2009) Multiple lineages of human breast cancer stem/progenitor cells identified by profiling with stem cell markers. PLoS One 4:e8377

46. Iinuma H, Watanabe T, Mimori K, Adachi M, Hayashi N, Tamura J, Matsuda K, Fukushima R, Okinaga K, Sasako M, Mori M (2011) Clinical significance of circulating tumor cells, including cancer stem-like cells, in peripheral blood for recurrence and prognosis in patients with Dukes' stage B and C colorectal cancer. J Clin Oncol 29:1547–1555
47. Jacobs JJ, Kieboom K, Marino S, DePinho RA, van Lohuizen M (1999) The oncogene and Polycomb-group gene bmi-1 regulates cell proliferation and senescence through the ink4a locus. Nature 397:164–168
48. Jaiswal S, Traver D, Miyamoto T, Akashi K, Lagasse E, Weissman IL (2003) Expression of BCR/ABL and BCL-2 in myeloid progenitors leads to myeloid leukemias. Proc Natl Acad Sci USA 100:10002–10007
49. Jamora C, DasGupta R, Kocieniewski P, Fuchs E (2003) Links between signal transduction, transcription and adhesion in epithelial bud development. Nature 422:317–322
50. Jia X, Li X, Xu Y, Zhang S, Mou W, Liu Y, Lv D, Liu CH, Tan X, Xiang R, Li N (2011) SOX2 promotes tumorigenesis and increases the anti-apoptotic property of human prostate cancer cell. J Mol Cell Biol 3(4):230–238
51. Jones RJ, Barber JP, Vala MS, Collector MI, Kaufmann SH, Ludeman SM, Colvin OM, Hilton J (1995) Assessment of aldehyde dehydrogenase in viable cells. Blood 85:2742–2746
52. Jordan CT, Upchurch D, Szilvassy SJ, Guzman ML, Howard DS, Pettigrew AL, Meyerrose T, Rossi R, Grimes B, Rizzieri DA, Luger SM, Phillips GL (2000) The interleukin-3 receptor alpha chain is a unique marker for human acute myelogenous leukemia stem cells. Leukemia 14:1777–1784
53. Karanu FN, Murdoch B, Gallacher L, Wu DM, Koremoto M, Sakano S, Bhatia M (2000) The notch ligand jagged-1 represents a novel growth factor of human hematopoietic stem cells. J Exp Med 192:1365–1372
54. Kashyap MP, Singh AK, Kumar V, Tripathi VK, Srivastava RK, Agrawal M, Khanna VK, Yadav S, Jain SK, Pant AB (2011) Monocrotophos induced apoptosis in PC12 cells: role of xenobiotic metabolizing cytochrome P450s. PLoS One 6:e17757
55. Kitamura H, Okudela K, Yazawa T, Sato H, Shimoyamada H (2009) Cancer stem cell: implications in cancer biology and therapy with special reference to lung cancer. Lung Cancer 66:275–281
56. Knudson AG Jr, Strong LC, Anderson DE (1973) Heredity and cancer in man. Prog Med Genet 9:113–158
57. Krause DS, Theise ND, Collector MI, Henegariu O, Hwang S, Gardner R, Neutzel S, Sharkis SJ (2001) Multi-organ, multi-lineage engraftment by a single bone marrow-derived stem cell. Cell 105:369–377
58. Lagasse E, Connors H, Al-Dhalimy M, Reitsma M, Dohse M, Osborne L, Wang X, Finegold M, Weissman IL, Grompe M (2000) Purified hematopoietic stem cells can differentiate into hepatocytes in vivo. Nat Med 6:1229–1234
59. Lanza F, Healy L, Sutherland DR (2001) Structural and functional features of the CD34 antigen: an update. J Biol Regul Homeost Agents 15:1–13
60. Leethanakul C, Patel V, Gillespie J, Pallente M, Ensley JF, Koontongkaew S, Liotta LA, Emmert-Buck M, Gutkind JS (2000) Distinct pattern of expression of differentiation and growth-related genes in squamous cell carcinomas of the head and neck revealed by the use of laser capture microdissection and cDNA arrays. Oncogene 19:3220–3224
61. Lessard J, Sauvageau G (2003) Bmi-1 determines the proliferative capacity of normal and leukaemic stem cells. Nature 423:255–260
62. Li L, Connelly MC, Wetmore C, Curran T, Morgan JI (2003) Mouse embryos cloned from brain tumors. Cancer Res 63:2733–2736
63. Li L, Borodyansky L, Yang Y (2009) Genomic instability en route to and from cancer stem cells. Cell Cycle 8:1000–1002
64. Lobo NA, Shimono Y, Qian D, Clarke MF (2007) The biology of cancer stem cells. Annu Rev Cell Dev Biol 23:675–699

65. Mezey E, Chandross KJ, Harta G, Maki RA, McKercher SR (2000) Turning blood into brain: cells bearing neuronal antigens generated in vivo from bone marrow. Science 290: 1779–1782
66. Morrison SJ, Qian D, Jerabek L, Thiel BA, Park IK, Ford PS, Kiel MJ, Schork NJ, Weissman IL, Clarke MF (2002) A genetic determinant that specifically regulates the frequency of hematopoietic stem cells. J Immunol 168:635–642
67. Muranyi AL, Dedhar S, Hogge DE (2010) Targeting integrin linked kinase and FMS-like tyrosine kinase-3 is cytotoxic to acute myeloid leukemia stem cells but spares normal progenitors. Leuk Res 34:1358–1365
68. Murga M, Jaco I, Fan Y, Soria R, Martinez-Pastor B, Cuadrado M, Yang SM, Blasco MA, Skoultchi AI, Fernandez-Capetillo O (2007) Global chromatin compaction limits the strength of the DNA damage response. J Cell Biol 178:1101–1108
69. Nusse R, Varmus HE (1982) Many tumors induced by the mouse mammary tumor virus contain a provirus integrated in the same region of the host genome. Cell 31:99–109
70. Pardal R, Clarke MF, Morrison SJ (2003) Applying the principles of stem-cell biology to cancer. Nat Rev Cancer 3:895–902
71. Park IK, Qian D, Kiel M, Becker MW, Pihalja M, Weissman IL, Morrison SJ, Clarke MF (2003) Bmi-1 is required for maintenance of adult self-renewing haematopoietic stem cells. Nature 423:302–305
72. Peault B, Weissman IL, Buckle AM, Tsukamoto A, Baum C (1993) Thy-1-expressing CD34+ human cells express multiple hematopoietic potentialities in vitro and in SCID-hu mice. Nouv Rev Fr Hematol 35:91–93
73. Perou CM, Sorlie T, Eisen MB, van de Rijn M, Jeffrey SS, Rees CA, Pollack JR, Ross DT, Johnsen H, Akslen LA, Fluge O, Pergamenschikov A, Williams C, Zhu SX, Lonning PE, Borresen-Dale AL, Brown PO, Botstein D (2000) Molecular portraits of human breast tumours. Nature 406:747–752
74. Peters R, Leyvraz S, Perey L (1998) Apoptotic regulation in primitive hematopoietic precursors. Blood 92:2041–2052
75. Petersen BE, Bowen WC, Patrene KD, Mars WM, Sullivan AK, Murase N, Boggs SS, Greenberger JS, Goff JP (1999) Bone marrow as a potential source of hepatic oval cells. Science 284:1168–1170
76. Phillips TM, McBride WH, Pajonk F (2006) The response of CD24(−/low)/CD44+ breast cancer-initiating cells to radiation. J Natl Cancer Inst 98:1777–1785
77. Ponti D, Costa A, Zaffaroni N, Pratesi G, Petrangolini G, Coradini D, Pilotti S, Pierotti MA, Daidone MG (2005) Isolation and in vitro propagation of tumorigenic breast cancer cells with stem/progenitor cell properties. Cancer Res 65:5506–5511
78. Prince ME, Sivanandan R, Kaczorowski A, Wolf GT, Kaplan MJ, Dalerba P, Weissman IL, Clarke MF, Ailles LE (2007) Identification of a subpopulation of cells with cancer stem cell properties in head and neck squamous cell carcinoma. Proc Natl Acad Sci USA 104:973–978
79. Reya T, O'Riordan M, Okamura R, Devaney E, Willert K, Nusse R, Grosschedl R (2000) Wnt signaling regulates B lymphocyte proliferation through a LEF-1 dependent mechanism. Immunity 13:15–24
80. Reya T, Morrison SJ, Clarke MF, Weissman IL (2001) Stem cells, cancer, and cancer stem cells. Nature 414:105–111
81. Reya T, Duncan AW, Ailles L, Domen J, Scherer DC, Willert K, Hintz L, Nusse R, Weissman IL (2003) A role for Wnt signalling in self-renewal of haematopoietic stem cells. Nature 423:409–414
82. Roy SK, Srivastava RK, Shankar S (2010) Inhibition of PI3K/AKT and MAPK/ERK pathways causes activation of FOXO transcription factor, leading to cell cycle arrest and apoptosis in pancreatic cancer. J Mol Signal 5:10
83. Sgroi DC, Teng S, Robinson G, LeVangie R, Hudson JR Jr, Elkahloun AG (1999) In vivo gene expression profile analysis of human breast cancer progression. Cancer Res 59:5656–5661
84. Shackleton M, Vaillant F, Simpson KJ, Stingl J, Smyth GK, Asselin-Labat ML, Wu L, Lindeman GJ, Visvader JE (2006) Generation of a functional mammary gland from a single stem cell. Nature 439:84–88

85. Shankar S, Chen Q, Srivastava RK (2008) Inhibition of PI3K/AKT and MEK/ERK pathways act synergistically to enhance antiangiogenic effects of EGCG through activation of FOXO transcription factor. J Mol Signal 3:7
86. Shankar S, Nall D, Tang SN, Meeker D, Passarini J, Sharma J, Srivastava RK (2011) Resveratrol inhibits pancreatic cancer stem cell characteristics in human and KrasG12D transgenic mice by inhibiting pluripotency maintaining factors and epithelial-mesenchymal transition. PLoS One 6:e16530
87. Shipitsin M, Campbell LL, Argani P, Weremowicz S, Bloushtain-Qimron N, Yao J, Nikolskaya T, Serebryiskaya T, Beroukhim R, Hu M, Halushka MK, Sukumar S, Parker LM, Anderson KS, Harris LN, Garber JE, Richardson AL, Schnitt SJ, Nikolsky Y, Gelman RS, Polyak K (2007) Molecular definition of breast tumor heterogeneity. Cancer Cell 11:259–273
88. Singh SK, Clarke ID, Terasaki M, Bonn VE, Hawkins C, Squire J, Dirks PB (2003) Identification of a cancer stem cell in human brain tumors. Cancer Res 63:5821–5828
89. Sleeman JP, Cremers N (2007) New concepts in breast cancer metastasis: tumor initiating cells and the microenvironment. Clin Exp Metastasis 24:707–715
90. Sophos NA, Vasiliou V (2003) Aldehyde dehydrogenase gene superfamily: the 2002 update. Chem Biol Interact 143–144:5–22
91. Stingl J, Eaves CJ, Kuusk U, Emerman JT (1998) Phenotypic and functional characterization in vitro of a multipotent epithelial cell present in the normal adult human breast. Differentiation 63:201–213
92. Sutherland DR, Keating A (1992) The CD34 antigen: structure, biology, and potential clinical applications. J Hematother 1:115–129
93. Taieb N, Maresca M, Guo XJ, Garmy N, Fantini J, Yahi N (2009) The first extracellular domain of the tumour stem cell marker CD133 contains an antigenic ganglioside-binding motif. Cancer Lett 278:164–173
94. Taipale J, Beachy PA (2001) The hedgehog and Wnt signalling pathways in cancer. Nature 411:349–354
95. Tang SN, Singh C, Nall D, Meeker D, Shankar S, Srivastava RK (2010) The dietary bioflavonoid quercetin synergizes with epigallocathechin gallate (EGCG) to inhibit prostate cancer stem cell characteristics, invasion, migration and epithelial-mesenchymal transition. J Mol Signal 5:14
96. Terskikh AV, Easterday MC, Li L, Hood L, Kornblum HI, Geschwind DH, Weissman IL (2001) From hematopoiesis to neuropoiesis: evidence of overlapping genetic programs. Proc Natl Acad Sci USA 98:7934–7939
97. Tu SM, Lin SH, Logothetis CJ (2002) Stem-cell origin of metastasis and heterogeneity in solid tumours. Lancet Oncol 3:508–513
98. Van Den Berg DJ, Sharma AK, Bruno E, Hoffman R (1998) Role of members of the Wnt gene family in human hematopoiesis. Blood 92:3189–3202
99. van der Lugt NM, Domen J, Linders K, van Roon M, Robanus-Maandag E, te Riele H, van der Valk M, Deschamps J, Sofroniew M, van Lohuizen M et al (1994) Posterior transformation, neurological abnormalities, and severe hematopoietic defects in mice with a targeted deletion of the bmi-1 proto-oncogene. Genes Dev 8:757–769
100. Varnum-Finney B, Xu L, Brashem-Stein C, Nourigat C, Flowers D, Bakkour S, Pear WS, Bernstein ID (2000) Pluripotent, cytokine-dependent, hematopoietic stem cells are immortalized by constitutive Notch1 signaling. Nat Med 6:1278–1281
101. Vermeulen L, De Sousa EMF, van der Heijden M, Cameron K, de Jong JH, Borovski T, Tuynman JB, Todaro M, Merz C, Rodermond H, Sprick MR, Kemper K, Richel DJ, Stassi G, Medema JP (2010) Wnt activity defines colon cancer stem cells and is regulated by the microenvironment. Nat Cell Biol 12:468–476
102. Wang J, Sakariassen PO, Tsinkalovsky O, Immervoll H, Boe SO, Svendsen A, Prestegarden L, Rosland G, Thorsen F, Stuhr L, Molven A, Bjerkvig R, Enger PO (2008) CD133 negative glioma cells form tumors in nude rats and give rise to CD133 positive cells. Int J Cancer 122:761–768
103. Wechsler-Reya RJ, Scott MP (1999) Control of neuronal precursor proliferation in the cerebellum by Sonic hedgehog. Neuron 22:103–114

104. Willert K, Brown JD, Danenberg E, Duncan AW, Weissman IL, Reya T, Yates JR 3rd, Nusse R (2003) Wnt proteins are lipid-modified and can act as stem cell growth factors. Nature 423:448–452
105. Woodward WA, Chen MS, Behbod F, Alfaro MP, Buchholz TA, Rosen JM (2007) WNT/beta-catenin mediates radiation resistance of mouse mammary progenitor cells. Proc Natl Acad Sci USA 104:618–623
106. Wright MH, Calcagno AM, Salcido CD, Carlson MD, Ambudkar SV, Varticovski L (2008) Brca1 breast tumors contain distinct CD44+/CD24− and CD133+ cells with cancer stem cell characteristics. Breast Cancer Res 10:R10
107. Yang YM, Chang JW (2008) Current status and issues in cancer stem cell study. Cancer Invest 26:741–755
108. Yin AH, Miraglia S, Zanjani ED, Almeida-Porada G, Ogawa M, Leary AG, Olweus J, Kearney J, Buck DW (1997) AC133, a novel marker for human hematopoietic stem and progenitor cells. Blood 90:5002–5012
109. Zhang Y, Kalderon D (2001) Hedgehog acts as a somatic stem cell factor in the Drosophila ovary. Nature 410:599–604
110. Zhou S, Schuetz JD, Bunting KD, Colapietro AM, Sampath J, Morris JJ, Lagutina I, Grosveld GC, Osawa M, Nakauchi H, Sorrentino BP (2001) The ABC transporter Bcrp1/ABCG2 is expressed in a wide variety of stem cells and is a molecular determinant of the side-population phenotype. Nat Med 7:1028–1034

Chapter 2
The Perspectives of Stem Cell-Based Therapy in Neurological Diseases*

Wojciech Maksymowicz, Joanna Wojtkiewicz, Hanna Kozłowska, Aleksandra Habich, and Wlodek Lopaczynski

Contents

* Dr. Lopaczynski contributed to this article in his personal capacity. The views expressed are his own and do not necessarily represent the views of the National Institutes of Health or the United States Government.

W. Maksymowicz • J. Wojtkiewicz • A. Habich
Stem Cell Research Laboratory, Department of Neurology and Neurosurgery,
Faculty of Medical Sciences, University of Warmia and Mazury,
Warszawska 30 Streat, 10-081 Olsztyn, Poland
e-mail: maksymowicz@interia.pl; joanna.wojtkiewicz@uwm.edu.pl;
olahabich@gmail.com

H. Kozłowska
Stem Cell Research Laboratory, Department of Neurology and Neurosurgery,
Faculty of Medical Sciences, University of Warmia and Mazury,
Warszawska 30 Streat, 10-081, Olsztyn, Poland

Stem Cell Bioengineering Laboratory, Neurorepair Department,
Mossakowski Medical Research Centre, Polish Academy of Sciences,
Pawinskiego 5 Str, 02-106 Warsaw, Poland
e-mail: hkozlowska@cmdik.pan.pl

W. Lopaczynski (✉)
National Cancer Institute, National Institutes of Health, Bethesda,
MD 20917, USA
e-mail: wlopaczynski@hotmail.com

R.K. Srivastava and S. Shankar (eds.), *Stem Cells and Human Diseases*,
DOI 10.1007/978-94-007-2801-1_2, © Springer Science+Business Media B.V. 2012

Abstract The impairment of function of Central Nervous System (CNS) due to the loss of nervous cells is the crucial feature of so called neurological degenerative diseases, including: Parkinson disease (PD), Alzheimer disease (AD), Amyotrophic Lateral Sclerosis (ALS) and Huntington Disease (HD). The social importance of treatment of those two first pathologies is increasing contemporary as a result of the aging of population. Multiple Sclerosis (MS) is another devastating neurological disease in which not only myelin sheet but also neuronal degeneration occurs (Brain 132(Pt 5):1175–1189, 2009). Until now there is no effective treatment, although during last decades the diagnostic possibilities dramatically improved. It is understandable that new opportunities of the use of stem cell progenitors of neurons are the topics in the developing research. There are also perspectives for implementation of the stem cells transplantation in the treatment of loss of neurons due to the brain or spinal cord damage, as a result of the stroke and mechanical injury. In human, the early transplantation stem cells trials present a huge variety of outcomes ranging from significant clinical benefit to worsening of symptoms with severe side effects. As the pathophysiology differs in PD, ALS, MS and stroke, different cell sources for transplantation might be required for optimal clinical improvement. Elementary examination is compulsory before stem cell transplantation therapy can become a realistic clinical treatment. Recently, the overall goal for many laboratories in their research became to understand the function of human brain stem cells and how they may play a role in the origin of brain tumors. Understanding the relationship between the genesis of brain tumors and the potential interventions using stem cells are of greatest importance and has been also recently a topic for many publications.

Keywords Neurodegenerative diseases • Restorative neurology and stem cells • Spinal cord and traumatic brain injury • Stroke • Brain tumor

2.1 Introduction to Stem Cell-Based Therapy of Neurological Diseases

Along with the discovery of stem cells, especially neural stem cells and their potential to reconstruct damaged brain tissue, new strategies for clinical treatment have appeared. However, despite discovery of neurogenesis in adult brain, the use of brain derived stem cells is still very limited. The lack of treatment based on transplantation is caused by specific factors, including brain complexity and disease-specific phenotypes that cannot be easily simulated in non-human systems. Moreover, the number of available cells and inconvenience of the isolation procedures of neural stem cells inspired the scientists to look for different sources of stem cells capable to specific neural differentiation.

A diversity of cell types has been considered as a good source for neural cell therapy. It is required that cells for transplantation should be available in large quantities as a homogenous population with low-immunogenicity, genetic stability and appropriate differentiation potential. Also, it is important that those cells should be permissible to cryopreservation. Thus, the potential cell sources for CNS transplantation include not only adult or fetal neural stem cells (NSCs) but also mesenchymal stem cells (MSCs) [1].

2.1.1 Source of Stem Cells for Neurological Repair

2.1.1.1 NSCs

Neural stem cells (NSCs) are the self-renewing, multipotent cells that generate the main phenotypes constituting complex structure of nervous system. Initially, existence of NSCs was only identified with embryonic tissue at the beginning of neural tube formation. Only since the last decade of the twentieth century, it is wildly known that neural stem cells reside in the periventricular regions and cerebral cortex during development and persist into adulthood in a number of sites, including the dentate *gyrus* of the hippocampus, *substantia nigra* and olfactory bulb [2]. NSCs can be isolated and propagated also from spinal cord of rodents and humans. *In vitro*, clones of single NSCs can be cultivated either as monolayers or as suspended, spherical structures called "neurospheres" under strictly definite culture conditions in the presence of epidermal growth factor (EGF) or basic fibroblast growth factor (bFGF). Cells derived from neurospheres retain their differentiation capacities and after neuromorphogenes stimulation give rise to neurons, astrocytes and oligodendrocytes. Because of the invasive isolation method and small number of NSCs from adult brain, many scientists focused on cells obtained from fetal brain. Despite of origin they are less ethically controversial than ESCs because they are generally isolated from aborted material. Fetal NSCs also have the capacity to differentiate into neurons, astrocytes and oligodendrocytes *in vitro*. What is important in a field of regenerative medicine, many reports have shown that NSCs isolated from fetal brain tissues after transplantation can survived as undifferentiated cells in ectopic perivascular niches in the inflamed CNS. These cells interact with CNS-infiltrating immune cells and the local microenvironment facilitates survival and proliferation of progenitor cells. However, the prolonged cell population of fetal NSCs conducts to an ever increasing glial differentiation system at the expense of neuronal differentiation, which meaningfully reduces their therapeutic treatment potential [3].

2.1.1.2 hNT Neurons

Cells derived by either genetic transformation or cultured embryonic and adult tissue offer a ready and unlimited source of cells, thereby eliminating ethical concerns in obtaining aborted fetal tissue. For example, LBS-Neurons (Layton Bioscience Inc., Sunnyvale, CA, USA) were produced from a NT2/D1 human precursor cell line and induced to differentiate into neurons by the addition of retinoic acid. This cell line was originally derived from a human testicular tumor more than 25 years ago [4] and the final product gives a neuronal cell population virtually indistinguishable from terminally differentiated postmitotic neurons [5]. However, malignant transformation following therapeutic transplantation of this cell type is a key concern for this approach.

The hNT cells improve functional recovery when transplanted into the ischemia-damaged striatum of rats, in a dose-dependent manner. Behavioral improvement persisted for up to 6 months after transplantation as long as the cells survived. However, survival of transplanted cells is not always correlated with functional recovery because there is no effect of these cells when transplanted into the ischemic cortex despite robust cell survival. This finding highlights the need to determine the parameters required for cell-enhanced functional recovery [6].

2.1.1.3 MSCs

Mesenchymal stem cells (MSCs) are the heterogeneous subset of stromal progenitors of mesodermal cells that originally have been isolated from bone marrow. Friedenstain et al. described them as clonal, plastic adherent cells from bone marrow capable of differentiating *in vitro* into marrow stromal cells, osteoblastic cells, chondrocytes, fat cells and myocytes, which are normally originated from mesenchymal stem cells [7]. Furthermore, their differentiation potential is much broader than it was initially accepted. They are also capable of differentiating into epithelial cells [8] and ectodermal origin neurons [9, 10]. Recent studies have indicated that MSCs can be isolated also from almost every connective tissue not only from human bone marrow but also from human umbilical cord blood, Wharton jelly and adipose tissue. Although the number of MSCs in the bone marrow is small, about five cells per every 1×10^6 mononuclear cells, they can be relatively easily isolated and propagated in culture as the fibroblast-like morphology adherent cells on plastic which can form colonies *in vitro*. Moreover, they have a great proliferation capacity *ex vivo* without losing their stem cell features [11]. However, the capacity of cells to proliferate and differentiate differs between isolates and it depends on the origin of tissue, isolation procedure and culture conditions. These differences reflect the highly heterogeneous nature of MSC population. The differentiation potential and accessibility of MSCs make them convenient for research leading to the possible clinical utilization in regenerative cell therapy.

Human bone marrow cells (HBMC), human umbilical cord blood (HUCB) and peripheral blood progenitor cells (PBPC) or adipose mesenchymal progenitor cells tissue are alternative sources of stem cells; their use carries minimal ethical unease when transplanted in an autologous manner. These types of cells have been reported to enhance recovery after stroke with intracerebral or intravascular delivery [12, 13]. Both HBMCs and HUCB cells target the ischemic border, as with NSCs, this is thought to be mediated by injury-induced chemokines [14]. However, very few transplanted cells are found in the brain, even when delivered intracerebrally and of these were expressed only a small percentage of neural markers. Approximately, 10–20% of human bone marrow-derived stem cells (HBMCs) are multipotent, with the remaining representing more differentiated committed cells [11]. Nevertheless, HBM cells improved outcome in experimental models of stroke [15] and preserved cognitive function, which also been observed with intravenous transplantation of MSCs into rat models of Middle Cerebral Artery occlusion (MCAo) [16]. Similarly, behavioral and neurological improvement has been demonstrated with intravenous infusion of $CD133^+$ cells in stroke rats [17]; the stem cell marker CD133 is a transmembrane cell surface antigen, which is specifically expressed on undifferentiated stem cells. In addition, they secrete trophic factors that enhance endogenous mechanisms of brain repair. That functional recovery is found often with very few transplanted cells in the brain suggests that the cells may exert an acute but persistent effect on the brain before they die; intravenously administered cells may not even need to enter the brain to elicit an effect but rather act in the periphery to increase trophic factor expression in the brain [18]. The trophic effects of HBMCs can be augmented by engineering them to overexpress trophic factors [19]. An advantage of a hematopoietic source of cells is that they avoid the ethical issues and tissue limitation associated with embryonic and fetal tissue. HBMCs and peripheral blood stem cells also offer the potential of autologous transplants, negating the need for immunosuppression regimens.

After transplantation of bone marrow derived mesenchymal stem cells into the lateral ventricles of a mouse, cells migrate toward forebrain and cerebellum, and differentiated into neurons and astrocytes [20]. Interestingly also cells injected intravenously migrate and differentiate toward neuronal lineages [21]. Moreover other transplantation studies have demonstrated that human MSCs generate a local immunosuppressive microenvironment by secreting cytokines and interfering with dendritic and T-cell function. Thus, naturally immunosuppresive nature of MSCs that is substantially increased by inflammatory cues provides a strong reason for the use of MSCs in therapeutic trials for neurological diseases of inflammatory origination.

2.1.1.4 Induced Pluripotent Stem (iPS) Cells

In 2006 Yamanaka and Takahashi first reprogrammed adult somatic cells so that they acquire characteristics very similar to ESCs [22]. The first reprogramming of differentiated human somatic cells was done in 2007 and 2008 [23]. Induced pluripotent stem cells were initially created by fibroblasts reprogramming required retroviral transfer of four transcription factors (c-myc, oct4, klf2, and sox2). Subsequent research indicated that iPS cells could be generated by non-viral transfection systems using only one transcription factor [24]. These findings are predominant for future potential clinical application in transplantation therapies. Moreover, such new technique allows the application of induced pluripotent stem cells from patient's own tissue and their subsequent differentiation into cell class lost by sporadic injury or neurodegenerative diseases. However, because of strong similarity to ESCs, the risk of tumor formation cannot be eliminated yet. In addition, the key challenge is to deal with the purity of the cell population, specifically looking at phenotypes and function of neurons to be used for therapy. Although the iPS personalize therapy is very promising, it may not be clinically accepted within the near future.

2.2 Stem Cell-Based Approach to Neurodegenerative Diseases—Parkinson Disease (PD)

The loss of over 50% of dopaminergic neurons localized in *substantia nigra* of the mesencephalon is the cause of the clinical manifestation of Parkinson Disease. Pharmaceutical substitutional treatment is effective usually only in the early stage of disease and play the symptomatic not causative role. Also neurosurgical procedures (stereotactic lesions and deep brain stimulation) are not truly curative type of treatment. Fetal cells transplantation for PD has initiated application cell therapy for neurodegenerative diseases [25]. Transplantation of fetal dopamine neurons into the striatum of Parkinson's disease patients can provide restoration of the dopaminergic neurons and motor deficits. The limited efficacy and ethical doubts related to neural transplantation from the fetal brain opened the way of the research looking for the alternative therapies. Main attention has been focused on the use of stem cells, which could develop into dopaminergic neurons.

Embryonic stem cells (ESC) have great potential for developing specialized cells like dopaminergic neurons, which have been commonly used in animal studies. Proper cell differentiation was possible on two ways: generation of embryonic bodies followed by selection and expansion of nestin-positive neural precursors and then differentiation into neural subtypes requiring multiple stages procedure and the use of number of media and culture conditions; and co-culture method based on induction of differentiation activity culturing cells on the stromal-derived feeder layer. Besides the technical problems, there are limitations of the possible ESC implementation in the clinical practice. However, there is concern about the ability of these immature cells to develop the teratogenic cancerous tissue; although the introductory procedure of *in vitro* differentiation prior to the transplantation can minimize the risk. Secondly, there is a second limitation concern about the ethical issues.

Fetal tissue stem cells, mostly derived from the bone marrow, have the high proliferative potential but lower risk of the development of tumors than embryonic stem cells [26]. They can be expanded and differentiated into dopaminergic neurons *in vitro* before implantation. In contrast to embryonic stem cells, there is the possibility of use of cells removed from the naturally death fetus, if close cooperation with obstetric departments is well organized.

Also, the human umbilical cord blood is rich in ontogenically young stem cells and their capacity for the neural differentiation was proved in animal models [27]. This source of the transplantation material was also used in clinics for bone marrow replacement and tumorogenic activity was not observed. The human umbilical cord mesenchymal stem cells from Wharton`s jelly were induced to differentiate *in vitro* into a population that contained dopaminergic neurons [28]. In addition, the human NSCs are highly expansive *ex vivo* and the protocols have been developed to enrich dopaminoergic (DA) neuron progenitor. The number of DA neurons produced is at most of 15% of the total cell population [29]. This approach appears to be not favorable for the development of PD therapies instead the future approaches aim towards the use of embryonic stem cell or even induced pluripotent stem cell (iPS)—derived dopamine neurons. Another source for cell transplantation are mesenchymal stem cells derived from bone marrow and adipose tissue, which can be converted into neural cell progenitors. The advantage of this kind of transplantation is the possibility of the use of expanded autogenic cells.

To date, only single reports on stem cell transplantation for the treatment of patients with Parkinson Disease have been published. Brazzini and coworkers described autologous implantation of stem cells into posterior region of the circle of Willis, with superselective arterial catheterization in 50 patients [21]. In mean follow-up of 7.4 months, patients showed the improvement. Another publication [30] presented prospective, uncontrolled, pilot study of single-dose, unilateral transplantation of autologous bone-marrow-derived mesenchymal stem cells. The cells were transplanted into the sublateral ventricular zone by stereotactic surgery. Three of seven patients showed a steady improvement in their "off/on" Unified Parkinson's Disease Rating Scale during the time of observation ranged from 10–36 months. Investigations did not show any complications; however, the time of observation and the type of study in this group of patients do not provide a proof for permit to conclude about the true efficacy of the method. The possible stem cell-based therapy can be performed by implanting stem cells modified to release growth factors for protection of neurons derived from stem cells. Another way could be the transplantation of stem cell-derived DA neuron precursors into the putamen for generation of new neurons, what would ameliorate disease-induced motor impairments (Fig. 2.1).

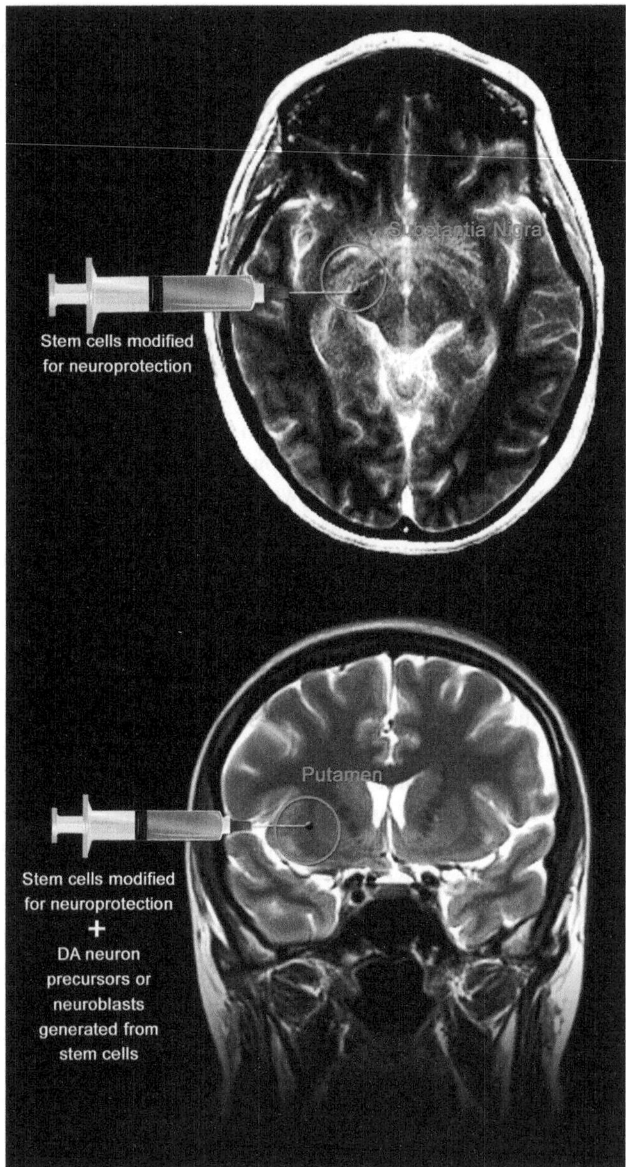

Fig. 2.1 The possible stem cell-based therapy of Parkinson Disease by implanting stem cells modified to release growth factors for protection of neurons derived from stem cells or by transplantation stem cell-derived DA neuron precursors into the putamen for generation of new neurons

Nonetheless, the time of translation of animal study into humans has come and the NIH just registered new trials:

- "To Study the Safety and Efficacy of Bone Marrow Derived Mesenchymal Stem Cells Transplant in Parkinson Disease" by Jaslok Hospital and Research Centre from India (Clinical trials. gov identifier NCT00976430)
- "Adult Neuronal Progenitor Stem Cell Project" by Rajavithi Hospital from Bangkok (Thailand) (Clinical trials. gov identifier NCT00927108)

2.3 Stem Cell-Based Approach to Neurodegenerative Diseases—Alzheimer Disease (AD)

Alzheimer Disease (AD) is progressive disorder of memory, cognitive functions and finally occurring dementia, caused by neuronal loss in amygdale, hippocampus, association neocortex and entorhinal cortex with basal forebrain cholinergic system. Pathological changes include abnormal intracellular protein aggregates (neurofibrillary tangles), composed of hyperphosphorylated forms of microtubule-associated protein tau, and extracellulary deposited protein, amyloid-β (Aβ), which is derived from amyloid precursor protein (APP). There is evidence that altered APP processing can lead to increased neurogenesis, but this change is not sufficient because of increasing neuronal dysfunction during the chronic disease. Amyloid deposits influence on the microgliosis and the inflammatory mechanism can be responsible for the impaired hippocampal neurogenesis [31].

There is no effective treatment significantly reversing neuro-degeneration and cognitive decline in Alzheimer's disease. It is stimulating the searching for new possible therapeutic methods including stem cells transplantation. However, in animal models the hypothesis that implanted stem cells could simply replace degenerated neurons has not been proved. For example, in mice model only small amount of transplanted bone marrow cells could be engrafted into affected tissue and they quickly disappeared. While some reports suggest the trans-differentiation of marrow-derived mesenchymal stem cells (BM-MSCs) into cells of neural lineages in cell replacement therapies, this is seen at low frequency *in vivo*, and it is, in fact, unlikely to be the predominant beneficial effect of BM-MSCs transplantation [32]. During this same experimental study evidence has emerged that BM-MSCs transplantation can reduce Aβ plaque formation, stimulating microglial activation that activated microglia leading to decreased Aβ deposits in the induced AD mouse model. In fact, BM-MSCs transplantation is able to reduce tau hyperphosphorylation. BM-MSCs treatment ameliorates spatial learning and memory impairments. Similar effect has been observed using human umbilical cord blood-derived mesenchymal stem cells (hUCB-MSCs). Keene and coworkers results provide a foundation for an adult stem cell-based therapy to suppress soluble Aβ-peptide and plaque accumulation in the cerebrum of patients with AD [33]. Also, the research performed by Ryu and coworkers demonstrated neuroprotective efficacy for NPCs in an animal model of AD [34]. There was evidenced that AD disturbs hippocampal neurogenesis and it is probably the pathomechanism of cognitive deficit of AD patients. Normalization of the formation and maturation of new hippocampal neurons by β-amyloid immunotherapy could have therapeutic potential. Stem cell-based therapy could also be used to prevent progression of the disease by transplanting modified stem cells to release growth factors. It could be supported by the infusion of compounds and/or antibodies to restore impaired hippocampal neurogenesis [35]. To date no clinical study on stem cell based therapy of AD patients has been published.

2.4 Stem Cell-Based Approach to Neurodegenerative Diseases—Amyotrophic Lateral Sclerosis (ALS)

Amyotrophic lateral sclerosis (ALS) is a neurodegenerative disease with still not fully discovered background. One strategy in treating ALS is to use stem cells in replacing lost spinal motor neurons and restore their functions [36]. Many research groups were successful in development of motor neurons from variety of stem cells, like mouse and human ESCs or NSCs, by initial induction of neuroepithelial differentiation followed by final motor neurons specification [36]. Those processes required specific signals and culture conditions including FGF2 stimulation, co-culture feeder cells presence and treatment with combination of neuromorphogenes. Finally, the genetic modifications were also used to increase the efficiency of spinal motor neuron generation [37]. Although those reports have shown the possibility to generate the motor neurons *in vitro* from variety of stem cells, it is still not sufficient for experimental treatment [38]. New neurons should integrate and cooperate with existing neural circuitries; thus, a more realistic strategy is based on neuroprotective capacity of transplanted stem cells on endogenous motor neurons through releasing of trophic factors that directly promote survival or decrease inflammation [39]. Since 2000, various types of stem cells therapies have been tested on transgenic ALS animals, including rodent bone marrow and mesenchymal stem cells, mouse neural stem cells, mouse olfactory bulb neural progenitor cells and fetal human neural stem and progenitor cells, human mesenchymal stem cells and human umbilical cord blood stem cells [40, 41].

The therapeutic potential of fetal human NSCs has been evaluated after injection into the lumbar region of presymptomatic ALS rats. This treatment, combined with FK-506 immunosuppression therapy, prolonged the lifespan and delayed the disease, probably by protection of endogenous motor neurons through the secretion of GDNF and BDNF. Interestingly, transplanted NSC did not differentiate toward motor neurons. The secretion of neurotrophic factors and neuroprotection capacity of grafted human NSCs probably underline the beneficial outcome in transplantation experiments. Comparable observation was also made for different stem cells types [42].

Human MSCs injected intravenously into irradiated mice not only delayed the disease onset and extended lifespan, but also prolonged motor performance and protected endogenous motor neurons [43]. What is interesting, the implanted cells

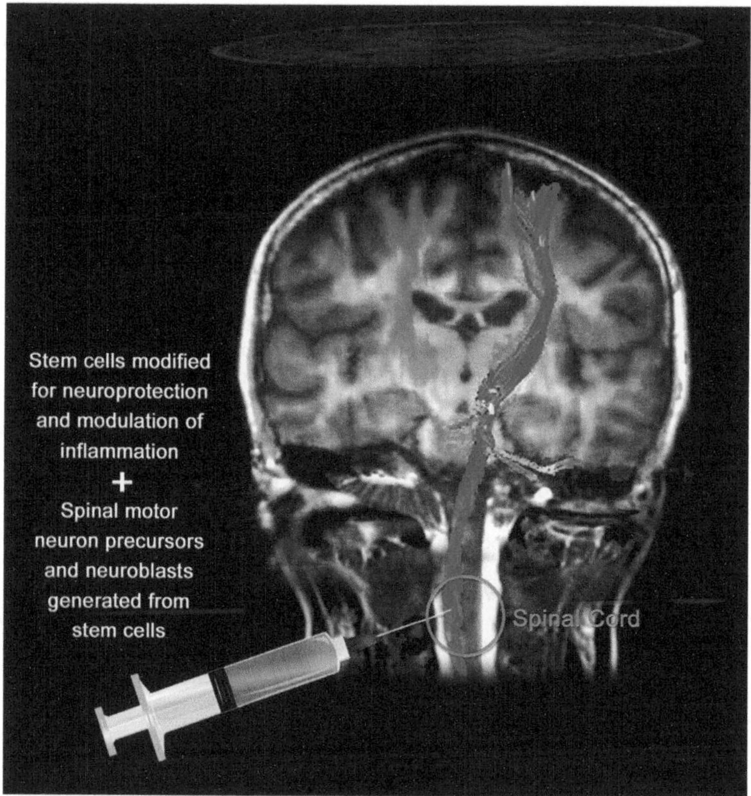

Fig. 2.2 The possible stem cell-based therapy of ALS

migrated to the brain and spinal cord, but only a few of them had characteristic neuronal and astroglial markers. Surprisingly, after transplantation of hMSCs intrathecally into the cisterna magna to presymptomatic ALS mice that kind of efficient effect was not observed. Although stem cells were delivered directly to the CNS, the migration to pathological areas and cell survival seemed to be less efficient compared to stem cells placed into the circulatory system [41]. These data could support the hypothesis that hMSCs transplantation also affect not through replacing motor neurons, but through neuroprotection and possibly direct neurotrophic support. Similar findings were observed in case of HUCBCs transplantation to presymptomatic irradiated ALS mice. Moreover, the effect was not only dependent from injection method but also dose dependent. The number of transplanted cells directly affected on the rate of the disease progression and lifespan extension of the animals [40]. Thus, probably the major beneficial effect on stem cells transplantation is the neuroprotection through an alteration of the ALS environment. Overall, using stem cells for delivery of trophic factors and neuroprotection to prevent disease progression seems a more achievable clinical goal than neuronal replacement (Fig. 2.2).

Table 2.1 Ongoing clinical trials in ALS from http://clinicaltrials.gov

Cell type	Title of clinical trial	NCTI ID
Human spinal cord derived NSC	A Phase l, Open-label, First in Human, Feasibility and Safety Study of Human Spinal Cord Derived Neural Stem Cell Transplantation for the Treatment of Amyotrophic Lateral Sclerosis	NCT01348451
BMSC	Phase I/II Clinical Trial on The Use of Autologous Bone Marrow Stem Cells in Amyotrophic Lateral Sclerosis (Extension CMN/ELA)	NCT01254539
Peripheral blood stem cell	A Pilot Study of High-Dose Immunosuppressive Therapy Using Carmustine, Etoposide, Cytarabine, and Melphalan (BEAM) + Thymoglobulin Followed by Syngeneic or Autologous Hematopoietic Cell Transplantation for Patients With Autoimmune Neurologic Diseases	NCT00716066
MSC	A Single Patient Treatment Protocol for Autologous Mesenchymal Stem Cell Intraspinal Therapy in Amyotrophic Lateral Sclerosis (ALS)	NCT01142856
BMSC	Phase I, Single Center, Prospective, Non-randomized, Open Label, Safety/Efficacy Study of the Infusion of Autologous Bone Marrow-derived Stem Cells, in Patients With Amyotrophic Lateral Sclerosis	NCT01082653
BMSC	An Open-label, Phase I/II Trial for Safety and Efficacy Study of Autologous Bone Marrow Derived Stem Cell Treatment in Amyotrophic Lateral Sclerosis	NCT01363401

Several clinical trials with stem cells transplantation have also been reported lately but no clinical benefits were observed. Although autologous peripheral blood cells injected intrathecally or mesenchymal stem cells injected intraspinally were well tolerated showing a few or no adverse effects [44]. Ongoing clinical trials in ALS patients are summarized in Table 2.1.

2.5 Stem Cell-Based Approach to Multiple Sclerosis (MS)

Multiple sclerosis (MS) is caused by the distruction of myelin in CNS. An important area of research is focused on finding the ways to identify the factors that lead to a failure of cells to produce myelin and enhance remyelination. Thus, transplantation of stem cells differentiating toward myelin producing offspring appears to be a promising therapeutic approach. The oligodendrocyte progenitor cells (OPCs) derived from ESCs and adult stem cells have been shown to myelinate dysmyelinated mouse brain and spinal cord after transplantation [45, 46]. However, there are major concerns about the inflammatory environment and multiple location of demyelinated MS sides across the brain [45].

In late 1970s, Blakemore in his studies presented that injection of exogenous myelinating cells in the rodent brain with demyelinated lesion could result in moderate remyelination [47]. However, further transplantation studies indicated that benefit effect of stem cells treatment is not connected mainly with their oligodendrocyte differentiation capacity but with the potential to induce local immunomodulation and neuroprotection [48]. Human bone marrow cells injected into dentate *gyrus* of *hippocampus* of immunodeficient mice substantially increased proliferation, migration and differentiation of endogenous neural stem cells [49]. Furthermore, intravenous injection of MSCs into mouse model of MS caused suppression of CNS inflammation through induction of T-cell anergy, stimulation brain parenchymal cells to express NGF and decrease of demyelination [50]. In addition, transplantation of MSCs also reduced gliotic scar formation, one of the major problems with spontaneous repair [49]. The effects induced by the human MSCs remain uncertain but probably could be basically explained by their secretion of chemokines, antioxidants, and growth factors such as NGF, BDNF and GDNF [51]. A stem cells therapy appears to be promising in treatment of MS because of their immunoprotective and immunomodulative properties. In 2011 Connick with his coworkers published the trial protocol for mesenchymal stem cells transplantation in multiple sclerosis (MSCIMS). In this study mesenchymal stem cells were successfully isolated, expanded and characterised *in vitro*. The goal of this trial was to evaluate the safety and feasibility of this method [52]. Other open stem cells clinical trials are summarized in Table 2.2.

2.6 Stem Cell-Based Approach to Stroke

Stroke is the third leading cause of death in industrialized countries and the most frequent cause of permanent disability in adults worldwide [53]. Previously known medically as a cerebrovascular accident (CVA), it is the rapidly developing loss of brain functions due to disturbance in blood supply to the brain. This can be due to ischemia (lack of blood flow) caused by blockage (thrombosis, arterial embolism), or a hemorrhage. Acute ischemic stroke accounts for about 85% of all cases while hemorrhagic stroke is responsible for almost 15% of all strokes [54]. The most common cause of stroke is the sudden occlusion of a blood vessel resulting in an almost immediate loss of oxygen and glucose to the cerebral tissue [55]. The deprivation of oxygen and glucose leads to several events including excitotoxity, oxidative stress and inflammation. It causes irreversible neuronal injury as well as oligodendrocytes, astrocytes and endothelial cells damage [35]. Moreover, stroke may also affect both white and grey matter and disrupt various anatomical pathways that need to be restored [56]. Widespread tissue damage throughout the central nervous system (CNS) has been shown to cause marked and multifarious functional impairments in the ischemic brain. Deficits can include partial paralysis, difficulties with memory, thinking, language, and movements. Functional recovery depends upon neuroplasticity enhanced by intensive rehabilitation [57] as well as the still-poorly-understood process of neurogenesis [58].

Table 2.2 Ongoing clinical trials in MS from http://clinicaltrials.gov

Cell type	Title of clinical trial	NCTI ID
UCMSC	Phase I/II Study of Umbilical Cord Mesenchymal Stem Cell Therapy for Patients With Progressive Multiple Sclerosis and Neuromyelitis Optica	NCT01364246
CD34 selected hematopoietic stem cell	Targeting Multiple Sclerosis as an Autoimmune Disease With Intensive Immunoablative Therapy and Immunological Reconstitution: A Potential Curative Therapy for Patients With a Predicted Poor Prognosis	NCT01099930
MSC	Multicenter Clinical Trial Phase I/II Randomized, Placebo-controlled Study to Evaluate Safety and Feasibility of Therapy With Two Different Doses of Autologous Mesenchymal Stem Cells in Patients With Secondary Progressive Multiple Sclerosis Who do Not Respond to Treatment	NCT01056471
HSC	Hematopoietic Stem Cell Therapy for Patients With Inflammatory Multiple Sclerosis Failing Interferon Therapy: A Phase II Multi-Center Trial	NCT00278655
MSC	A Phase I Study to Assess the Feasibility, Safety, and Tolerability of Autologous Mesenchymal Stem Cell Transplantation in Patients With Relapsing Forms of Multiple Sclerosis	NCT00813969
HSC	A Phase II Study of High-Dose Immunosuppressive Therapy (HDIT) Using Carmustine, Etoposide, Cytarabine, and Melphalan (BEAM) and Thymoglobulin, and Autologous CD34+ Hematopoietic Stem Cell Transplant (HCT) for the Treatment of Poor Prognosis Multiple Sclerosis	NCT00288626
HSC	Hematopoietic Stem Cell Therapy for Patients With Inflammatory Multiple Sclerosis Failing Interferon: A Randomized Study	NCT00273364
HSC	Autologous Hematopoietic Stem Cell Transplant in Patients With Neuromyelitis Optica	NCT01339455

2.6.1 General Issues in Developing Stem Cells Transplantation Therapy for Stroke

The survival of the stem cells grafted after cerebral ischemia for the most part has been reported to be minimal. Kim and co-workers reported a variable graft survival without detailed quantification of 3 weeks after human neural progenitor cells transplantation into basal ganglia after MCAO in rats [59]. Also, some studies show that nearly 40% of neural stem/progenitor cells transplanted in the striatum or cortex in MCAO models survive [6]. However, Hicks and colleagues showed that only about 1% of the transplanted cells survived 2 months after the transplantation, and that half of the transplanted animals had surviving human cells. The trypsination of hNPC neurospheres used these investigators may have impaired the capacity of the cells to integrate into the tissue, even though they continued to grow normally after replating *in vitro* [13]. As the cerebral environment after ischemia might not support stem cell survival (e.g. formation of glial scar tissue) [59], the poor survival of the grafts is a significant limitation in experimental stroke studies, and more attention should be given to studies trying to improve this issue. Transplanted cells may reduce inflammation [78], enhance endogenous neurogenesis and angiogenesis, stimulate release of neurotrophic factors and increase brain plasticity [13].

2.6.1.1 Potential Mechanisms of Transplanted Cell–Mediated Recovery

Motor, sensory and cognitive dysfunctions are common results of stroke, and cause considerable disability and social distress. Restoration of movement and motor function forms the focus of rehabilitation based on physical therapy. Recovery of motor and cognitive function occurs to a variable degree through a number of pathways [60]:

- Unmasking—recruitment of existing but latent connections
- Sprouting—development of new neural connections (including synaptogenesis)
- Long-term potentiation—enhancement of memory and learning
- Resolution of diaschisis (remote functional depression)
- Neurogenesis—replacing lost neurones

Additionally, animal models of recovery of function after stroke reveal that repetitive practice or exercise can evoke endogenous neurogenesis and the expression of signaling molecules, i.e., brain-derived neurotrophic factor—BDNF, vascular endothelial growth factor—VEGF, and glial cell-derived neurotrophic factor, which then promote neuronal repair, enhancing learning and memory [25].

There is increasing interest in the use of interventions that might augment these normal restorative events after stroke, either pharmacologically with drugs such as amphetamine [61] or, more recently, through the use of stem cells. Advances in molecular biology are creating opportunities to use cellular therapies to enhance neuronal regeneration in adjunct to a neuro-rehabilitation program. However, the mechanism by which stem cells may improve recovery is still poorly understood. There are two theories:

- Neuroprotection—preventing damaged neurons undergoing cell death in the acute phase of cerebral ischaemia
- Neurorepair—the repair of broken neuronal networks in the chronic phase of cerebral ischaemia.

When considering how stem cells promote recovery, there is probably some overlap between these two groups. Understanding how transplanted cells affect the brain, and vice versa, in model systems is important before proceeding to the clinical trials. Various mechanisms may be responsible for the transplanted cell–mediated effect. Thus, the connection of stimulating endogenous neurogenesis with cytokines in stem cell therapy will likely have a great impact for future therapy of patients with stroke.

2.6.1.2 Integration into the Host Circuitry

The attraction of using neural progenitor cells is their potential to replace lost circuitry and thus to have prolonged beneficial effects. However, evidence for this is rather limited [62, 63]. Expression of synaptic proteins by transplanted cells has been reported when rat NPCs were used in a model of global ischemia [63] and with hNT neurons in a model of traumatic brain injury (not reported in the ischemic brain), [64] indicating the potential of these cells to form synaptic contacts. Electron microscopy studies revealed that human NPCs formed synapses with host circuits in the ischemic brain [62]. Specific neuronal subpopulations may be required for functional integration. For example, striatal neuronal progenitors (e.g. from the lateral ganglionic eminence) may be necessary to repair damaged striatal circuitry; this has yet to be explored in the ischemic brain [12]. The small numbers of transplanted cells surviving in the post-stroke brain reported in some studies may pose a major problem if integration into the host circuitry is necessary for improved neurological outcome.

2.6.1.3 Transplanted Cells Reduce Death of Host Cells

Acute delivery of cells often reduces lesion size and inhibits apoptosis in the ischemic *penumbra*, suggesting that enhanced recovery results from neuroprotection. A myriad of cell types elicit this effect. Common to all is the secretion of trophic factors, such as vascular endothelial growth factor, fibroblast growth factor, glial cell-derived neurotrophic factor, and brain-derived neurotrophic factor, which are likely to contribute to this neuroprotective mechanism [18, 19].

2.6.1.4 Induction of Host Brain Plasticity

An increase in endogenous brain plasticity and motor remapping after ischemia is postulated to underlie the spontaneous recovery seen after a stroke [65]. Such plasticity events include an increase in afferent and efferent connections between the site of injury and both adjacent and contralateral brain regions, restoration of local synaptic activity by synaptogenesis, and probably strengthening of existing synapses as well as activation of silent synapses. Cell transplantation may enhance these endogenous repair mechanisms. Human cord blood cells in the ischemic cortex increased sprouting of nerve fibers from the contralateral to the ischemic hemisphere [66]. Shen et al. [67] reported increased synaptophysin expression in the penumbra after intravenous delivery of human bone marrow stromal cells.

2.6.1.5 Increased Neovascularization

Increased vascularization in the penumbra within a few days after stroke is associated with neurological recovery and offers another potential target for cell therapy [68].

Transplanted cell–induced blood vessel formation has been reported with bone marrow stromal cells [67], neural stem cells [69], and cells from human cord blood and peripheral blood [70]. Direct incorporation of the transplanted cells into the new blood vessels has been observed in some cases. An indirect cell-induced effect on blood vessel formation is also likely; human bone marrow stromal cells promoted angiogenesis in the ischemic border by increasing endogenous levels of the angiogenic factor vascular endothelial growth factor [67]. Transplanted cells have been reported to increase endogenous levels of other factors (BDNF, SDF-1, and FGF) that could induce proliferation of existing vascular endothelial cells (angiogenesis) and mobilization with homing of endogenous endothelial progenitors (vasculogenesis).

2.6.1.6 Attenuation of Inflammation

An intriguing potential repair mechanism is the ability of transplanted cells to attenuate the stroke-induced inflammatory/immune response. Intravenous injection of HUCBCs reduced leukocyte infiltration into the brain [54], although it is not clear whether this is a direct effect on the inflammatory response or a secondary effect attributable to reduction in infarct size. It seems paradoxical that a xenotransplant would inhibit the immune response. However, there is evidence in the literature that stem cells can directly inhibit T-cell activation [54].

2.6.1.7 Recruitment of Endogenous Progenitors

Endogenous neurogenesis is increased after stroke. This function has not been determined yet, but may signify a natural repair mechanism of brain that could potentially be further enhanced by transplanted cells, which was evidenced with cord blood cells and bone marrow cells [68]. In addition to local effects on the damaged tissue, transplanted cells could potentially recruit different progenitor cell types from other tissues. As mentioned above, they could mobilize endogenous endothelial progenitors into circulation to enhance vascularization. The issue whether the cell transplantation enhances endogenous hematopoietic or mesenchymal cell mobilization after ischemia has not been determined yet. With all the aforementioned mechanisms, whether the phenomenon measured is a cause or a secondary effect of cell-enhanced functional recovery needs to be determined.

2.6.2 *Animal Models of Stem Cells-Based Therapy of Stroke*

The clinical translation of stem cells-based therapies requires preclinical experimentation to assess their feasibility, safety, and efficacy. Depending on the stage of therapy development, different animal species are tested before clinical translation.

Most studies use rodent models of disease to assess stem cells properties. Nonhuman primates are the next essential platform for assessing first-in-class and invasive therapies for neurological diseases, because their brain size and complexity allow for the evaluation of surgical targets and cell distribution [71]. Animal models demonstrate that ischemic stroke is associated with differentiation of cells into neurons phenotypically similar to those lost in the ischemic lesion [3], findings suggesting that adult brain has capacity for self repair. As observed in rodents, subventricular and hippocampal cells from adult human brain can be expanded *in vitro*, differentiate into all three neural cell lineages (neuronal, astrocytic and oligodendroglial) and can improve functional recovery when administered intravenously into a rat model of stroke [14].

2.6.2.1 Classification of Stroke Models According to Different Etiologies

The procedures undertaken in animal models of stroke, including non-human primates, intend to provoke pathophysiological states that are similar to those of human stroke to study basic processes or potential therapeutic interventions in this disease. The aim is to extend the knowledge on and/or improvement of medical treatment of human stroke.

The term stroke subsumes cerebrovascular disorders of different etiologies, featuring diverse pathophysiological processes. Thus, for each stroke etiology one or more animal models have been developed:

- Animal models of ischemic stroke
- Animal models of intracerebral hemorrhage
- Animal models of subarachnoid hemorrhage and cerebral vasospasm
- Animal models of sinus vein thrombosis

Models of Ischemic Stroke

Animal models of ischemic stroke are based on the protocols to induce cerebral ischemia with the goal to study basic processes or potential therapeutic interventions and extend the pathophysiological knowledge on and/or improvement of medical treatment of human ischemic stroke (Table 2.3). Ischemic stroke has a complex pathophysiology involving the interplay of many different cells and tissues such as neurons, glia, endothelium, and the immune system. These events cannot be mimicked satisfactorily in vitro yet. Thus a large portion of stroke research is conducted on animals. Several models in different species are currently known to produce cerebral ischemia. Global ischemia models, both complete and incomplete, tend to be easier to perform; however, they are less immediately relevant to human stroke than the focal stroke models, because global ischemia is not a common feature of human stroke. Also, in various settings global ischemia is relevant, e.g., in global anoxic brain damage due to cardiac arrest. Different species also vary in their susceptibility

Table 2.3 Some of the mechanisms of ischemia

Ischemia			
Complete global	Incomplete global	Focal cerebral	Multifocal cerebral
Decapitation	Hemorrhage or hypotension	Middle cerebral artery occlusion	Blood clot embolization
Aorta/vena cava occlusion	Hypoxic ischemia	Endothelin-1-induced constriction of arteries and veins	Microsphere embolization
External neck tourniquet or cuff	Intracranial hypertension and common carotid artery occlusion	Spontaneous brain infarction (in spontaneously hypertensive rats)	Photothrombosis
Cardiac arrest	Two-vessel occlusion and hypotension	Macrosphere embolization	
	Four-vessel occlusion		
	Unilateral common carotid artery occlusion (in some species only)		

to the various types of ischemic insults, e.g. gerbils. They do not have a Circle of Willis and stroke can be induced by common carotid artery occlusion alone.

Focal Cerebral Ischemia

Models of focal cerebral ischemia are divided into techniques including reperfusion of the ischemic tissue (transient focal cerebral ischemia) and those without reperfusion (permanent focal cerebral ischemia). Focal cortical lesion can be triggered by distal ligation of the Middle Cerebral Artery (MCA) after temporal craniotomy [72]. However, on account of the rather complicated surgery, this experimental model is often eschewed in favor of the induction of cortical lesions photo-chemically [73] or by using endothelin-1 [74]. While these models are technically much easier to handle, they produce a kind of stroke mainly limited to the cortex. However, the cortical stroke is rather rarely observed in clinical practice on account of the wide network of vessels on the brain surface. Additionally, the clinical and behavioral symptoms of the cortical stroke are not sufficient to justify its implementation as a model for experimental treatment. In turn, while proximal endovascular occlusion of the MCA (MCAO model) is most frequently employed by researchers as a clinically relevant model of stroke, it causes huge brain damage within the cortical, sub-cortical and brain stem areas, followed by severe symptoms mainly related to space-occupying brain edema [75]. To obtain an ischemia limited to deep-brain structures in rodents, the usage of chemical substance is necessary. Intraparenchymal application of the Na/K ATPase pump inhibitor, ouabain, produces a lacunar deep-brain, stroke-like lesion, which is detectable behaviorally, notwithstanding the relatively limited brain damage [76]. The effect of ouabain administration is based on the crucial role of energy-dependent ionic pump for the functioning of excitable cells via membrane depolarization. Thus, a chemical blockade of the pump brings about metabolic and structural brain changes that mimic ischemic injury. Also, when Veldhuis compared MR images of focal ischemia and ouabain-induced brain lesion, he found the same pattern and evolution of brain changes [77]. Ouabain-induced lesion was then successfully used in neuroprotection studies [78] (Fig. 2.4).

2.6.2.2 Systematic Identification of Transplant Parameters for Stroke

Despite many animal studies showing that cell transplantation can improve recovery from stroke, the variables responsible for the success of these therapies are largely unknown. Researchers have used different cell types transplanted at different times after stroke and in different locations and they have used different behavioral tests to assess the efficacy of the transplant. However, the optimal conditions for cell transplant therapy after stroke are not known.

Timing of Transplantation

The optimal time to transplantation after a stroke is also unknown. The brain environment dramatically changes over time after ischemia. In the acute phase, there is an increase in excitatory amino acid release, peri-infarct depolarization, and reactive oxygen species release. This is followed by an inflammatory/immune response and cell death, which, in the penumbra, can last up to several weeks. Brain repair and plasticity after the acute phase take place over several weeks to months. The optimal timing of delivery will depend on the used cell type and their mechanism of action. If a treatment strategy focuses on neuroprotective mechanisms, acute delivery of the cells will be critical. If the cells act to enhance endogenous repair mechanisms (e.g., plasticity, angiogenesis, and neurogenesis) or require these events in order to survive and integrate, then early delivery would be pertinent because these events are most prevalent in the first 2–3 weeks after ischemia [65]. If cell survival is important, then transplanting late, after inflammation has subsided, could be beneficial. However, a systematic analysis of transplantation timing and its effect on functional recovery has not been established. The literature reports a wide range of stroke to transplantation intervals. Many studies demonstrating functional recovery report transplantation within the first 3 days after ischemia. However, cell-enhanced recovery has been reported with chronic delivery of cells even at 1 month after ischemia [79]. The optimal approach in the clinical setting has yet to be defined.

Lesion Localization and Size

Lesion location and size are important factors in determining patient suitability for cell therapy. The majority of preclinical studies show cell-enhanced recovery after striatal lesions [12, 13], although cell-induced improvements with cortical lesions are also reported [70]. However, not all studies find that cell therapy is effective [80]. Two research groups report that neural progenitor cells (NPCs) improve recovery, but only if combined with enriched housing [81], and very little effect of hNT cells in cortical stroke was found [6] despite multiple studies showing efficacy of the same cells with striatal stroke [79]. To date, most experimental studies showing cell-enhanced recovery used a stroke model that damages the striatum (with some damage to the cortex), and the cells are often delivered into the striatum. Similarly, Makinen et al. [82] found no behavioral improvement after transplantation of human umbilical cord blood stem cells while other studies using similar cells, stroke model, and timing of transplantation reported recovery [18, 78]. However, more discussion of inclusion/exclusion criteria is required in the field [83]. The recovery from cortical damage may be more complex than from striatal damage; it may simply be that the infarct associated with cortical stroke model is too large and too many essential connections are severed to make repair feasible. Precise anatomic location of the lesion and its functional implication, as well as lesion size, will be critical determinants to define the target population for the potential clinical trials.

Route and Site of Cell Delivery

Functional recovery has been reported with intracerebral, intravascular, and intracerebroventricular delivery of cells, but the best route of injection is not established yet. Intracerebral delivery results in more transplanted cells targeting the lesion in the brain compared to other delivery routes [4, 12, 13]. However, that intravascular delivery may be more appropriate for larger lesions as it could lead to wider distribution of cells around the ischemic area. Many studies using systemic delivered cells find significant functional recovery with very few [13] or sometimes no cells [18] entering the brain. Even with intracerebral delivery, proximity of the graft to the lesion may not be important: Modo et al. [84] found equal functional recovery when cells were grafted in the ipsi- or contralesional hemispheres, and the recent work from our laboratory revealed that the hNPCs exerted their major effect 1 week before they migrated to the lesion. Thus, the need for transplanted cells to be near the lesion, or even to be in brain, requires further investigation. In addition, each route of delivery has safety issues; intravascular delivery is less invasive than injection into the brain but raises concerns of cells sticking together creating microemboli, and cells homing to other organs. Intraarterial (intracarotid) administration is preferable for intravenous infusion, allowing first pass delivery and better targeting of cells to the brain [85] and fewer cells found in other organs [13]. However, Bang et al. [86] reported no adverse effects of intravenous infusion of MSCs in their clinical trial; however, this observation was made on only five patients. Intraparenchymal transplantation avoids this biodistribution issue, but is more invasive and often results in a physical mass of cells, which itself could disrupt the healthy tissue. The Phase I and II clinical trials with hNT cells transplanted into several striatal sites surrounding the lesion reported some adverse events in 4 out of 30 patients including a seizure and subdural hemorrhage; however, there is concern about whether the transplantation surgery contributed to these events [87, 88]. Progenitor or neural stem cells grafted into parenchyma (intracerebrally) surrounding the lesion or delivered intravenously survive, differentiate, and can enhance functional recovery [14]. Transplantation into the lesion cavity has been investigated by other research groups [89]. The cavity presents a hostile inflammatory environment that lacks trophic support. As proximity to blood vessels and the extent of inflammation can influence graft survival [90]. It is probable that cells will need to be encapsulated or delivered within a scaffold to facilitate their survival in the cavity. This strategy raises issues such as biocompatibility of the matrix material with the patient and the transplanted cells. Intracisternal and intraventricular delivery routes have been also tested [91].

In summary, the optimum route of human stem cell delivery has not been determined but will ultimately depend on the timing of delivery, the cell type used, and their mechanism of action.

2.6.3 Clinical Trials of Cell Transplantation for Stroke Treatment

The growing amount of preclinical data showing the potential benefit associated with neural stem/progenitor cell therapy warrants the development of well controlled clinical trials to investigate therapeutic safety and efficacy for central nervous system insults such as ischemic stroke. Cell-based therapies have emerged as a novel and highly promising investigational approach to enhance recovery after stroke in animal models. The encouraging preliminary results have led several investigative teams to launch clinical trials evaluating the safety of cell-based therapies in stroke patients (Table 2.4). The safety, feasibility, ideal cell type, optimal dosage, and most favorable delivery method of cells are currently unknown (Fig. 2.3).

Preliminary clinical trials investigating the role of cell therapeutics for ischemic stroke have been limited to date and powered only to evaluate safety. The majority of these studies are restricting enrollment to patients with MCA infarct territory and do not assess the role of these cells in other areas of the brain. No optimal method of delivery has been established, and it is still unclear whether intravenous, intra-arterial, or other approaches may be safer and lead to better outcomes. Additionally, the studies employ different outcome measures limiting the ability to compare results among trials (Table 2.5). While these preliminary studies have yielded some data to support the safety of cellular transplantation, additional trials need to be completed prior to controlled multicenter trials [92].

These human trials imply that stem cell treatment in ischemic brain is practicable; however, the limited clinical trial data have been obtained providing a little consistent evidence of clinical benefit [93]. Nonetheless, the previous studies determining the optimal cell population and the protocol of administration are really needed to improve the outcome of cell therapy for stroke. On the other hand, the optimal time of post-ischaemic transplantation has not yet been designated, as activated microglia can limit survival of transplanted cells.

Table 2.4 Completed clinical stem cells trials for stroke

Study (year)	Cell type	Source	Patient (active/control)	Administration	Stroke type	Location	Stroke age	Comments
Kondziolka et al. (2000)	hNT	Immortalized neuronal cell line	12/0	IC (×1) Stereotactic transplantation into region of the stroke	Basal ganglia infarct	Striatum	6 months to 6 years	Feasibility proven. No effect on fuctional outcome. PET showed increased metabolic activity Adverse effect in 2 patients, unclear if related to transplant surgery
Kondziolka et al. (2005)	hNT	Immortalized neuronal cell line	14/4	IC (×1) Stereotactic transplantation into region of the stroke	Basal ganglia infarct	Striatum	12 months to 6 years	Feasibility proven. No effect on functional outcome. PET showed increased metabolic activity Adverse effect in 2 patients, unclear if related to transplant surgery
Savitz et al. (2005)	LGE	Xeno/swine	5 (of planned 12)/0	IC (×1) Stereotactic transplantation into region of the stroke	MCA infarct	Striatum	3 months to 10 years	Study stopped early after 2 serious adverse event Terminated by the FDA due to possible side effects
Bang et al. (2005)	BMSC	Autologous	5/25	IV (×2) Intravenous	MCA infarct	Striatum + cortex	4–5 weeks, 7–9 weeks	Questionable study quality, e.g. 10 patients lost to follow-up. Feasibility proven. No surgery-related adverse effects

Abbreviations: *hNT* human immature neurons generated from the immortalized NT2 cell line derived from human teratocarcinoma, *LGE* Lateral Ganglionic Eminence, *BMSC* Bone Marrow-derived Stem Cells, *IC* intracerebral, *IV* intravenous, *MCA* middle cerebral artery, *FDA* Food and Drug Administration

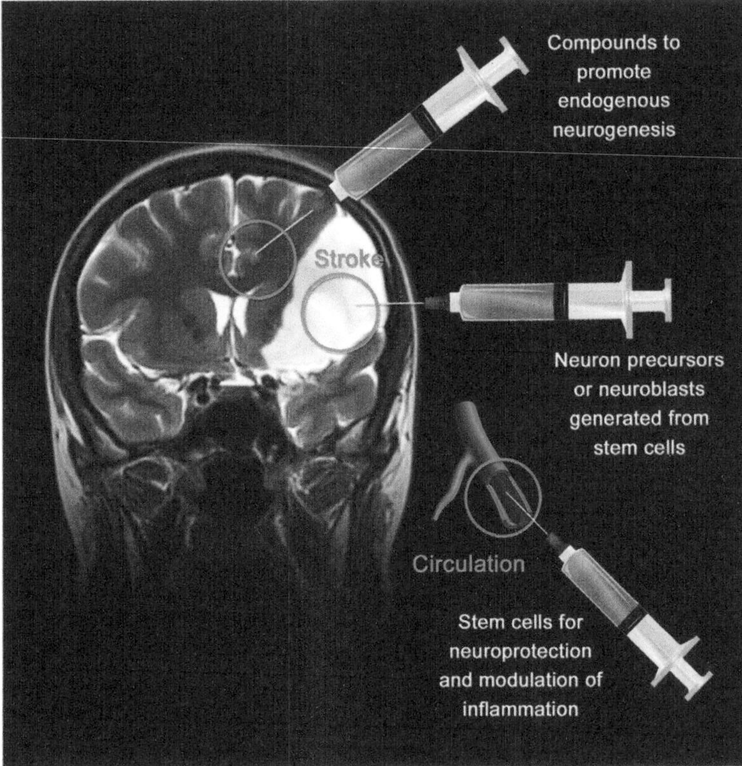

Fig. 2.3 The possible delivery methods for stem cell transplantation in stroke treatment

2.7 Stem Cell-Based Approach to Traumatic Brain Injury (TBI)

Traumatic brain injury (TBI) can occur in different stages of the severity and complexity. In more serious cases the pathological changes can be generally similar like after the haemorrhagic stroke; however, other complications can be observed even fare from the place of impact as a result of big destroying energy due to acceleration or deceleration of the head. Clinical examples of the recovery after serious TBI with the functional improvement even after long time after the incidence, gave the impulse to search for the role of different factors responsible for the recovery, including the neurogenesis [94]. Important observations have been made using the animal models of TBI. Following TBI, stem and progenitor cells located in subgranular zone (SGZ) of the lateral ventricles and the SVZ of hippocampal dentate gyrus become activated [95]. Interventions stimulating hippocampal neurogenesis appear to improve cognitive recovery after experimental TBI [96]. The studies on effect of transplantation of neural progenitor cells (NPCs) derived from mouse brain and injected into the striatum of mice 1 week following unilateral cortical impact

Table 2.5 Current clinical cell transplantation trials for stroke

Clinical identifier, clinical phase and country	Study design	Cell type	Patient	Administration	Inclusion criteria	Outcomes	Time windows
NCT00473057, Phase I, Brazil (Rio de Janeiro)	2 arms (non randomized: 10 IA/5IV	Autologous BMMCs	15	IA/IV	MCA stroke, 18–75 yo; NIHSS: 4–20	Safety	3 h to 90 days
NCT00859014, Phase I, USA (Huoston)	Single arm	Autologous BMMCs	10	IV	MCA stroke, 18–80 yo; NIHSS: 6–20	Safety and feasibility	24–72 h
NCT00535197, Phase I/II, UK (London)	Single arm	Autologous $CD34^+$ BMC	10	IA	MCA stroke, 30–80 yo; Severe stroke conforming to the TACS phenotype (weakness, homonymous, hemianopia and a focal cognitive deficit)	Safety and tolerability	7 days
NCT00875654, Phase II, France (Grenoble)	Randomized (Control versus treatment groups)	Autologous bone marrow derived progenitor cells	30	IV	Carotid territory stroke, 18–65 yo, NIHSS: more then 2	Feasibility and tolerablity	6 weeks
NTC00950521, Phase II, China (Taiwan)	Randomized (Cell infusion versus conventional treatment)	Autologous peripheral blood $CD34^+$ cells	30	IC	Stable deficit s hemiplegia, 35–70 yo, NIHSS: 9–20	Safety and eficacy	6 months to 5 years

Abbreviations: *IA* intra-arterial, *IV* intravenous, *IC* intracerebral, *BMMC* Bone Marrow Mononuclear Cells, *BMC* Bone marrow cells, *MCA* middle cerebral artery, *NIHSS* National Institutes of Health Stroke Scale, *TACS* total anterior circulation stroke injury, demonstrated that cells survived in the host brain up to 14 months, migrated to the site of injury and enhanced motor and cognitive recovery [97]. The environment associated with brain lesion in rat model of TBI can significantly modulate the phenotype and migratory patterns of the implanted NSCs [98]. In later work, the findings support the idea that mechanisms other than the replacement of damaged neurons or glia, such as NSC-induced increases in protective neurotrophic factors, may be responsible for the functional recovery in the model of TBI [99]. The intravenously transplanted HUCB cells engrafted in parenchyma of brain lesion in a rat model of TBI and decrease in neurological damage was observed [100]. The avoidance of secondary brain injury after TBI is main topic of research interest and combined treatments possibly can provide the best results. These potential combination include agents cytokines, pharmaceuticals or stem cells, and additionally physical or electric stimulation [101]. The first single center clinical study was performed on seven patients after TBI using autologous MSCs isolated from bone marrow, expanded in culture and administered partly directly to the injured region or partly intravenously. The safety of the method was main aim of that trial but neurological function improvement was also observed at 6 months after the procedure [102]. A study of autologous stem cell treatment in children with TBI was performed in the University of Texas Medical School at Houston (Clinical trials. gov identifier: NCT00254722). The purpose of the study was to determine if bone marrow progenitor cell (BMPC) autologous transplantation in children after isolated traumatic brain injury is safe and will improve functional outcome. There was observed limited rate of cell engraftment; however, the study also showed that implanted cells interact with immunologic cells located in organ systems distant to the CNS, thereby, altering the systemic immunologic/inflammatory response, the observed post-injury pro-inflammatory environment, and leading to neuroprotection [103].

2.8 Stem Cell-Based Approach to Spinal Cord Injury (SCI)

Severe spinal cord injury (SCI) is resulting in loss of motor function, sensation and autonomic control below the level of impact. The treatment of SCI is still limited to surgical decompression of the spinal cord and stabilization of fractured spine. This can give the chance for natural recovery supported by the complex, rather than only symptomatic, treatment and rehabilitation; however, the patients usually persist in chronic disabling state. There is a need of more active methods of treatment improving the ability of functional recovery. It justified the very intensive research including the efficacy of stem cell transplantation. Researchers are focused to explain pathophysiology of the processes in the injured spinal cord repair, including:

- Protecting surviving nerve cells from further damage
- Replacing damaged neurons
- Stimulating the regrowth of axons and targeting their connections appropriately
- Rebuilding neural circuits to restore body functions

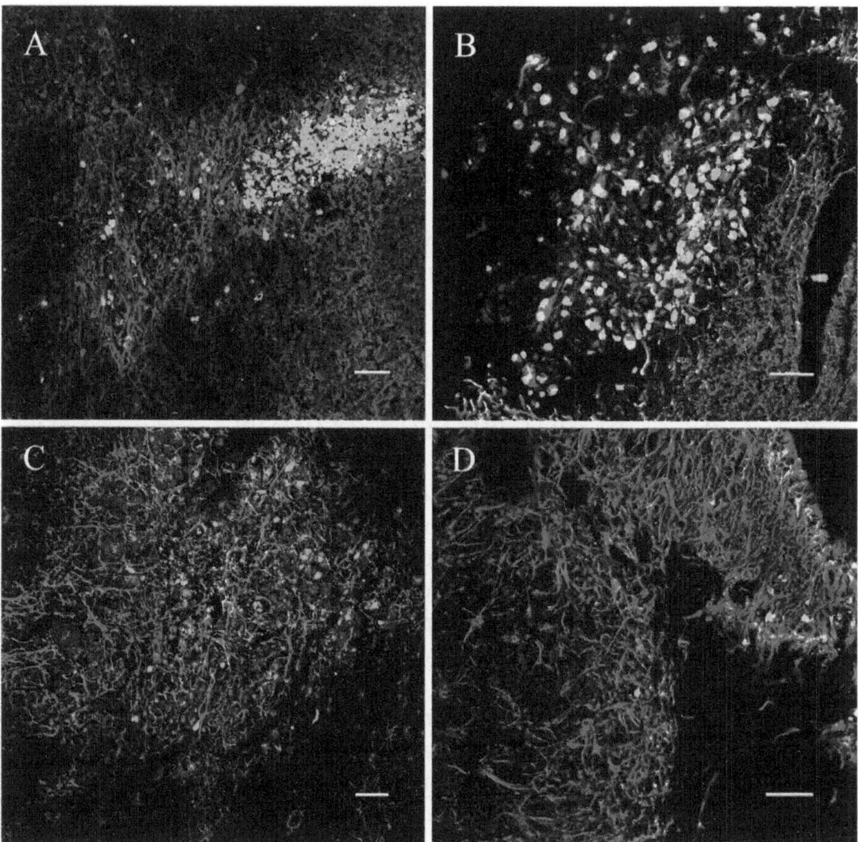

Fig. 2.4 Immunohistochemical staining of HUCB-NSCs transplanted into lesioned rat forbrain. (**a**) Migration CMFDA-labeled HUCB-NSCs (*green*) surrounding NF-200 reactive host neuron (*red*) 3 days after cells injection. Co-staining of these cells with neuron specific anti-NF-200 antibody (*yellow*) can be found in minimal number of cells. (**b**) 7 days after transplantation HUCB-NSCs stained with human nuclei marker NuMA (*green*) co-expressing neuronal marker NF-200 (*red* cytoplasm and protrusions) in the infarct area. (**c**) 7 days after transplantation HUCB-NSCs stained with human nuclei marker NuMA (*green*) in close vicinity of GFAP reactive astrocytes (*red*) (**d**) 30 days after transplantation glial scar around lesion stained with GFAP (*red*) harbors a few of NuMA –positive nuclei (*green*). Scale bar – 50 μm

The animal models of different cells transplantation after the experimental SCI, including NSCs, ESCs, MSCs, iPSCs, but also Olfactory Ensheating cells (OECs), Schwann Cells (SC) and macrophages, showed the possibility of functional improvement. Human NSCs transplantation in mouse model of SCI generated neurons and oligodendrocytes resulted the locomotor function recovery [104]. Similar study performed on marmoset (belonging to primates), proved this observation. Grafted NSCs migrated up to 7 mm far from the lesion epicenter [105]. Human NSCs transplanted into injured rat spinal cord can differentiate into neurons that form axons and synapses to establish contacts with host motor neurons [106].

SCI provokes also endogenous precursors to differentiate and migrate toward the spinal lesion. It was showed using an example of ependymal cells lining the central canal [107] and spinal parenchymal progenitors, which gave rise to oligodendrocytes [108]. The implantation of stem cell-derived cells that remyelinate can be important goal of SCI treatment because demyelination contributes to loss of function after injury. Transplantation of human ESCs into shiverer mouse model of dysmyelination resulted in integration, differentiation into oligodendrocytes, and compact myelin formation [46]. Also, Schwan cells can be derived from BMCs *in vitro*, and then implanted into the injured rat spinal cord reducing the size of the posttraumatic cystic cavity and promoting axonal regeneration and sparing, which results in functional recovery [109]. Co-transplantation of the NSCs and OECs into partially transected spinal cord of rats might have a synergistic effect on promoting neural regeneration and improving locomotor functions. The results were significantly better than single graft of either NSCs or OECs [110]. In addition, the MSCs can play the neuroprotective role which is shown using the example of the gene-modified human MSCs secreting brain derived neurotrophic factor (BDNF). Their transplantation resulted in structural changes in spinal cord, which were associated with improved functional outcome in the acute SCI rat model [111]. The need of support and guiding of growing cells, especially in the model of spinal cord transection, focused the researchers on the appropriate biomaterials (as the scaffolds) for the cells and neurotropic substances. Different polymers formed as tubules or fibers were developed, but results are still not satisfactory [112]. Therefore, combining complementary strategies might be required to advance stem cells treatment of SCI to the clinical stage. Dramatic situation of patients in poor neurological state after injury put pressure on researchers to begin the clinical studies. The possible mechanism of stem cells-based therapy of patients after SCI include: replacement of damaged or dead motor neurons, promote the remyelination and modification of release different factors, which could counteract detrimental inflammation.

Eleven clinical projects were completed and published since 2006, all of them outside of the USA. Over 500 patients were treated in many countries. The results are difficult to compare because the studies were not randomized, seldom used control groups, the ways of cell delivery and the time of intervention were different. The benefit was reported from open-label trials where patients have also been subjected to intensive rehabilitation. In most cases autologous BMSCs were introduced and no tumor formation was diagnosed as a result of stem cells transplantation [113–118]. Consequently, serious complications of the procedures were mostly reported and the improvement of the patients' neurological state was described in different ways; in some attempts no improvement was observed [119]. In other papers, the percentage of improved patients run from 29.5% [120] to 66.7% [121]. The largest material of 297 patients was presented by Indian researchers, and their clinical results were provided in 32.6% neurologically improved patients [122]. The post-transplantation improvement of somato-sensory evoked potentials (SSEP) was reported in 66.7% patients (two publications) [123]. Summarized data of completed clinical trials of the use of stem cells transplantation for the treatment of patients after SCI are presented in Table 2.6.

Table 2.6 Completed clinical stem cell transplantation trials for the SCI patients

Study (year)	Cell type	Source	Patient (active/control)	Administration	Location of injury	Injury age	Observation (months)	Outcone
Callera et al. (2006)	BMSC	Autologous	10/0	ITh	7 paraplegia 3 tetraplegia	3 years (mean)	3	Feasibility proven. No effect on functional outcome. No adverse effects
Moviglia et al. (2006)	BMSC	Autologous	2/0	ITh	C2, Th6	8 and 30 months	3–6	Feasibility proven. Neurological recovery. No adverse effects
Sykov'a et al. (2006)	BMSC	Autologous	20/0	IA (6 patients) IV (14 patients)	C4 to Th11	Subacute (8 patients) Chronic (12 patients	12	Improvement in SSEP (66.7%) No adverse effects
Yoon et al. (2007)	BMSC	Autologous	35/13	IS	C (23 patients) Th, ThL (12 patients)	3 groups: <2 weeks, 2–8 weeks, >8 weeks	3 month to 10 years	Neurological improvement only in acute (29.5%) and subacute group (33.3%). Transient neurolog. deterioration in 1 case, fewer in 22 cases, abdominal discomfort in 7 cases, neutopathic pain in 7 cases
Chernykh et al. (2007)	BMSC	Autologous	18/+	IS IV	6 paraplegia 12 tetraplegia	Chronic (mean 36.4 months)	Mean 9.4 ± 4.6	Neurological improvement (66.7%)
Deda et al. (2008)	BMSC	Autologous	9/0	IS ITh IV	C3 to Th11	Chronic	12	Neurological improvement. No adverse effects

(continued)

Table 2.6 (continued)

Study (year)	Cell type	Source	Patient (active/control)	Administration	Location of injury	Injury age	Observation (months)	Outcone
Pal et al. (2009)	BMSC	Autologous	30/0	ITh	C4 to Th10	1 group-<6 months 2 group>6 months	12–36	No Neurological improvement No adverse effects
Cristante et al. (2009)	BMSC	Autologous	39/0	IA	Cervical and thoracic	Chronic (>2 years)	30	Improvement in SEEP (66.7%) No adverse effects
Kumar et al. (2009)	BMSC	Autologous	297/33 nontraumatic	ITh	215 paraplegia 49 tetraplegia 33 nontraumatic	Chronic	3	Neurological improvement (32.6%) No serious adverse effects
Kishk et al. (2010)	BMSC	Autologous	44/20	ITh	Multilevel group	Chronic (mean 3. 6 years)	18	No significant between-group improvement (higher percentage of the MSC group increased motor score by 1–2 points and changed from ASIA A to B). Neuropathic pain in 24 out of the 43 patients, 1-encephalomyelitis
Ra et al. (2011)	hAdMSC	Autologous	8/0	IV	7 tetraplegia 1 paraplegia	1.07–7.88 years	3	1 patientout of 4 with ASIA grade A improved to ASIA grade C. Motor score improvement was shown in 3 patients

Abbreviations: *BMSC* Bone Marrow-derived Stem Cells, *hAdMSC* human Adipose tissue-derived Mesenhymal Stem Cells, *IS* intraspinal, *IV* intravenous, *IA* intraarterial, *ITh* intrathecal, *C* cervical, *Th* thoracic, *THL* thoraco-lumbar junction, *SSEP* somatosensory evoked potentials

2.9 Stem Cells Research as the Way to Restorative Neurology

The understanding of functional recovery on a cellular level is crucial to optimize circuit reconstruction and progenitor cell guidance. The stem cell research becomes the basis of restorative neurology. This term was put forward by Dimitrijevic in 1980-ties [124] and it was defined as a branch of the neurological sciences that applied not destructive procedures, which could improve impaired function of nervous system. The neurorestorative methods modulate abnormal neurocontrol according to underlying mechanisms and "sleeping" residual functions, which cannot be clinically recognized. They include cell transplants, but also neurostimulation, neuromodulation, and pharmaceuticals. In twenty-first century this idea develops in wide range.

2.10 Stem Cells and Brain Tumor

Although tumorigenecity of stem cells implanted to the brain has been mentioned in several places of this chapter, the overall goal for many laboratories in their research is to understand the function of human brain stem cells and how they may play a role in the origin of brain tumors. Understanding the relationship between this area of research and the genesis of brain tumors is of greatest importance and has been recently a topic for many publications to ultimately aim for the ability to manipulate the microenvironment of neural stem cells in order to find better therapies to fight brain cancer, which is still an underdeveloped area of research. However, this particular area of stem cell research requires a separate monography; thus, we would like to address only a few but the most important topics of interest in this chapter.

Several concepts for the origin of brain cancer have been proposed in recent years. There is some evidence that brain tumors may arise from progenitor cells that display deviant differentiation with some type of steam cell niches. Consequently, brain tumors appear to contain stem-like cancer cells that are required to propagate tumor growth [125]. Neural stem cells in the brain have been shown to be cells of origin of certain brain cancers, most notably astrocytomas and medulloblastoma. In particular, in a mouse model, the targeting of genetic modifications for astrocytoma-relevant tumor suppressors to neural stem cells causes malignant astrocytoma to arise, thereby suggesting that astrocytoma is derived from neural stem cells [126]. However, it remains to be determined whether this important finding is reproducible in humans. Also, recently, Wang et al. [127] demonstrated that a subpopulation of endothelial cells within glioblastomas possesses the same somatic mutations identified within tumor cells, such as amplification of EGFR and chromosome 7 rearrangements. In additon, they demonstrated that the stem-cell-like CD133(+) fraction includes a subset of vascular endothelial-cadherin (CD144)-expressing cells that show characteristics of endothelial progenitors capable of maturation into endothelial cells, which explains, at least partially, the mechanism how normal neural stem cells become malignant.

Interestingly, NSCs have the ability to function as cell carriers for targeted delivery of an oncolytic adenovirus because of their inherent tumor-tropic migratory activity [128]. It was reported that delivery of CRAD-S-pk7, a glioma restricted adenovirus, inhibits tumor growth and increases median survival by 50% in an orthopic xenograph model of human glioma [128].

A recently identified approach for the delivery of antibodies to brain tumor uses stem cells as antibody delivery vehicles [121, 129, 130]. There are well documented strategies for enhansing antibody delivery to the brain, including: methods for intracranial injection of antibodies, disruption of the blood-brain barrier, antibody cationization, nanoparticles and liposomes, and engineered antibody fragments. However, the antibody-secreting NSCs are able to cross the blood-brain barrier and deliver the therapeutic effects on ectopic human cancer xenografts in the mouse brain [131].

For many patients diagnosed with brain cancer each year, more than half that number die and the identification of successful treatment remains a challenging goal. Some pioneering research using stem cells may be changing that. Recently, CBS News correspondent, Ben Tracy, reported that a woman in California made medical history as the first human being to have stem cells injected into her brain to try to cure brain tumor, glioblastoma. Karen Aboody (City of Hope, CA) and her investigative team injected ten million of neural stem cells secreting protease into the patient's brain with the hope is that this new type of treatment will kill the tumor and leave healthy brain tissue alone.

References

1. Teo AK, Vallier L (2010) Emerging use of stem cells in regenerative medicine. Biochem J 428(1):11–23
2. Richardson RM, Holloway KL, Bullock MR, Broaddus WC, Fillmore HL (2006) Isolation of neuronal progenitor cells from the adult human neocortex. Acta Neurochir (Wien) 148(7):773–777
3. Andersson ER, Lendahl U (2009) Regenerative medicine: a 2009 overview. J Intern Med 266(4):303–310
4. Andres RH, Choi R, Steinberg GK, Guzman R (2008) Potential of adult neural stem cells in stroke therapy. Regen Med 3(6):893–905
5. Hara K, Yasuhara T, Maki M et al (2008) Neural progenitor NT2N cell lines from teratocarcinoma for transplantation therapy in stroke. Prog Neurobiol 85(3):318–334
6. Bliss TM, Kelly S, Shah AK et al (2006) Transplantation of hNT neurons into the ischemic cortex: cell survival and effect on sensorimotor behavior. J Neurosci Res 83(6):1004–1014
7. Friedenstein AJ, Chailakhyan RK, Latsinik NV, Panasyuk AF, Keiliss-Borok IV (1974) Stromal cells responsible for transferring the microenvironment of the hemopoietic tissues. Cloning in vitro and retransplantation in vivo. Transplantation 17(4):331–340
8. Ma J, Shen Z, Zhang Q, Zhu T, Yao K (2011) The effect of siRNA-VEGF on the growth of REC in retinal pigment epithelial cell and retinal endothelial cell co-culture system. Yan Ke Xue Bao 26(2):20–27
9. Sanchez-Ramos J, Song S, Cardozo-Pelaez F et al (2000) Adult bone marrow stromal cells differentiate into neural cells in vitro. Exp Neurol 164(2):247–256
10. Woodbury D, Schwarz EJ, Prockop DJ, Black IB (2000) Adult rat and human bone marrow stromal cells differentiate into neurons. J Neurosci Res 61(4):364–370

11. England T, Martin P, Bath PM (2009) Stem cells for enhancing recovery after stroke: a review. Int J Stroke 4(2):101–110
12. Bliss T, Guzman R, Daadi M, Steinberg GK (2007) Cell transplantation therapy for stroke. Stroke 38(2 Suppl):817–826
13. Hicks A, Jolkkonen J (2009) Challenges and possibilities of intravascular cell therapy in stroke. Acta Neurobiol Exp (Wars) 69(1):1–11
14. Gornicka-Pawlak E, Janowski M, Habich A et al (2011) Systemic treatment of focal brain injury in the rat by human umbilical cord blood cells being at different level of neural commitment. Acta Neurobiol Exp (Wars) 71(1):46–64
15. Nagai A, Kim WK, Lee HJ et al (2007) Multilineage potential of stable human mesenchymal stem cell line derived from fetal marrow. PLoS One 2(12):e1272
16. Sokolova IB, Fedotova OR, Zin'kova NN, Kruglyakov PV, Polyntsev DG (2006) Effect of mesenchymal stem cell transplantation on cognitive functions in rats with ischemic stroke. Bull Exp Biol Med 142(4):511–514
17. Borlongan CV, Evans A, Yu G, Hess DC (2005) Limitations of intravenous human bone marrow CD133+ cell grafts in stroke rats. Brain Res 1048(1–2):116–122
18. Borlongan CV, Hadman M, Sanberg CD, Sanberg PR (2004) Central nervous system entry of peripherally injected umbilical cord blood cells is not required for neuroprotection in stroke. Stroke 35(10):2385–2389
19. Kurozumi K, Nakamura K, Tamiya T et al (2005) Mesenchymal stem cells that produce neurotrophic factors reduce ischemic damage in the rat middle cerebral artery occlusion model. Mol Ther 11(1):96–104
20. Kopen GC, Prockop DJ, Phinney DG (1999) Marrow stromal cells migrate throughout forebrain and cerebellum, and they differentiate into astrocytes after injection into neonatal mouse brains. Proc Natl Acad Sci USA 96(19):10711–10716
21. Brazzini A, Cantella R, De la CA et al (2010) Intraarterial autologous implantation of adult stem cells for patients with Parkinson disease. J Vasc Interv Radiol 21(4):443–451
22. Takahashi K, Yamanaka S (2006) Induction of pluripotent stem cells from mouse embryonic and adult fibroblast cultures by defined factors. Cell 126(4):663–676
23. Nakagawa M, Koyanagi M, Tanabe K et al (2008) Generation of induced pluripotent stem cells without Myc from mouse and human fibroblasts. Nat Biotechnol 26(1):101–106
24. Kim D, Kim CH, Moon JI et al (2009) Generation of human induced pluripotent stem cells by direct delivery of reprogramming proteins. Cell Stem Cell 4(6):472–476
25. Ekonomou A, Ballard CG, Pathmanaban ON et al (2010) Increased neural progenitors in vascular dementia. Neurobiol Aging 32(12):2152–61
26. Michejda M (2004) Which stem cells should be used for transplantation? Fetal Diagn Ther 19(1):2–8
27. Habich A, Jurga M, Markiewicz I, Lukomska B, Bany-Laszewicz U, Domanska-Janik K (2006) Early appearance of stem/progenitor cells with neural-like characteristics in human cord blood mononuclear fraction cultured in vitro. Exp Hematol 34(7):914–925
28. Fu YS, Cheng YC, Lin MY et al (2006) Conversion of human umbilical cord mesenchymal stem cells in Wharton's jelly to dopaminergic neurons in vitro: potential therapeutic application for Parkinsonism. Stem Cells 24(1):115–124
29. Deierborg T, Soulet D, Roybon L, Hall V, Brundin P (2008) Emerging restorative treatments for Parkinson's disease. Prog Neurobiol 85(4):407–432
30. Venkataramana NK, Kumar SK, Balaraju S et al (2010) Open-labeled study of unilateral autologous bone-marrow-derived mesenchymal stem cell transplantation in Parkinson's disease. Transl Res 155(2):62–70
31. Zhang C, McNeil E, Dressler L, Siman R (2007) Long-lasting impairment in hippocampal neurogenesis associated with amyloid deposition in a knock-in mouse model of familial Alzheimer's disease. Exp Neurol 204(1):77–87
32. Lee JK, Jin HK, Endo S, Schuchman EH, Carter JE, Bae JS (2010) Intracerebral transplantation of bone marrow-derived mesenchymal stem cells reduces amyloid-beta deposition and rescues memory deficits in Alzheimer's disease mice by modulation of immune responses. Stem Cells 28(2):329–343

33. Keene CD, Chang RC, Lopez-Yglesias AH et al (2010) Suppressed accumulation of cerebral amyloid beta peptides in aged transgenic Alzheimer's disease mice by transplantation with wild-type or prostaglandin E2 receptor subtype 2-null bone marrow. Am J Pathol 177(1):346–354
34. Ryu JK, Cho T, Wang YT, McLarnon JG (2009) Neural progenitor cells attenuate inflammatory reactivity and neuronal loss in an animal model of inflamed AD brain. J Neuroinflammation 6:39
35. Lindvall O, Kokaia Z (2010) Stem cells in human neurodegenerative disorders–time for clinical translation? J Clin Invest 120(1):29–40
36. Thonhoff JR, Ojeda L, Wu P (2009) Stem cell-derived motor neurons: applications and challenges in amyotrophic lateral sclerosis. Curr Stem Cell Res Ther 4(3):178–199
37. Bohl D, Liu S, Blanchard S, Hocquemiller M, Haase G, Heard JM (2008) Directed evolution of motor neurons from genetically engineered neural precursors. Stem Cells 26(10):2564–2575
38. Mitne-Neto M, Machado-Costa M, Marchetto MC et al (2011) Downregulation of VAPB expression in motor neurons derived from induced pluripotent stem cells of ALS8 patients. Hum Mol Genet 20(18):3642–52
39. Karussis D, Karageorgiou C, Vaknin-Dembinsky A et al (2010) Safety and immunological effects of mesenchymal stem cell transplantation in patients with multiple sclerosis and amyotrophic lateral sclerosis. Arch Neurol 67(10):1187–1194
40. Garbuzova-Davis S, Sanberg CD, Kuzmin-Nichols N et al (2008) Human umbilical cord blood treatment in a mouse model of ALS: optimization of cell dose. PLoS One 3(6):e2494
41. Habisch HJ, Janowski M, Binder D et al (2007) Intrathecal application of neuroectodermally converted stem cells into a mouse model of ALS: limited intraparenchymal migration and survival narrows therapeutic effects. J Neural Transm 114(11):1395–1406
42. Yan J, Xu L, Welsh AM et al (2006) Combined immunosuppressive agents or CD4 antibodies prolong survival of human neural stem cell grafts and improve disease outcomes in amyotrophic lateral sclerosis transgenic mice. Stem Cells 24(8):1976–1985
43. Zhao CP, Zhang C, Zhou SN et al (2007) Human mesenchymal stromal cells ameliorate the phenotype of SOD1-G93A ALS mice. Cytotherapy 9(5):414–426
44. Mazzini L, Ferrero I, Luparello V et al (2010) Mesenchymal stem cell transplantation in amyotrophic lateral sclerosis: a phase I clinical trial. Exp Neurol 223(1):229–237
45. Brustle O, Jones KN, Learish RD et al (1999) Embryonic stem cell-derived glial precursors: a source of myelinating transplants. Science 285(5428):754–756
46. Nistor GI, Totoiu MO, Haque N, Carpenter MK, Keirstead HS (2005) Human embryonic stem cells differentiate into oligodendrocytes in high purity and myelinate after spinal cord transplantation. Glia 49(3):385–396
47. Blakemore WF (1977) Remyelination of CNS axons by Schwann cells transplanted from the sciatic nerve. Nature 266(5597):68–69
48. Einstein O, Fainstein N, Vaknin I et al (2007) Neural precursors attenuate autoimmune encephalomyelitis by peripheral immunosuppression. Ann Neurol 61(3):209–218
49. Bai L, Lennon DP, Eaton V et al (2009) Human bone marrow-derived mesenchymal stem cells induce Th2-polarized immune response and promote endogenous repair in animal models of multiple sclerosis. Glia 57(11):1192–1203
50. Zhang J, Li Y, Lu M et al (2006) Bone marrow stromal cells reduce axonal loss in experimental autoimmune encephalomyelitis mice. J Neurosci Res 84(3):587–595
51. Chen Q, Long Y, Yuan X et al (2005) Protective effects of bone marrow stromal cell transplantation in injured rodent brain: synthesis of neurotrophic factors. J Neurosci Res 80(5):611–619
52. Connick P, Kolappan M, Patani R et al (2011) The mesenchymal stem cells in multiple sclerosis (MSCIMS) trial protocol and baseline cohort characteristics: an open-label pre-test: post-test study with blinded outcome assessments. Trials 12:62
53. Luo Y (2011) Cell-based therapy for stroke. J Neural Transm 118(1):61–74
54. Jablonska A, Lukomska B (2011) Stroke induced brain changes: implications for stem cell transplantation. Acta Neurobiol Exp (Wars) 71(1):74–85

55. Espinoza-Rojo M, Iturralde-Rodriguez KI, Chanez-Cardenas ME, Ruiz-Tachiquin ME, Aguilera P (2010) Glucose transporters regulation on ischemic brain: possible role as therapeutic target. Cent Nerv Syst Agents Med Chem 10(4):317–325
56. Locatelli F, Bersano A, Ballabio E et al (2009) Stem cell therapy in stroke. Cell Mol Life Sci 66(5):757–772
57. Biernaskie J, Corbett D (2001) Enriched rehabilitative training promotes improved forelimb motor function and enhanced dendritic growth after focal ischemic injury. J Neurosci 21(14):5272–5280
58. Lichtenwalner RJ, Parent JM (2006) Adult neurogenesis and the ischemic forebrain. J Cereb Blood Flow Metab 26(1):1–20
59. Kim DY, Park SH, Lee SU et al (2007) Effect of human embryonic stem cell-derived neuronal precursor cell transplantation into the cerebral infarct model of rat with exercise. Neurosci Res 58(2):164–175
60. Savitz SI, Dinsmore J, Wu J, Henderson GV, Stieg P, Caplan LR (2005) Neurotransplantation of fetal porcine cells in patients with basal ganglia infarcts: a preliminary safety and feasibility study. Cerebrovasc Dis 20(2):101–107
61. Jagasia R, Song H, Gage FH, Lie DC (2006) New regulators in adult neurogenesis and their potential role for repair. Trends Mol Med 12(9):400–405
62. Ishibashi S, Sakaguchi M, Kuroiwa T et al (2004) Human neural stem/progenitor cells, expanded in long-term neurosphere culture, promote functional recovery after focal ischemia in Mongolian gerbils. J Neurosci Res 78(2):215–223
63. Toda H, Takahashi J, Iwakami N et al (2001) Grafting neural stem cells improved the impaired spatial recognition in ischemic rats. Neurosci Lett 316(1):9–12
64. Zhang J, Li Y, Chen J et al (2005) Human bone marrow stromal cell treatment improves neurological functional recovery in EAE mice. Exp Neurol 195(1):16–26
65. Carmichael ST (2006) Cellular and molecular mechanisms of neural repair after stroke: making waves. Ann Neurol 59(5):735–742
66. Xiao J, Nan Z, Motooka Y, Low WC (2005) Transplantation of a novel cell line population of umbilical cord blood stem cells ameliorates neurological deficits associated with ischemic brain injury. Stem Cells Dev 14(6):722–733
67. Shen LH, Li Y, Chen J et al (2006) Intracarotid transplantation of bone marrow stromal cells increases axon-myelin remodeling after stroke. Neuroscience 137(2):393–399
68. Chen J, Zhang ZG, Li Y et al (2003) Intravenous administration of human bone marrow stromal cells induces angiogenesis in the ischemic boundary zone after stroke in rats. Circ Res 92(6):692–699
69. Jiang Q, Zhang ZG, Ding GL et al (2005) Investigation of neural progenitor cell induced angiogenesis after embolic stroke in rat using MRI. Neuroimage 28(3):698–707
70. Shyu WC, Lin SZ, Chiang MF, Su CY, Li H (2006) Intracerebral peripheral blood stem cell (CD34+) implantation induces neuroplasticity by enhancing beta1 integrin-mediated angiogenesis in chronic stroke rats. J Neurosci 26(13):3444–3453
71. Capitanio JP, Emborg ME (2008) Contributions of non-human primates to neuroscience research. Lancet 371(9618):1126–1135
72. Risedal A, Mattsson B, Dahlqvist P, Nordborg C, Olsson T, Johansson BB (2002) Environmental influences on functional outcome after a cortical infarct in the rat. Brain Res Bull 58(3):315–321
73. Kozlowska H, Jablonka J, Janowski M, Jurga M, Kossut M, Domanska-Janik K (2007) Transplantation of a novel human cord blood-derived neural-like stem cell line in a rat model of cortical infarct. Stem Cells Dev 16(3):481–488
74. Horie N, Maag AL, Hamilton SA, Shichinohe H, Bliss TM, Steinberg GK (2008) Mouse model of focal cerebral ischemia using endothelin-1. J Neurosci Meth 173(2):286–290
75. Hofmeijer J, Veldhuis WB, Schepers J et al (2004) The time course of ischemic damage and cerebral perfusion in a rat model of space-occupying cerebral infarction. Brain Res 1013(1):74–82

76. Janowski M, Gornicka-Pawlak E, Kozlowska H, Domanska-Janik K, Gielecki J, Lukomska B (2008) Structural and functional characteristic of a model for deep-seated lacunar infarct in rats. J Neurol Sci 273(1–2):40–48
77. Veldhuis WB, van der Stelt M, Delmas F, Gillet B et al (2003) In vivo excitotoxicity induced by ouabain, a Na+/K+-ATPase inhibitor. J Cereb Blood Flow Metab 23(1):62–74
78. Vendrame M, Cassady J, Newcomb J et al (2004) Infusion of human umbilical cord blood cells in a rat model of stroke dose-dependently rescues behavioral deficits and reduces infarct volume. Stroke 35(10):2390–2395
79. Saporta S, Borlongan CV, Sanberg PR (1999) Neural transplantation of human neuroterato-carcinoma (hNT) neurons into ischemic rats. A quantitative dose–response analysis of cell survival and behavioral recovery. Neuroscience 91(2):519–525
80. Hicks AU, MacLellan CL, Chernenko GA, Corbett D (2008) Long-term assessment of enriched housing and subventricular zone derived cell transplantation after focal ischemia in rats. Brain Res 1231:103–112
81. Hicks AU, Hewlett K, Windle V et al (2007) Enriched environment enhances transplanted subventricular zone stem cell migration and functional recovery after stroke. Neuroscience 146(1):31–40
82. Makinen S, Kekarainen T, Nystedt J et al (2006) Human umbilical cord blood cells do not improve sensorimotor or cognitive outcome following transient middle cerebral artery occlusion in rats. Brain Res 1123(1):207–215
83. Dirnagl U (2006) Bench to bedside: the quest for quality in experimental stroke research. J Cereb Blood Flow Metab 26(12):1465–1478
84. Modo M, Stroemer RP, Tang E, Patel S, Hodges H (2002) Effects of implantation site of stem cell grafts on behavioral recovery from stroke damage. Stroke 33(9):2270–2278
85. Fischer JM, Bramow S, Dal-Bianco A et al (2009) The relation between inflammation and neurodegeneration in multiple sclerosis brains. Brain 132(Pt 5):1175–1189
86. Bang OY, Lee JS, Lee PH, Lee G (2005) Autologous mesenchymal stem cell transplantation in stroke patients. Ann Neurol 57(6):874–882
87. Kondziolka D, Wechsler L, Goldstein S, Meltzer C et al (2000) Transplantation of cultured human neuronal cells for patients with stroke. Neurology 55(4):565–569
88. Kondziolka D, Steinberg GK, Wechsler L et al (2005) Neurotransplantation for patients with subcortical motor stroke: a phase 2 randomized trial. J Neurosurg 103(1):38–45
89. Bible E, Chau DY, Alexander MR, Price J, Shakesheff KM, Modo M (2009) The support of neural stem cells transplanted into stroke-induced brain cavities by PLGA particles. Biomaterials 30(16):2985–2994
90. Kelly S, Bliss TM, Shah AK et al (2004) Transplanted human fetal neural stem cells survive, migrate, and differentiate in ischemic rat cerebral cortex. Proc Natl Acad Sci USA 101(32):11839–11844
91. Li L, Jiang Q, Zhang L et al (2006) Ischemic cerebral tissue response to subventricular zone cell transplantation measured by iterative self-organizing data analysis technique algorithm. J Cereb Blood Flow Metab 26(11):1366–1377
92. Walker PA, Harting MT, Shah SK et al (2010) Progenitor cell therapy for the treatment of central nervous system injury: a review of the state of current clinical trials. Stem Cells Int 2010:369578
93. Banerjee S, Williamson D, Habib N, Gordon M, Chataway J (2011) Human stem cell therapy in ischaemic stroke: a review. Age Ageing 40(1):7–13
94. Ewing-Cobbs L, Barnes MA, Fletcher JM (2003) Early brain injury in children: development and reorganization of cognitive function. Dev Neuropsychol 24(2–3):669–704
95. Richardson RM, Sun D, Bullock MR (2007) Neurogenesis after traumatic brain injury. Neurosurg Clin N Am 18(1):169–181, xi
96. Kernie SG, Parent JM (2010) Forebrain neurogenesis after focal ischemic and traumatic brain injury. Neurobiol Dis 37(2):267–274
97. Shear DA, Tate MC, Archer DR et al (2004) Neural progenitor cell transplants promote long-term functional recovery after traumatic brain injury. Brain Res 1026(1):11–22

98. Boockvar JA, Schouten J, Royo N et al (2005) Experimental traumatic brain injury modulates the survival, migration, and terminal phenotype of transplanted epidermal growth factor receptor-activated neural stem cells. Neurosurgery 56(1):163–171
99. Shear DA, Tate CC, Tate MC et al (2011) Stem cell survival and functional outcome after traumatic brain injury is dependent on transplant timing and location. Restor Neurol Neurosci 29(4):215–225
100. Lu D, Sanberg PR, Mahmood A et al (2002) Intravenous administration of human umbilical cord blood reduces neurological deficit in the rat after traumatic brain injury. Cell Transplant 11(3):275–281
101. Xiong Y, Mahmood A, Chopp M (2009) Emerging treatments for traumatic brain injury. Expert Opin Emerg Drugs 14(1):67–84
102. Zhang ZX, Guan LX, Zhang K, Zhang Q, Dai LJ (2008) A combined procedure to deliver autologous mesenchymal stromal cells to patients with traumatic brain injury. Cytotherapy 10(2):134–139
103. Walker PA, Letourneau PA, Bedi S, Shah SK, Jimenez F, Charles S Jr - Jr CS (2011) Progenitor cells as remote "bioreactors": neuroprotection via modulation of the systemic inflammatory response. World J Stem Cells 3(2):9–18
104. Cummings BJ, Uchida N, Tamaki SJ et al (2005) Human neural stem cells differentiate and promote locomotor recovery in spinal cord-injured mice. Proc Natl Acad Sci USA 102(39):14069–14074
105. Nakamura M, Toyama Y, Okano H (2005) Transplantation of neural stem cells for spinal cord injury. Rinsho Shinkeigaku 45(11):874–876
106. Yan J, Xu L, Welsh AM et al (2007) Extensive neuronal differentiation of human neural stem cell grafts in adult rat spinal cord. PLoS Med 4(2):e39
107. Meletis K, Barnabe-Heider F, Carlen M et al (2008) Spinal cord injury reveals multilineage differentiation of ependymal cells. PLoS Biol 6(7):e182
108. Ohori Y, Yamamoto S, Nagao M et al (2006) Growth factor treatment and genetic manipulation stimulate neurogenesis and oligodendrogenesis by endogenous neural progenitors in the injured adult spinal cord. J Neurosci 26(46):11948–11960
109. Someya Y, Koda M, Dezawa M et al (2008) Reduction of cystic cavity, promotion of axonal regeneration and sparing, and functional recovery with transplanted bone marrow stromal cell-derived Schwann cells after contusion injury to the adult rat spinal cord. J Neurosurg Spine 9(6):600–610
110. Wang G, Ao Q, Gong K, Zuo H, Gong Y, Zhang X (2010) Synergistic effect of neural stem cells and olfactory ensheathing cells on repair of adult rat spinal cord injury. Cell Transplant 19(10):1325–1337
111. Sasaki M, Radtke C, Tan AM et al (2009) BDNF-hypersecreting human mesenchymal stem cells promote functional recovery, axonal sprouting, and protection of corticospinal neurons after spinal cord injury. J Neurosci 29(47):14932–14941
112. Hyun JK, Kim HW (2010) Clinical and experimental advances in regeneration of spinal cord injury. J Tissue Eng 2010:650857
113. Moviglia GA, Fernandez VR, Brizuela JA et al (2006) Combined protocol of cell therapy for chronic spinal cord injury. Report on the electrical and functional recovery of two patients. Cytotherapy 8:202–209
114. Deda H, Inci MC, Kurekci AE et al (2008) Treatment of chronic spinal cord injured patients with autologous bone marrow-derived hematopoietic stem cell transplantation: 1-year follow-up. Cytotherapy 10:565-574
115. Cristante AF, Barros-Filho TE, Tatsui N (2009) Stem cells in the treatment of chronic spinal cord injury: evaluation of somatosensitive evoked potentials in 39 patients. Spinal Cord 47:733–738
116. Pal R, Venkataramana NK, Bansal A et al (2009) Ex vivo-expanded autologous bone marrow-derived mesenchymal stromal cells in human spinal cord injury/paraplegia: a pilot clinical study. Cytotherapy 11(7):897–911

117. Kishk NA, Gabr H, Hamdy S et al(2010) Case control series of intrathecal autologous bone marrow mesenchymal stem cell therapy for chronic spinal cord injury. Neurorehabil. Neural Repair 24: 702–708
118. Ra JC, Shin IS, Kim SH et al (2011) Safety of intravenous infusion of human adipose tissue-derived mesenchymal stem cells in animals and humans. Stem Cells Dev 20:1297–1308
119. Callera F, do Nascimento RX (2006) Delivery of autologous bone marrow precursor cells into the spinal cord via lumbar puncture technique in patients with spinal cord injury: a preliminary safety study. Exp Hematol 34(2):130–131
120. Yoon SH, Shim YS, Park YH et al (2007) Complete spinal cord injury treatment using autologous bone marrow cell transplantation and bone marrow stimulation with granulocyte macrophage-colony stimulating factor: phase I/II clinical trial. Stem Cells 25(8):2066–2073
121. Chernykh ER, Stupak VV, Muradov GM et al (2007) Application of autologous bone marrow stem cells in the therapy of spinal cord injury patients. Bull Exp Biol Med 143(4):543–547
122. Kumar AA, Kumar SR, Narayanan R, Arul K, Baskaran M (2009) Autologous bone marrow derived mononuclear cell therapy for spinal cord injury: a phase I/II clinical safety and primary efficacy data. Exp Clin Transplant 7(4):241–248
123. Sykova E, Homola A, Mazanec R et al (2006) Autologous bone marrow transplantation in patients with subacute and chronic spinal cord injury. Cell Transplant 15:675–687
124. Dimitrijevic MR (1989) Restorative neurology of head injury. J Neurotrauma 6(1):25–29
125. Germano I, Swiss V, Casaccia P (2010) Primary brain tumors, neural stem cell, and brain tumor cancer cells: where is the link? Neuropharmacology 58(6):903–910
126. Lee JS, Lee HJ, Moon BH, Song SH et al (2011) Generation of cancerous neural stem cells forming glial tumor by oncogenic stimulation. Stem Cell Rev DOI: 10.1007/s12015-011-9280-4 (Abstract)
127. Wang R, Chadalavada K, Wilshire J, Kowalik U et al (2010) Glioblastoma stem-like cells give rise to tumour endothelium. Nature 468(7325):829–833
128. Ahmed AU, Thaci B, Alexiades NG, Han Y et al (2011) Neural stem cell-based cell carriers enhance therapeutic efficacy of an oncolytic adenovirus in an orthotopic mouse model of human glioblastoma. Mol Ther 19(9):1714–26
129. Frank RT, Edmiston M, Kendall SE (2009) Neural stem cells as a novel platform for tumor-specific delivery of therapeutic antibodies. PLoS One 4(12):e8314
130. Frank RT, Najbauer J, Aboody KS (2010) Concise review: stem cells as an emerging platform for antibody therapy of cancer. Stem Cells 28(11):2084–2087
131. Frank RT, Aboody KS, Najbauer J (2011) Strategies for enhancing antibody delivery to the brain. Biochim Biophys Acta 1816(2):191–198

Chapter 3
Stem Cell-Based Therapy for Lysosomal Storage Diseases

Brittni A. Scruggs, Xiujuan Zhang, Jeffrey M. Gimble, and Bruce A. Bunnell

Contents

B.A. Scruggs
Center for Stem Cell Research and Regenerative Medicine, School of Medicine, Tulane University, 1430 Tulane Ave, SL-99 New Orleans, LA, USA

Department of Pharmacology, School of Medicine, Tulane University, New Orleans, LA, USA

X. Zhang
Center for Stem Cell Research and Regenerative Medicine, School of Medicine, Tulane University, 1430 Tulane Ave, SL-99 New Orleans, LA, USA

J.M. Gimble
Stem Cell Laboratory, Pennington Biomedical Research Center, Louisiana State University System, Baton Rouge, LA, USA

R.K. Srivastava and S. Shankar (eds.), *Stem Cells and Human Diseases*,
DOI 10.1007/978-94-007-2801-1_3, © Springer Science+Business Media B.V. 2012

Abstract The family of lysosomal storage diseases (LSDs) affects 1 out of every 5,000–7,000 live births, and most affected LSD patients experience a rapidly progressive course after disease onset due to dysfunction of the lysosome. The brain is particularly sensitive to the abnormal storage of lysosomal aggregates, and lysosomal dysfunction in neurons and other cell types leads to the progressive degeneration of the central nervous system (CNS) in as many as 75% of all LSDs. Transplantation of stem cells has been studied for several LSDs due to the tremendous promise of these cells to deliver therapeutic gene products, sequester toxic compounds (e.g., accumulated metabolites), replace affected cells through engraftment and differentiation, and decrease inflammation in neurodegenerative disease states. This chapter describes the LSDs that have been tested using stem cell transplantation (SCT), and a review of the current state of LSD research is presented and highlights clinical improvements that have been reported.

Keywords Lysosomal storage disease • Hematopoietic stem cell transplantation • Stem cell-based therapy • Embryonic stem cells • Neural stem cells • Mesenchymal stem cells

3.1 Pathogenesis and Disease Presentation of Lysosomal Storage Diseases

Comprised of over 40 clinically distinct diseases, the family of lysosomal storage diseases (LSDs) affects 1 out of every 5,000–7,000 live births, and collectively, this heterogeneous group represents one of the most common inherited diseases in pediatric populations [1–3]. Although asymptomatic at birth, most affected LSD patients experience a rapidly progressive course after disease onset due to dysfunction of the lysosome, which is the organelle responsible for the degradation and/or metabolism of glycolipids, glycoproteins, glycosaminoglycans (GAGs), oligosaccharides, proteoglycans and sphingolipids, among other substrates [2, 4]. LSDs are most commonly caused by a specific mutation in a gene encoding for a catabolic enzyme, usually a specific lysosomal acid hydrolase, and such genetic aberrations lead to the accumulation of the enzyme's substrate due to the inability of the macromolecules to escape from the lysosome [4, 5].

B.A. Bunnell, Ph.D. (✉)
Center for Stem Cell Research and Regenerative Medicine, School of Medicine, Tulane University, 1430 Tulane Ave, SL-99 New Orleans, LA, USA

Department of Pharmacology, School of Medicine, Tulane University, New Orleans, LA, USA

Division of Regenerative Medicine, Tulane National Primate Research Center, Covington, LA, USA
e-mail: bbunnell@tulane

There are at least 50 known lysosomal acid hydrolase enzymes that act on these macromolecules, and in healthy cells, the degradation products are not stored in the lysosome but instead passively diffuse out of the organelle to be recycled in a variety of biosynthesis reactions. Other causes of LSDs include genetic defects in post-translational modification and/or cellular targeting of lysosomal enzymes or mutations in non-enzymatic proteins (e.g., integral lysosomal membrane proteins) [6]. All causative mutations and defects result in lysosomal distension, which induces mechanical stress and cell changes that often result in cell death [3]. Secondary effects, including oxidative stress, abnormal lipid trafficking, and increased inflammation of the affected tissue, occur as a result of cellular malfunction and/or apoptotic death [1, 6, 7]. Because most lysosomal enzymes are ubiquitously expressed, the genetic defects of LSDs have devastating effects in multiple organ systems [1].

LSDs are often classified according to either the type of accumulated substrate present in the cellular vacuoles (e.g., sphingolipidoses or mucopolysaccharidoses) or presence of demyelination in the nervous systems (e.g., leukodystrophies) [6]. For any given LSD, affected patients may have different mutations of the same enzyme, and the severity of the mutation directly relates to the lysosomal enzymatic activity and thus disease presentation. In general, mutations causing low residual enzyme activity (e.g., nonsense mutations that cause protein truncation) will result in the most severe, early onset variants of LSDs (e.g., infantile type), whereas genetic defects that result in a moderate reduction of enzymatic activity will more likely result in a late onset form (e.g., juvenile or adult types) of the same disease [4]. Thus, there is marked variability of clinical presentation within the family of LSDs due to the differences in the severity of enzymatic mutations and substrate accumulation rates [2].

Although each specific disease has a unique presentation, storage disease patients commonly exhibit symptoms of abnormal skeletal growth, hepatosplenomegaly, and cardiac disease [1]. Additionally, the brain is particularly sensitive to the abnormal storage of lysosomal aggregates, and lysosomal dysfunction in neurons and other cell types in the brain leads to apoptotic cell death, which instigates subsequent brain inflammation [7]. The massive loss of neurons and myelin producing cells results in severe central nervous system (CNS) abnormalities [7]. Neurodegeneration occurs in as many as 75% of LSDs, making the progressive degeneration of the CNS a fundamental clinical problem in this family of diseases. Affected patients often present with cognitive delay, visual or auditory defects, and seizures as a result of the loss of their neuromotor and neurophysiological ability [1, 8].

It is known that 2–8% of all synthesized lysosomal enzymes are not transported directly to the lysosome for immediate use but instead are released through exocytosis to be taken up by adjacent cells [1, 4, 5, 9]. This cross-correction mechanism is the basis for multiple current therapeutic approaches for LSDs, including enzyme replacement therapy (ERT), gene therapy using viral vectors, and stem cell transplantation (SCT). Effective treatment for LSDs will require persistent widespread correction of the underlying pathologies involved in these globally affected organ systems [10], and enzymatic cross-correction using these techniques provides long-term sustained delivery of enzyme to affected tissue for use in substrate catabolism. Released lysosomal enzymes are transported and incorporated into the lysosome's interior as a result of the interaction between the mannose-6-phosphate (M6P) moieties found on the proteins and the specific M6P receptor on the lysosomal membrane. Introduction of low amounts of enzyme (5–10% of normal levels), especially when administered into a depot organ (e.g., liver), has been shown to decrease substrate material, reduce pathology, and alleviate symptoms in several LSDs [11]. Also, over-expression of these enzymes has never been documented to cause deleterious effects, suggesting that precise transcriptional regulation is not likely to be necessary when applying these treatments to LSD patients [1].

3.2 Stem Cell Biology and Mechanisms of Repair

The severity and progressive nature of LSDs, in addition to the lack of effective treatment options, highlight the need for further development of innovative therapeutic approaches. Stem cell-based therapy has shown tremendous promise in preclinical trials of many different LSDs, and the mechanisms of repair and therapeutic benefits will be discussed herein for a variety of stem cell types (Fig. 3.1).

Stem cells (SCs) can be derived from a variety of tissue sources, including but not limited to post-implantation embryos, amniotic fluid, placental cord blood, skin, bone

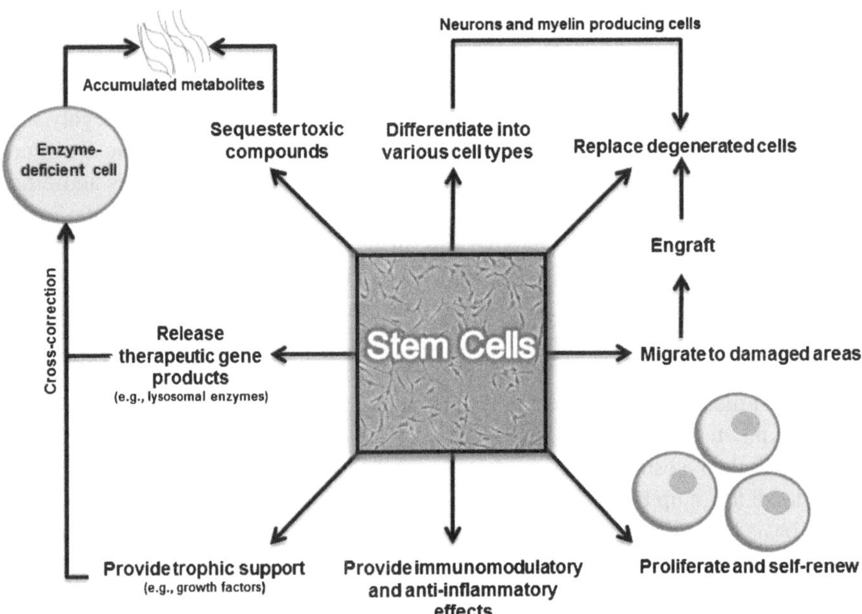

Fig. 3.1 Mechanisms of repair involved in stem cell therapy for lysosomal storage diseases

marrow, adipose and neural tissue; all such cells exhibit properties of "stemness," such that they have the ability to proliferate, self-renew, and differentiate into a wide range of cell types [7, 12]. The main type of mammalian pluripotent stem cell is the embryonic stem cell (ESC), which has unlimited self-renewal capabilities and is highly amenable to genetic modification (e.g., viral vector transduction) [11]. Neural stem cells (NSCs) are defined as CNS progenitor cells that are capable of self-renewal and differentiation into neurons and glial cells; these cells can be extracted directly from fetal neural tissue or indirectly derived from ESCs [13]. Likewise, mesenchymal stem cells (MSCs) are defined as progenitor cells derived from bone marrow, adipose tissue, or other organs that are capable of self-renewal and differentiation into mesenchymal lineages. It is known that ESCs are more versatile and plastic than both NSCs and their adult stem cell counterparts; however, adult MSCs are very promising tools for cell therapy due to their accessibility in adult populations [12]. Recent stem cell advances have led to the development of induced pluripotent stem cells (iPSCs), which are also derived from adult tissue but are capable of differentiating into most cell types. Such technology involves the introduction of embryogenesis-related genes to adult somatic cells, such as skin fibroblasts [7, 12].

Transplantation of SCs has been studied for a wide variety of diseases due to the tremendous promise of these cell-based paracrine factories to deliver therapeutic gene products, sequester toxic compounds (e.g., accumulated metabolites), replace affected cells through engraftment and differentiation, and decrease inflammation in neurodegenerative disease states [14–16]. Specifically for LSD therapy, genetic modification using transduction by recombinant virus vectors (e.g., lentiviral or AAV) has allowed researchers to implant SCs of various types that overexpress the lysosomal enzyme that is either missing or defective in affected cells; additionally, many SCs produce sufficient endogenous lysosomal enzyme without any genetic modification to have a beneficial effect through cross-correction in LSD animal models [3].

For the past three decades, hematopoietic stem cell transplantation (HSCT) has been routinely performed for a variety of LSDs in an attempt to repopulate recipient hematopoietic compartments with healthy donor cells capable of demonstrating cross-correction of the missing functional enzyme [17]. Over 1,000 patients with storage disease have already been treated with intravenous HSCT, and it has been demonstrated that donor myeloid progenitors and monocytes migrate to various organ systems, including the CNS, and replace affected cells [18, 19]. Thus, HSCT restores a critical scavenger function for the catabolism of the storage material [19]. Specifically in the brain tissue, these donor progenitor cells have been shown to engraft in the recipient tissue and differentiate into microglial cells that stably produce lysosomal enzymes for uptake by neighboring affected cells, such as oligodendrocytes and neurons [20, 21]. The ability of hematopoietic cells to migrate, cross the blood–brain barrier (BBB), provide enzyme, replace affected cells, alleviate symptoms, and increase lifespan in several LSDs has challenged researchers in recent years to develop cell-based therapies that incorporate these mechanisms of repair without the added risks associated with donor mismatch and immunological rejection.

ESCs serve as an attractive source of cells for transplantation due to their capability to differentiate into all tissues of an organism, their ability to be easily cultivated *in vitro*, and their ability to provide enzyme delivery in storage disorders [22]. However, the risk of teratoma formation in recipients of ESC transplantation should be weighed carefully against the benefits provided by these cells. Robinson et al. reported that 20% of all ESC-implanted animals were found to have formed teratomas [22]. Similarly, induced pluripotent stem cells are prone to uncontrolled proliferation due to the genetic manipulation (e.g., retroviral transduction) necessary to produce such cells, but advances in cell cultivation and purification are likely to allow for future autologous transplantation of these iPSCs without the high risk of cancer formation [11].

Most prevalent in neonate central nervous systems, NSCs can also be derived from the ventricular subependyma region of the adult CNS [23]. As multipotent cells, NSCs have the ability to differentiate into neurons, oligodendrocytes, and astrocytes, which comprise the three major neuroepithelial-derived brain cell types, yet NSC transplantation does not show signs of tumorigenicity [7, 24]. Glial progenitor cells (GPC) are derived either from NSCs or directly from tissue as primitive neural precursors, and they also have the ability to generate new myelinogenic oligodendrocytes [23], making NSCs and GPCs promising therapeutic vectors for the restoration of myelin. A small number of GPCs may be sufficient to rectify enzymatic deficiencies [24]; however, data suggest that NSCs should be transduced using an appropriate viral vector encoding the specific enzyme of interest in order to achieve sufficient endogenous enzyme levels for cross-correction [3].

Mesenchymal stem cells, also known as multipotent stromal cells (MSCs), include adipose-derived (ASCs) and bone marrow-derived stromal cells (BMSCs) and are capable of self-renewal and differentiation into a wide range of cell types (e.g., osteocytes, adipocytes, chondrocytes, etc.), thus making them a promising therapeutic candidate for restoration of damaged tissue [25–28]. In addition to being accessible for harvesting in clinical settings and easily expanded in culture, MSCs are known to express several lysosomal enzymes, modulate immune reactions *in vitro*, and do not elicit an immunological response *in vivo* [26, 29–32]. Homing to damaged tissue, these cells undergo systemic migration and, unlike ESCs, are not prone to tumor formation. Non-hematopoietic adult stem cells, although restricted in their differentiation potential, can be exploited without ethical constraints and can be transplanted into patients as autografts or as allografts without severe immunologic rejection, which is a major problem and limitation of other cell-based transplantations (e.g., HSCT) [24, 33].

Previous studies have shown that MSCs administered through intravenous (IV) and intracerebroventricular (ICV) routes migrate to areas of CNS inflammation, providing evidence for the potential neuroprotective (e.g., neural differentiation and remyelination) and immunomodulatory effects of MSCs in neurodegenerative diseases [34, 35]. MSCs have also been shown to promote cellular proliferation within the brain after traumatic brain injuries, secrete a variety of cytokines and growth factors that have paracrine activities, and promote differentiation of the host's stem cells [35–39]. Evidence of the suppression of T-lymphocyte proliferation by allogeneic bone marrow-derived MSCs highlights these cells as a potentially safer option for treatment of LSDs than the current treatment with HSCT.

Multiple research teams in recent years have reported that neurons and glial cells have been successfully generated in affected LSD rodent brains after transplantation of ESCs, MSCs, or NSCs; directed differentiation of SCs before transplantation has been proposed as a means of enhancing the production of certain trophic factors and providing cell replacement. Clonally derived immortalized human NSCs, referred to as F3 NSCs, have been implanted through ICV administration in murine models of LSDs, stroke, Parkinson's disease, and Huntington's disease [13]. For all disease states studied, these cells were capable of migrating inside the CNS, integrating in brain parenchyma, and repairing damaged tissue through the replacement of degenerated cells and release of neurotrophic factors. Additionally, SCs are capable of modulating the inflammatory environment in neurodegenerative diseases by providing trophic support to synapses and scavenging toxic metabolites [7, 13]. Such results highlight the powerful therapeutic potential stem cells possess for alleviating the severity of neuronopathic LSDs.

A variety of stem cells hold significant promise as vectors for enzyme delivery in LSDs, especially in view of their extensive migratory capacity throughout a transplanted individual [12]. *In vitro* and *in vivo* tools for determination of stem cell safety and efficacy must be further developed, and non-invasive transplantation techniques must be actively pursued to circumvent problems associated with the poor regenerative capacity of the brain [3]. The implementation of SCT as primary treatment in LSDs will require the extensive evaluation of data in preclinical and phase I/II trials to assess the most appropriate cell type for each individual storage disease.

Delivery of therapeutic product(s) to the entire CNS may be required for successful treatment of globally expressed, heritable disorders like LSDs, and currently, there are three main routes of administration to the CNS: intraparenchymal, intraventricular, and intrathecal [15]. Transplanted stem cells have been reported to migrate throughout the recipient and cross the blood brain barrier (BBB) when administered systemically, and recent publications highlight the ability of rodent MSCs to travel to the brain when administered by an intranasal route [24]. Such non-conventional routes of administration for SCT may provide therapeutic benefit without exposing the patient to the high risks associated with invasive surgical procedures. Increasing the permeability of the BBB to enhance the engraftment in the brain of transplanted stem cells should also be investigated, especially in light of the newly developed technology that incorporates ultrasound and MRI guided microbubble delivery to the BBB [24].

Disease severity and phenotype should be assessed when determining SCT eligibility, especially considering that HSCT is effective only in certain LSDs if given in the asymptomatic period (i.e., before neuro-cognitive damage occurs) [17, 40]. Unfortunately, this asymptomatic period is often in the first sixth months of an affected patient's life [2]. This highlights the need for widespread newborn screening; however, even with prenatal screening testing widely available, there are only a few states (e.g., New York) that mandate newborn screening of specific

LSDs due to the very limited treatment options available, the poor prognosis of even the earliest treated patients, and the inability to predict phenotype with certainty [17, 41, 42].

3.3 Specific LSDs and Current Stem Cell Research

The following discussion highlights the current findings in stem cell research as it relates to several LSDs that have shown significant improvement after SCT in preclinical or clinical trials (Table 3.1). The LSDs that have been tested using SCT are described, and a review of the current state of LSD research is presented to highlight any clinical improvement resulting from SCT in animal models (Table 3.2). A large number of naturally occurring and genetically modified animal models of neuronopathic LSDs have been utilized, including large (e.g., cat, sheep, rhesus macaques, etc.) and small (e.g., mouse) animal species (Table 3.1) [6]. Disease associated improvements subsequent to SCT indicate a promising future for such cells as a novel therapeutic for patients affected with LSDs.

3.3.1 Alpha-Mannosidosis

There are over 40 known mutations of alpha-D-mannosidase (MANB) that result in alpha-mannosidosis, an autosomal recessive LSD with a wide range of clinical phenotypes. The MANB enzyme serves an essential role as a lysosomal hydrolase in glycoprotein degradation, and its deficiency leads to multisystemic accumulation of mannose-rich oligosaccharides before birth and throughout a patient's lifetime, especially in the CNS [22, 100]. This disease can clinically present as a mild (type I), moderate (type II) or severe form (type III). The most severe form can manifest as prenatal loss, but often type III affected infants develop coarse facial features, impaired hearing, corneal clouding, hepatosplenomegaly and skeletal abnormalities before dying prematurely as a result of primary CNS disease (e.g., ataxia) and/or recurrent infection. Serious bacterial infections are common in these patients due to the decreased ability of leukocytes to undergo intracellular killing [100].

Preclinical trials that test cell-based therapies for alpha-mannosidosis are ongoing as a result of excellent cat and guinea pig animal models. These animals have similar pathological and clinical disease presentation to that of affected human patients [22]. In the feline model, results show that HSCT can prevent neurologic deterioration if administered in asymptomatic animals, and this procedure increases the levels of MANB enzymatic activity in the brain of transplanted cats [20, 24]. This is a remarkable finding in light of the massive accumulation of oligosaccharides in neurons and glial cells of affected animals [55]. The guinea pig has been used to assess the therapeutic effects of ESC transplantation. Partially differentiated mouse ESCs were implanted into the brains of neonatal guinea pigs in an attempt

Table 3.1 Enzyme deficiencies and substrate accumulation in various lysosomal storage diseases. Listed are the most commonly used animal models to test stem cell-based therapy for LSDs

Storage disease	Enzyme deficiency	Accumulated substrate	Animal models
Alpha-mannosidosis	Alpha-D-mannosidase	Oligosaccharides	Cat, guinea pig [22]
Aspartylglucosaminuria	Aspartylglucosaminidase	Asn-GlcNAc	–
Fabry disease	Alpha-galactosidase A	Globotriaosylceramide (Gb3)	Mouse (*galactosidase A-KO*) [43]
Farber disease	Acid ceramidase	Ceramides	–
Gaucher disease	Glucocerebrosidase (GCase)	Glucosylceramide	Mouse (*GCase-KO*) [44]
Krabbe disease	Galactosylceramidase	Galactosylceramide	Mouse, rhesus macaque [45, 46]
Metachromatic leukodystrophy	Arylsulfatase A (ASA)	Galactosyl-3-sulfate ceramide	Mouse (*ASA-KO*) [11, 18]
MPS IH (Hurler's disease)	Alpha-L-iduronidase	Heparan and dermatan sulfate	Cat, dog [47]
MPS III (Sanfilippo syndrome)	GAG degrading enzymes	Heparan sulfate	Mouse (*KO*), dog [48]
MPS VII (Sly syndrome)	Beta-glucuronidase	Various mucopolysaccharides	Mouse (*KO*), cat, dog [5]
Niemann-Pick disease (types A, B)	Acid sphingomyelinase (ASM)	Sphingomyelin	Mouse (*ASM-KO*) [49, 50]
Niemann-Pick disease (type C)	Sphingomyelin phosphodiesterase	Sphingomyelin	Mouse, cat [51, 52]
Neuronal ceroid lipofuscinosis	Palmitoyl protein thioesterase (PPT)	Lipofuscin material	Mouse (*PPT1-KO*) [53]
Sandhoff disease	Beta-hexosaminidase (A and B)	Ganglioside	Mouse (*HexB-KO*) [54]
Tay Sachs disease	Beta-hexosaminidase A	Ganglioside	Mouse (*HexA-KO*) [54]

Table 3.2 Preclinical and clinical stem cell (SC) therapy for various lysosomal storage diseases

Storage disease	Preclinical and clinical cell-based therapies
Alpha-mannosidosis	Feline: HSCT [20, 24], monocytes [41]
	Guinea pig: partially differentiated mouse ESCs [22]
	Human patients: peripheral blood-derived stem cells [40], **HSCT** [55]
Aspartylglucosaminuria	Human patients: HSCT (limited success after symptomatic) [56–60]
Fabry disease	Mouse: **HSCT** [61], **transduced HSCs** [62–64], **autologous HSCs** [65]
Farber disease	Type 1 human patients: HSCT (limited effect on CNS symptoms) [66]
	Types 2 and 3 human patients: **HSCT** [67–69]
Gaucher disease	Mouse: human transduced bone marrow-derived HSCs [44, 70–73]
	Types 1 and 3 human patients: **HSCT** (improved somatic manifestations) [74–78]
Krabbe disease	Mouse: BMSCs and ASCs [29, 79], NSCs [80, 81], HSCT [82, 83]
	Human patients: **HSCT** (limited effect on CNS symptoms) [29, 84, 85]
Metachromatic leukodystrophy	Mouse: human ESCs, transduced OPCs, and GPCs [11], transduced HSCs [86]
	Human patients: **HSCT** (not effective in infantile type), UCB-derived HSCs [9, 21, 23], transfected CD34 cells (phase I trial in progress) [21]
MPS IH (Hurler's disease)	Human patients: **HSCT** [5, 87]
MPS III (Sanfilippo syndrome)	Mouse: UCB-derived HSCs [88], transduced mouse GPCs and HSCT (only limited success) [48]
MPS VII (Sly syndrome)	Mouse: transduced MSCs (BMSCs) [33], **transduced NSCs and NPCs** [7, 12, 89]
Niemann-Pick disease (A, B)	Mouse: HSCT, transduced MSCs, UCB-derived HSCs, NSCs [90–94]
	Type B human patients: HSCT [95, 96]
Niemann-Pick disease (C)	Mouse: BMSCs, ASCs, UCB-derived HSCs, NSCs [97, 98]
Neuronal ceroid lipofuscinosis	Mouse: **human CNS-derived stem cells** [53]
	Human patients: human CNS stem cells (phase I/II clinical trial for INCL patients) [53, 99]
Sandhoff disease	Mouse: **mouse NSCs** [98]
Tay Sachs disease	Mouse: mouse NSCs [12]
	Human patients: HSCT (only limited success) [17]

The most successful stem cell transplantations in preclinical and clinical trials are listed for each LSD, separated by species tested. All SCT that resulted in complete reversal of lysosomal storage and/or impressive outcomes are listed in **Boldface** type. Outcomes of each procedure listed are described in detail in the LSD subsections of Sect. 3.3

to provide a source of MANB in the CNS, and cells were able to extensively migrate throughout the brain, differentiate into neural lineages, and survive for up to 8 weeks post-transplantation [22]. However, high cell doses (e.g., 10^5 cells per hemisphere) resulted in teratoma formation in most of the transplanted animals [22]. These preclinical results highlight the importance of developing more stringent differentiation protocols to circumvent problems associated with pluripotent cell contamination.

Sun and colleagues have developed a retrovirus vector that expresses high levels of MANB in fibroblasts derived from affected cats and human patients [101]. Using gene therapy methods, mouse D3 ESCs and various adult stem cells have been successfully modified to overexpress human lysosomal alpha-mannosidase [102]. Although no clinical trials have implanted these modified cells, increasing the dose of the enzyme through over-expression would be expected to increase the therapeutic benefit of SCT. An interesting study using the feline model of this LSD demonstrated the feasibility of directly transplanting affected animals *in utero* with monocytic cells derived from normal marrow. This was the first description of an *in utero* transplantation in a cat model and marked the first *in utero* transplantation of cells other than HSCs. This strategy has the potential to prevent morbidities of LSD affected humans before a definitive treatment (e.g., SCT) can be provided [55].

Cell-based therapies have already been applied in clinical settings for several MANB deficient patients. The increase of MANB enzyme levels and the decrease of mannose-rich oligosaccharide urinary excretion have been reported for a young boy 13 months after receiving peripheral blood-derived stem cells from his HLA-identical mother [100]. Retrospective analysis of 17 HSC transplantations provides evidence of cell engraftment in all 15 of the survivors. Although moderate development progress was documented for all patients, eight developed GVHD and three underwent subsequent transplantation due to graft rejection [103]. Allogeneic bone marrow-derived stem cell transplantations have shown reduced pathology and clinical improvements in three out of four treated patients; specifically, adaptive skills and verbal memory were shown to improve after this procedure [20]. Other beneficial effects included the resolution of hepatomegaly and improved hearing in two of the patients [20, 100]. These results suggest that HSCT can halt the progression of alpha-mannosidosis, but based on preclinical studies, it is unlikely that such marked neuro-cognitive improvements would be demonstrated in patients with advanced stages of the disease.

3.3.2 *Aspartylglucosaminuria*

Aspartylglucosaminuria (AGU), a rare disease mainly found in Finland, is the most common LSD involving dysfunctional glycoprotein metabolism. AGU is caused by the defective activity of aspartylglucosaminidase, which leads to the accumulation of the Asn-GlcNAc linkage unit in tissue lysosomes [56, 104]. The clinical hallmarks of AGU involve mental retardation, skeletal abnormalities, and coarse facial features [57, 58]. The patients usually develop normally during infancy, mental retardation becomes worse during adolescence, and their lifespan is usually less than 50 years [59].

Numerous HSC transplantations have been performed on AGU patients [59, 60, 105–107]. While some reported promising results, many commentaries have indicated that there is continued failure. Studies have indicated that the timing of transplantation is very important and that AGU patients should receive transplantation as newborns before the presentation of symptoms.

3.3.3 Fabry Disease

Fabry disease, an X-linked LSD, is caused by the deficiency of α-galactosidase A (α-Gal A) and the subsequent systemic accumulation of sphingoglycolipid ceramidetrihexoside, also known as globotriaosylceramide (Gb3), predominantly in the lysosomes of vascular endothelial cells; this substrate accumulation often leads to renal failure, cardiovascular and cerebrovascular disease [61–63, 65, 108, 109]. A mouse model of Fabry disease has been generated by homologous recombination of a mutated 129/SvJ-derived genomic Gla clone in J1 ESCs, which were obtained from the 129 S4/SvJae strain [43]. Although the mice appear clinically normal at 10 weeks of age, the ultrastructure and lipid analyses reveal the accumulation of Gb3 in the liver and kidney, indicating that the pathophysiology in the mouse model is similar to that of patients with Fabry disease. The model has been widely used as a tool to assess treatment options for affected patients.

HSC transplantation from wild type mice has been performed on the mutant mice, and it was found that the α-Gal A activity was increased in the liver, kidney and heart with a concomitant reduction of Gb3 levels 6 months after the BMT [61]. α-Gal A deficient bone marrow cells were transduced with retrovirus or lentivirus encoding α-Gal A and transplanted into the α-Gal A deficient mice [62–64], and autologous transplantations have also been investigated under multiple reduced-intensity conditioning regimens [65]. HSCT using transduced cells led to long-term enzyme correction and Gb3 lipid reduction in multiple organs of the Fabry mice, and the outcome was enhanced if the functionally transduced cells were selectively enriched prior to transplantation. The minimum requirement of donor cells or gene corrected cells to reduce Gb3 was also investigated [108]. The normal bone marrow cells were mixed with α-Gal A deficient bone marrow cells at various ratios and were transplanted into Fabry mice, and the study showed that a 30% gene correction might be sufficient to reverse disease manifestation in Fabry disease.

3.3.4 Farber Disease

Farber lipogranulomatosis, which was first described by Farber in 1957 [110] and later described in detail by Moser and Chen [111], is a rare LSD caused by mutation of the *ASAH1* gene, which encodes for acid ceramidase. Accumulation of ceramide in cells and tissues [67, 112, 113] leads to several phenotypes (types 1–7) of Farber disease, differing in severity and organ involvement, which can include the central nervous system, heart, lung and lymph nodes. Although the phenotype varies, most patients present characteristic symptoms like progressive hoarseness, subcutaneous granulomas, and painful swollen joints [67, 112, 113].

HSCT was first performed in 1989 [114] and 2000 [66] on patients with Farber disease type 1, which mainly involves the CNS. While the non-neurological manifestations (e.g., granuloma formation) resolved, both patients died of progressive neurological disease. Since 2001, several patients with Farber disease types 2 and 3, which do not involve the CNS, have received HSCT, and all transplanted patients have demonstrated improvement, including improved joint mobility and regression of granulomas [67–69]. HSCT seems ineffective for Farber patients with CNS involvement, especially because ceramide neurotoxicity may not be reversible by BMT. However, BMT data are very encouraging for patients with types 2 or 3 Farber disease without CNS involvement.

3.3.5 Gaucher Disease

First described by Fillip Gaucher in 1882, Gaucher disease is the most prevalent LSD and is caused by mutations in the structural gene that encodes for glucocerebrosidase. Deficiency of this lysosomal enzyme results in the build-up of glucosylceramide in macrophages throughout the reticuloendothelial system [70, 115–117]. The clinical symptoms of Gaucher disease include hepatosplenomegaly, hypersplenism and lytic bone lesions. Based on the CNS involvement, Gaucher disease is classified into three types: type 1 is the most common type and does not involve CNS; type 2 is a rare, infantile form with severe neurological symptoms that often leads to death by the age of 2; and type 3 patients develop slowly progressive neurological symptoms and often live to be 40 years [70, 115–117].

Substrate accumulation in affected macrophages is responsible for the pathology and symptoms displayed in Gaucher disease, and since macrophages are derived from bone marrow, BMT has routinely been used to replace the defective macrophages with new marrow-derived macrophages containing functional glucocerebrosidase enzyme [74–78]. BMT has shown beneficial effects in types 1 and 3 Gaucher disease affected patients who received and survived the transplantation procedure. The somatic symptoms were greatly alleviated if the BMT was given before the development of irreversible skeletal and organ changes [74–78].

Transplantation of HSCs transduced with retroviral and lentiviral vectors containing human glucocerebrosidase into the murine model of Gaucher disease also has been widely investigated [44, 70–73]. These studies showed that HSCs could be efficiently transduced if the HSCs were prestimulated by cytokines and growth factors [44, 70–72], and glucocerebrosidase transduced HSCs could generate higher enzyme activity in the murine models.

3.3.6 Krabbe Disease

Krabbe disease (KD), also known as globoid cell leukodystrophy (GLD), is an inherited, fatal neurodegenerative disorder that results from the deficiency of galactosylceramidase (GALC), a lysosomal enzyme responsible for the degradation of glycolipids such as galactosylceramide (GalCer) and galactosphingosine (psychosine) during active myelination [45, 118–121]. With an incidence of 1 in 100,000 live births, Krabbe disease is characterized by demyelination both in the central and peripheral nervous systems with rapid neurological deterioration and death. Based on the onset and severity, Krabbe disease is clinically divided into infantile type, juvenile type and adult type [122]. The infantile variant of GLD is most common with symptomatic onset between 3 and 6 months of age, and symptoms include vomiting, feeding difficulty, extreme irritability, complete blindness, deafness, spastic paralysis, extreme emaciation and dementia leading to death by 18 months [41, 45, 123–126].

Animal models, including the dog, sheep, rhesus monkey and mouse, are extensively being used to develop safe and effective treatments for GLD [45, 46]. The twitcher mouse model is an authentic model of human GLD that developed through spontaneous mutation of the GALC gene ($GALC^{twi}$) at the Jackson Laboratory in 1976 [125, 127]. Studied mainly in the C57Bl/6 genetic background, the twitcher mouse presents at postnatal day (PND) 14–15 with symptoms of motor function deterioration, weight reduction, and decreased activity; this onset of symptoms correlates well with the onset of the active myelination period [46]. Further disease progression leads to the twitcher mouse developing hind limb weakness, kyphosis of the spine, and severe wasting and twitching. If left untreated, affected twitcher mice will experience rapid motor and neurological deterioration and death by PND40 [45, 127]. Pathological features are similar to those found in the affected human, including central and peripheral demyelination, astrocytic gliosis, psychosine-induced apoptosis of oligodendrocytes, and increased brain inflammation, thus highlighting the usefulness of this model to study human GLD [46, 127, 128].

The twitcher mouse model has been treated using HSCT [82, 83], combination of HSCT and gene therapy [129, 130], bone-marrow and adipose-derived mesenchymal stem cells (BMSCs, ASCs) transplantation [29, 79], and HSCT with ERT [80, 81]. These treatments prolonged the lifespan of the treated mice; however, the treated mice died with symptoms of progressive neurological degeneration, indistinguishable from those untreated twitchers. The ICV transplantation of BMSCs and ASCs significantly decreased the brain inflammation of twitcher mice, although they only modestly improved the motor function and the lifespan of the affected mice [79].

Currently, there is no cure for GLD. The only known treatment is HSCT, such as bone marrow or umbilical cord blood HSCT, after myeloablative chemotherapy. HSCT has been shown to increase GALC levels and slow the progression of GLD [29, 84, 85], and it has the potential to provide life-long gene correction as a result of the self-renewing properties of the donor myeloid progenitor cells [5]. However, such transplantation is not effective in patients with overt neurological symptoms or aggressive forms of the disease (e.g., early-onset infantile type) [19], and the majority of presymptomatic infants transplanted still develop severe disabilities, including motor and language deterioration [42, 84, 131], and the treatment of GLD with bone marrow transplantation is currently limited by the unavailability of donors.

3.3.7 *Metachromatic Leukodystrophy*

Leukodystrophies, such as GLD and Metachromatic leukodystrophy (MLD), are characterized as LSDs that involve neurological dysfunction typically as a direct result of myelin absence or loss due to the death of neurons and/or myelin producing cells in the CNS. There is no cure for any leukodystrophy, and only supportive therapies (e.g., seizure control) are currently available for the mitigation of the disease severity. The main goal in developing cell-based therapy for leukodystrophies has been to replace myelinogenic cells using cells capable of differentiating into oligodendrocytes and also to transplant non-oligodendrocyte cells capable of providing enzyme through cross-correction [17, 23].

Metachromatic leukodystrophy (MLD) is an autosomal recessive inherited LSD due to the deficiency of arylsulfatase A (ASA), an enzyme that catalyzes the degradation of a major myelin sphingolipid, galactosyl-3-sulfate ceramide [4, 11, 21]. The non-metabolized sulfatide molecules accumulate in metachromatic granules of oligodendrocytes and Schwann cells mainly, but also in macrophages, neurons, and various visceral organs (e.g., kidney, liver and pancreas) [11, 17]. The high concentrations of sulfatide substrate in myelin lead to symptoms associated with progressive demyelination of the CNS and PNS, including ataxia, seizures, dementia, spastic quadriparesis, deafness, and possible blindness from optic atrophy [11, 21]. The four disease variants of MLD include the late infantile, early juvenile, late juvenile, and adult forms, and patients with infantile and juvenile MLD present with a more severe phenotype due to their possession of one or two null mutations in the *ASA* gene [9, 21].

There is no naturally occurring animal model of MLD; however, an ASA knockout (KO) mouse has been generated and various preclinical trials have been performed using stem cell-based therapies [11, 18]. Specifically, human ESCs have been genetically modified to overexpress ASA 30% more than cells transfected with a control vector, and implantation of these cells into the brain of ASA-deficient mice has shown a dramatic reduction of sulfatide deposits in areas surrounding the hASA-positive engrafted cells [11]. Other cell types, such as primary oligodendrocyte progenitor (OPCs) and glial progenitor cells (GPCs), have also corrected sulfatide storage in these ASA-deficient mice after brain implantation [11]. However, to enable a global distribution of stem cells or progenitor cells, multiple transplantation procedures using high numbers of cells in the neonatal animal will likely be necessary to globally correct the metabolic defect. In one recent study, HoxB4 overexpressing bone marrow-derived HSCs were transplanted in an attempt to increase the number of cells capable of enzyme delivery in the brain of these mice; transplanted animals had improved behavior test performance (e.g., balance beam walking) and decreased sulfatide accumulation in brain tissue [86]. HoxB4 transduction increases the self-renewal and regeneration of HSCs, but the possibility of these animals developing myeloid leukemia long-term is a serious consequence that should be studied extensively before clinical application [86].

The promise of cell-based therapy for MLD is evident from preclinical trials, but it should be noted that the ASA deficient mouse has only a mild phenotype with scarce demyelination [18]. MLD therapy is currently limited to symptomatic treatment in human patients, and use of HSCT in MLD is contradicted unless performed in affected newborns as a part of approved clinical trials [11]. MSC and HSC grafts have proven ineffective in correcting CNS manifestations in late infantile MLD or in any MLD patient with overt neurological symptoms [9, 21, 23]. Preclinical success using hASA-transduced HSCs and primary cells suggests that autologous HSCT after genetic modification may be an effective source of ASA for these patients without the side effects associated with allogeneic HSCT [21]. A phase I/II clinical trial to test autologous ASA-transfected CD34 cells for MLD treatment is in the final process of being approved [21].

An alternative approach to using bone marrow-derived HSCs is the transplantation of umbilical cord blood (UCB) in MLD affected children. Umbilical cord blood stem cell transplantation (UCBSCT) in three siblings improved nerve conduction and neurophysiological abilities and showed total resolution of signal abnormalities on neuroimaging for the two youngest siblings [9]. These results highlight UCBSCT as a better option than bone marrow HSCT for MLD patients [9, 17]. All clinical studies using cell-based therapy for MLD have shown that transplantation is ineffective for symptomatic disease, thus indicating that the absence of symptoms based on a pre-transplantation neurological examination may be the most significant predictor of transplantation outcome in these individuals [9].

3.3.8 Mucopolysaccharidosis

The group of LSDs called mucopolysaccharidosis (MPS) is categorized into seven distinct diseases: MPS types I–VII. All types except MPS II are inherited as autosomal recessive disorders resulting from the defective catabolism of glycosylaminoglycans (GAGs), which are long, unbranched polysaccharides that consist of repeating disaccharide units [88]. Such molecules are precursor components of connective tissue; therefore, the accumulation of GAGs leads to clinical consequences in the connective tissue of skin, heart valves, airway, and the skeleton of affected MPS patients [132]. Additionally, GAGs accumulate in neuronal cells, and patients often develop severe, progressive neurological manifestations, such as behavioral disturbances, dementia, and decline of all motor function [88]. Although stem cell therapy has been tested in both preclinical and clinical settings for every MPS type, only MPS type IH (Hurler's disease), MPS type III (Sanfilippo syndrome), and MPS type VII (Sly syndrome) will be discussed herein.

The first HSCT on an LSD patient was performed in 1981 by Hobbs in an attempt to treat MPS IH [88]. Since then, most clinical experience in stem cell treatment for MPS has been gained with MPS type IH, which is due to an alpha-L-iduronidase deficiency [133]. Those affected patients who receive HSCT before 2 years usually develop normal hearing and speech that is appropriate for their developmental stage. In addition to improved cognitive and neurological function, HSCT has resulted in decreased GAG excretion, resolution of hepatosplenomegaly, reduced corneal clouding, and improved obstructive airway symptoms in MPS type I patients [5, 87]. The therapeutic potential of combining ERT and HSCT has been explored, and it seems that ERT administered before HSCT helps reduce respiratory complications after transplantation [134]. Overall, impressive results have been obtained in children with MPS I after stem cell-based therapy.

HSCT may dramatically alter the course of many mucopolysaccharidoses, such as MPS I, but this has not been seen in every type. MPS III, the most common type of MPS, results in the accumulation of heparan sulfate due to the defective nature of one of four lysosomal enzymes required to degrade this specific GAG [88]. Although mild somatic symptoms occur in MPS III affected individuals, this disease is characterized by severe neurological deterioration that leads to progressive dementia. Inflammatory responses from neuronal cell death most likely contribute to the pathophysiology of this MPS, making anti-inflammatory adjuvant therapy a promising agent for combination with other procedures, such as SCT [134]. However, to date, HSCT has not been shown to be an effective treatment for MPS III, even when administered prior to symptom onset [134].

Interestingly, UCB-derived HSCs have been administered both through ICV and IV routes in MPS III patients, and these patients did show improvement in cognitive function [88]. Such results indicate that human UCB may be a better source of HSCs than bone marrow for treating MPS types involving severe neurological consequences, perhaps because this source contains a higher number of HSCs capable of developing into neural cells [88]. Robinson and colleagues recently developed sulfamidase-overexpressing GPCs derived from ESCs for the treatment of MPS III mice. Intraparenchymal injections proved successful for long-term persistence (12 weeks), and the lack of teratoma formation indicated that this newly designed protocol for the selective expansion of GPCs may be sufficient for eliminating residual pluripotent cells capable of undergoing cancerous proliferation [48].

MPS VII is a rare form of MPS caused by the deficiency of beta-glucuronidase; however, SCT has been extensively investigated for this type due to the availability of a NOD/SCID/MPS VII mouse model [5, 33]. An almost complete normalization of GAG storage has been documented in this mouse model after transplantation of BMSCs overexpressing beta-glucuronidase. Such cells engrafted throughout the brain and persisted for up to 4 months after transplantation [33]. Similarly, beta-glucuronidase-transduced NSCs and neural progenitor cells (NPCs) have been injected through ICV administration into presymptomatic MPS VII mice, and results demonstrated that both cell types had extensive migratory capacities and the ability to reduce lysosomal storage throughout the brain [7, 12, 89].

3.3.9 Niemann-Pick Disease

Niemann-Pick disease (NPD), a classic autosomal recessive genetic disease, is classified into a subgroup of LSDs called sphingolipidoses or lipid storage disorders. Now with a better understanding of the molecular genetics of the disease, NPD has been subdivided into two groups: acid sphingomyelinase (ASM) deficiencies (NPD types A and B), which are due to sphingomyelin phosphodiesterase 1 *(SMPD1)* mutations, and Niemann-Pick type C (NPC), which results in the defective trafficking of low-density lipoprotein (LDL) derived cholesterol due to *NPC1* or *NPC2* mutations [135, 136]. Type A NPD, also known as infantile NPD, is the most common type and is a severe, neuronopathic disorder with which children rarely live beyond the age of 3 [90, 91, 137, 138]. Type B is a milder, non-neurologic form that occurs in late childhood or adolescence [137]. NPC is a neuro-visceral condition with an extremely heterogeneous age of onset ranging from perinatal to adulthood, and the lifespan varies, too, with most deaths occurring between 10 and 25 years of age [135, 136, 139–142].

There are no naturally occurring animal models with type A and B NPD; however, two mouse models have been generated through gene targeting by two independent groups [49, 50]. These two ASM knockout (ASMKO) mouse models have slightly different phenotypes, but both resemble the human forms clinically, biochemically and pathologically. Extensive research has been conducted on naturally occurring murine and feline models of NPC with spontaneous NPC1 mutations, which biochemically, clinically and morphologically resemble the human NPC disease [51, 52].

The Schuchman group has investigated SCT for the treatment of ASMKO mice [90–94]. Direct bone marrow transplantation (BMT) [90], hematopoietic stem cell therapy (HSCT) [91] and intracerebral transplantation of MSCs transduced with a retroviral vector over-expressing ASM [92] have been used to treat ASMKO mice independently, and the results demonstrate that all three methods could improve the lifespan and cerebellar function of the treated animals. BMSCs [94, 143–145], ASCs [146], human umbilical cord blood-derived MSCs [147], and NSCs have been investigated for therapeutic effects on mice with NPC [97, 98]. Upon transplantation into the cerebellum of NPC mice, MSCs developed into electrically active Purkinje neurons; the functional synaptic transmission demonstrated in the neurons suggests that adult SCs may have the potential to promote neuronal networks [94]. Such findings shed light on how SCs would uniquely benefit neuronopathic LSD patients, whose affected neurons undergo apoptosis or necrosis during advanced disease stages.

The combination of HSCT and intracerebral MSC transplantation was also investigated for NPC, and it was shown that the combined procedure leads to a more improved clinical and pathological outcome compared to the individual transplant procedures [93]. However, although the histology phenotype was improved in the reticuloendothelial organs and the onset of ataxia was delayed, none of the therapeutic effects persisted, and all treated mice eventually developed neurological symptoms and died prematurely. This research demonstrated that HSCT and MSC transplantation might be viable options for type B NPD, which is non-neurological, but not sufficient therapy for type A NPD. Additionally, mouse and human NSCs have been used to treat ASMKO mice, and the data showed that there was no obvious improvement in the behavioral disorder, although there was a dramatic decrease in neuronal and glial vacuolization and cholesterol accumulation in parts of the brain [148].

Clinically, there is no effective treatment for type A NPD. One prenatally diagnosed girl with type A NPD received umbilical cord blood stem cell transplantation (UCBSCT) when she was 3 months old, but there was only minimal overall benefit from the UCBSCT [138]. HSCT has proven more effective for type B NPD. HSCT was clinically performed on two affected girls with type B NPD [95, 96], and while the initial transplantation failed on both of them, the symptoms were highly improved after their second transplantation, although they suffered from GVHD.

3.3.10 Neuronal Ceroid Lipofuscinosis (Batten Disease)

Neuronal ceroid lipofuscinoses (NCLs) are the most common inherited pediatric neurodegenerative disorder [99], and this family, which is comprised of over ten distinct LSDs, is among the most common progressive encephalopathies of childhood [99, 149]. With more than 269 causal mutations in eight known genes, NCLs are a result of deficiencies in the lysosomal enzyme, palmitoyl protein thioesterase-1 (PPT1) [53, 149]. All NCL variants involve the loss of neurons in the cortical regions of the cerebrum and cerebellum due to the accumulation of lipofuscin-like material in neural cells. Disease presentation often includes loss of vision, decreased cognitive and motor skills, seizures, and premature death [149].

Tamaki and colleagues have generated a mouse model for infantile NCL (INCL) on the NOD/SCID background. Promising preclinical trials have been performed on these mice, including the transplantation of human CNS stem cells that had been grown as neurospheres. This ICV treatment provided SCs capable of migrating, differentiating into neurons and myelin producing cells, and releasing the PPT1 enzyme through cross-correction. This group provided convincing evidence for the reduction of accumulated lipofuscin material in multiple affected organs and a decreased rate of motor function decline compared to control mice [53]. Due to these preclinical results, a phase I trial is currently treating infantile NCL (INCL) patients with human CNS–derived stem cells. To date, six patients have each been treated with one billion SCs using intraparenchymal and cerebral ventricular administration. This trial has thus far demonstrated the acceptable safety profile of SC transplantation in INCL patients, although continued monitoring of these patients will have to be performed to assess long-term side effects of this procedure [53, 99].

3.3.11 *Sandhoff Disease and Tay Sachs Disease*

LSDs that are caused by genetic defects in the degradation of sphingolipids are referred to as sphingolipidoses; two sphingolipidoses are Sandhoff and Tay Sachs diseases. As is the case for most LSDs, decreased catabolic hydrolase activity constitutes the main deficiency for this subgroup of diseases [4]. Both Sandhoff and Tay Sachs diseases result from the deficiency of beta-hexosaminidase, and patients with these diseases are clinically indistinguishable. These are progressive disorders, and patients usually develop severe mental retardation and motor dysfunction before dying prematurely, often in infancy [98].

NSCs have been the primary stem cell used in the mouse models of Sandhoff and Tay Sachs diseases, most likely due to their potential to alleviate the severe CNS dysfunction (mainly cerebellar pathology) that contributes to the symptomatology of these sphingolipidoses [98]. Primary and secondary NSCs from fetal tissue and ESCs, respectively, have been transplanted into neonatal mice that model Sandhoff disease, and these cells prolonged life, improved motor function, sustained coat condition, reduced ganglioside storage, and reduced brain inflammation (e.g., activation of microglial cells) [98]. When these cells were transplanted at two different time points through intraparenchymal and ICV routes, the results proved even more promising, with extension of lifespan and further delayed symptom onset [98]. A separate study provided evidence that NSCs prolonged lifespan and improved motor function in adult Sandhoff mice even when transplantation occurred after the mice were symptomatic [150]. Similarly, human immortalized NSCs in a mouse model of Tay Sachs disease have been shown to reduce ganglioside accumulation [12]. The majority of affected patients have responded poorly to HSCT, and NSCs may serve as alternative delivery vectors for beta-hexosaminidase to treat patients with Sandhoff and Tay Sachs diseases [17].

3.4 Limitations and Complications of Cell-Based Therapy for LSDs

As previously mentioned, HSCT has been used extensively as a therapy for LSDs, and survival data indicate that HSC engraftment can prolong lifespan by two decades in a limited number of LSD patients [151]. Although marked clinical improvement has been documented in some patients, the majority of HSCT cases result in only slight mitigation of disease severity [134]. In individuals who need transplantation frequently, HSCT has a high mortality rate and risk for graft failure and graft versus host disease (GVHD), a serious and often fatal complication resulting from the immunological attack of foreign T cells against the patient [152–155]. The significant risk of GVHD, the harsh conditioning regimens, and the high mortality rates associated with HSCT (reported between 10% and 25%) all have helped form the current recommendation for physicians that HSCT should be administered to only those LSDs that show a clear beneficial response and for which enzyme replacement therapy is not standard of care [1, 2]. As recommended by Peters and colleagues, HSCT should be considered as the primary treatment for the following LSDs: MPS Types IH, VI, and VII, globoid cell leukodystrophy, metachromatic leukodystrophy, alpha-fucosidosis, alpha-mannosidosis, Gaucher disease, and Niemann-Pick disease type B [2]. It is likely that the progression rate of most storage diseases is too rapid for the transplanted cells to efficiently engraft and replace affected cells throughout the patient [21].

Several studies have indicated that both bone marrow and umbilical cord blood-derived HSCT have succeeded in stabilizing neurodegeneration associated with LSDs, potentially as a result of transplanted cells migrating into the CNS to reconstitute therapeutic levels of lysosomal enzymes in the affected brain [9]. Umbilical cord blood is an alternative tissue source for obtaining hematopoietic cells, and compared to bone marrow-derived hematopoietic cells, cord blood-derived HSCs have better engraftment rates, decreased frequency of mixed chimaerism, lowered incidence of rejection and transmission of CMV, and can be transplanted rapidly upon diagnosis [3, 156]. In several studies, bone marrow obtained from affected LSD patients has been modified to overexpress specific lysosomal enzymes for subsequent autologous transplantation; such a procedure, now in phase I/II clinical trials, eliminates the need for dangerous myeloablation treatment, eliminates the risk of GVHD, and overcomes the limitation of potential donor availability [3, 6].

Therapeutic efficacy of cell-based therapy depends on disease-specific parameters, such as enzyme secretion rates, and also on the timing of administration, severity of disease causing mutations, and nature of the accumulated substrate. Optimization of SCT (e.g., administration route, timing, cell type, adjuvant therapy, etc.) should be investigated in human clinical trials focused on safety and efficacy; however, clinical trial design for LSDs can be limited in power and sample size due to the small number of LSD patients available for testing. Additionally, clinical trials of rare neurodegenerative diseases often enroll patients in advanced stages due to this limitation in sample size, and such patients often have co-morbidities and irreversible, end-stage organ damage that confound results. Specifically, the inflammatory state and gliotic changes that accompany late stage LSD often impede the engraftment of stem cells, limiting the therapeutic potential of cell-based treatment [15].

There are multiple secondary effects, including altered cellular signaling and increased inflammation, that result from the single gene deficiency of LSDs, and these mechanisms of pathogenesis increase the complexity associated with designing a single therapeutic agent [1]; thus, there is not one type of stem cell that can be generalized as a potential vector for enzyme delivery for all LSDs [11]. The complexity of cellular and tissue changes in LSDs highlights the potential for combining other treatment modalities with SCT to receive synergistic or additive therapeutic effects upon transplantation. This approach of combining various therapeutic strategies utilizes the regenerative capacity of the stem cell, especially in advanced stages when neural pathways are lost, while also targeting a secondary consequence of disease with transient therapies, such as ERT, substrate reduction, or anti-inflammatory pharmacological agents [134].

3.5 Discussion: The Potential of Stem Cell Therapy for LSDs

The long-term efficacy of SCT needs to be further evaluated to determine whether multiple transplantations should be performed in LSD patients, and the effects of successive transplantations should also be evaluated to ensure that no immune challenges result from a second allograft [15]. Reproducing preclinical results of NPCs in LSD animal models, a recent phase I clinical trial for the use of human NPCs in children with a storage disease (INCL) demonstrated safety, engraftment, and long-term survival of transplanted cells [6]. Future trials using NPCs engineered to express high levels of therapeutic enzyme would be useful for comparison to similar preclinical trials [6]. Other considerations of SCT include the ideal donor tissue source due to each type of stem cell having their own unique advantages and disadvantages. With the accessibility of adult stem cells and recent advances in using these cells both *in vitro* and *in vivo*, it is unlikely that ESCs will be the best source of SCs for transplantation in LSD patient due to their potential for neoplastic transformation and the ethical limitations associated with using embryonic or fetal tissues to obtain stem cells. Somatic stem cells can be extensively propagated *in vitro* while maintaining a stable profile in regards to karyotyping and molecular profiling; these cells have been implanted after long-term culture and do not result in tumor formation [7, 24].

Stem cell therapy for LSDs is a tremendous undertaking in scientific, regulatory and ethical terms. The results of recent research, many of which are discussed in this chapter, suggest that stem cells offer the greatest potential for effective treatment of LSDs, both those with and without neurological complications [15]. Upon transplantation in LSD patients, especially when distributed throughout the ventricular system in the CNS, most types of stem cells are capable of engrafting in affected tissue to provide neuroprotective and immunomodulatory activity, differentiation and replacement for degenerated cells, and delivery of missing lysosomal enzymes while maintaining low immunogenicity and stable profiles [6].

References

1. Hawkins-Salsbury JA, Reddy AS, Sands MS (2011) Combination therapies for lysosomal storage disease: is the whole greater than the sum of its parts? Hum Mol Genet 20(R1):R54–60. PMCID: 3095053
2. Heese BA (2008) Current strategies in the management of lysosomal storage diseases. Semin Pediatr Neurol 15(3):119–126
3. Hodges BL, Cheng SH (2006) Cell and gene-based therapies for the lysosomal storage diseases. Curr Gene Ther 6(2):227–241
4. Eckhardt M (2010) Pathology and current treatment of neurodegenerative sphingolipidoses. Neuromolecular Med 12(4):362–382
5. Beck M (2010) Therapy for lysosomal storage disorders. IUBMB Life 62(1):33–40
6. Gritti A (2011) Gene therapy for lysosomal storage disorders. Expert Opin Biol Ther 11(9):1153–1167

7. de Filippis L (2011) Neural stem cell-mediated therapy for rare brain diseases: perspectives in the near future for LSDs and MNDs. Histol Histopathol 26(8):1093–1109
8. Prasad VK, Kurtzberg J (2010) Cord blood and bone marrow transplantation in inherited metabolic diseases: scientific basis, current status and future directions. Br J Haematol 148(3):356–372
9. Pierson TM, Bonnemann CG, Finkel RS, Bunin N, Tennekoon GI (2008) Umbilical cord blood transplantation for juvenile metachromatic leukodystrophy. Ann Neurol 64(5): 583–587. PMCID: 2605197
10. Proia RL, Wu YP (2004) Blood to brain to the rescue. J Clin Invest 113(8):1108–1110. PMCID: 385411
11. Klein D, Schmandt T, Muth-Kohne E, Perez-Bouza A, Segschneider M, Gieselmann V et al (2006) Embryonic stem cell-based reduction of central nervous system sulfatide storage in an animal model of metachromatic leukodystrophy. Gene Ther 13(24):1686–1695
12. Kim SU, de Vellis J (2009) Stem cell-based cell therapy in neurological diseases: a review. J Neurosci Res 87(10):2183–2200
13. Kim SU (2007) Genetically engineered human neural stem cells for brain repair in neurological diseases. Brain Dev 29(4):193–201
14. Lee JK, Jin HK, Endo S, Schuchman EH, Carter JE, Bae JS (2010) Intracerebral transplantation of bone marrow-derived mesenchymal stem cells reduces amyloid-beta deposition and rescues memory deficits in Alzheimer's disease mice by modulation of immune responses. Stem Cells 28(2):329–343
15. Selden NR, Guillaume DJ, Steiner RD, Huhn SL (2008) Cellular therapy for childhood neurodegenerative disease. Part II: clinical trial design and implementation. Neurosurg Focus 24(3–4):E23
16. Guillaume DJ, Huhn SL, Selden NR, Steiner RD (2008) Cellular therapy for childhood neurodegenerative disease. Part I: rationale and preclinical studies. Neurosurg Focus 24(3–4):E22
17. Orchard PJ, Tolar J (2010) Transplant outcomes in leukodystrophies. Semin Hematol 47(1):70–78
18. Sevin C, Cartier-Lacave N, Aubourg P (2009) Gene therapy in metachromatic leukodystrophy. Int J Clin Pharmacol Ther 47(Suppl 1):S128–S131
19. Rovelli AM (2008) The controversial and changing role of haematopoietic cell transplantation for lysosomal storage disorders: an update. Bone Marrow Transplant 41(Suppl 2):S87–S89
20. Grewal SS, Shapiro EG, Krivit W, Charnas L, Lockman LA, Delaney KA et al (2004) Effective treatment of alpha-mannosidosis by allogeneic hematopoietic stem cell transplantation. J Pediatr 144(5):569–573
21. Biffi A, Lucchini G, Rovelli A, Sessa M (2008) Metachromatic leukodystrophy: an overview of current and prospective treatments. Bone Marrow Transplant 42(Suppl 2):S2–S6
22. Robinson AJ, Meedeniya AC, Hemsley KM, Auclair D, Crawley AC, Hopwood JJ (2005) Survival and engraftment of mouse embryonic stem cell-derived implants in the guinea pig brain. Neurosci Res 53(2):161–168
23. Goldman SA, Windrem MS (2006) Cell replacement therapy in neurological disease. Philos Trans R Soc Lond B Biol Sci 361(1473):1463–1475. PMCID: 1664668
24. Shihabuddin LS, Aubert I (2010) Stem cell transplantation for neurometabolic and neurodegenerative diseases. Neuropharmacology 58(6):845–854
25. Fink DW Jr (2009) FDA regulation of stem cell-based products. Science 324(5935):1662–1663
26. Chen X, Armstrong MA, Li G (2006) Mesenchymal stem cells in immunoregulation. Immunol Cell Biol 84(5):413–421
27. Dwyer RM, Kerin MJ (2010) Mesenchymal stem cells and cancer: tumor-specific delivery vehicles or therapeutic targets? Hum Gene Ther 21(11):1506–1512
28. Hardy SA, Maltman DJ, Przyborski SA (2008) Mesenchymal stem cells as mediators of neural differentiation. Curr Stem Cell Res Ther 3(1):43–52
29. Croitoru-Lamoury J, Williams KR, Lamoury FM, Veas LA, Ajami B, Taylor RM et al (2006) Neural transplantation of human MSC and NT2 cells in the twitcher mouse model. Cytotherapy 8(5):445–458

30. Gimble JM, Katz AJ, Bunnell BA (2007) Adipose-derived stem cells for regenerative medicine. Circ Res 100(9):1249–1260
31. Koc ON, Peters C, Aubourg P, Raghavan S, Dyhouse S, DeGasperi R et al (1999) Bone marrow-derived mesenchymal stem cells remain host-derived despite successful hematopoietic engraftment after allogeneic transplantation in patients with lysosomal and peroxisomal storage diseases. Exp Hematol 27(11):1675–1681
32. Giordano A, Galderisi U, Marino IR (2007) From the laboratory bench to the patient's bedside: an update on clinical trials with mesenchymal stem cells. J Cell Physiol 211(1):27–35
33. Meyerrose T, Olson S, Pontow S, Kalomoiris S, Jung Y, Annett G et al (2010) Mesenchymal stem cells for the sustained in vivo delivery of bioactive factors. Adv Drug Deliv Rev 62(12):1167–1174
34. Kassis I, Grigoriadis N, Gowda-Kurkalli B, Mizrachi-Kol R, Ben-Hur T, Slavin S et al (2008) Neuroprotection and immunomodulation with mesenchymal stem cells in chronic experimental autoimmune encephalomyelitis. Archives of neurology 65(6):753–761
35. Uccelli A, Benvenuto F, Laroni A, Giunti D (2011) Neuroprotective features of mesenchymal stem cells. Best Pract Res Clin Haematol 24(1):59–64
36. Mahmood A, Lu D, Chopp M (2004) Marrow stromal cell transplantation after traumatic brain injury promotes cellular proliferation within the brain. Neurosurgery 55(5):1185–1193
37. Ricks DM, Kutner R, Zhang XY, Welsh DA, Reiser J (2008) Optimized lentiviral transduction of mouse bone marrow-derived mesenchymal stem cells. Stem Cells Dev 17(3):441–450
38. Munoz JR, Stoutenger BR, Robinson AP, Spees JL, Prockop DJ (2005) Human stem/progenitor cells from bone marrow promote neurogenesis of endogenous neural stem cells in the hippocampus of mice. Proc Natl Acad Sci USA 102(50):18171–18176. PMCID: 1312406
39. Caplan AI, Dennis JE (2006) Mesenchymal stem cells as trophic mediators. J Cell Biochem 98(5):1076–1084
40. Soni S (2007) Allogeneic stem cell transplantation for genetic disorders. J Ky Med Assoc 105(1):12–16
41. Duffner PK, Caggana M, Orsini JJ, Wenger DA, Patterson MC, Crosley CJ et al (2009) Newborn screening for Krabbe disease: the New York State model. Pediatr Neurol 40(4):245–252; discussion 53–55
42. Duffner PK, Caviness VS Jr, Erbe RW, Patterson MC, Schultz KR, Wenger DA et al (2009) The long-term outcomes of presymptomatic infants transplanted for Krabbe disease: report of the workshop held on July 11 and 12, 2008, Holiday Valley, New York. Genet Med 11(6):450–454
43. Ohshima T, Murray GJ, Swaim WD, Longenecker G, Quirk JM, Cardarelli CO et al (1997) alpha-Galactosidase A deficient mice: a model of Fabry disease. Proc Natl Acad Sci USA 94(6):2540–2544. PMCID: PMC20124
44. Enquist IB, Nilsson E, Ooka A, Månsson JE, Olsson K, Ehinger M et al (2006) Effective cell and gene therapy in a murine model of Gaucher disease. Proc Natl Acad Sci USA 103(37):13819–13824. PMCID: PMC1564262
45. Kolodny EH (1996) Globoid leukodystrophy. Elsevier, Amsterdam
46. Suzuki K, Suzuki K (1995) The twitcher mouse: a model for Krabbe disease and for experimental therapies. Brain pathol 5(3):249–258
47. Metcalf JA, Ma X, Linders B, Wu S, Schambach A, Ohlemiller KK et al (2010) A self-inactivating gamma-retroviral vector reduces manifestations of mucopolysaccharidosis I in mice. Mol Ther 18(2):334–342. PMCID: 2839301
48. Robinson AJ, Zhao G, Rathjen J, Rathjen PD, Hutchinson RG, Eyre HJ et al (2010) Embryonic stem cell-derived glial precursors as a vehicle for sulfamidase production in the MPS-IIIA mouse brain. Cell Transplant 19(8):985–998
49. Horinouchi K, Erlich S, Perl DP, Ferlinz K, Bisgaier CL, Sandhoff K et al (1995) Acid sphingomyelinase deficient mice: a model of types A and B Niemann-Pick disease. Nat Genet 10(3):288–293
50. Otterbach B, Stoffel W (1995) Acid sphingomyelinase-deficient mice mimic the neurovisceral form of human lysosomal storage disease (Niemann-Pick disease). Cell 81(7):1053–1061

51. Pentchev PG, Gal AE, Booth AD, Omodeo-Sale F, Fouks J, Neumeyer BA et al (1980) A lysosomal storage disorder in mice characterized by a dual deficiency of sphingomyelinase and glucocerebrosidase. Biochim Biophys Acta 619(3):669–679
52. Somers KL, Royals MA, Carstea ED, Rafi MA, Wenger DA, Thrall MA (2003) Mutation analysis of feline Niemann-Pick C1 disease. Mol Genet Metab 79(2):99–103
53. Tamaki SJ, Jacobs Y, Dohse M, Capela A, Cooper JD, Reitsma M et al (2009) Neuroprotection of host cells by human central nervous system stem cells in a mouse model of infantile neuronal ceroid lipofuscinosis. Cell Stem Cell 5(3):310–319
54. Sango K, Yamanaka S, Hoffmann A, Okuda Y, Grinberg A, Westphal H et al (1995) Mouse models of Tay-Sachs and Sandhoff diseases differ in neurologic phenotype and ganglioside metabolism. Nat Genet 11(2):170–176
55. Abkowitz JL, Sabo KM, Yang Z, Vite CH, Shields LE, Haskins ME (2009) In utero transplantation of monocytic cells in cats with alpha-mannosidosis. Transplantation 88(3):323–329. PMCID: 2742773
56. Kaartinen V, Mononen I, Voncken JW, Noronkoski T, Gonzalez-Gomez I, Heisterkamp N et al (1996) A mouse model for the human lysosomal disease aspartylglycosaminuria. Nat Med 2(12):1375–1378
57. Autio S, Visakorpi JK, Järvinen H (1973) Aspartylglycosaminuria (AGU). Further aspects on its clinical picture, mode of inheritance and epidemiology based on a series of 57 patients. Ann Clin Res 5(3):149–155
58. Määttä A, Järveläinen HT, Nelimarkka LO, Penttinen RP (1994) Fibroblast expression of collagens and proteoglycans is altered in aspartylglucosaminuria, a lysosomal storage disease. Biochim Biophys Acta 1225(3):264–270
59. Arvio M, Sauna-Aho O, Peippo M (2001) Bone marrow transplantation for aspartylglucosaminuria: follow-up study of transplanted and non-transplanted patients. J Pediatr 138(2):288–290
60. Autti T, Santavuori P, Raininko R, Renlund M, Rapola J, Saarinen-Pihkala U (1997) Bone-marrow transplantation in aspartylglucosaminuria. Lancet 349(9062):1366–1367
61. Ohshima T, Schiffmann R, Murray GJ, Kopp J, Quirk JM, Stahl S et al (1999) Aging accentuates and bone marrow transplantation ameliorates metabolic defects in Fabry disease mice. Proc Natl Acad Sci USA 96(11):6423–6427. PMCID: PMC26897
62. Yoshimitsu M, Higuchi K, Ramsubir S, Nonaka T, Rasaiah VI, Siatskas C et al (2007) Efficient correction of Fabry mice and patient cells mediated by lentiviral transduction of hematopoietic stem/progenitor cells. Gene Ther 14(3):256–265
63. Takenaka T, Murray GJ, Qin G, Quirk JM, Ohshima T, Qasba P et al (2000) Long-term enzyme correction and lipid reduction in multiple organs of primary and secondary transplanted Fabry mice receiving transduced bone marrow cells. Proc Natl Acad Sci USA 97(13):7515–7520. PMCID: PMC16577
64. Takenaka T, Qin G, Brady RO, Medin JA (1999) Circulating alpha-galactosidase A derived from transduced bone marrow cells: relevance for corrective gene transfer for Fabry disease. Hum Gene Ther 10(12):1931–1939
65. Liang SB, Yoshimitsu M, Poeppl A, Rasaiah VI, Cai J, Fowler DH et al (2007) Multiple reduced-intensity conditioning regimens facilitate correction of Fabry mice after transplantation of transduced cells. Mol Ther 15(3):618–627
66. Yeager AM, Uhas KA, Coles CD, Davis PC, Krause WL, Moser HW (2000) Bone marrow transplantation for infantile ceramidase deficiency (Farber disease). Bone Marrow Transplant 26(3):357–363
67. Ehlert K, Frosch M, Fehse N, Zander A, Roth J, Vormoor J (2007) Farber disease: clinical presentation, pathogenesis and a new approach to treatment. Pediatr Rheumatol Online J 5:15. PMCID: PMC1920510
68. Vormoor J, Ehlert K, Groll AH, Koch HG, Frosch M, Roth J (2004) Successful hematopoietic stem cell transplantation in Farber disease. J Pediatr 144(1):132–134
69. Ehlert K, Roth J, Frosch M, Fehse N, Zander N, Vormoor J (2006) Farber's disease without central nervous system involvement: bone-marrow transplantation provides a promising new approach. Ann Rheum Dis 65(12):1665–1666. PMCID: PMC1798467

70. Ohashi T, Boggs S, Robbins P, Bahnson A, Patrene K, Wei FS et al (1992) Efficient transfer and sustained high expression of the human glucocerebrosidase gene in mice and their functional macrophages following transplantation of bone marrow transduced by a retroviral vector. Proc Natl Acad Sci USA 89(23):11332–11336. PMCID: PMC50544
71. Havenga M, Valerio D, Hoogerbrugge P, Es H (1999) In vivo methotrexate selection of murine hemopoietic cells transduced with a retroviral vector for Gaucher disease. Gene Ther 6(10):1661–1669
72. Kim EY, Hong YB, Lai Z, Cho YH, Brady RO, Jung SC (2005) Long-term expression of the human glucocerebrosidase gene in vivo after transplantation of bone-marrow-derived cells transformed with a lentivirus vector. J Gene Med 7(7):878–887
73. Medin JA, Migita M, Pawliuk R, Jacobson S, Amiri M, Kluepfel-Stahl S et al (1996) A bicistronic therapeutic retroviral vector enables sorting of transduced CD34+ cells and corrects the enzyme deficiency in cells from Gaucher patients. Blood 87(5):1754–1762
74. Hobbs JR, Jones KH, Shaw PJ, Lindsay I, Hancock M (1987) Beneficial effect of pretransplant splenectomy on displacement bone marrow transplantation for Gaucher's syndrome. Lancet 1(8542):1111–1115
75. Ringdén O, Groth CG, Erikson A, Bäckman L, Granqvist S, Månsson JE et al (1988) Long-term follow-up of the first successful bone marrow transplantation in Gaucher disease. Transplantation 46(1):66–70
76. Erikson A, Groth CG, Månsson JE, Percy A, Ringdén O, Svennerholm L (1990) Clinical and biochemical outcome of marrow transplantation for Gaucher disease of the Norrbottnian type. Acta Paediatr Scand 79(6–7):680–685
77. Rappeport JM, Barranger JA, Ginns EI (1986) Bone marrow transplantation in Gaucher disease. Birth Defects Orig Artic Ser 22(1):101–109
78. Tsai P, Lipton JM, Sahdev I, Najfeld V, Rankin LR, Slyper AH et al (1992) Allogenic bone marrow transplantation in severe Gaucher disease. Pediatr Res 31(5):503–507
79. Ripoll CB, Flaat M, Klopf-Eiermann J, Fisher-Perkins JM, Trygg CB, Scruggs BA et al (2011) Mesenchymal-lineage stem cells have pronounced anti-inflammatory effects in the twitcher mouse model of Krabbe's disease. Stem Cells 29(1):67–77
80. Strazza M, Luddi A, Carbone M, Rafi MA, Costantino-Ceccarini E, Wenger DA (2009) Significant correction of pathology in brains of twitcher mice following injection of genetically modified mouse neural progenitor cells. Mol Genet Metab 97(1):27–34
81. Pellegatta S, Tunici P, Poliani PL, Dolcetta D, Cajola L, Colombelli C et al (2006) The therapeutic potential of neural stem/progenitor cells in murine globoid cell leukodystrophy is conditioned by macrophage/microglia activation. Neurobiol Dis 21(2):314–323
82. Yagi T, McMahon EJ, Takikita S, Mohri I, Matsushima GK, Suzuki K (2004) Fate of donor hematopoietic cells in demyelinating mutant mouse, twitcher, following transplantation of GFP+ bone marrow cells. Neurobiol Dis 16(1):98–109
83. Luzi P, Rafi MA, Zaka M, Rao HZ, Curtis M, Vanier MT et al (2005) Biochemical and pathological evaluation of long-lived mice with globoid cell leukodystrophy after bone marrow transplantation. Mol Genet Metab 86(1–2):150–159
84. McGraw P, Liang L, Escolar M, Mukundan S, Kurtzberg J, Provenzale JM (2005) Krabbe disease treated with hematopoietic stem cell transplantation: serial assessment of anisotropy measurements–initial experience. Radiology 236(1):221–230
85. Shapiro EG, Lockman LA, Balthazor M, Krivit W (1995) Neuropsychological outcomes of several storage diseases with and without bone marrow transplantation. J Inherit Metab Dis 18(4):413–429
86. Miyake N, Miyake K, Karlsson S, Shimada T (2010) Successful treatment of metachromatic leukodystrophy using bone marrow transplantation of HoxB4 overexpressing cells. Mol Ther 18(7):1373–1378. PMCID: 2911255
87. Peters C, Steward CG (2003) Hematopoietic cell transplantation for inherited metabolic diseases: an overview of outcomes and practice guidelines. Bone Marrow Transplant 31(4):229–239

88. de Ruijter J, Valstar MJ, Wijburg FA (2011) Mucopolysaccharidosis type III (Sanfilippo Syndrome): emerging treatment strategies. Curr Pharm Biotechnol 12(6):923–930
89. Snyder EY, Taylor RM, Wolfe JH (1995) Neural progenitor cell engraftment corrects lysosomal storage throughout the MPS VII mouse brain. Nature 374(6520):367–370
90. Miranda SR, Erlich S, Friedrich VL, Haskins ME, Gatt S, Schuchman EH (1998) Biochemical, pathological, and clinical response to transplantation of normal bone marrow cells into acid sphingomyelinase-deficient mice. Transplantation 65(7):884–892
91. Miranda SR, Erlich S, Friedrich VL, Gatt S, Schuchman EH (2000) Hematopoietic stem cell gene therapy leads to marked visceral organ improvements and a delayed onset of neurological abnormalities in the acid sphingomyelinase deficient mouse model of Niemann-Pick disease. Gene Ther 7(20):1768–1776
92. Jin HK, Carter JE, Huntley GW, Schuchman EH (2002) Intracerebral transplantation of mesenchymal stem cells into acid sphingomyelinase-deficient mice delays the onset of neurological abnormalities and extends their life span. J Clin Invest 109(9):1183–1191. PMCID: PMC150966
93. Jin HK, Schuchman EH (2003) Ex vivo gene therapy using bone marrow-derived cells: combined effects of intracerebral and intravenous transplantation in a mouse model of Niemann-Pick disease. Mol Ther 8(6):876–885
94. Bae JS, Han HS, Youn DH, Carter JE, Modo M, Schuchman EH et al (2007) Bone marrow-derived mesenchymal stem cells promote neuronal networks with functional synaptic transmission after transplantation into mice with neurodegeneration. Stem Cells 25(5):1307–1316
95. Schneiderman J, Thormann K, Charrow J, Kletzel M (2007) Correction of enzyme levels with allogeneic hematopoeitic progenitor cell transplantation in Niemann-Pick type B. Pediatr Blood Cancer 49(7):987–989
96. Shah AJ, Kapoor N, Crooks GM, Parkman R, Weinberg KI, Wilson K et al (2005) Successful hematopoietic stem cell transplantation for Niemann-Pick disease type B. Pediatrics 116(4):1022–1025
97. Ahmad I, Hunter RE, Flax JD, Snyder EY, Erickson RP (2007) Neural stem cell implantation extends life in Niemann-Pick C1 mice. J Appl Genet 48(3):269–272
98. Lee JM, Bae JS, Jin HK (2010) Intracerebellar transplantation of neural stem cells into mice with neurodegeneration improves neuronal networks with functional synaptic transmission. J Vet Med Sci 72(8):999–1009
99. Wong AM, Rahim AA, Waddington SN, Cooper JD (2010) Current therapies for the soluble lysosomal forms of neuronal ceroid lipofuscinosis. Biochem Soc Trans 38(6):1484–1488
100. Malm D, Nilssen O (2008) Alpha-mannosidosis. Orphanet J Rare Dis 3:21. PMCID: 2515294
101. Sun H, Yang M, Haskins ME, Patterson DF, Wolfe JH (1999) Retrovirus vector-mediated correction and cross-correction of lysosomal alpha-mannosidase deficiency in human and feline fibroblasts. Hum Gene Ther 10(8):1311–1319
102. Robinson AJ, Crawley AC, Hopwood JJ (2005) Over-expression of human lysosomal alpha-mannosidase in mouse embryonic stem cells. Mol Genet Metab 85(3):203–212
103. Mynarek M, Tolar J, Albert MH, Escolar ML, Boelens JJ, Cowan MJ et al (2011) Allogeneic hematopoietic SCT for alpha-mannosidosis: an analysis of 17 patients. Bone Marrow Transplant doi:10.1038/bmt.2011.99
104. Aronson NN (1999) Aspartylglycosaminuria: biochemistry and molecular biology. Biochim Biophys Acta 1455(2–3):139–154
105. Ringdén O, Remberger M, Svahn BM, Barkholt L, Mattsson J, Aschan J et al (2006) Allogeneic hematopoietic stem cell transplantation for inherited disorders: experience in a single center. Transplantation 81(5):718–725
106. Autti T, Rapola J, Santavuori P, Raininko R, Renlund M, Liukkonen E et al (1999) Bone marrow transplantation in aspartylglucosaminuria–histopathological and MRI study. Neuropediatrics 30(6):283–288

107. Malm G, Månsson JE, Winiarski J, Mosskin M, Ringdén O (2004) Five-year follow-up of two siblings with aspartylglucosaminuria undergoing allogeneic stem-cell transplantation from unrelated donors. Transplantation 78(3):415–419
108. Yokoi T, Kobayashi H, Shimada Y, Eto Y, Ishige N, Kitagawa T et al (2011) Minimum requirement of donor cells to reduce the glycolipid storage following bone marrow transplantation in a murine model of Fabry disease. J Gene Med 13(5):262–268
109. Noben-Trauth K, Neely H, Brady RO (2007) Normal hearing in alpha-galactosidase A-deficient mice, the mouse model for Fabry disease. Hear Res 234(1–2):10–14
110. Farber S, Cohen J, Uzman LL (1957) Lipogranulomatosis; a new lipo-glycoprotein storage disease. J Mt Sinai Hosp N Y 24(6):816–837
111. Moser H, Chen W (1983) Ceramidase deficiency: Farber's lipogranulomatosis. In: Stanbury JB, Wyngaarden JB, Fredrickson DS, Goldstein JL, Brown MS (eds) The metabolic basis of inherited disease, 5th edn. McGraw-Hill, New York, pp 820–830
112. Jameson RA, Holt PJ, Keen JH (1987) Farber's disease (lysosomal acid ceramidase deficiency). Ann Rheum Dis 46(7):559–561. PMCID: PMC1002193
113. Ramsubir S, Nonaka T, Girbés CB, Carpentier S, Levade T, Medin JA (2008) In vivo delivery of human acid ceramidase via cord blood transplantation and direct injection of lentivirus as novel treatment approaches for Farber disease. Mol Genet Metab 95(3):133–141. PMCID: PMC2614354
114. Souillet G, Guiband P, Fensom A, Maire I, Zabot M (1989) Outcome of displacement bone marrow transplantation in Farber's disease: a report of case. In: Hobbs J (ed) Correction of certain genetic disease by transplantation. CoGENT, London, pp 137–141
115. Dunbar C, Kohn D (1996) Retroviral mediated transfer of the cDNA for human glucocerebrosidase into hematopoietic stem cells of patients with Gaucher disease. A phase I study. Hum Gene Ther 7(2):231–253
116. Cox TM (2010) Gaucher disease: clinical profile and therapeutic developments. Biologics 4:299–313. PMCID: PMC3010821
117. Barranger JA, Rice EO, Dunigan J, Sansieri C, Takiyama N, Beeler M et al (1997) Gaucher's disease: studies of gene transfer to haematopoietic cells. Baillieres Clin Haematol 10(4):765–778
118. Suzuki K, Suzuki Y (1970) Globoid cell leucodystrophy (Krabbe's disease): deficiency of galactocerebroside beta-galactosidase. Proc Natl Acad Sci USA 66(2):302–309. PMCID: 283044
119. Rafi MA, Zhi Rao H, Passini MA, Curtis M, Vanier MT, Zaka M et al (2005) AAV-mediated expression of galactocerebrosidase in brain results in attenuated symptoms and extended life span in murine models of globoid cell leukodystrophy. Mol Ther 11(5):734–744
120. Sakai N (2009) Pathogenesis of leukodystrophy for Krabbe disease: molecular mechanism and clinical treatment. Brain Dev 31(7):485–487
121. Svennerholm L, Vanier MT, Mansson JE (1980) Krabbe disease: a galactosylsphingosine (psychosine) lipidosis. J Lipid Res 21(1):53–64
122. Pannuzzo G, Cardile V, Costantino-Ceccarini E, Alvares E, Mazzone D, Perciavalle V (2010) A galactose-free diet enriched in soy isoflavones and antioxidants results in delayed onset of symptoms of Krabbe disease in twitcher mice. Mol Genet Metab 100(3):234–240
123. Krabbe K (1916) A new familial, infantile form of diffuse brain-sclerosis. Brain 39(1):74–114
124. Borda JT, Alvarez X, Mohan M, Ratterree MS, Phillippi-Falkenstein K, Lackner AA et al (2008) Clinical and immunopathologic alterations in rhesus macaques affected with globoid cell leukodystrophy. Am J Pathol 172(1):98–111. PMCID: 2189619
125. Wenger DA, Rafi MA, Luzi P, Datto J, Costantino-Ceccarini E (2000) Krabbe disease: genetic aspects and progress toward therapy. Mol Genet Metab 70(1):1–9
126. Lin D, Donsante A, Macauley S, Levy B, Vogler C, Sands MS (2007) Central nervous system-directed AAV2/5-mediated gene therapy synergizes with bone marrow transplantation in the murine model of globoid-cell leukodystrophy. Mol Ther 15(1):44–52
127. Kobayashi T, Yamanaka T, Jacobs JM, Teixeira F, Suzuki K (1980) The Twitcher mouse: an enzymatically authentic model of human globoid cell leukodystrophy (Krabbe disease). Brain Res 202(2):479–483

128. Wenger DA (2000) Murine, canine and non-human primate models of Krabbe disease. Mol Med Today 6(11):449–451
129. Galbiati F, Givogri MI, Cantuti L, Rosas AL, Cao H, van Breemen R et al (2009) Combined hematopoietic and lentiviral gene-transfer therapies in newborn Twitcher mice reveal contemporaneous neurodegeneration and demyelination in Krabbe disease. J Neurosci Res 87(8):1748–1759
130. Gentner B, Visigalli I, Hiramatsu H, Lechman E, Ungari S, Giustacchini A et al (2010) Identification of hematopoietic stem cell-specific miRNAs enables gene therapy of globoid cell leukodystrophy. Sci Transl Med 2(58):58ra84
131. Escolar ML, Poe MD, Provenzale JM, Richards KC, Allison J, Wood S et al (2005) Transplantation of umbilical-cord blood in babies with infantile Krabbe's disease. N Engl J Med 352(20):2069–2081
132. Turbeville S, Nicely H, Rizzo JD, Pedersen TL, Orchard PJ, Horwitz ME et al (2011) Clinical outcomes following hematopoietic stem cell transplantation for the treatment of mucopolysaccharidosis VI. Mol Genet Metab 102(2):111–115
133. Aldenhoven M, Boelens JJ, de Koning TJ (2008) The clinical outcome of Hurler syndrome after stem cell transplantation. Biol Blood Marrow Transplant 14(5):485–498
134. Schiffmann R (2010) Therapeutic approaches for neuronopathic lysosomal storage disorders. J Inherit Metab Dis 33(4):373–379
135. Kolodny EH (2000) Niemann-Pick disease. Curr Opin Hematol 7(1):48–52
136. Vanier MT (2010) Niemann-Pick disease type C. Orphanet J Rare Dis 5:16. PMCID: PMC2902432
137. Schuchman EH (1999) Hematopoietic stem cell gene therapy for Niemann-Pick disease and other lysosomal storage diseases. Chem Phys Lipids 102(1–2):179–188
138. Morel CF, Gassas A, Doyle J, Clarke JT (2007) Unsuccessful treatment attempt: cord blood stem cell transplantation in a patient with Niemann-Pick disease type A. J Inherit Metab Dis 30(6):987
139. Spiegel R, Raas-Rothschild A, Reish O, Regev M, Meiner V, Bargal R et al (2009) The clinical spectrum of fetal Niemann-Pick type C. Am J Med Genet A 149A(3):446–450
140. Wraith JE, Baumgartner MR, Bembi B, Covanis A, Levade T, Mengel E et al (2009) Recommendations on the diagnosis and management of Niemann-Pick disease type C. Mol Genet Metab 98(1–2):152–165
141. Vanier MT, Wenger DA, Comly ME, Rousson R, Brady RO, Pentchev PG (1988) Niemann-Pick disease group C: clinical variability and diagnosis based on defective cholesterol esterification. A collaborative study on 70 patients. Clin Genet 33(5):331–348
142. Trendelenburg G, Vanier MT, Maza S, Millat G, Bohner G, Munz DL et al (2006) Niemann-Pick type C disease in a 68-year-old patient. J Neurol Neurosurg Psychiatry 77(8):997–998. PMCID: PMC2077625
143. Lee H, Lee JK, Min WK, Bae JH, He X, Schuchman EH et al (2010) Bone marrow-derived mesenchymal stem cells prevent the loss of Niemann-Pick type C mouse Purkinje neurons by correcting sphingolipid metabolism and increasing sphingosine-1-phosphate. Stem Cells 28(4):821–831
144. Bae JS, Furuya S, Ahn SJ, Yi SJ, Hirabayashi Y, Jin HK (2005) Neuroglial activation in Niemann-Pick Type C mice is suppressed by intracerebral transplantation of bone marrow-derived mesenchymal stem cells. Neurosci Lett 381(3):234–236
145. Bae JS, Furuya S, Shinoda Y, Endo S, Schuchman EH, Hirabayashi Y et al (2005) Neurodegeneration augments the ability of bone marrow-derived mesenchymal stem cells to fuse with Purkinje neurons in Niemann-Pick type C mice. Hum Gene Ther 16(8):1006–1011
146. Bae JS, Carter JE, Jin HK (2010) Adipose tissue-derived stem cells rescue Purkinje neurons and alleviate inflammatory responses in Niemann-Pick disease type C mice. Cell Tissue Res 340(2):357–369
147. Lee H, Bae JS, Jin HK (2010) Human umbilical cord blood-derived mesenchymal stem cells improve neurological abnormalities of Niemann-Pick type C mouse by modulation of neuroinflammatory condition. J Vet Med Sci 72(6):709–717

148. Sidman RL, Li J, Stewart GR, Clarke J, Yang W, Snyder EY et al (2007) Injection of mouse and human neural stem cells into neonatal Niemann-Pick A model mice. Brain Res 1140:195–204
149. Kohan R, Cismondi IA, Oller-Ramirez AM, Guelbert N, Anzolini TV, Alonso G et al (2011) Therapeutic approaches to the challenge of neuronal ceroid lipofuscinoses. Curr Pharm Biotechnol 12(6):867–883
150. Jeyakumar M, Lee JP, Sibson NR, Lowe JP, Stuckey DJ, Tester K et al (2009) Neural stem cell transplantation benefits a monogenic neurometabolic disorder during the symptomatic phase of disease. Stem Cells 27(9):2362–2370
151. Krivit W, Peters C, Shapiro EG (1999) Bone marrow transplantation as effective treatment of central nervous system disease in globoid cell leukodystrophy, metachromatic leukodystrophy, adrenoleukodystrophy, mannosidosis, fucosidosis, aspartylglucosaminuria, Hurler, Maroteaux-Lamy, and Sly syndromes, and Gaucher disease type III. Curr Opin Neurol 12(2):167–176
152. Di Nicola M, Carlo-Stella C, Magni M, Milanesi M, Longoni PD, Matteucci P et al (2002) Human bone marrow stromal cells suppress T-lymphocyte proliferation induced by cellular or nonspecific mitogenic stimuli. Blood 99(10):3838–3843
153. Djouad F, Charbonnier LM, Bouffi C, Louis-Plence P, Bony C, Apparailly F et al (2007) Mesenchymal stem cells inhibit the differentiation of dendritic cells through an interleukin-6-dependent mechanism. Stem Cells 25(8):2025–2032
154. Krampera M, Cosmi L, Angeli R, Pasini A, Liotta F, Andreini A et al (2006) Role for interferon-gamma in the immunomodulatory activity of human bone marrow mesenchymal stem cells. Stem Cells 24(2):386–398
155. Selmani Z, Naji A, Zidi I, Favier B, Gaiffe E, Obert L et al (2008) Human leukocyte antigen-G5 secretion by human mesenchymal stem cells is required to suppress T lymphocyte and natural killer function and to induce CD4+CD25highFOXP3+ regulatory T cells. Stem Cells 26(1):212–222
156. Martin PL, Carter SL, Kernan NA, Sahdev I, Wall D, Pietryga D et al (2006) Results of the cord blood transplantation study (COBLT): outcomes of unrelated donor umbilical cord blood transplantation in pediatric patients with lysosomal and peroxisomal storage diseases. Biol Blood Marrow Transplant 12(2):184–194

Chapter 4
Drosophila Germline Stem Cells

Yalan Xing and Willis X. Li

Contents

Abstract Adult stem cells are capable of generating differentiated cell types for maintaining tissue homeostasis. They usually reside in a defined locale called 'stem-cell niche', which supports and maintains the stem-cell fate. Within the niche, cell-cell interactions between stem cells, interactions between stem cells and neighboring differentiated cells, as well as interactions between stem cells and exterior factors, such as adhesion molecules, extracellular matrix components, growth factors, cytokines, have all been found to be important for maintaining the stem cell fate. In addition, adult stem cells and their niche may induce each other during development and reciprocally signal to maintain each other during adulthood. *Drosophila* ovaries and testes have been used as excellent models for understanding the mechanisms underlying adult stem cell maintenance. Studies of the *Drosophila* model have provided insights into mammalian stem cell maintenance. This chapter summarizes the advances gained in studies of *Drosophila* male and female germline stem cells.

Y. Xing
Department of Biochemistry, Institute for Stem Cell and Regenerative Medicine, University of Washington, Seattle, WA 98109, USA
e-mail: yalan_xing@hotmail.com

W.X. Li (✉)
Department of Medicine, University of California San Diego, La Jolla, CA 92093, USA
e-mail: willisli@ucsd.edu

R.K. Srivastava and S. Shankar (eds.), *Stem Cells and Human Diseases*,
DOI 10.1007/978-94-007-2801-1_4,

Keywords *Drosophila* • Germline stem cells • JAK • STAT

4.1 Introduction

Adult stem cells, compared to embryonic stem (ES) cells, are multipotent and capable of generating one or several differentiated cell types for maintaining the homeostasis and normal functions of various tissues [36]. These lifetime dividing stem cells usually reside in a defined locale called 'stem-cell niche', which is composed of localized signaling cells and an extracellular matrix to support and maintain the stem-cell fate and prevent their differentiation [33, 41]. Multiple factors are essential for regulating stem cell characteristics within the niche: cell-cell interactions between stem cells, interactions between stem cells and neighboring differentiated cells, as well as interactions between stem cells and exterior factors, such as adhesion molecules, extracellular matrix components, growth factors, cytokines, etc. Adult stem cells and their niche may induce each other during development and reciprocally signal to maintain each other during adulthood.

As the adult stem cells that continuously produce gametes throughout lifetime, germline stem cells (GSCs) are essential for the continuation of species and inheritance of genomic information through generations. Therefore, studies revealing the mechanisms of how GSCs maintain their population, and how they properly divide and differentiate, are fundamentally and practically necessary. Compared to mammalian gonads, *Drosophila* ovaries and testes provide excellent models to elucidate the adult stem cell mystery, with the advantages of having identifiable markers for various cell types, simple anatomy, and most importantly, powerful genetic tools. A thorough understanding of *Drosophila* GSCs' self-renewal and their regulatory mechanisms should provide insightful guidance to mammalian stem cell studies. In this chapter, we will summarize and compare the latest knowledge and theory from both male and female *Drosophila* GSCs, which may contribute to our better understanding of human adult stem cells and diseases.

4.2 Establishment of the *Drosophila* Germline Stem Cell (GSC) Niche

Lineage analysis in many organisms has demonstrated that the mature GSCs originate from precursors named primordial germ cells (PGCs), and that GSC development is highly similar between vertebrates and invertebrates [38]. In *Drosophila*, PGCs form at the posterior pole of the embryo, migrate through the hindgut epithelium after gastrulation, and eventually reach somatic gonadal precursors (SGPs) in parasegment 10 [9]. The difference between male and female gonad development appears as early as the gonad coalescence; male-specific SGPs join the posterior of the male gonad but die by apoptosis in females [12]. Meanwhile, the SGPs start sending out sex-specific signals to the germ cells and instructing them to express sexually dimorphic genes [47]. It has been shown that functional male hub cells differentiate from SGPs during the final stage of embryogenesis (stage 17) [30]; PGCs first become GSCs, marked by the transcription factor *escargot* (*esg*) [42], in the developing male gonad at the transition from embryo to larval stage [39]. In contrast, female GSCs are formed from PGCs at the anterior of the gonad, at the transition from larval to pupal stage [2, 16, 53].

4.3 *Drosophila* Adult Male Germline Stem Cells, the Niche, and the Regulatory Mechanisms

Drosophila spermatogenesis initiates from the male germline stem cells and their niche located at the apical tip of *Drosophila* testis. Consisting of a cluster of post-mitotic somatic cells named "hub cells", 6–12 GSCs, each flanked by a pair of somatic stem cells called cyst stem cells (CySCs), the male stem cell niche forms a tightly packed "Rosetta" pattern (Fig. 4.1) [18]. The somatic 'niche' maintains the GSC population by direct attachment to them via adherens junctions and by short-range signals to prevent them from differentiation [18, 51]. A GSC divides asymmetrically to produce two daughter cells, with the mitotic spindle orienting perpendicularly to the hub. Of the two daughter cells, the one retaining contact with the hub maintains stem cell identity, while the one displaced away from the hub experiences a weaker niche signal and initiates differentiation as a gonialblast (GB), which will then go through four synchronous mitotic divisions and enter meiosis to become 16-cell spermatocytes [14, 18, 31].

Extensive studies have been done to reveal the molecular mechanism of how GSCs are maintained within the niche, which ensures fertility throughout an animal's life span. The primary signaling pathway controlling male GSC self-renewal is the Janus Kinase/Signal Transducer and Activators of Transcription (JAK/STAT) pathway. The cytokine-like ligand Unpaired (Upd) is secreted by hub cells and binds to the transmemberane receptor Domeless (Dom). In this fashion, Upd activates the JAK/STAT pathway in the neighboring GSCs and somatic stem cells (SSCs) to specify stem cell self-renewal [24, 31, 44]. Restoration of JAK/STAT signaling in the prematurely differentiated cysts due to *Stat* loss-of-function mutation can initiate their de-differentiation and repopulate the niche with new germline cells with stem cell identity [3]. Although the direct STAT target genes responsible for GSCs maintenance remains obscure, evidence from studying somatic stem cells (SSCs) has shown that a transcriptional repressor *zfh-1*, a presumptive target of JAK/STAT, is necessary and sufficient for SSC maintenance, and indirectly regulates GSC self-renewal [31]. A recent study exploring niche signaling has suggested a novel role of Suppressor of cytokine signaling 36E (SOCS36E), a JAK-STAT signaling target as well as inhibitor, in preventing SSC over-proliferation within the niche, which will indirectly affect GSC population through cell-cell competition [22].

Other than Upd/JAK/STAT, it has been shown that BMP signal is also necessary for proper GSC self-renewal. Ligands of the BMP signaling pathway, gbb and dpp,

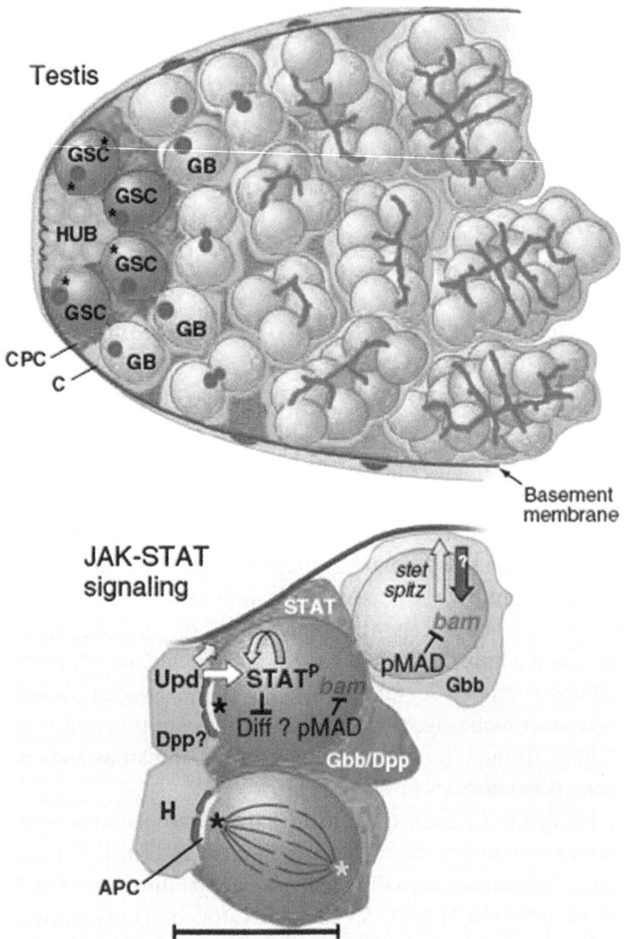

Fig. 4.1 (With permission, from Fuller and Spradling [15]). See text and ref [15] for detail

are expressed in hub and SSCs, where they activate BMP signaling in GSCs and prevent their differentiation at least partially through repression of the differentiation factor bag-of-marbles (bam) [4, 23, 37]. EGFR signal from SSCs, on the other hand, has been shown to regulate gonialblast differentiation and proliferation, rather than directly regulate GSCs [25].

Nutrient availability is an essential factor to instruct animals to adjust their metabolism in response to limited resources so that tissue homeostasis is maintained. Latest findings from *Drosophila* male and female GSC models, as well as intestinal stem cell biology, have demonstrated the sensitivity of stem cells to nutrient changes and the signaling pathway that is involved: Insulin signaling [13, 20, 35, 52]. Insulin receptor (InR) is cell-autonomously required for male GSC maintenance; and the GSC population will be reduced to a smaller number upon nutrient stress, then restored to normal level when re-fed [35]. In addition, a novel role for insulin signaling is revealed in promoting GSC cell cycle progression through the G2/M phase, leading to accelerated cell proliferation [45]. These findings all together indicate a theory that adult stem cells, especially GSCs, can accordingly reduce their population and metabolism upon nutrient stress to maintain tissue homeostasis, and repopulate afterwards.

4.4 *Drosophila* Female Germline Stem Cells, the Niche, and Regulatory Mechanisms

Drosophila female and male GSCs, as well as their niche, share high anatomic and functional similarity, yet with considerable differences [15]. *Drosophila* ovaries are made up of 14–16 ovarioles with the germaria as the anteriormost structure. Two to three GSCs, marked by spherical fusomes, locate at the apical tip of each germarium, anchored through DE-cadherin-mediated cell adhesion to the somatic niche composed of terminal filament (TF) and cap cells (Fig. 4.2) [32]. The direct attachments to cap cells, as well as the presence of germ cell-specific organelle fusome, are mostly widely used criteria for GSC identification [34, 50]. Similar to male GSCs and many other adult stem cell systems, *Drosophila* female GSCs divide asymmetrically with their spindles oriented perpendicular to the somatic niche, producing one daughter cell that remains in the niche with stem cell identity, and the other daughter cell sent away from the niche, which becomes a differentiating cystoblast (CB). After four synchronized divisions with incomplete cytoplasm separation, a cystoblast becomes a cyst with 16 interconnected cells, marked by highly branched fusome [10, 48].

One major difference between the two niches is that self-renewal of female GSCs is primarily regulated by bone morphogenetic protein (BMP) signaling from the niche, mediated by the ligands decapentaplegic (*dpp*) and glassbottomed boat (*gbb*) expressed in niche cells [40, 49]. BMP signaling activates cytoplasmic Mad and Medea, the *Drosophila* Smads, in GSCs, which form complex and silence the transcription of *bam* gene, the key differentiation factor that is normally turned off in GSCs [5, 40]. Other extrinsic signals, like Piwi and Yb, are also expressed in the somatic niche and are essential for female GSC maintenance, through hedgehog (hh) signaling [7, 8, 26, 27]. A study suggests that JAK/STAT signaling also plays a critical role in the female GSC niche function, but acting within the somatic escort cell lineage rather than directly in GSCs [11].

In addition to the extrinsic regulatory signaling from the niche, GSC self-renewal is also mediated by self-renewal and differentiation-promoting intrinsic factors. Multiple studies have demonstrated that Pumilio (Pum)/Nanos (Nos) complex-mediated and microRNA-mediated translational repression is essential for GSC maintenance and division, respectively through repression of *bam* expression in GSC and down-regulation of cyclin-dependent kinase inhibitor Dacapo [17, 19, 43]. Female GSC self-renewal and division are also tightly regulated by cell cycle factors.

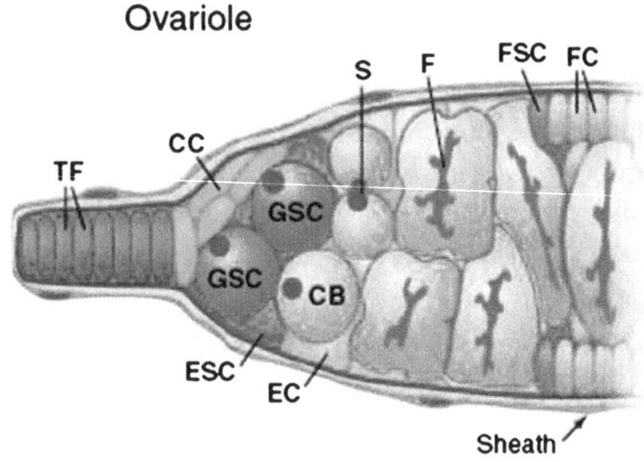

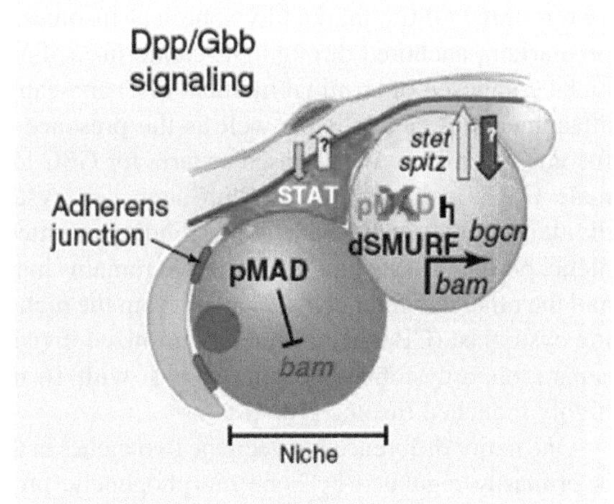

Fig. 4.2 (With permission, from Fuller and Spradling [15]). See text and ref [15] for detail

Emerging evidence has shown that an E2-ubiquitin-ligase dependent degradation of Cyclin A is required for GSC maintenance; while Cyclin B is specifically essential for normal GSC division, and possibly for GSC self-renewal [6, 46].

The roles of insulin signaling and steroid hormones in GSCs are better characterized in female niche. Insulin signaling is directly required for somatic niche maintenance rather than GSCs, but consequently leading to a GSC loss [20]. The division of GSCs and germline cysts, meanwhile, is immediately affected by diet status, partially via insulin signaling [21, 29]. Ecdysone, the only major steroid hormone in *Drosophila*, has been implicated to be functional in multiple cell types of the niche, including modulating the somatic niche size, affecting GSC proliferation through chromatin remodeling factors, and regulating GSC and cystoblast differentiation through the TGF-b pathway [1, 28]. Taken together, various classes of extrinsic and intrinsic signals, as well as cell-cell interactions, cooperatively function to maintain a proper GSC population and division rate, while specifying self-renewal of GSCs and differentiation of cystoblasts.

References

1. Ables ET, Drummond-Barbosa D (2010) The steroid hormone ecdysone functions with intrinsic chromatin remodeling factors to control female germline stem cells in Drosophila. Cell Stem Cell 7(5):581–592
2. Asaoka M, Lin H (2004) Germline stem cells in the Drosophila ovary descend from pole cells in the anterior region of the embryonic gonad. Development 131(20):5079–5089
3. Brawley C, Matunis E (2004) Regeneration of male germline stem cells by spermatogonial dedifferentiation in vivo. Science 304(5675):1331–1334
4. Bunt SM, Hime GR (2004) Ectopic activation of Dpp signalling in the male Drosophila germline inhibits germ cell differentiation. Genesis 39(2):84–93
5. Chen D, McKearin D (2003) Dpp signaling silences bam transcription directly to establish asymmetric divisions of germline stem cells. Curr Biol 13(20):1786–1791
6. Chen D, Wang Q et al (2009) Effete-mediated degradation of Cyclin A is essential for the maintenance of germline stem cells in Drosophila. Development 136(24):4133–4142
7. Cox DN, Chao A et al (1998) A novel class of evolutionarily conserved genes defined by piwi are essential for stem cell self-renewal. Genes Dev 12(23):3715–3727
8. Cox DN, Chao A et al (2000) Piwi encodes a nucleoplasmic factor whose activity modulates the number and division rate of germline stem cells. Development 127(3):503–514
9. Dansereau DA, Lasko P (2008) The development of germline stem cells in Drosophila. Methods Mol Biol 450:3–26
10. de Cuevas M, Lilly MA et al (1997) Germline cyst formation in Drosophila. Annu Rev Genet 31:405–428
11. Decotto E, Spradling AC (2005) The Drosophila ovarian and testis stem cell niches: similar somatic stem cells and signals. Dev Cell 9(4):501–510
12. DeFalco TJ, Verney G et al (2003) Sex-specific apoptosis regulates sexual dimorphism in the Drosophila embryonic gonad. Dev Cell 5(2):205–216
13. Drummond-Barbosa D, Spradling AC (2001) Stem cells and their progeny respond to nutritional changes during Drosophila oogenesis. Dev Biol 231(1):265–278
14. Fuller MT (1993) Development of Drosophila melanogaster, vol 1. Cold Spring Habor Laboratory Press, New York, pp 71–147
15. Fuller MT, Spradling AC (2007) Male and female Drosophila germline stem cells: two versions of immortality. Science 316(5823):402–404
16. Gilboa L, Forbes A et al (2003) Germ line stem cell differentiation in Drosophila requires gap junctions and proceeds via an intermediate state. Development 130(26):6625–6634
17. Gilboa L, Lehmann R (2004) Repression of primordial germ cell differentiation parallels germ line stem cell maintenance. Curr Biol 14(11):981–986
18. Hardy RW, Tokuyasu KT et al (1979) The germinal proliferation center in the testis of Drosophila melanogaster. J Ultrastruct Res 69(2):180–190
19. Hatfield SD, Shcherbata HR et al (2005) Stem cell division is regulated by the microRNA pathway. Nature 435(7044):974–978
20. Hsu HJ, Drummond-Barbosa D (2009) Insulin levels control female germline stem cell maintenance via the niche in Drosophila. Proc Natl Acad Sci USA 106(4):1117–1121
21. Hsu HJ, LaFever L et al (2008) Diet controls normal and tumorous germline stem cells via insulin-dependent and -independent mechanisms in Drosophila. Dev Biol 313(2):700–712

22. Issigonis M, Tulina N et al (2009) JAK-STAT signal inhibition regulates competition in the Drosophila testis stem cell niche. Science 326(5949):153–156
23. Kawase E, Wong MD et al (2004) Gbb/Bmp signaling is essential for maintaining germline stem cells and for repressing bam transcription in the Drosophila testis. Development 131(6):1365–1375
24. Kiger AA, Jones DL et al (2001) Stem cell self-renewal specified by JAK-STAT activation in response to a support cell cue. Science 294(5551):2542–2545
25. Kiger AA, White-Cooper H et al (2000) Somatic support cells restrict germline stem cell self-renewal and promote differentiation. Nature 407(6805):750–754
26. King FJ, Lin H (1999) Somatic signaling mediated by fs(1)Yb is essential for germline stem cell maintenance during Drosophila oogenesis. Development 126(9):1833–1844
27. King FJ, Szakmary A et al (2001) Yb modulates the divisions of both germline and somatic stem cells through piwi- and hh-mediated mechanisms in the Drosophila ovary. Mol Cell 7(3):497–508
28. Konig A, Yatsenko AS et al (2011) Ecdysteroids affect Drosophila ovarian stem cell niche formation and early germline differentiation. EMBO J 30(8):1549–1562
29. LaFever L, Drummond-Barbosa D (2005) Direct control of germline stem cell division and cyst growth by neural insulin in Drosophila. Science 309(5737):1071–1073
30. Le Bras S, Van Doren M (2006) Development of the male germline stem cell niche in Drosophila. Dev Biol 294(1):92–103
31. Leatherman JL, Dinardo S (2008) Zfh-1 controls somatic stem cell self-renewal in the Drosophila testis and nonautonomously influences germline stem cell self-renewal. Cell Stem Cell 3(1):44–54
32. Lin H (1997) The tao of stem cells in the germline. Annu Rev Genet 31:455–491
33. Lin H (2002) The stem-cell niche theory: lessons from flies. Nat Rev Genet 3(12):931–940
34. Lin H, Yue L et al (1994) The Drosophila fusome, a germline-specific organelle, contains membrane skeletal proteins and functions in cyst formation. Development 120(4):947–956
35. McLeod CJ, Wang L et al (2010) Stem cell dynamics in response to nutrient availability. Curr Biol 20(23):2100–2105
36. Morrison SJ, Shah NM et al (1997) Regulatory mechanisms in stem cell biology. Cell 88(3):287–298
37. Schulz C, Kiger AA et al (2004) A misexpression screen reveals effects of bag-of-marbles and TGF beta class signaling on the Drosophila male germ-line stem cell lineage. Genetics 167(2):707–723
38. Seydoux G, Braun RE (2006) Pathway to totipotency: lessons from germ cells. Cell 127(5):891–904
39. Sheng XR, Posenau T et al (2009) Jak-STAT regulation of male germline stem cell establishment during Drosophila embryogenesis. Dev Biol 334(2):335–344
40. Song X, Wong MD et al (2004) Bmp signals from niche cells directly repress transcription of a differentiation-promoting gene, bag of marbles, in germline stem cells in the Drosophila ovary. Development 131(6):1353–1364
41. Spradling A, Drummond-Barbosa D et al (2001) Stem cells find their niche. Nature 414(6859):98–104
42. Streit A, Bernasconi L et al (2002) mgm 1, the earliest sex-specific germline marker in Drosophila, reflects expression of the gene esg in male stem cells. Int J Dev Biol 46(1):159–166
43. Szakmary A, Cox DN et al (2005) Regulatory relationship among piwi, pumilio, and bag-of-marbles in Drosophila germline stem cell self-renewal and differentiation. Curr Biol 15(2):171–178
44. Tulina N, Matunis E (2001) Control of stem cell self-renewal in Drosophila spermatogenesis by JAK-STAT signaling. Science 294(5551):2546–2549
45. Ueishi S, Shimizu H et al (2009) Male germline stem cell division and spermatocyte growth require insulin signaling in Drosophila. Cell Struct Funct 34(1):61–69

46. Wang Z, Lin H (2005) The division of Drosophila germline stem cells and their precursors requires a specific cyclin. Curr Biol 15(4):328–333
47. Wawersik M, Milutinovich A et al (2005) Somatic control of germline sexual development is mediated by the JAK/STAT pathway. Nature 436(7050):563–567
48. Xie T (2008) Germline stem cell niches. In: Lin H, Donahoe P (eds) StemBook. StemBook ed: The stem cell research community. doi/103824/stembook.1.23.1. Available at: http://www.stembook.org. Accessed 30 Sept 2008
49. Xie T, Spradling AC (1998) Decapentaplegic is essential for the maintenance and division of germline stem cells in the Drosophila ovary. Cell 94(2):251–260
50. Xie T, Spradling AC (2000) A niche maintaining germ line stem cells in the Drosophila ovary. Science 290(5490):328–330
51. Yamashita YM, Jones DL et al (2003) Orientation of asymmetric stem cell division by the APC tumor suppressor and centrosome. Science 301(5639):1547–1550
52. Yu JY, Reynolds SH et al (2009) Dicer-1-dependent Dacapo suppression acts downstream of insulin receptor in regulating cell division of Drosophila germline stem cells. Development 136(9):1497–1507
53. Zhu CH, Xie T (2003) Clonal expansion of ovarian germline stem cells during niche formation in Drosophila. Development 130(12):2579–2588

Chapter 5
Mesenchymal Stem Cells in Hematopoietic Stem Cell Transplantation

Peiman Hematti, Jaehyup Kim, and Minoo Battiwalla

Contents

Abstract Hematopoietic stem cell (HSC) transplantation, the only form of stem cell therapy in routine clinical practice, provides a curative option for treatment of a wide variety of malignant and non-malignant disorders. However, this potentially

---

P. Hematti, MD (✉)
Department of Medicine, School of Medicine and Public Health,
University of Wisconsin, Madison, WI, USA

University of Wisconsin Carbone Cancer Center, Madison, WI, USA

WIMR 4033, 1111 Highland Avenue, Madison, WI 53705, USA
e-mail: pxh@medicine.wisc.edu

J. Kim, MD
Department of Medicine, School of Medicine and Public Health,
University of Wisconsin, Madison, WI, USA

M. Battiwalla, MD, MS
Section of Allogeneic Stem Cell Transplantation, Hematology Branch,
National Heart, Lung and Blood Institute Division of Intramural Research,
National Institutes of Health, Bethesda, MD, USA

R.K. Srivastava and S. Shankar (eds.), *Stem Cells and Human Diseases*,
DOI 10.1007/978-94-007-2801-1_5,

life-saving procedure is often associated with significant morbidity and mortality mainly due to graft versus host disease (GVHD). Mesenchymal stem cells (MSCs) are another type of stem cells present in bone marrow. These cells not only provide the supportive microenvironmental niche for hematopoietic stem cells but are also capable of differentiating into various cell types of mesenchymal origin, such as bone, fat, and cartilage. *In vitro* and *in vivo* data suggest that MSCs have low inherent immunogenicity, and modulate immunological responses through interactions with a wide range of innate and adaptive immune cells. MSCs participate in tissue regenerative processes through their diverse biological properties such as paracrine effects through growth factor and cytokine secretion. MSCs derived from bone marrow have been investigated extensively in the context of HSC transplantation, for promotion of engraftment, prevention of GVHD or treatment of GVHD or other complications associated with transplantation. Clinical studies of use of MSCs in HSC transplantation have paved the way for use of MSCs in a wide variety of other clinical indications.

Keywords Mesenchymal stem cells • Hematopoietic stem cells • Bone marrow transplant • Graft versus host disease

5.1 Introduction

Allogeneic hematopoietic stem cell (HSC) transplantation is a high-risk procedure indicated not only for the treatment of various hematologic malignancies but also non-malignant conditions such as bone marrow failure states and immunodeficiency syndromes. Allogeneic HSC transplantation utilizes a preparative conditioning regimen prior to donor stem cell infusion to optimally cytoreduce the malignancy (in case of hematological malignancies) and to make immunologic space so the host can accept the graft (donor stem cells). The donor HSC graft eventually replaces the recipient's hematopoietic and immune systems. The traditional source of hematopoietic stem cells has been bone marrow harvest; nowadays, peripheral blood progenitor cell and umbilical cord grafts are increasing in popularity. Despite decades of progress HSC transplantation remains a high-risk procedure with significant nonrelapse morbidity and mortality related to the conditioning regimen related toxicity, graft failure, infectious complications and graft-versus-host disease (Fig. 5.1). Lethal organ injury can result from the combination of uncontrolled inflammation, drug side effects and infections. While these complications have been reduced in recent years, there is still great room for improvement. There is critical need for nontoxic treatments that will reduce inflammation, permit tissue and organ regeneration and prevent relapse. In this chapter, we discuss how the use of mesenchymal stem cells (MSCs) could provide novel options for reducing the morbidity and mortality associated with HSC transplantation. This could potentially allow use of HSC transplantation for treatment of a wider variety of disorders. Also, our clinical experience in using MSCs in the field of HSC transplantation has paved the way for use of MSCs for treatment of a wide variety of other disorders.

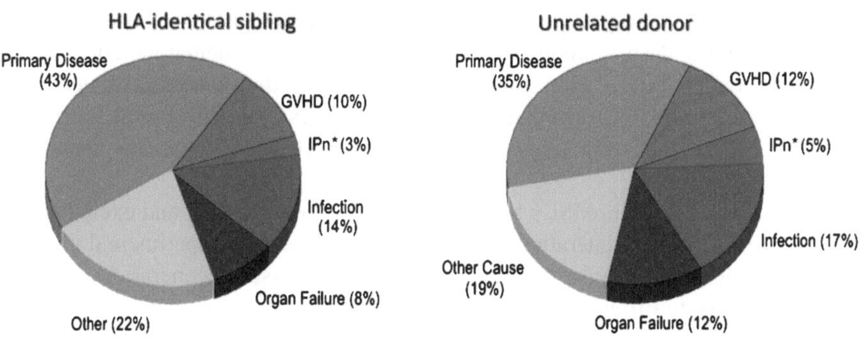

Fig. 5.1 Causes of mortality after allogeneic HSCT (Reproduced with permission from Center for International Blood and Marrow Transplant Research Center)

5.2 Mesenchymal Stem Cells

Bone marrow (BM) stromal cells were first reported by Friedenstein et al. as an adherent, fibroblast-like population derived from adult rodent BM that were capable of regenerating rudimentary bone *in vivo* capable of supporting hematopoiesis [16]. These cells comprise a very small population (<0.1%) of adult BM cells and are thought to be a major component of the supportive BM stromal niche for HSCs; however, some recent studies favor osteoblasts, a progeny of MSCs, as the main contributor to the HSC niche [72]. Another prominent characteristic of these cells is their capability to differentiate into other cells of mesenchymal lineage including bone, cartilage and fat; and thus commonly termed mesenchymal stem cells (MSCs) [7]. MSCs from BM are most commonly isolated by plating aspirated mononuclear cells in culture plates, followed by serial passage of the adherent cells. Due to lack of reliable markers for their direct isolation from BM samples it must be noted that even after three decades since their original description, there is still much controversy about their exact anatomical location inside BM, true physiological role, and their multilineage differentiation [21]. To better reflect the true identity of these cells the term multipotent mesenchymal stromal cells (with the same acronym as MSCs) has been proposed as the preferred terminology and specific combination of cell surface markers recommended for their definition [13].

MSCs are part of the bone marrow stromal microenvironment and their fate during HSC transplantation is of keen interest to physicians who use HSC transplantation for treatment of malignant or non-malignant disorders. Recipient MSCs seem to survive myeloablative conditioning regimens and the majority of studies have shown MSCs to be predominantly recipient-derived after bone marrow transplantation [2, 30, 58]. Thus, although donor MSCs comprises a constituent of marrow grafts, and at least theoretically they should be carried within the graft, they do not replace recipient MSCs despite complete engraftment of a new hematopoietic system. Interestingly, this lack of engraftment does not obviate their potential clinical utilities. Indeed, recent studies suggest that huge numbers of *ex vivo* culture expanded MSCs are capable of mediating clinically favorable effects without any meaningful level of engraftment.

Over the last decade, MSCs have generated a lot of attention and excitement in the field of cellular regenerative medicine. There are ample experimental and pre-clinical evidence that make these cells attractive for tissue regeneration including their capability to differentiate into a variety of cell types, their ability to support and stimulate proliferation and survival of resident progenitor cells, their tropism for migration into sites of injury or inflammation after intravenous infusion, and their therapeutic tendency to promote recovery of damaged tissues through secretion of a variety of cytokines and chemokines [8, 11, 17, 23, 36, 45, 53, 55]. Additionally, these cells possess many immunomodulatory and anti-inflammatory properties which could prove to be beneficial in many pathological conditions [38, 56, 69]. For example, MSCs have been shown to avoid inducing lymphocyte proliferation in vitro and escape being targeted by cytotoxic T cells or NK cells [26, 41, 57, 67]. Those properties of immunological stealth enable MSCs to be transplanted over major histocompatibility complex barriers in humans [35]. Capitalizing on these immunological characteristics, many studies involving human subjects have used "off-the-shelf " *ex vivo* culture expanded BM-derived-MSCs from HLA-mismatched "third party" donors, and to a more limited extent, MSCs derived from other tissues such as fat. Several phase I-II trials for a variety of non-hematological indications including treatment of patients with metachromatic leukodystrophy and Hurler's disease [27], osteogenesis imperfecta [20], myocardial infarction [9, 64], amyotrophic lateral sclerosis [10, 43, 44], stroke [4], Crohn's disease [66], diabetes mellitus [1] and refractory wounds [73] have been performed so far [18]. However, the focus of this review will be on the use of MSCs in HSC transplantation, based on their numerous biological properties and functions relevant to HSC transplantation biology (Fig. 5.2) [5].

5.3 MSCs to Enhance Engraftment After HSCT

Primary graft failure is a feared complication of allogeneic, and to a lesser extent autologous, HSCT which can be caused by failure to engraft or rejection of transplanted cells. The incidence of graft failure is typically <5% in fully ablative (when high dose chemotherapy or irradiation is used to wipe out the recipient BM) HLA-matched transplants but is higher in patients who receive less intense conditioning,

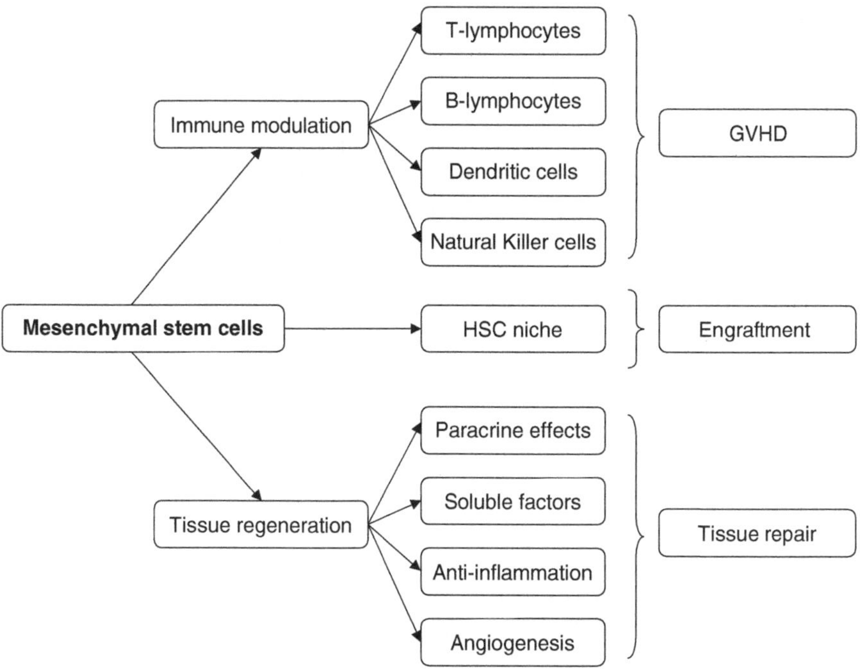

Fig. 5.2 Mechanisms of action and potential role of mesenchymal stem cells in hematopoietic stem cell transplantation. Reproduced with Permission from Battiwalla M and Hematti P (2009) Mesenchymal stem cells in hematopoietic stem cell transplantation. Cytotherapy 11:503–515

with HLA-mismatched grafts or in recipients who are HLA allo-immunized due to immunologic resistance against donor hematopoietic progenitors. Since MSCs are believed to support hematopoiesis, animal studies have been conducted in which co-transplantation of MSCs with HSCs showed improved engraftment of the latter [6, 51]. Subsequently, the original studies on use of MSCs were conducted to investigate their potential to improve engraftment of HSC transplants.

The first clinical trial to show feasibility of collection and, *ex vivo* culture-expansion, and safety of intravenous infusion of human BM-derived MSCs was performed by Lazarus et al. [32]. In that trial, MSCs were expanded from 10 ml BM aspirates from 23 patients with hematologic malignancies while in complete remission. After several weeks of *ex vivo* expansion, autologous MSCs were re-infused intravenously into 15 patients without any adverse reactions. This first-in-human study proved the feasibility of expanding MSCs from a small volume of marrow and their safety in infusion; however, this study was not intended to investigate the potential of MSCs in promoting hematopoietic engraftment as no simultaneous HSC transplantation was done.

The same group then conducted a phase I-II clinical trial with the goal to further determine not only the feasibility and safety but also the hematopoietic supportive effects of culture-expanded autologous MSCs. For this study autologous MSCs were infused at the time of autologous HSC transplantation into 28 breast cancer patients who received high-dose chemotherapy [28] to test the hypothesis that autologous MSCs would accelerate recovery of hematopoiesis. In this study MSCs were culture-expanded up to passages 2–6 from a small BM aspirate, and were infused at a dose of $1–2.2 \times 10^6$ MSCs/Kg. Again, infusion of autologous MSCs were not associated with toxicity and hematopoietic recovery was prompt. However, due to non-randomized nature of this study, definite conclusions regarding the effect of MSCs on HSC engraftment could not be made [28]. Nevertheless, these studies provided reassuring evidence that infusion of culture expanded MSCs is a very safe procedure.

The Lazarus group next investigated the potential of MSCs in an allogeneic HSC transplantation setting. In that study 46 recipient of HLA-identical sibling HSC transplantation for various hematological malignancies received an infusion of culture-expanded donor MSCs 4 h before infusion of HSCs [33]. In this study, no infusion-related toxicities, ectopic tissue formation, or increase in the incidence or severity of GVHD was reported. However, compared to historical controls, hematopoietic engraftment occurred at a similar pace. Importantly, incidences of acute and chronic GVHD in this study were similar to other comparable studies. It is notable that for the first time the authors suggested that culture-expanded MSCs from unrelated healthy HLA unmatched third-party donors could be potentially served as a universal donor product, a pioneering vision that was realized later by other investigators.

Since these original attempts, the use of allogeneic or autologous *ex vivo* culture expanded BM-derived MSCs co-transplanted with HSCs for promotion of HSC engraftment have been reported by other investigators. In one study, MSC transplantation was used to prevent repeat graft rejection/failure after HSC re-transplantation in three patients and to enhance hematopoietic engraftment after allogeneic HSC transplantation in four patients [39]. The source of HSCs in this study included three HLA-matched siblings, three unrelated donors and one cord blood unit. MSC donors were HLA-matched siblings in three cases and haplo-identical donors in the remaining four cases. Engraftment of neutrophils (≥500/μL) was achieved at a median of 12 days, and platelets (≥30,000/μL) at a median of 12 days. In summary, co-transplantation of MSCs resulted in rapid engraftment and 100% donor chimerism, even in three patients who received re-transplantation for previous graft failure/rejections, a group with high risk for rejection of HSCs. In another study by Ball et al., 14 children undergoing haplo-identical HSC transplantation with G-CSF mobilized CD34+ cells were co-transplanted with MSCs from the same donors. All the patients showed appropriate hematopoietic engraftment without any adverse reactions, a favorable result compared to a graft failure rate of 15% in 47 historic controls. This suggests that MSCs have the potential capability to reduce the risk of graft failure in haplo-identical HSC transplant recipients [3]. The authors proposed that the reduction of graft failure could be explained by the immunosuppressive effect of MSCs on residual allo-reactive recipient T lymphocytes which survived the preparative regimen. In another phase I-II trial by MacMillan et al. the potential of MSCs from haplo-identical parental donors to accelerate hematopoietic recovery added to a single unit of umbilical cord blood transplantation was tested. However,

7 out of 15 enrolled patients did not receive MSC infusions for a variety of reasons. Eight patients received MSCs on the day of infusion of umbilical cord blood and three patients received a repeat dose infused on day 21. Patients who received MSC transfusion achieved neutrophil engraftment (>500/μL) at a median of 19 days and 75% chance of platelet engraftment (first day of a seven consecutive day of having a platelet count of >50,000/μL) at a median of 53 days. Furthermore, with a median follow-up of 6.8 years, five patients remained alive and disease free at the time of report [42].

5.4 MSCs for Prevention and/or Treatment of Graft Versus Host Disease

One of the most intriguing properties of *ex vivo* culture expanded MSCs is their ability to modulate a wide variety of immune responses either through cell-cell contact or secretion of paracrine factors. These immunomodulatory effects involves interaction with a broad range of immune cells including T-lymphocytes, B-lymphocytes, natural killer and dendritic cells [47, 50, 62, 65]. Based on these immune properties it was suggested that MSCs could be potentially useful for amelioration of GVHD, a common adverse effect after allogeneic HSC transplantation [12, 29]. Steroid-refractory GVHD has an extremely poor prognosis with no satisfactory treatment options available [15]. In a seminal study Le Blanc et al. were the first to investigate the potential of MSCs for the treatment of refractory GVHD [37]. A 9-year-old boy who received HSC transplantation from a matched unrelated donor for treatment of leukemia developed severe acute GVHD of the gut and liver unresponsive to all types of immunosuppression including high dose methylprednisolone, infliximab and daclizumab. Haplo-identical MSCs were generated from the patient's mother, who was not the original donor of HSCs. After 3 weeks of *ex vivo* culture expansion, infusion of one dose of MSCs resulted in an impressive control of the GVHD symptoms. Infusion of a second dose of MSCs generated from the same batch of MSCs was also effective in treating a recurrent episode of GVHD that happened months later. Following this encouraging case, the same group embarked on a larger study in which MSCs were given to eight patients with steroid-refractory grades III-IV GVHD and one with extensive chronic GVHD [60]. These MSCs were derived from HLA-identical siblings (2), haplo-identical family donors (6), and unrelated mismatched donors (4). Acute toxicities after MSC infusions were absent and acute GVHD completely resolved in six of eight patients, although two other patients did not respond to MSC treatment and died soon. At the time of publication five patients were alive between 2 months and 3 years after transplantation. This survival rate was significantly superior to a control group of 16 patients with refractory acute gut GVHD who did not receive MSCs. Later, the same group reported the result of administration of BM-derived MSCs that were culture expanded according to the European Group for Blood and Marrow Transplantation *ex vivo* expansion procedure. In this multicenter phase II trial, a total of 55 patients with steroid refractory GVHD received a median dose of 1.4×10^6 MSCs (range $0.4–9 \times 10^6$)/kg derived from multiple donor sources; interestingly, some patient even received MSCs from multiple donors. Complete responses occurred in 30 patients and 53% of the complete responders were still alive after 2 years while non-responders showed a survival rate of 16% [34]. Generation of donor specific MSCs is time consuming, costly, and in many occasions impractical due to the urgent nature of the need for its use; so these issues add to the importance of these studies that showed GVHD responses were independent of the source of MSCs. Thus, these pioneering clinical trials ushered an era of use of MSCs derived from third party unrelated donors not only in numerous more GVHD trials but also many trials in which MSCs were tested in the context of other disorders.

The largest clinical trials performed with MSCs have been two Phase III double-blind, placebo-controlled randomized trial evaluating (Prochymal™) as a first-line treatment for acute GVHD or for treatment of refractory acute GVHD. The final results of these industry sponsored trials are eagerly awaited. However, use of (Prochymal™) in pediatric patients with severe refractory acute graft-versus-host disease in a compassionate study showed promising results [54]. In this study 7 out of 12 patients had complete response, two showed partial response, and three had mixed response and complete resolution of GI symptoms occurred in nine patients. If approved, this will be the first stem cell product on market.

Recently, two groups reported the safety and potential efficacy of MSCs in the setting of chronic GVHD [71, 74]. Again, the small size and non-randomized nature of these studies preclude any firm conclusion on the potential efficacy of MSCs in this condition it re-affirms the safety of MSCs even in situations that there are chronic multi-organ injuries. Clearly, randomized clinical trials to test the efficacy of MSCs in chronic GVHD are warranted, as long as there are no other viable therapeutic options.

5.5 MSCs for Tissue Repair After HSC Transplantation

While MSCs have been used widely for immunomodulatory function or bone marrow microenvironment reconstitution, another avenue for clinical use of MSCs is to utilize tissue regenerative properties of MSCs in treating other complications associated with HSC transplantation. For example, Ringden et al. treated ten patients who had a variety of tissue/organ toxicities following allogeneic HSC transplantation, including seven with hemorrhagic cystitis, two with pneumomediastinum and one with perforated colon and peritonitis with MSCs [61]. After MSC infusion, severe hemorrhagic cystitis resolved in five patients, reduction in transfusion requirements was achieved in two patients, although one died of multiorgan failure and one from progressive GVHD. In two patients, pneumomediastinum resolved after MSC infusions. In the case of a patient with steroid-refractory GVHD of the gut who developed perforated diverticulitis and peritonitis, MSC infusion was used as the patient refused surgery. She recovered after MSC infusion but later developed recurrent peritonitis. Although she responded a second time, she eventually died of disseminated fungal infection. Like other pioneering studies on use of MSCs in HSC transplantation these studies could pave the way for use of MSCs in a wider range of sever tissue injuries such as refractory Crohn's disease [14]

5.6 Practical Issues with the Use of MSCs

5.6.1 Effect of Culture Methodology

One of major concerns regarding the use of MSC is enormous heterogeneity in the production methodologies for MSCs as currently there is no standard method to culture and expand those cells [19, 48, 63]. Bone marrow MSCs are usually derived from small volume (25–50 ml) BM aspirates. Since MSCs are present in the mononuclear cell (MNC) fraction, the MNC fraction is usually, but not always, separated using density gradient centrifugation followed by culturing cells in media supplemented with fetal bovine serum (FBS). Twenty-four hours to several days later, nonadherent cells are removed and the adherent cells are fed with culture media changes every 3–4 days. When the cells reach near confluence, they are trypsinized and passaged into new culture plates. There will be cells other than MSCs at low passages but after a few passages there will be a homogenous population of cells resembling fibroblasts. Nevertheless, despite their seemingly similar morphologies, the ultimate identity and phenotypes of such cells could depend on numerous factors such as: the starting source of the cells, age of the donor, density of the cells plated, and type of media and serum used. For example, the use of FBS generates concern about the potential risk of transmitting animal diseases of yet unknown origin and also for potential reaction of recipients to antigens of animal origin. Serum free media, autologous serum, fresh frozen plasma, and human platelet lysates have been investigated as an alternative, but this just adds to potentially more heterogeneity in the final delivered product [31, 40, 46].

In a trial performed by Von Bonin et al., 13 patients with steroid-refractory acute GVHD received MSCs that were expanded in culture medium containing platelet lysate and derived from unrelated HLA disparate donors [70]. Two of these patients showed complete response and did not require further escalation of their immunosuppressive medications after the first dose of MSCs. Eleven patients received additional immunosuppressive therapy along with more MSC infusions, with five of them showing response after 28 days. After a median follow-up of 257 days four patients (31%) were alive. Lower response rate in this cohort compared to that of other reports could potentially be attributable to several factors including the use of platelet lysate-containing medium, difference in number of MSC infusions, cell dose, or other factors such as the interval between development of GVHD and the first dose of MSCs. Indeed, lack of uniformity in these parameters is a recurring theme that makes comparison between different studies problematic. On the other hand, Perez-Simon et al. performed a clinical trial involving bone marrow MSCs expanded with autologous serum [52]. Among ten patients treated for acute GVHD, one patient achieved complete remission, six patients achieved partial remission and three patients did not respond. In the group with chronic GVHD, total of eight patients have received MSC treatment, achieving complete remission in one, partial remission in three and four patients did not show response. This trial showed feasibility of expanding MSCs using donor serum which eliminates worries about using FBS for culturing MSCs in vitro. However, these variations in the culture methodologies could easily affect the immunomodulatory properties of MSCs and thus confound the interpretation of results of different studies.

5.6.2 Optimal Dose

The use of MSCs for enhancing HSC engraftment and mitigating side effects associated with HSC transplantation is being actively researched by numerous groups around the world. However, it is not clear how many cells need to be transplanted to achieve clinical efficacy, how the cells should be used and what would be the optimal schedule for using MSCs. A recent report by Kebriaei et al. compared high versus low dose MSCs [24]. In this study, doses of either 2 or 8×10^6 MSCs/kg for grades II-IV acute GVHD were equivalent in safety and efficacy. Furthermore, use of eight million MSC/Kg in pediatric patients has also been shown to be safe in kids [54]. The optimal dose and schedule may also differ by indication while the potency might differ by laboratory.

5.6.3 Optimal Source

Another issue about the application of MSCs in clinical settings is whether MSCs should be from single donor or multiple donors can be pooled. There are suggestions that an improved response rate in mixed lymphocyte culture might result in greater benefit from pooling MSCs [63]. Ringden and Le Blanc also reported a case series where pooled MSCs were used for treatment of a patient with severe hemorrhage after HSC transplantation [59]. Clinical utility of pooled MSCs further enhance the notion of the unique immunomodulatory characteristics enjoyed by MSCs.

Also, numerous logistical concerns are relevant to clinical translation. Even in an ideal situation with minimal delay, it takes about 4 weeks to expand autologous MSCs and this has prevented many trial subjects to receive target dose of MSCs. In this respect, the finding that third party MSCs are equally effective in treating GVHD and safe to administer, suggest that use of third party MSCs could be the solution of choice in indications requiring urgent intervention.

5.6.4 Tumorigenic Potential

In an open-label randomized trial by Ning et al. involving patients who received HLA-matched sibling HSC transplantation for treatment of hematologic malignancies, 10 patients received MSC transfusion while 15 patients served as control [49]. In this study, the prevalence of grades II-IV acute GVHD was far lower in MSC group (11.1%) compared to control where eight patients (53.3%) developed acute GVHD. One of the interesting findings was that there was more relapse in the MSC group where six relapsed (60.0%) in contrast with three patients (20.0%) in the non-MSC group. Three-year disease-free survival was also lower in the MSC treated group at 30.0% compared to the non-treated group at 66.7%. The authors suggested that co-transplantation of MSCs and HSCs may prevent GVHD, but could be also associated with a higher rate of relapse. Although tumor promoting effects of MSCs have been shown animal models [22, 25] it is re-assuring that this worrisome and concerning trend has not been reported by others and it is not clear what factors might have caused this. Durability of the infused MSCs is also a matter of great debate [68]. Although MSC engraftment level is now believed to be only transient and at a low level it must be considered that MSCs might exert their beneficial effects through mechanisms unrelated to engraftment, such as paracrine effects, prior to their demise. This short lifespan of MSCs post-transplant reduces their tumorigenic potential.

5.7 Conclusions

MSCs were originally isolated from BM and, due to their presumed roles in hematopoiesis, HSC transplant physicians were the first to use them in clinical trials. Not surprisingly, MSCs have been most extensively studied in the context of HSC transplantation for promotion of engraftment, prevention and/or treatment of GVHD and for treatment of organ and tissue toxicity after HSC transplantation. These attempts at capitalizing on the unique properties of MSCs have demonstrated safety and allayed fears about the generation of ectopic tissue. However, the final place of MSCs in HSC transplantation remains to be fully defined. For example, evidence of MSCs boosting engraftment of HSC is inconclusive. Addressing this uncertainty will require large randomized trials in populations vulnerable to engraftment failure (such as cord blood transplants) which is undoubtedly a daunting task. Similarly, despite reassuring safety results reported in GVHD trials, small sample size and non-randomized nature limit conclusive determination of clinical benefit. However, ease of production, apparent lack of immunogenicity, and impressive safety records of MSCs warrant their continued investigation not only in the HSC transplant arena but also other disciplines of medicine. Carefully designed clinical trials with sufficient size will be necessary to define the optimum dose, schedule, source, route of administration, and specific indications for MSC.

References

1. Abdi R, Fiorina P, Adra CN, Atkinson M, Sayegh MH (2008) Immunomodulation by mesenchymal stem cells: a potential therapeutic strategy for type 1 diabetes. Diabetes 57:1759–1767
2. Awaya N, Rupert K, Bryant E, Torok-Storb B (2002) Failure of adult marrow-derived stem cells to generate marrow stroma after successful hematopoietic stem cell transplantation. Exp Hematol 30:937–942
3. Ball LM, Bernardo ME, Roelofs H, Lankester A, Cometa A, Egeler RM, Locatelli F, Fibbe WE (2007) Cotransplantation of ex vivo expanded mesenchymal stem cells accelerates lymphocyte recovery and may reduce the risk of graft failure in haploidentical hematopoietic stem-cell transplantation. Blood 110:2764–2767
4. Bang OY, Lee JS, Lee PH, Lee G (2005) Autologous mesenchymal stem cell transplantation in stroke patients. Ann Neurol 57:874–882
5. Battiwalla M, Hematti P (2009) Mesenchymal stem cells in hematopoietic stem cell transplantation. Cytotherapy 11:503–515
6. Bensidhoum M, Chapel A, Francois S, Demarquay C, Mazurier C, Fouillard L, Bouchet S, Bertho JM, Gourmelon P, Aigueperse J, Charbord P, Gorin NC, Thierry D, Lopez M (2004) Homing of in vitro expanded Stro-1− or Stro-1+ human mesenchymal stem cells into the NOD/SCID mouse and their role in supporting human CD34 cell engraftment. Blood 103:3313–3319
7. Caplan AI (1991) Mesenchymal stem cells. J Orthop Res 9:641–650
8. Caplan AI (2007) Adult mesenchymal stem cells for tissue engineering versus regenerative medicine. J Cell Physiol 213:341–347
9. Chen SL, Fang WW, Ye F, Liu YH, Qian J, Shan SJ, Zhang JJ, Chunhua RZ, Liao LM, Lin S, Sun JP (2004) Effect on left ventricular function of intracoronary transplantation of autologous bone marrow mesenchymal stem cell in patients with acute myocardial infarction. Am J Cardiol 94:92–95
10. Deda H, Inci MC, Kurekci AE, Sav A, Kayihan K, Ozgun E, Ustunsoy GE, Kocabay S (2009) Treatment of amyotrophic lateral sclerosis patients by autologous bone marrow-derived hematopoietic stem cell transplantation: a 1-year follow-up. Cytotherapy 11:18–25
11. Devine SM, Cobbs C, Jennings M, Bartholomew A, Hoffman R (2003) Mesenchymal stem cells distribute to a wide range of tissues following systemic infusion into nonhuman primates. Blood 101:2999–3001
12. Devine SM, Hoffman R (2000) Role of mesenchymal stem cells in hematopoietic stem cell transplantation. Curr Opin Hematol 7:358–363
13. Dominici M, Le BK, Mueller I, Slaper-Cortenbach I, Marini F, Krause D, Deans R, Keating A, Prockop D, Horwitz E (2006) Minimal criteria for defining multipotent mesenchymal stromal cells. The International Society for Cellular Therapy position statement. Cytotherapy 8:315–317
14. Duijvestein M, Vos AC, Roelofs H, Wildenberg ME, Wendrich BB, Verspaget HW, Kooy-Winkelaar EM, Koning F, Zwaginga JJ, Fidder HH, Verhaar AP, Fibbe WE, van den Brink GR, Hommes DW (2010) Autologous bone marrow-derived mesenchymal stromal cell treatment for refractory luminal Crohn's disease: results of a phase I study. Gut 59:1662–1669
15. Ferrara JL, Levine JE, Reddy P, Holler E (2009) Graft-versus-host disease. Lancet 373(9674):1550–1561
16. Friedenstein AJ, Petrakova KV, Kurolesova AI, Frolova GP (1968) Heterotopic of bone marrow. Analysis of precursor cells for osteogenic and hematopoietic tissues. Transplantation 6:230–247
17. Gao J, Dennis JE, Muzic RF, Lundberg M, Caplan AI (2001) The dynamic in vivo distribution of bone marrow-derived mesenchymal stem cells after infusion. Cells Tissues Organs 169:12–20
18. Giordano A, Galderisi U, Marino IR (2007) From the laboratory bench to the patient's bedside: an update on clinical trials with mesenchymal stem cells. J Cell Physiol 211:27–35

19. Haack-Sorensen M, Bindslev L, Mortensen S, Friis T, Kastrup J (2007) The influence of freezing and storage on the characteristics and functions of human mesenchymal stromal cells isolated for clinical use. Cytotherapy 9:328–337
20. Horwitz EM, Prockop DJ, Fitzpatrick LA, Koo WW, Gordon PL, Neel M, Sussman M, Orchard P, Marx JC, Pyeritz RE, Brenner MK (1999) Transplantability and therapeutic effects of bone marrow-derived mesenchymal cells in children with osteogenesis imperfecta. Nat Med 5:309–313
21. Javazon EH, Beggs KJ, Flake AW (2004) Mesenchymal stem cells: paradoxes of passaging. Exp Hematol 32:414–425
22. Karnoub AE, Dash AB, Vo AP, Sullivan A, Brooks MW, Bell GW, Richardson AL, Polyak K, Tubo R, Weinberg RA (2007) Mesenchymal stem cells within tumour stroma promote breast cancer metastasis. Nature 449:557–563
23. Keating A (2006) Mesenchymal stromal cells. Curr Opin Hematol 13:419–425
24. Kebriaei P, Robinson S (2011) Treatment of graft-versus-host-disease with mesenchymal stromal cells. Cytotherapy 13:262–268
25. Kidd S, Spaeth E, Klopp A, Andreeff M, Hall B, Marini FC (2008) The (in) auspicious role of mesenchymal stromal cells in cancer: be it friend or foe. Cytotherapy 10:657–667
26. Klyushnenkova E, Mosca JD, Zernetkina V, Majumdar MK, Beggs KJ, Simonetti DW, Deans RJ, McIntosh KR (2005) T cell responses to allogeneic human mesenchymal stem cells: immunogenicity, tolerance, and suppression. J Biomed Sci 12:47–57
27. Koc ON, Day J, Nieder M, Gerson SL, Lazarus HM, Krivit W (2002) Allogeneic mesenchymal stem cell infusion for treatment of metachromatic leukodystrophy (MLD) and Hurler syndrome (MPS-IH). Bone Marrow Transplant 30:215–222
28. Koc ON, Gerson SL, Cooper BW, Dyhouse SM, Haynesworth SE, Caplan AI, Lazarus HM (2000) Rapid hematopoietic recovery after coinfusion of autologous-blood stem cells and culture-expanded marrow mesenchymal stem cells in advanced breast cancer patients receiving high-dose chemotherapy. J Clin Oncol 18:307–316
29. Koc ON, Lazarus HM (2001) Mesenchymal stem cells: heading into the clinic. Bone Marrow Transplant 27:235–239
30. Koc ON, Peters C, Aubourg P, Raghavan S, Dyhouse S, DeGasperi R, Kolodny EH, Yoseph YB, Gerson SL, Lazarus HM, Caplan AI, Watkins PA, Krivit W (1999) Bone marrow-derived mesenchymal stem cells remain host-derived despite successful hematopoietic engraftment after allogeneic transplantation in patients with lysosomal and peroxisomal storage diseases. Exp Hematol 27:1675–1681
31. Lange C, Cakiroglu F, Spiess AN, Cappallo-Obermann H, Dierlamm J, Zander AR (2007) Accelerated and safe expansion of human mesenchymal stromal cells in animal serum-free medium for transplantation and regenerative medicine. J Cell Physiol 213:18–26
32. Lazarus HM, Haynesworth SE, Gerson SL, Rosenthal NS, Caplan AI (1995) Ex vivo expansion and subsequent infusion of human bone marrow-derived stromal progenitor cells (mesenchymal progenitor cells): implications for therapeutic use. Bone Marrow Transplant 16:557–564
33. Lazarus HM, Koc ON, Devine SM, Curtin P, Maziarz RT, Holland HK, Shpall EJ, McCarthy P, Atkinson K, Cooper BW, Gerson SL, Laughlin MJ, Loberiza FR Jr, Moseley AB, Bacigalupo A (2005) Cotransplantation of HLA-identical sibling culture-expanded mesenchymal stem cells and hematopoietic stem cells in hematologic malignancy patients. Biol Blood Marrow Transplant 11:389–398
34. Le Blanc K, Frassoni F, Ball L, Locatelli F, Roelofs H, Lewis I, Lanino E, Sundberg B, Bernardo ME, Remberger M, Dini G, Egeler RM, Bacigalupo A, Fibbe W, Ringden O (2008) Mesenchymal stem cells for treatment of steroid-resistant, severe, acute graft-versus-host disease: a phase II study. Lancet 371:1579–1586
35. Le Blanc K, Gotherstrom C, Ringden O, Hassan M, McMahon R, Horwitz E, Anneren G, Axelsson O, Nunn J, Ewald U, Norden-Lindeberg S, Jansson M, Dalton A, Astrom E, Westgren M (2005) Fetal mesenchymal stem-cell engraftment in bone after in utero transplantation in a patient with severe osteogenesis imperfecta. Transplantation 79:1607–1614
36. Le Blanc K, Pittenger M (2005) Mesenchymal stem cells: progress toward promise. Cytotherapy 7:36–45

37. Le Blanc K, Rasmusson I, Sundberg B, Gotherstrom C, Hassan M, Uzunel M, Ringden O (2004) Treatment of severe acute graft-versus-host disease with third party haploidentical mesenchymal stem cells. Lancet 363:1439–1441
38. Le Blanc K, Ringden O (2005) Immunobiology of human mesenchymal stem cells and future use in hematopoietic stem cell transplantation. Biol Blood Marrow Transplant 11:321–334
39. Le Blanc K, Samuelsson H, Gustafsson B, Remberger M, Sundberg B, Arvidson J, Ljungman P, Lonnies H, Nava S, Ringden O (2007) Transplantation of mesenchymal stem cells to enhance engraftment of hematopoietic stem cells. Leukemia 21:1733–1738
40. Le Blanc K, Samuelsson H, Lonnies L, Sundin M, Ringden O (2007) Generation of immunosuppressive mesenchymal stem cells in allogeneic human serum. Transplantation 84:1055–1059
41. Le Blanc K, Tammik C, Rosendahl K, Zetterberg E, Ringden O (2003) HLA expression and immunologic properties of differentiated and undifferentiated mesenchymal stem cells. Exp Hematol 31:890–896
42. Macmillan ML, Blazar BR, DeFor TE, Wagner JE (2009) Transplantation of ex-vivo culture-expanded parental haploidentical mesenchymal stem cells to promote engraftment in pediatric recipients of unrelated donor umbilical cord blood: results of a phase I-II clinical trial. Bone Marrow Transplant 43:447–454
43. Martinez HR, Gonzalez-Garza MT, Moreno-Cuevas JE, Caro E, Gutierrez-Jimenez E, Segura JJ (2009) Stem-cell transplantation into the frontal motor cortex in amyotrophic lateral sclerosis patients. Cytotherapy 11:26–34
44. Mazzini L, Mareschi K, Ferrero I, Vassallo E, Oliveri G, Boccaletti R, Testa L, Livigni S, Fagioli F (2006) Autologous mesenchymal stem cells: clinical applications in amyotrophic lateral sclerosis. Neurol Res 28:523–526
45. Mouiseddine M, Francois S, Semont A, Sache A, Allenet B, Mathieu N, Frick J, Thierry D, Chapel A (2007) Human mesenchymal stem cells home specifically to radiation-injured tissues in a non-obese diabetes/severe combined immunodeficiency mouse model. Br J Radiol 80(Spec No 1):S49–S55
46. Muller I, Kordowich S, Holzwarth C, Spano C, Isensee G, Staiber A, Viebahn S, Gieseke F, Langer H, Gawaz MP, Horwitz EM, Conte P, Handgretinger R, Dominici M (2006) Animal serum-free culture conditions for isolation and expansion of multipotent mesenchymal stromal cells from human BM. Cytotherapy 8:437–444
47. Nauta AJ, Fibbe WE (2007) Immunomodulatory properties of mesenchymal stromal cells. Blood 110(10):3499–3506
48. Neuhuber B, Swanger SA, Howard L, Mackay A, Fischer I (2008) Effects of plating density and culture time on bone marrow stromal cell characteristics. Exp Hematol 36:1176–1185
49. Ning H, Yang F, Jiang M, Hu L, Feng K, Zhang J, Yu Z, Li B, Xu C, Li Y, Wang J, Hu J, Lou X, Chen H (2008) The correlation between cotransplantation of mesenchymal stem cells and higher recurrence rate in hematologic malignancy patients: outcome of a pilot clinical study. Leukemia 22:593–599
50. Noel D, Djouad F, Bouffi C, Mrugala D, Jorgensen C (2007) Multipotent mesenchymal stromal cells and immune tolerance. Leuk Lymphoma 48:1283–1289
51. Noort WA, Kruisselbrink AB, in't Anker PS, Kruger M, van Bezooijen RL, de Paus RA, Heemskerk MH, Lowik CW, Falkenburg JH, Willemze R, Fibbe WE (2002) Mesenchymal stem cells promote engraftment of human umbilical cord blood-derived CD34(+) cells in NOD/SCID mice. Exp Hematol 30:870–878
52. Perez-Simon JA, Lopez-Villar O, Andreu EJ, Rifon J, Muntion S, Campelo MD, Sanchez-Guijo FM, Martinez C, Valcarcel D, Canizo CD (2011) Mesenchymal stem cells expanded in vitro with human serum for the treatment of acute and chronic graft-versus-host disease: results of a phase I/II clinical trial. Haematologica 96:1072–1076
53. Pittenger MF, Mackay AM, Beck SC, Jaiswal RK, Douglas R, Mosca JD, Moorman MA, Simonetti DW, Craig S, Marshak DR (1999) Multilineage potential of adult human mesenchymal stem cells. Science 284:143–147
54. Prasad VK, Lucas KG, Kleiner GI, Talano JA, Jacobsohn D, Broadwater G, Monroy R, Kurtzberg J (2011) Efficacy and safety of ex vivo cultured adult human mesenchymal stem

cells (Prochymal) in pediatric patients with severe refractory acute graft-versus-host disease in a compassionate use study. Biol Blood Marrow Transplant 17:534–541
55. Prockop DJ (2007) "Stemness" does not explain the repair of many tissues by mesenchymal stem/multipotent stromal cells (MSCs). Clin Pharmacol Ther 82:241–243
56. Prockop DJ, Olson SD (2007) Clinical trials with adult stem/progenitor cells for tissue repair: let's not overlook some essential precautions. Blood 109:3147–3151
57. Rasmusson I, Ringden O, Sundberg B, Le Blanc K (2003) Mesenchymal stem cells inhibit the formation of cytotoxic T lymphocytes, but not activated cytotoxic T lymphocytes or natural killer cells. Transplantation 76:1208–1213
58. Rieger K, Marinets O, Fietz T, Korper S, Sommer D, Mucke C, Reufi B, Blau WI, Thiel E, Knauf WU (2005) Mesenchymal stem cells remain of host origin even a long time after allogeneic peripheral blood stem cell or bone marrow transplantation. Exp Hematol 33:605–611
59. Ringden O, Leblanc K (2011) Pooled MSCs for treatment of severe hemorrhage. Bone Marrow Transplant 46(8):1158–1160
60. Ringden O, Uzunel M, Rasmusson I, Remberger M, Sundberg B, Lonnies H, Marschall HU, Dlugosz A, Szakos A, Hassan Z, Omazic B, Aschan J, Barkholt L, Le BK (2006) Mesenchymal stem cells for treatment of therapy-resistant graft-versus-host disease. Transplantation 81:1390–1397
61. Ringden O, Uzunel M, Sundberg B, Lonnies L, Nava S, Gustafsson J, Henningsohn L, Le Blanc K (2007) Tissue repair using allogeneic mesenchymal stem cells for hemorrhagic cystitis, pneumomediastinum and perforated colon. Leukemia 21:2271–2276
62. Ryan JM, Barry FP, Murphy JM, Mahon BP (2005) Mesenchymal stem cells avoid allogeneic rejection. J Inflamm (Lond) 2:8
63. Samuelsson H, Ringden O, Lonnies H, Le Blanc K (2009) Optimizing in vitro conditions for immunomodulation and expansion of mesenchymal stromal cells. Cytotherapy 11:129–136
64. Schuleri KH, Boyle AJ, Hare JM (2007) Mesenchymal stem cells for cardiac regenerative therapy. Handb Exp Pharmacol 2007(180):195–218
65. Stagg J, Galipeau J (2007) Immune plasticity of bone marrow-derived mesenchymal stromal cells. Handb Exp Pharmacol 2007(180):45–66
66. Taupin P (2006) OTI-010 Osiris therapeutics/JCR pharmaceuticals. Curr Opin Investig Drugs 7:473–481
67. Tse WT, Pendleton JD, Beyer WM, Egalka MC, Guinan EC (2003) Suppression of allogeneic T-cell proliferation by human marrow stromal cells: implications in transplantation. Transplantation 75:389–397
68. Uccelli A, Moretta L, Pistoia V (2008) Mesenchymal stem cells in health and disease. Nat Rev Immunol 8:726–736
69. Uccelli A, Pistoia V, Moretta L (2007) Mesenchymal stem cells: a new strategy for immunosuppression? Trends Immunol 28:219–226
70. von Bonin M, Stolzel F, Goedecke A, Richter K, Wuschek N, Holig K, Platzbecker U, Illmer T, Schaich M, Schetelig J, Kiani A, Ordemann R, Ehninger G, Schmitz M, Bornhauser M (2009) Treatment of refractory acute GVHD with third-party MSC expanded in platelet lysate-containing medium. Bone Marrow Transplant 43:245–251
71. Weng JY, Du X, Geng SX, Peng YW, Wang Z, Lu ZS, Wu SJ, Luo CW, Guo R, Ling W, Deng CX, Liao PJ, Xiang AP (2010) Mesenchymal stem cell as salvage treatment for refractory chronic GVHD. Bone Marrow Transplant 45:1732–1740
72. Wu JY, Scadden DT, Kronenberg HM (2009) Role of the osteoblast lineage in the bone marrow hematopoietic niches. J Bone Miner Res 24:759–764
73. Yoshikawa T, Mitsuno H, Nonaka I, Sen Y, Kawanishi K, Inada Y, Takakura Y, Okuchi K, Nonomura A (2008) Wound therapy by marrow mesenchymal cell transplantation. Plast Reconstr Surg 121:860–877
74. Zhou H, Guo M, Bian C, Sun Z, Yang Z, Zeng Y, Ai H, Zhao RC (2010) Efficacy of bone marrow-derived mesenchymal stem cells in the treatment of sclerodermatous chronic graft-versus-host disease: clinical report. Biol Blood Marrow Transplant 16:403–412

Chapter 6
Cancer Stem Cells in Solid Tumors, Markers and Therapy

Ortiz-Sánchez Elizabeth, González-Montoya José Luis, Langley Elizabeth, and García-Carrancá Alejandro

Contents

O.-S. Elizabeth (✉) • G.-M. José Luis • L. Elizabeth • G.-C. Alejandro
Instituto Nacional de Cancerología SS, Av. San Fernando 22,
Mexico City 14080, Mexico
e-mail: elinfkb@yahoo.com.mx

R.K. Srivastava and S. Shankar (eds.), *Stem Cells and Human Diseases*,
DOI 10.1007/978-94-007-2801-1_6,

Abstract Cancer is a multi-factorial disease related with a high number of deaths in the world. There are three cancer models which try to explain the origin and behavior of most known cancers: The stochastic, hierarchical and phenotype plasticity models, which are not exclusive of one an other. However, the most common cancers known until now follow the hierarchical model. In this model only cancer stem cells and some early progenitor cells have the capacity to initiate tumor growth. In this chapter, we focus on the cancers that follow this hierarchical model and the current therapeutic strategies designed to eliminate cancer stem cells in order to improve patient health.

Keywords Cancer • Cancer stem cells • Cancer stem cell targets • Cancer stem cell therapy • Solid tumors

6.1 Introduction: Clonal Evolution, Cancer Stem Cells and Phenotype Plasticity Models

Cancer is an important human public health problem which has long been an important focus of scientific study. Carcinogenesis has been explained, in part, by clonal evolution or a stochastic model in which all cells present in a tumor would be able to originate, develop, and maintain tumor growth, including the capacity for metastasis (Fig. 6.1). The stochastic model considers that all genetic aberrations in cancer cells give them some advantages, which will be maintained in all subsequent cells [3, 4].

The stochastic model is based on the hallmarks of cancer. The mutations involved during early and late-onset cancer [5] are initiated at different times [6], including cancer initiation via DNA damage response and DNA repair, p53-mediated senescence and the accumulation of mutations in relevant genes [7]. Recent analyses show that human cancers, as well as cancer cell lines, exhibit extensive and highly localized mutations following DNA damage and repair responses. It is known that chromothripsis, single chromosome mutations that are developed in 2–3% of human cancers, drive the relevant mutations [8]. This stochastic model was considered for a long period of time; however, approximately 30 years ago, another model was developed to understand the aggressive cancer disease. This model, named hierarchy model, involves highly organized cell compartments of which the top one is represented by the "cancer stem cells" (CSC) or the "Tumor initiation cells" (TIC). Uncountable research groups have demonstrated the presence of CSCs in tumors, including solid tumors that have the ability to originate and maintain tumor growth. Additionally, there are many groups who are designing experiments to determine if stem cells can acquire important mutations that would then promote the origin of cancer stem cells or if CSCs can be originated by precancerous progenitor cells that then become CSCs. Furthermore, hypoxia and microenvironment play an important role in CSC generation. In the 1970s, spleen colony formation tests helped to evaluate

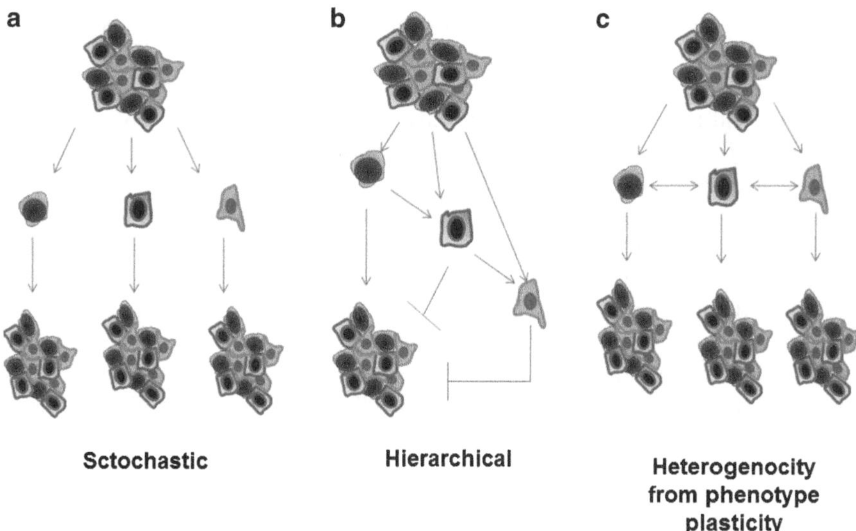

Fig. 6.1 Cancer origin models. (**a**) Stochastic model: cells that compose the tumor with different phenotypes exhibit similar potential to proliferate and capacity to induce tumor growth in xenotransplants. In addition, the phenotypes of these cells are irreversible. (**b**) Hierarchical model: only a small un-differentiated subpopulation represented in a tumor has the capability to induce tumor growth. These cells are named Cancer Stem Cells; they have similar characteristics to their normal counterparts like self-renewal, high potential for differentiation, and these cells exhibit chemoresistance and radioresistance to conventional treatments. The non-tumorigenic cells lack the ability to induce tumor growth and have lost the capability of differentiation. (**c**) Phenotype plasticity model: most of the cells that compose a tumor have the capability to induce tumor growth. These cells, with specific phenotypes, can induce tumor growth with a heterogenic phenotype. In contrast to the stochastic model, this phenotype, related with specific epigenetic changes, can be reversible. The original phenotype of the cells that induce tumor growth is heterogenic and epigenetically different compared to the cells that the tumor originates from (This Figure was based on references 1 and 2)

the variable capacity of leukemia cells to induce tumor growth in mice. These observations suggest that cancer is another stem cell human disease, causing research in this area to move at great speed.

As with their normal counterparts, it is known that some cancer stem cells express relevant protein regulators including p53, Wnt, Notch, and Hedgehog (Hg) signaling pathways, as well as c-Myc transcription profiles, similar to embryonic stem cells [9]. Cancer stem cells from solid tumors were described for the first time in breast cancer [5], moving on to liver, ovarian, prostate, head and neck, colon, and brain carcinomas [10–14].

The specific microenvironment of cancer stem cells is known as a niche, in which CSCs get special and specific signals to maintain and promote most of the properties of CSCs, including epigenetic modifications [15] as well as protection from apoptosis signals [16].

The term, "stem cells", includes embryonic, germinal and somatic stem cells, which are related to their respective potential for differentiation. Therefore, cancer diseases include these stem cell categories, in which the CSCs have the ability to initiate and maintain tumor growth. Like normal stem cells, the CSCs have the capability of self-renewal, and are characterized by their ability to generate functional cells in their particular tissues [17]. However, the CSCs lack the capacity to control cell numbers, losing the homeostasis function of normal stem cells. There is a small number of CSCs in tumors; however it depends on the type of tumor as well as specific tumor aggressiveness. Clinically, CSCs are chemo-resistant and radio-resistant making ordinary cancer therapies inefficient for eliminating tumor growth and relapse, which is not good for patients with these diseases.

It is known that before the advent of cancer lesions, precancerous lesions are present. These pre-stage precancerous stem cells show some stem-like properties such as multi-differentiation and the capability for tumor growth in immune deficient mice. Through epigenetic changes induced by microenvironment, these precancerous stem-like cells become cancer stem cells [18]. Chen et al., have defined these as precancerous stem cells, since they are present in premalignant lesions [18]. Through various steps in carcinogenesis, the normal stem cells are transformed into precancerous and then primary cancer stem cells [19]. Furthermore, these precancerous stem cells can be mutated in genes that encode proteins related with stemness such as OCT3/4, KLF-4, MYC, etc., to gain self-renewal abilities, multi-differentiation and resistance to chemo- and radiotherapies [20]. Finally, precancerous stem cells can be dependent on their microenvironment to generate malignant or benign tumors. There are some examples that support the presence of precancerous stem cells. In breast carinogenesis, it is possible to identify the precancerous and cancer stem cells based on their localization. The precancerous stem cells are present in ductal carcinoma *in situ* (DCIS), which is a pre-malignant lesion that precedes invasive ductal carcinoma (IDC). In DICS, precancerous stem cells are confined within the duct, which can be subjected to hypoxia and other stimuli from the microenvironment leading to primary cancer stem cells. Differences in gene expression in DCIS compared to IDC can be used to distinguish the genotypes exhibited for precancerous versus cancer stem cells [18].

Regarding the origin of cancer stem cells, it has been demonstrated that Nkx3-1 positive luminal epithelial cells (considered as prostate stem cells) are mutated in the tumor suppressor gene PTEN, leading to prostate carcinoma [19]. Additionally, Lgr5 or promonin1 positive subpopulations have been considered as intestinal stem cells, which can be mutated for the activation of endogenous Wnt signaling, leading to transformations of intestinal stem cells to cancer stem cells that support malignant intestinal disease [20, 21]. Yang et al., reported that $CD90^+$, but not $CD90^-$, liver cancer cells are able to form tumors. Also, $CD90^+CD44^+$ subpopulations have an important capacity for tumorigenesis and metastases compared to $CD90^+CD44^-$ cells. Moreover, there is a higher proportion of $CD90^+CD44^+$ cells in metastases than in primary cancer tissue samples [6].

Metastasis is a process that involves progressive tumor growth, vascularization, invasion, detachment, embolization, survival in the circulation, arrest, extravasations, evasion of immune defenses, and progressive expansive growth [4]. The cancer stem cell model is closely related with metastases. Mesenchymal stem cells can also experience malignant transformation, losing their differentiation potential, increasing telomerase activity, blocking senescence, and acquiring the ability for anchorage-independent growth. Additionally, these transformed mesenquimal stem cells are able to induce tumorigenesis in mice [22]. The migration capacity of cancer stem cells is still limited, due to a lack of specific migration markers. However, it is known that the epithelial-mesenchymal transition plays an important role in the migration of primary CSCs. Gene expression analysis has shown that CD44^{high}CD24^{low} subpopulations also express high levels of mesenchymal markers such as N-cadherin, Vimentin, Fibronectin, Zeb2, Foxc2, Snail, Slug, Twist1, and Twist2 and have low levels of E-cadherin. Breast cancer stem cells that express Snail and Twist are more efficient at inducing tumor growth in immune-deficient mice in which the CD44^{high} and CD24^{low} subpopulations are elevated [3].

However, not all cancer stem cells are able to migrate. CD133^{+}CXCR4^{+} subsets are considered to be cancer stem cells with the capacity to migrate and induce pancreatic tumor growth, in contrast CD133^{+}CXCR4^{-} are also cancer stem cells but without the migration capacity [7].

An additional feature of CSCs is their resistance to traditional chemotherapy and radiotherapy through several mechanisms such as blocking apoptosis pathways and the expression of ABC pumps (ATP-binding cassette superfamily) that are related to drug efflux. There is a subpopulation of human colon cancer stem cells that is resistant to the popular chemotherapeutic agents oxaliplatin and 5-fluorouracil (5-FU) because of the autocrine expression of intherleukin-4 (IL-4) which induces anti-apoptotic proteins cFLIP, Bcl-xL, and PED. Interestingly, the combination of an antagonist of IL-4 with oxaliplatin or 5-FU has cytotoxic activity against these cancer stem cells *in vitro* and *in vivo* ([8].

Conversely, in the ABC family, the ABCG2 has been considered as a marker of cancer stem cells [13]. Nevertheless, using an ABCG2 antagonist it is possible to observe partial inhibition of Side population (SP) cells in the sample. This partial effect could be explained by the expression of other ABC proteins on these cells [10].

Regarding the radioresistance property of cancer stem cells, it is known that these particular stem cells contain high levels of antioxidant defense systems, thus decreasing DNA damage after ionizing radiation, compared to non-tumor cells. The antixodant state depends on the reactive oxygen species (ROS) levels in the cancer stem cells, promoting resistance to radiotherapy [11]. Furthermore, in CD133^{+} glioma stem cells, the expression of autophagy related proteins LC3, ATG5, and ATG12 was increased in response to γ-radiation. Nevertheless, the mechanisms of radioresistance in cancer stem cells remain unclear [14].

In other human stem cell diseases, benign tumor stem cells have been isolated from stem-like cells from pituitary adenoma, which share some stem cell features such as self-renewal, multi-lineage differentiation and neurosphere formation capacity. Additionally, pituitary adenoma stem cells express anti-apoptotic proteins as well as pituitary progenitor markers in addition to stronger resistance to chemotherapy compared to differentiated pituitary adenoma cells. Finally, these stem cells have the capability to induce benign tumor growth by xenotransplantation [23], demonstrating for the first time, the existence of benign tumor stem cells.

There is a large group of solid tumors that follow this CSC model. However, recent research from Dr Sean J Morrison's Group has demonstrated that not all tumors follow this hierarchy model. There are an important number of studies that have demonstrated that cells from non-tumorigenic/non-leukemogeneic cancer cell populations, occasionally have the ability to induce tumor growth under permissive conditions for tumorigenesis through a small number of CSCs [24–27]. Non-tumorigenic cells, under CSC models, have irreversibly lost tumorigenic capacity, or only retain this capacity in rare circumstances. Currently, the CSC and clonal evolution models emphasize the role of irreversible epigenetic and genetic changes which are related to heterogeneity among cancer cells [1].

Recent studies on cancer cell lines suggest that some phenotypic and functional features of tumorigenic cells can be reversibly turned on and off [28–31]. These observations have motivated some researchers to elucidate if these reversible changes can be observed in primary cancers from patients and whether many or few cells in these samples undergo such modifications. This is where Dr. Morrison and his lab said: "If most cells in a cancer can reversibly gain and lose competence to form a tumor then this is a transient state rather than hierarchically determined attribute possessed only by rare cancer stem cells [1]."

Studies have shown that tumorigenic melanoma cells can be identified by the expression of ABCB5 [32] or CD271 [33] of which 1:5,000 to 1:1,090,000 cells are able to induce tumors in xenotransplant models. Variations in the tumorigenic capacity of melanoma cells are highly assay-dependent [1]. Thus, Quinatana et al., have made changes in assay conditions to increase the frequency of tumorigenic melanoma cells to 1 in 4 cells obtained from samples of patients with primary cutaneous or metastasic melanomas that are able to form tumors into NOD/SCID IL2Rγnull (NSG) mice by xenotransplantation [1]. These amazing results demonstrate that many melanoma cells from primary mouse tumors [34] or from human cancer cell lines [30] are capable of forming tumors independently of primary cancer tissues and cancer cells [1].

Results from Morrison's Lab demonstrated that $CD133^{+}$ and $CD133^{-}$ melanoma cells have the capability to form tumors which exhibit similar heterogeneity in CD133 expression, suggesting that CD133 is reversibly expressed by tumorigenic melanoma cells rather than being a distinguishing characteristic for cells at different levels of a hierarchy [35].

In recent studies, Quintana and et al., were unable to identify specific markers related to the tumorigenic capacity of melanoma cells, since similar growth rates were observed for all cells. Therefore, this tumorigenic ability is not restricted to a small population of melanoma cells, but interestingly this ability is widely shared among phenotypically diverse cells. Additionally, these distinct melanoma cells form tumors that recapitulate the phenotypic diversity of the tumor from which they were derived, suggesting that these tumorigenic melanoma cells undergo reversible changes in marker expression, *in vivo*. These results demonstrate that melanoma cells do not follow the hierarchy model of non clonal evolution, in which the genetic changes are irreversible. These studies predict that, in some cancers such as melanoma, many cells will be able to form phenotypically diverse tumors without hierarchical organization.

All studies that attempt to identify tumorigenic cells from human cancer cells must be studied in highly immunocompromised mice and the results extrapolated to contribute to our knowledge about the cancer cells and related with the disease in patients.

The contribution of these studies is related to the heterogeneity in melanomas obtained from patients in which reversible changes can be observed and a broad range of markers turn on and off in lineages of tumorigenic cells. The term "phenotypic plasticity" was introduced to explain this behavior, which is contrary to the CSC and clonal evolution models where heterogeneity is attributed to irreversible epigenetic and genetic changes. Conversely, the phenotype plasticity model (Fig. 6.1) can be an independent source of heterogeneity in several cancers [1].

Taken together, this emphasizes the importance of knowing the specific physiology of the many cancer stem cells to develop new and effective strategies for eliminating this population in order to increase the survival of patients with cancer.

6.2 Common CSC of Solid Tumors

In several biomedical areas, cancer stem cells have become an important and relevant field of research, in order to determine the characteristics of the tumorigenic cells responsible for human cancer diseases. The following paragraphs show some examples of human tumors that follow the cancer stem cell model. Although leukemia is the first cancer disease to demonstrate the presence of cancer stem cells and/or tumor initiation cells in hematopoietic malignances, we will focus on solid tumors, among which breast cancer stem cells were the first CSCs from solid tumors to be studied (Table 6.1).

6.2.1 Breast Cancer Stem Cells

Flow cytometry assays of leukemia cells from primary cell cultures demonstrated that the small population of $CD34^{+}CD38^{-}$ leukemia cells has the capability to induce and maintain tumor growth. This work made the cancer stem cell model real, for the first time. Al-Hajj et al., also demonstrated by xenotransplant in mice, that the

Table 6.1 Specific markers for cancer stem cells

Marker	Type of cancer	Reference
$CD24^-CD44^+Lin^-$	Breast cancer	[3]
$ALDH1^+$	Breast cancer	[38, 39]
$a2\beta1^{hi}CD133_+$	Prostate cancer	[40]
$Lin^-Sca\text{-}1^+CD49f^{high}$	Prostate cancer	[10]
$CD44^+$	HNSCC	[42]
$ALDH1^+$	HNSCC	[43]
$CD90^+$	Liver cancer	[44]
$CD133^+$	Liver cancer	[45]
$EpCAM^+$	Liver cancer	[46]
$CD105^+$	Renal carcinomas	[38]
$Lgr5^+$	Intestinal cancer	[47]
$CD133^+$	Intestinal cancer	[48]
$CD44^+CD24^+ESA^+$	Pancreatic cancer	[49]
$CD133^+$	Pancreatic cancer	[50]
$ALDH1^+$	Bladder cancer	[51]
$CD133^+$	Colon cancer	[52, 53]
$Lgr5^+$	Colon cancer	[47]
$ALDH1^+$	Colon cancer	[54]
$CD133^+$	Endometrial tumors	[55]
$CD44^+CD117^+$	Ovarian cancer	[38]
$CD133^+$	Ovarian cancer	[39]
$SP\text{-}C^+CCA+$	Lung cancer	[40]
$CD133^+$	Lung cancer	[40]
$ALDH1^+$	Lung cancer	[41]
$Oct\text{-}4^+$	Osteosarcoma	[42]
$SSEA\text{-}1^+$	Glioblastoma	[43]
$A2B5^+$	Glioblastoma	[44]
$CD133^+$	Brain tumors	[45]
$CD133^+$	Hepatocellular cancer	[38, 46]

$CD44^+CD24^-$ are the surface markers that identify human tumorigenic breast cancer stem cells [56]. After these pioneer studies, many researchers have determined the presence of cancer stem cells by analyzing surface proteins. The cancer stem cell markers known at this time are included in Table 6.1.

Using mammoshpere formation as a tool, breast cancer stem cells with $CD49f^+$, and $CD44^{high}$ $CD24^{low/negative}$ markers, as well as increased ALDH1 activity have been identified [26]. ALDH1 is a detoxifying enzyme that plays an important role in stem cell differentiation [27]. As mentioned previously, CD44, a membrane receptor, is involved in cell adhesion, motility, and metastasis. CD44 has been used to enrich and isolate breast cancer stem cells [27]. The $CD44^{high}CD24^{low}$ phenotype in human breast tumors has been related to basal-like tumors and has also been associated with BRCA1 gene heredity, in addition to the expression of epidermal growth factor receptor (EGFR) combined with low expression of estrogen receptor (ER), progesterone receptor (PgR), and HER-2 [1]. The presence of basal-like tumors is commonly related to a poorer prognosis. Interestingly, the CD44^{high}CD24^{low} phenotype has been found in lower proportions in luminal type HER-2 positive tumors, independent of ER status.

Signaling pathways such as Notch, Hedgehog (Hg), and Wnt/β-catenin play a role in organogenesis and embryogenesis and are also related with the self-renewal and differentiation capacities of stem cells. Therefore, these pathways are also important for the self-renewal of breast cancer stem cells. It has been demonstrated that Notch-4 protein is over-expressed in breast cancer, promoting tumor growth *in vivo*. In contrast, it can inhibit proliferation of normal mammary epithelial cells [57]. Using mammospheres as tools for the enrichment of CSCs, an increase in Notch protein expression was observed, suggesting a relevant role for the Notch pathway in the generation and maintenance of breast cancer stem cells [58]. In the same way, the β-catenin/Wnt pathway is also deregulated in hematopoietic malignant diseases as well in breast cancer cells [2, 59].

6.2.2 Liver Cancer Stem Cells

Hepatocellular carcinoma has been considered an aggressive malignant disease. The Wnt pathway is closely related with the self-renewal feature of stem cells and is also important for liver development [60], suggesting a role for this pathway in liver cancer stem cells generation. The bipotential liver progenitor cells present in this tissue could be candidate target cells to explain the origin of these cancer stem cells and their ability to raise hepatocarcinomas and cholangiocarcinomas [61]. Immunotype profiling shows that 28–50% of cells that compose the hepatocarcinoma tissue express progenitor markers such as cytokeratin 7 (CK7) and CK19, which have been considered markers for poor prognosis and high recurrence after surgical treatment [62]. Furthermore, the presence of components of the TGF-β signaling pathway, TBRII and ELF, in human hepatic stem cells has been related to hepatocarcinogenesis. Heterozygous elf$^{+/-}$ mice spontaneously develop liver tumors, in which the lesions include early centrilobular steatosis, proceeding to poorly differentiated carcinoma. A study of 10 patients with hepatocarcinoma showed the presence of small clusters of 3–4 cells strongly positive to OCT4 expression, which are negative for TBRII and ELF. Interestingly, this phenotype is not observed in normal liver from biopsies of regenerating organs. These observations suggest that STAT3^{+}/OCT4^{+} cells negative for proteins belonging to the TGF-β pathway are able to generate hepatocarcinoma lesions and could represent the liver cancer progenitor-stem cells [63]. Recent studies show that liver cancer stem cells isolated from alpha-fetoprotein (AFP)-positive hepatocarcinoma cells have a EpCAM (CD326)$^{+}$ phenotype. These EpCAM^{+}AFP^{+} cancer stem cells showed overexpression of the Wnt signaling pathway [64] related with high tumorigenicity and metastases.

6.2.3 Ovarian Cancer Stem Cells

Normal stem ovarian cells are considered as multi-potent adult stem cells because of the ability they have to regenerate epithelium after ovulatory rupture. These stem cells undergo asymmetric division to give rise to a daughter cell programmed for terminal differentiation and an undifferentiated self-copy. Constant asymmetric self-renewal can lead to the acquisition of mutations over time, promoting the generation of ovarian cancer stem cells and thus malignant progression. The ROS generated by the ovulatory process [65] can also contribute to stem cell malignant transformation through the induction of DNA strand breaks.

Due to the multipotential capacity of ovarian stem cells, it is possible to find sex-cord and germinal cell tumors (GCTs). These tumors can be found along the middle of the body, along the migration route of primordial germ cells (PGCs) during embryogenesis [66]. Histopathologic studies distinguish them as (1) Teratomas and yolk sac tumors, (2) seminomas/dysgerminomas/germinomas of the testis and ovary, mediastinum, and midline of the brain, (3) ovarian dermoid cysts, and (4) spermatocytic seminomas of elderly testis [7].

Analysis of SP cells has been used to isolate a subpopulation enriched in cancer stem cells such as ovarian cancer stem cells with the ability to induce tumor growth. These specific CSCs provide a model of cancer metastases since they have the ability to colonize, expand and differentiate into heterogeneous tumor phenotypes like primary tumors. In this tumor model, there is a similar gene expression signature between primary and metastatic tumors because both of them are related to the same ovarian cancer stem cell. In ovarian tumors, histological heterogeneity can be found in tumors [12, 15, 24]. Antibodies against CD44 and CD117 proteins can be used to identify a subpopulation with the capacity for anchorage-independent growth and to induce tumor growth in mice. These cell populations are also chemo-resistant [16]. Zhang et al, showed that the ovarian cancer initiating cells overexpress ABCG2 proteins and thus show high resistance to chemotherapeutic drugs such as cisplatin and paclitaxel, suggesting a possible role for these cells in ovarian cancer resistance to chemotherapy [25].

There is a close relationship between testicular GCT and embryonic stem cells because of the specific expression of pluripotent genes such as OCT3/4 [67], a necessary transcription factor for the self-renewal of ES [68]. Additionally, there is expression of other genes such as c-Kit, Nanog and AP-2γ. It has been suggested that the overexpression of c-Kit in dysgerminomas is due to spontaneous gene mutations occurring before oocytes enter meiosis to increase the proliferation of undifferentiated germ cells [69]. Immunohistochemical studies of stem cell-related markers could be a diagnostic tool to evaluate premalignant germ-cell lesions and can be used for targeted therapy.

6.2.4 Head and Neck Cancer Stem Cells

Head and neck squamous cell carcinoma (HNSCC) is the sixth cancer-related mortality in the world. The combination of chemotherapy and radiotherapy has resulted in modest benefits for patients with this disease [70]. CD133 is a putative marker for CSCs in tumors and in HNSCC cell lines. HNSCC cell line $CD133^+$ stem cells have shown increased clonogenicity, tumor sphere formation and more importantly, $CD133^+$ HNSCC stem cells are better able to promote tumor growth compared with $CD133^-$ cells [71]. There is some evidence showing that $CD44^{high}$ cells isolated from HNCSS tumors have a high capacity to form tumors in mice compared to $CD44^{low}$ cells [72]. These tumors recapitulate the original tumor's cellular heterogeneity and can be serially passaged. Another important tool is the evaluation of $ALDH^{high}$ activity which enriches cells with characteristics of cancer stem cells. Clay et al, demonstrated that 500 $ALDH^{high}$ cancer cells can induce tumor growth in mice, which is tenfold less cells than $CD44^+$ cells isolated from HNSCC tumors, suggesting that ALDH activity, $CD44^{high}$ can define HNSCC stem cells [73]. The spheres generated from HNSCC samples enrich CSCs identified by $CD44^{high}$ [72], OCT3/4, Nanog, Nestin, and $CD133^{high}$ markers which exhibit an increase in tumorigenicity in xenografted mice [74, 75].

6.2.5 Oral Squamous Cell Carcinoma Stem Cells

Patients with oral squamous cell carcinoma (OSCC) or basal cell cancer (BCC) of the head and neck represent the majority of oncological clinical cases. It is known that these cancers are linked to the cancer stem cell model; however, currently there is not enough data to get successful isolation of specific CSCs from BCC and OSCC. OSCC is derived from malignant keratinocytes in which the malignant stem cells can be found in the basal stratum of the oral and dermal epithelium. Immunohistochemical studies have shown that these cancer stem cells are located adjacent to the dermal papillas and the basal membrane.

Risk factors for OSCC include smoking and alcoholic beverages, as well as genetic and epidemiological markers. OSCC development follows a series of more or less defined intermediate stages, through the accumulation of molecular changes that promote the progression of pre-cancerous leukoplakia to pre-invasive (erythroplakia) and finally invasive cancer with outspread and severe local destruction [76, 77]. There are several ongoing trials for therapies against this kind of cancer, without showing significant progress, from a combination of surgeries and radiothearpy with chemothearpy. In addition, the protocols or common therapeutic strategies have not been standardized. The identification of relevant cellular or molecular bio-markers involved in oral pre-cancerous lesions and OSCC still unclear.

Prolonged sunlight and/or UV-light exposure can promote BCC, the most common form of skin cancer, specifically from basal epithelial cells. BCC carcinogenesis is composed by several steps and its development is slower than melanoma. BCC has been considered to be an epidermal stem cell-derived carcinoma of the skin. An increase in the activation of Sonic hedgehog (SHH) and Wnt pathways, as well as Notch1 and Notch2 expression, were observed in BBC [78]. However, this cancer has a slight metastatic potential, therefore is considered a "semi-malignant" tumor [7].

6.2.6 Pancreatic Cancer Stem Cells

Pancreatic cancer is the fourth leading cause of cancer death in the United States. The American Cancer Society estimated that 42,470 new cases could be diagnosed with pancreatic cancer [79]. This malignancy is a rapidly invasive, metastatic tumor that is resistant to standard therapies [80, 81]. Pancreatic ductal adenocarcinoma (PDAC) is an aggressive cancer in which putative pancreatic cancer cells have been identified by the expression of $CD44^{+}CD24^{high}ESA^{+}$. Subcutaneous injection of 100 cells with this phenotype is able to induce tumor growth in immune-compromised mice. However, via intraperitoneal injection, tumor growth needs 500 $CD44^{+}CD24^{+}ESA^{+}$ cells, and 5,000 cells in an orthotopic model, compared to $CD44^{-}CD24^{-}ESA^{-}$ cells. For that reason, it is not clear if pancreatic cancer cells follow the hierarchical model [82]. However, these observations can be explained by the clonal evolution model. It has been considered that clonal evolution and CSC models are not exclusive. Taken together, the origin of this type cancer still unknown, perhaps pancreatic cancer is closed to the phenotype plasticity model, proposed by Dr. Morrison, that we were talking about before.

In addition, some reports have tried to isolate pancreatic cancer stem cells by sphere formation assays, suggesting that a rare population $CD133^{+}$ $CXCR4^{+}$ can be considered as cancer stem cells with highly tumorigenic and metastatic capacities [50].

6.2.7 Colon Cancer Stem Cells

The second leading cause of death by cancer in the United States is colorectal cancer [83], which affects around 146, 970 individuals per year [84, 85]. Polyps are early lesions that can develop into colorectal cancer if these polyps persist for a long time [85]. The colon cancer stem cells are characterized as $ALDH^{+}$ $CD133^{+}$ cells which express high levels of STAT3, a protein that can be used for target therapy [87]. CD44 is a glycoprotein that has been used as a CSC marker in several tumors including colon carcinoma [88]. The colon cancer cells sorted by $CD44^{+}CD133^{+}$ cells have high tumorigenic capacity compared to CD44 negative colon cancer cells [88, 89]. In addition to CD44, the mesenchymal stem cell marker CD166 (ALCAM) is also expressed in colon cancer stem cells [88]. Finally, CD24 and CD29 have also been identified as colon cancer stem cells in human and mouse colorectal carcinomas [90]. These studies have provided a good battery of markers to isolate colon cancer stem cells [91].

6.2.8 Lung Cancer Stem Cells

The side population (SP) cell assay is another tool to identify and isolate possible populations enriched in cancer stem cells and early progenitor cells. This tool has been used to isolate several CSCs, including lung cancer stem cells. These specific cells exhibit properties of cancer stem cells, such as clonogenic proliferation, sphere tumor growth, and 3D matrigel invasion [92]. Furthermore, $CD133^{+}$ cell populations from human lung cancer specimens of human non-small cell lung cancer (NSCLC) and small cell lung carcinoma (SCLC) showed CSC properties. These lung cancer stem cells isolated using a SP assay are able to induce tumor growth in immune-deficient mice [93]. The $CD133^{+}$ lung cancer stem cells are chemo-resistant and showed OCT3/4 expression, supporting self-renewal of these CSCs [22]. This CD133 marker has been tested in 60 human lung cancer samples demonstrating the accurate use of CD133 as positive marker for these specific cancer cells [94]. The K-Ras oncogenic pathway plays an important role in the physiology of lung cancer stem cells related with self-renewal and proliferation. Several groups have observed that the PI3K/PTEN pathway is also needed for the self-renewal of human NSCLC ([22, 95]).

6.3 Clinical Development in Anti-cancer Stem Cells Therapies

One of the most promising features of cancer stem cells (CSC), since the early beginning of their discovery, is to target CSC with therapeutic approaches that might represent excellent strategies to improve clinical cancer therapy. It is believed that this population is the major component responsible for sustained tumor growth and metastasis. Therefore, novel therapeutic strategies must be developed to identify and kill this cancer stem cell population (Fig. 6.2). Once the stem cells can be identified, we should be able to direct therapeutic targets against proteins expressed by these cancer stem cells [2].

Normal stem cells tend to be more resistant to chemotherapeutics than mature cell types from the same tissues [2]. Cancer stem cells have inherited or acquired some of their related mechanisms such as quiescence, expression of ABC pumps that promote drug efflux [96, 97], high expression of anti-apoptotic proteins, and resistance to DNA damage [98]. These are actually the key points of study in CSC related therapy [10, 17].

Current therapies have been developed against the bulk tumor population, being able to shrink tumors due to the cancer cell's limited proliferative potential. Despite the fact that available drugs used for therapy can shrink tumors, these effects are usually transient. One reason for the failure of these treatments is the acquisition of drug resistance by the cancer cells as they evolve. Another possibility could be that existing therapies fail to kill cancer stem cells effectively. Based on the observation of stem cell mechanisms [75, 97–100] one might predict that these CSC would be more resistant to chemotherapeutic agents than other tumor cells with limited proliferative potential. Thus, therapies that are more specifically directed against cancer stem cells would prove more effective to treat and eradicate metastatic cancer and avoid relapse [13].

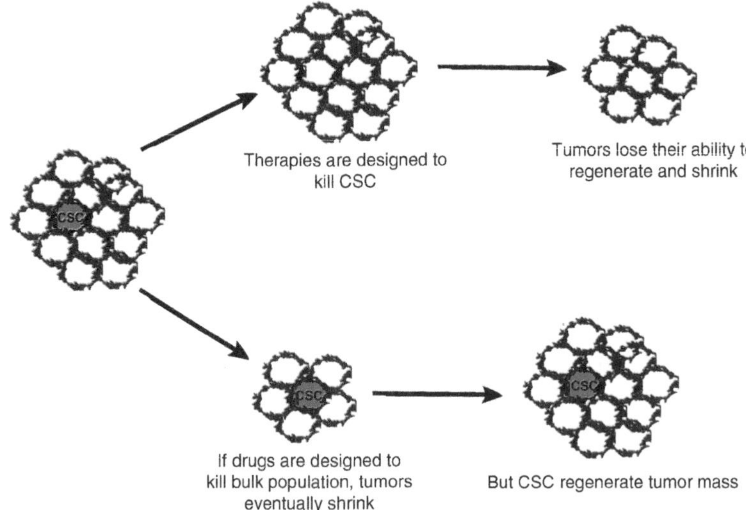

Fig. 6.2 Therapies designed to kill bulk population of tumors might shrink the tumor leading to apparent response (*lower arrow*), but presence of CSC in tumors will repopulate tumors. If therapies are designed to eradicate CSC (*upper arrow*) tumors lose their ability to regenerate and eventually shrink

The CSC model proposes that signaling pathways associated with stem cells could be targeted to enhance therapy. The CSC hypothesis suggests that the most effective therapeutics would target survival or self-renewal pathways specific to CSCs compared to their normal tissue counterparts. New drugs should be aimed at killing or inactivating CSCs and not normal stem cells. Obviously, care should be taken when making molecular comparisons between CSCs and normal tissue stem cells; however, promising data exist suggesting that CSC-specific pathways known to be important in oncogenesis and oncologic therapy are also implicated in stem cell self-renewal [101–103].

The main goal of this section is to review some examples and implications of therapy aimed at CSCs. First, we will look at the different pathways that are more commonly targeted for CSC therapy. Then we will review the classical chemo- and radiotherapies as well as different immunotherapeutic approaches. The focus will be on the advances and achievements in different solid and hematopoietic cancers. We will also give an insight of other useful strategies for CSC therapy, such as virotherapy, epigenetic therapy, and mesenchymal stem cell therapy, as well as other future emerging technologies for diagnosis and treatment in CSC directed therapy.

6.3.1 Pathways Targeted as Treatment Opportunities (Notch, Sonic/Hedgehog, Wnt/β-Catenin, HMGA2, Bcl-2, and Bmi-1)

The process of self-renewal in stem cells is tightly controlled and governed by both signals from the stem cell niche as well as regulated control of key developmental pathways such as the Notch, Sonic/Hedgehog, Wnt/β-catenin, HMGA2, Bcl-2, and Bmi-1 signaling pathways, among others. CSC self-renewal is subjected to same pathways involved in normal stem cell renewal; however, the deregulation of these key pathways also plays a role in tumorigenesis [7, 59, 79, 82, 101–107]. Elucidation of the pathways that regulate self-renewal in breast stem cells has led to a clearer picture of how deregulation of these pathways may lead to carcinogenesis in this model. Furthermore, these pathways may provide targets for breast cancer prevention and therapy.

There is growing evidence suggesting that key oncogenic pathways known to be deregulated in breast cancer also regulate stem-cell behavior and there is hope that such CSC-specific pathways can be successfully targeted for therapy. Among these targets are included Wnt5, Notch, Hg, BMI1, PTEN and BMP as well as the pro-survival transcription factor NF-κB [59, 101, 105, 108].

Future efforts to develop drugs that target these pathways in combination with analysis of their effects on CSC populations in human tumors, holds significant promise for developing CSC-directed therapies.

6.3.2 The HMGA2 Pathway

The HMGA2 gene is suggested to control growth, proliferation, and differentiation. Apparently, it plays a role in modulating macromolecular complexes that are involved in many biological processes, including binding directly to the DNA and aiding in the regulation of many genes that have also been implicated in survival and self-renewal of cancer stem cells. Its expression during embryogenesis suggests that it is important for development. In cancer, it correlates with the presence of metastases and reduced survival, and has been found overexpressed in pancreatic and lung carcinomas [109].

6.3.3 The Bmi-1 Pathway

The polycomb gene Bmi-1 plays an active role in signaling in carcinogenesis and has an effect on cancer stem cell self-renewal. It was identified as a player in promoting the generation of lymphomas. This demonstrates that Bmi-1 plays a role in cancer development. Additionally, it has been found to be activated in human breast stem cells ($CD44^{+}CD24^{-/low}Lin^{-}$). It was also found to mediate the mammosphere-initiating cell number and mammosphere size, supporting a role in the regulation of self-renewal of normal and tumorigenic human mammary stem cells. Therefore, deregulation of the Bmi-1 pathway within cancer stem cells may be associated with the acquisition of self-renewal properties. It regulates, along with Hedhog, the self-renewal of normal and malignant human mammary stem cells [105, 108, 110].

6.3.4 The Bcl-2 Pathway

Bcl-2 is an anti-apoptotic protein, which has been found over-expressed in many cancers, leading to prevention of apoptosis. This Bcl-2 effect causes increased number of stem cells *in vivo*. The key role in cancer stem cells is to prolong CSC lifespan by avoiding death and contributing to resistance [111].

6.3.5 The Wnt/β-Catenin Pathway

The Wnt/β-catenin pathway plays an important role in the regulation of hematopoietic stem cell self-renewal. However, the role of Wnt signaling in lung epithelial stem cells is less well understood. The prospect of Wnt signaling as a driver of lung tumorigenesis and stem cell self-renewal make the Wnt signaling pathway an appealing target for therapy; however, further studies will be necessary to define the context of activated Wnt signaling in lung cancer stem cells, as well as in different types of lung cancer.

6.3.6 The Hh Signaling Pathway

The Hh signaling pathway is activated when one of the three extracellular Hh ligands (three in mammals), sonic hedgehog, desert hedgehog, and Indian hedgehog, binds to and inactivates its receptor, patched (PTCH). The Hh signaling pathway is a key developmental pathway required for proper embryogenesis. In the developing lungs, activated Hh signaling is involved in pulmonary cell fate determination and branching morphogenesis. Aberrations in expression and activation of this pathway lead to deformations in development as well as contributing to tumorigenesis [79]. The Hh pathway has been demonstrated to play a role in tumorigenesis and is activated in primary glioblastoma and glioma cell lines. Treatment of neurosphere cultures with cyclopamine (an inhibitor of the Hh signaling pathway), inhibited sphere formation and enhanced radiation treatment and temozolomide chemotherapy, suggesting that inhibiting Hh will increase the efficiency of current glioblastoma therapies by targeting the CSC population [102].

6.3.7 The Notch Signaling Pathway

The Notch signaling pathway is involved in cell fate determination, organogenesis, and tissue homeostasis. Notch-mediated cell-cell interactions dictate the preservation or differentiation of stem cells by binding Notch ligands to membrane receptors on adjacent cells. In the developing lung, Notch signaling appears to be required for determining proximal and distal lung epithelial cell fates. In lung cancer, elevated Notch signaling transcripts have been described in non-squamous cell lung carcinoma, however the role of Notch in tumor maintenance remains poorly understood [79]. Notch receptors, their ligands, and their downstream targets are commonly overexpressed in glioma cell lines and primary glioblastoma samples. Additionally, Notch is involved in stem cell functions and is deregulated in glioblastoma. This makes it a promising target for directed therapy, especially since Notch can be inhibited at multiple steps of Notch signaling. The most common method of Notch inhibition in basic research, as well as in current Phase I and Phase II clinical trials, is via small molecule inhibitors of γ-secretase. When γ-secretase inhibitors (GSIs) are utilized, the Notch receptor is not cleaved and remains bound to the cellular membrane. This halts the Notch signaling pathway, because the intracellular domain fails to translocate into the nucleus. It has been suggested that an active Notch pathway is required to maintain the CSC population of glioblastomas and may be a promising target for therapy [102, 106, 107]. Furthermore, it is frequently deregulated in invasive breast cancer. It was demonstrated that Notch signaling can act on mammary stem cells to promote self-renewal and on early progenitor cells to promote their proliferation. It is suggested that atypical Notch signaling could lead to deregulation of the self-renewal properties of cancer stem cells. [59, 103, 108]

6.3.8 Insights into Diverse Therapeutic Approaches

Solid tumors are mainly treated by surgery or chemo-radiation treatment and solid tumors that have metastasized are, by definition, incurable. If tumor growth and re-growth after therapy is a property of CSCs, the response of these cells to radiation or chemotherapy is a critical factor to achieve curability. While unusual cancers, like testicular cancers, can be cured at advanced stages with conventional therapy, therapeutic resistance is common among most advanced cancers. Many mechanisms may contribute to the development of therapeutic resistance, including the stochastic selection of resistant genetic subclones, microenvironmental factors (hypoxia, lactic acidosis, etc.), and extrinsic cellular factors. Although the chemo- and radio-resistance of CSCs presents a therapeutic challenge, their similarity to normal stem cells may, at the same time, provide a therapeutic target. It is intriguing to ask whether it might be possible to develop radio/chemo-sensitizers that will preferentially sensitize CSCs and not normal stem cells (Table 6.2).

One example of chemo-resistance of CSC is ABCG2, a member of the ATP-binding cassette superfamily and one type of multidrug resistant proteins, which can pump chemo-therapy drugs out of the cell. Despite targeted therapy with ABCG2 antagonists, one can only partially inhibit the growth of SP cells and cancer stem cells. This may be because cancer stem cells express other drug resistant proteins such as ABCB1 [116, 117].

Radio-resistance in cancer stem cells is associated with the levels of reactive oxygen species (ROS). In human and mouse breast cancer stem cells, lower levels of reactive oxygen species (ROS) have been found compared to their non-tumorigenic progeny. Moreover, human cancer stem cells contain higher levels of antioxidant defense systems and develop less DNA damage after ionizing radiation, compared with non-tumor cells. Therefore, the heterogeneity of ROS levels in cancer stem cell subsets might contribute to their radio-resistance [117, 118]. Oxygen has long been known to be one of the most potent radio-sensitizing agents. Tumors contain areas of low oxygen tension and cells residing in these areas are considered to be relatively protected from radiation. Consequently, considerable effort has been made to overcome tumor hypoxia in order to improve radiation treatment results [119].

6.3.9 Therapy for Hematopoietic Malignancies

Leukemic Stem Cells (LCS) exhibit significantly higher daunorubicin and mitoxantrone efflux than bulk tumor components. It has been shown that both malignant and physiologic $CD34^{+}CD38^{-}$ hematopoietic cells exhibit reduced *in vitro* sensitivity to daunorubicin. This preferential chemoresistance has been associated with increased transcript levels of multi-drug resistance-associated proteins and reduced Fas-induced apoptosis. In chronic myeloid leukemia (CML), resistance to the ABL tyrosine kinase inhibitor imatinib is responsible for the failure of the chemotherapy. AML monoclonal antibody-based strategies, targeting aberrant expression of CD123, the α chain of the IL-3 receptor, preferentially inhibit leukemic stem cells [76]. Cytosine arabinoside (Ara-C) treatment of human AML-bearing mice revealed that LSC are recruited to the osteoblast niche within the bone marrow, by a process termed homing, and remain protected from Ara-C-induced apoptosis [13]. Chronic myeloid leukemia (CML) is sustained by leukemic stem cells that are relatively resistant to the drug imatinib. Additionally, leukemia-initiating cells in T-lineage acute lymphoblastic leukemia (T-ALL) persist following dexamethasone treatment in high-risk T-ALL. The survival mechanisms (such as dormancy) of these stem-like cells have not been directly explored. Ishikawa and his colleagues, provided evidence for the presence of quiescent AML stem cells within the bone marrow of

Table 6.2 Therapeutic approaches in CSC. Most of them are targeted to same pathways as normal stem cell and lack specificity. Combined immunotherapy and chemotherapy is often used

Cancer	Agent	Target	Effect	Reference
AML	Monoclonal antibodies	Aberrant expression of CD123	Preferentially inhibit leukemic stem cells	[76]
	Asparthenolide	Inhibit NF-κB signaling	Inhibit CSC rather than normal stem cells	[76]
CML	Nilotinib and dasatinib	Inhibit BCR–ABL harboring additional point mutations	Produce responses in patients with imatinib-resistant CML	[76]
Multiple mieloma	Glucocorticoids, alkylators, lenalidomide bortezomib	May be mediated by similar processes that protect normal stem cells	Debulking of the tumor	[82]
T-ALL	Granulocyte colony–stimulating factor	Broke the quiescent state of the cell	Induce cell cycle entry and increase the sensitivity for chemotherapy	[112]
Glioblastoma	BMPs Suberoylanilide hydroxamic acid	Self-renewal pathways of normal stem cells	Reduced cell proliferation in vitro and induced differentiation into astrocytes	[12]
Brain malignancies	Debromohymenialdisine	$CD133^+$ and $CD133^-$ cells Chk1 and Chk2 kinase inhibitor	Are rendered less resistant	[113] [113]
	Focused beam radiation delivered with the gamma knife	Whole brain	Minimize radiation to the normal brain	[114]
Medulloblastoma	Cyclopamine	Hedgehog signaling	Inhibited growth	[113]

(continued)

Table 6.2 (continued)

Cancer	Agent	Target	Effect	Reference
Breast cancer	Metformin	Four genetically different types of breast cancer	Inhibit cellular transformation and selectively kill cancer stem cells	[84]
	Lapatinib	EGFR and HER-2/neu (ErbB-2) dual-tyrosine kinase inhibitor	Decreases in the percentage of CD44+/CD24low	[115]
	Pheophorbide and chemotherapy	Specific probe for ABCG2	Increase the efficiency of chemotherapeutic drugs to kill cancer stem cells	[59]
	Gamma secretase inhibitors	Notch signaling	Improves the therapeutic outcome in advanced breast cancer	[13]
Hepatocellular carcinoma	Blocking IL-6	IL-6 signaling pathway	Could provide powerful new therapeutic approaches to HCC	[64]
Ovarian carcinoma	Retinoic acid N-t-boc-Daidzein	Differentiation Ovarian cancer stem cells	Daidzein inhibits cell growth and decrease CSC viability	[64]

mice transplanted with human AML cells. The quiescent state of these cells can be broken: granulocyte colony–stimulating factor treatment induces cell cycle entry and increases sensitivity of the AML stem cells to chemotherapy [112].

In AML, small molecules that inhibit NF-κB signaling, such asparthenolide and its derivatives, seem to primarily inhibit CSCs rather than normal stem cells. The second generation BCR–ABL inhibitors, nilotinib and dasatinib, have been approved for use in CML. Although these agents can produce responses in patients with imatinib-resistant CML, through the ability to inhibit BCR–ABL harboring additional point mutations, in vitro studies suggest that they are not able to fully to eliminate CML stem cells [76].

In multiple Myeloma, studies have found that circulating clonotypic B cells may persist following systemic treatment. Treatments with glucocorticoids, alkylators, the thalidomide analogue lenalidomide, and the proteosome inhibitor bortezomib, that are currently used in the treatment of myeloma patients, produce debunking of the tumor but lack curative potential as single agents. Therefore, it is thought that the drug resistance exhibited by myeloma stem cells may be mediated by similar processes that protect normal stem cells [82].

6.3.10 Therapy for Solid Tumors

Radiation and cytotoxic chemotherapies (particularly the oral methylator, temozolomide) remain the main treatment strategies for brain cancer treatment [107]. Radio-resistance of CD133$^+$ cells in glioma was one of the first reported. This resistance was attributed to constitutive activation of the DNA repair checkpoint and inhibition of the corresponding kinase in radio-sensitized CD133$^+$ cells [120]. In human glioblastoma xenografts, the fraction of CD133$^+$ CSC was found enriched after ionizing radiation. The radio-resistance of brain CSCs was mechanistically linked to the preferential activation of the DNA damage repair pathway and to significantly reduced rates of apoptosis through the involvement of DNA checkpoint kinases. CD133$^+$ glioblastoma cells also exhibit lower rates of apoptosis in response to chemotherapeutic agents compared to their CD133$^-$ counterparts [113]. It has been shown that treatment of CD133$^+$ CSCs, derived from human GBMs, with bone morphogenic proteins (BMPs) reduced cell proliferation in vitro and induced differentiation into astrocytes. Other strategies employ drugs that target posttranslational modifiers such as histone deacetylases (HDACs). Suberoylanilide hydroxamic acid and other HDAC inhibitors are currently in phase 1 clinical trials. Signaling pathways regulating self-renewal of normal stem cells (e.g., polycomb gene Bmi-1, Notch, Wnt, and Hedgehog (Hh)) may be possible targets. Treatment of mice with the Hh pathway inhibitor cyclopamine inhibited the growth of medulloblastoma. Similarly, inhibition of Notch with specific gamma secretase inhibitors attenuated CSC self-renewal and tumor growth. Studies showed that both CD133$^+$ and CD133$^-$ cells could be rendered less resistant by treatment with the Chk1 and Chk2 kinase inhibitor, debromohymenialdisine. Radiation toxicity that could ensue from lack of tumor specificity is a major concern. There has already been a move from whole brain radiation to more focused beam radiation delivered with the gamma knife to minimize radiation of the normal brain [114]. An interesting hypothesis postulated that the tumor-tropic property of stem cells or progenitor cells could be exploited to selectively deliver a therapeutic gene to metastatic solid tumors. They injected HB1. F3.C1 cells transduced to express an enzyme that efficiently activates the anti-cancer prodrug CPT-11 intravenously into mice bearing disseminated neuroblastoma tumors. The HB1.F3.C1 cells migrated selectively to tumor sites regardless of the size or anatomical location of the tumors. The treated mice, showed tumor-free survival in 100% of the mice for 6 months [121] .

Radio-resistance of breast CSCs is contrary to glioma. Breast CSC produce less reactive oxygen species in response to radiation. Breast cancer cell line-derived spheroids enriched for $CD44^{+}CD24^{-/low}$ subsets exhibited increased in vitro resistance to ionizing radiation as well as reduced levels of reactive oxygen species (ROS) and DNA double strand breaks after radiation exposure, compared to adherent monolayer cultures [59, 115, 118, 122]. Low doses of metformin, a standard drug for diabetes, inhibit cellular transformation and selectively kill cancer stem cells in four genetically different types of breast cancer. The combination of metformin and doxorubicin, kills both cancer stem cells and non-stem cancer cells in culture. Furthermore, this combinatorial therapy reduces tumor mass and prevents relapse much more effectively than either drug alone in a xenograft mouse model [84]. Treatment of patients with HER-2-positive tumors with lapatinib, an EGFR and HER-2/*neu* (ErbB-2) dual-tyrosine kinase inhibitor, resulted in a nonstatistically significant decrease in the percentage of the $CD44^{+}/CD24^{low}$ population and in the ability for self-renewal as assessed by mammosphere formation; however, lapatinib does not show an increase in tumorigenic cells [115]. The combined use of ABC transporter inhibitors such as Pheophorbide, a chlorophyll catabolite that is a specific probe for ABCG2, and chemotherapy could be used to increase the efficiency of chemotherapeutic drugs to kill cancer stem cells [59]. Clinical trials utilizing gamma secretase inhibitors to target Notch in combination with chemotherapy for women with advanced breast cancer are being initiated, as well as use of cyclopamine analogs and other Hh inhibitors. Such trials will directly test the hypothesis that targeting breast cancer stem cells improves the therapeutic outcome in these women [13].

In hepatocellular carcinoma (HCC), hepatic stem cells that have lost TBRII and ELF, and have an activated IL-6 signaling pathway, suggesting that HCC may be induced by IL-6 activation within the setting of loss of the TGF-β signaling pathway. This makes IL-6 an important therapeutic target for HCC therapy. Some successful strategies involve blocking IL-6 signaling, which in effect limits self-renewal. Besides, differences in posttranscriptional and posttranslational modifications in normal and cancerous cells reveal that different splicing variants of the adhesion receptor, CD44, are differentially expressed so therapeutic strategies can focus on targeting this cell surface marker. Differentiation therapy in HCC is an attractive strategy but will require a major effort to define the liver cancer stem cells and their differentiation pathways. Targeting pathways that lead to the self-renewal and proliferating properties of cancer stem cells such as Hedgehog, Wnt, Notch, and IL-6, and using differentiation therapies, such as retinoic acid, and targeting the microenvironment of the stem cell niche could all provide powerful new therapeutic approaches to HCC [63].

In head and neck squamous cell carcinoma (HNSCC), CSCs were made more chemo-sensitive via knockdown of Bmi-1 [67]. Moreover, knockdown of CD44 increased the sensitivity of HNSCC cells to cisplatin, indicating that this CSC marker may be involved in mediating the response of these cells to chemotherapy [75].

In human malignant melanoma CSCs, an ABCB5$^+$ subset was shown to possess increased resistance to the chemotherapeutic agent doxorubicin as a result of diminished drug accumulation. Doxorubicin resistance could be reversed through monoclonal antibody-mediated inhibition of ABCB5-dependent cellular drug efflux, resulting in preferential chemo-sensitization of the more resistant CSC subset. Additionally, ABCB5 gene silencing through siRNA treatment significantly increased the sensitivity of human melanoma cells to doxorubicin, to 5-fluorouracil, and to camptothecin. Killing of melanoma CSCs through their prospective identifier, ABCB5, could reduce experimental tumor initiation and growth *in vivo* [13, 123].

Ovarian cancer stem cells (OCSCs) are thought to be able to regenerate the tumor following chemotherapy. Treatment with increasing concentrations of N-t-boc-Daidzein inhibits cell growth and decreases cell viability of OCSCs and may provide vital aide in improving overall survival in patients with epithelial ovarian cancer [64].

In colorectal carcinoma, autocrine production of IL-4 by CD133$^+$ colon cancer stem-like cells has been reported. Though their cell isolates were more resistant to the chemotherapeutic agents, 5-fluorouracil or oxaliplatin, the ability of these agents to decrease tumorigenic growth was significantly increased when the cells were first treated with antibodies to IL-4. In xenografts, the addition of anti-IL-4 antibodies significantly reduced tumor growth after chemotherapy. *In vivo*, this enhancement was expressed as a substantial slowing of growth, and normal growth resumed without the chemotherapeutic agent. It can be concluded that IL-4 exerts a protective effect on the CD133$^+$ stem-like colon cancer isolates [124].

An interesting point of view about CSC therapy is stated by Jones [99], comparing this kind therapy to a Dandelion. It is observed in multiple myeloma that anti-myeloma agents bortezomib and lenalidomide display significant activity against myeloma plasma cells, in vitro and clinically, but exhibited little activity against myeloma CSCs *in vitro*. Conversely, rituximab and alemtuzumab eliminated myeloma CSCs *in vitro*, but had no activity against myeloma plasma cells (CD20$^-$ and CD52$^-$) [125]. The former agents attack could be seen as mowing a dandelion and the latter as attacking the root of the dandelion.

6.3.11 Other Therapeutic Approaches

An innovative approach compared a hypogravity environment obtained using a Hydrodynamic Focusing Bioreactor (HFB) to a rotary cell culture system (RCCS).

The HFB greatly sensitized the diverse CD133$^+$ cancer cells lines used, which are normally resistant to chemo treatment. After exposure to hypogravity they become susceptible to various chemotherapeutic agents, proposing a less toxic and more effective chemotherapeutic treatment that could be used for patients [126].

Combined epigenetic and chemical therapies have been used for chemo-resistant cancer stem cells in advanced gastric adenocarcinoma. They use antisense molecular targeting to eradicate cancer stem cells, as well as docetaxel. Tumor cells that over-expressed bcl-2 and CSCs were obtained from chemo-resistant patients with metastatic advanced gastric adenocarcinoma and were orthotopically transplanted into immune-deficient mice. The animals were treated with pegylated liposomes composed of phospholipids with high transition temperature entrapping docetaxel in the acyl chains, and encapsulating a 21-base pair of siRNA strand targeted to Msi1 in the liposomes [127].

Mesenchymal stem cell (MSC) lines are also being used as therapeutic cell transplanting reagents (cytoreagents). Gene-modified MSCs are useful as therapeutic tools for brain tissue damage and malignant brain neoplasms [128]. Gene therapy employing MSCs as a tissue-protecting and targeting cytoreagent could be a promising approach [129].

Another interesting approach is *in vivo* multicolor photoacoustic (PA) flow cytometry. It has been used for ultra-sensitive molecular detection of the CD44$^+$ circulating tumor cells (CTCs) in a mouse model of human breast cancer. The idea was to target CTCs with a stem-like phenotype, which are naturally shed from parent tumors. This is carried out with functionalized gold and magnetic nanoparticles. This novel, noninvasive platform, integrates multispectral PA detection and photothermal therapy with a potential for multiplex targeting of many cancer biomarkers using multicolor nanoparticles. This may help clinicians detect circulating stem cells before they migrate to foreign tissues and improve diagnosis as well as treatment [130].

6.4 Tools for the Study of CSC

There are several tools that have been used to identify and isolate cancer stem cells or at least to get a population enriched in CSCs and/or early progenitors. The majority of the papers reviewed in this text are based on similar assays. These assays include:

(a) Immunotyping. It is necessary to determine the specific markers for each type of cancer stem cell obtained from different tissue tumor samples, in order to use them as target proteins to directly eliminate only the cancer stem cells desired. These specificity markers make it possible to design the best anticancer therapy, with less side effects for the patients. Elucidating the candidate proteins for CSC markers is not easy since it is tumor dependent. In some tumors, CSC markers can be determined by the close relationship with their normal counterparts; however, it is not a rule for all cancers. Flow cytometry assay is a useful tool for screening a big battery of markers on cancer cells. This tool simplifies the determination of possible CSC markers.

(b) Sphere formation assay. A property of CSCs is their ability to grow in suspension with specific serum-free media plus at least epidermal growth factor (EGF) and basic fibroblast growth factor (FGF-b). However, the sphere growth enriches the number of cancer stem cells and also their early and late progenitors, compared to traditional culture. In addition, the number of spheres formed is related with the stemness capacity of the cancer cells studied. Currently, several studies on CSCs are based on cells obtained after sphere tumor culture. These conditions induce sphere formation, which has to have a considerable size. It is necessary to keep in mind that only a few cells of these spheres are clearly undifferentiated [106].

(c) Colony forming assay. Similar to sphere formation, CSCs have the capability to grow on soft agar, generating colonies which are also enriched in cancer cells with stem features. The number of colonies generated correlates with the number of possible CSCs. In theory, each colony comes from a single cancer stem cell.

(d) Side population assay. This is a useful tool for isolating CSCs based on exclusion of Hoescht 33342 dye through ABC pumps. It has been demonstrated that the cells sorted as SP cells, show several characteristics of stem cells. However, it is important to emphasize that this sub-population is enriched in CSCs and early progenitors. Furthermore, SP cells are tested using colony forming assays, sphere tumor formation, and in combination with immunotyping.

(e) ALDH activity. It is known that there is an increase in ALDH activity on CSCs. This property has been used by the company Stem Cell Technologies (Canada), which introduced the Aldefluor kit, to sort a population with high tumorigenic capability yielding enrichment of CSCs and their progenitors. This assay is also relevant to identify possible CSCs in combination with the assays described previously.

(f) Signaling pathway identification. There are a large group of proteins belonging to specific signal transduction pathways that are related with self-renewal, proliferation and differentiation of CSCs. These proteins can be evaluated by flow cytometry, immunofluorescence, immunohistochemistry, western blot and proteomics. To confirm the crucial role of proteins from specific signaling pathways in the tumorigenic capacity of specific CSC groups, it is necessary to use specific inhibitors.

(g) Microarrays to explore gene expression, proteomics, transcriptomics and metabolomics are molecular biology tools that can help to screen possible candidates to target specific cancer stem cells.

(h) The gold standard to confirm the presence of cancer stem cells in a sample is to demonstrate their capacity to induce tumor growth in animal models. The mice model depends, in part, on the cancer cells tested, and as we observed before, the route of cell administration is also decisive to determine the capacity of candidate CSCs to induce tumor growth. In addition, it is also extremely important to determine the right immune-deficient mice to confirm the presence or absence of cancer stem cells in tumors. As we saw in studies on melanoma, the use of accurate immune-deficient mice demonstrated the phenotypic plasticity of these cancer cells [1].

6.5 Concluding Remarks

Although a wide group of cancer stem cells have been identified, the origin of these CSCs still unknown. We don't know if the primary CSCs are the same in patients with same type of tumor and we have not elucidated a way to distinguish normal from cancer stem cells . In order to create cancer therapies to decrease and eliminate tumor growth, scientists have focused on determining the physiology of cancer stem cells compared with their normal counterparts. Cellular signal transduction pathways and epigenomics [104] have become relevant tools to figure out specific and differential cancer stem cell markers in comparison with normal stem cells, which can then help to design new specific drugs and strategies that can lend a hand to the physician in order to improve the patient's chances for cancer-free survival.

Acknowledgments We thank CONACyT for the fellowship to support the Ph.D. student José Luis González-Montoya. Finally, we thank Instituto Nacional de Cancerología for supporting my professional development.

References

1. Quintana E, Shackleton M, Foster HR, Fullen DR, Sabel MS, Johnson TM, Morrison SJ (2010) Phenotypic heterogeneity among tumorigenic melanoma cells from patients that is reversible and not hierarchically organized. Cancer Cell 18:510–523
2. Reya T, Morrison SJ, Clarke MF, Weissman IL (2001) Stem cells, cancer, and cancer stem cells. Nature 414:105–111
3. Collado M, Blasco MA, Serrano M (2007) Cellular senescence in cancer and aging. Cell 130:223–233
4. Levine B, Kroemer G (2008) Autophagy in the pathogenesis of disease. Cell 132:27–42
5. Al-Hajj M, Wicha MS, Benito-Hernandez A, Morrison SJ, Clarke MF (2003) Prospective identification of tumorigenic breast cancer cells. Proc Natl Acad Sci USA 100:3983–3988
6. Bozic I, Antal T, Ohtsuki H, Carter H, Kim D, Chen S, Karchin R, Kinzler KW, Vogelstein B, Nowak MA (2010) Accumulation of driver and passenger mutations during tumor progression. Proc Natl Acad Sci USA 107:18545–18550
7. Hombach-Klonisch S, Paranjothy T, Wiechec E, Pocar P, Mustafa T, Seifert A, Zahl C, Gerlach KL, Biermann K, Steger K, Hoang-Vu C, Schulze-Osthoff K, Los M (2008) Cancer stem cells as targets for cancer therapy: selected cancers as examples. Arch Immunol Ther Exp 56:165–180
8. Stephens PJ, Greenman CD, Fu B, Yang F, Bignell GR, Mudie LJ, Pleasance ED, Lau KW, Beare D, Stebbings LA, McLaren S, Lin ML, McBride DJ, Varela I, Nik-Zainal S, Leroy C, Jia M, Menzies A, Butler AP, Teague JW, Quail MA, Burton J, Swerdlow H, Carter NP, Morsberger LA, Iacobuzio-Donahue C, Follows GA, Green AR, Flanagan AM, Stratton MR,

Futreal PA, Campbell PJ (2011) Massive genomic rearrangement acquired in a single catastrophic event during cancer development. Cell 144:27–40
9. Bhagwandin VJ, Shay JW (2009) Pancreatic cancer stem cells: fact or fiction? Biochim Biophys Acta 1792:248–259
10. Jones RJ, Matsui W (2007) Cancer stem cells: from bench to bedside. Biol Blood Marrow Transplant 13:47–52
11. Lawson DA, Witte ON (2007) Stem cells in prostate cancer initiation and progression. J Clin Invest 117:2044–2050
12. Ponnusamy MP, Batra SK (2008) Ovarian cancer: emerging concept on cancer stem cells. J Ovarian Res 1:4
13. Schatton T, Frank NY, Frank MH (2009) Identification and targeting of cancer stem cells. Bioessays 31:1038–1049
14. Sell S, Leffert HL (2008) Liver cancer stem cells. J Clin Oncol 26:2800–2805
15. Clarke MF, Fuller M (2006) Stem cells and cancer: two faces of eve. Cell 124:1111–1115
16. Idikio HA (2011) Human cancer classification: a systems biology- based model integrating morphology, cancer stem cells, proteomics, and genomics. J Cancer 2:107–115
17. Sagar J, Chaib B, Sales K, Winslet M, Seifalian A (2007) Role of stem cells in cancer therapy and cancer stem cells: a review. Cancer Cell Int 7:9
18. Castro NP, Osorio CA, Torres C, Bastos EP, Mourao-Neto M, Soares FA, Brentani HP, Carraro DM (2008) Evidence that molecular changes in cells occur before morphological alterations during the progression of breast ductal carcinoma. Breast Cancer Res 10:R87
19. Wang X, Kruithof-de Julio M, Economides KD, Walker D, Yu H, Halili MV, Hu YP, Price SM, Abate-Shen C, Shen MM (2009) A luminal epithelial stem cell that is a cell of origin for prostate cancer. Nature 461:495–500
20. Barker N, Ridgway RA, van Es JH, van de Wetering M, Begthel H, van den Born M, Danenberg E, Clarke AR, Sansom OJ, Clevers H (2009) Crypt stem cells as the cells-of-origin of intestinal cancer. Nature 457:608–611
21. Rosland GV, Svendsen A, Torsvik A, Sobala E, McCormack E, Immervoll H, Mysliwietz J, Tonn JC, Goldbrunner R, Lonning PE, Bjerkvig R, Schichor C (2009) Long-term cultures of bone marrow-derived human mesenchymal stem cells frequently undergo spontaneous malignant transformation. Cancer Res 69:5331–5339
22. Kratz JR, Yagui-Beltran A, Jablons DM (2010) Cancer stem cells in lung tumorigenesis. Ann Thorac Surg 89:S2090–S2095
23. Xu Q et al (2009) Isolation of tumour stem-like cells from benign tumours. Br J Cancer 101(2):303–311
24. Oravecz-Wilson KI, Philips ST, Yilmaz OH, Ames HM, Li L, Crawford BD, Gauvin AM, Lucas PC, Sitwala K, Downing JR, Morrison SJ, Ross TS (2009) Persistence of leukemia-initiating cells in a conditional knockin model of an imatinib-responsive myeloproliferative disorder. Cancer Cell 16:137–148
25. Read TA, Fogarty MP, Markant SL, McLendon RE, Wei Z, Ellison DW, Febbo PG, Wechsler-Reya RJ (2009) Identification of CD15 as a marker for tumor-propagating cells in a mouse model of medulloblastoma. Cancer Cell 15:135–147
26. Ricci-Vitiani L, Lombardi DG, Pilozzi E, Biffoni M, Todaro M, Peschle C, De Maria R (2007) Identification and expansion of human colon-cancer-initiating cells. Nature 445:111–115
27. Singh SK, Hawkins C, Clarke ID, Squire JA, Bayani J, Hide T, Henkelman RM, Cusimano MD, Dirks PB (2004) Identification of human brain tumour initiating cells. Nature 432:396–401
28. Mani SA, Guo W, Liao MJ, Eaton EN, Ayyanan A, Zhou AY, Brooks M, Reinhard F, Zhang CC, Shipitsin M, Campbell LL, Polyak K, Brisken C, Yang J, Weinberg RA (2008) The epithelial-mesenchymal transition generates cells with properties of stem cells. Cell 133:704–715

29. Pinner S, Jordan P, Sharrock K, Bazley L, Collinson L, Marais R, Bonvin E, Goding C, Sahai E (2009) Intravital imaging reveals transient changes in pigment production and Brn2 expression during metastatic melanoma dissemination. Cancer Res 69:7969–7977
30. Roesch A, Fukunaga-Kalabis M, Schmidt EC, Zabierowski SE, Brafford PA, Vultur A, Basu D, Gimotty P, Vogt T, Herlyn M (2010) A temporarily distinct subpopulation of slow-cycling melanoma cells is required for continuous tumor growth. Cell 141:583–594
31. Sharma SV, Lee DY, Li B, Quinlan MP, Takahashi F, Maheswaran S, McDermott U, Azizian N, Zou L, Fischbach MA, Wong KK, Brandstetter K, Wittner B, Ramaswamy S, Classon M, Settleman J (2010) A chromatin-mediated reversible drug-tolerant state in cancer cell subpopulations. Cell 141:69–80
32. Schatton T, Murphy GF, Frank NY, Yamaura K, Waaga-Gasser AM, Gasser M, Zhan Q, Jordan S, Duncan LM, Weishaupt C, Fuhlbrigge RC, Kupper TS, Sayegh MH, Frank MH (2008) Identification of cells initiating human melanomas. Nature 451:345–349
33. Boiko AD, Razorenova OV, van de Rijn M, Swetter SM, Johnson DL, Ly DP, Butler PD, Yang GP, Joshua B, Kaplan MJ, Longaker MT, Weissman IL (2010) Human melanoma-initiating cells express neural crest nerve growth factor receptor CD271. Nature 466:133–137
34. Held MA, Curley DP, Dankort D, McMahon M, Muthusamy V, Bosenberg MW (2010) Characterization of melanoma cells capable of propagating tumors from a single cell. Cancer Res 70:388–397
35. Li HZ, Yi TB, Wu ZY (2008) Suspension culture combined with chemotherapeutic agents for sorting of breast cancer stem cells. BMC Cancer 8:135
36. Al-Hajj M, Wicha M, Ito-Hernandez A, Morrison S, Clarke M (2003) Prospective identification of tumorigenic breast cancer cells. Proc Natl Acad Sci USA 100:3983–3988
37. Bauerschmitz GJ, Ranki T, Kangasniemi L, Ribacka C, Eriksson M, Porten M, Herrmann I, Ristimaki A, Virkkunen P, Tarkkanen M, Hakkarainen T, Kanerva A, Rein D, Pesonen S, Hemminki A (2008) Tissue-specific promoters active in CD44+CD24-/low breast cancer cells. Cancer Res 68:5533–5539
38. Charafe-Jauffret E, Ginestier C, Iovino F, Tarpin C, Diebel M, Esterni B, Houvenaeghel G, Extra JM, Bertucci F, Jacquemier J, Xerri L, Dontu G, Stassi G, Xiao Y, Barsky SH, Birnbaum D, Viens P, Wicha MS (2010) Aldehyde dehydrogenase 1-positive cancer stem cells mediate metastasis and poor clinical outcome in inflammatory breast cancer. Clin Cancer Res 16:45–55
39. Ginestier C, Hur MH, Charafe-Jauffret E, Monville F, Dutcher J, Brown M, Jacquemier J, Viens P, Kleer CG, Liu S, Schott A, Hayes D, Birnbaum D, Wicha MS, Dontu G (2007) ALDH1 is a marker of normal and malignant human mammary stem cells and a predictor of poor clinical outcome. Cell Stem Cell 1:555–567
40. Collins AT, Berry PA, Hyde C, Stower MJ, Maitland NJ (2005) Prospective identification of tumorigenic prostate cancer stem cells. Cancer Res 65:10946–10951
41. Mulholland DJ, Xin L, Morim A, Lawson D, Witte O, Wu H (2009) Lin-Sca-1+CD49fhigh stem/progenitors are tumor-initiating cells in the Pten-null prostate cancer model. Cancer Res 69:8555–8562
42. Shimono Y, Zabala M, Cho RW, Lobo N, Dalerba P, Qian D, Diehn M, Liu H, Panula SP, Chiao E, Dirbas FM, Somlo G, Pera RA, Lao K, Clarke MF (2009) Downregulation of miRNA-200c links breast cancer stem cells with normal stem cells. Cell 138:592–603
43. Wodarz A, Gonzalez C (2006) Connecting cancer to the asymmetric division of stem cells. Cell 124:1121–1123
44. Yang ZF, Ho DW, Ng MN, Lau CK, Yu WC, Ngai P, Chu PW, Lam CT, Poon RT, Fan ST (2008) Significance of CD90+ cancer stem cells in human liver cancer. Cancer Cell 13:153–166
45. Suetsugu A, Nagaki M, Aoki H, Motohashi T, Kunisada T, Moriwaki H (2006) Characterization of CD133+ hepatocellular carcinoma cells as cancer stem/progenitor cells. Biochem Biophys Res Commun 351:820–824
46. Terris B, Cavard C, Perret C (2010) EpCAM, a new marker for cancer stem cells in hepatocellular carcinoma. J Hepatol 52:280–281

47. Barker N, van Es JH, Kuipers J, Kujala P, van den Born M, Cozijnsen M, Haegebarth A, Korving J, Begthel H, Peters PJ, Clevers H (2007) Identification of stem cells in small intestine and colon by marker gene Lgr5. Nature 449:1003–1007
48. Di Tomaso T, Mazzoleni S, Wang E, Sovena G, Clavenna D, Franzin A, Mortini P, Ferrone S, Doglioni C, Marincola FM, Galli R, Parmiani G, Maccalli C (2010) Immunobiological characterization of cancer stem cells isolated from glioblastoma patients. Clin Cancer Res 16:800–813
49. Li Y, Zhang T, Korkaya H, Liu S, Lee HF, Newman B, Yu Y, Clouthier SG, Schwartz SJ, Wicha MS, Sun D (2010) Sulforaphane, a dietary component of broccoli/broccoli sprouts, inhibits breast cancer stem cells. Clin Cancer Res 16:2580–2590
50. Dey M, Ulasov IV, Lesniak MS (2010) Virotherapy against malignant glioma stem cells. Cancer Lett 289:1–10
51. Pannuti A, Foreman K, Rizzo P, Osipo C, Golde T, Osborne B, Miele L (2010) Targeting Notch to target cancer stem cells. Clin Cancer Res 16:3141–3152
52. Obrien CA, Pollett A, Gallinger S, Dick JE (2007) A human colon cancer cell capable of initiating tumour growth in immunodeficient mice. Nature 445:106–110
53. Ricci-Vitiani L (2007) Identification and expansion of human colon-cancer-initiating cells. Nature 445:111–115
54. Wei J, Barr J, Kong LY, Wang Y, Wu A, Sharma AK, Gumin J, Henry V, Colman H, Sawaya R, Lang FF, Heimberger AB (2010) Glioma-associated cancer-initiating cells induce immunosuppression. Clin Cancer Res 16:461–473
55. Rutella S, Bonanno G, Procoli A, Mariotti A, Corallo M, Prisco MG, Eramo A, Napoletano C, Gallo D, Perillo A, Nuti M, Pierelli L, Testa U, Scambia G, Ferrandina G (2009) Cells with characteristics of cancer stem/progenitor cells express the CD133 antigen in human endometrial tumors. Clin Cancer Res 15:4299–4311
56. Deonarain MP, Kousparou CA, Epenetos AA (2009) Antibodies targeting cancer stem cells: a new paradigm in immunotherapy? MAbs 1:12–25
57. Soriano JV, Uyttendaele H, Kitajewski J, Montesano R (2000) Expression of an activated Notch4(int-3) oncoprotein disrupts morphogenesis and induces an invasive phenotype in mammary epithelial cells in vitro. Int J Cancer 86:652–659
58. Dontu G, Al-Hajj M, Abdallah WM, Clarke MF, Wicha MS (2003) Stem cells in normal breast development and breast cancer. Cell Prolif 36(Suppl 1):59–72
59. Morrison BJ, Schmidt CW, Lakhani SR, Reynolds BA, Lopez JA (2008) Breast cancer stem cells: implications for therapy of breast cancer. Breast Cancer Res 10:210
60. Jerry DJ, Tao L, Yan H (2008) Regulation of cancer stem cells by p53. Breast Cancer Res 10:304
61. Theise ND, Saxena R, Portmann BC, Thung SN, Yee H, Chiriboga L, Kumar A, Crawford JM (1999) The canals of Hering and hepatic stem cells in humans. Hepatology 30:1425–1433
62. Chiba T, Kita K, Zheng YW, Yokosuka O, Saisho H, Iwama A, Nakauchi H, Taniguchi H (2006) Side population purified from hepatocellular carcinoma cells harbors cancer stem cell-like properties. Hepatology 44:240–251
63. Mishra L, Banker T, Murray J, Byers S, Thenappan A, He AR, Shetty K, Johnson L, Reddy EP (2009) Liver stem cells and hepatocellular carcinoma. Hepatology 49:318–329
64. Green JM, Alvero AB, Kohen F, Mor G (2009) 7-(O)-Carboxymethyl daidzein conjugated to N-t-Boc-hexylenediamine: a novel compound capable of inducing cell death in epithelial ovarian cancer stem cells. Cancer Biol Ther 8:1747–1753
65. Ness RB, Cottreau C (1999) Possible role of ovarian epithelial inflammation in ovarian cancer. J Natl Cancer Inst 91:1459–1467
66. Anderson R, Copeland TK, Scholer H, Heasman J, Wylie C (2000) The onset of germ cell migration in the mouse embryo. Mech Dev 91:61–68
67. Salcido CD, Larochelle A, Taylor BJ, Dunbar CE, Varticovski L (2010) Molecular characterisation of side population cells with cancer stem cell-like characteristics in small-cell lung cancer. Br J Cancer 102:1636–1644
68. Niwa H, Miyazaki J, Smith AG (2000) Quantitative expression of Oct-3/4 defines differentiation, dedifferentiation or self-renewal of ES cells. Nat Genet 24:372–376

69. Hoei-Hansen CE, Kraggerud SM, Abeler VM, Kaern J, Rajpert-De Meyts E, Lothe RA (2007) Ovarian dysgerminomas are characterised by frequent KIT mutations and abundant expression of pluripotency markers. Mol Cancer 6:12
70. Vermorken JB, Remenar E, van Herpen C, Gorlia T, Mesia R, Degardin M, Stewart JS, Jelic S, Betka J, Preiss JH, van den Weyngaert D, Awada A, Cupissol D, Kienzer HR, Rey A, Desaunois I, Bernier J, Lefebvre JL (2007) Cisplatin, fluorouracil, and docetaxel in unresectable head and neck cancer. N Engl J Med 357:1695–1704
71. Zhang Q, Shi S, Yen Y, Brown J, Ta JQ, Le AD (2010) A subpopulation of CD133(+) cancer stem-like cells characterized in human oral squamous cell carcinoma confer resistance to chemotherapy. Cancer Lett 289:151–160
72. Okamoto A, Chikamatsu K, Sakakura K, Hatsushika K, Takahashi G, Masuyama K (2009) Expansion and characterization of cancer stem-like cells in squamous cell carcinoma of the head and neck. Oral Oncol 45:633–639
73. Clay MR, Tabor M, Owen JH, Carey TE, Bradford CR, Wolf GT, Wicha MS, Prince ME (2010) Single-marker identification of head and neck squamous cell carcinoma cancer stem cells with aldehyde dehydrogenase. Head Neck 32:1195–1201
74. Chiou SH, Yu CC, Huang CY, Lin SC, Liu CJ, Tsai TH, Chou SH, Chien CS, Ku HH, Lo JF (2008) Positive correlations of Oct-4 and Nanog in oral cancer stem-like cells and high-grade oral squamous cell carcinoma. Clin Cancer Res 14:4085–4095
75. Monroe MM, Anderson EC, Clayburgh DR, Wong MH (2011) Cancer stem cells in head and neck squamous cell carcinoma. J Oncol 2011:762–780
76. Lin T, Jones RJ, Matsui W (2009) Cancer stem cells: relevance to SCT. Bone Marrow Transplant 43:517–523
77. Neville BW, Day TA (2002) Oral cancer and precancerous lesions. CA Cancer J Clin 52:195–215
78. Okuyama R, Tagami H, Aiba S (2008) Notch signaling: its role in epidermal homeostasis and in the pathogenesis of skin diseases. J Dermatol Sci 49:187–194
79. Sullivan JP, Minna JD, Shay JW (2010) Evidence for self-renewing lung cancer stem cells and their implications in tumor initiation, progression, and targeted therapy. Cancer Metastasis Rev 29:61–72
80. Bao S, Wu Q, Li Z, Sathornsumetee S, Wang H, McLendon RE, Hjelmeland AB, Rich JN (2008) Targeting cancer stem cells through L1CAM suppresses glioma growth. Cancer Res 68:6043–6048
81. Real FX (2003) A "catastrophic hypothesis" for pancreas cancer progression. Gastroenterology 124:1958–1964
82. Ghosh N, Matsui W (2009) Cancer stem cells in multiple myeloma. Cancer Lett 277:1–7
83. Wilkinson N, Scott-Conner CE (2008) Surgical therapy for colorectal adenocarcinoma. Gastroenterol Clin North Am 37:253–267, ix
84. Hirsch HA, Iliopoulos D, Tsichlis PN, Struhl K (2009) Metformin selectively targets cancer stem cells, and acts together with chemotherapy to block tumor growth and prolong remission. Cancer Res 69:7507–7511
85. Jemal A, Siegel R, Ward E, Hao Y, Xu J, Thun MJ (2009) Cancer statistics, 2009. CA Cancer J Clin 59:225–249
86. Subramaniam D, Ramalingam S, Houchen CW, Anant S (2010) Cancer stem cells: a novel paradigm for cancer prevention and treatment. Mini Rev Med Chem 10:359–371
87. Lin L, Liu Y, Li H, Li PK, Fuchs J, Shibata H, Iwabuchi Y, Lin J (2011) Targeting colon cancer stem cells using a new curcumin analogue, GO-Y030. Br J Cancer 105:212–220
88. Dalerba P, Dylla SJ, Park IK, Liu R, Wang X, Cho RW, Hoey T, Gurney A, Huang EH, Simeone DM, Shelton AA, Parmiani G, Castelli C, Clarke MF (2007) Phenotypic characterization of human colorectal cancer stem cells. Proc Natl Acad Sci USA 104: 10158–10163
89. Du L, Wang H, He L, Zhang J, Ni B, Wang X, Jin H, Cahuzac N, Mehrpour M, Lu Y, Chen Q (2008) CD44 is of functional importance for colorectal cancer stem cells. Clin Cancer Res 14:6751–6760

90. Vermeulen L, Todaro M, de Sousa MF, Sprick MR, Kemper K, Perez Alea M, Richel DJ, Stassi G, Medema JP (2008) Single-cell cloning of colon cancer stem cells reveals a multi-lineage differentiation capacity. Proc Natl Acad Sci USA 105:13427–13432
91. Kemper K, Grandela C, Medema JP (2010) Molecular identification and targeting of colorectal cancer stem cells. Oncotarget 1:387–395
92. Ho MM, Ng AV, Lam S, Hung JY (2007) Side population in human lung cancer cell lines and tumors is enriched with stem-like cancer cells. Cancer Res 67:4827–4833
93. Eramo A, Lotti F, Sette G, Pilozzi E, Biffoni M, Di Virgilio A, Conticello C, Ruco L, Peschle C, De Maria R (2008) Identification and expansion of the tumorigenic lung cancer stem cell population. Cell Death Differ 15:504–514
94. Bertolini G, Roz L, Perego P, Tortoreto M, Fontanella E, Gatti L, Pratesi G, Fabbri A, Andriani F, Tinelli S, Roz E, Caserini R, Lo Vullo S, Camerini T, Mariani L, Delia D, Calabro E, Pastorino U, Sozzi G (2009) Highly tumorigenic lung cancer CD133+ cells display stem-like features and are spared by cisplatin treatment. Proc Natl Acad Sci USA 106:16281–16286
95. Yang Y, Iwanaga K, Raso MG, Wislez M, Hanna AE, Wieder ED, Molldrem JJ, Wistuba II, Powis G, Demayo FJ, Kim CF, Kurie JM (2008) Phosphatidylinositol 3-kinase mediates bronchioalveolar stem cell expansion in mouse models of oncogenic K-ras-induced lung cancer. PLoS One 3:e2220
96. Gottesman MM, Fojo T, Bates SE (2002) Multidrug resistance in cancer: role of ATP-dependent transporters. Nat Rev Cancer 2:48–58
97. Zhou S, Schuetz JD, Bunting KD, Colapietro AM, Sampath J, Morris JJ, Lagutina I, Grosveld GC, Osawa M, Nakauchi H, Sorrentino BP (2001) The ABC transporter Bcrp1/ABCG2 is expressed in a wide variety of stem cells and is a molecular determinant of the side-population phenotype. Nat Med 7:1028–1034
98. Zhou BB, Zhang H, Damelin M, Geles KG, Grindley JC, Dirks PB (2009) Tumour-initiating cells: challenges and opportunities for anticancer drug discovery. Nat Rev Drug Discov 8:806–823
99. Jones RJ (2009) Cancer stem cells-clinical relevance. J Mol Med (Berl) 87:1105–1110
100. Vaish M (2007) Mismatch repair deficiencies transforming stem cells into cancer stem cells and therapeutic implications. Mol Cancer 6:26
101. Diehn M, Cho RW, Clarke MF (2009) Therapeutic implications of the cancer stem cell hypothesis. Semin Radiat Oncol 19:78–86
102. Gilbert CA, Ross AH (2009) Cancer stem cells: cell culture, markers, and targets for new therapies. J Cell Biochem 108:1031–1038
103. Regenbrecht CR, Lehrach H, Adjaye J (2008) Stemming cancer: functional genomics of cancer stem cells in solid tumors. Stem Cell Rev 4:319–328
104. DeSano JT, Xu L (2009) MicroRNA regulation of cancer stem cells and therapeutic implications. AAPS J 11:682–692
105. Kakarala M, Wicha MS (2008) Implications of the cancer stem-cell hypothesis for breast cancer prevention and therapy. J Clin Oncol 26:2813–2820
106. Keith B, Simon MC (2007) Hypoxia-inducible factors, stem cells, and cancer. Cell 129:465–472
107. Rich JN (2008) The implications of the cancer stem cell hypothesis for neuro-oncology and neurology. Future Neurol 3:265–273
108. Dontu G, Liu S, Wicha MS (2005) Stem cells in mammary development and carcinogenesis: implications for prevention and treatment. Stem Cell Rev 1:207–213
109. Fusco A, Fedele M (2007) Roles of HMGA proteins in cancer. Nat Rev Cancer 7:899–910
110. Liu S, Dontu G, Mantle ID, Patel S, Ahn NS, Jackson KW, Suri P, Wicha MS (2006) Hedgehog signaling and Bmi-1 regulate self-renewal of normal and malignant human mammary stem cells. Cancer Res 66:6063–6071
111. Domen J, Cheshier SH, Weissman IL (2000) The role of apoptosis in the regulation of hematopoietic stem cells: overexpression of Bcl-2 increases both their number and repopulation potential. J Exp Med 191:253–264
112. Ishikawa F, Yoshida S, Saito Y, Hijikata A, Kitamura H, Tanaka S, Nakamura R, Tanaka T, Tomiyama H, Saito N, Fukata M, Miyamoto T, Lyons B, Ohshima K, Uchida N, Taniguchi S,

Ohara O, Akashi K, Harada M, Shultz LD (2007) Chemotherapy-resistant human AML stem cells home to and engraft within the bone-marrow endosteal region. Nat Biotechnol 25:1315–1321

113. Liu G, Yuan X, Zeng Z, Tunici P, Ng H, Abdulkadir IR, Lu L, Irvin D, Black KL, Yu JS (2006) Analysis of gene expression and chemoresistance of CD133+ cancer stem cells in glioblastoma. Mol Cancer 5:67
114. Sakariassen PØ, Immervoll H, Chekenya M (2007) Cancer stem cells as mediators of treatment resistance in brain tumors: status and controversies. Neoplasia 9:882–892
115. Li X (2008) Intrinsic resistance of tumorigenic breast cancer cells to chemotherapy. J Natl Cancer Inst 100:672–679
116. Ding XW, Wu JH, Jiang CP (2010) ABCG2: a potential marker of stem cells and novel target in stem cell and cancer therapy. Life Sci 86:631–637
117. Liu HG, Chen C, Yang H, Pan YF, Zhang XH (2011) Cancer stem cell subsets and their relationships. J Transl Med 9:50
118. Diehn M, Cho RW, Lobo NA, Kalisky T, Dorie MJ, Kulp AN, Qian D, Lam JS, Ailles LE, Wong M, Joshua B, Kaplan MJ, Wapnir I, Dirbas FM, Somlo G, Garberoglio C, Paz B, Shen J, Lau SK, Quake SR, Brown JM, Weissman IL, Clarke MF (2009) Association of reactive oxygen species levels and radioresistance in cancer stem cells. Nature 458:780–783
119. Vlashi E, McBride WH, Pajonk F (2009) Radiation responses of cancer stem cells. J Cell Biochem 108:339–342
120. Bao S, Wu Q, McLendon RE, Hao Y, Shi Q, Hjelmeland AB, Dewhirst MW, Bigner DD, Rich JN (2006) Glioma stem cells promote radioresistance by preferential activation of the DNA damage response. Nature 444:756–760
121. Aboody KS, Bush RA, Garcia E, Metz MZ, Najbauer J, Justus KA, Phelps DA, Remack JS, Yoon KJ, Gillespie S, Kim SU, Glackin CA, Potter PM, Danks MK (2006) Development of a tumor-selective approach to treat metastatic cancer. PLoS One 1:e23
122. Charafe-Jauffret E, Monville F, Ginestier C, Dontu G, Birnbaum D, Wicha MS (2008) Cancer stem cells in breast: current opinion and future challenges. Pathobiology 75:75–84
123. Frank NY, Margaryan A, Huang Y, Schatton T, Waaga-Gasser AM, Gasser M, Sayegh MH, Sadee W, Frank MH (2005) ABCB5-mediated doxorubicin transport and chemoresistance in human malignant melanoma. Cancer Res 65:4320–4333
124. Huang EH, Wicha MS (2008) Colon cancer stem cells: implications for prevention and therapy. Trends Mol Med 14:503–509
125. Matsui W, Wang Q, Barber JP, Brennan S, Smith BD, Borrello I, McNiece I, Lin L, Ambinder RF, Peacock C, Watkins DN, Huff CA, Jones RJ (2008) Clonogenic multiple myeloma progenitors, stem cell properties, and drug resistance. Cancer Res 68:190–197
126. Giannios et al (2010) Gastrointest Cancer Res (Suppl 2):S14–S15
127. Kelly SE, Di Benedetto A, Greco A, Howard CM, Sollars VE, Primerano DA, Valluri JV, Claudio PP (2010) Rapid selection and proliferation of CD133+ cells from cancer cell lines: chemotherapeutic implications. PLoS One 5:e10035
128. Nakamizo A, Marini F, Amano T, Khan A, Studeny M, Gumin J, Chen J, Hentschel S, Vecil G, Dembinski J, Andreeff M, Lang F (2005) Human bone marrow-derived mesenchymal stem cells in the treatment of gliomas. Cancer Res 65:3307–3318
129. Hamada H, Kobune M, Nakamura K, Kawano Y, Kato K, Honmou O, Houkin K, Matsunaga T, Niitsu Y (2005) Mesenchymal stem cells (MSC) as therapeutic cytoreagents for gene therapy. Cancer Sci 96:149–156
130. Galanzha EI, Kim JW, Zharov VP (2009) Nanotechnology-based molecular photoacoustic and photothermal flow cytometry platform for in-vivo detection and killing of circulating cancer stem cells. J Biophotonics 2:725–735

Chapter 7
Cancer Stem Cells: Paradigm Shifting or Perishing Concept?

Senthil K. Pazhanisamy and Keith Syson Chan

Contents

S.K. Pazhanisamy, Ph.D.
Scott Department of Urology, Baylor College of Medicine, One Baylor Plaza, BCM 380, 77030 Houston, TX, USA

K.S. Chan, Ph.D. (✉)
Scott Department of Urology, Baylor College of Medicine, One Baylor Plaza, BCM 380, 77030 Houston, TX, USA

Department of Molecular & Cellular Biology, Baylor College of Medicine, One Baylor Plaza, BCM 380, 77030 Houston, TX, USA
e-mail: kc1@bcm.edu

R.K. Srivastava and S. Shankar (eds.), *Stem Cells and Human Diseases*,
DOI 10.1007/978-94-007-2801-1_7,

Abstract There has been tremendous progress in our understanding of intratumoral heterogeneity over the last decade. One emerging concept postulates that tumors are analogous to self-renewing adult tissues, and are maintained by a unique subpopulation of stem cells. Cancer stem cells are best defined functionally by their enriched tumorigenic properties, with distinct self-renewal and differentiation potential to regenerate the cellular heterogeneity of original patient tumors in immunocompromised mice. Cancer stem cells have attracted enormous attention in cancer research, and exploring their biology will provide tremendous opportunities to develop more effective therapeutic strategies against tumors. In this review, we summarize the historical perspective leading to the prospective isolation of CSCs, their intriguing biological characteristics and clinical implications with emphasis on solid tumors. We also discuss our perspectives on the controversies and possible misconcepts of the cancer stem cell model.

Keywords Cancer stem cells • Hierarchical model • Plasticity • Molecular heterogeneity

7.1 Intra-tumoral Heterogeneity of Human Cancer

One major characteristic of human cancers is that they are composed of tumor cells with profound heterogeneity in cellular morphology, biological properties and karyotype. Since early nineteenth century, modernization and application of microscopy by Rudolf Virchow allowed for the detailed pathological evaluation of cancer tissues. This led to the findings that significant differences exist in the appearance of cancer and its surrounding normal tissues. Pathological analyses in the 1980s led to the early concept of intratumoral heterogeneity—cancer cells with different cell sizes, variation in nuclear feature and differentiation morphology exist within an individual tumor [1]. Importantly, intratumoral heterogeneity can be also characterized by their difference in biological properties such as anchorage-independent growth in soft agar [2], proliferative capacities [3], expression of markers that define different stages of differentiation [4], and metastatic potential [5].

7.2 Cancer Stem Cell Concept to Explain Intra-tumoral Heterogeneity

7.2.1 Human Tumor Stem Cell Assays

In 1977, Hamburger and Salmon et al. were among the first to experimentally describe "tumor stem cells" in a variety of tumor types from patient-derived tumor tissues. Their findings revealed a rare subpopulation of tumor cells possess the unique ability to grow in then a newly established two-layer soft agar assay [2]. Utilization of this tumor stem cell assay led to a series of *in vitro* preclinical studies and eventually prospective clinical trials [6] to predict chemo and radiation sensitivity in patients based on drug sensitivity in this assay, with remarkable true positive and false negative rate for prospective prediction of response or lack of response of individual patients to these cytotoxic therapy [6]. Although their definition of "tumor stem cell colonies" may be better defined as colonies with anchorage-independent growth properties in modern terms, these early observations clearly supported the existence of a rare tumorigenic or transforming subpopulation within primary tumors.

7.2.2 Teratocarcinoma as an Early Model for Cancer Stem Cells

Mouse teratocarcinomas (or embryonal carcinomas) were long described with the capacities to maintain undifferentiated teratocarcinoma cells and differentiated derivatives *in vivo* [7]. Undifferentiated embryonic carcinoma cells could be distinguished from the differentiated somatic cell progenies based on the expression level of enzyme alkaline phosphatase, anywhere from 5- to 100-fold difference [7]. Remarkably, after more than 200 generations of transplanting teratocarcinoma cells as malignant tumors in nude mice, these cells retained their pluripotent stem cell property. When injected into blastocysts for generation of chimeric mice, these malignant teratocarcinoma cells can form various developmentally unrelated normal somatic and germ-line tissues [8]. Further lineage-tracing experiments by cloning single teratocarcinoma cells in genetically marked blastocysts and following the fate of these tumor cells by analyses of mutation-carrying cells revealed similar results, that these malignant teratocarcinoma cells can form developmentally unrelated normal somatic and germ-line tissues and the majority of these chimeric mice were free of teratomas [9]. Monoclonal antibodies were generated to isolate subpopulation of embryonal carcinoma cells (or modern term cancer stem cells) in comparison to differentiated derivatives [10]. These early studies supported the existence of cancer stem cells within teratocarcinomas, and that they may be epigenetically modulated by the surrounding microenvironment to determine their fate decisions to self-renew or differentiate into normal or malignant tissues.

7.2.3 *Prospective Isolation of Cancer Stem Cells from Patient Tumors*

The cancer stem cell model hypothesizes that a tumorigenic stem cell is able to self-renew and differentiate into phenotypically diverse tumor cells that explain the intratumoral heterogeneity of cancer (Fig. 7.1a). Development of fluorescence activated cell sorting (FACS) by Herzenberg and colleagues [11] in Stanford University and the invention of monoclonal antibodies allowed for the application of these technologies by Irving Weissman and colleagues to isolate the first mouse [12, 13] and human hematopoietic stem cells (HSCs) [14]. These become fundamental techniques and instrument in the stem cell field for the prospective isolation of viable stem cells from other adult tissues, and eventually cancer stem cells from multiple tissue types. Establishment of a unique *in vivo* bone marrow reconstitution assay from Weissman's group allowed for the experimental interrogation of HSC biology *in vivo*, which set the gold standard for the prospective isolation of normal and cancer stem cells from other tissue types. John Dick and colleagues utilized similar approaches to isolate the first leukemic stem cells from human acute myeloid leukemia (AML). Their landmark study demonstrated that human AML leukemia cells with the Lineage-CD34+CD38− antigen profile possess the unique ability to produce colony-forming progenitors *in vitro*, after engraftment in a severe combined immune-deficient (SCID) mouse [15]. Importantly, they also demonstrated that human AML are hierarchically organized, leukemic stem cells can serially transplant (self-renewal capacity) and recapitulate the original bulk tumor in immunocompromised mice; these properties are limited or absent in the rest majority of differentiated tumor cells. Michael Clarke's group was the first to apply this principle to isolate cancer stem cells from solid tumor; using primary patient tumors and ascites from mammary tumors, they found that breast CSCs were enriched in Lin-/CD44+/CD24− subpopulation [16]. Since then, various cancer attributes such as xenotransplantation assays, sphere-forming ability, and exclusion of the Hoechst dye were used to schematically isolate CSC subpopulation from other solid tumor types. Thus far, a large body of evidence supported the existence of cancer stem cells in multiple solid tumors; these include brain [17], colon [18], head and neck [19], bladder [20], prostate [21], melanoma [22], pancreatic [23] and lung cancer [24].

7.2.3.1 *In Vivo* Xenotransplantation Assay and *In Vitro* Stem Cell Assays

Cancer stem cells are best defined functionally by their enriched tumorigenic properties, with distinct self-renewal and differentiation potential to regenerate the cellular heterogeneity of original patient tumors. The current gold standard for assaying CSC remained to be the xenotransplantation assay using immunocompromised mice [25]. Primary tumors, xenografts or immortalized cell lines are usually subfractionated by FACS or magnetic beads separation and the relative tumorigenic potential between subpopulations is assayed by their latency and xenograft take rate. The ability of CSCs to self-renew is examined by serial transplantation into a second recipient

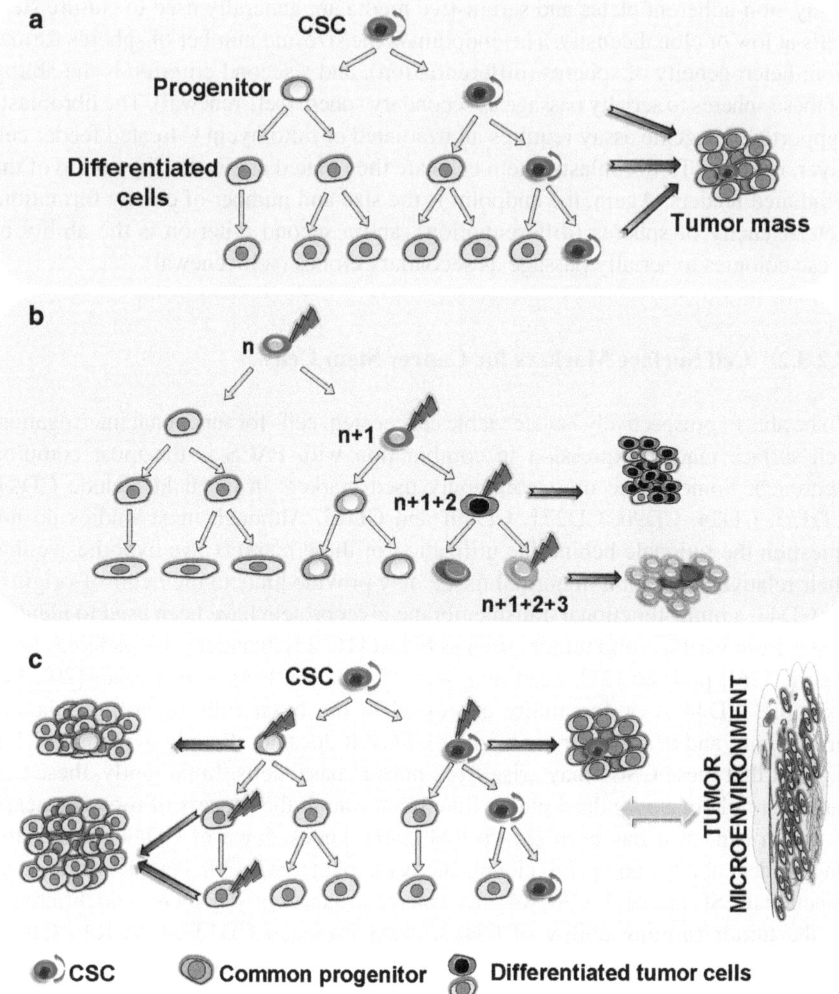

Fig. 7.1 Models explaining intratumoural heterogeneity. (**a**) *Cancer Stem Cell (CSC) model.* Only CSCs, with exquisite self-renewal and differentiation potential, can continuously generate common and more differentiated progenitor cells, thereby ultimately reconstitute a heterogeneous tumor. (**b**) *Clonal evolution model.* Normal cells accumulate sufficient hits to become neoplastic, only one variant clone have the selective advantage to survive under evolutionary selection. (**c**) *Integrated model.* CSCs primarily drive tumor growth, however, multiple clones may existence during the course of cancer development by acquisition of additional alterations

(similar to that in assaying hematopoietic stem cells), and their ability to differentiate is examined whether a pure CSC population can regenerate the heterogeneity of original tumor by analyzing by flow cytometry or immunohistochemistry [25].

Several *in vitro* assays have been adapted from assaying neural or skin stem cells, and are now routinely used for assaying cancer stem cells; these include the sphere-forming assay and the fibroblasts supported clonogenic assay. In the sphere-forming assay, non-adherent plates and serum-free media are generally used to culture stem cells at low or clonal density. The endpoint is the size and number of spheres formation; heterogeneity of spheres (differentiation), and a second criterion is the ability of these spheres to serially passage as secondary sphere (self-renewal). The fibroblasts supported clongenic assay requires an irradiated or mitomycin C treated feeder cell layer, usually 3T3 fibroblasts. Stem cells are then plated at low density on top of the irradiated feeders. Again, the endpoint is the size and number of colony formation; heterogeneity of spheres (differentiation), and a second criterion is the ability of these colonies to serially passage as secondary clones (self-renewal).

7.2.3.2 Cell Surface Markers for Cancer Stem Cells

To be able to prospectively isolate viable cancer stem cells for functional interrogation, cell surface marker expression in combination with FACS is the most common approach. Some of the most commonly used markers in the field include CD44, CD133, CD24, CD90, CD271, CD49f and CD13. Although most studies do not question the rationale behind the utilization of these markers, we hypothesize that their relative expression in normal tissue may provide hints to their cell-of-origin.

CD44, a multi-functional transmembrane glycoprotein have been used to identify CSCs from various solid tumors such as breast [16, 23], bladder [20], pancreas [23], gastric [26], prostate [27], head and neck [28], colon [18], and ovarian [29]. For example, CD44 is preferentially expressed in the basal cells of normal bladder urothelium, and is a marker for bladder CSCs; it does not directly demonstrate but suggest that these CSCs may arise from normal basal cells. Importantly, these cell surface markers may indeed play a functional role in the process of tumorigenesis. In colon cancer, it has been shown that stable knock down of CD44 reduces the engraftment ability and *in vitro* clonogenicity of colon CSCs [30]. Another example of functional CSC marker is CXCR4; in a pancreatic cancer model, there is no difference in the tumor forming ability of CD133+CXCR4+ and CD133+CXCR4− CSCs. Interestingly, only CD133+CXCR4+ CSCs can form distal metastasis, suggestive of a functional role for CXCR4 in the metastatic process [31].

7.2.3.3 Cytokeratin Markers

Cytokeratins, intermediate filaments that form the major components of cytoskeleton in the cytoplasm of epithelial cells, are classically used to characterize epithelial cells from different stages of cellular differentiation. There are two types of cytokeratins, the acidic type I and the basic type II cytokeratins, which are usually found in heterodimers. The expression of these cytokeratins depends on the type of epithelium and their expression also alters during the course of normal epithelial cell differentiation [32]. One of the most well characterized patterns of cytokeratin expression is in mouse skin, where CK15 expression is restricted to the bulge region multipotent stem cells, wheras its expression is lost in epidermal basal cells and replaced by CK5.

With further differentiation down the path, CK5 is replaced by CK10 and eventually by the terminal differentiation marker involucrin. In the bladder urothelium, others and we have demonstrated that CK14 and CK5 are expressed in urothelial basal cells, CK8 and CK18 are expressed in intermediate cells and CK20 are expressed in terminally differentiated umbrella cells. We have applied these cytokeratins to characterize cancer stem cells that are being prospectively isolated, as supporting evidence for their differentiation status [20]. Although cytokeratins are intracellular proteins and cannot be utilized in FACS for prospective isolation of CSCs, they are extremely valuable as supporting evidence when additional new cell surface markers are being identified to subfranctionate populations of stem, progenitor and differentiated cells in solid cancers [19, 20].

7.2.3.4 Side Population and Aldehyde Dehydrogenase (ALDH)

Although cell surface markers are widely used for prospective isolation of CSCs, another commonly used method to isolate stem cells and CSCs is the Side Population (SP) discrimination assay. This method is based on the unique properties of cell subpopulation such as stem cells that expresses higher level of the ATP-binding cassette (ABC) family of transporters. Several family members, such as ABCG2 (breast cancer resistance protein, BRCP1), and ABCB1 (MDR1, or P-glycoprotein) have been implicated to rapidly efflux the dye Hoschst 33342 and responsible for this SP phenotype. SP cells have been identified in a number of cancer types, including prostate [33], bladder [34], breast [35], and lung [24]. In general, these SP cells were reported to contain the functional criteria of cancer stem cells: enrichment for tumorigenecity, ability to give rise to both SP and non-SP cells. Another property commonly used to identify normal and cancer stem cells is the enzyme Aldehyde Dehydrogenase (ALDH). It was first demonstrated in normal and cancer stem cells from breast that cells express ALDH1 contain functional properties of stem cells and this marker is a predictor for poor clinical outcome for breast cancer patients [36]. Since then, this markers have been found to isolate tumor cell with CSC properties in glioblastoma [37], lung [38] liver [39] and a number of other cancers.

7.3 Molecular Heterogeneity in Self-renewal Pathways

Embryonic and tissue adult stem cells intrinsically activate certain signaling pathways to maintain their stemness or self-renewal, and these pathways are downregulated upon differentiation. A handful of which includes the Wnt/β-catenin, Sonic Hedgehog (SHH), Notch, and Signal transducer and activation of transcription 3 (Stat3). All of these described pathways are also activated and implicated in various solid cancers, suggestive of their roles in cancer stem cell self-renewal. However, it is not our intention to summarize in detail the self-renewal pathways involved in normal and cancer stem cells in this chapter. Numerous studies often implicate the role of an individual self-renewal pathway to cancer stem cells from a particular cancer type, misleading readers to believe that it represents a predominant pathway in all cancer specimens. However, results from our laboratory revealed that the activation of these self-renewal pathways in human cancer are extremely heterogeneous, only a subset of patient correlate to activation of a particular pathway, e.g. <5% β-catenin, ~40% Stat3 etc. in human bladder cancer. In fact, the cancer stem cell marker CD44 was only shown to be expressed in ~40.4% of human bladder cancer that we analyzed. Interestingly, when we first submitted our manuscript for publication, one of the reviewers' comments was "since CD44 and these self-renewal proteins are expressed at such low percentage of human bladder cancer, the significance of these markers are not relevant at all". In this section, we would like to highlight that in human cancer specimens, heterogeneity in active pathways among patients is rather common, reflecting the complexity of human cancer development.

7.3.1 Wnt/β-catenin Signaling

Wnt canonical pathway is normally in a state of default repression, under the repression of the GSK-3, axin and APC complex which targets β-catenin for ubiquitination degradation. Upon Wnt binding to the Frizzled receptors and co-receptors, β-catenin degradation is blocked, stabilized, and translocated into the nucleus, where it heterodimerizes with Tcf-LEF to drive transcription. β-catenin has originally been proposed to mark intestinal stem cell in the crypt. This pathway was long known to play an important role in familial adenomatous polyposis of colon, due to common germline mutation of APC. However, in non-familial colorectal cancer, nuclear active form of β-catenin was only evident in 39% (21/54) and 26.9% (32/118) of all cases analyzed [40]. In breast cancer and non-small cell lung cancer, it was demonstrated that nuclear β-catenin was found in only 60% (74/123) [41] and 7% (16/261) [42] respectively, supporting our observation that activation of self-renewal pathways are heterogeneous among patients. In human chronic myelogenous leukemia, nuclear β-catenin was demonstrated to confer self-renewal of granulocyte–macrophage progenitors, as demonstrated by their ability to replate in colony-forming assay *in vitro* [43]. In mouse model of skin cancer, genetic deletion of β-catenin not only depleted CSCs but also abrogated tumor regression, indicating that nuclear β-catenin is crucial for maintenance of malignant squamous cell carcinomas. Collectively, these results revealed an important role of β-catenin signaling in maintaining the self-renewal of CSCs in both leukemia and solid cancer.

7.3.2 Sonic Hedgehog (SHH) Signaling

SHH is a ligand that binds to the Patched-1 (PTCH1) receptor. Without SHH, PTCH1 is a repressor that inhibits Smoothened (SMO), which is essential in activating the downstream Gli family of transcription factors. SHH pathway has been implicated in a wide variety of human tumors including glioblastoma, breast cancer, pancreatic adenocarcinoma, multiple myeloma, and chronic myeloid leukemia (CML) [44–46]. Hh pathway has been reported in maintaining the self-renewal and tumorigenic potential of cancer stem cells. For instance, depletion of Smo gene disrupts SHh signaling and consequently diminishes the self-renewal in leukemic stem cells [46]. Similarly, SHh ligand activation or GLI1 inhibition upregulates another stemness gene BMI-1 and consequently mediates self-renewal and tumorigenic potential of mammary cancer stem cells [47]. Conversely, cyclopamine mediated inhibition of SMO activation led to loss of tumorigenicity in glioblastoma cells [48]. Intriguingly, Hh signaling mediates crosstalks among CSCs, tumor cells and the microenvironment in multiple myeloma model, and its inhibition led to loss of CSC self-renewal [45]. Moreover, inhibition of Hh pathway diminished EMT, invasion, and metastasis potential in pancreatic cancer [49]. Collectively, active SHH signaling seemed to confer CSC self-renewal in several cancer types.

7.3.3 Notch Signaling

Notch pathway is instrumental to maintain various developmental processes, such as cell fate specification, and self-renewal of stem cells. Notch receptor is a transmembrane protein with a large extracellular domain. Upon activation through direct binding to ligand such as Delta and Jagged, proteolytic cleavage of the extracellular domain occurs, which releases the intracellular domain that translocates into nucleus to activate gene transcription. Aberrant Notch pathway signaling has been implicated in multiple cancer types, including T cell acute lymphoblastic leukaemia [50], breast [51], cervix [52, 53], prostate tumors [53] and pancreatic cancers [54]. A role for Notch in breast and glioma CSCs is well established, it is reported that Notch inhibitor diminished serial passaging of secondary mammosphere formation in breast cancer specimens, and the ability to form multilineage mammospheres [55]. Interestingly, co-operation of the Notch and EGFR/Erbb2 signaling seemed to be important for conferring maintenance of breast cancer stem cells [56]. In glioblastomas and medulloblastomas, inhibition of Notch led to depletion of CSCs [57], and sensitized CSC to radiation-induced apoptosis, but not bulk tumor cells [58]. Collectively, these results established a role for Notch signaling in maintaining CSC self-renewal and intrinsic radiation resistance.

7.3.4 Stat3 Signaling

Signal transducer and activator of transcription 3 (Stat3) is a latent transcription factor that localizes in the cytoplasm in an in active form. Upon receptor or non-receptor tyrosine kinase activation, Stat3 rapidly homodimerizes or heterodimerizes with Stat1. This dimerization masks the nuclear export signal, and enables its translocation into the nucleus to activate gene transcription. Stat3 was demonstrated to maintain self-renewal of mouse embryonic stem cell [59] and was shown to be constitutively activated in a wide spectrum of epithelial cancers. We have previously demonstrated that Stat3 plays multiple roles during different stages of tumor development in mouse skin model [60, 61]. Recently, its role on cancer stem cells have been explored in breast cancer and glioblastoma [62]. In breast cancer, a large scale loss-of-function screen led to the finding that a number of proteins (e.g. IL-6) when inhibited, reduced Stat3 activity in CD44+CD24– breast CSCs. Inhibition of the JAK/Stat3 pathway in breast CSCs led to a reduction in xenograft tumor growth [62]. In glioblastomas, Stat3 was demonstrated to be upregulated by the non-receptor kinase bone marrow X-linked (BMX) in glioblastoma stem cells [63]. Knock down of BMX inhibited Stat3 activity, and repressed expression of other stem cell-related transcription factor e.g. Sox2, Oct4. Interestingly, BMX seemed to be exclusively important to maintain self-renewal of glioblastoma stem cells but not that in normal neural progenitor cells. This is important, since Stat3 was previously established to be important for glioblastoma self-renewal [64, 65], but this transcription factor is more ubiquitously expressed. Identification of BMX as Stat3's upstream activator provided a more specific approach to knock down this pathway in glioblastoma cancer stem cells [63]. Collectively, these results link a role for Stat3 in the self-renewal of CSCs. Since Stat3 also mediates other important properties (e.g. proliferation, anti-apoptosis) of cancer cells, targeted therapy to Stat3 will likely affect both CSCs and downstream differentiated tumor cells.

7.4 Cancer Stem Cell Niche

7.4.1 Normal Stem Cell Niche

The early concept of a stem cell "niche" was proposed by Schofield in 1978, when he explored the microenvironment in modulating hematopoietic stem cells [66]. The niche was then thought to be a physical microenvironment for harboring or anchoring of stem cells to maintain their stemness. One of the most well established stem cell niche is in the Drosophila testis; a cluster of 10–12 cells form the "hub" [67], which are in direct contact with basement membrane distally and form a "niche" for germline stem cells (GSCs) and progenitor somatic stem cells (SSCs) [67, 68]. The hub cells secret paracrine signals to regulate surrounding GSCs and SSCs, and these stem cells differentiate as they are further away from the hub niche [68]. In adult epithelial tissues, the bulge region stem cells and the dynamic interaction of its dermal papilla niche in skin is rather intriguing. During catagen phase (degrowth) of the hair cycle, the lower follicle degenerates, allowing the dermal papilla niche cells to come in contact with bulge stem cells [69]. Paracrine signals from dermal papilla stimulate the next round of anagen (hair growth), when early progenitors from the bulge stem cells re-initiate hair shaft and differentiation [69]. Interestingly, recently lineage-tracing experiments revealed that these early progenitors home back to the niche after hair growth stops, retains stemness and become the primary stem cells for the next hair cycle [70].

7.4.2 *Cancer Stem Cell Niche*

Similar to normal stem cells, cancer stem cells are proposed to be nourished by a physical or physiological niche (Fig. 7.1c). Indeed, the concept of tumor micro-environment has long been established. Stephen Paget hypothesized in 1889 that tumor cells ("seeds") preferentially metastasize into permissive microenvironment ("fertile soil")—the famous Seed and Soil Hypothesis [71]. This seminal finding suggested the intricate reciprocal relationship between tumor cells and their stromal mileu [71]. Despite this early finding, the microenvironment was considered to provide passive influence on the tumor progression. Recent studies now demonstrate an active interplay between cancer cells and the tumor microenvironment. The tumor microenvironment is characterized by both acellular and cellular components [72]. Non-cellular components include extracellular matrixes such as collagen and tenascin. Major cell components in the stroma include fibroblasts, endothelial cells, pericytes, infiltrating inflammatory and immune cells [72]. It is likely the cancer stem cell niche is more dynamic than that of a normal stem cell niche, and may comprise multiple components depending on the state of cancer progression.

7.4.2.1 Perivascular Niche

Brain cancer stem cells were found to be in close proximity to tumor vasculature and vascular endothelial cells within the stroma to maintain their self-renewal and tumorigenecity [73]. More intricate association was evident as Bevacizumab, a VEGF neutralizing antibody, drastically suppressed the CSC pool self-renewal and subsequently tumor growth via depletion of blood vessels and angiogenesis [73]. In another study, targeting the vascular niche by anti-angiogenic therapy in combination with chemotherapy showed much better efficacy in eradicating tumor growth than either treatment alone. These evidences support the concept of targeting the perivascular niche, which likely lead to loss of CSC stemness and intrinsic drug resistance, and therefore sensitize CSCs to chemotherapy [74]. Further examination of the perivascular niche revealed that paracrine signals from endothelial cells mediate Notch signaling in CSCs, which protected glioblastoma CSCs from radiation-induced damage [75]. These results clearly establish a beneficial role for endothelial cells in the niche to maintain stemness of brain CSCs. Reciprocally, it has been shown that glioma cells when co-cultured with endothelial cells, can rescue them from hypoxia-induced apoptosis [76]. Altogether, these findings shed light on a dynamic interplay between brain CSCs and its niche endothelial cells, which reciprocally protect each other from therapeutic or physiological stress such as hypoxia (Fig. 7.2).

7.4.2.2 Hypoxic Niche

Hypoxia is a common feature of solid cancer when the demand of oxygen supply exceeds the capacity of tumor vasculature. It is an independent prognostic factor for advanced disease progression and poor survival. Tumor cells respond to hypoxia by

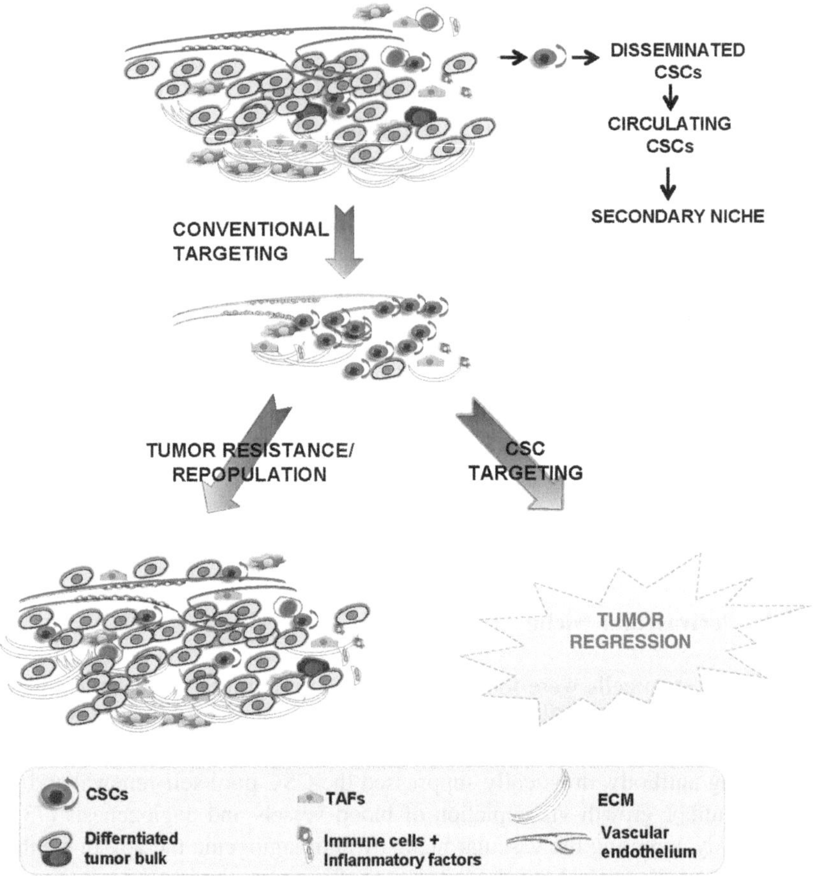

Fig. 7.2 CSCs in metastatic cascade, therapeutic resistance and its niche. Model summarizing the putative role of CSCs in the initiation of metastatic cascade, and its interaction with the cancer stem cell niche that induces therapeutic resistance. Novel therapies targeting both the CSCs and its niche will likely sensitize CSCs to cytotoxic therapy, CSC exhaustion and regression of tumor

upregulating Hypoxia-inducible factors (HIFs) in attempt to restore oxygen homeostasis, through induction of glycolysis and angiogenesis; HIFs also induce reduction in proliferation and increase in apoptosis of tumor cells, likely contributing to the selection of tumor colonies with selective growth advantage. Interestingly, increased expression of HIF-1α and HIF-2α was evident in neuroblastoma [77, 78] and glioma stem cells [79]. HIF-2α was particularly important in maintaining the undifferentiated state of CSCs as evident by upregulation of neural crest stem cell markers [78], and to maintain the self-renewal of CSCs *in vitro* and to support xenograft growth *in vivo* [79]. Based on these findings and those discussed in perivascular niche section, it is likely that hypoxia induces a positive feedback loop from CSCs that protects endothelial cells in the perivascular niche, which in turn initiate a vicious cycle of positive feedback between CSCs and the endothelial cells to mediate CSC self-renewal and therapeutic resistance (Fig. 7.2).

7.4.2.3 Cancer Stem Cells Form Its Own Niche

Neural stem cells are known to be multipotent, that can give rise to neurons, astrocytes, oligodendrocytes, glial cells, and to a smaller extent, endothelial cells [80]. Remarkably, glioblastoma stem cells were currently found to retain properties of normal neural stem cells, and can differentiate into phenotypic and functional vascular endothelial cells both *in vitro* and *in vivo* [81]. The proportion of CSCs derived vascular endothelial cells can range from 20% to 90% and carry the same genomic alteration as tumor cells, supporting their clonal origin [81]. This striking finding possibly sheds light to the complex dynamics between CSCs and their different niches; subpopulation of metastatic CSCs are highly migratory. It is reasonable to rationalize that during metastatic colonization, this intriguing niche-forming ability of CSCs likely play a major role in establishing their initial environment at a foreign site.

7.5 Clinical Significance of Cancer Stem Cells

7.5.1 Cancer Stem Cells in Metastasis and Epithelial Mesenchymal Transition (EMT)

Metastasis, spread of cancer cells from primary site to distal sites as secondary malignant growth, is responsible for over 90% of deaths in solid cancer patients. During metastasis cascade, tumor cells in primary site acquire a number of timely and spatially well-orchestrated events: gain of migratory of invasive ability via a process known as epithelial mesenchymal transition or through other mechanisms, intravasate into blood or lymphatic system, survive in the vasculature, extravasate to preferred tissues, adapt to the new microenvironment of distant tissues, and subsequent clonogenic growth in the distant site [82, 83]. Since CSCs are the tumor-initiating cells at the primary site, it is conceivable that disseminating cells may be CSCs and eventually metastatic seeding cells for metastatic formation (Fig. 7.2).

Indeed, Balic et al. demonstrated that disseminated cancer cells found in the bone marrow of early-stage breast cancer patients are enriched for phenotypic CD44+CD24− cancer stem cells [84]. These results support the hypothesis that a subpopulation of cancer stem cells may be the metastatic forming cells that intravasate into the vasculature and eventually become overt metastasis (Fig. 7.2). A number of prior studies have implicated the chemokine receptor CXCR4 in bone marrow homing and metastatic spread. Interestingly, several investigators have described metastatic subpopulation of cancer stem cells in different tissue types. In pancreatic cancer, it has been shown that CXCR4 can subfractionate CD133+ cancer stem cells into CD133+CXCR4+ and CD133+CXCR4− subpopulations [31]. There is no difference in the ability of these two populations to initiate primary xenografts; however, CD133+CXCR4+ pancreatic tumor cells can initiate metastasis formation in distal sites [31]. These results suggest that CXCR4 is not necessary for primary tumor growth, but is important for metastatic spread.

Independently, it has been shown that in colorectal cancer, primary tumors with distal metastasis consistently express the glycoprotein CD26 in comparison to non-metastatic tumors [85]. The presence of CD26+ colorectal tumor cells is significantly associated with poor tumor differentiation and microscopic vascular invasion. Remarkably, CD26+ CSCs express phenotypic EMT markers such as loss of E-cadherin, concomitant upregulation of N-cadherin, Twist and Slug, and are more migratory and invasive in transwell migration and matrigel invasion assays [85]. When co-transplanted with intestinal fibroblasts, these CD26+ CSCs readily form liver metastasis [85]. Collectively, these results provide evidence for a subpopulation of cancer stem cells in the initiation of metastatic cascade in several cancer types (Fig. 7.2).

7.5.2 Resistance to Radiation and Chemotherapy

Undoubtedly, one of the major obstacles to the eradication of advanced stage cancer is drug resistance. Since cancers contain a heterogeneous population of cells, certain fraction of tumor cells either acquire selective advantage or possess cell intrinsic properties to survive following cytotoxic therapies. A large body of evidence demonstrates a variety of mechanisms to therapeutic resistance: elevated DNA damage response and repair, altered cell cycle checkpoint control, cellular quiescence, acquired anti-apoptotic resistance, and overexpression of multi-drug transporters to facilitate drug efflux [86].

Based on the cancer stem cell hypothesis, cancer stem cells intrinsically adopt selective advantages that allow them to repopulate residue tumors following cytotoxic treatment (Fig. 7.2). One of the mechanisms for intrinsic drug resistance of CSCs is that they harbor ATP-binding cassette (ABC) family transporters to efflux intracellular drug to escape cytotoxic toxicity. These ABC transporters effectively efflux a number of commonly used chemotherapeutic agents such as vinblastine, doxorubicin, daunorubicin, actinomycin-D, and paclitaxel. [87]. Independently, it was demonstrated that in breast patients that received neoadjuvant chemotherapy, the percentage of phenotypic CSC significantly increased after therapy [88]. Interestingly, breast cancer patients who received the EGFR/HER2 inhibitor lapatinib showed a decrease in phenotypic CSC and mammosphere forming efficiency [88], suggestive of the EGFR/HER2 signaling in conferring chemoresistance and stemness.

Separately, in response to radiation therapy, brain CSCs were found to have elevated DNA damage response, could avoid accumulation of DNA damage or possess intrinsic anti-apoptotic mechanisms that all contributed to their therapeutic resistance. In response to radiation, glioma stem cells and non-stem cells accumulated the same amount of DNA damage, however glioma stem cells preferentially activated the DNA damage checkpoint kinases Chk1 and Chk2 [89]. Specific blockade of these Chk kinases removed the radioresistance properties of CD133+ glioma stem cells [89]. Interestingly, studies also outline that more pronounced ATM-Chk2-p53 DNA damage responses were observed in clinical specimens obtained from advanced stages human tumors from the urinary bladder, breast, lung and colon [90]. It is therefore reasonable to speculate that advanced stage cancer may harbor a higher fraction of CSCs and subsequent radioresistance.

Breast cancer stem cells, similar to normal mammary stem cells, intrinsically contain lower level of reactive oxygen species (ROS) via upregulation of free radical scavenging systems [91]. This markedly increases the viability of breast CSCs following radiation, since ROS is a critical mediator for ionizing radiation induced killing of cancer cells. Phillips et al. showed similar results in CSCs from two immortalized breast cancer cell lines, and proposed Notch1 signal in mediating this radiation resistance [92]. Independently, the Wnt/beta-catenin pathway was shown to confer radiation resistance in mammary progenitor cells and breast CSCs [93], possibly through upregulation of an anti-apoptotic protein, survivin [93, 94].

7.5.3 Cancer Stem Cell Gene-Signature and Prognosis Significance

Global gene expression profiling of solid cancer has led to major success in the sub-classification of cancer in molecular subtypes and the concept of personalized medicine. There was high expectation that oncologists can utilize genomic and molecular information generated from patient cancer tissues, to assist them in tailoring treatment options that can favor individual patient's clinical outcome. Recently, the US Food and Drug Administration (FDA) has approved the first clinical use of a gene expression test, Onco*type* DX, which is a 21-gene signature that was shown to predict breast cancer patients' risk of recurrence [95] and their benefit from chemotherapy [96]. Since cancer stem cells are the drivers for tumor initiation, it is reasonable to rationalize that global gene expression profile of CSCs likely reveal novel insights into prognosis that are linked to its biology. Liu et al. derived a 186-gene signature by comparing breast CSCs and normal breast epithelium, and found that this signature can predict overall survival and metastasis-free survival in breast cancer [97]. This gene signature is also associated with prognosis in other cancer types, including neuroblastoma, lung cancer, and prostate cancer [97], suggesting overlapping signaling pathways in driving tumorigenesis in different cancer types. Our laboratory has reported a bladder CSC gene-signature by comparing CSCs and paired non-CSCs within the same cancer specimens. This gene-signature is able to statistically and effectively segregate muscle-invasive cancer from non-invasive cancer [20]. More importantly, the bladder CSC signature can predict overall survival of non-invasive bladder cancer [20]. Collectively, these results support the notion that genetic information from a subpopulation of cancer cells indeed shed light to the overall biology, in particular invasive and metastasis properties of solid cancer.

Independently, numerous studies have reported substantial clinical relevance in using immunohistochemistry or other methods to correlate CSC frequency in primary tumor sections or micrometastasis with clinical outcome. In head and neck cancer, high frequency of the Lin-CD44+ CSCs correlate with known poor prognostic factors such as advanced T stage classification and recurrence [98]. In oral and hypopharynx cancer, the CSC marker CD44 significantly associates with decrease 5-year survival [99]. In glioblastomas, co-expression of the CSC marker CD133 and proliferation marker Ki67 are significant independent prognostic factors for overall and disease-free survival [100]. In breast cancer, the expression of ALDH/CD44+ CSCs in bone marrow micrometastasis significant predict high risk cancer patients [101]. These are only isolated examples of a large series of studies revealing the clinical relevance of CSCs to clinical prognosis. Collectively, these findings point to various biological roles of CSCs in invasion, tumor progression, and metastasis, which are major contributing factors to overall survival.

7.5.4 Targeted Therapies Against Cancer Stem Cells

Since cancer stem cells were demonstrated to effectively escape cytotoxic therapies by various mechanisms, identification of novel therapies directed against CSCs may be useful to sensitize them to cytotoxic or other conventional therapies. A number of studies have aimed at targeting specific self-renewal pathways in different cancer types, e.g. Notch signaling by γ-secretase inhibitors, Hedgehog pathway by cyclopamine and Stat3 signaling by upstream inhibition or small molecule inhibitors. When targeting CSC self-renewal pathways in combination with chemo or radiation, these studies all demonstrated favorable treatment outcome, in comparison to single treatment control alone. Targeting of CSCs led to differentiation, that decreases intrinsic resistance and therefore sensitize CSCs to cytotoxic therapies.

Further, a number of studies have employed small to large scale screening of small molecule inhibitors or natural extracts to identify novel targets for CSCs. One study utilized the in vitro properties of normal neural spheres to screen chemicals that can inhibit its proliferation and also brain cancer stem cells in vitro; interestingly, a compound affect the neurotransmission pathways was identified [102]. Independently, other group took advantage of the EMT properties of CSC and performed their first screen (>16,000 compounds) on immortalized breast epithelial cells that were induced to undergo EMT by knockdown of E-cadherin. The compound Salinomycin was identified with a unique efficacy against CSCs by decrease its proportion by 20-fold in comparison to control, in contrast chemotherapies increased the proportion of phenotypic CSCs. Salinomycin was also effective in decreasing tumorspheres *in vitro*, xenograft tumor growth and metastatic nodule formation *in vivo* [103]. Recently, a small kinase inhibitor screen revealed the compound BI 2536, that inhibit polo-like kinase 1, has good cytotoxic effect against neuroblastoma stem cells, while it is relatively non-cytotoxic to normal pediatric neural stem cells [104]. Another approach exploits the physical properties to target CSCs, it was found that short-term local hyperthermia could effectively target CSC and sensitize them for ionizing radiation. Intravenous delivery of optically activated gold nanoshells in combination to radiation treatment induced differentiation of CSCs, and showed inhibitory effect on xenograft tumorigenecity [105].

Since CSCs are in dynamic interaction with its niche, we hypothesize that in addition to direct targeting of CSCs, combinatory targeting of signals from cancer stem cell niche and factors from CSCs will likely disrupt the interaction in both compartments and will be able to eradicate the most primitive CSCs in the niche.

7.6 Controversies in the Cancer Stem Cell Model

7.6.1 Frequency of Cancer Stem Cells, Genotype and Cell-of-Origin

A number of studies posed challenges to the cancer stem cell model. One of them utilized oncogene-driven mouse pre-B/B cell lymphomas, thymic lymphomas and *PU.1*$^{-/-}$ tumor suppressor driven leukemia to investigate the frequencies of cancer stem cells or tumor-initiating cells that can transplant cancer in congenic mice. Depending on individual tumor model, the authors found that rare unfractionated tumor cells, down to a single-cell up to ten-cell level could transplant cancer [106]. Their conclusion was that in the majority of cancer stem cell studies, the utilization of immunocompromised mice may undermine the effect of immune microenvironment and therefore the frequency and even the very existence of cancer stem cells is now debatable [106]. We will be very cautious about these interpretations, since it should be noted that different tumor models have their own strengths and weaknesses. For instance, in tumor models where specific oncogenes were driven to a certain tissue specific expression, especially when tumor incidences occur rapidly; it is likely that individual tumor cells are genetically identical (without sufficient time for accumulating other alterations unless genetic instability is involved) and their expression are driven to the same cell types. It is unclear how such tumor models are analogous to human cancer development, where intra-tumoral heterogeneity is clearly observed. These models may be relevant to investigate the role of specific oncogenes in mediating tumor development, however the relevance of using such tumor models in the investigation of cancer stem cells remained debatable. Further, the genes involved in these tumor models will definitely influence the tumor phenotype and tumor-initiating cell frequencies. For instance PU.1, a transcription factor of the ETS family, is clearly shown to be up-regulated in response to all-trans retinoic acid-induced leukemic differentiation. And overexpression of PU.1 can reverse differentiation block of leukemic blasts [107, 108]. These results clearly pointed to an essential role for PU.1 deficiency to induce block of differentiation during leukemogenesis. Given that, high frequencies of tumor-initiating cell will not be too surprising in *PU.1*$^{-/-}$ tumor suppressor driven leukemia model, where block of differentiation is evident.

Based on these findings, we rationalize that the type of genetic or epigenetic alterations, and the target cell expression are crucial determinants affecting the frequencies and immunophenotype of cancer stem cells. Indeed, in a study analyzing lung tumors-induced from three different genetic mouse models (EGFR mutant, Kras, Kras;p53$^{-/-}$); the authors found that Sca1-negative cells are the tumor-initiating cells for the EGFR mutant model, while all the cells can initiate tumors in Kras model, Sca1-positive cells are the tumor-initiating cells in the Kras;p53$^{-/-}$ model [109]. This study supported our hypothesis that genotype driving the cancer can determine the frequency and immunophenotype of cancer stem cells. Separately, it has been demonstrated in poorly-differentiated or high grade (G3) human breast cancer, the frequency of phenotypic cancer stem cells are much higher in comparison to lower grade (G1) breast cancer [110]. Collectively, these findings did not pose strong opposition, but indeed, support the cancer stem cell model, that the frequency of cancer stem cells is associated with the genotype driving cancer, the target cell expression and pathological grade of cancer.

7.6.2 Validity of Xenograft Transplantation Using Immunocompromised Mice for Cancer Stem Cell Research

It remained a central debate whether it is relevant to employ immunocompromised mice as a model to study human cancer stem cells, since the immune microenvironment is well established to play a major role in early stage cancer, in which cancer has to go through the "Three Es" of cancer immunoediting—elimination, equilibrium and escape [111]. Another series of landmark studies, primarily focusing on human melanoma, sparked the continuity of this heated debate in the field. Despite the original finding that melanoma-initiating cells can be prospectively isolated based on the expression of ABCB5 [112], Quintana E et al. demonstrated that depending on the severity of immunocompromised mouse models and the type of vehicles (e.g. matrigel) being used for tumor cell engraftment, the frequencies of melanoma-initiating cells can significantly vary. They concluded that single melanoma cell could engraft as xenografts given the appropriate environment, regardless of their immunophenotype. Interestingly, Boiko et al. later demonstrated that, even with the most severed immunocompromised mice available, and/or using a more relevant humanized model with human skin graft, CD271-positive melanoma cells recapitulate all the properties of cancer stem cells. Quintana E et al. later demonstrated that only 28% of single melanoma cells directly from patient could engraft as xenografts. This suggested that only a subpopulation of melanoma cells are tumor-initiating; however, after screening a long list of cell surface markers including ABCB5 and CD271, although heterogeneity in the expression of these markers were found, but they found no evidence in the functional difference in the relative ability of melanoma subpopulations to form xenograft tumors. These results are in direct contradiction to that reported by Boiko et al. and Schatton et al. Even more interestingly, Quintana E et al. found that melanoma cell are rather plastic in their immunophenotype, CD271+ cells can form negative cells and vice versa. Most recently, Civenni G et al. reported that in less severe immunocompromised mouse models [e.g. nude or nonobese diabetic/severe combined immunodeficient (NOD/SCID) mice], only CD271+ cells could initiate xenograft; while in more severely immunocompromised mouse models (e.g. NOD/SCID/IL2rg null), both CD271+ and CD271− cells can initiate xenografts. More importantly, they found that in patient specimens, the positive expression of CD271 significantly correlated with worse patient survival outcome. Separately, Boiko et al. also demonstrated that CD271+ cells lack the expression of current immunotherapy targets TYR, MART1 and MAGE. It is clear that the existence of melanoma-initiating cells remains controversial, however it is also evident that the biology of CD271+ melanoma cells seemed unique and may reveal new insight to the therapeutic targeting of melanoma. It is obviously not ideal to utilize immunocompromised mice for cancer research, since it isolates the significance and interaction of the immune microenvironment in cancer development. Development of humanized mouse models with intact humanized immune system may be more appropriate for studying melanoma stem cells; however, as previously discussed, individual tumor model has its own strengths and limitations depending on the biological questions to be asked. Other technical issues may need to be pondered, e.g. the tumor dissociation protocol, that may involve the timing of trypsin digestion and cleavage of cell surface receptors; this may lead to false negativity of a certain cell surface receptor when analyzed by FACS. Also, the environment and conditioning of immunocompromised mice for tumor injection may play a role, e.g. the type of matrigel or growth factors, or humanization of stromal environment has been well known to influence xenograft engraftments.

7.6.3 Plasticity, Spontaneous Conversion, or Reprogramming of Cancer Stem Cells

Recently, Yamanaka and colleagues revolutionized stem cell biology by demonstrating that four transcription factors can induce the generation of pluripotent stem cells from terminally differentiated or adult somatic cells [113]; these cells are known as induced pluripotent stem cells (iPS cells) [113]. This raises a possibility that during cancer development, instead of the conventional view that requires accumulation of a series of genetic and epigenetic alterations, cancer may result from reprogramming given the precise combination of genetic and epigenetic events. Indeed, c-myc, one of the four transcription factors originally described in the generation of iPS cells is a potent oncogene which is also described to repress differentiation program via histone modification mechanisms [114].

It was demonstrated that in immortalized human mammary epithelial cells, induction of epithelial-mesenchymal transition by transcription factors (e.g. Snail, Twist) and cytokines (e.g. TGF-β) led to a phenotypic stem cell phenotype and higher efficiency in mammospheres formation [115]. Interestingly, steady state normal mammary stem cells endogenously express higher level of EMT markers. Of relevance to cancer is that mammary epithelial cells overexpressing these EMT

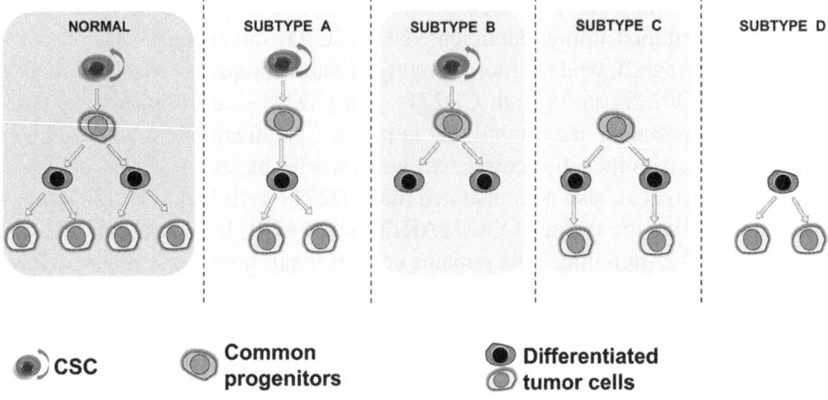

Fig. 7.3 Hypothetical model of cancer stem cells and multiple cancer subtypes. Normal stem cells give rise to common progenitor that eventually become more committed differentiated cells. *Subtype A*. Cancer contains a full hierarchical organization of stem, progenitor and differentiated cells. *Subtype B*. An alteration induces blocks of differentiation of the cancer stem cells and allows the expansion of cancer stem cell pool. *Subtype C*. Cancer lacks the stem cell compartment, which contains the progenitor and differentiated cells only. *Subtype D*. Cancer contains only differentiated cells

transcription factors, when assayed, form colonies in soft agar and xenograft tumors more efficiently [115]. These results led to the fundamental question whether cancer stem cell properties is an intrinsic stable phenotype, or other biological-driven mechanisms (e.g. EMT) can induce the cancer stem cell phenotype from non-stem cells. Another group utilized the H3K4 demethylase JARD1B as a marker and found significant overlap of JARD1B with melanoma label-retaining cells [116]. With a GFP reporter construct to trace and visualize viable JARD1B+ melanoma cells, they found that these cells recapitulate all the functional properties of cancer stem cells [116]. Knockdown of this gene led to an initial acceleration of xenograft tumor growth, followed by exhaustion upon serial passaging [116]. However, surprising finding revealed that expression of JARID1B seemed to be very dynamic, JARID1B− cells were able to give rise to JARID1B+ cells, although the majority of cells remained JARID1B− [116]. This led to the view that in certain cancer type, such as melanoma cells, whether their phenotype is more dynamically regulated than we have expected. In contrast, unpublished results from our laboratory revealed a hierarchical organization of stem, progenitor and differentiated cancer cells in human bladder cancer specimens, which can be prospectively isolated and characterized. Even more interestingly, not all bladder cancers contain all cellular compartments (Fig. 7.3); using these newly identified markers defining different stages of differentiation, we found at least several different subtypes of bladder cancers (Fig. 7.3). Certain cancer contains a full hierarchical organization of stem, progenitor and differentiated cells (Fig. 7.3, Subtype A); certain cancer contains only the stem and progenitor cells, with a potential block of differentiation (Fig. 7.3, Subtype B); certain cancer lacks the stem cell compartment, which contains the progenitor and differentiated cells only (Fig. 7.3, Subtype C); certain cancer contains only differentiated cells (Fig. 7.3, Subtype D). We hypothesize that there will likely be more subtypes of bladder cancer, depending on the type of genotype driving their development. It is often misunderstood that the cancer stem cell concept attempts to over simply the reality of cancer development, which is not entirely true in our opinion.

7.7 Concluding Remarks

Paradigm shifting concept always generates controversies. Remarkably, after more than 40 years since the cancer stem cell concept was originally proposed, with mounting clinical evidence revealing unique biological relevance of these cells in therapeutic resistance and other processes such as metastasis, some cancer researchers consider the very existence of these CSCs a myth or misconception. It is possible that certain cancer types follow the cancer stem cell model and some do not. To view this heated controversy from a different perspective; it is clear that most, if not all, epithelial tissues contain normal stem, progenitor and differentiated cells; and cancer arises from this hierarchical organization of cells. It will be reasonable to rationalize that in any type of cancers with a normal counterpart of stem cell hierarchy, the resulting cancers likely retain a certain feature of normal hierarchical organization, and unquestionably less defined as that in the normal tissue counterpart due to additional genetic and epigenetic alterations. In our opinion, the cancer stem cell concept does not exclude the co-existence of clonal evolution; instead, preliminary results from our laboratory and other published literatures suggest that these two concepts are not mutually exclusive [117, 118] (Fig. 7.1c). The ultimate proof or disproof of the cancer stem cell concept likely lies in the clinical relevance of these tumor cell subpopulation, their prognostic value, and whether targeted therapies toward CSS will improve patient survival and revolutionize future anti-cancer therapies.

References

1. Heppner GH (1984) Tumor heterogeneity. Cancer Res 44(6):2259–2265
2. Hamburger AW, Salmon SE (1977) Primary bioassay of human tumor stem cells. Science 197(4302):461–463
3. Riley RS (1992) Cellular proliferation markers in the evaluation of human cancer. Clin Lab Med 12(2):163–199
4. Corson JM (1986) Keratin protein immunohistochemistry in surgical pathology practice. Pathol Annu 21(Pt 2):47–81
5. Fidler IJ (1978) Tumor heterogeneity and the biology of cancer invasion and metastasis. Cancer Res 38(9):2651–2660
6. Von Hoff DD et al (1983) Prospective clinical trial of a human tumor cloning system. Cancer Res 43(4):1926–1931

7. Berstine EG et al (1973) Alkaline phosphatase activity in mouse teratoma. Proc Natl Acad Sci USA 70(12):3899–3903
8. Mintz B, Illmensee K (1975) Normal genetically mosaic mice produced from malignant teratocarcinoma cells. Proc Natl Acad Sci USA 72(9):3585–3589
9. Illmensee K, Mintz B (1976) Totipotency and normal differentiation of single teratocarcinoma cells cloned by injection into blastocysts. Proc Natl Acad Sci USA 73(2):549–553
10. Stern PL et al (1978) Monoclonal antibodies as probes for differentiation and tumor-associated antigens: a Forssman specificity on teratocarcinoma stem cells. Cell 14(4):775–783
11. Herzenberg LA, Sweet RG (1976) Fluorescence-activated cell sorting. Sci Am 234(3):108–117
12. Spangrude GJ, Heimfeld S, Weissman IL (1988) Purification and characterization of mouse hematopoietic stem cells. Science 241(4861):58–62
13. Smith LG, Weissman IL, Heimfeld S (1991) Clonal analysis of hematopoietic stem-cell differentiation in vivo. Proc Natl Acad Sci USA 88(7):2788–2792
14. Baum CM et al (1992) Isolation of a candidate human hematopoietic stem-cell population. Proc Natl Acad Sci USA 89(7):2804–2808
15. Lapidot T et al (1994) A cell initiating human acute myeloid leukaemia after transplantation into SCID mice. Nature 367(6464):645–648
16. Al-Hajj M et al (2003) Prospective identification of tumorigenic breast cancer cells. Proc Natl Acad Sci USA 100(7):3983–3988
17. Singh SK et al (2004) Identification of human brain tumour initiating cells. Nature 432(7015): 396–401
18. Dalerba P et al (2007) Phenotypic characterization of human colorectal cancer stem cells. Proc Natl Acad Sci USA 104(24):10158–10163
19. Prince ME et al (2007) Identification of a subpopulation of cells with cancer stem cell properties in head and neck squamous cell carcinoma. Proc Natl Acad Sci USA 104(3): 973–978
20. Chan KS et al (2009) Identification, molecular characterization, clinical prognosis, and therapeutic targeting of human bladder tumor-initiating cells. Proc Natl Acad Sci USA 106(33):14016–14021
21. Patrawala L et al (2006) Highly purified CD44+ prostate cancer cells from xenograft human tumors are enriched in tumorigenic and metastatic progenitor cells. Oncogene 25(12):1696–1708
22. Boiko AD et al (2010) Human melanoma-initiating cells express neural crest nerve growth factor receptor CD271. Nature 466(7302):133–137
23. Li C et al (2007) Identification of pancreatic cancer stem cells. Cancer Res 67(3):1030–1037
24. Ho MM et al (2007) Side population in human lung cancer cell lines and tumors is enriched with stem-like cancer cells. Cancer Res 67(10):4827–4833
25. Clarke MF et al (2006) Cancer stem cells—perspectives on current status and future directions: AACR workshop on cancer stem cells. Cancer Res 66(19):9339–9344
26. Takaishi S et al (2009) Identification of gastric cancer stem cells using the cell surface marker CD44. Stem Cells 27(5):1006–1020
27. Hurt EM et al (2008) CD44+ CD24(−) prostate cells are early cancer progenitor/stem cells that provide a model for patients with poor prognosis. Br J Cancer 98(4):756–765
28. Prince ME, Ailles LE (2008) Cancer stem cells in head and neck squamous cell cancer. J Clin Oncol 26(17):2871–2875
29. Zhang S et al (2008) Identification and characterization of ovarian cancer-initiating cells from primary human tumors. Cancer Res 68(11):4311–4320
30. Du L et al (2008) CD44 is of functional importance for colorectal cancer stem cells. Clin Cancer Res 14(21):6751–6760
31. Hermann PC et al (2007) Distinct populations of cancer stem cells determine tumor growth and metastatic activity in human pancreatic cancer. Cell Stem Cell 1(3):313–323
32. Fuchs E, Coulombe PA (1992) Of mice and men: genetic skin diseases of keratin. Cell 69(6):899–902

33. Patrawala L et al (2005) Side population is enriched in tumorigenic, stem-like cancer cells, whereas ABCG2+ and ABCG2- cancer cells are similarly tumorigenic. Cancer Res 65(14):6207–6219
34. She JJ et al (2008) Identification of side population cells from bladder cancer cells by DyeCycle Violet staining. Cancer Biol Ther 7(10):1663–1668
35. Hiraga T, Ito S, Nakamura H (2011) Side population in MDA-MB-231 human breast cancer cells exhibits cancer stem cell-like properties without higher bone-metastatic potential. Oncol Rep 25(1):289–296
36. Ginestier C et al (2007) ALDH1 is a marker of normal and malignant human mammary stem cells and a predictor of poor clinical outcome. Cell Stem Cell 1(5):555–567
37. Rasper M et al (2010) Aldehyde dehydrogenase 1 positive glioblastoma cells show brain tumor stem cell capacity. Neuro Oncol 12(10):1024–1033
38. Jiang F et al (2009) Aldehyde dehydrogenase 1 is a tumor stem cell-associated marker in lung cancer. Mol Cancer Res 7(3):330–338
39. Ma S et al (2008) Aldehyde dehydrogenase discriminates the CD133 liver cancer stem cell populations. Mol Cancer Res 6(7):1146–1153
40. Kobayashi M et al (2000) Nuclear translocation of beta-catenin in colorectal cancer. Br J Cancer 82(10):1689–1693
41. Lin SY et al (2000) Beta-catenin, a novel prognostic marker for breast cancer: its roles in cyclin D1 expression and cancer progression. Proc Natl Acad Sci USA 97(8):4262–4266
42. Pirinen RT et al (2001) Reduced expression of alpha-catenin, beta-catenin, and gamma-catenin is associated with high cell proliferative activity and poor differentiation in non-small cell lung cancer. J Clin Pathol 54(5):391–395
43. Jamieson CH et al (2004) Granulocyte-macrophage progenitors as candidate leukemic stem cells in blast-crisis CML. N Engl J Med 351(7):657–667
44. Jiang J, Hui CC (2008) Hedgehog signaling in development and cancer. Dev Cell 15(6):801–812
45. Merchant AA, Matsui W (2010) Targeting Hedgehog—a cancer stem cell pathway. Clin Cancer Res 16(12):3130–3140
46. Zhao C et al (2007) Loss of beta-catenin impairs the renewal of normal and CML stem cells in vivo. Cancer Cell 12(6):528–541
47. Liu S et al (2006) Hedgehog signaling and Bmi-1 regulate self-renewal of normal and malignant human mammary stem cells. Cancer Res 66(12):6063–6071
48. Clement V et al (2007) HEDGEHOG-GLI1 signaling regulates human glioma growth, cancer stem cell self-renewal, and tumorigenicity. Curr Biol 17(2):165–172
49. Feldmann G et al (2008) An orally bioavailable small-molecule inhibitor of Hedgehog signaling inhibits tumor initiation and metastasis in pancreatic cancer. Mol Cancer Ther 7(9):2725–2735
50. Ellisen LW et al (1991) TAN-1, the human homolog of the Drosophila notch gene, is broken by chromosomal translocations in T lymphoblastic neoplasms. Cell 66(4):649–661
51. Weijzen S et al (2002) Activation of Notch-1 signaling maintains the neoplastic phenotype in human Ras-transformed cells. Nat Med 8(9):979–986
52. Zagouras P et al (1995) Alterations in Notch signaling in neoplastic lesions of the human cervix. Proc Natl Acad Sci USA 92(14):6414–6418
53. Santagata S et al (2004) JAGGED1 expression is associated with prostate cancer metastasis and recurrence. Cancer Res 64(19):6854–6857
54. Miyamoto Y et al (2003) Notch mediates TGF alpha-induced changes in epithelial differentiation during pancreatic tumorigenesis. Cancer Cell 3(6):565–576
55. Farnie G et al (2007) Novel cell culture technique for primary ductal carcinoma in situ: role of Notch and epidermal growth factor receptor signaling pathways. J Natl Cancer Inst 99(8):616–627
56. Osipo C et al (2008) ErbB-2 inhibition activates Notch-1 and sensitizes breast cancer cells to a gamma-secretase inhibitor. Oncogene 27(37):5019–5032
57. Fan X et al (2010) NOTCH pathway blockade depletes CD133-positive glioblastoma cells and inhibits growth of tumor neurospheres and xenografts. Stem Cells 28(1):5–16

58. Wang J et al (2010) Notch promotes radioresistance of glioma stem cells. Stem Cells 28(1):17–28
59. Niwa H et al (1998) Self-renewal of pluripotent embryonic stem cells is mediated via activation of STAT3. Genes Dev 12(13):2048–2060
60. Chan KS et al (2004) Disruption of Stat3 reveals a critical role in both the initiation and the promotion stages of epithelial carcinogenesis. J Clin Invest 114(5):720–728
61. Chan KS et al (2008) Forced expression of a constitutively active form of Stat3 in mouse epidermis enhances malignant progression of skin tumors induced by two-stage carcinogenesis. Oncogene 27(8):1087–1094
62. Marotta LL et al (2011) The JAK2/STAT3 signaling pathway is required for growth of CD44+CD24- stem cell-like breast cancer cells in human tumors. J Clin Invest 121(7): 2723–2735
63. Guryanova OA et al (2011) Nonreceptor tyrosine kinase BMX maintains self-renewal and tumorigenic potential of glioblastoma stem cells by activating STAT3. Cancer Cell 19(4):498–511
64. Villalva C et al (2011) STAT3 is essential for the maintenance of neurosphere-initiating tumor cells in patients with glioblastomas: a potential for targeted therapy? Int J Cancer 128(4):826–838
65. Sherry MM et al (2009) STAT3 is required for proliferation and maintenance of multipotency in glioblastoma stem cells. Stem Cells 27(10):2383–2392
66. Schofield R (1978) The relationship between the spleen colony-forming cell and the haemopoietic stem cell. Blood Cells 4(1–2):7–25
67. Kiger AA, White-Cooper H, Fuller MT (2000) Somatic support cells restrict germline stem cell self-renewal and promote differentiation. Nature 407(6805):750–754
68. Tran J, Brenner TJ, DiNardo S (2000) Somatic control over the germline stem cell lineage during Drosophila spermatogenesis. Nature 407(6805):754–757
69. Fuchs E, Tumbar T, Guasch G (2004) Socializing with the neighbors: stem cells and their niche. Cell 116(6):769–778
70. Hsu YC, Pasolli HA, Fuchs E (2011) Dynamics between stem cells, niche, and progeny in the hair follicle. Cell 144(1):92–105
71. Paget S (1989) The distribution of secondary growths in cancer of the breast. 1889. Cancer Metastasis Rev 8(2):98–101
72. Pietras K, Ostman A (2010) Hallmarks of cancer: interactions with the tumor stroma. Exp Cell Res 316(8):1324–1331
73. Calabrese C et al (2007) A perivascular niche for brain tumor stem cells. Cancer Cell 11(1):69–82
74. Folkins C et al (2007) Anticancer therapies combining antiangiogenic and tumor cell cytotoxic effects reduce the tumor stem-like cell fraction in glioma xenograft tumors. Cancer Res 67(8):3560–3564
75. Hovinga KE et al (2010) Inhibition of notch signaling in glioblastoma targets cancer stem cells via an endothelial cell intermediate. Stem Cells 28(6):1019–1029
76. Ezhilarasan R et al (2007) Glioma cells suppress hypoxia-induced endothelial cell apoptosis and promote the angiogenic process. Int J Oncol 30(3):701–707
77. Pietras A et al (2008) High levels of HIF-2alpha highlight an immature neural crest-like neuroblastoma cell cohort located in a perivascular niche. J Pathol 214(4):482–488
78. Pietras A et al (2009) HIF-2alpha maintains an undifferentiated state in neural crest-like human neuroblastoma tumor-initiating cells. Proc Natl Acad Sci USA 106(39):16805–16810
79. Li Z et al (2009) Hypoxia-inducible factors regulate tumorigenic capacity of glioma stem cells. Cancer Cell 15(6):501–513
80. Wurmser AE et al (2004) Cell fusion-independent differentiation of neural stem cells to the endothelial lineage. Nature 430(6997):350–356
81. Ricci-Vitiani L et al (2010) Tumour vascularization via endothelial differentiation of glioblastoma stem-like cells. Nature 468(7325):824–828
82. Chiang AC, Massague J (2008) Molecular basis of metastasis. N Engl J Med 359(26): 2814–2823

83. Shibue T, Weinberg RA (2011) Metastatic colonization: settlement, adaptation and propagation of tumor cells in a foreign tissue environment. Semin Cancer Biol 21(2):99–106
84. Balic M et al (2006) Most early disseminated cancer cells detected in bone marrow of breast cancer patients have a putative breast cancer stem cell phenotype. Clin Cancer Res 12(19): 5615–5621
85. Pang R et al (2010) A subpopulation of CD26+ cancer stem cells with metastatic capacity in human colorectal cancer. Cell Stem Cell 6(6):603–615
86. Dean M, Fojo T, Bates S (2005) Tumour stem cells and drug resistance. Nat Rev Cancer 5(4):275–284
87. Ambudkar SV et al (1999) Biochemical, cellular, and pharmacological aspects of the multidrug transporter. Annu Rev Pharmacol Toxicol 39:361–398
88. Li X et al (2008) Intrinsic resistance of tumorigenic breast cancer cells to chemotherapy. J Natl Cancer Inst 100(9):672–679
89. Bao S et al (2006) Glioma stem cells promote radioresistance by preferential activation of the DNA damage response. Nature 444(7120):756–760
90. Bartkova J et al (2005) DNA damage response as a candidate anti-cancer barrier in early human tumorigenesis. Nature 434(7035):864–870
91. Diehn M et al (2009) Association of reactive oxygen species levels and radioresistance in cancer stem cells. Nature 458(7239):780–783
92. Phillips TM, McBride WH, Pajonk F (2006) The response of CD24(–/low)/CD44+ breast cancer-initiating cells to radiation. J Natl Cancer Inst 98(24):1777–1785
93. Woodward WA et al (2007) WNT/beta-catenin mediates radiation resistance of mouse mammary progenitor cells. Proc Natl Acad Sci USA 104(2):618–623
94. Chen MS et al (2007) Wnt/beta-catenin mediates radiation resistance of Sca1+ progenitors in an immortalized mammary gland cell line. J Cell Sci 120(Pt 3):468–477
95. Paik S et al (2004) A multigene assay to predict recurrence of tamoxifen-treated, node-negative breast cancer. N Engl J Med 351(27):2817–2826
96. Paik S et al (2006) Gene expression and benefit of chemotherapy in women with node-negative, estrogen receptor-positive breast cancer. J Clin Oncol 24(23):3726–3734
97. Liu R et al (2007) The prognostic role of a gene signature from tumorigenic breast-cancer cells. N Engl J Med 356(3):217–226
98. Joshua B et al (2011) Frequency of cells expressing CD44, a head and neck cancer stem cell marker: correlation with tumor aggressiveness. Head Neck Feb 14
99. Kokko LL et al (2011) Significance of site-specific prognosis of cancer stem cell marker CD44 in head and neck squamous-cell carcinoma. Oral Oncol 47(6):510–516
100. Pallini R et al (2008) Cancer stem cell analysis and clinical outcome in patients with glioblastoma multiforme. Clin Cancer Res 14(24):8205–8212
101. Reuben JM et al (2011) Primary breast cancer patients with high risk clinicopathologic features have high percentages of bone marrow epithelial cells with ALDH activity and CD44(+)CD24(lo) cancer stem cell phenotype. Eur J Cancer 47(10):1527–1536
102. Diamandis P et al (2007) Chemical genetics reveals a complex functional ground state of neural stem cells. Nat Chem Biol 3(5):268–273
103. Gupta PB et al (2009) Identification of selective inhibitors of cancer stem cells by high-throughput screening. Cell 138(4):645–659
104. Grinshtein N et al (2011) Small molecule kinase inhibitor screen identifies polo-like kinase 1 as a target for neuroblastoma tumor-initiating cells. Cancer Res 71(4):1385–1395
105. Atkinson RL et al (2010) Thermal enhancement with optically activated gold nanoshells sensitizes breast cancer stem cells to radiation therapy. Sci Transl Med 2(55):55ra79
106. Kelly PN et al (2007) Tumor growth need not be driven by rare cancer stem cells. Science 317(5836):337
107. Durual S et al (2007) Lentiviral PU.1 overexpression restores differentiation in myeloid leukemic blasts. Leukemia 21(5):1050–1059
108. Cook WD et al (2004) PU.1 is a suppressor of myeloid leukemia, inactivated in mice by gene deletion and mutation of its DNA binding domain. Blood 104(12):3437–3444

109. Curtis SJ et al (2010) Primary tumor genotype is an important determinant in identification of lung cancer propagating cells. Cell Stem Cell 7(1):127–133
110. Pece S et al (2010) Biological and molecular heterogeneity of breast cancers correlates with their cancer stem cell content. Cell 140(1):62–73
111. Dunn GP, Old LJ, Schreiber RD (2004) The three Es of cancer immunoediting. Annu Rev Immunol 22:329–360
112. Schatton T et al (2008) Identification of cells initiating human melanomas. Nature 451(7176): 345–349
113. Takahashi K, Yamanaka S (2006) Induction of pluripotent stem cells from mouse embryonic and adult fibroblast cultures by defined factors. Cell 126(4):663–676
114. Smith KN, Singh AM, Dalton S (2010) Myc represses primitive endoderm differentiation in pluripotent stem cells. Cell Stem Cell 7(3):343–354
115. Mani SA et al (2008) The epithelial-mesenchymal transition generates cells with properties of stem cells. Cell 133(4):704–715
116. Roesch A et al (2010) A temporarily distinct subpopulation of slow-cycling melanoma cells is required for continuous tumor growth. Cell 141(4):583–594
117. Siegmund KD et al (2009) Inferring clonal expansion and cancer stem cell dynamics from DNA methylation patterns in colorectal cancers. Proc Natl Acad Sci USA 106(12):4828–4833
118. Odoux C et al (2008) A stochastic model for cancer stem cell origin in metastatic colon cancer. Cancer Res 68(17):6932–6941

Chapter 8
Genomics of Prostate Cancer

Kern Rei Chng, Shin Chet Chuah, and Edwin Cheung

Contents

Abstract Prostate cancer is one of the most common types of cancer afflicting the male population. Although prostate cancer is initially slow-growing and regresses upon androgen ablation, the disease is capable of transiting into an aggressive and metastatic form that is hormone refractory. Studies have attributed alterations in the cancer genome and transcriptome as being integral to this transition process. With

K.R. Chng • S.C. Chuah
Cancer Biology and Pharmacology, Genome Institute of Singapore, A*STAR (Agency for Science, Technology and Research), 60 Biopolis Street, #02-01 Genome, Singapore 138672, Singapore

E. Cheung (✉)
Cancer Biology and Pharmacology, Genome Institute of Singapore, A*STAR (Agency for Science, Technology and Research), 60 Biopolis Street, #02-01 Genome, Singapore 138672, Singapore

Department of Biochemistry, Yong Loo Lin School of Medicine, National University of Singapore, Singapore 117597, Singapore

School of Biological Sciences, Nanyang Technological University, Singapore 637551, Singapore
e-mail: cheungcwe@gis.a-star.edu.sg

R.K. Srivastava and S. Shankar (eds.), *Stem Cells and Human Diseases*,
DOI 10.1007/978-94-007-2801-1_8,

the aim of developing alternative therapeutic strategies for advanced prostate cancer, research efforts are now directed towards a more comprehensive understanding of prostate cancer genomics. Herein, we review the progress made recently in prostate cancer genomics research with a focus on the application of next generation sequencing technologies.

Keywords Prostate cancer • Genomics • Androgen receptor • Transcription factor • Next-generation sequencing

8.1 Introduction

Prostate cancer is the most commonly diagnosed cancer in the male population, accounting for about one in four newly identified cancers each year. It is responsible for the second highest number of cancer-related fatalities in men in the United States, with an estimated death toll of 32,000 in 2010 alone [1]. According to data provided by the American Cancer Society, it is estimated that one in six men will be diagnosed with prostate cancer in their lifetime and, among those who are diagnosed, one in six will die of the disease [1]. Furthermore, with reference to statistics obtained from the Surveillance, Epidemiology, End Results (SEER) database, the 5-year survival rate of prostate cancer patients after diagnosis at localized and regional stages is 100%. However, this number dropped drastically to 28.8% in patients whose cancer have metastasized [2].

For most cases, prostate cancer is detected during routine medical checkups either through prostate-specific antigen (PSA) screening or a digital rectal exam. Those who screened positive will undergo a core needle biopsy for further diagnosis. Clinicians generally assess the development of the tumor based on both Gleason scoring and TNM (Tumor, Lymph Nodes and Metastases) Tumor Staging systems. The Gleason scoring system is a powerful prognostic indicator based on examination of the tissue morphology. It relies on the extent of loss of normal glandular tissue architecture, and the overall Gleason score takes into account the sum of the two most predominant tumor patterns [3]. The TNM Tumor Staging system has its basis on evaluating the extent of the tumor (T), and assigns different stages/ sub-stages to cases where cancer cells have spread to the lymph nodes (N) as well as distant organs (M) where metastases have occurred. However, both grading systems only aid in determining a patient's prognosis, and are not able to account for the heterogeneity of biological and morphological behaviors of prostate cancers.

The development of prostate cancer is largely dependent on the presence of androgens for cellular growth and proliferation. Major treatments of clinically localized prostate cancer include radiation therapy and radical prostatectomy, which involves surgical removal of the prostate together with neighboring tissues [1]. Regional lymph nodes may also be removed to determine if metastases has occurred. In cases of patients with androgen-dependent prostate cancer, androgen deprivation therapy through surgical castration or pharmacological intervention using luteinizing hormone-releasing hormone (LHRH) agonists is generally recommended along with external beam radiation therapy [1]. This approach generally results in the initial regression of the prostate tumor, but most prostate cancers eventually recur and become resistant to androgen-deprivation therapy. The median survival of these patients is about 1–2 years [4]. Ninety percentage of patients who eventually die of the cancer have skeletal metastases, and this condition is the most significant cause of morbidity and mortality in prostate cancer [5, 6]. Currently, no curative modalities have been found for metastatic, androgen-independent forms of prostate cancer. Hence, it is of critical importance that we elucidate the mechanisms governing prostate cancer and develop a greater understanding of the progression of this disease to improve therapeutic outcome.

8.2 The Role of Androgen Receptor in Prostate Cancer

The pivotal discovery of the role of androgen in regulating prostate cancer by Huggins and Hodges in 1941 culminated in the development of androgen ablation therapy for advanced prostate cancer [7]. In the landmark paper, it was reported that marked reductions in androgen levels by castration resulted in the regression of metastatic prostate carcinoma. Conversely, administration of exogenous androgen stimulated the growth of prostate tumor [7].

Androgen receptor (AR) is a ligand-dependent transcription factor from the nuclear hormone receptor superfamily. Unliganded AR is maintained in the cytoplasm of prostate cells by binding to heatshock (Hsp) proteins such as Hsp90. Upon androgen/dihydrotestosterone (DHT) binding to AR, AR dissociates from the heatshock proteins and translocates into the nucleus. Following this, AR undergoes homodimerization and subsequent binding to specific DNA sequences called AR response elements (AREs). AR then recruits a myriad of coregulators as well as general transcription factors to orchestrate the transcriptional regulation of its target genes.

Most prostate cancers express AR, and a reduction in the levels of AR protein decreases prostate cancer growth in both androgen-dependent and -independent model systems. In a study by Eder and colleagues, silencing of AR expression in LNCaP (an androgen-sensitive AR-expressing prostate cancer cell line) was accompanied by a notable inhibition on cellular growth as well as a reduction of PSA secretion [8]. In the same study, by utilizing a mouse xenograft model in which LNCaP cells were implanted, the authors demonstrated a significant retardation in tumor growth when AR expression was abrogated [8]. In addition to being involved in androgen-dependent prostate cancer, several groups have also established that castration-recurrent tumors continue to rely on the AR signaling pathway [9–12]. In androgen-starved conditions, it was demonstrated that AR is reactivated and has a role in both prostate tumor growth and survival. When AR was silenced in the androgen-independent AR-positive prostate cancer cell line C4-2 in hormone-stripped conditions, there was a corresponding reduction in cellular proliferation and PSA expression. Thus, despite the ability to grow in an androgen-depleted medium, prostate cancer cells remain AR-dependent [9]. Additionally, in a xenograft mouse model, it was observed that AR was transcriptionally reactivated as the cancer transitioned from androgen-dependent to recurrent growth [12]. Similar observations were made earlier by Pound and colleagues who examined data from a clinical cohort and observed an increase in serum PSA levels in castration-recurrent patients, hence implicating AR reactivation in recurrent prostate cancer [13].

AR overexpression is reportedly the single most consistent event during the progression from androgen sensitivity to androgen independence in this disease. This has been validated in different experimental models of androgen refractory prostate cancer [14]. From *in vivo* xenograft work, it was reported that the consistent increase in AR levels is correlated with the development of resistance to anti-androgen therapy [14]. Indeed, Chen and co-workers further demonstrated that AR is required for the progression of prostate cancer growth from a hormone-sensitive to a hormone-refractory stage. Furthermore, it was reported that AR antagonists such as bicalutamide and cyproterone acetate surprisingly displayed agonistic behavior in LAPC4 and LNCaP cells (both androgen sensitive prostate cancer cell lines) with elevated AR levels. High levels of AR may hence convert the actions of an antagonist to a weak agonist, resulting in the activation of androgen-sensitive genes in hormone-refractory prostate cancer. More specifically, treating prostate cancer cells with bicalutamide resulted in an increase in AR protein, and this was accompanied by an increase in both PSA transcript levels as well as secreted PSA [14]. This is likely because of differential recruitment of cofactors to the promoters of AR target genes due to aberrant AR levels. As such, it is of paramount importance that we thoroughly examine the AR signaling pathway, starting off by identifying coregulators that participate in mediating AR action so as to obtain a more comprehensive understanding of the disease state.

8.3 The Transcriptional Cofactors of Androgen Receptor

Upon androgen stimulation, AR assembles a series of coregulators on the cis-regulatory regions of its target genes to initiate transcription. As AR is involved in the regulation of both androgen-dependent and androgen-independent prostate cancers, it is of great interest to identify the coregulators that participate in these processes. Hence, there has been an intense effort in this field by researchers to identify the various coregulators that work together with AR. These factors could eventually serve as potential therapeutic candidates to better block AR action, and hence, bring about a regression in prostate cancer development. These factors may represent the most direct and promising targets for therapeutic development as they are less ubiquitous than upstream signaling molecules and reside at a focal point in deregulated pathways [15]. Information obtained from The Nuclear Receptor Signaling Atlas (NURSA) indicated that approximately 260 coregulators have been identified to interact with nuclear receptors. Coregulators are typically made up of two groups of proteins that serve different functions, namely co-activators and co-repressors which act to enhance, or repress the transcription of target genes, respectively.

Classical cofactors of AR, such as members of the SRC family (SRC1, SRC2 and SRC3), were initially identified on the basis of transcriptional interference in transient transfection experiments [16, 17]. To determine the recruitment of AR and its cofactors to regions near selected androgen regulated genes such as prostate-specific antigen (PSA), researchers typically use chromatin immunoprecipitation (ChIP) assays. However, this method is limited in its approach as it only allows the targeted identification of known sites. In recent years, the advent of newer and higher throughput technology such as ChIP-Seq and ChIP-chip allows for the unbiased simultaneous identification of the binding sites of AR and its cofactors in a genome-wide manner. The ChIP-Seq approach incorporates the ChIP process together with next-generation sequencing technology. ChIP-chip, on the other hand, integrates ChIP together with tiled DNA oligonucleotide microarray analysis to allow for a genome-wide scale view of DNA-protein interactions.

Using the latter approach, Wang and colleagues mapped binding sites of AR on chromosomes 21 and 22 in LNCaP cells. In addition to identifying 90 AR binding sites on both chromosomes, they found the majority of these binding sites were positioned far from androgen-regulated genes, suggesting a potential enhancer role for the ARs found at these distal locations. Furthermore, they also reported significant enrichment of the DNA recognition motifs for three transcription factors, namely FoxA1, Gata2 and Oct1 [18]. Forkhead box A1 (FoxA1), a member of the Forkhead family of winged-helix transcription factors, plays a role in the development and differentiation of several major organs such as liver, prostate and mammary glands [19]. FoxA1 has been shown to physically interact with AR in prostate cancer cells [18, 20]. Moreover, Gao and colleagues showed that FoxA1 binds immediately adjacent to the AREs near two AR-regulated genes, including the rat probasin (PB) gene promoter and the enhancer of the prostate-specific antigen (PSA) gene [20]. Taken together, these observations point to a common functional mechanism for FoxA1 in governing prostate-specific responses. Moreover, as FoxA1 is present on the chromatin at the basal state, it is suggested to serve as a pioneer factor and functions to initiate a cascade of events that culminates in transcriptional activation upon androgen stimulation [20]. This work was subsequently substantiated by Lupien and co-workers who carried out a genomic scale study of FoxA1 binding on chromosomes 8, 11 and 12 using the ChIP-chip approach. In their work, they report FoxA1 occupy more than 60% of AR-binding sites and is required to drive the AR transcriptional response in prostate cancer cells [21].

Gata2, a member of the GATA family of zinc-finger transcription factors, appears to also function as a pioneer factor, similar to FoxA1, as its presence on chromatin is a prerequisite for AR recruitment. Gata2 functions likely by decompacting and opening up the chromatin to facilitate AR binding [18]. In contrast, Oct1, a member of the POU transcription factor family is not required for AR binding. This suggests that Oct1 has a role after Gata2 action occurring in conjunction with AR. Functionally, both Gata2 and Oct1 are required for proper androgen-dependent transcription and in promoting cellular proliferation [18, 22, 23].

In addition to the above co-factors, the consensus binding sequences for the ETS (avian erythroblastosis virus E26 homologue) family of transcription factors were found enriched in 70% of AR-associated promoters in LNCaP cells [24]. Specifically, the sequence motif of ETS1 was found to be most prevalent among a panel of AR ChIP-chip promoters. In subsequent functional studies, ETS1 was shown to bind in an androgen-dependent manner to a subset of AR-bound promoters including CCNG2 and UNQ9419 [24]. In similar studies but using ChIP-Seq, the ETS motif was found to be highly enriched in AR binding sites of VCaP cells [25]. Remarkably, ERG, another member of the ETS family, was shown co-localized to approximately half of the AR binding sites and bound to over 90% of AR occupied genes. Surprisingly, in contrast to ETS1, ERG appears to potentiate prostate cancer dedifferentiation by acting as an AR co-repressor.

8.4 Genomic Integrity of Prostate Cancer

Genome instability is one of the major hallmarks of cancer. Indeed, genomic mutations such as point mutations, insertions, deletions, gene amplifications and chromosomal translocations are widely present in prostate cancer. A clear understanding of how these genomic aberrations drive tumor progression would provide us with fundamental insights on prostate cancer pathogenesis. This knowledge would also be extremely helpful in the development of prostate cancer therapeutics especially in the field of personalized medicine. Cytogenetic and molecular techniques such as Spectral Karyotyping (SKY) [26, 27], Array Comparative Genomic Hybridization (Array CGH) [28, 29] and Fluorescence In-situ Hybridization (FISH) [30, 31] are commonly used for the detection and analysis of prostate cancer-specific genome alterations. Using these techniques, several chromosomal abnormalities underpinning prostate cancer initiation and progression have been identified. For instance, the PTEN tumour suppressor gene was found to be frequently inactivated by loss of heterozygosity or somatic mutations [32, 33], while genomic amplification of the AR gene was reported to contribute to hormone refractory prostate cancer [14, 34, 35]. More recently, a large proportion of prostate cancers were found to harbor recurrent ETS gene fusions arising from structural chromosomal rearrangements [36, 37]. In most of the ETS fusion positive prostate cancers, the regulatory region of androgen responsive gene TMPRSS2 was fused upstream to ETS transcription factors such as ERG or ETV1. This process renders the expression of ETS transcription factors androgen responsive and overexpressed in prostate cancers. This seminal finding has drastically altered our view on the types of genome aberration that occurs in prostate cancers. Although common in haematological sarcomas [38], chromosomal translocations have previously not been discovered in solid epithelial tumors. Recently, several studies suggest ETS transcription factors to be major drivers of prostate cancer progression [39, 40]. These examples therefore highlight the importance and impact of genomic studies in prostate cancer research.

Although traditional molecular cytogenetic techniques have been applied in the study and analysis of prostate cancer genetic anomalies with some success, their applications as screening and discovery tools for chromosomal aberrations are largely limited with respect to their relative low resolution and/or low throughput. For instance, a molecular cytogenetic technique like FISH is more suited as a targeted approach rather than for high throughput screening purposes. Indeed, a shift from the conventional molecular cytogenetics based experimental approach to a bioinformatics based strategy resulted in the landmark discovery of recurrent ETS fusion genes [36]. This finding highlights the need to explore alternative approaches and strategies in the screen for novel genomic anomalies. Recent advancements that has accelerated and lowered the cost of sequencing by means of parallel sequencing (next-generation sequencing) could be exploited for this very purpose. The advent of the next-generation sequencing technologies (i.e. RNA-Seq, DNA-PET) brings much promise for breakthroughs in the field of cancer genomics as a screening tool of extremely high resolution and throughput. In spite of all its potentials, there are still substantial challenges for cancer genome sequencing in general. An excellent review of these methodological considerations for next-generation sequencing technologies is available elsewhere [41] and will not be discussed in further detail in this chapter. For this section, we will instead limit our discussion by focusing on genomic aberrations associated with prostate cancer. In addition, we will review the contributions and what future prospects will hold for the utility, application and integration of conventional molecular cytogenetic techniques, novel bioinformatic analysis along with the next-generation sequencing (NGS) technologies in the field of prostate cancer genomics.

8.4.1 Point Mutations and DNA Copy Number Changes

Nucleotide changes in the coding sequence of a gene can drastically impact the function of the translated protein. For example, a single point mutation in the catalytic domain of an enzyme can enhance or abolish its activity, or alter the structure of the protein domain such that it confers a different binding specificity for its interacting partner. Consequently, tumors usually harbor genes containing various different point mutations that give rise to abnormally functioning proteins, resulting in tumor cells with selective advantages for proliferation, survival, metastasis, and drug resistance. Previously, the identification of novel point mutations in genes required a gene targeted approach which required the amplification of a genomic region and subsequent sequencing [42–44]. Although this gene targeted approach has been successful in elucidating point mutations in tumor suppressor genes such as PTEN and in oncogenes such as EGFR and AR [43–46], the potential of this strategy is severely limited due to its laborious and low throughput nature. Consequently, point mutation analysis for prostate cancer is limited to a small subset of specifically chosen target genes that are of strong biological interest. To circumvent this limitation, one group used next-generation sequencing technology to sequence and construct genomic maps of seven prostate tumors with patient-matched normal samples [47]. The high sequence coverage of the genomic maps offered an unbiased view into the genomic landscape of prostate cancers for the very first time [47]. Using this approach, several novel (e.g. SPTA1, CHD1, CHD5, HDAC9, HSPA2, HSPA5, HSP90AB1, PRKCI and Dicer, etc.) and previously known genes (e.g. PTEN and SPOP) with coding mutations in at least one tumor (out of seven) were identified. In another study, using exon sequencing, 138 genes were examined in 80 tumors [48]. In this study, only the AR gene was found to harbor mutations in more than three different samples. However, somatic mutations were also found in genes associated with the AR pathway such as NCOA2, NCOR2, NRIP1, TNK2 and EP300. Although the study found mutations in commonly mutated oncogenes (i.e. PIK3CA, KRAS and BRAF, IDH2), they occurred rarely in the samples tested. TP53, an important tumor suppressor widely mutated in most cancers, was only found to harbor missense mutations in only 2 out of the 80 tumors while no point mutations in PTEN were detected. From these high resolution, high throughput and wide sequence coverage studies, it appears that the prostate cancer genomes are generally characterized by a relatively low rate of point mutations in genes but with a high rate of chromosomal rearrangements. Indeed, a recent study using targeted mutagenesis analysis on metastatic prostate cancers via NGS technologies also painted a similar genomic landscape of prostate cancer cells [49].

DNA copy number alterations (CNAs) in the form of amplification and deletion of chromosomal segments occur commonly in cancers through a variety of mechanisms [50]. This change in copy number could facilitate carcinogenesis by providing an effective mean for cancer cells to bypass transcriptional regulatory mechanisms and produce abnormal levels of gene transcripts from gene loci residing in the affected chromosomal regions. For instance, by gaining several copies of the same genes through chromosomal amplification, cancer cells can acquire the potential to exceed the maximum transcript production threshold of the amplified genes by several folds. With deletion of chromosomal segments, the maximum production threshold of the deleted genes is completely abolished if two loci are both deleted or reduced by 50% if one locus is deleted. Furthermore, in comparison to point mutations, CNAs can affect a large genomic region and thus is able to impact the expression of multiple genes simultaneously. Consequently, amplification of oncogenes and deletion of tumor suppressors are regarded as potential causative factors underlying tumor formation and progression.

In the field of prostate cancer genomics, massive efforts have been channeled into examining the role of CNAs in prostate cancer pathologies. For example, using a cytogenetic method, several CNA hotspots in prostate cancers were identified [51–53]. Loss of 7q and 10q were two of the earliest CNAs detected in prostate cancer by chromosomal banding techniques (i.e. G-banding) [54, 55]. Loss of 10q was reported again in a later study utilizing a DNA probe hybridization approach in prostate carcinomas [56]. This same study also discovered a loss of heterozygosity at 16q in prostate cancers. With the advent of molecular cytogenetics such as FISH and array

CGH, these techniques were frequently applied to study CNAs in prostate cancers [51]. These techniques provided several advantages over the classical cytogenetics approach in the study of solid tumors [57, 58]. Several CNAs which were infrequently reported by conventional cytogenetics were detected by molecular cytogenetic techniques. For instance, aberrations in chromosome 8 that were rarely detected by conventional cytogenetics [53] were demonstrated successfully by several studies that utilized FISH [59, 60]. In addition to the detection of novel aberrations, previously identified CNAs of prostate cancer cells were largely recapitulated and further refined to a better resolution through defining minimally overlapping regions (MORs) with improved molecular cytogenetic techniques. This is exemplified by recent studies utilizing array CGH on clinical prostate cancers [61, 62]. A higher quality CNA data and an improved bioinformatic analysis methodology would allow for deeper and more insightful biological interpretation. Through evaluating the high resolution CNA data obtained from metastatic prostate cancers, two recent studies were able to argue for a monoclonal origin of metastatic prostate cancer [61, 63].

Additionally, we would like to specifically highlight an extensive study that was carried out with 218 prostate tumor samples for CNAs analysis using array CGH [48]. The large number of samples analyzed in the study allows for observations of high statistical significance. Again, similar to previous findings, the most common deletion found was loss of 8p. Deletions targeting 10q23.31 (PTEN), 13q14.2 (RB1) and 17p31.1 (TP53) were also detected. Furthermore, the study also found deletions between 21q22.2–3 (ERG and TMPRSS2) and 12p13.31–p12.3 (ETV6, CDKN1B and DUSP16). The more frequently occurring amplified loci were 8q24.21 (MYC) and a novel amplification of the NCOA2 gene on 8q13.3. On the other hand, the common amplification of AR (Xq12) was only detected in metastatic prostate cancers. Less commonly found amplifications occurred on discontinuous region of 7q, including the BRAF and EZH2 loci and 5p13.3–p13.1 (AMACR, RICTOR, and SKP2).

As highlighted, PTEN and AR are two genes that are frequently affected by either somatic mutations and/or CNAs in prostate cancers. Indeed, multiple studies have demonstrated their importance in prostate cancer progression [14, 64] and their value as prognostic biomarkers of prostate cancers [65–67].

8.4.1.1 Mutations in Phosphatase and Tensin Homolog (PTEN)

Phosphatase and tensin homolog (PTEN) is a tumor suppressor that functions as a protein phosphatase. The main role of PTEN is to convert phosphatidylinositol-3,4,5-trisphosphate (PIP_3) to phosphatidylinositol-4,5-bisphosphate (PIP_2) [68, 69]. Loss of PTEN leads to a massive accumulation of PIP_3 which acts as the signal for aberrant activation of downstream signaling components such as the serine/threonine protein (Akt) kinase [68, 69]. A constitutive active PI3K/Akt signaling pathway will then result in uncontrolled cell growth and survival [68, 69].

Point mutations and frame shifts that disrupt the coding region of PTEN gene, leading to a loss of PTEN protein function have been detected in prostate cancers but are relatively rare [47, 70]. Instead, the more common loss of PTEN in prostate cancer occurs through deletion of the gene [48, 49]. In fact, PTEN was regarded as the main target for inactivation in the frequent loss of 10q23 in prostate cancer [71]. Several murine PTEN prostate cancer models have been developed to show the importance of PTEN loss in prostate cancer development [72, 73]. Even though both hemizygous and homozygous deletion of PTEN are present in prostate cancer, primary prostate cancer commonly exhibit hemizygous deletions while prostate cancer with homozygous deletions of PTEN are more likely to be metastatic and hormone refractory [67, 74]. This clinical observation suggests a strong support for the inversely correlated relationship between PTEN dose and prostate cancer progression [64].

8.4.1.2 Mutations in the Androgen Receptor

As discussed above, AR plays a major role in calibrating the transcriptional output of genes that regulate the normal development and maintenance of the prostate [75]. AR is also a key player in the development and progression of prostate cancer by driving proliferation and survival in both androgen dependent [76] and hormone refractory prostate cancers [77]. Due to its importance in prostate cancer, mutations in AR tend to exert profound effects on prostate cancer pathology and prognosis. For instance, the mutation status of AR largely determines the outcome of anti-androgens therapy in prostate cancers. A single point mutation in the ligand binding domain of AR (sense codon 868 (Thr to Ala)) can alter the ligand specificity of AR [78, 79]. This mutated AR can be activated by other steroids such as estrogens and anti-androgens including flutamide, cyproterone acetate and nilutamide [78, 79]. Thus, mutation of AR renders prostate cancer resistance to certain forms of androgen ablation therapy (i.e. the use of flutamide as an AR antagonist). It is also interesting to note that this specific AR mutation does not convert all wild-type AR antagonists into agonists of the mutated AR. Casodex, for instance, still act as an antagonist of the mutated AR, thus supporting the idea that different antagonists are likely to exert their effects on AR through unique mechanisms.

The progression of hormone refractory prostate cancer has been attributed to increased AR levels [14]. High levels of AR resulting from amplification of the AR gene locus in prostate cancer have been associated with endocrine treatment failure [65]. Indeed, recent studies that profiled clinical prostate cancers detected amplification of the AR gene locus in castration resistant metastatic prostate tumors [48, 49]. In spite of all the mounting evidence that prostate cancers with an AR amplified gene locus are clinically associated with subversion of androgen ablation treatment, the low frequency of AR amplifications in advanced prostate cancers render this mechanism inapplicable for the majority of hormone refractory prostate cancers [14, 65]. In some prostate cancers, even though the AR gene itself does not carry any mutations, the androgen signaling pathway was still altered to favor cancer progression. For example, mutations in other members of the androgen signaling pathway (i.e. co-factors of AR) are present in some prostate cancers and thus can deregulate the androgen signaling pathway [48].

8.4.2 *Fusion Genes*

Chromosomal rearrangements such as inversion, translocation and interstitial deletion can result in the formation of fusion genes (Fig. 8.1). Fusion gene transcripts can also be formed from read through events (Fig. 8.1) during transcription [80]. The resulting fusion gene may give rise to fusion gene products that have distinct (but likely to be related) functions from the two parent fusion partners that it was generated from. The novel function acquired by the fusion gene product is likely to be disruptive to normal cellular processes and hence contributes to cancer progression. The BCR-ABL fusion protein associated with chronic myelogenous leukemia and the EWS-FLI fusion protein associated with Ewing sarcoma are two such examples. Another kind of gene fusion involves foreign transcriptional regulatory elements (i.e. promoter) fusing with another gene (tumor suppressor or oncogene) resulting in the deregulated expression of that gene as its native regulatory elements would have been displaced or disrupted by the fusion. An example of this type of fusion is the TMPRSS2-ERG fusion gene in prostate cancer.

The landmark discovery of recurrent fusion genes in prostate cancers occurred in 2005 [36]. For this discovery, the group formulated a novel bioinformatic approach termed Cancer Outlier Profile Analysis (COPA). This analysis was implemented on clinical studies available in the Oncomine database (www.oncomine.com) to detect outlier gene expression in a specific subset of prostate cancers. Consequently, a subset of prostate cancers was found to be ETS-fusion positive. Currently, with the advent of NGS, fusion genes in cancers are being detected in a high-throughput manner using various NGS platforms. Paired end tag, long read and short read sequencing technologies have been applied on both the cancer transcriptome and genome for the discovery of gene fusions as drivers of neoplasm. The clinical relevance to cancer therapy resulting from the discovery such gene fusions is exemplified by the development of imatinib, an inhibitor of tyrosine kinase activity of ABL [81].

8.4.2.1 Detection of Gene Fusions Using Transcriptome Sequencing

Through an integrative analysis of both long and short transcriptome sequencing reads generated in VCaP cells, LNCaP cells, and other prostate tumor tissues, Maher and colleagues discovered numerous novel fusion genes in prostate cancers [80]. Apart from the 'rediscovery' of the TMPRSS2-ERG fusion in VCaP cells, the authors managed to discover several novel fusion transcripts in both VCaP and LNCaP cells. Specifically, in VCaP cells, they found fusions between exon 1 of USP10 with exon 3 of ZDHHC7, exon 10 of RC3H2 with exon 20 of RGS3, exon 8 and exon 9 of HJURP with exon 2 of EIF4E2 and exon 25 of INPP4A. For LNCaP cells, they detected a fusion between exon 11 of MIPOL1 with the last exon of DGKB, resulting in ETV1 insertion into an intron of MIPOL1. In prostate tumor samples, they found novel fusions such as between exon 9 of STRN4 with exon2 of

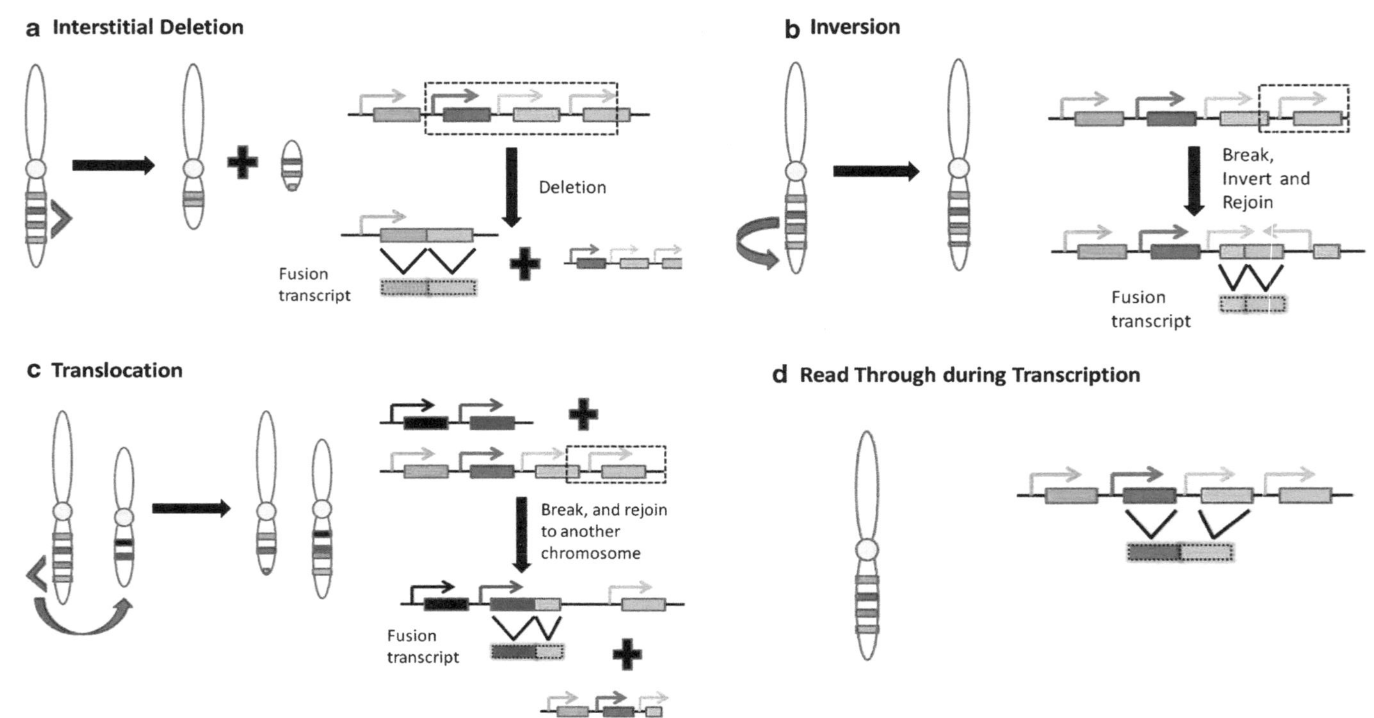

Fig. 8.1 Different forms of fusion transcripts generated in prostate cancers

GPSN2 and exon1 of LMAN2 with exon 2 of AP3S1. Besides fusion transcripts arising from chromosomal rearrangements, the authors also detected chimeric transcripts between neighboring genes generated from transcriptional read-through events. Unlike, their counterpart fusion transcriptions which are largely cancer cell specific, the chimeric transcripts from read-through events are distributed more broadly across both normal and cancerous cells. However, there are still recurrent read-through chimera transcripts that occur mainly in malignant prostate cancers. For example, the read-through chimera SLC45A3-ELK4 (exon 4 of SLC45A3 with exon 2 of ELK4), identified in several metastatic prostate cancers and in LNCaP cells is one such representative. This fusion is suggested to collaborate with ETV1 for the progression of prostate cancers with SLC45A3- ELK4 fusion and genomic ETV1 aberrations.

Following this study, several other studies have also attempted to describe novel chimera transcripts in prostate cancer cell lines and clinical samples using NGS technologies. In one work, the authors made use of paired-end transcriptome sequencing technology to catalogue fusion transcripts in VCaP cells [82]. Paired-end sequencing offer several advantages over single tag sequencing approach in that it increases tag mapping specificity and the sequencing coverage by specifying the start and ends of a transcript [82, 83]. Importantly, it also reduces the requirement for sequencing tags that map exactly across the fusion junction for calling a chimeric transcript. By leveraging on the strengths of paired-end transcriptome sequencing in detecting gene fusions, the study was able to discover several novel chimeric transcripts in addition to previously identified fusion transcripts. Two examples are the fusion between ZDHHC7 with ABCB9 and TIA1 with DIRC2. Furthermore, this study managed to discover novel instances of ETS gene fusions (HERPUD1-ERG and AX747630-ETV1) in two prostate cancer samples that were previously reported to lack ETS fusions, further demonstrating the superiority associated with paired-end transcriptome sequencing. It was also found that in comparison to the single read approach, paired-end transcriptome sequencing technology is better suited to detect transcript chimeras of adjacent genes that are usually transcribed in different orientations. To facilitate the identification of fusion transcripts using paired-end RNA sequencing, the bioinformatic framework FusionSeq was established [84]. FusionSeq was successfully applied for the analysis of the paired end transcriptome sequencing data of 25 human prostate cancer samples to identify novel cancer specific gene fusions [85]. Of interest, this work brought about the discovery of several novel non-ETS fusion genes that are present in ETS-fusion positive prostate cancers.

The development of NGS technologies has largely helped to circumvent the bottleneck of chimeric transcript detection by identifying chromosomal rearrangements of the cancer genome in a high throughput and efficient manner. The main challenge now is to formulate strategies that could help to effectively distinguish driver structural variations from a large pool of passenger genomic aberrations. One straightforward way to solve this problem is through analyzing a large set of tumor samples for recurrent fusions. However, this is a relatively laborious approach. In additional, variables such tumor selection and sampling in such studies could confound the described frequency of a particular rearrangement [86]. Hence, a more general approach may be preferred. To this end, an integrative approach utilizing gene ontology and protein interactome data was used to derive the 'ConSig' score for the ranking the biological relevance of gene fusions detected in cancer genome sequencing studies [87].

8.4.2.2 Formation of Fusion Genes

Since the discovery of recurrent gene fusions in prostate cancer, there has been intense interest on how such fusions are formed. Studies have since showed that the fusion gene TMPRSS2-ERG could arise as a result of chromosomal looping during AR mediated transcription [88, 89]. Interestingly, rearrangement breakpoints were reported to be enriched for AR and ERG binding sites in ERG fusion positive prostate cancers [47]. These results could potentially explain how recurrent fusion genes arise and highlight a previously unappreciated role of androgen receptor in generating chromosomal aberrations through the mechanics of transcription. However, the key question on whether this phenomenon is mediated by other transcription factors and if it extends to the formation of other non-ETS fusion genes through specific chromosomal looping during transcription still remains. Recent genomic technologies such as ChIA-PET [90] and Hi-C [91] that were originally developed for studying the 3D conformation of chromatin during transcription, if integrated with DNA and RNA paired end sequencing techniques, may offer tremendous opportunities for exploring this gap in knowledge.

8.4.3 The Epigenome of Prostate Cancer

Drastic epigenetic modifications are known to occur in cancers, severely distorting the transcriptome output in the process and ultimately leading to the disruption of vital cellular process that maintain cell functions [92, 93]. In cancers, tumor suppressors are frequently silenced through epigenetic mechanisms including but not limited to DNA and histone hypermethylation [94]. Consequently, these cancer associated epigenetic modifications are deemed as potential cancer biomarkers and targets for therapeutic interventions [95].

The epigenetic alterations associated with prostate cancer progression have always been an area of intense research. For example, several studies revealed prostate cancers harbor aberrant epigenetic profiles in histone modifications and DNA methylation [96, 97], and as a result of these abnormal epigenetic modifications, the transcriptome profiles of prostate cells were altered to one that promotes cancer progression. Specifically, the promoters of tumor suppressors were frequently found hypermethylated, resulting in their low expression levels [96, 98]. On the contrary, DNA hypomethylation was suggested to occur late in prostate cancer progression leading to the upregulation of genes that contribute to metastasis [99].

Histone modifications also play a huge role in determining the transcriptional output of prostate cancer cells [100]. For instance, H3K4 methylation marks are enriched at enhancers of AR and associated with AR-mediated transcriptional activation [21], while H3K9 methylation marks are repressive marks on AR enhancers and are reduced upon AR activation [101, 102]. Specific examples include H3K4 methylation of the AR enhancer of the proto-oncogene UBE2C which enhanced AR binding and activation of transcription [103], while H3K9 methylation of the AR enhancer of PSA and NKX3.1 resulted in the reduced transcription of these genes [101]. Trimethylation of H3K27 is another repressive mark that contributes to suppression of transcriptional activities. In prostate cancer, genes enriched in H3K27me3 are generally repressed and associated with metastasis [104]. Apart from methylation, acetylation and deacetylation of the histone tails are two more examples of common histone modifications. Importantly, histone modifications serve as epigenetic rheostat for co-coordinating dynamic transcriptional activities [105]. Finally, given the massive influence of the epigenome on the profile of gene expression, it was not surprising that specific patterns of epigenetic modifications in the prostate cancer cells were found to be predictive of the progression and prognosis of the disease [106].

8.4.3.1 Epigenetic Regulators

Epigenetic changes are mediated by enzymes that catalyze the chemical modification of DNA and histones. In concordance with the abnormal epigenetic profile in prostate cancers, the enzymes that catalyze these epigenetic alterations are frequently aberrantly expressed. A decreased in level of H3K9 methylation is detected in prostate cancers when compared to their non-malignant counterparts [107]. Correspondingly, the histone lysine demethylase JHDM2A which demethylates H3K9 during AR mediated transcription [101] were found to be increased in prostate cancers when compared to benign prostate hyperplasia [108]. Similarly, the expression of another histone lysine demethylase, LSD1, which also demethylates H3K9me1 and H3K9me2 [102] was associated with prostate cancer recurrence [109]. The polycomb repressor EZH2 which catalyses the H3K27 trimethylation is over-expressed in metastatic prostate cancers [110] and is known to mediate a polycomb repression signature in prostate cancers [25, 104]. The regulation of histone acetylation and deacetylation is also exploited as a strategy for driving a cancer-promoting gene expression profile in prostate cancers. Consequently, histone deacetylases (HDACs) are frequently over-expressed in prostate cancers and are suggested as predictors of poor disease prognosis [111, 112]. Recently a large scale study was carried out to obtain the promoter methylation profile of 86 benign and 95 cancerous prostate samples [113]. In general, the study detected 5912 CpG regions with a significant rise in DNA methylation and 2151 CpG regions with a significant reduction in DNA methylation [113]. Furthermore, the authors identified 69 methylation changes that correlated with prostate cancer recurrence [113]. The mechanism underlying these methylation alterations was attributed to the increase in the expression levels of DNA methyltransferase, DMNT3A2 and DMNT3B, in prostate tumors [113].

Apart from the epigenetic modifiers, it is also essential to identify other players that are required for modifying the epigenome of prostate cancers. The interplay between the transcription factors and the epigenetic modifiers may be crucial for cancer progression. For instance, AR recruits histone lysine demethylase, JHDM2A [101], and histone acetylase, p300 [114], to activate transcription. Even though it has been shown that pioneering transcription factors such as PU.1 and FoxA1 are important in establishing epigenetic signatures in other cell types [115, 116], there is no direct proof that pioneering factors are capable of doing so in prostate cancers. However, FoxA1 was shown to be associated with H3K4me1 and H3K4me2 in prostate cancers [21]. Interestingly, while it has been shown that the expression of the ETS fusion gene, ERG, is correlated with histone deacetylases (HDACs) [117] and polycomb histone deacetylase EZH2 [25], it is not known if they could collaborate with each other to promote cancer progression. If such cross-talk exists, it would be interesting and important to investigate the mechanism of this co-operation and the function outcome associated with it.

8.5 The Future of Prostate Cancer Genomics

Cancer is a disease of the genome. Genome instability of the cell leads to the loss of genomic integrity and results in an aberrant transcriptome and proteome that distort cellular processes such as cell cycle and apoptosis. The properties of the carcinoma will be a direct manifestation of the type of acquired genomic lesions. Consequently, cancer patients with differing genomic profiles will display dissimilar disease prognosis and exhibit diverse responses to disparate therapeutic strategies. A comprehensive understanding of the cancer genome will contribute much to the successful formulation of a tailored therapy for the patient. Indeed, the characterization of breast cancer subtypes have resulted in the development and application of stratified medicine for breast cancer therapeutics [118]. This is already the first step towards the development of personalized medicine for breast cancer treatments. In contrast, prostate cancer genomics research has not enabled the stratification of distinct prostate cancer subtypes until the recent discovery of ERG fusion positive prostate cancer subtype [36, 37]. Even though that there are reports suggesting that ERG fusion positive prostate cancers are more sensitive to HDAC inhibitors [119], the reasons behind this phenomenon are not clear. Importantly, there is a lack of highly reliable and specific biomarkers for prostate cancer diagnosis and risk stratification. PSA, the most widely used biomarker in prostate cancer detection has several limitations [120]. As a biomarker for prostate cancers, PSA was shown to exhibit a modest pooled sensitivity of 72.1%, a pooled specificity of 93.2% and a relatively low pooled positive predictive value of 25.1% [121]. Furthermore, it is not possible to accurately distinguish indolent forms of prostate cancer that is better off (for the patients) if left untreated from the aggressive ones that need prompt treatment

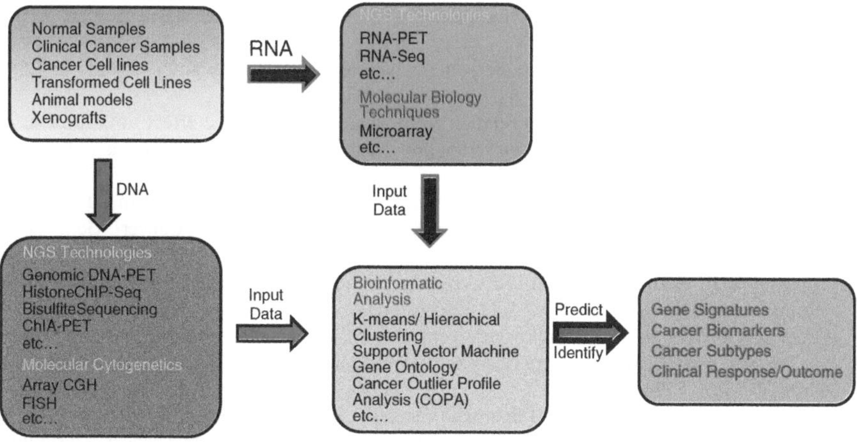

Fig. 8.2 Flow chart illustrating a typical cancer genomic profiling workflow

through the monitoring of PSA levels. As a result, patients who are over-diagnosed and over-treated for prostate cancers would have to endure side effects arising from non-beneficial therapies [122]. Strikingly, a landmark clinical trial suggests that while PSA increased prostate cancer detection rate, it did not significantly improve the survival rate of the patients [123]. No doubt, the challenges involved in identifying better biomarkers and in delineating specific subtypes of prostate cancers that display similar prognosis are great, but it is not insurmountable. With improvements in genomic profiling technologies, computational powers and bioinformatic algorithms, recent studies have achieved some significant progress [36, 47, 48, 87]. Indeed, the application of a more innovative and integrative bioinformatic analysis approach on the massive amounts of high quality genomic data generated by NGS technologies appears to hold the future of prostate cancer genomics research in achieving the ultimate goal (Fig. 8.2).

Acknowledgments This work was supported by the Biomedical Research Council/Science and Engineering Research Council of A*STAR (Agency for Science and Technology), Singapore.

References

1. American Cancer Society (2010) Cancer facts & figures. American Cancer Society, Atlanta
2. Howlader N, Noone AM, Krapcho M, Neyman N, Aminou R, Waldron W, Altekruse SF, Kosary CL, Ruhl J, Tatalovich Z, Cho H, Mariotto A, Eisner MP, Lewis DR, Chen HS, Feuer EJ, Cronin KA, Edwards BK (eds) (2011) SEER cancer statistics review, 1975–2008. National Cancer Institute, Bethesda
3. Wang R, Tomlins SA, Chinnaiyan AM (2009) Androgen regulation of prostate cancer gene fusions. In: Androgen action in prostate cancer. Springer, New York, pp 701–721
4. Lassi K, Dawson NA (2009) Emerging therapies in castrate-resistant prostate cancer. Curr Opin Oncol 21(3):260–265

5. Singh AS, Figg WD (2005) In vivo models of prostate cancer metastasis to bone. J Urol 174(3):820–826
6. Bubendorf L et al (2000) Metastatic patterns of prostate cancer: an autopsy study of 1,589 patients. Hum Pathol 31(5):578–583
7. Huggins C, Hodges CV (1972) Studies on prostatic cancer. I. The effect of castration, of estrogen and androgen injection on serum phosphatases in metastatic carcinoma of the prostate. CA Cancer J Clin 22(4):232–240
8. Eder IE et al (2002) Inhibition of LNCaP prostate tumor growth in vivo by an antisense oligonucleotide directed against the human androgen receptor. Cancer Gene Ther 9(2):117–125
9. Agoulnik IU et al (2005) Role of SRC-1 in the promotion of prostate cancer cell growth and tumor progression. Cancer Res 65(17):7959–7967
10. Gregory CW et al (1998) Androgen receptor expression in androgen-independent prostate cancer is associated with increased expression of androgen-regulated genes. Cancer Res 58(24):5718–5724
11. Hara T et al (2003) Enhanced androgen receptor signaling correlates with the androgen-refractory growth in a newly established MDA PCa 2b-hr human prostate cancer cell subline. Cancer Res 63(17):5622–5628
12. Zhang L et al (2003) Interrogating androgen receptor function in recurrent prostate cancer. Cancer Res 63(15):4552–4560
13. Pound CR et al (1999) Natural history of progression after PSA elevation following radical prostatectomy. JAMA 281(17):1591–1597
14. Chen CD et al (2004) Molecular determinants of resistance to antiandrogen therapy. Nat Med 10(1):33–39
15. Koehler AN (2010) A complex task? Direct modulation of transcription factors with small molecules. Curr Opin Chem Biol 14(3):331–340
16. Bocquel MT et al (1989) The contribution of the N- and C-terminal regions of steroid receptors to activation of transcription is both receptor and cell-specific. Nucleic Acids Res 17(7):2581–2595
17. Meyer ME et al (1989) Steroid hormone receptors compete for factors that mediate their enhancer function. Cell 57(3):433–442
18. Wang Q et al (2007) A hierarchical network of transcription factors governs androgen receptor-dependent prostate cancer growth. Mol Cell 27(3):380–392
19. Friedman JR, Kaestner KH (2006) The Foxa family of transcription factors in development and metabolism. Cell Mol Life Sci 63(19–20):2317–2328
20. Gao N et al (2003) The role of hepatocyte nuclear factor-3 alpha (Forkhead Box A1) and androgen receptor in transcriptional regulation of prostatic genes. Mol Endocrinol 17(8):1484–1507
21. Lupien M et al (2008) FoxA1 translates epigenetic signatures into enhancer-driven lineage-specific transcription. Cell 132(6):958–970
22. Ewen ME (2000) Where the cell cycle and histones meet. Genes Dev 14(18):2265–2270
23. Tsai FY, Orkin SH (1997) Transcription factor GATA-2 is required for proliferation/survival of early hematopoietic cells and mast cell formation, but not for erythroid and myeloid terminal differentiation. Blood 89(10):3636–3643
24. Massie CE et al (2007) New androgen receptor genomic targets show an interaction with the ETS1 transcription factor. EMBO Rep 8(9):871–878
25. Yu J et al (2010) An integrated network of androgen receptor, polycomb, and TMPRSS2-ERG gene fusions in prostate cancer progression. Cancer Cell 17(5):443–454
26. van Bokhoven A et al (2003) Spectral karyotype (SKY) analysis of human prostate carcinoma cell lines. Prostate 57(3):226–244
27. Beheshti B et al (2001) Evidence of chromosomal instability in prostate cancer determined by spectral karyotyping (SKY) and interphase fish analysis. Neoplasia 3(1):62–69
28. Ishkanian AS et al (2009) High-resolution array CGH identifies novel regions of genomic alteration in intermediate-risk prostate cancer. Prostate 69(10):1091–1100

29. Brookman-Amissah N et al (2005) Genome-wide screening for genetic changes in a matched pair of benign and prostate cancer cell lines using array CGH. Prostate Cancer Prostatic Dis 8(4):335–343
30. Celep F et al (2003) Detection of chromosomal aberrations in prostate cancer by fluorescence in situ hybridization (FISH). Eur Urol 44(6):666–671
31. Liu HL et al (2001) Detection of low level HER-2/neu gene amplification in prostate cancer by fluorescence in situ hybridization. Cancer J 7(5):395–403
32. Trybus TM et al (1996) Distinct areas of allelic loss on chromosomal regions 10p and 10q in human prostate cancer. Cancer Res 56(10):2263–2267
33. Pesche S et al (1998) PTEN/MMAC1/TEP1 involvement in primary prostate cancers. Oncogene 16(22):2879–2883
34. Palmberg C et al (1997) Androgen receptor gene amplification in a recurrent prostate cancer after monotherapy with the nonsteroidal potent antiandrogen Casodex (bicalutamide) with a subsequent favorable response to maximal androgen blockade. Eur Urol 31(2):216–219
35. Linja MJ et al (2001) Amplification and overexpression of androgen receptor gene in hormone-refractory prostate cancer. Cancer Res 61(9):3550–3555
36. Tomlins SA et al (2005) Recurrent fusion of TMPRSS2 and ETS transcription factor genes in prostate cancer. Science 310(5748):644–648
37. Tomlins SA et al (2007) Distinct classes of chromosomal rearrangements create oncogenic ETS gene fusions in prostate cancer. Nature 448(7153):595–599
38. Rowley JD (1973) Letter: a new consistent chromosomal abnormality in chronic myelogenous leukaemia identified by quinacrine fluorescence and Giemsa staining. Nature 243(5405):290–293
39. Tomlins SA et al (2008) Role of the TMPRSS2-ERG gene fusion in prostate cancer. Neoplasia 10(2):177–188
40. Shin S et al (2009) Induction of prostatic intraepithelial neoplasia and modulation of androgen receptor by ETS variant 1/ETS-related protein 81. Cancer Res 69(20):8102–8110
41. Meyerson M, Gabriel S, Getz G (2010) Advances in understanding cancer genomes through second-generation sequencing. Nat Rev Genet 11(10):685–696
42. Stapleton AM et al (1997) Primary human prostate cancer cells harboring p53 mutations are clonally expanded in metastases. Clin Cancer Res 3(8):1389–1397
43. Douglas DA et al (2006) Novel mutations of epidermal growth factor receptor in localized prostate cancer. Front Biosci 11:2518–2525
44. Newmark JR et al (1992) Androgen receptor gene mutations in human prostate cancer. Proc Natl Acad Sci USA 89(14):6319–6323
45. Peraldo-Neia C et al (2011) Epidermal Growth Factor Receptor (EGFR) mutation analysis, gene expression profiling and EGFR protein expression in primary prostate cancer. BMC Cancer 11:31
46. Suzuki H et al (1998) Interfocal heterogeneity of PTEN/MMAC1 gene alterations in multiple metastatic prostate cancer tissues. Cancer Res 58(2):204–209
47. Berger MF et al (2011) The genomic complexity of primary human prostate cancer. Nature 470(7333):214–220
48. Taylor BS et al (2010) Integrative genomic profiling of human prostate cancer. Cancer Cell 18(1):11–22
49. Robbins CM et al (2011) Copy number and targeted mutational analysis reveals novel somatic events in metastatic prostate tumors. Genome Res 21(1):47–55
50. Hastings PJ et al (2009) Mechanisms of change in gene copy number. Nat Rev Genet 10(8):551–564
51. Nupponen NN, Visakorpi T (2000) Molecular cytogenetics of prostate cancer. Microsc Res Tech 51(5):456–463
52. Brothman AR (2002) Cytogenetics and molecular genetics of cancer of the prostate. Am J Med Genet 115(3):150–156
53. Brothman AR (1997) Cytogenetic studies in prostate cancer: are we making progress? Cancer Genet Cytogenet 95(1):116–121

54. Atkin NB, Baker MC (1985) Chromosome 10 deletion in carcinoma of the prostate. N Engl J Med 312(5):315
55. Lundgren R et al (1988) Multiple structural chromosome rearrangements, including del(7q) and del(10q), in an adenocarcinoma of the prostate. Cancer Genet Cytogenet 35(1):103–108
56. Carter BS et al (1990) Allelic loss of chromosomes 16q and 10q in human prostate cancer. Proc Natl Acad Sci USA 87(22):8751–8755
57. Vorsanova SG, Yurov YB, Iourov IY (2010) Human interphase chromosomes: a review of available molecular cytogenetic technologies. Mol Cytogenet 3:1
58. Persson K et al (1999) Chromosomal aberrations in breast cancer: a comparison between cytogenetics and comparative genomic hybridization. Genes Chromosomes Cancer 25(2):115–122
59. Deubler DA et al (1997) Allelic loss detected on chromosomes 8, 10, and 17 by fluorescence in situ hybridization using single-copy P1 probes on isolated nuclei from paraffin-embedded prostate tumors. Am J Pathol 150(3):841–850
60. Brothman AR et al (1992) Analysis of prostatic tumor cultures using fluorescence in-situ hybridization (FISH). Cancer Genet Cytogenet 62(2):180–185
61. Holcomb IN et al (2009) Comparative analyses of chromosome alterations in soft-tissue metastases within and across patients with castration-resistant prostate cancer. Cancer Res 69(19):7793–7802
62. Kim JH et al (2007) Integrative analysis of genomic aberrations associated with prostate cancer progression. Cancer Res 67(17):8229–8239
63. Liu W et al (2009) Copy number analysis indicates monoclonal origin of lethal metastatic prostate cancer. Nat Med 15(5):559–565
64. Trotman LC et al (2003) Pten dose dictates cancer progression in the prostate. PLoS Biol 1(3):E59
65. Visakorpi T et al (1995) In vivo amplification of the androgen receptor gene and progression of human prostate cancer. Nat Genet 9(4):401–406
66. DeMarzo AM et al (2003) Pathological and molecular aspects of prostate cancer. Lancet 361(9361):955–964
67. Yoshimoto M et al (2007) FISH analysis of 107 prostate cancers shows that PTEN genomic deletion is associated with poor clinical outcome. Br J Cancer 97(5):678–685
68. Stambolic V et al (1998) Negative regulation of PKB/Akt-dependent cell survival by the tumor suppressor PTEN. Cell 95(1):29–39
69. Maehama T, Dixon JE (1998) The tumor suppressor, PTEN/MMAC1, dephosphorylates the lipid second messenger, phosphatidylinositol 3,4,5-trisphosphate. J Biol Chem 273(22):13375–13378
70. Feilotter HE et al (1998) Analysis of PTEN and the 10q23 region in primary prostate carcinomas. Oncogene 16(13):1743–1748
71. Cairns P et al (1997) Frequent inactivation of PTEN/MMAC1 in primary prostate cancer. Cancer Res 57(22):4997–5000
72. Wang S et al (2003) Prostate-specific deletion of the murine Pten tumor suppressor gene leads to metastatic prostate cancer. Cancer Cell 4(3):209–221
73. Kwabi-Addo B et al (2001) Haploinsufficiency of the Pten tumor suppressor gene promotes prostate cancer progression. Proc Natl Acad Sci USA 98(20):11563–11568
74. Sircar K et al (2009) PTEN genomic deletion is associated with p-Akt and AR signalling in poorer outcome, hormone refractory prostate cancer. J Pathol 218(4):505–513
75. Meeks JJ, Schaeffer EM (2011) Genetic regulation of prostate development. J Androl 32(3):210–217
76. Heinlein CA, Chang C (2004) Androgen receptor in prostate cancer. Endocr Rev 25(2):276–308
77. Buchanan G et al (2001) Contribution of the androgen receptor to prostate cancer predisposition and progression. Cancer Metastasis Rev 20(3–4):207–223
78. Veldscholte J et al (1992) Anti-androgens and the mutated androgen receptor of LNCaP cells: differential effects on binding affinity, heat-shock protein interaction, and transcription activation. Biochemistry 31(8):2393–2399

79. Veldscholte J et al (1990) A mutation in the ligand binding domain of the androgen receptor of human LNCaP cells affects steroid binding characteristics and response to anti-androgens. Biochem Biophys Res Commun 173(2):534–540
80. Maher CA et al (2009) Transcriptome sequencing to detect gene fusions in cancer. Nature 458(7234):97–101
81. Druker BJ et al (1996) Effects of a selective inhibitor of the Abl tyrosine kinase on the growth of Bcr-Abl positive cells. Nat Med 2(5):561–566
82. Maher CA et al (2009) Chimeric transcript discovery by paired-end transcriptome sequencing. Proc Natl Acad Sci USA 106(30):12353–12358
83. Fullwood MJ et al (2009) Next-generation DNA sequencing of paired-end tags (PET) for transcriptome and genome analyses. Genome Res 19(4):521–532
84. Sboner A et al (2010) FusionSeq: a modular framework for finding gene fusions by analyzing paired-end RNA-sequencing data. Genome Biol 11(10):R104
85. Pflueger D et al (2011) Discovery of non-ETS gene fusions in human prostate cancer using next-generation RNA sequencing. Genome Res 21(1):56–67
86. Inaki K et al (2011) Transcriptional consequences of genomic structural aberrations in breast cancer. Genome Res 21(5):676–687
87. Wang XS et al (2009) An integrative approach to reveal driver gene fusions from paired-end sequencing data in cancer. Nat Biotechnol 27(11):1005–1011
88. Mani RS et al (2009) Induced chromosomal proximity and gene fusions in prostate cancer. Science 326(5957):1230
89. Lin C et al (2009) Nuclear receptor-induced chromosomal proximity and DNA breaks underlie specific translocations in cancer. Cell 139(6):1069–1083
90. Fullwood MJ et al (2009) An oestrogen-receptor-alpha-bound human chromatin interactome. Nature 462(7269):58–64
91. Lieberman-Aiden E et al (2009) Comprehensive mapping of long-range interactions reveals folding principles of the human genome. Science 326(5950):289–293
92. Feinberg AP, Tycko B (2004) The history of cancer epigenetics. Nat Rev Cancer 4(2):143–153
93. Fraga MF et al (2005) Loss of acetylation at Lys16 and trimethylation at Lys20 of histone H4 is a common hallmark of human cancer. Nat Genet 37(4):391–400
94. Herman JG, Baylin SB (2003) Gene silencing in cancer in association with promoter hypermethylation. N Engl J Med 349(21):2042–2054
95. Perry AS et al (2010) The epigenome as a therapeutic target in prostate cancer. Nat Rev Urol 7(12):668–680
96. Kang GH et al (2004) Aberrant CpG island hypermethylation of multiple genes in prostate cancer and prostatic intraepithelial neoplasia. J Pathol 202(2):233–240
97. Coolen MW et al (2010) Consolidation of the cancer genome into domains of repressive chromatin by long-range epigenetic silencing (LRES) reduces transcriptional plasticity. Nat Cell Biol 12(3):235–246
98. Maruyama R et al (2002) Aberrant promoter methylation profile of prostate cancers and its relationship to clinicopathological features. Clin Cancer Res 8(2):514–519
99. Yegnasubramanian S et al (2008) DNA hypomethylation arises later in prostate cancer progression than CpG island hypermethylation and contributes to metastatic tumor heterogeneity. Cancer Res 68(21):8954–8967
100. Chen Z et al (2010) Histone modifications and chromatin organization in prostate cancer. Epigenomics 2(4):551–560
101. Yamane K et al (2006) JHDM2A, a JmjC-containing H3K9 demethylase, facilitates transcription activation by androgen receptor. Cell 125(3):483–495
102. Metzger E et al (2005) LSD1 demethylates repressive histone marks to promote androgen-receptor-dependent transcription. Nature 437(7057):436–439
103. Wang Q et al (2009) Androgen receptor regulates a distinct transcription program in androgen-independent prostate cancer. Cell 138(2):245–256
104. Yu J et al (2007) A polycomb repression signature in metastatic prostate cancer predicts cancer outcome. Cancer Res 67(22):10657–10663

105. Clayton AL, Hazzalin CA, Mahadevan LC (2006) Enhanced histone acetylation and transcription: a dynamic perspective. Mol Cell 23(3):289–296
106. Seligson DB et al (2005) Global histone modification patterns predict risk of prostate cancer recurrence. Nature 435(7046):1262–1266
107. Ellinger J et al (2010) Global levels of histone modifications predict prostate cancer recurrence. Prostate 70(1):61–69
108. Suikki HE et al (2010) Genetic alterations and changes in expression of histone demethylases in prostate cancer. Prostate 70(8):889–898
109. Kahl P et al (2006) Androgen receptor coactivators lysine-specific histone demethylase 1 and four and a half LIM domain protein 2 predict risk of prostate cancer recurrence. Cancer Res 66(23):11341–11347
110. Varambally S et al (2002) The polycomb group protein EZH2 is involved in progression of prostate cancer. Nature 419(6907):624–629
111. Weichert W et al (2008) Histone deacetylases 1, 2 and 3 are highly expressed in prostate cancer and HDAC2 expression is associated with shorter PSA relapse time after radical prostatectomy. Br J Cancer 98(3):604–610
112. Halkidou K et al (2004) Upregulation and nuclear recruitment of HDAC1 in hormone refractory prostate cancer. Prostate 59(2):177–189
113. Kobayashi Y et al (2011) DNA methylation profiling reveals novel biomarkers and important roles for DNA methyltransferases in prostate cancer. Genome Res 21(7):1017–1027
114. Huang ZQ et al (2003) A role for cofactor-cofactor and cofactor-histone interactions in targeting p300, SWI/SNF and mediator for transcription. EMBO J 22(9):2146–2155
115. Serandour AA et al (2011) Epigenetic switch involved in activation of pioneer factor FOXA1-dependent enhancers. Genome Res 21(4):555–565
116. Heinz S et al (2010) Simple combinations of lineage-determining transcription factors prime cis-regulatory elements required for macrophage and B cell identities. Mol Cell 38(4):576–589
117. Iljin K et al (2006) TMPRSS2 fusions with oncogenic ETS factors in prostate cancer involve unbalanced genomic rearrangements and are associated with HDAC1 and epigenetic reprogramming. Cancer Res 66(21):10242–10246
118. Olopade OI et al (2008) Advances in breast cancer: pathways to personalized medicine. Clin Cancer Res 14(24):7988–7999
119. Bjorkman M et al (2008) Defining the molecular action of HDAC inhibitors and synergism with androgen deprivation in ERG-positive prostate cancer. Int J Cancer 123(12):2774–2781
120. Wright JL, Lange PH (2007) Newer potential biomarkers in prostate cancer. Rev Urol 9(4):207–213
121. Mistry K, Cable G (2003) Meta-analysis of prostate-specific antigen and digital rectal examination as screening tests for prostate carcinoma. J Am Board Fam Pract 16(2):95–101
122. Bangma CH, Roemeling S, Schroder FH (2007) Overdiagnosis and overtreatment of early detected prostate cancer. World J Urol 25(1):3–9
123. Andriole GL et al (2009) Mortality results from a randomized prostate-cancer screening trial. N Engl J Med 360(13):1310–1319

Chapter 9
Ischemia, Reactive Radicals, Redox Signaling and Hematopoietic Stem Cells

Suman Kanji, Vincent J. Pompili, and Hiranmoy Das

Contents

Abstract Ischemia is a medical condition generated by inadequate or no blood flow to the organs such as heart, brain, limbs and kidney, contributes to the pathophysiology of several diseases such as myocardial infarction, cerebral ischemia (stroke), peripheral vascular insufficiency and hypovolemic shock etc. As a result of ischemic insult, cascades of metabolic and ultra structural changes occur at the cellular level resulted in irreversible tissue injury and cell death. Several pharmacological interventions such as antioxidant therapy, anti-inflammatory

S. Kanji • V.J. Pompili • H. Das Ph.D. (✉)
Cardiovascular Stem Cell Research Laboratory, The Dorothy M. Davis Heart and Lung Research Institute, The Ohio State University Medical Center, 460 W. 12th Avenue, BRT 382, Columbus, OH 43210, USA
e-mail: hiranmoy.das@osumc.edu

R.K. Srivastava and S. Shankar (eds.), *Stem Cells and Human Diseases*,
DOI 10.1007/978-94-007-2801-1_9,

therapy along with reperfusion strategies have been tried for several decades to protect the affected tissue from necrosis and cell death after ischemic insult. But, none of these interventions had proven to be cure for ischemia. Thus, novel therapeutic approach such as stem cell becomes very popular in the field of regenerative medicine to treat post ischemic conditions and to improve patient lifestyle as an alternate. Hematopoietic stem and progenitor cells from different sources showed promise to the ischemic therapy. Human umbilical cord blood-derived stem cells are one of the most promising cells that have shown beneficial effects in treating ischemic disorders in pre-clinical models and some clinical trials. However, several hurdles such as adequate supply of stem cells, suitable route of administration, need to be addressed to use as a regular therapeutic regiment for ischemic patients in the clinic.

Keywords Hematopoietic stem cells • Redox signaling • Stem cell therapy • Leukemia

9.1 Introduction

The term ischemia originated from Greek word "ischaimía" where "isch" means restriction and "haema" means blood. In medical science, ischemia can be described as an inadequate or no supply of blood to a part of the body due to total constriction or partial blockage of blood vessels supplying it. Ischemia is a condition rather than a disease. In ischemic condition, due to inadequate supply of blood flow to the organs such as heart, brain, limbs etc. become necrotic. Ischemia is developed due to the occlusion of coronary artery or cerebral artery or peripheral artery with atherosclerotic plug. Thus, ischemia contributes to the pathophysiology of many conditions, such as myocardial infarction, cerebral ischemia (stroke), peripheral vascular insufficiency and hypovolemic shock etc. A variety of cellular metabolic and ultra structural changes occur as a result of prolonged ischemia. Thus, restoration of blood flow and revascularization of ischemic tissue is the primary aim to prevent irreversible cellular injury, and salvaging the viable tissue at the ischemic zone. However, restoration of blood flow to these ischemic organs mostly augments tissue injury that produced by ischemia alone. Various thrombolytic agents or surgical techniques are used to dissolve or remove the clot and to restore blood flow to ischemic tissues. It was evident that pathological changes of injury after 3 h of intestinal ischemia followed by 1 h of reperfusion are far worse than the changes observed after 4 h of ischemia alone [74]. Thus, reperfusion acts as two-edged sword, which have no alternative to salvage the viable tissue but post perfusion injury is a big concern for cellular damage. As a result, different tissue protective agents have been tried to prevent these damages after revascularization but none has become a standard regular clinical regimen. Thus, ischemia of different vital and non-vital organs becomes a primary concern for the clinicians as well as for the researchers due to its multimodal complexity of cellular events and lack of suitable pharmacological therapy. However, regenerative therapy by using stem cell is getting serious attention for treating the ischemia related disorders to regenerate affected tissue, improve affected organ function, reduce over all mortality and improve lifestyle of the patients. Thus, stem cell therapy, especially hematopoietic stem cell, may be a suitable candidate for treating ischemia related disorders as evident by several preclinical and clinical studies [76, 86, 88].

9.2 Development of Ischemic Injury

Prolonged ischemia mediates a variety of cellular metabolic and ultra structural changes such as, altered membrane potential, altered ion-distribution, cellular swelling, cytoskeletal disorganization, decreased phosphocreatine, decreased glutathione and acidosis (Fig. 9.1). During ischemic events, deficiency of oxygen occurs to the tissues due to loss of blood supply or inadequate blood supply. With the lack of oxygen, mitochondrial oxidative phosphorylation is hindered and resulted to failure in

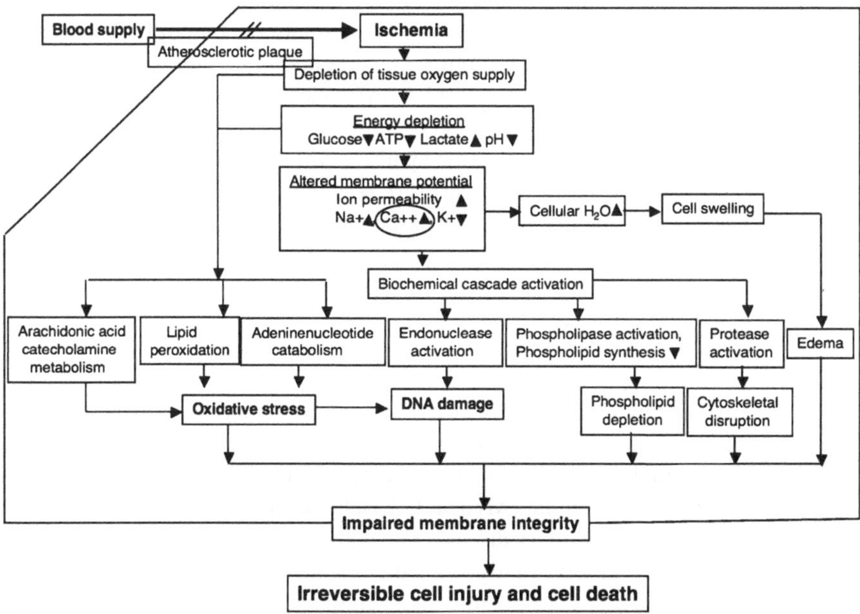

Fig. 9.1 Schematics of cascade of various molecular events leading to irreversible ischemic cellular injury and cell death. Inadequate blood supply to the tissue leads to inadequate production of ATP. Reduced number of ATP resulted to imbalance in sodium and potassium gradients and calcium overload. Higher concentration of calcium causes activation of phospholipases and proteases, leads to cellular ultra structural changes. Accumulation of lactate, hydrogen ion and reduced nicotinamide adenine dinucleotide inhibits ATP production. Adenine nucleosides and bases generate free radicals via the xanthine oxidase reactions generate oxidative stress and causes DNA damage. As a result of increased intracellular catabolites, cell-swelling and causes rupture of cell membrane, which leads to cellular damage and cell death

producing energy-rich phosphates such as, adenosine 5′-triphosphate (ATP) and phosphocreatine, which are essential for energy metabolism. As a result, anaerobic glycolysis takes place for ATP production leads to accumulation of hydrogen ions and lactate resulted to intracellular acidosis and inhibition of glycolysis. Mitochondrial and cellular membrane structure altered, ATP-dependent ionic pump function is disrupted, favoring the entry of calcium, sodium, and water into the cell. As a result of accumulated metabolites and inorganic phosphates in the ischemic tissue, an increased K^+ efflux along with osmotic load takes place. An increase in cytosolic Ca^{2+} triggers the activation of proteases and phospholipases resulted to cytoskeletal damage and impaired balance in membrane phospholipids (PL). The alteration of lipids due to ischemic event is a combined incidence of increased degradation of PL with the release of free fatty acids (FFA), lysophospholipids (LPL) and decreased synthesis of phospholipids. The accumulation of amphipathic lipids resulted in altered membrane fluidity. Free radicals are also generated in ischemia, partly from the excess electrons (e^-) produced by lipid peroxidation in oxygen-deprived mitochondria. Free radicals are also derived from the metabolism of arachidonic acid and catecholamine. Moreover, during ischemia catabolism of adenine nucleotide results in intracellular accumulation of hypoxanthine, which is subsequently converted into toxic reactive oxygen species (ROS) upon the reintroduction of molecular oxygen. In the endothelium, ischemia promotes expression of certain proinflammatory factors (e.g., leukocyte adhesion molecules, cytokines) and bioactive agents (e.g., endothelin, thromboxane A2), while repressing other cells to produce protective molecules (e.g., constitutive nitric oxide synthase, thrombomodulin) and bioactive agents (e.g., prostacyclin, nitric oxide) [20, 67]. Thus, ischemia also induces a proinflammatory condition that increases tissue vulnerability.

9.3 Types of Ischemia

There are various types of ischemia, depending to the organs such as heart, brain, limbs, and kidney experiencing the ischemic insult. Although the nature of ischemia may differ based on the organ involved but their post ischemic sequences of cellular events are more or less similar. Myocardial ischemic injury develops as clinical manifestations of coronary artery disease. The pathogenesis of myocardial ischemic injury demonstrated experimentally with the coronary occlusion model in dog, where irreversible myocardial ischemic injury initiated after 20 min of occlusion in the subendocardium region and progressed into the subepicardium of the ischemic myocardial bed-at-risk [47]. The anatomical and biochemical substrates for ischemic myocardial injury and death are well characterized [17]. Normal cerebral blood flow (CBF) in man is typically in the range of 45–50 ml/min/100 g between a mean arterial pressure (MAP) of 60 and 130 mm of Hg [26]. When CBF falls below 20–30 ml/min/100 g, marked disturbances in brain metabolism begin to occur, such as water and shifts in electrolyte and the regional areas of the cerebral cortex experiences failure in perfusion. At blood flow rates below 10 ml/min/100 g, sudden depolarization of the neurons occurs with a rapid loss of intracellular potassium [40]. Early observations on the mechanisms of ischemic injury focused on relatively simple biochemical and physiological changes resulted from interruption of circulation. For example, loss of high-energy compounds [78], acidosis due to anaerobic generation of lactate [77], and no re-flow due to swelling of astrocytes with compression of brain capillaries [6]. Subsequent finding has shown the problem to be far more complex than previously thought, involved interactions of many factors [54]. Like heart and brain, peripheral vasculatures are also vulnerable for the development of ischemia. In conditions like peripheral artery disease, it has been found that when arterial obstruction exceeds 50% of the internal lumen, approximately 20–25% patients will require revascularization and fewer than 5% will lead to critical limb ischemia [46]. Skeletal muscles, predominant tissues in the limb are affected due to the ischemic arteries. Physiological and anatomical studies showed that irreversible muscle cell damage starts after 3 h of ischemia and is nearly completed in 6 h of ischemia [43].

9.4 Factors Involved in Ischemic Disorders

Various factors are responsible for the development of ischemic conditions including systemic arterial hypertension, left ventricular hypertrophy, hyperlipidemia, atherosclerosis, diabetes, insulin resistance, heart failure, aging and stress. These factors frequently co-exist with other disease states.

9.4.1 Hyperlipidemia

Hyperlipidemia is considered as an independent risk factor for the development of ischemic heart disease including myocardial infarction. Epidemiological studies showed that there is a strong relationship between the elevation of serum total cholesterol level and the morbidity and mortality of myocardial infarction [31]. Total cholesterol primarily consists of the low-density lipoprotein (LDL cholesterol) and high-density lipoprotein (HDL cholesterol), which have opposite associations with ischemic heart disease (IHD). Results from randomized trials suggest that treatment with a LDL-cholesterol lowering agent like statin, reduces the incidence of IHD significantly [9]. These trials have also demonstrated a substantial reduction in the incidence of ischemic stroke (without increase the incidence of hemorrhagic stroke). However, several trials contrast the epidemiological studies of direct relationship between blood cholesterol and stroke [99]. Recent, large studies confirmed the positive correlation of total cholesterol and stroke. After meta-analysis of almost 900,000 individuals in 61 prospective observational collaborative studies, it was shown that there was an incidence of 55,000 vascular deaths in 12 million individuals per years of follow-up. These studies have reliably characterized the age-specific associations of total cholesterol with IHD, stroke, and other vascular mortality, and have assessed the quantitative and qualitative relevance of other risk factors to these associations [64].

9.4.2 Diabetes

Diabetes is of two types, type 1 (insulin-dependent) and type 2 (non-insulin-dependent). It has been found from different epidemiological studies and clinical trials that both type 1 and type 2 diabetic patients are prone to developing ischemic heart disease, including acute myocardial infarction and post-ischemic complications [97]. Ischemic heart disease accounts for more than 50% of deaths in diabetic patients [52]. It has also been observed that mortality due to acute myocardial infarction is almost doubled in diabetic patients compared to non-diabetics [1]. The frequency of hospitalization also increased in myocardial infarction patient when accompanied with diabetes [30]. Thus, treatment of patients with diabetes who have underlying ischemic heart disease is challenging because of its complex pathophysiological condition [56]. Although, preclinical studies of myocardial ischemia in diabetic animal models are controversial and inconclusive, large-scale human studies have shown that both type 1 and type 2 diabetes increase the ischemic cardiac injury and increase overall cardiovascular risk [45]. In a large clinical trial, it was shown that diabetes is independently associated with 30-day or 1-year mortality after acute coronary syndrome while receiving thrombolytic treatment [28]. These data suggest that diabetes is very closely associated with ischemic heart disease and increase severity and mortality of myocardial infarction patient.

9.4.3 Mental Stress

Mental stress induces myocardial ischemia in almost one third to one half of patients with cardiovascular associated diseases (CAD). Ischemia can be induced by behavioral challenges that we encounter in everyday life but not by extremely severe emotional stress [39]. Mental stress-induced ischemia is typically without pain, and occurs at lower levels of oxygen demand than physical exercise. Stress-induced hemodynamic changes are associated with the increase in systemic vascular resistance, coronary artery vasoconstriction, microvascular changes, and may all contribute to the development of ischemia. No pharmacological interventions have shown consistency in blocking mental stress-induced ischemia [79]. There is also variability in results for the development of cardiovascular ischemic disorders in response to mental stress. Limited evidences claim that stress-induced responses can predict adverse coronary outcomes as an independent risk factor. Mental stress testing may provide a tool of evaluating the role of emotional factors in acute coronary syndromes and sudden cardiac death [39]. After summarization of 34 different correlative studies of myocardial ischemia and mental stress, it has been shown that mental stress induces transient myocardial ischemia in about 30% of CAD patients using electrocardiographic analysis, 37–41% of patients with CAD based on decreased ejection fraction or wall motion abnormalities, and 75% of patients with CAD have perfusion abnormalities measured by scintigraphy or PET scanning. Whereas, Rates of experimental stress induced in patients without CAD or in healthy volunteers are in the range of 16% (VEST studies) to 21% by radio nuclei ventriculography analysis [85]. In these experimental set-up differences in the sensitivity of the method of assessing myocardial function is responsible for some of the variation in rates of experimental stress. Variability between studies using the same assessment technique is a result of other factors such as, the selection of patient, severity of CAD, cardiac history, age, nature of stressors etc. Some mental stress tests are more potent than others in stimulating myocardial ischemia. Hence, mental stress testing is not yet an integral part of the routine clinical investigation of patients with suspected or proven CAD.

9.4.4 Aging

Ischemic cardiovascular diseases related mortality is significantly higher in elderly individuals than in young adults. Impaired diastolic and systolic functions are the features of the aged heart. Oxidative stress is one of the major factors contributing to the alterations in cardiac function [23]. Several mechanisms have been proposed for the increased oxidative stress in aged hearts, such as upregulation of the angiotensin II type 1 receptors and subsequent activation of NADPH oxidases [71]. Increased cardiac monoamine oxidase-A activity, increased mitochondrial free oxy-radical formation and decreased mitochondrial oxidative defense are other proposed contributing factors to ischemic cardiovascular disorders in aged heart [82]. In healthy heart, within the mitochondrial matrix, reactive species are eliminated by super oxide dismutase (SOD) enzyme together with catalase and glutathione redox system [63]. In aged heart, production of these antioxidant enzymes is impaired resulted to oxidative stress [50] and become more vulnerable to ischemic insult. Ultimately, overall increased oxidative stress causes protein, lipid, and DNA oxidation contributing to contractile failure of heart [53]. It has also observed that Cytochrome c release from mitochondria in the left ventricle increases with age leading to programmed cell death and contributes to a reduction in cardiomyocyte number and an increase in the extent of fibrosis [12]. The aged myocardium showed a reduced tolerance to ischemia mediated cardiac injury [2]. In experimental model using C57/BL6 mice the myocardial tolerance to ischemia shown to decrease at 12 months of age [94]. In clinical study, similar results were also found, supporting the idea of reduced tolerance toward ischemic injury in aged hearts. Retrospective analysis of the TIMI-4B trial revealed that patients with 60 years and older age had a higher rate of death compared with younger patients due to ischemic cardiovascular events [57].

9.5 Role of Reactive Radicals in Ischemia

In ischemic condition reactive radicals such as reactive oxygen species (ROS) and reactive nitrogen species (RNS) are generated. ROS is a collective name for a group of oxygen-containing molecules such as superoxide anion radical ($O_2^{\cdot-}$), hydroxyl radical (OH·) and hydrogen peroxide (H_2O_2). ROS plays a critical role in tissue injury that follows ischemia and reperfusion [102]. After reperfusion the exposure of O_2 to the ischemic tissues is a critical event for the generation of ROS [5]. However, it is also evident that the generation of ROS occurs during ischemia that damages the tissues such as cardiomyocytes in the heart before reperfusion process [15]. During ischemia, redox-reduced electron carriers in the respiratory chain directly transfer electrons to the residual molecular O_2 and generate superoxide [42]. Reactive nitrogen species (RNS) include nitrogen dioxide radicals (NO2·), peroxynitrite (ONOO-), and nitroxyl (HNO), which are derived from NO· by oxidation or reduction processes. A number of defense mechanisms are involved to counteract the accumulation of reactive species radicals. These defense mechanisms include enzymatic scavengers such as catalase, glutathione peroxidase and superoxide dismutase (SOD). During ischemia and reperfusion there is an imbalance between formation of ROS, RNS and antioxidant defense mechanism and as a result oxidative stress is generated within the ischemic tissue. Oxidative stress is implicated in a wide variety of heart and vascular diseases, such as myocardial ischemia/reperfusion injury and atherosclerosis [62].

In ischemic tissue, ROS randomly react with DNA, protein and lipid molecules. ROS oxidize DNA bases and form 8-hydroxyguanosine (8-OHG) leading to DNA strand breakage [100]. ROS also reacts with amino acid side chains of proteins and form protein carbonyl, methionine sulfide etc. [89]. When ROS reacts with lipids, lipid peroxidation takes place in the ischemic tissue as a result of peroxyl and alkoxyl radical formations [4]. In case of reperfusion of ischemic tissues, generation of ROS or RNS occur resulted to massive oxidative stress in ischemic tissues leading to tissue necrosis [33]. After reperfusion of ischemic tissues $O_2^{\cdot-}$ react with NO· and forms $ONOO^-$ (peroxynitrite). As a result, the level of NO· is reduced and $ONOO^-$ level is increased in the ischemic-reperfused tissue and contributes to the tissue damage especially in myocardium [32, 72]. Thus, high concentration of $ONOO^-$ is considered to be highly toxic to the cell especially to the vulnerable ischemic tissue [72]. Peroxynitrite cytotoxicity is a result of its reaction with proteins (on tyrosine and thiol groups), lipids, and DNA [11, 18]. $ONOO^-$ inactivates prostacyclin synthase by tyrosine nitration, contributing to the pathophysiology of the cardiovascular system [101]. Excessive reactive species are also considered deleterious to the cell function, especially in mitochondria, where they act as inducers of mitochondrial permeability transition pore (mPTP) opening. The mPTP is a nonspecific mega channel of the mitochondrial membrane. Prolonged opening of these channels during the reperfusion of blood in the ischemic tissue promotes cell death especially in cardiomyocytes [66].

9.6 Redox Signaling in Ischemia

Unlike oxidative stress, low levels of reactive species act as secondary messengers and modulate signaling pathways by covalent modification of target molecules referred to as redox signaling. Both ROS and RNS are not destructive in physiological condition rather contribute to the signaling of mitochondria. In redox signaling, proteins may undergo reversible chemical changes in response to changes in local redox potential. Thus, transmission and amplification of signals generally involves posttranslational redox protein modifications in presence of reactive species. The reactive species-mediated molecular signals like modification of redox state transduce into the cell organelles and the nucleus, which resulted as modification functions such as resistance to stress, senescence, and programmed cell death in physiological or pathophysiological conditions. For example, low level of reactive oxygen species found to be cardioprotective in ischemia reperfusion injury [49]. Posttranslational redox modifications of myocardial proteins, such as cysteine and methionine thiol oxidation, proline and arginine hydroxylation, and tyrosine nitration, affect the conformation, stability, and activity of diverse receptors, ion transporters, kinases, phosphatases, caspases, transcription factors, and structural or contractile proteins [41]. In oxidative state, s-nitrosylation of cysteine residues of protein occurs by incorporation of a $NO^{\cdot}$ moiety to a sulfur atom. In ischemic and hypoxic condition, the redox state and ultrastructure of cysteine residue under low oxygen pressure may determine the occurrence of s-nitrosylation of a particular protein [80]. Cysteine residues within proteins can adopt multiple oxidation states and can react with reactive species to yield a number of species such as disulfide cross linking, S-nitrosylated proteins, and the formation of mixed disulfides with glutathione [41]. S-nitrosylation and glutathionylation of proteins can be considered pivotal redox signaling in ischemic condition. Proteins phosphatases such as PTP1, PTEN, caspase-3, and STAT3 are widely modulated by such thiol-redox switching between reduced and oxidized states [29]. In cardiac tissue, redox modulated proteins such as protein kinase A (PKA), protein kinase G (PKG), the ryanodine receptor (RyR) [16, 96] and several others affect the cardiac function.

9.7 Current Therapy for Ischemia Related Disorders

Many therapeutic strategies have been used to treat and prevent ischemia-induced cellular damage in heart and brain. There are several therapeutic or preventive medicines are available in the market but none of them have really shown a cure for ischemia-mediated cell injury (Table 9.1). However, the primary aim of treating myocardial or cerebral ischemia is to restore the circulation at the ischemic zone by dissolving clot with the use of medicine (fibrinolysis) or other surgical procedures (percutaneous coronary intervention, and emergency coronary artery bypass grafting).

Table 9.1 Current pharmacological interventions for ischemic disorders and their mechanisms of action

Current pharmacological interventions for ischemia	
Therapeutic interventions	Mechanisms of action
Fibrinolysis: Recombinant tissue plasminogen activator	Restores blood circulation to the affected organs
Surgical procedure: Percutaneous coronary intervention Coronary artery bypass grafting	To restore blood circulation to the ischemic organs
Antioxidant therapy: Superoxide dismutase Catalase Mannitol Allopurinol Vitamin E N-acetylcysteine Iron chelating agents	To prevent or attenuate the extent of cellular damage resulted from ischemia
Anti-complement treatment: C3, C5 inhibitors	To inhibit complement cascades
Anti-inflammatory therapy: Interleukin-1 receptor antagonist Antitumor necrosis Factor-α antibody Leukotriene B4 antagonist	To inhibit or attenuate effects of inflammatory mediators and inflammatory responses

There are several anti-inflammatory and anti-oxidant therapy are available along with reperfusion strategy. In the absence of reperfusion, no intervention is able to limit development of infarction (in heart) or development of core surrounded by penumbral region, in case of stroke lesion. Thus, reperfusion and revascularization therapies are the primary aim of salvaging viable tissue at the ischemic zone [81]. In myocardial infarction, ischemic preconditioning has become beneficial as a non-pharmaceutical intervention. Ischemic preconditioning refers to the phenomenon by which exposure of cardiac tissue to a brief period of ischemia protects them from the harmful effects of prolonged ischemia and reperfusion in myocardial infarction or coronary artery disease. It has been shown that preconditioning improves ventricular function, decrease neutrophil accumulation in the myocardium and apoptosis after ischemic reperfusion in preclinical models [48]. In clinical studies, ischemic preconditioning also have a protective effect on recovery of right ventricular contractility in patients who had coronary artery bypass grafting [95]. Antioxidant therapy, such as the use of superoxide dismutase, catalase, mannitol, allopurinol, vitamin E, N-acetylcysteine, iron chelating compounds have been found efficacious in preventing or attenuating cardiac and cerebral ischemia as well as reperfusion injury in experimental animal studies [59]. In clinical studies beneficial effects of superoxide dismutase have been demonstrated in cardiac ischemic reperfusion injuries [27]. However, the use of these antioxidants in clinical practice for cardiac ischemia or reperfusion injuries, yet to be established. Many compounds that were claimed to be free radical scavengers in cerebral ischemia have been tried in different clinical trials with limited success [35]. The limited success of anti-oxidant therapy in ischemic and reperfusion injury is probably due to insufficient correlation between the preclinical and clinical conditions [34]. Anti-complement treatment (C3, C5 inhibitor) has found to be effective in both preclinical and clinical studies of myocardial ischemia reperfusion injuries, yet to be established. In ischemic condition and also ischemia-reperfusion, leukocyte activation is mediated by the release of inflammatory mediators such as histamine, platelet activation factor, leukotriene B4, and tumor necrosis factor α [73]. In this pathophysiological condition, leukocyte activation can be attenuated by the inhibition of inflammatory mediator release or receptor blocking using therapeutic agents such as soluble interleukin-1 receptor antagonists, anti–tumor necrosis factor α antibodies, or platelet activation factor, leukotriene B4 antagonists etc. However, these anti-leukocyte strategies have received limited attention in the clinical setting irrespective of their effectiveness in animal models of ischemia and reperfusion [22]. Dissolving the blood clot in the artery of the affected brain after acute ischemic stroke, can be treated. Numerous attempts have been made to find the best thrombolytic agent. The use of intravenous (IV) streptokinase in treating ischemic stroke demonstrated an unacceptably high risk of fatal intracranial haemorrhages and death. The efficacy of recombinant tissue plasminogen activator (rt-PA) has found to be very effective in treating acute ischemic stroke. In clinical practice, the treatment with rt-PA of acute ischemic stroke within a 3 h time window found to be very effective [60].

9.8 Stem Cell Therapy for Ischemia

Cardiomyocytes undergo necrosis or death during ischemia and also after reperfusion in a complicated manner. In search of cardioprotective therapeutics, several interventions were found, such as Na^+ -H^+ exchange inhibitors, activation of kinases, perfusion with erythropoietin, inhibition of protein kinase C (PKC)-δ, inhibitors of the mitochondrial permeability transition pore (MPT), inhibition of glycogen synthase kinase (GSK)-3β [36–38, 44], and other interventions protect heart in cardiac ischemic episode. However, none of these interventions have found to be effective in myocardial ischemia during clinical trials [14]. Hence, none of them are used in regular clinical practice. Cell-based therapy is a currently considered as a promising approach for the cardiac repair in patients with ischemic heart resulted from coronary artery disease. In preclinical and early clinical studies, investigators have preliminary evidence showing that stem cell therapy can safely and effectively improve myocardial perfusion and left ventricular function. Stem cell therapy also decreases left ventricular remodeling in cases of myocardial infarction (MI) and may alleviate symptoms and prevent cardiac enlargement in chronic ischemic heart disease. Different types of stem cells have shown promises in functional improvement of heart after myocardial ischemia, such as bone marrow [76], endothelial progenitor cells (EPC) [8], mesenchymal stem cells (MSC), embryonic stem cells, cardiac stem cells [75].

In cerebral ischemia, various cells of central nervous system will undergo apoptosis and necrosis. Although, various drugs have shown promises as neuroprotective agent in preclinical and clinical conditions, none of them are used for regular clinical practice [98]. However, surgical decompression can reduce mortality in severe cases [90], the only currently available intervention to reduce the size of the infarct is recombinant tissue plasminogen activator (t-PA). Innovative approaches, such as stem cell transplantation offers a promising new therapeutic avenue for cerebral ischemic insult not only to prevent damage as a conventional therapeutic strategy but also regenerate or repair the injured brain after ischemic episode. Stem cell transplantation has shown much promise in experimental models of cerebral ischemia with a diverse array of stem cells obtained from bone marrow, blood or central nervous system, reported to enhance functional recovery and reduction of affected area after ischemic insult [13]. Several phase I and II clinical trials were performed by using a cell line of immature neurons (hNT) derived from a human teratocarcinoma and autologous MSCs [93]. The aim of these studies was the safety and feasibility of cell transplantation therapy. No cell-related adverse effects were reported with the hNT [58] or MSC transplants [10].

9.9 Hematopoietic Stem Cell Therapy for Ischemia

Hematopoietic stem cells (HSC) transplantation has been most established regenerative therapy in various malignant and non-malignant blood disorders. Its role in cardiac regeneration in ischemic events is under extensive investigation. Hematopoietic stem and progenitor cells are obtained from several sources, such as bone marrow, umbilical cord blood and peripheral blood. Multiple pre-clinical and clinical studies have claimed the beneficial effects of these cells in ischemic conditions of various organ systems. Bone marrow derived stem and progenitor cells, such as endothelial progenitor cells (EPC), HSC and hematopoietic progenitor cells (HPC) can restore tissue vascularization after ischemic events in limbs [7], myocardium, endothelium [68], retina [70], brain [21], resulted to regeneration and tissue repair of ischemic organs (Fig. 9.2). In preclinical studies different populations of stem and progenitor cells including CD34+ obtained from bone marrow contribute to myocardial regeneration and revascularization after cardiac ischemic events [69]. Autologous bone marrow cells contribute to neoangiogenesis in the ischemic myocardium in preclinical rodent and porcine models [51]. Several clinical studies also support the preclinical observations of neovascularization. In human subjects, improvement in cardiac function has been observed when autologous bone marrow or peripheral stem cell is injected directly into damaged myocardium after ischemic event [88]. Transplantation of purified populations of autologous bone marrow derived CD133+ cells also improves cardiac function in acute myocardial ischemic

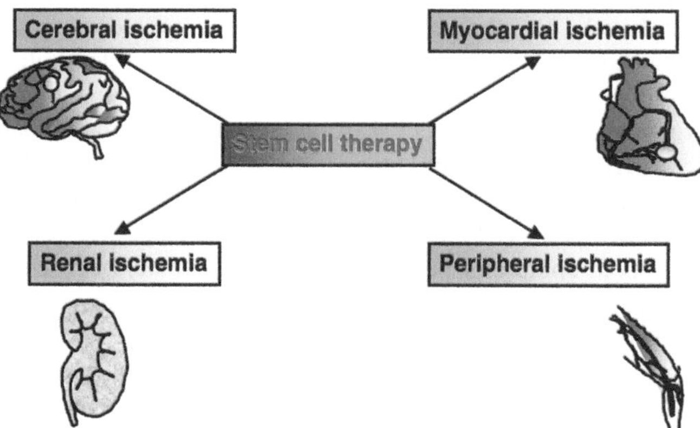

Fig. 9.2 Stem cell therapy for various ischemic disorders. Stem cell therapy helps to regenerate damaged organs or helps to improve functionality of the organs like heart, brain, kidney and peripheral regions after ischemic insult

patients [3, 83]. All these evidences encourage the application of cell therapy in cardiac regeneration after ischemic event. Cord blood-derived stem and progenitor cells also have shown improved cardiac functions in sheep and porcine cardiac ischemic model [55]. Cord blood-derived CD133+ cells also enhance function and repair of myocardial infarction in athymic nude rats [91]. Thus, cord blood-derived stem and progenitor cells are a great resource and have potentials in transplantation therapy of myocardial ischemia.

In cerebral ischemia, hematopoietic stem and progenitor cells from various sources were found to be useful for transplantation. Several phase I and II clinical trials are underway around the world for treating cerebral ischemia by using CD34+ cells obtained from autologous bone marrow, peripheral blood. Human umbilical cord blood (HUCB) is enriched with CD34+ stem cells and can be a suitable agent for stem cell therapy in cerebral ischemia. Several preclinical studies have demonstrated its suitability of treating cerebral ischemia [21]. Different *in vitro* experiments also demonstrate the HUCB stem cells were able to form neuronal cells [19]. Thus, HUCB is having multifunctional biological therapy. The cells modify more than one physiological process to induce recovery from ischemic damage, such as neurogenesis, angiogenesis, immunomodulation and induction of host cell plasticity [21, 86, 92].

9.10 *Ex-vivo* Expansion of Hematopoietic Stem Cell and Its Implication in Ischemic Therapy

Cord blood is an attractive source for tissue regeneration due to its easy access and potential use in non-hematopoietic diseases [91]. Among several advantages of cord blood, the collection of stem cells from cord blood of a placental umbilical cord is much simpler than the complex collection from bone marrow or peripheral sources. Furthermore, an advantage of cord blood stem cells over adult stem cells is that they possess a primitive ontogeny and have not been exposed to immunologic challenges. It has been found that there is a lower risk of acute graft-versus-host-disease (GVHD) and chronic GVHD among the cord blood transplanted patients [84]. But the number of HSCs obtained from single cord is not sufficient for clinical application. Therefore, there is a dire need for a suitable method, by which HSCs can be expanded many folds without compromising their phenotype and stem cell characteristics. Ex-vivo expansion of HSC became very important in regenerative therapy for hematopoietic and ischemic disorders. In ex-vivo expansion, *in vivo* environment of bone marrow is simulated by using feeder-stromal layer, and by supplying cocktail of growth factors and cytokines to maintain microenvironment signaling cascades. In various studies it was shown that cocktail of cytokines and growth factors such as interleukin (IL)-3, IL-6, IL-11, thrombopoietin (TPO), fetal liver tyrosine kinase 3 (Flt3) ligand, stem cell factor (SCF) in serum free media increase the expansion of HSCs many folds with in a short period of time [65]. However, feeder layer was replaced currently by non-feeder expansion technology due to the complication generated from feeder layer. The chemical nature of different stromal layer culture is defined and as a result it is very challenging to maintain a balance between self renewal and differentiation in presence of stromal layer. Among several techniques of ex-vivo expansion, nanofiber expansion technology became promising for HSC expansion, where nanofiber is used as physical support combining with cytokines to achieve large number of HSCs. Our lab has successfully expanded CD133+/CD34+ cells from freshly isolated human cord blood by using PES nanofiber-scaffolds along with serum-free media containing TPO, Flt3, IL3, low density lipoprotein (LDL), and SCF [24]. The therapeutic potential of these expanded cells are promising for the ischemic conditions. It was found that after transplantation of expanded stem cell to the myocardial infarcted rat, an elevated heart function and neovascularization was observed compare to control group. Moreover, nanofiber expanded mononuclear cells were functionally more efficient than that of freshly isolated cord blood cells in terms of functional recovery and formation of neovascularization in ischemic heart [25]. The functional superiority of nanofiber expanded cells have been partially explained by the higher CXCR4 expression compared to freshly isolated cells [24]. Other group has also showed improved cardiac function in murine model of myocardial infarction after transplantation of expanded cells obtained from different ex-vivo expansion technique [87]. These preclinical evidences encourage using these cells in future clinical setting.

9.11 Conclusions

Ischemia is a complex pathophysiological condition, especially, in the heart and brain it is complicated with the reperfusion, which is mostly followed the ischemic event. The developmental process of myocardial and cerebral ischemic injury is complex and multifaceted, thereby difficult to target for therapeutic intervention pharmacologically. Enormous amount of tissue is damaged in the organs after ischemic insults. Thus, novel approach such as stem cell therapy could be an alternative to prevent tissue destruction and to restore damaged tissue through regeneration. However, several hurdles need to overcome for stem cell transplantation after development of ischemic conditions. In myocardial ischemia, it has been observed that the formation of edema and dilation of hemorrhage prone vessels were associated with the heterogeneous nature of stem cell transplantation [61]. Thus, identification and selection of appropriate homogeneous population of stem cells for transplantation in ischemic condition will provide a future successful intervention. Moreover, the local introduction of stem cells into the affected tissue is inappropriate due to the destruction of blood vessels in the ischemic tissue bed. Thus, Route of delivery of stem cells might have a critical role in the success of tissue revascularization. And finally, development of appropriate ex-vivo expansion protocols will provide large number of clinical grade homogeneous transplantable stem cells for effective therapeutic application.

Acknowledgements This work was supported in part by National Institutes of Health grants, K01 AR054114 (NIAMS), SBIR R44 HL092706-01 (NHLBI), R21 CA143787 (NCI) and The Ohio State University start-up fund. The funders had no role in study design, data collection and analysis, decision to publish or preparation of the manuscript.

References

1. Abbott RD, Donahue RP, Kannel WB, Wilson PW (1988) The impact of diabetes on survival following myocardial infarction in men vs women. The Framingham study. JAMA 260(23):3456–3460
2. Abete P, Cioppa A, Calabrese C, Pascucci I, Cacciatore F, Napoli C, Carnovale V, Ferrara N, Rengo F (1999) Ischemic threshold and myocardial stunning in the aging heart. Exp Gerontol 34(7):875–884
3. Adler DS, Lazarus H, Nair R, Goldberg JL, Greco NJ, Lassar T, Laughlin MJ, Das H, Pompili VJ (2011) Safety and efficacy of bone marrow-derived autologous CD133+ stem cell therapy. Front Biosci (Elite Ed) 3:506–514
4. Ambrosio G, Flaherty JT, Duilio C, Tritto I, Santoro G, Elia PP, Condorelli M, Chiariello M (1991) Oxygen radicals generated at reflow induce peroxidation of membrane lipids in reperfused hearts. J Clin Invest 87(6):2056–2066
5. Ambrosio G, Zweier JL, Duilio C, Kuppusamy P, Santoro G, Elia PP, Tritto I, Cirillo P, Condorelli M, Chiariello M et al (1993) Evidence that mitochondrial respiration is a source of potentially toxic oxygen free radicals in intact rabbit hearts subjected to ischemia and reflow. J Biol Chem 268(25):18532–18541
6. Ames A 3rd, Wright RL, Kowada M, Thurston JM, Majno G (1968) Cerebral ischemia. II. The no-reflow phenomenon. Am J Pathol 52(2):437–453
7. Asahara T, Masuda H, Takahashi T, Kalka C, Pastore C, Silver M, Kearne M, Magner M, Isner JM (1999) Bone marrow origin of endothelial progenitor cells responsible for postnatal vasculogenesis in physiological and pathological neovascularization. Circ Res 85(3):221–228
8. Asahara T, Murohara T, Sullivan A, Silver M, van der Zee R, Li T, Witzenbichler B, Schatteman G, Isner JM (1997) Isolation of putative progenitor endothelial cells for angiogenesis. Science 275(5302):964–967

9. Baigent C, Keech A, Kearney PM, Blackwell L, Buck G, Pollicino C, Kirby A, Sourjina T, Peto R, Collins R, Simes R (2005) Efficacy and safety of cholesterol-lowering treatment: prospective meta-analysis of data from 90,056 participants in 14 randomised trials of statins. Lancet 366(9493):1267–1278
10. Bang OY, Lee JS, Lee PH, Lee G (2005) Autologous mesenchymal stem cell transplantation in stroke patients. Ann Neurol 57(6):874–882
11. Beckman JS, Beckman TW, Chen J, Marshall PA, Freeman BA (1990) Apparent hydroxyl radical production by peroxynitrite: implications for endothelial injury from nitric oxide and superoxide. Proc Natl Acad Sci USA 87(4):1620–1624
12. Bernecker OY, Huq F, Heist EK, Podesser BK, Hajjar RJ (2003) Apoptosis in heart failure and the senescent heart. Cardiovasc Toxicol 3(3):183–190
13. Bliss T, Guzman R, Daadi M, Steinberg GK (2007) Cell transplantation therapy for stroke. Stroke 38(2 Suppl):817–826
14. Bolli R, Becker L, Gross G, Mentzer R Jr, Balshaw D, Lathrop DA (2004) Myocardial protection at a crossroads: the need for translation into clinical therapy. Circ Res 95(2):125–134
15. Bolli R, Patel BS, Jeroudi MO, Lai EK, McCay PB (1988) Demonstration of free radical generation in "stunned" myocardium of intact dogs with the use of the spin trap alpha-phenyl N-tert-butyl nitrone. J Clin Invest 82(2):476–485
16. Brennan JP, Bardswell SC, Burgoyne JR, Fuller W, Schroder E, Wait R, Begum S, Kentish JC, Eaton P (2006) Oxidant-induced activation of type I protein kinase A is mediated by RI subunit interprotein disulfide bond formation. J Biol Chem 281(31):21827–21836
17. Buja LM (1998) Modulation of the myocardial response to ischemia. Lab Invest 78(11): 1345–1373
18. Burwell LS, Brookes PS (2008) Mitochondria as a target for the cardioprotective effects of nitric oxide in ischemia-reperfusion injury. Antioxid Redox Signal 10(3):579–599
19. Buzanska L, Jurga M, Stachowiak EK, Stachowiak MK, Domanska-Janik K (2006) Neural stem-like cell line derived from a nonhematopoietic population of human umbilical cord blood. Stem Cells Dev 15(3):391–406
20. Carden DL, Granger DN (2000) Pathophysiology of ischaemia-reperfusion injury. J Pathol 190(3):255–266
21. Chen J, Li Y, Wang L, Zhang Z, Lu D, Lu M, Chopp M (2001) Therapeutic benefit of intravenous administration of bone marrow stromal cells after cerebral ischemia in rats. Stroke 32(4):1005–1011
22. Chiang N, Gronert K, Clish CB, O'Brien JA, Freeman MW, Serhan CN (1999) Leukotriene B4 receptor transgenic mice reveal novel protective roles for lipoxins and aspirin-triggered lipoxins in reperfusion. J Clin Invest 104(3):309–316
23. Csiszar A, Pacher P, Kaley G, Ungvari Z (2005) Role of oxidative and nitrosative stress, longevity genes and poly(ADP-ribose) polymerase in cardiovascular dysfunction associated with aging. Curr Vasc Pharmacol 3(3):285–291
24. Das H, Abdulhameed N, Joseph M, Sakthivel R, Mao HQ, Pompili VJ (2009) Ex vivo nanofiber expansion and genetic modification of human cord blood-derived progenitor/stem cells enhances vasculogenesis. Cell Transplant 18(3):305–318
25. Das H, George JC, Joseph M, Das M, Abdulhameed N, Blitz A, Khan M, Sakthivel R, Mao HQ, Hoit BD, Kuppusamy P, Pompili VJ (2009) Stem cell therapy with overexpressed VEGF and PDGF genes improves cardiac function in a rat infarct model. PLoS One 4(10):e7325
26. Dearden NM (1985) Ischaemic brain. Lancet 2(8449):255–259
27. Dhalla NS, Elmoselhi AB, Hata T, Makino N (2000) Status of myocardial antioxidants in ischemia-reperfusion injury. Cardiovasc Res 47(3):446–456
28. Donahoe SM, Stewart GC, McCabe CH, Mohanavelu S, Murphy SA, Cannon CP, Antman EM (2007) Diabetes and mortality following acute coronary syndromes. JAMA 298(7):765–775
29. Eaton P (2006) Protein thiol oxidation in health and disease: techniques for measuring disulfides and related modifications in complex protein mixtures. Free Radic Biol Med 40(11):1889–1899

30. Fang J, Alderman MH (2006) Impact of the increasing burden of diabetes on acute myocardial infarction in New York City: 1990–2000. Diabetes 55(3):768–773
31. Fang J, Mensah GA, Alderman MH, Croft JB (2006) Trends in acute myocardial infarction complicated by cardiogenic shock, 1979–2003, United States. Am Heart J 152(6):1035–1041
32. Ferdinandy P, Schulz R (2003) Nitric oxide, superoxide, and peroxynitrite in myocardial ischaemia-reperfusion injury and preconditioning. Br J Pharmacol 138(4):532–543
33. Ferdinandy P, Schulz R, Baxter GF (2007) Interaction of cardiovascular risk factors with myocardial ischemia/reperfusion injury, preconditioning, and postconditioning. Pharmacol Rev 59(4):418–458
34. Feuerstein GZ, Chavez J (2009) Translational medicine for stroke drug discovery: the pharmaceutical industry perspective. Stroke 40(3 Suppl):S121–S125
35. Feuerstein GZ, Zaleska MM, Krams M, Wang X, Day M, Rutkowski JL, Finklestein SP, Pangalos MN, Poole M, Stiles GL, Ruffolo RR, Walsh FL (2008) Missing steps in the STAIR case: a translational medicine perspective on the development of NXY-059 for treatment of acute ischemic stroke. J Cereb Blood Flow Metab 28(1):217–219
36. Gross ER, Hsu AK, Gross GJ (2004) Opioid-induced cardioprotection occurs via glycogen synthase kinase beta inhibition during reperfusion in intact rat hearts. Circ Res 94(7): 960–966
37. Hanlon PR, Fu P, Wright GL, Steenbergen C, Arcasoy MO, Murphy E (2005) Mechanisms of erythropoietin-mediated cardioprotection during ischemia-reperfusion injury: role of protein kinase C and phosphatidylinositol 3-kinase signaling. FASEB J 19(10):1323–1325
38. Hausenloy DJ, Duchen MR, Yellon DM (2003) Inhibiting mitochondrial permeability transition pore opening at reperfusion protects against ischaemia-reperfusion injury. Cardiovasc Res 60(3):617–625
39. Hemingway H, Malik M, Marmot M (2001) Social and psychosocial influences on sudden cardiac death, ventricular arrhythmia and cardiac autonomic function. Eur Heart J 22(13): 1082–1101
40. Hertz L (1981) Features of astrocytic function apparently involved in the response of central nervous tissue to ischemia-hypoxia. J Cereb Blood Flow Metab 1(2):143–153
41. Hess DT, Matsumoto A, Nudelman R, Stamler JS (2001) S-nitrosylation: spectrum and specificity. Nat Cell Biol 3(2):E46–E49
42. Hess ML, Manson NH (1984) Molecular oxygen: friend and foe. The role of the oxygen free radical system in the calcium paradox, the oxygen paradox and ischemia/reperfusion injury. J Mol Cell Cardiol 16(11):969–985
43. Hickey MJ, Hurley JV, Angel MF, O'Brien BM (1992) The response of the rabbit rectus femoris muscle to ischemia and reperfusion. J Surg Res 53(4):369–377
44. Inagaki K, Chen L, Ikeno F, Lee FH, Imahashi K, Bouley DM, Rezaee M, Yock PG, Murphy E, Mochly-Rosen D (2003) Inhibition of delta-protein kinase C protects against reperfusion injury of the ischemic heart in vivo. Circulation 108(19):2304–2307
45. Jaffe JR, Nag SS, Landsman PB, Alexander CM (2006) Reassessment of cardiovascular risk in diabetes. Curr Opin Lipidol 17(6):644–652
46. Jelnes R, Gaardsting O, Hougaard Jensen K, Baekgaard N, Tonnesen KH, Schroeder T (1986) Fate in intermittent claudication: outcome and risk factors. Br Med J (Clin Res Ed) 293(6555):1137–1140
47. Jennings RB, Sommers HM, Smyth GA, Flack HA, Linn H (1960) Myocardial necrosis induced by temporary occlusion of a coronary artery in the dog. Arch Pathol 70:68–78
48. Jerome SN, Akimitsu T, Gute DC, Korthuis RJ (1995) Ischemic preconditioning attenuates capillary no-reflow induced by prolonged ischemia and reperfusion. Am J Physiol 268 (5 Pt 2):H2063–H2067
49. Jones DP (2006) Disruption of mitochondrial redox circuitry in oxidative stress. Chem Biol Interact 163(1–2):38–53
50. Judge S, Jang YM, Smith A, Hagen T, Leeuwenburgh C (2005) Age-associated increases in oxidative stress and antioxidant enzyme activities in cardiac interfibrillar mitochondria: implications for the mitochondrial theory of aging. FASEB J 19(3):419–421

51. Kamihata H, Matsubara H, Nishiue T, Fujiyama S, Tsutsumi Y, Ozono R, Masaki H, Mori Y, Iba O, Tateishi E, Kosaki A, Shintani S, Murohara T, Imaizumi T, Iwasaka T (2001) Implantation of bone marrow mononuclear cells into ischemic myocardium enhances collateral perfusion and regional function via side supply of angioblasts, angiogenic ligands, and cytokines. Circulation 104(9):1046–1052
52. Kannel WB, McGee DL (1979) Diabetes and cardiovascular disease. The Framingham study. JAMA 241(19):2035–2038
53. Kanski J, Behring A, Pelling J, Schoneich C (2005) Proteomic identification of 3-nitrotyrosine-containing rat cardiac proteins: effects of biological aging. Am J Physiol Heart Circ Physiol 288(1):H371–H381
54. Kaplan J, Dimlich RV, Biros MH, Hedges J (1987) Mechanisms of ischemic cerebral injury. Resuscitation 15(3):149–169
55. Kim BO, Tian H, Prasongsukarn K, Wu J, Angoulvant D, Wnendt S, Muhs A, Spitkovsky D, Li RK (2005) Cell transplantation improves ventricular function after a myocardial infarction: a preclinical study of human unrestricted somatic stem cells in a porcine model. Circulation 112(9 Suppl):I96–I104
56. Klein L, Gheorghiade M (2004) Management of the patient with diabetes mellitus and myocardial infarction: clinical trials update. Am J Med 116(Suppl 5A):47S–63S
57. Kloner RA, Przyklenk K, Shook T, Cannon CP (1998) Protection conferred by preinfarct angina is manifest in the aged heart: evidence from the TIMI 4 trial. J Thromb Thrombolysis 6(2):89–92
58. Kondziolka D, Steinberg GK, Wechsler L, Meltzer CC, Elder E, Gebel J, Decesare S, Jovin T, Zafonte R, Lebowitz J, Flickinger JC, Tong D, Marks MP, Jamieson C, Luu D, Bell-Stephens T, Teraoka J (2005) Neurotransplantation for patients with subcortical motor stroke: a phase 2 randomized trial. J Neurosurg 103(1):38–45
59. Kunz A, Park L, Abe T, Gallo EF, Anrather J, Zhou P, Iadecola C (2007) Neurovascular protection by ischemic tolerance: role of nitric oxide and reactive oxygen species. J Neurosci 27(27):7083–7093
60. Kwiatkowski TG, Libman RB, Frankel M, Tilley BC, Morgenstern LB, Lu M, Broderick JP, Lewandowski CA, Marler JR, Levine SR, Brott T (1999) Effects of tissue plasminogen activator for acute ischemic stroke at one year. National Institute of Neurological Disorders and Stroke Recombinant Tissue Plasminogen Activator Stroke Study Group. N Engl J Med 340(23):1781–1787
61. Lee RJ, Springer ML, Blanco-Bose WE, Shaw R, Ursell PC, Blau HM (2000) VEGF gene delivery to myocardium: deleterious effects of unregulated expression. Circulation 102(8):898–901
62. Lefer DJ, Granger DN (2000) Oxidative stress and cardiac disease. Am J Med 109(4):315–323
63. Lesnefsky EJ, Moghaddas S, Tandler B, Kerner J, Hoppel CL (2001) Mitochondrial dysfunction in cardiac disease: ischemia–reperfusion, aging, and heart failure. J Mol Cell Cardiol 33(6):1065–1089
64. Lewington S, Whitlock G, Clarke R, Sherliker P, Emberson J, Halsey J, Qizilbash N, Peto R, Collins R (2007) Blood cholesterol and vascular mortality by age, sex, and blood pressure: a meta-analysis of individual data from 61 prospective studies with 55,000 vascular deaths. Lancet 370(9602):1829–1839
65. Li Y, Ma T, Kniss DA, Yang ST, Lasky LC (2001) Human cord cell hematopoiesis in three-dimensional nonwoven fibrous matrices: in vitro simulation of the marrow microenvironment. J Hematother Stem Cell Res 10(3):355–368
66. Lim SY, Davidson SM, Hausenloy DJ, Yellon DM (2007) Preconditioning and postconditioning: the essential role of the mitochondrial permeability transition pore. Cardiovasc Res 75(3):530–535
67. Maxwell SR, Lip GY (1997) Reperfusion injury: a review of the pathophysiology, clinical manifestations and therapeutic options. Int J Cardiol 58(2):95–117

68. Orlic D, Kajstura J, Chimenti S, Jakoniuk I, Anderson SM, Li B, Pickel J, McKay R, Nadal-Ginard B, Bodine DM, Leri A, Anversa P (2001) Bone marrow cells regenerate infarcted myocardium. Nature 410(6829):701–705
69. Orlic D, Kajstura J, Chimenti S, Limana F, Jakoniuk I, Quaini F, Nadal-Ginard B, Bodine DM, Leri A, Anversa P (2001) Mobilized bone marrow cells repair the infarcted heart, improving function and survival. Proc Natl Acad Sci USA 98(18):10344–10349
70. Otani A, Kinder K, Ewalt K, Otero FJ, Schimmel P, Friedlander M (2002) Bone marrow-derived stem cells target retinal astrocytes and can promote or inhibit retinal angiogenesis. Nat Med 8(9):1004–1010
71. Oudot A, Martin C, Busseuil D, Vergely C, Demaison L, Rochette L (2006) NADPH oxidases are in part responsible for increased cardiovascular superoxide production during aging. Free Radic Biol Med 40(12):2214–2222
72. Pacher P, Beckman JS, Liaudet L (2007) Nitric oxide and peroxynitrite in health and disease. Physiol Rev 87(1):315–424
73. Panes J, Perry M, Granger DN (1999) Leukocyte-endothelial cell adhesion: avenues for therapeutic intervention. Br J Pharmacol 126(3):537–550
74. Parks DA, Granger DN (1986) Contributions of ischemia and reperfusion to mucosal lesion formation. Am J Physiol 250(6 Pt 1):G749–G753
75. Passier R, van Laake LW, Mummery CL (2008) Stem-cell-based therapy and lessons from the heart. Nature 453(7193):322–329
76. Rafii S, Lyden D (2003) Therapeutic stem and progenitor cell transplantation for organ vascularization and regeneration. Nat Med 9(6):702–712
77. Rehncrona S (1985) Brain acidosis. Ann Emerg Med 14(8):770–776
78. Reichelt KL (1968) The chemical basis for the intolerance of the brain to anoxia. Acta Anaesthesiol Scand Suppl 29:35–46
79. Rozanski A, Blumenthal JA, Kaplan J (1999) Impact of psychological factors on the pathogenesis of cardiovascular disease and implications for therapy. Circulation 99(16):2192–2217
80. Saini HK, Machackova J, Dhalla NS (2004) Role of reactive oxygen species in ischemic preconditioning of subcellular organelles in the heart. Antioxid Redox Signal 6(2):393–404
81. Simoons ML, Boersma E, Maas AC, Deckers JW (1997) Management of myocardial infarction: the proper priorities. Eur Heart J 18(6):896–899
82. Sivonova M, Tatarkova Z, Durackova Z, Dobrota D, Lehotsky J, Matakova T, Kaplan P (2007) Relationship between antioxidant potential and oxidative damage to lipids, proteins and DNA in aged rats. Physiol Res 56(6):757–764
83. Stamm C, Westphal B, Kleine HD, Petzsch M, Kittner C, Klinge H, Schumichen C, Nienaber CA, Freund M, Steinhoff G (2003) Autologous bone-marrow stem-cell transplantation for myocardial regeneration. Lancet 361(9351):45–46
84. Stanevsky A, Goldstein G, Nagler A (2009) Umbilical cord blood transplantation: pros, cons and beyond. Blood Rev 23(5):199–204
85. Strike PC, Steptoe A (2003) Systematic review of mental stress-induced myocardial ischaemia. Eur Heart J 24(8):690–703
86. Taguchi A, Soma T, Tanaka H, Kanda T, Nishimura H, Yoshikawa H, Tsukamoto Y, Iso H, Fujimori Y, Stern DM, Naritomi H, Matsuyama T (2004) Administration of CD34+ cells after stroke enhances neurogenesis via angiogenesis in a mouse model. J Clin Invest 114(3):330–338
87. Templin C, Kotlarz D, Faulhaber J, Schnabel S, Grote K, Salguero G, Luchtefeld M, Hiller KH, Jakob P, Naim HY, Schieffer B, Hilfiker-Kleiner D, Landmesser U, Limbourg FP, Drexler H (2008) Ex vivo expanded hematopoietic progenitor cells improve cardiac function after myocardial infarction: role of beta-catenin transduction and cell dose. J Mol Cell Cardiol 45(3):394–403
88. Tse HF, Kwong YL, Chan JK, Lo G, Ho CL, Lau CP (2003) Angiogenesis in ischaemic myocardium by intramyocardial autologous bone marrow mononuclear cell implantation. Lancet 361(9351):47–49

89. Turko IV, Murad F (2002) Protein nitration in cardiovascular diseases. Pharmacol Rev 54(4):619–634
90. Vahedi K, Hofmeijer J, Juettler E, Vicaut E, George B, Algra A, Amelink GJ, Schmiedeck P, Schwab S, Rothwell PM, Bousser MG, van der Worp HB, Hacke W (2007) Early decompressive surgery in malignant infarction of the middle cerebral artery: a pooled analysis of three randomised controlled trials. Lancet Neurol 6(3):215–222
91. van de Ven C, Collins D, Bradley MB, Morris E, Cairo MS (2007) The potential of umbilical cord blood multipotent stem cells for nonhematopoietic tissue and cell regeneration. Exp Hematol 35(12):1753–1765
92. Vendrame M, Gemma C, de Mesquita D, Collier L, Bickford PC, Sanberg CD, Sanberg PR, Pennypacker KR, Willing AE (2005) Anti-inflammatory effects of human cord blood cells in a rat model of stroke. Stem Cells Dev 14(5):595–604
93. Wechsler LR (2009) Clinical trials of stroke therapy: which cells, which patients? Stroke 40 (3 Suppl):S149–S151
94. Willems L, Zatta A, Holmgren K, Ashton KJ, Headrick JP (2005) Age-related changes in ischemic tolerance in male and female mouse hearts. J Mol Cell Cardiol 38(2):245–256
95. Wu ZK, Tarkka MR, Pehkonen E, Kaukinen L, Honkonen EL, Kaukinen S (2000) Beneficial effects of ischemic preconditioning on right ventricular function after coronary artery bypass grafting. Ann Thorac Surg 70(5):1551–1557
96. Xu L, Eu JP, Meissner G, Stamler JS (1998) Activation of the cardiac calcium release channel (ryanodine receptor) by poly-S-nitrosylation. Science 279(5348):234–237
97. Zairis MN, Lyras AG, Makrygiannis SS, Psarogianni PK, Adamopoulou EN, Handanis SM, Papantonakos A, Argyrakis SK, Prekates AA, Foussas SG (2004) Type 2 diabetes and intravenous thrombolysis outcome in the setting of ST elevation myocardial infarction. Diabetes Care 27(4):967–971
98. Zhang RL, Zhang ZG, Chopp M (2005) Neurogenesis in the adult ischemic brain: generation, migration, survival, and restorative therapy. Neuroscientist 11(5):408–416
99. Zhang X, Patel A, Horibe H, Wu Z, Barzi F, Rodgers A, MacMahon S, Woodward M (2003) Cholesterol, coronary heart disease, and stroke in the Asia Pacific region. Int J Epidemiol 32(4):563–572
100. Zorov DB, Juhaszova M, Sollott SJ (2006) Mitochondrial ROS-induced ROS release: an update and review. Biochim Biophys Acta 1757(5–6):509–517
101. Zou M, Martin C, Ullrich V (1997) Tyrosine nitration as a mechanism of selective inactivation of prostacyclin synthase by peroxynitrite. Biol Chem 378(7):707–713
102. Zweier JL, Flaherty JT, Weisfeldt ML (1987) Direct measurement of free radical generation following reperfusion of ischemic myocardium. Proc Natl Acad Sci USA 84(5):1404–1407

Chapter 10
Cancer Stem Cell Models and Role in Drug Discovery

Rohit Duggal, Boris Minev, Angelo Vescovi, and Aladar Szalay

Contents

Abstract With the cementing of the cancer stem cell (CSC) concept, cancer biology and cancer drug discovery has attained a new avenue to approach cancer from. Studying the hierarchy of tumor tissue organization and how to inhibit the cell that resides at the very top of this hierarchy has opened up a new branch of tumor biology

R. Duggal • B. Minev • A. Szalay (✉)
Genelux Corporation, 3030 Bunker Hill Street, Suite 310, San Diego, CA 92109, USA
e-mail: rohit.duggal@genelux.com; aaszalay@genelux.com

A. Vescovi
Department of Biotechnology and Biosciences, University of Milano-Bicocca, Piazza della Scienza 2, I-20126 Milan, Italy

R.K. Srivastava and S. Shankar (eds.), *Stem Cells and Human Diseases*,
DOI 10.1007/978-94-007-2801-1_10, © Springer Science+Business Media B.V. 2012

and given the opportunity to develop novel inhibitors that target cancer. With the discovery of CSCs in majority of cancer indications there seems to be a universal applicability of the concept. However, the CSC field is still at an early fledgling state and a lot more needs to be done in terms of understanding their emergence, maintenance, role in metastasis and determining the architecture of the tumor.

Keywords Cancer stem cells • CSC detection • Cancer drug discovery • Oncolytic therapy • Glioblastoma

10.1 Introduction to Cancer Stem Cell Concept

The emergence of a cancer stem cell (CSC) concept has if not revolutionized but certainly altered views about the origin(s) of cancer and what the new anti cancer modalities should target. In this concept, the CSC behaves like the "queen bee" in a bee hive and the drones in the hive are analogous to the bulk tumor cells in a tumor. The CSC, may or may not have a stem cell origin but functionally is similar to a stem cell. The main properties of CSCs as identified by a distinguished group of CSC scientists after the AACR workshop in 2006 [4] are the ability to initiate and maintain a tumor including the CSC compartment (self renewal) and generation of differentiated progeny that make up the bulk of the tumor. This makes the CSC at the apex of neoplastic transformation where its unique stem cell properties of self renewal and multipotency enables it to initiate, fuel and sustain tumor growth (Fig. 10.1). This hierarchical model is supported by the fact that significant heterogeneity is observed in tumors and develops the notion that tumor initiation and

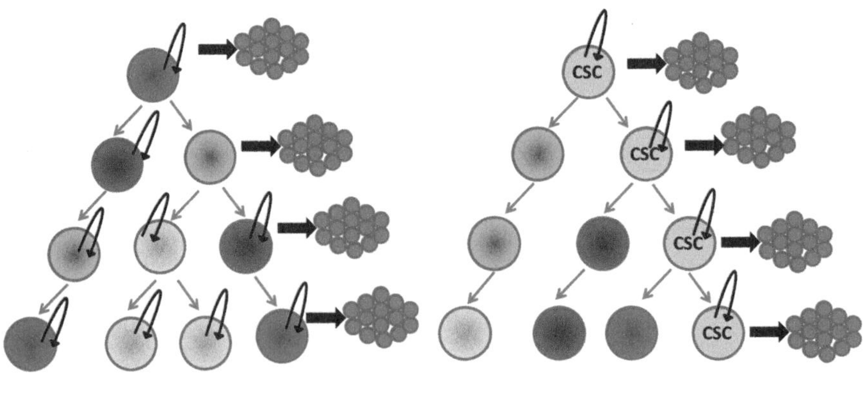

Fig. 10.1 Cartoons depicting the two models for tumor formation [30]. The stochastic model indicates that each tumor cell in the tumor has the ability to initiate and maintain a tumor. On the other hand the hierarchical model shows that the cancer stem cell (*CSC*) resides at the apex of the transformation process and due to its multipotent nature differentiates into the final cell lineages that comprise the tumor. Furthermore, the CSC goes through an asymmetric cell division that replicates more of itself (self renewal) as well as cells of other lineages

maintenance properties reside in a subset of cells within the tumor that make up the CSC population. This is in contrast to the stochastic model of tumor growth where every cell in the tumor is capable of initiating, advancing and maintaining a tumor (Fig. 10.1) [30].

Self-renewal is an important property of CSCs and in embryonic or the further differentiated pluripotent stem cells consists of two events: (a) Formation of a new stem cell and (b) formation of a progenitor cell that can undergo proliferation and differentiation but not further self-renewal [53]. Abnormal regulation of the self-renewal may lead to malignancy [15]. In malignancies, unregulated self-renewal is the cause of uncontrolled proliferation of cells. It is hypothesized that there exist in vivo a few cancer stem cells that cause de novo proliferation even in the presence of therapies that target and kill most of the existing cancer cells [53].

In order to demonstrate a stem cell origin of cancer, we will need to understand the early formation of CSC. Mutations in the genes that cause self-renewal like STAT5A, and those involved in multi-lineage CD34+ progenitor stem cell differentiation such as c-mpl [11] may be inducers of CSC, as well as mutations in JAK2 and c-mpl that induce myeloproliferative disorders [9]. Constitutive activation of STAT5 is associated with several leukemias, chromosomal aberrations, and the transformation of hematopoietic cells to cytokine independent growth [49].

10.2 Recent Advances in Cancer Stem Cell Field

The original study by John Dick and colleagues that used immunodeficient mice to xenograft tumorous cells was a seminal study [43]. These researchers found that most subtypes of acute myeloid leukemia could be implanted in these mice, but found heterogeneity within these tumors. Only one in a million tumor cells could initiate tumors, thereby this capability lying in only a subset of tumorous cells. In case of solid tumors, the ground breaking work was carried out by Clarke and coworkers in 2003 [1]. They established the tumor initiating capability to reside in a subset of cells in breast tumors. This was followed by identification of CSCs in brain tumors [8, 38]. Very interestingly it was demonstrated that the GBM CSCs are multipotent and could be maintained as spheroids in vitro almost indefinitely without significant change in properties [8]. CSCs have also been identified now in colon cancer, pancreatic cancer, liver cancer, ovarian cancer, melanoma and thyroid cancer [20, 26, 33, 36, 42, 50, 51].

10.2.1 Use of Functional Properties and Not Marker Based CSC Discovery

Even though the first major breakthroughs in CSC identification came through the use of cell surface markers, recently marker based CSC identification is not considered the best way of separating CSCs from other tumor cells. The reasons being two fold, firstly functional identification of CSCs is more meaningful and secondly even if CSCs identified by marker based means are established to be functional, it is possible that the markers might be biased for identification of a subset of the CSC population. Hence, relying solely on functional assay based CSC identification might be a suitable approach for CSC isolation since it would rely on functional properties of CSCs and prevent a bias from entering in the way the CSCs were selected from tumor tissue. This has been greatly enhanced by the development of a neurosphere assay [44]. Using this assay CSCs can be maintained as cell lines without significant alteration of genomic or RNA signatures [7, 19, 21].

10.2.2 Attempts to Improve CSC Detection by Using a More Sensitive Assay

Sean Morrison's laboratory successfully demonstrated the use of highly immunocompromised mice for improving the frequency of detection of CSCs [28]. Using a variant of the commonly used NOD/SCID mouse model that is deficient in the interleukin-2 gamma receptor, NSG, that further debilitates the immune system by the inability to make natural killer cells, these researchers showed that in melanoma, 25% of unsorted single cells could generate tumors upon transplantation in NSG mice. Therefore, the use of NSG mice could further assist in CSC detection. Indeed, Ishizawa et al. [14] found a ten fold increase in CSC frequency detection by use of NSG mice. However, unlike in melanoma the frequency of detection was still much more rare in the tumors from other cancer indications that were examined in this study.

10.2.3 Induction and Characterization of Human CSCs In Vivo Using the SCID-hu Thy/Liv Mouse Model System

We reported recently for the first time the utilization of the SClDhu Thy/Liv chimeric small animal model system for induction of human cancer stem cells and their early detection [52]. This model system allows long-term systemic human T-cell reconstitution in vivo, and also provides both human antigen-presenting cells (APCs) and effector cells for induction of anti-tumor immune responses. We were able to generate human hematopoietic cancer stem cells (HCSC) using the SClD-hu Thy/Liv system, and confirmed the expression of both the CD34 marker and two human tumor antigens in these cells. Importantly, we observed an enhanced expression of several embryonic stem cell markers in the HCSC, as well as morphological appearance typical for undifferentiated cells, suggesting the acquisition of highly malignant aggressive properties. Specifically, we demonstrated a successful induction of CD34+ hematopoietic cancer stem cells (HCSC) in the SCID-hu Thy/Liv implants following injection of human melanoma cells (Mel-624). A cell-cell fusion or endocytosis in the SCID-hu ThylLiv implants between the injected Mel-624 cancer cells and the human CD34+ cells of the implants may *occur* in the generation of HCSC. This may also include exchange of genetic material and/or metabolic cooperation due to these mel-624 cell and CD34+ cell contact and interactions. Therefore, these findings are an important step in our understanding of the mechanisms involved in the early formation of the cancer stem cells, and in demonstrating the stem cell origin of cancer. Since human antigen presenting cells and cytotoxic T lymphocytes (CTL) are in the blood circulation of these animals, this *NOD/SCID-hu* model system enables testing of the effects of the human CTL-directed killing of the HCSC versus original cancer cells in vivo.

10.2.4 CSCs Contribution to Endothelium in Tumors

An interesting observation that was published simultaneously last year was the capability of CSCs to generate the vasculature of tumors [34, 47]. This was unprecedented since CSCs were thought to contribute to generation of only bulk tumor cells and not contribute to the stroma. Working with glioblastoma CSCs these researchers demonstrated that the CSCs could be differentiated into endothelium cells both in vitro and in vivo. Furthermore, Ricci-Vitiani et al. [34] demonstrated the importance of the CSC lines to authentically mimic glioblastoma upon xenotransplantation. The generation of vasculature was driven by human CSCs as opposed to transplanting in mice of traditionally serum grown glioma cell lines such as U-87 where the vasculature is contributed by mouse endothelium cells.

10.3 Approaches for Targeting Cancer Stem Cells in a Drug Discovery Setting

Initial efforts for targeting CSCs involved targeting pathways that are involved in development that are thought to be active in undifferentiated and primitive cells, namely the Wnt-beta catenin, Notch and the Hedgehog pathways [24, 35, 41]. Limited success has been achieved targeting these pathways using small and large molecule inhibitors [12, 13, 45]. With the advent of the development of authentic CSC models that can be utilized in vitro and in vivo besides providing model systems to test the activity of the above mentioned pathway inhibitors, the CSC model systems themselves can be utilized for identification of new targets and inhibitors. Utilization of siRNA screens for new target identification could be an approach that would work by using robust CSC in vitro systems. These systems could also be directly used for identification of small or large molecule inhibitors. Another approach could entail using the CSC in vitro systems for generating antibodies against the cells followed by identification of inhibitor or detection antibodies. The in vivo CSC models tend to reproduce the disease similar to that seen in patients in terms of histopathology [8] and spread of disease (unpublished data). Furthermore, the expression and genomic signatures of CSC systems is similar to that of patient tumor samples [21]. Therefore, they potentially provide more clinically relevant model systems for testing inhibitors. Finding new targets and inhibitors using these approaches could potentially inhibit angiogenesis driven by CSCs. Therefore the potential effect of a CSC inhibitor could be more substantial than previously anticipated.

The SCID-hu Thy/Liv chimeric small animal model system [40] appears to be ideally suited for induction of human cancer stem cells and their early detection. This model system consists of a functional human hematopoietic organ that maintains hematopoiesis and allows the reconstitution of human CD34+ progenitor hematopoietic stem cells (HSC) in a fresh stromal microenvironment, or in the host circulatory system [23, 40]. A NOD/SCID-hu Thy/Liv system has been used to generate humanized immune responses in vivo utilizing the host vascular and cytokine physiology [8]. Importantly, this model system allows long-term systemic human T-cell reconstitution *in vivo.* These humanized mice are also highly reconstituted with human B cells, monocyte/macrophages, and CD123+ and CD11c+dendritic cells. Thus, this model provides both human antigen presenting cells (APCs) and effector cells for induction of anti-tumor immune responses [23]. We have utilized this model system to induce and characterize human CSC in vivo as potential targets for immune-based therapies.

10.4 Use of Oncolytic Viruses as a Viable Option for Targeting Cancer Stem Cells and Treating Cancer

Oncolytic viruses are not affected by mechanisms attributed to generate cancer resistance against chemotherapeutic agents and radiation modalities that are thought to reside in CSCs [5]. Furthermore, there is a lack of precedence for robust and validated CSC systems being tested for targeting extensively with oncolytic viruses, especially with oncolytic vaccinia viruses. CSCs display potential resistance to infection (replication) by oncolytic viruses engineered for an attenuated phenotype. Also an elevated interferon (IFN) response in CSCs relative to bulk tumor cells is thought to decrease sensitivity to oncolytic virus infection [5].

10.4.1 Oncolytic Herpes Simplex Viruses (HSV) and CSCs

In case of HSV, several mutants, deletion of neurovirulence factor, ICP34.5 and early expression of U_s11 gene (G47Δ) that compensates for ICP34.5 deletion have been used for infecting potential CSC systems, mainly connected with brain cancers. A variant that conditionally expresses ICP34.5 under the transcriptional control of nestin, a gene expressed in undifferentiated brain cells, rQNestin34.5, has also been claimed to grow in primary glioma cells and provides significant survivability in glioma animal models [17]. The primary glioma cells in this study were grown in the presence of fetal bovine serum that significantly reduces the primary tumor signature and the presence of undifferentiated cells. Furthermore, studies for addressing safety in terms of infection and lytic activity in normal neural stem cells (that express nestin in copious quantities) were not carried out. Therefore, the overall utility of the rQNestin34.5 HSV variant at this point in the glioma setting is questionable. In another interesting study, the mutant G47Δ, demonstrated CSClytic activity by reducing growth of GBM CSC cultures albeit to a reduced extent compared to wild type or another mutant virus [46]. A single intratumoral dose of G47Δ reduced intracranial GBM tumor formation in immunocompromised mice. Yet another study by Otsuki et al. [27] demonstrated the utility of valporic acid (VPA), an histone deacetylase (HDAC) inhibitor, as an agent that would target the IFN response in cells upon infection with oncolytic HSV variants. The HDAC inhibitors increased virus titer and cell killing in response to rQNestin34.5 HSV variant, even in primary glioma cultures grown under spheroid formation conditions. IFN beta treated glioma cells showed significant resistance to viral infection which was counteracted by treating with VPA. The variant rQNestin34.5 was less sensitive to IFN beta treatment in glioma lines compared to in the context of primary glioma cells. This and other evidence from breast cancer initiating cells [5] indicate a heightened IFN response in CSCs upon viral infection. Furthermore, some primary neuroblastoma CSC cultures have been found to be resistant to oncolytic HSV infection [5]. Again leading to the observation of a possibility of encountering reduced levels of infection, possibly resistance at both cell entry and replication stages, by oncolytic viruses in CSCs.

10.4.2 Oncolytic Adenoviruses (Ad) and CSCs

Studies of oncolytic Ad and CSCs have described the use of mutants [31, 32, 37], firstly in E1A (Δ24, deletion of 24 bp to prevent Rb binding; [48]) to prevent infection in normal cells and capsid mutants that would facilitate uptake and infection in cancer cells. Two commonly used capsid mutants that help in this tropism modification are chimeric capsid mutants with Ad 3, Ad5/3-Δ24 that uses the Ad serotype 3 receptor [18] and Ad5.pk7-Δ24 that uses heparin sulfate proteoglycans for cell entry [29]. Both types of double mutant Ads were effective in killing $CD44^{high}$ and $CD24^{low}$ populations of breast cancer cells from fresh or briefly cultured pleural effusions [6, 25]. The $CD44^{high}$ and $CD24^{low}$ population of breast cancer cells were shown to comprise of breast CSCs [1]. However, these studies have been found to be difficult to reproduce and solely marker based identification of CSCs is not a recognized concept in the field. Nevertheless it is safe to say that the oncolytic Ad mutants used by Eriksson et al. [6] were able to destroy $CD44^{high}$ and $CD24^{low}$ sorted and unsorted cells that resulted in abrogating tumor formation and increased animal survival. However, oncolytic Ad viruses were found to reduce the putative breast CSC population only to a small extent [2] and did not completely remove tumors. Therefore, lower efficiency of infection in vivo and the possibility of not targeting other tumor initiating cells exists that needs to be addressed with newer studies comprised of a non biased breast CSC population and possibly different viruses.

A brain CSC study involving oncolytic Ad has shown successful infection of primary GBM spheroid cultures followed by improvement in survival of glioma bearing mice [16]. These researchers found evidence of autophagy related processes responsible for cell death upon viral infection. Another set of researchers have found other Ad serotypes such as Ad16 and chimpanzee Ad CV23 to infect low passage GBM cultures as well as $CD133^{+}$ and $CD133^{-}$ cells freshly isolated from glioma patients [39]. The infection efficiency was significantly better for these isolates compared to the Ad5 serotype, especially in early passages of these cultures.

10.4.3 Other Oncolytic Viruses and CSCs

Reovirus causes mild respiratory and gastrointestinal diseases in humans and has the potential to be used as an oncolytic virus in its natural state. Marcato et al. [22] found tumor growth inhibition of primary breast cancer sample induced tumor formation in immunocompromised mice upon intratumoral administration of reovirus. They found both putative breast CSCs and non breast CSCs to be killed by apoptosis by viral infection. Reovirus is particularly useful in treating cancers with dysregulation due to Ras signaling since aberrant Ras signaling facilitates various steps in the reovirus life cycle. Primary CSCs from GBM have been shown to have genetic lesions that would result in dysregulated Ras signaling that could be a target indication for using reovirus oncolytic viruses.

There are numerous reports on infection of cancer cells with vesicular stomatitis virus (VSV). However, CSC infection studies have only been recently initiated with oncolytic VSVs. One study describes resistance of primary neuroblastoma CSC cultures to infection by a VSV mutant virus [5], possibly due to increased sensitivity of this virus to an enhanced IFN response considered to be prevalent in CSC cultures.

10.4.4 Oncolytic Poxviruses and CSCs

There are few literature reports of studies involving CSCs and oncolytic poxviruses. Vaccinia virus has been found to not infect all primary hematolymphoid cells [3, 10]. Therefore, there could be a tropism issue associated with infection of primary cells by vaccinia virus that could be accentuated upon using attenuated mutants used for oncolytic therapy. However, some other poxviruses, such as myxoma virus has been shown to readily infect primary neuroblastoma CSCs. Therefore, it would

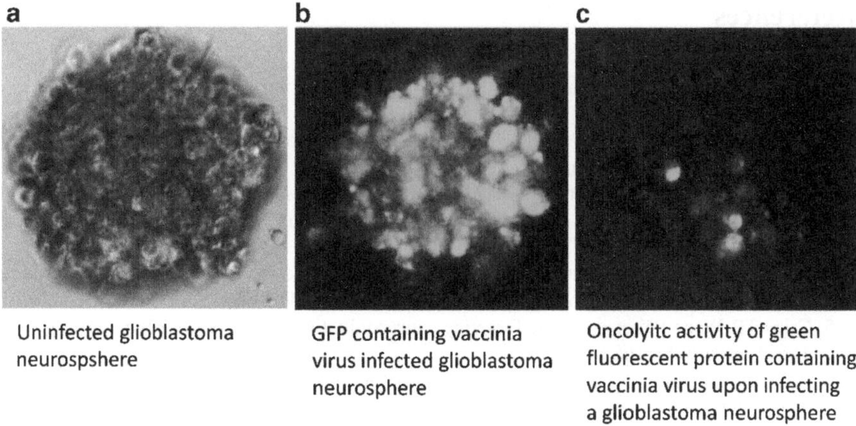

a
b
c
Uninfected glioblastoma neurospshere
GFP containing vaccinia virus infected glioblastoma neurosphere
Oncolyitc activity of green fluorescent protein containing vaccinia virus upon infecting a glioblastoma neurosphere

Fig. 10.2 Micrographs showing (**a**) A glioblastoma neurosphere with several CSCs after propagation in specific stem cell medium. (**b**) A glowing glioblastoma neurosphere established after infecting a single-cell suspension of glioblastoma CSCs with a green fluorescent protein (*GFP*) containing vaccinia virus. (**c**) A glioblastoma neurosphere undergoing oncolysis upon infecting with a GFP containing vaccinia virus

be very interesting to test oncolytic vaccinia viruses against bonafide CSC preparations to determine susceptibility to infection. Our preliminary observations using authentic glioblastoma CSC systems indicate that vaccinia virus, indeed, does infect CSCs (Fig. 10.2). However, the extent of infection remains to be determined and requires further experimentation and analyses.

10.5 Implications for Targeting Cancer Stem Cells and Monitoring Effects in the Clinic

With CSCs comprising a small population of the tumor there is a concern that the effect of CSC specific inhibitors might not be visible in vivo. Furthermore this could be reflected in the clinic where the outcome might not register as suitable patient response as evaluated by classical Response Evaluation Criteria In Solid Tumors (RECIST). However, testing the effects of CSC specific inhibitors in diseases such as glioblastoma where the tumor is comprised of a larger proportion of CSCs, the effect might be more noticeable in a preclinical setting. Glioblastoma is also an aggressive disease with a median survival duration of 14 months. The benefit, if any, could be readily visible for a CSC specific inhibitor in terms of increased survival if traditional RECIST end points cannot be evaluated. Oncolytic viruses on the other hand, if found to target CSCs could have the ability to register suitable RECIST end points due to their ability to target bulk tumor cells as well and consequently increase the chances of observing suitable tumor regression.

References

1. Al-Hajj M, Wicha MS et al (2003) Prospective identification of tumorigenic breast cancer cells. Proc Natl Acad Sci USA 100(7):3983–3988
2. Bauerschmitz GJ, Ranki T et al (2008) Tissue-specific promoters active in CD44+CD24-/low breast cancer cells. Cancer Res 68(14):5533–5539
3. Chahroudi A, Chavan R et al (2005) Vaccinia virus tropism for primary hematolymphoid cells is determined by restricted expression of a unique virus receptor. J Virol 79(16):10397–10407
4. Clarke MF, Dick JE et al (2006) Cancer stem cells—perspectives on current status and future directions: AACR Workshop on cancer stem cells. Cancer Res 66(19):9339–9344
5. Cripe TP, Wang PY et al (2009) Targeting cancer-initiating cells with oncolytic viruses. Mol Ther 17(10):1677–1682
6. Eriksson M, Guse K et al (2007) Oncolytic adenoviruses kill breast cancer initiating CD44+CD24-/low cells. Mol Ther 15(12):2088–2093
7. Ernst A, Hofmann S et al (2009) Genomic and expression profiling of glioblastoma stem cell-like spheroid cultures identifies novel tumor-relevant genes associated with survival. Clin Cancer Res 15(21):6541–6550
8. Galli R, Binda E et al (2004) Isolation and characterization of tumorigenic, stem-like neural precursors from human glioblastoma. Cancer Res 64(19):7011–7021
9. Gibson SE, Schade AE et al (2008) Phospho-STAT5 expression pattern with the MPL W515L mutation is similar to that seen in chronic myeloproliferative disorders with JAK2 V617F. Hum Pathol 39(7):1111–1114
10. Guo ZS, Thorne SH et al (2008) Oncolytic virotherapy: molecular targets in tumor-selective replication and carrier cell-mediated delivery of oncolytic viruses. Biochim Biophys Acta 1785(2):217–231
11. Hitchcock IS, Chen MM et al (2008) YRRL motifs in the cytoplasmic domain of the thrombopoietin receptor regulate receptor internalization and degradation. Blood 112(6):2222–2231
12. Hoey T, Yen WC et al (2009) DLL4 blockade inhibits tumor growth and reduces tumor-initiating cell frequency. Cell Stem Cell 5(2):168–177
13. Huang SM, Mishina YM et al (2009) Tankyrase inhibition stabilizes axin and antagonizes Wnt signalling. Nature 461(7264):614–620
14. Ishizawa K, Rasheed ZA et al (2010) Tumor-initiating cells are rare in many human tumors. Cell Stem Cell 7(3):279–282
15. Jamieson CH, Weissman IL et al (2004) Chronic versus acute myelogenous leukemia: a question of self-renewal. Cancer Cell 6(6):531–533
16. Jiang H, Gomez-Manzano C et al (2007) Examination of the therapeutic potential of Delta-24-RGD in brain tumor stem cells: role of autophagic cell death. J Natl Cancer Inst 99(18):1410–1414
17. Kambara H, Okano H et al (2005) An oncolytic HSV-1 mutant expressing ICP34.5 under control of a nestin promoter increases survival of animals even when symptomatic from a brain tumor. Cancer Res 65(7):2832–2839
18. Kanerva A, Zinn KR et al (2003) Enhanced therapeutic efficacy for ovarian cancer with a serotype 3 receptor-targeted oncolytic adenovirus. Mol Ther 8(3):449–458
19. Lee J, Kotliarova S et al (2006) Tumor stem cells derived from glioblastomas cultured in bFGF and EGF more closely mirror the phenotype and genotype of primary tumors than do serum-cultured cell lines. Cancer Cell 9(5):391–403
20. Li C, Heidt DG et al (2007) Identification of pancreatic cancer stem cells. Cancer Res 67(3):1030–1037
21. Li A, Walling J et al (2008) Genomic changes and gene expression profiles reveal that established glioma cell lines are poorly representative of primary human gliomas. Mol Cancer Res 6(1):21–30

22. Marcato P, Dean CA et al (2009) Oncolytic reovirus effectively targets breast cancer stem cells. Mol Ther 17(6):972–979
23. Melkus MW, Estes JD et al (2006) Humanized mice mount specific adaptive and innate immune responses to EBV and TSST-1. Nat Med 12(11):1316–1322
24. Mullendore ME, Koorstra JB et al (2009) Ligand-dependent Notch signaling is involved in tumor initiation and tumor maintenance in pancreatic cancer. Clin Cancer Res 15(7):2291–2301
25. Nguyen NP, Almeida FS et al (2010) Molecular biology of breast cancer stem cells: potential clinical applications. Cancer Treat Rev 36(6):485–491
26. O'Brien CA, Pollett A et al (2007) A human colon cancer cell capable of initiating tumour growth in immunodeficient mice. Nature 445(7123):106–110
27. Otsuki A, Patel A et al (2008) Histone deacetylase inhibitors augment antitumor efficacy of herpes-based oncolytic viruses. Mol Ther 16(9):1546–1555
28. Quintana E, Shackleton M et al (2008) Efficient tumour formation by single human melanoma cells. Nature 456(7222):593–598
29. Ranki T, Kanerva A et al (2007) A heparan sulfate-targeted conditionally replicative adenovirus, Ad5.pk7-Delta24, for the treatment of advanced breast cancer. Gene Ther 14(1):58–67
30. Reya T, Morrison SJ et al (2001) Stem cells, cancer, and cancer stem cells. Nature 414(6859):105–111
31. Ribacka C, Hemminki A (2008) Virotherapy as an approach against cancer stem cells. Curr Gene Ther 8(2):88–96
32. Ribacka C, Pesonen S et al (2008) Cancer, stem cells, and oncolytic viruses. Ann Med 40(7):496–505
33. Ricci-Vitiani L, Lombardi DG et al (2007) Identification and expansion of human colon-cancer-initiating cells. Nature 445(7123):111–115
34. Ricci-Vitiani L, Pallini R et al (2010) Tumour vascularization via endothelial differentiation of glioblastoma stem-like cells. Nature 468(7325):824–828
35. Scales SJ, de Sauvage FJ (2009) Mechanisms of Hedgehog pathway activation in cancer and implications for therapy. Trends Pharmacol Sci 30(6):303–312
36. Schatton T, Murphy GF et al (2008) Identification of cells initiating human melanomas. Nature 451(7176):345–349
37. Short JJ, Curiel DT (2009) Oncolytic adenoviruses targeted to cancer stem cells. Mol Cancer Ther 8(8):2096–2102
38. Singh SK, Hawkins C et al (2004) Identification of human brain tumour initiating cells. Nature 432(7015):396–401
39. Skog J, Edlund K et al (2007) Adenoviruses 16 and CV23 efficiently transduce human low-passage brain tumor and cancer stem cells. Mol Ther 15(12):2140–2145
40. Sundell IB, Koka PS (2006) Chimeric SCID-hu model as a human hematopoietic stem cell host that recapitulates the effects of HIV-1 on bone marrow progenitors in infected patients. J Stem Cells 1(4):283–300
41. Teo JL, Kahn M (2010) The Wnt signaling pathway in cellular proliferation and differentiation: a tale of two coactivators. Adv Drug Deliv Rev 62(12):1149–1155
42. Todaro M, Iovino F et al (2010) Tumorigenic and metastatic activity of human thyroid cancer stem cells. Cancer Res 70(21):8874–8885
43. Uckun FM, Sather H et al (1995) Leukemic cell growth in SCID mice as a predictor of relapse in high-risk B-lineage acute lymphoblastic leukemia. Blood 85(4):873–878
44. Vescovi AL, Parati EA et al (1999) Isolation and cloning of multipotential stem cells from the embryonic human CNS and establishment of transplantable human neural stem cell lines by epigenetic stimulation. Exp Neurol 156(1):71–83
45. Von Hoff DD, LoRusso PM et al (2009) Inhibition of the hedgehog pathway in advanced basal-cell carcinoma. N Engl J Med 361(12):1164–1172
46. Wakimoto H, Kesari S et al (2009) Human glioblastoma-derived cancer stem cells: establishment of invasive glioma models and treatment with oncolytic herpes simplex virus vectors. Cancer Res 69(8):3472–3481

47. Wang R, Chadalavada K et al (2010) Glioblastoma stem-like cells give rise to tumour endothelium. Nature 468(7325):829–833
48. Whyte P, Buchkovich KJ et al (1988) Association between an oncogene and an anti-oncogene: the adenovirus E1A proteins bind to the retinoblastoma gene product. Nature 334(6178):124–129
49. Wierenga AT, Vellenga E et al (2008) Maximal STAT5-induced proliferation and self-renewal at intermediate STAT5 activity levels. Mol Cell Biol 28(21):6668–6680
50. Yang ZF, Ho DW et al (2008) Significance of CD90+ cancer stem cells in human liver cancer. Cancer Cell 13(2):153–166
51. Zhang S, Balch C et al (2008) Identification and characterization of ovarian cancer-initiating cells from primary human tumors. Cancer Res 68(11):4311–4320
52. Zhang M, Dias P et al (2010) Induction characterization and targeting of human hematopoietic cancer stem cells. J Stem Cells 5(1):1–7
53. Zou GM (2007) Cancer stem cells in leukemia, recent advances. J Cell Physiol 213(2):440–444

Chapter 11
Origins of Metastasis-Initiating Cells

Sara M. Nolte and Sheila K. Singh

Contents

Abstract Current models of primary cancers suggest that tumour formation and growth is due to a rare subpopulation of stem cell-like tumour-initiating cells. These cells have the capability of self-renewal and the ability to form all cell types of the heterogenous tumour. Due to the nature of the metastatic process, where only a small number of primary cells are capable of successfully forming a metastasis, it is

S.M. Nolte
McMaster Stem Cell and Cancer Research Institute, 1280 Main Street West, Hamilton ON, L8S 4K1, Canada

S.K. Singh (✉)
McMaster Children's Hospital, 1200 Main Street West., Room 4E5, Hamilton, ON L8N 3Z5, Canada

Department of Surgery, McMaster Stem Cell and Cancer Research Institute, Hamilton, ON, Canada
e-mail: ssingh@mcmaster.ca

R.K. Srivastava and S. Shankar (eds.), *Stem Cells and Human Diseases*, DOI 10.1007/978-94-007-2801-1_11, © Springer Science+Business Media B.V. 2012

suggestive that metastases may also form from a rare tumour-initiating population. Currently, the existence and origin of these putative metastasis-initiating cells is unclear. Here we aim to discuss current evidence for such a metastasis-initiating cell population, and the potential models for the origin of these cells. The therapeutic implications of targeting chemo- and radioresistent primary tumour-initiating cells may also apply to the treatment of metastatic disease.

Keywords Cancer stem cell • Epithelial-mesenchymal transition • Metastasis • Metastasis-initiating cell • Tumour-initiating cell

List of Abbreviations

ALDH1	Aldehyde dehydrogenase 1
BBB	Blood–brain-barrier
Bmi1	B-cell-specific Moloney murine leukemia virus insertion site-1
COX2	Cyclooxygenase 2
CSC	Cancer stem cell
EGF	Epidermal growth factor
EMT	Epithelial-mesenchymal transition
EpCAM	Epithelial cell adhesion molecule (a.k.a ESA)
ESA	Epithelial-specific antigen (a.k.a. EpCAM)
FoxM1	Forkhead box M1
HIF1	Hypoxia-inducible factor-1
HNSCC	Head and neck squamous cell carcinoma
HOXB9	Homeobox B9
HSC	Hematopoietic stem cell
ID1	Inhibitor of DNA binding 1
IL11 / IL8	Interleukin-11 / Interleukin-8
LAC	Lung adenocarcinoma
LEF1	Lymphoid enhancer binding factor 1
LWS	Lung cancer WNT gene set
MIC	Metastasis-initiating cell
MMP	Matrix metalloproteinase
MV	Microvesicle
NOD-SCID	Non-obese diabetic severe combined immunodeficient
ST6GALNAC5	α2,6-sialytransferase
TGFβ	Transforming growth factor β
TIC	Tumour-initiating cell
TNFα	Tumour necrosis factor α
VEGF	Vascular endothelial growth factor

11.1 Making a Case for Metastasis-Initiating Cells

11.1.1 The Metastatic Process

Metastasis, the spread of a primary tumour to additional tissues in the body, is an extremely inefficient, multi-step process (Fig. 11.1). In order to form a tumour in a secondary location, the metastatic cell must first escape the primary tumour, then intravasate into the circulation. Once in the circulation, the cell must survive host immunological defences and shearing forces. The cell then arrests in a secondary location, and extravasates into the tissue stroma. Most often this secondary site is the first capillary bed encountered [1, 2], but occasionally, homing mechanisms direct the cell to a specific secondary tissue [1]. Once in the secondary location, the cell must initiate growth of a micrometastasis. This in turn develops into a macrometastasis, with angiogensis occurring to supply the growing tumour with nutrients. Finally, the tumour becomes well-vascularized and clinically-detectable. Studies have shown that while close to 90% of cells that have escaped the primary tumour are capable of completing all the steps up to and including extravasation, only 2 and 0.02% of these cells can develop micro- and macrometastases, respectively [3].

The ability of a small subset of primary tumour cells to produce macrometastases is reminiscent of the tumour-initiating cells (TICs) or cancer stem cells (CSCs) hypothesized to form primary tumours (Table 11.1). These TICs are considered "stem cell-like cells" due to their ability to undergo self-renewal and differentiation programs, producing all cell types of the tumour. The term is strictly functional, and does not imply that TICs originate from a normal stem cell population; the cell of

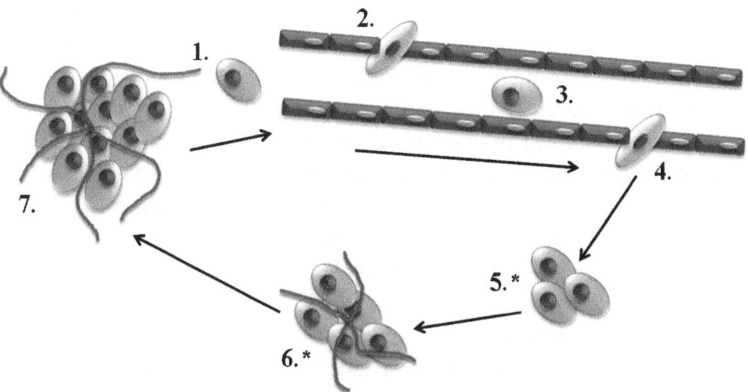

Fig. 11.1 Metastasis is an extremely inefficient multi-step process. In order to form a tumour in a secondary location, the metastatic cell must complete several steps: *1* escape of the primary tumour, *2* intravasation into the circulation, *3* survival of hostile circulation environment, *4* arrest in a secondary location and extravasation into the tissue stroma, *5* initiation of a micrometastasis, *6* angiogenesis occurs to supply developing macrometastasis, and *7* development of a well-vascularized and clinically-detectable tumour. *Denotes particularly inefficient stages in the metastatic process

Table 11.1 Markers used to identify and characterize tumour-initiating cell (TIC) populations in human primary solid tumours

Tumour type	TIC population	Reference
Brain	CD133$^+$	[4, 5]
	CD15$^+$	[6]
Breast	CD44$^{+/hi}$/CD24$^{-/lo}$	[7]
	Aldefluor$^+$	[8]
Colon	CD133$^+$	[9, 10]
	Aldefluor$^+$	[11]
Lung	CD133$^+$	[12]
Melanoma	ABCB5$^+$	[13]
Head & neck	Aldefluor$^+$	[14]
Pancreatic	CD44$^+$/CD24$^+$/EpCAM$^+$	[15]

origin may also be a more mature cell or progenitor that has acquired the ability of self-renewal [16]. Like the subset of cells forming metastases, primary TICs are most often a rare subpopulation of cells [17], with some exceptions [18]. Based on the functional definition of a TIC, cells capable of forming a metastasis could be putatively considered to be metastasis-initiating cells, or MICs.

11.1.2 Potential Models for the Origin of Metastasis-Initiating Cells

Similar to their primary tumour counterparts, the origin of MICs has yet to be determined. Several possible models exist (Fig. 11.2). First, metastases could be formed by the very TICs that have initiated the growth of the primary tumour. This would suggest that primary TICs have some sort of inherent metastatic ability. Primary TICs may also acquire metastatic properties by an epithelial-mesenchymal transition (EMT), through which they become more motile and invasive. Alternatively, some primary non-TICs may be inherently metastatic, and acquire tumour-initiating capabilities potentially through an EMT or once in the appropriate secondary microenvironment. Signalling instigated either by the primary tumour or host may also induce site-specific homing, tumour-initiation, and/or proliferation of migration cells.

It is quite possible that not all types of primary tumours leading to metastatic disease have the same origin of MICs. Clinically, systemic involvement often presents differently in various types of primary cancers. For example, lung cancer metastases often occur within several months to a few years of initial diagnosis; whereas, breast cancer patients tend to relapse with metastatic disease several years to decades after supposed eradication of the primary tumour [19]. Due to the differences in the timing of clinical presentation, this might suggest that lung cancer-derived MICs originate from cells that already possess some tumour-initiating and/or metastasis machinery, while breast cancer-derived MICs may require priming of the secondary niche and supportive signalling.

In any case, it is likely a combination of several models that leads to the origin of the MIC, with some models being more successful than others, depending on the

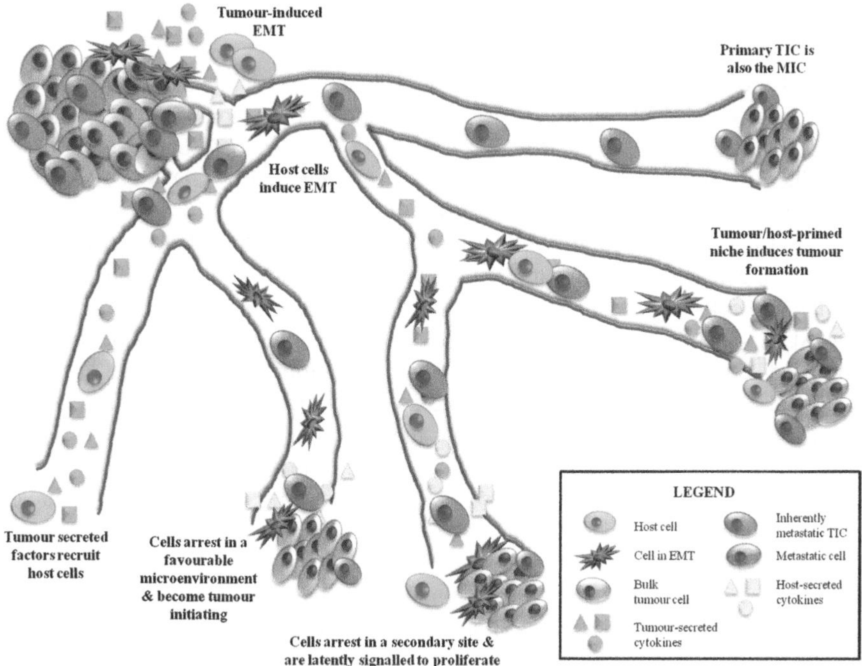

Fig. 11.2 Potential models of origin for metastasis initiation and formation: development of metastasis-initiating cells (*MICs*). Metastasis formation could arise from primary tumour-initiating cells (*TICs*) which are inherently metastatic, where the MIC and TIC populations are one in the same. TICs may also obtain invasive and migratory characteristics through an epithelial-mesenchymal transition (*EMT*) induced by the primary tumour and/or surrounding host cells. Alternatively, inherently metastatic cells may acquire tumour initiating capabilities through an EMT, once exposed to a suitable microenvironment, or when arrested in a niche primed by host and/or tumour-secreted cytokines. Cells may also arrest in a secondary location and remain indolent until they receive signals to proliferate. It is most likely that a combination of these possible models is occurring simultaneously, and that the success of each model is dependent upon the type of primary tumour

primary tumour and secondary site involved. As described in this chapter, a distinction between possible models becomes less clear, suggesting that the underlying mechanism of metastasis initiation has yet to be determined.

11.2 Inherent Metastatic Characteristics of Primary Tumours

11.2.1 Migration and Invasion of Primary Tumour-Initiating Cells

The study of metastasis has been focussed on breast cancer, and much of the research linking TICs to metastasis is modelled in breast cancer. Several studies have used $CD44^+/CD24^-$ with/without $Aldefluor^+$ populations to select breast TICs, and have found these populations of cells to be intrinsically migratory and invasive *in vitro*, and metastatic *in vivo* [20–24].

Croker et al. used several breast TIC markers to select stem cell-like subpopulations from several human breast cancer cell lines. Populations that were $CD44^{+}$/$CD24^{-}$/$Aldefluor^{hi}$ or $CD44^{+}$/$CD133^{+}$/$Aldefluor^{hi}$ demonstrated enhanced abilities of adhesion, migration, and invasion *in vitro*, compared to $CD44^{-/lo}$/$Aldeflour^{lo}$ controls [20]. Immunodeficient mice injected with either $CD44^{+}$/$CD24^{-}$/$Aldefluor^{hi}$ or $CD44^{+}$/$CD133^{+}$/$Aldefluor^{hi}$ cells had significantly higher incidence of metastasis to the lung and developed larger metastases than mice injected with $CD44^{-/lo}$/ $Aldeflour^{lo}$ cells [20]. Similarly, in a study of inflammatory breast cancer, MARY-X and SUM149 cell lines were sorted based on their Aldefluor activity. $Aldefluor^{+}$ and unsorted cells demonstrated tumour initiation capabilities *in vivo*, while $Aldefluor^{-}$ cells did not [21]. $Aldefluor^{+}$ TICs were more invasive *in vitro*; demonstrated an ability to initiate spontaneous, systemic metastases in non-obese diabetic severe combined immunodeficient (NOD-SCID) mice; and correlated with decreased overall survival and an increased probability of metastasis in inflammatory breast cancer patients, as compared to unsorted and $Aldefluor^{-}$ cells [21].

Furthermore, Sheridan et al. showed that breast cancer cell lines with a $CD44^{+}$/ $CD24^{-/lo}$ percentage of over 30% consistently expressed pro-invasive genes (e.g. *CXCR4, MMP1, osteopontin*) and had a higher capacity for invasion *in vitro* [23]. However, when intracardiac injection of $CD44^{+}$/$CD24^{-/lo}$ cells was performed in nude mice, a limited number of metastases formed. Interestingly, these *in vivo* metastases were formed by $CD44^{+}$/$CD24^{-/lo}$ populations isolated from breast cancer cell lines derived from metastatic sites, rather than the primary breast tumour [23]. This suggests that while the primary TIC population may possess a higher level of invasiveness, it is not sufficient to form metastases *in vivo* [23].

Another group used an orthotopic xenograft model of human breast cancer in NOD-SCID mice, to show that use of an EGF (epidermal growth factor) gradient caused injected cells to be more invasive and migratory [24]. While not prospectively examining the formation of metastases by the $CD44^{+}$ population, analysis of the invasive portions compared to the tumour bulk by flow cytometry showed that the invasive portions were significantly enriched for the breast TIC marker CD44 [24]. In addition, this model allowed for the observation of spontaneous metastases to the lung, where the lung metastases were found to possess a $CD44^{+}$ population similar to that of the primary tumour. When the $CD44^{+}$ cells from the metastases were injected into the mammary fat pad, they were able to recapitulate the primary tumour [24]. From these findings, it may be inferred that metastases do retain the TIC populations from the primary breast tumour (i.e. $CD44^{+}$/$CD24^{-/lo}$ cells) and that these populations are capable of recapitulating the primary tumour; however, it appears as though only a subset of the primary TICs are capable of forming a metastasis.

In a model of head and neck squamous cell carcinoma (HNSCC) it was also shown that $CD44^{high}$ TICs were more migratory *in vitro* and formed more lung metastases when injected via tail vein into NOD-SCID mice, compared to $CD44^{low}$ non-TICs [25]. However, $CD44^{low}$ cells demonstrated a trend towards being more invasive *in vitro* than CD44^{high} cells, but this trend was not significant [25]. A study examining CD44^{+}/CD24^{-}status in paraffin-embedded breast cancer patient samples also found a lack of correlation between the TIC population and event-free or overall patient survival; however, they did show that the majority of patients with metastases had more than 10% of CD44^{+}/CD24^{-} cells [26].

Together, these studies confirm that a link exists between stem cell-like populations and metastatic ability. What is not clear is whether this ability is truly inherent, or whether these cells acquire metastatic capability over time.

11.2.2 Organ-Specific Metastasis Gene Signatures in Primary Tumours

While not studying a TIC population, the Massague group has identified several gene signatures for organ-specific metastasis of breast cancer [27–29]. A list of 54 genes was identified as being upregulated in lung metastases from the breast, and a subset of these genes correlated with lung metastasis in clinical samples [27]. While the majority of the genes seemed to be important only for colonization of the lung, the transcription factor *ID1* (*inhibitor of DNA binding 1*) was found to be important for both growth of the primary tumour and metastasis to the lung [27]. Several genes were also identified as being overexpressed in bone metastases from breast cancer. These genes have specific roles in homing (*CXCR4*), osteolysis (*interleukin 11, IL11; osteopontin*), and invasion (*matrix metalloproteinase 1, MMP1*), and when expressed in combination, significantly increased the ability of previously non-metastatic cells to successfully metastasize to bone [29]. This breast-to-bone signature was found to be independent of another signature predicting prognosis and tumorigenicity of breast cancer [29, 30].

Another recent study by the Massague group has identified several genes that mediate breast cancer metastasis to the brain. By using genomic expression analysis to compare *in vivo*-selected brain metastasis cell lines (MDA231-BrM2; CN34-BrM2) to the corresponding parental lines (MDA231; CN34-BrM2), 17 genes were identified as correlating with brain relapse of breast cancer patients [31]. Several genes overlapped with the previously identified lung metastasis signature [32], including *COX2* (*cyclooxygenase 2*), *MMP1*, and other genes implicated in adhesion, migration, and invasion [31]. Knockdown of these genes reduced the migration and invasiveness of MDA231-BrM2 cells. Furthermore, *ST6GALNAC5* (α2,6-sialyltransferase), a gene normally restricted to the brain [33], was identified as being required to the formation of brain metastases *in vivo* and for the crossing of an *in vitro* blood–brain-barrier (BBB) [31].

Using *in vivo* selection, metastatic derivatives of the PC9 and H2030 lung adenocarcinoma cell lines were obtained, and consistently metastasized to the brain [34]. A set of Wnt3a-regulated genes (lung cancer WNT gene set; LWS) were found to predict for metastasis to the brain. The LWS signature did not correspond with the signature previously determined for brain metastasis by the Massague group [31].

Among these LWS genes were *LEF1* (*lymphoid enhancer binding factor 1*) and *HOXB9* (*homeobox B9*), which in addition to being required for brain colonization [34], they have also been shown to promote the metastasis of breast cancer to the lung through acquisition of mesenchymal characteristics [35].

Through these gene signature studies, it is clear that primary tumours express genes that are useful for metastasis and secondary site-specific colonization. It is possible that these gene signatures may contain putative MIC markers, yet separation of functionally distinct populations has yet to be shown. While important for metastasis, these genes are not necessarily essential for primary tumour formation. No relationship of the metastasis gene signatures with a primary TIC population has been examined, nor how they are related temporally to primary tumour initiation and growth. Answering these questions may help further elucidate their role in secondary tumour development.

11.3 The Epithelial-Mesenchymal Transition

11.3.1 Acquired Metastatic Properties of Primary Tumour-Initiating Cells

The epithelial-mesenchymal transition (EMT) is an important process in embryonic development, allowing for increased plasticity of cells as epithelial cells lose their characteristics and acquire those of mesenchymal lineages [36–38]. It is becoming increasingly apparent that EMT is involved in metastasis: cells of the primary tumour may undergo EMT, becoming more motile and invasive, allowing them to complete the initial steps of metastasis [38–42].

A variety of signalling cascades widely known to play roles in tumorigenesis are also implicated in the process of EMT: Wnt, Notch, and transforming growth factor β (TGFβ) signalling [36, 37, 43, 44]. These signalling cascades ultimately result in the upregulation of the EMT promoters *Snail1, Snail2/Slug,* and *Twist1,* where *Twist1* induces the expression of transcription factors *Snail1* and *Snail2/Slug*, which in turn suppress *E-cadherin* expression, allowing cells to lose their epithelial phenotype (Fig. 11.3) [45, 46]. Genes upregulated by these EMT promoters are responsible for disruption of tight junctions, cytoskeleton remodelling, and reorganization of the extra-cellular matrix. Other regulators of EMT include hypoxia (hypoxia-inducible factor (HIF-1) signalling), epithelial-stromal cell interactions, and several microRNAs from the miR200 family [36, 37, 44]. Of particular interest are the recent links between TICs and EMT, where TIC populations have been shown to express higher levels of EMT markers than the bulk population [47–51].

Much of the research regarding the association and mechanism of EMT in TICs has been done in HNSCC models. In a study examining several HNSCC cell lines, Aldefluor^{+}/CD44^{+}/CD24^{-} status was found to be indicative of invasion and expression of EMT markers (*Snail1, Twist1, Snail2/Slug*) [47]. Cell lines with higher

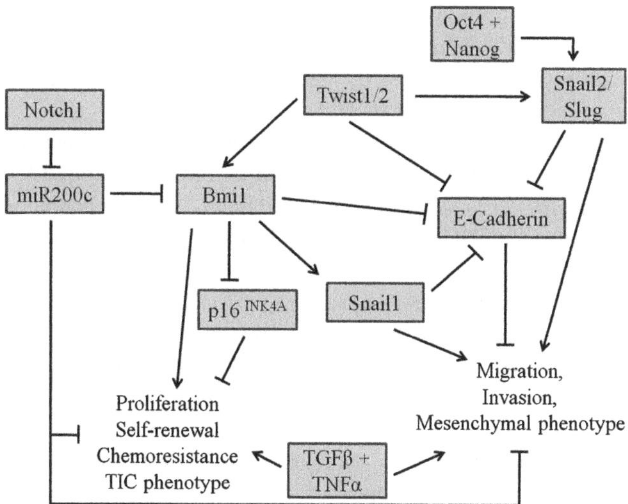

Fig. 11.3 Regulators of tumour-initiating cell (*TIC*) populations are implicated to also have a role in the epithelial-mesenchymal transition (*EMT*). Bmi1 is a known oncogene, and has been found to regulate self-renewal, leading to an increase in TIC populations. Recently Bmi1 has been shown to be regulated by Twist1, and to also regulate genes downstream of Twist1 in the EMT pathway. Other stem cell-related genes and signalling pathways (Oct4, Nanog, and Notch) have also been implicated in this process. Thus, EMT allows for stem cell-like populations to possess mesenchymal cell properties

levels of vimentin and α-smooth muscle actin, markers of myofibroblasts, an important component of tumour stroma [47]. Similarly, naturally present $CD44^{+}/CD24^{low}$ cells in normal and tumorigenic breast tissue were found to have higher expression of EMT markers and genes than their non-stem cell counterparts [48].

In oral and cutaneous squamous cell carcinomas, it was found that the $CD44^{hi}$ CSC population could be further fractionated by epithelial-specific antigen (ESA) [52]. The $CD44^{hi}/ESA^{lo}$ cells also expressed significantly higher levels of EMT markers, which corresponded with an elongated fibroblast-like morphology [52]. Both $CD44^{hi}$ populations had similar orthotopic tumour formation and were capable of generating heterogenous tumours; however, only the $CD44^{hi}/ESA^{lo}$ cells showed any lymph node infiltration [52]. Through clonal expansion and analysis, all $CD44^{hi}/ESA^{hi}$ clones were found to be bipotent, producing both ESA^{hi} and ESA^{lo} cells; only a small subset of $CD44^{hi}/ESA^{lo}$ that were also aldehyde dehydrogenase 1 positive ($ALDH1^{+}$) were found to be bipotent [52]. These results suggest that CSCs that may have undergone EMT retain their stem-like properties of self-renewal and multi-lineage differentiation.

Another group demonstrated that overexpression of *FoxM1* (*forkhead box M1*), a transcription factor often overexpressed in solid tumours [53], in AsPC-1 pancreatic cancer cells led to the acquisition of a mesenchymal-like cell morphology and upregulation of EMT transcription factors; the change in phenotype correlated with an increase in migration, self-renewal, and drug resistance [54]. All of these findings suggest a model where EMT and tumour-initiation are interrelated; however, they fail to address the causation or mechanism by which the two phenomena are related.

Several groups have made an effort to determine a genetic mechanism for the link between TICs and EMT. Aldefluor$^+$ HNSCC TICs were found to overexpress *Bmi1* (*B-cell-specific Moloney murine virus insertion site 1*), a potent oncogene and regulator of stemness in TICs [55, 56], which subsequently led to higher levels of *Snail1* [49]. Knockdown of *Bmi1* in Aldefluor$^+$ cells led to a decrease in *Snail* and *ALDH1* expression; whereas, overexpression of *Bmi1* in Aldefluor$^-$ cells led to an increase in *Snail* and *ALDH1* expression, while enhancing tumorigenicity, metastasis, and radioresistance *in vivo* [49]. Similarly, Lo et al. found *Bmi1* to be highly expressed in Aldefluor$^+$/CD44$^+$ HNSCC cells, while miR200c, a suppressor of *Bmi1*[57] and EMT-mediated metastasis [58], was found to have lower expression levels [50]. Knockdown of *Bmi1* or overexpression of miR200c reduced properties of EMT, TIC ability, and the formation of lung metastases [50]. Overexpression of miR200c also lead to the sensitization of previously radioresistent Aldefluor$^+$/CD44$^+$ cells [50]. In another study of breast cancer, co-overexpression of *Bmi1* with *H-RAS* in MCF10A human mammary epithelial cells led to the cells forming spontaneous metastases more frequently than MCF10A$^{H\text{-}RAS}$ cells when injected orthotopically into *scid* mice [59].

Roles for *Bmi1* in EMT have been found in other metastatic cancers. Overexpression of *Bmi1* in human nasopharyngeal epithelial cells induced a morphological change, causing the epithelial cells to resemble fibroblasts and acquire increased motility. Silencing of *Bmi1* reduced transformation and metastatic abilities [60]. High levels of Bmi1 were predictive of high grade melanoma with metastatic disease [61].

Other stemness genes, such as *Oct4* [62, 63] and *Nanog* [63, 64], have been linked to EMT. Multi-dimensional scaling analysis of CD133$^+$ and CD133$^-$ lung adenocarcinoma (LAC) cells, LAC metastases, and primary tumours showed that CD133$^+$ cells and metastases had similar gene signatures, and both had significantly higher levels of *Oct4* and *Nanog* than CD133$^-$ cells and the primary tumours [65]. When *Oct4* and *Nanog* were overexpressed in the A549 LAC cell line, A549$^{Oct4/Nanog}$ cells had higher levels of genes implicated in EMT, and a higher frequency of metastasis *in vivo* than control [65].

Bao et al. showed that overexpressing *Notch1*, where the Notch pathway has a role in CSC maintenance [66, 67], in the AsPC-1 pancreatic cancer cell line resulted in an increase of CSC markers EpCAM (epithelial cell adhesion molecule) and CD44, while at the same time inducing a mesenchymal-like morphology, EMT genes, and increased migration [51]. This corresponded with a decrease of several microRNAs in the miR200 family [51], which have been shown to be negative regulators of EMT and CSC genes [57, 68, 69]. Taken together, these findings and those regarding *Bmi1* (Fig. 11.3), suggest that TICs may be intrinsically primed to acquire metastatic abilities through EMT, further suggesting that TICs are not always natively metastatic.

11.3.2 Acquired Stem-Cell Like Properties of Migratory Cells

While it should come as no surprise that stem cell-like cells possess the ability to undergo an EMT, it is surprising to see that EMT can induce the stem-cell like property of self-renewal, and an increase in the TIC population. Mani et al. showed that by overexpressing the EMT transcription factors *Snail* and *Twist1* in human mammary epithelial cells, the cells not only acquired a mesenchymal phenotype, but had also acquired the breast stem cell markers $CD44^{+}/CD24^{low}$, accompanied by increased self-renewal and the ability to differentiate into multiple breast lineages [48].

When colorectal cancer cell lines and clinical samples were cultured as colonospheres, CSC marker expression (CD44, CD24, ALDH1, etc.) was enhanced, and transcriptome analysis showed that *Snail1, IL-8,* and *VEGF* (vascular endothelial growth factor) were significantly upregulated in colonospheres, compared to adherent models [70]. The $CD44^{+}$ populations were found to have much higher levels of *Snail1* than $CD44^{-}$ cells, and knockdown of *Snail1* resulted in a loss of CD44 expression, suggesting that *Snail1* has a regulatory role over this CSC marker [70]. Overexpression of *Snail1* also led to an increase in self-renewal, as demonstrated by sphere formation, as well as increased tumorigenecity and chemoresistance *in vivo* [70], further suggesting a role for EMT regulators in acquiring stemness properties.

Though *Bmi1* was found to regulate EMT genes, other EMT transcription factors are also able to regulate *Bmi1* expression. *Twist1* overexpression consistently led to the upregulation of *Bmi1* in several human HNSCC cell lines; furthermore, both *Twist1* and *Bmi1* were found to be upregulated in $Aldeflour^{+}/CD44^{+}$ TICs [71]. Overexpression of *Twist1* or *Bmi1* was also correlated with increased radioresistence and tumour formation *in vivo* [71]. *Bmi1* and *Twist1* were found to be mutually essential in inducing EMT and tumour-initiation capabilities, and cooperatively bind and repress the *E-cadherin* and $p16^{INK4A}$ promoters [71]. These findings suggest that *Bmi1* plays a regulatory role in the initiation of EMT signalling while expanding the TIC compartment.

Twist has also been shown to regulate breast CSCs by increasing the percentage of $CD44^{+}/CD24^{-}$ cells through direct suppression of the CD24 promoter [72]. In addition to increases in the percentage of $CD44^{+}/CD24^{-}$ cells in MCF7 and MCF10A cell lines, these populations were also able to self-renew, produced both $CD44^{+}/CD24^{-}$ and $CD44^{+}/CD24^{+}$ cells, and were more tumorigenic *in vivo* [72]. They were also shown to have increased ALDH and ATP-binding cassette transporter activity through the Aldefluor and Hoechst-efflux assays, respectively [72].

Twist2 has also been demonstrated to induce EMT in MCF7 and MCF10A cells, while also increasing the stemness of these two cell lines. Overexpression of *Twist2* resulted in increased migration, colony formation, and tumour formation *in vivo* [73]. Additionally, the $CD44^{high}/CD24^{low}$ population increased from 2.43 and 7.5% to 15.04 and 35.65% in MCF7 and MCF10A cells, respectively [73]. This was accompanied by increased self-renewal, and an increased expression of the stemness genes *Bmi1* and *Sox2* [73]. These results clearly demonstrate a direct regulation of stemness by EMT pathways, further suggesting that migratory cells may acquire TIC capabilities.

Other methods of inducing EMT, such as TGFβ and tumour necrosis factor α (TNFα) have also been found to lead to an increase in cancer stem cell-like characteristics. Using mouse mammary carcinoma cells which had undergone TGFβ/TNFα-induced EMT, Asiedu et al. found that the cells became more efficient at sphere-formation, were highly $CD44^{+}/CD24^{low/-}$, and also had a greater tumour formation capacity at very low cell numbers [74]. Similar results were shown in human MCF10A cells [74]. Through various mechanisms of EMT-induction, it is becoming apparent that cells undergoing EMT are also able to acquire stem-like characteristics (Fig. 11.3), potentially becoming MICs.

11.4 Host-Tumour Interactions

11.4.1 Tumour-Induced Host Involvement

It is also possible that the primary tumour induces changes in the host, allowing the host to become more permissive to tumour seeding, enhancing the growth of tumours in secondary locations. While the following reports have not been executed in metastatic models, they provide a very plausible framework for how high grade primary cancers may interact with the host environment to allow for metastatic spread.

McAllister et al. demonstrated that otherwise indolent breast tumours (transformed human mammary epithelial HMLER-HR cells; "responders") acquired tumour-initiating capacity when injected contralaterally with "instigator" human breast cancer cell lines (BPLER, MDA-MB-231) into the mammary fat pads of nude mice [75]. This was found to be due to instigator-secreted osteopontin, leading to the recruitment of host bone marrow cells to the injection sites of both instigator and responder cells [75]. Knockdown of *osteopontin* prevented such recruitment and subsequent responder growth, along with development of lung metastases [75]. Instigator-conditioned bone marrow was found to be sufficient to promote growth of the responder cells. From this, it was concluded that highly malignant tumours can recruit host bone marrow cells, which in turn aid in the development of a tumour-initiating-niche for indolent tumour cells [75]. In a subsequent study, the bone marrow cells were found to be secreting granulin, which acted upon mammary fibroblasts, leading to an increase in inflammatory and chemotactic cytokines secreted by the fibroblasts. This suggests that the modification of fibroblasts is responsible for such a change in the tumour-initiating-niche [76].

Microvesicles (MVs) released from renal CSCs, identified as CD105 (also known as endoglin) positive, were found to induce human umbilical vein endothelial cells to form capillary-like structures on Matrigel *in vitro,* where those derived from $CD105^{-}$ non-CSC cells were not [77]. When NOD-SCID mice were pre-treated with $CD104^{+}$ MVs, a greater number of lung metastases formed with higher efficiency than $CD105^{-}$ MVs or unsorted MVs [77]. It was found that the lung epithelium of

CD105[+] MV pre-treated mice had higher levels of MMP9, VEGF, and MMP2 [77], suggesting that primary tumour CSCs are indirectly modifying the host niche by promoting matrix degradation and angiogenesis, allowing for circulating tumour cells to seed a secondary location and induce tumour formation.

As a whole, these reports of primary tumours interacting with and affecting the host environment on a systemic scale may indicate that circulating tumour cells are not required to possess a full complement of inherent tumour-initiating, invasion, and angiogenesis machinery. Cells escaping the primary tumour may travel to and remain latent in a secondary location until they receive the appropriate tumour-induced support from the host. When this occurs on a systemic scale, it greatly increases the chances of developing multiple macrometastases in several locations, a hallmark of metastatic disease, in which case, identifying and targeting the MIC population may prove to be more difficult than anticipated.

11.4.2 Metastatic Cell-Niche Interactions

Migratory cells are not solely dependent upon primary tumour interactions with the host, but are also capable of forming unique relationships with the secondary microenvironment. For instance, metastatic prostate cancer to the bone appears to compete with host hematopoietic stem cells (HSCs) native to the bone marrow niche. When mice were pre-injected subcutaneously with human prostate cancer cells, engraftment of HSCs during a bone marrow transplant was significantly reduced, compared to non-tumour controls [78]. This suggested that the primary tumour was shedding cells, and these were occupying the HSC niche. Further imaging showed that this was indeed the case, and that competition for the niche was dependent on the number of prostate cancer cells injected [78]. Similar results were found when human HSC cells were used [78].

Other studies have examined how CSC populations may be more capable of adapting to a new niche. Pommier et al. showed that breast TICs from a human breast cancer cell line were significantly more successful at forming brain tumours than non-TICs when injected intracranially into nude rats [79]. This suggests that inherent TIC properties are important in niche adaptation. However, it is not clear whether these primary TIC populations are the same cells expressing the genes implicated in brain metastasis, as identified by Bos et al. [31]. Pommier's findings also demonstrate that the brain microenvironment is not sufficient for inducing tumour-initiation capabilities in non-TICs; tumour cells must be inherently tumour-initiating. However, there is evidence suggesting that primary tumour cells, once they have formed metastases in the brain, are capable of activating the surrounding astrocytes, which in turn increase tumour growth [80, 81]. This further suggests that tumour-initiating capacity is linked to adapting to a new microenvironment, but is not necessarily required for sustained tumour growth.

The gene signature studies previously described indicate that migratory and metastatic cells appear to have acquired the expression of a subset of genes that allows for better niche adaptation and survival [27–29, 31, 32]. However, due to *in vivo* selection, it is difficult to say whether these gene signatures are present in a subset of cells of the primary tumour, which then home to their specific niche, or if it is an artefact of selection. Interestingly, overexpression of *Bmi1* has been shown to enhance the growth of breast cancer cells in the brain when injected intracranially [59]. This is supported by several studies implicating *Bmi1* in the initiation of primary brain tumours [82, 83], indicating it is important for tumour formation in the brain. It is clear that cells shed from the primary tumour do possess machinery to enhance metastatic growth in a particular environment; whether these genes are sufficient for metastasis initiation remains to be seen.

11.5 Origins of Metastasis: Impacts on Therapeutic Developments

The cancer stem cell model also stipulates that TICs evade current therapies [16, 84, 85], suggesting TICs are responsible for relapse and allow for metastasis. This would also suggest that putative MICs are also resistant to conventional chemotherapy and radiation. Therefore, future therapeutic developments must target the MIC subpopulation in addition to the bulk tumour. Current efforts to target TICs, and potentially MICs, include induction of differentiation, resulting in a loss of self-renewal and increased susceptibility to bulk tumour therapy [86–89]. Targeting of the EMT pathway to prevent acquisition of metastatic capabilities is also under study [44, 89]. Interfering with homing mechanisms through chemotactic axes, such as the CXCR4/CXCL12 axis, is also showing promise in reducing migration and infiltration in primary brain tumours [90, 91], and impairment of metastasis formation in breast cancer [92, 93]. Ideal therapeutic targets are those with roles in multiple pathways integral to the MIC population, such as *Bmi1* in self-renewal, proliferation, and EMT pathways [49, 71, 82]. An important consideration in targeting MICs is that resistance and protection may be conferred by the metastatic niche. For example, brain metastases are protected by the BBB, and recruit astrocytes into the tumour microenvironment [80, 81]. As such, a prophylactic approach is needed, using MIC targets to prevent metastasis and maintain a localized disease state.

Metastasis remains a prevalent and serious clinical problem. The TIC model of primary cancers could very well be a suitable model for explaining the occurrence of metastases. Currently, some evidence suggests that primary TICs may be inherently metastatic, while other studies suggest tumour-initiating and/or metastatic capabilities may be acquired through EMT. While it may seem as though these findings are contradictory, it is quite possible that both are true, and that we are simply observing the same population of cells at different points in time, something that would require lineage-tracing to clarify. Additionally, primary tumour-host and migratory cell-niche interactions suggest a more dynamic and complex method of metastasis initiation. Regardless of origin, it is clear that putative MICs are potential targets for treating metastatic disease. Further research combining the ideas of CSCs and gene signatures, lineage tracing with EMT, and interference with host and niche interactions is necessary to determine the existence and origins of a putative MIC population. This study may reveal genes and signalling pathways implicated in metastasis, which could allow for targeting of organ-specific or systemic metastasis. In doing so, patient survival could be markedly extended, as blocking the metastatic process would transform a uniformly fatal disease into a locally controlled, and eminently more treatable one.

References

1. Fidler IJ (2001) Seed and soil revisited: contribution of the organ microenvironment to cancer metastasis. Surg Oncol Clin N Am 10(2):257–269, vii–viiii
2. Croker AK, Allan AL (2008) Cancer stem cells: implications for the progression and treatment of metastatic disease. J Cell Mol Med 12(2):374–390
3. Luzzi KJ et al (1998) Multistep nature of metastatic inefficiency: dormancy of solitary cells after successful extravasation and limited survival of early micrometastases. Am J Pathol 153(3):865–873
4. Singh SK et al (2003) Identification of a cancer stem cell in human brain tumors. Cancer Res 63(18):5821–5828
5. Singh SK et al (2004) Identification of human brain tumour initiating cells. Nature 432(7015):396–401
6. Mao XG et al (2009) Brain tumor stem-like cells identified by neural stem cell marker CD15. Transl Oncol 2(4):247–257
7. Al-Hajj M et al (2003) Prospective identification of tumorigenic breast cancer cells. Proc Natl Acad Sci USA 100(7):3983–3988
8. Ginestier C et al (2007) ALDH1 is a marker of normal and malignant human mammary stem cells and a predictor of poor clinical outcome. Cell Stem Cell 1(5):555–567
9. O'Brien CA et al (2007) A human colon cancer cell capable of initiating tumour growth in immunodeficient mice. Nature 445(7123):106–110
10. Ricci-Vitiani L et al (2007) Identification and expansion of human colon-cancer-initiating cells. Nature 445(7123):111–115
11. Huang EH et al (2009) Aldehyde dehydrogenase 1 is a marker for normal and malignant human colonic stem cells (SC) and tracks SC overpopulation during colon tumorigenesis. Cancer Res 69(8):3382–3389
12. Eramo A et al (2008) Identification and expansion of the tumorigenic lung cancer stem cell population. Cell Death Differ 15(3):504–514
13. Schatton T et al (2008) Identification of cells initiating human melanomas. Nature 451(7176): 345–349
14. Chen YC et al (2009) Aldehyde dehydrogenase 1 is a putative marker for cancer stem cells in head and neck squamous cancer. Biochem Biophys Res Commun 385(3):307–313
15. Li C et al (2007) Identification of pancreatic cancer stem cells. Cancer Res 67(3):1030–1037
16. Clarke MF et al (2006) Cancer stem cells–perspectives on current status and future directions: AACR Workshop on cancer stem cells. Cancer Res 66(19):9339–9344
17. Ishizawa K et al (2010) Tumor-initiating cells are rare in many human tumors. Cell Stem Cell 7(3):279–282
18. Quintana E et al (2008) Efficient tumour formation by single human melanoma cells. Nature 456(7222):593–598
19. Schouten LJ et al (2002) Incidence of brain metastases in a cohort of patients with carcinoma of the breast, colon, kidney, and lung and melanoma. Cancer 94(10):2698–2705

20. Croker AK et al (2009) High aldehyde dehydrogenase and expression of cancer stem cell markers selects for breast cancer cells with enhanced malignant and metastatic ability. J Cell Mol Med 13(8B):2236–2252
21. Charafe-Jauffret E et al (2010) Aldehyde dehydrogenase 1-positive cancer stem cells mediate metastasis and poor clinical outcome in inflammatory breast cancer. Clin Cancer Res 16(1):45–55
22. Marcato P et al (2011) Aldehyde dehydrogenase activity of breast cancer stem cells is primarily due to isoform ALDH1A3 and its expression is predictive of metastasis. Stem Cells 29(1):32–45
23. Sheridan C et al (2006) CD44+/CD24- breast cancer cells exhibit enhanced invasive properties: an early step necessary for metastasis. Breast Cancer Res 8(5):R59
24. Liu H et al (2010) Cancer stem cells from human breast tumors are involved in spontaneous metastases in orthotopic mouse models. Proc Natl Acad Sci USA 107(42):18115–18120
25. Davis SJ et al (2010) Metastatic potential of cancer stem cells in head and neck squamous cell carcinoma. Arch Otolaryngol Head Neck Surg 136(12):1260–1266
26. Abraham BK et al (2005) Prevalence of CD44+/CD24-/low cells in breast cancer may not be associated with clinical outcome but may favor distant metastasis. Clin Cancer Res 11(3):1154–1159
27. Minn AJ et al (2005) Genes that mediate breast cancer metastasis to lung. Nature 436(7050): 518–524
28. Minn AJ et al (2005) Distinct organ-specific metastatic potential of individual breast cancer cells and primary tumors. J Clin Invest 115(1):44–55
29. Kang Y et al (2003) A multigenic program mediating breast cancer metastasis to bone. Cancer Cell 3(6):537–549
30. van't Veer LJ et al (2002) Gene expression profiling predicts clinical outcome of breast cancer. Nature 415(6871):530–536
31. Bos PD et al (2009) Genes that mediate breast cancer metastasis to the brain. Nature 459(7249):1005–1009
32. Minn AJ et al (2007) Lung metastasis genes couple breast tumor size and metastatic spread. Proc Natl Acad Sci USA 104(16):6740–6745
33. Okajima T et al (1999) Molecular cloning of brain-specific GD1alpha synthase (ST6GalNAc V) containing CAG/Glutamine repeats. J Biol Chem 274(43):30557–30562
34. Nguyen DX et al (2009) WNT/TCF signaling through LEF1 and HOXB9 mediates lung adenocarcinoma metastasis. Cell 138(1):51–62
35. Hayashida T et al (2010) HOXB9, a gene overexpressed in breast cancer, promotes tumorigenicity and lung metastasis. Proc Natl Acad Sci USA 107(3):1100–1105
36. Polyak K, Weinberg RA (2009) Transitions between epithelial and mesenchymal states: acquisition of malignant and stem cell traits. Nat Rev Cancer 9(4):265–273
37. Kalluri R, Weinberg RA (2009) The basics of epithelial-mesenchymal transition. J Clin Invest 119(6):1420–1428
38. Baum B, Settleman J, Quinlan MP (2008) Transitions between epithelial and mesenchymal states in development and disease. Semin Cell Dev Biol 19(3):294–308
39. Kang Y, Massague J (2004) Epithelial-mesenchymal transitions: twist in development and metastasis. Cell 118(3):277–279
40. Yang J, Weinberg RA (2008) Epithelial-mesenchymal transition: at the crossroads of development and tumor metastasis. Dev Cell 14(6):818–829
41. Thiery JP (2002) Epithelial-mesenchymal transitions in tumour progression. Nat Rev Cancer 2(6):442–454
42. Turley EA et al (2008) Mechanisms of disease: epithelial-mesenchymal transition–does cellular plasticity fuel neoplastic progression? Nat Clin Pract Oncol 5(5):280–290
43. Xu J, Lamouille S, Derynck R (2009) TGF-beta-induced epithelial to mesenchymal transition. Cell Res 19(2):156–172
44. Roussos ET et al (2010) AACR special conference on epithelial-mesenchymal transition and cancer progression and treatment. Cancer Res 70(19):7360–7364

45. Cano A et al (2000) The transcription factor snail controls epithelial-mesenchymal transitions by repressing E-cadherin expression. Nat Cell Biol 2(2):76–83
46. Casas E et al (2011) Snail2 is an essential mediator of Twist1-induced epithelial mesenchymal transition and metastasis. Cancer Res 71(1):245–254
47. Chen C et al (2011) Evidence for epithelial-mesenchymal transition in cancer stem cells of head and neck squamous cell carcinoma. PLoS One 6(1):e16466
48. Mani SA et al (2008) The epithelial-mesenchymal transition generates cells with properties of stem cells. Cell 133(4):704–715
49. Yu CC et al (2011) Bmi-1 regulates snail expression and promotes metastasis ability in head and neck aquamous cancer-derived ALDH1 positive cells. J Oncol 2011. doi:10.1155/2011/609259
50. Lo WL et al (2011) MicroRNA-200c attenuates tumour growth and metastasis of presumptive head and neck squamous cell carcinoma stem cells. J Pathol 223(4):482–495
51. Bao B et al (2011) Notch-1 induces epithelial-mesenchymal transition consistent with cancer stem cell phenotype in pancreatic cancer cells. Cancer Lett 307(1):26–36
52. Biddle A et al (2011) Cancer stem cells in squamous cell carcinoma switch between two distinct phenotypes that are preferentially migratory or proliferative. Cancer Res 71(15):5317–5326
53. Pilarsky C et al (2004) Identification and validation of commonly overexpressed genes in solid tumors by comparison of microarray data. Neoplasia 6(6):744–750
54. Bao B et al (2011) Over-expression of FoxM1 leads to epithelial-mesenchymal transition and cancer stem cell phenotype in pancreatic cancer cells. J Cell Biochem 112(9):2296–2306
55. Jiang L, Li J, Song L (2009) Bmi-1, stem cells and cancer. Acta Biochim Biophys Sin (Shanghai) 41(7):527–534
56. Park IK, Morrison SJ, Clarke MF (2004) Bmi1, stem cells, and senescence regulation. J Clin Invest 113(2):175–179
57. Shimono Y et al (2009) Downregulation of miRNA-200c links breast cancer stem cells with normal stem cells. Cell 138(3):592–603
58. Burk U et al (2008) A reciprocal repression between ZEB1 and members of the miR-200 family promotes EMT and invasion in cancer cells. EMBO Rep 9(6):582–589
59. Hoenerhoff MJ et al (2009) BMI1 cooperates with H-RAS to induce an aggressive breast cancer phenotype with brain metastases. Oncogene 28(34):3022–3032
60. Song LB et al (2009) The polycomb group protein Bmi-1 represses the tumor suppressor PTEN and induces epithelial-mesenchymal transition in human nasopharyngeal epithelial cells. J Clin Invest 119(12):3626–3636
61. Mihic-Probst D et al (2007) Consistent expression of the stem cell renewal factor BMI-1 in primary and metastatic melanoma. Int J Cancer 121(8):1764–1770
62. Nichols J et al (1998) Formation of pluripotent stem cells in the mammalian embryo depends on the POU transcription factor Oct4. Cell 95(3):379–391
63. Park IH et al (2008) Reprogramming of human somatic cells to pluripotency with defined factors. Nature 451(7175):141–146
64. Chambers I et al (2003) Functional expression cloning of Nanog, a pluripotency sustaining factor in embryonic stem cells. Cell 113(5):643–655
65. Chiou SH et al (2010) Coexpression of Oct4 and Nanog enhances malignancy in lung adenocarcinoma by inducing cancer stem cell-like properties and epithelial-mesenchymal transdifferentiation. Cancer Res 70(24):10433–10444
66. Ying M et al (2011) Regulation of glioblastoma stem cells by retinoic acid: role for Notch pathway inhibition. Oncogene 30:3454–3467
67. Sullivan JP et al (2010) Aldehyde dehydrogenase activity selects for lung adenocarcinoma stem cells dependent on notch signaling. Cancer Res 70(23):9937–9948
68. Korpal M et al (2008) The miR-200 family inhibits epithelial-mesenchymal transition and cancer cell migration by direct targeting of E-cadherin transcriptional repressors ZEB1 and ZEB2. J Biol Chem 283(22):14910–14914
69. Park SM et al (2008) The miR-200 family determines the epithelial phenotype of cancer cells by targeting the E-cadherin repressors ZEB1 and ZEB2. Genes Dev 22(7): 894–907

70. Hwang WL et al (2011) SNAIL regulates interleukin-8 expression, stem cell-like activity, and tumorigenicity of human colorectal carcinoma cells. Gastroenterology 141(1):279–291
71. Yang MH et al (2010) Bmi1 is essential in Twist1-induced epithelial-mesenchymal transition. Nat Cell Biol 12(10):982–992
72. Vesuna F et al (2009) Twist modulates breast cancer stem cells by transcriptional regulation of CD24 expression. Neoplasia 11(12):1318–1328
73. Fang X et al (2011) Twist2 contributes to breast cancer progression by promoting an epithelial-mesenchymal transition and cancer stem-like cell self-renewal. Oncogene 30(47):4707–4720
74. Asiedu MK et al (2011) TGF{beta}/TNF{alpha}-mediated epithelial-mesenchymal transition generates breast cancer stem cells with a Claudin-low phenotype. Cancer Res 71(13): 4707–4719
75. McAllister SS et al (2008) Systemic endocrine instigation of indolent tumor growth requires osteopontin. Cell 133(6):994–1005
76. Elkabets M et al (2011) Human tumors instigate granulin-expressing hematopoietic cells that promote malignancy by activating stromal fibroblasts in mice. J Clin Invest 121(2):784–799
77. Grange C et al (2011) Microvesicles released from human renal cancer stem cells stimulate angiogenesis and formation of lung pre-metastatic niche. Cancer Res 71(15):5346–5356
78. Shiozawa Y et al (2011) Human prostate cancer metastases target the hematopoietic stem cell niche to establish footholds in mouse bone marrow. J Clin Invest 121(4):1298–1312
79. Pommier SJ et al (2010) Characterizing the HER2/neu status and metastatic potential of breast cancer stem/progenitor cells. Ann Surg Oncol 17(2):613–623
80. Seike T et al (2011) Interaction between lung cancer cells and astrocytes via specific inflammatory cytokines in the microenvironment of brain metastasis. Clin Exp Metastasis 28(1):13–25
81. Arshad F et al (2010) Blood–brain barrier integrity and breast cancer metastasis to the brain. Pathol Res Int 2011:920509
82. Abdouh M et al (2009) BMI1 sustains human glioblastoma multiforme stem cell renewal. J Neurosci 29(28):8884–8896
83. Michael LE et al (2008) Bmi1 is required for Hedgehog pathway-driven medulloblastoma expansion. Neoplasia 10(12):1343–1349, 5p following 1349
84. Dean M, Fojo T, Bates S (2005) Tumour stem cells and drug resistance. Nat Rev Cancer 5(4):275–284
85. Shervington A, Lu C (2008) Expression of multidrug resistance genes in normal and cancer stem cells. Cancer Invest 26(5):535–542
86. Korur S et al (2009) GSK3beta regulates differentiation and growth arrest in glioblastoma. PLoS One 4(10):e7443
87. Li Y et al (2010) Sulforaphane, a dietary component of broccoli/broccoli sprouts, inhibits breast cancer stem cells. Clin Cancer Res 16(9):2580–2590
88. Srivastava RK et al (2011) Sulforaphane synergizes with quercetin to inhibit self-renewal capacity of pancreatic cancer stem cells. Front Biosci (Elite Ed) 3:515–528
89. Mimeault M, Batra SK (2010) New promising drug targets in cancer- and metastasis-initiating cells. Drug Discov Today 15(9–10):354–364
90. Rubin JB et al (2003) A small-molecule antagonist of CXCR4 inhibits intracranial growth of primary brain tumors. Proc Natl Acad Sci USA 100(23):13513–13518
91. Terasaki M et al (2011) CXCL12/CXCR4 signaling in malignant brain tumors: a potential pharmacological therapeutic target. Brain Tumor Pathol 28(2):89–97
92. Dewan MZ et al (2006) Stromal cell-derived factor-1 and CXCR4 receptor interaction in tumor growth and metastasis of breast cancer. Biomed Pharmacother 60(6):273–276
93. Liang Z et al (2004) Inhibition of breast cancer metastasis by selective synthetic polypeptide against CXCR4. Cancer Res 64(12):4302–4308

Chapter 12
The Reduction of Callus Formation During Bone Regeneration by BMP-2 and Human Adipose Derived Stem Cells

Claudia Keibl and Martijn van Griensven

Contents

Abstract Reconstructive medicine, trauma surgery and orthopaedics show an enormous increase in numbers of patients with degenerative diseases. Thus the demand for new therapeutic approaches is continuously growing. New technologies, the so called "Tissue Engineering" offers with the combination of the three components: cells, growth factor and matrix, new promising technologies.

In our studies, a 2 mm transcortical non-critical size drill hole in the middle of the femur shaft of male rats was applied as a small defect model that was used as a screening model for bone regeneration as well as the in vivo bone healing stimulation when the growth factor BMP-2 was embedded together with ASCs in a locally-applied fibrin matrix. After relatively short periods of times (2 and 4 weeks) our small animal model demonstrated that it is possible to get information about the osteogenetic

C. Keibl, DVM (✉) • M. van Griensven
Ludwig Boltzmann Institute for Experimental and Clinical Traumatology,
Research Center of the Austrian Worker's Compensation Board,
Austrian Cluster for Tissue Regeneration, Donaueschingenstrasse 13,
A-1200 Vienna, Austria
e-mail: claudia.keibl@trauma.lbg.ac.at

R.K. Srivastava and S. Shankar (eds.), *Stem Cells and Human Diseases*,
DOI 10.1007/978-94-007-2801-1_12, © Springer Science+Business Media B.V. 2012

potential and bone regeneration with little effort (no osteosynthesis). The most significant result of our scientific project with the help of micro-computer tomography and descriptive histology analysis is the fact that the combination of ASCs + BMP-2 in a fibrin matrix significantly reduces the callus reaction after 2 weeks. ASCs embedded alone in the fibrin matrix did not cause an increased bone regeneration. Consequently these stem cells rather prevented the osteoinductive reaction of BMP-2 and thereby less callus formation could be analysed.

Keywords BMP-2 • Bone regeneration • Drill hole model • Fibrin • Human adipose derived stem cells

12.1 Introduction

Due to the fact that the number of patients with degenerative diseases has steadily grown, the demand for new therapeutic approaches to treat bone defects and fractures has tremendously increased in trauma surgery and orthopaedics. The promising new technology "Tissue Engineering" is composed of three components: cells, matrix and growth factors. Main efforts are targeted at improving and accelerating recovery, especially of long bone fractures, and reducing the risk of delayed bone healing or pseudarthrosis.

Adult human stem cells (ASCs) isolated from fat tissue can differentiate into osteoblasts in an osteogenic surrounding. BMP-2 (bone morphogenetic protein-2) accelerates respectively initiates this differentiation. Fibrin, a matrix that promotes wound healing, is a promising carrier for ASCs and BMP-2. BMP-2 in combination with ASCs not only rebuilds the bone structure in a defect from the inside out, but it also expedites this process.

Traumatic bone defect fractures and also fractures caused by diseases such as osteoporosis show an enormous medical and economic problem in the trauma surgery and orthopaedics particularly with regard to a continuously growing older population [1].

Blood loss, injury of blood vessels and nerves are problems caused by trauma and operations and correlate with a number of complications and with increased length of treatment. A lack of vascularization and injury of the soft tissue cause a delayed fracture healing [2, 3].

So far the gold standard, autologous spongiosa or osteoconductive bone substitutes, has been used for bone regeneration to treat bone defects. New and innovative treatment is necessary in order to approve and accelerate the healing process in trauma surgery and orthopaedics [4].

12.2 Bone

Bone is the hardest tissue of the body and the most important orthetic and guarding part of the skeleton.

There are different types of bones: long bones (tubular bone) such as humera and femora; short bones such as vertebral body; plate bones such as shoulder and costal bone.

Histologically it is distinguished between woven bone (bone and collagen fiber are irregularly distributed) and the lamellar bone (regularly repeated structures with a higher number of bone cells and a lower number of minerals) [5].

The reaction of a bone on trauma depends on various factors, such as proliferation of bone stem cells, osteoid production, mineralization and modelling. The natural healing process depends on the age of the patient, the extent of the injury of muscles and bone structures, local pathological conditions, products of inflammation and the type of bone [6].

Fracture is defined as a united division of bone caused by direct (trauma) or indirect (lever force), pathological as well as spontaneous fracture.

Tscherne et al. distinguish between open and closed fractures. Certain clinical signs of fracture are deformation, crepitation, abnorm movement and free bone components [7]. Uncertain signs of fracture are dolor, rubor, haematoma, functio laesa and tumor.

Diagnostic ways to exclude a fracture comprise x-ray on two levels—considering the adjoining structures.

Local results of fracture are injury of the soft tissue, injury of internal organs, tendineum, nerves and big vessels [8].

The bone rebuilding in the fracture area is caused by the periost and endost. The surplus tissue is called *callus* [5].

The directives of Lorenz Boehler in fracture treatment are still valid [9]: reposition, retention and rehabilitation.

Aim of fracture healing is quickest and without complications remodelling of a damaged bone [8, 10].

Possible therapeutic options are biological (allogene, autogene, xenogene or syngene) or synthetic (polymere, ceramics) implants [8, 11, 12].

Indications for osteosynthesis, e.g. distraction-osteogenesis, are generally painful, long in duration and high in costs [8, 13].

The mechanical problem, the insufficient maximum stress capacity with regard to all bone substitutes mentioned above in order to guarantee stability without osteosynthesis, seem to be unsolved so far [8].

12.3 Tissue Engineering

New promising technologies, the so called "Tissue Engineering" are combining three components: cells, matrix and growth factors which open new possibilities for the therapy of bone defects [14].

"Tissue Engineering" means the reproduction of parts of tissues aiming at the acceleration of healing processes (open wounds, fractures) and also means the regeneration of injured tissue as well as the replacement of destroyed tissue (burned skin).

Charles und Josef Vacanti, Patrick, Mikos, Langer und McIntire started with this promising technologies in the mid-1980s [12, 15].

"Tissue Engineering" is carried out "in vivo" and also "in vitro". The real challenge of this technology is to find the three correspondent components which interact positively taking into account the cell to cell contacts and the mechanical power [15].

In our studies ASC (adipose derived stem cells) as cell-compartment, fibrin as matrix and BMP-2 (bone morphogenetic protein-2) as growth factor were applied following the principles of "Tissue Engineering".

12.4 Stem Cells

Stem cells are often called "human supplies" which have the capacity to differentiate in various tissue and cell types, e.g. nerves, cartilage, bone or skin-cells. They represent a most potential factor in medical research. Stem cells are usually unspecialized cells located in so called stem cell niches and proliferate and differentiate into specialized cell types caused by e.g. diseases and trauma.

Serious diseases which cannot be healed so far, e.g. spinal cord injuries, complicated fractures or myocardial infarct are hoped to be healed efficiently by means of stem cell research [16].

Ross Harrison developed the basic principles for the cell culture in 1907. American research scientists discovered stem cells in the bone marrow of mice in 1963. The first successful stem cell therapy of a blood cancer patient by means of bone marrow transplantation was reported in 1969 [17]. In 1981 embryonic stem cells of mice were cultivated in vitro for the first time.

In 1988 the first transplantation of cord blood of an anaemia patient was successfully carried out. In 1998 the first embryonic stem cells in vitro were effectively reproduced. In 1999 nerve cells were developed embryonic from stem cells of mice. In 2000 brain stem cells of mice could be transplanted into different areas (heart, lung, liver, kidney and nerve tissue for the first time). In 2003 degenerative human intervertebral disks were treated with bone marrow stem cells. In 2005, a British research team discovered stem cells in the cord blood, which are similar in their qualities to those of the embryonic stem cells [18].

There are two different types of stem cells: embryonic and adult (haematopoetic, mesenchymal and many others) cells.

Embryonic stem cells are omni/pluripotent. They can be obtained from the inner cells of a 5 day old embryo (blastocyst). During this procedure the blastocyst are destroyed. For this reason embryonic stem cells are critically discussed and their extraction strictly prohibited in many countries.

Adult stem cells are multipotent. They can be found in many different areas of the adult body (liver, bone marrow, brain, fat tissue). Adult stem cells are able to regenerate new cells of another tissue, which is called plasticity [19].

By means of additional factors (e.g.: vitamins, dexamethason, growth factors and others) the adult stem cells can be animated in vitro to differentiate into specialized cell types.

Adult stem cells can be extracted of the bone marrow (haematopoetic and mesenchymal stem cells) by means of puncture of the iliac crest, furthermore of the cord blood (haematopoetic stem cells) after omphalotomy or by means of liposuction (mesenchymal stem cells). The advantage of the use of these autologous cells is the absent immunogenicity.

Fat stem cells are mesenchymal stem cells. They can differentiate into bone, muscle, nerves, cartilage and tendon [20–22].

Fat stem cells are a resource which is easy to be obtained [19], they are advantageous with regard to expenses ("one-step procedure": "Tissue Engineering" in connection of an operation of a trauma patient) and what is more, there is an enormous amount of it in each of us! Fat tissue contains a number of different cells (e.g.: adipocytes, blood cells) and among them they also contain cells which have the capacity to regenerate themselves with the possibility to differentiate into different lines [22, 23].

In 2004 German surgeons succeeded in replacing missing parts of the skull bone of a 7 year old girl by means of transplantation of a mixture of her own fat stem cells, fibrin and single pieces of her pelvis [24].

12.5 Fibrin

In 1909 Bergel reported about the wound-healing-supporting capacity of fibrin. He described fibrin as scentless, not caustic, not toxic, alleviative against germs, promoting in building callus and a physiological wound healing preparation [25].

In 1975 Kuderna und Matras were responsible for the clinical application of the adhesive sealing for nerve anastomosis [26].

In 1978 the first commercial fibrin glue was produced from human plasma.

Fibrin is produced by Baxter-Immuno AG in Vienna, Austria for all over the world. In 1998, this fibrin was established under the brand name Tisseel/Tissucol in America.

Fibrin plays an essential part during the natural wound healing, hemostasis, and tissue occlusion, and is the end-product of the coagulation cascade. Fibrin glue is equivalent to the last phase of the blood coagulation and is based upon the transformation from fibrinogen to fibrin [27].

The result of mixing fibrinogen and thrombin is a fibrin clot. The addition of factor XIII and Ca^{2+} leads to cross-linking, after this to fibrinolysis and in the end to the wound healing [28].

The formation of fibrin is locally evoked as a result of fibrin glue [29].

This two-component-glue, used in a double syringe, the so called duploject (Baxter AG, Vienna), applied in an equal amount, contains on the one hand a high concentrated, viscous fibrinogen/fibronektin-dilution, factor XIII and aprotinin and on the other hand a dilution consisting of diluted thrombin and calciumchloride. All components mentioned above are derived from human plasma except aprotinin [27, 30].

Due to the fact that the protection against infections is of high importance, fibrin is very often and multivarious in use.

Fibrin is used as a reproducible matrix [31].

Fibrin is also applied in trauma surgery, dentistry, surgery, and many other medical areas [31].

In 1976, Bösch et al. showed in an experimental research while using fibrin that bone regeneration is accelerated because of better vascularisation, perfect local hemostasis and good glue in place of the bone transplant [32].

Guéhennec et al. reported about the usage of fibrin as a widely-used matrix for bone regeneration with the positive effect of being not cytotoxic, being completely resorbable and being not immunologically reactive. The easy application in clinical use and the possibility in adding cells and bio-ceramic materials made fibrin ideal for the reconstructive trauma surgery [27].

Fibrin is used as a bio-matrix and scaffold for stem cells, plasmids and growth factors [26].

Fibrin was used as a matrix in which the fat stem cells with or without BMP-2 were embedded.

12.6 BMP-2 (Bone Morphogenetic Protein 2)

The so called bone morphogenetic protein (BMP), which is a bone-forming activity, was discovered by M.R. Urist in 1965 [33].

BMPs are signalproteins and members of the transforming growth factor (TGF)-β superfamily. They are also well known to be osteoinductive [34]. BMPs influence local target cells in the direction of cell proliferation and cell differentiation. They are also responsible for the direct differentiation of mesenchymal stem cells in osteoblasts, they accelerate the bone healing process and improve the osseointegrity [35–37].

The production of recombinant human BMP (rh-BMP) is possible due to gene-technology, mainly in *E. coli* and CHO-cells and is locally applied on a collagen fleece in clinical routine, e.g. with tibial non-unions and fractures of vertebral bodies. In our drill hole model BMP-2 was injected in the femur of rats.

Rh-BMPs play a significant role in the bone healing process due to the stimulation of the differentiation of stem cells in the osteogenic direction [38, 39]. BMP-2 is commercially available for clinical use.

BMP-2 is a potential bone growth factor. On the one hand it may induce a surplus of callus and on the other hand bone may induce bone formation in muscles. Possible side effects like local erythema, tumour, heterotropic ossification, surplus callus and immune-reaction may develop [2].

Einhorn reported about a small number of infections, intra-operative bleeding and an accelerated wound healing in connection with open tibial shaft fractures [40].

The selection of the appropriate matrix, the kind of application and the individual dosage are very important for prospective research and clinical studies [2, 40, 41].

The aim of our studies was to find out the influence of human adult fat stem cells and BMP-2 on bone regeneration. Both components, human adult fat stem cells and BMP-2, embedded in fibrin as a matrix, were analysed in our scientific research project, in a 2 mm transcortical rat femur drill hole model, which served as a small non-critical size defect model for fracture simulation in two different periods (in an early healing phase after 2 weeks and in a late healing phase after 4 weeks).

12.7 Materials, Methods and Results

12.7.1 Materials

Fifty male Sprague–Dawley rats (Animal Research Laboratories, Himberg, Austria), weighing between 350 and 450 g were kept under standard laboratory conditions.

The rats were caged in macrolon-cages type IV (three rats per cage) with food (Ssniff, Germany) and water access *ad libitum*, with adequate beeding and enrichment. All animals were allowed to acclimatize to their environment for at least 7 days before the onset of the experiment. Experiments were approved by the animal protocol review board of the City Government of Vienna, Austria in accordance with the Guide for the Care and Use of Laboratory Animals as defined by the National Institute of Health (MA58/05766/2007/8).

Healthy rats were randomised and divided into five groups (n = 10 each): Control; Fibrin;

Fibrin + human adult fat stem cells (ASC); Fibrin + ASC + BMP-2; Fibrin + BMP-2.

Anaesthesia was induced by intramuscular injection of a mixture of 110 mg/kg ketaminhydrochlorid (ketamidor, Richter Pharma AG, Wels, Austria) and 12 mg/kg xylazin (rompun 2%, Bayer AG, Vienna, Austria). Due to species-specific risk of decreasing intestinal motility and hypoglycaemia rats were not starved prior to anaesthesia. The operated rats received subcutaneously 4 mg/kg carprofen (rimadyl, Pfizer Corporation GmbH, Vienna, Austria) once a day, for 4 days, as an analgetic therapy and preoperatively a 2 mL liquid ringer-solution-depot mixed with 0.3 mL butafosan (catosal, Bayer Health Care Austria GmbH, Vienna, Austria) subcutaneously. According to group allocation a bilateral 2 mm transcortical (non-critical size bone defect) drill hole in the middle of the femur shaft of the rats was filled with 0.2 mL filling compound and in the control group no filling was carried out (Fig. 12.1a–d).

In order to exclude fractures, latero-lateral and dorso-ventral X-rays were performed on day 7, 14, 21 under short inhalation (3 L/min air and 2 vol.% isoflurane (forane, Abbott GmbH, Vienna, Austria)) anaesthesia. Two animals showed a fracture: in the fibrin group on day 9 and in the control group on day 2. These rats were euthanised and excluded from further analysis.

Two and four weeks after surgery the rats were sacrificed in general anaesthesia (see above) by an overdose of thiopental-sodium (120 mg/kg, thiopental sandoz, Sandoz GmbH, Vienna, Austria) by intracardiac injection. The femora were harvested and stored in 4% formaldehyde solution (VWR International, Vienna, Austria) for 1 week for further analysis (μCT and descriptive histological analysis).

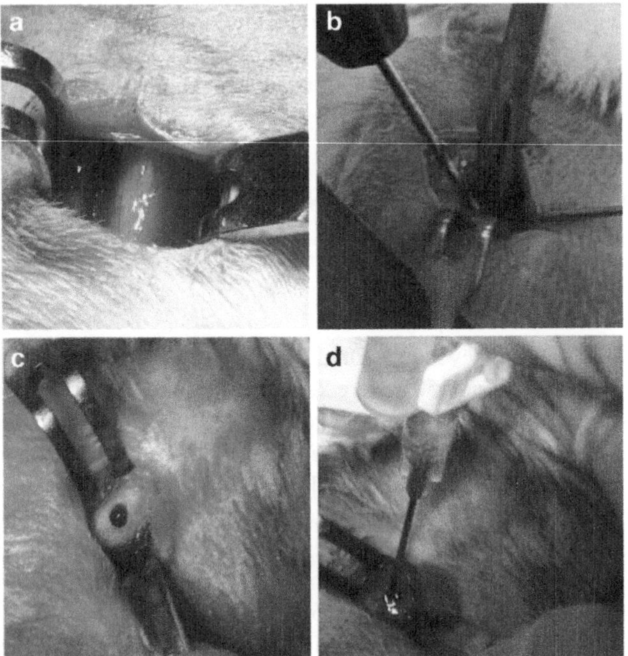

Fig. 12.1 Drill hole model: (**a**) surgical approach; (**b**) drilling; (**c**) 2 mm drill hole; (**d**) filling with test substance (From Keibl et al. [42] with permission)

12.7.2 Methods and Results

12.7.2.1 Cell Culture

During outpatient tumescence liposuction, subcutaneous adipose tissue was obtained under local anaesthesia mainly from female patients (IRB consent obtained).

ASCs were isolated and cultured under sterile conditions in DMEM-low glucose/HAM's F-12 supplemented with 2 mM L-glutamine, 100 U/mL penicillin, 10% fetal calf serum (FCS, PAA, Pasching, Austria), 0.1 mg/mL streptomycin and 1 ng/mL recombinant human basic fibroblast growth factor (rhFGF, R&D Systems, Minneapolis, USA) at 37°C, 95% air humidity and 5% CO_2 to a subconfluent state [43].

Cells in passage 3 and 4 were applied embedded in fibrin at 2×10^5 cells/200 μl clot. Fibrin (Tisseel, Baxter AG, Vienna, Austria) was used in 1 mL duploject syringes (500 U/mL thrombin solved in 40 mmol calcium chloride and 100 mg/mL fibrinogen solved in 3,000 KIU/mL aprotinin). 10 μg BMP-2 (Induct OS, Wyeth Europe LTD, Berkshire, UK) was mixed with 1 mL of the thrombin component for the BMP-2 groups. A cell pellet (for the ASC groups) of 2×10^6 cells was dissolved in 1 mL fibrinogen. Thus, no dilution effect of adding the cells was present.

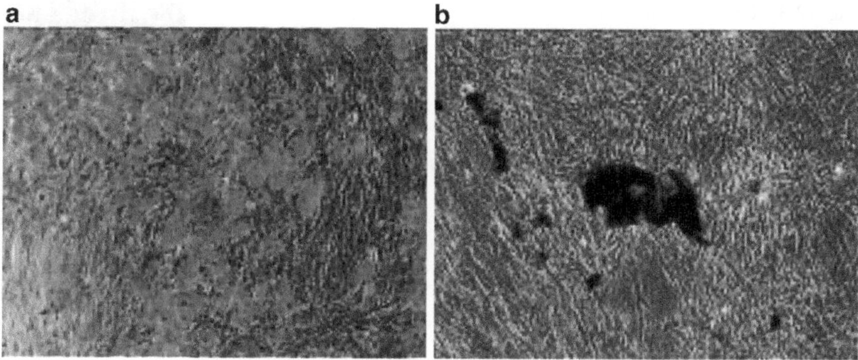

Fig. 12.2 (**a**) adipogenic differentiation, Oil Red staining, 2 × magnification; (**b**) osteogenic differentiation, Von Kossa staining, 2 × magnification

ASCs are routinely checked on the expression of the mesenchymal stem cell marker (CD29, CD44, CD73, CD90, CD105) and on the absence of CD14, CD34, CD45 and HLA-DR by means of flow cytometry. In order to characterize the ASCs with regard to their stem cell-quality they were differentiated along the adipogenic and osteogenic lineage. The verification of adipogenous differentiation was carried out by staining the intracellular lipid accumulation with Oil Red (Sudan III) (Fig. 12.2a). Twenty four hours afterwards (after implanting ASCs) the verification of osteogenous differentiation followed by staining the cells with "*Von Kossa*" (Fig. 12.2b).

12.7.2.2 Micro Computer Tomography (μCT) Analysis and Results

3-D reconstructions of all explanted femora were subjected to Micro Computer Tomography analysis (μCT) (μCT 20, Scanco Medical AG, Bassersdorf, Switzerland).

The μCT analyses were carried out by scanning (20 h for one femur), picture processing and "μCT evaluation program" (regions of interest, bone volume/tissue volume (BV/TV) and trabecular thickness). Callus was determined by measuring the borders of the drill hole as well as the height at the centre. All callus' measurements were performed in the middle plane of the femur at the defect area.

Three regions of interest were defined [44] (Fig. 12.3):

M1: periosteal callus area; M2: cortical area; M3: medullary area. M1 was used to evaluate the callus formation of the specimens. In region M2 and M3 two parameters: bone volume/tissue volume (BV/TV) and trabecular thickness (tb.th., only in section M3) were used to study the bone regeneration within the defect. Both regions of interest had a cylindrical shape. M2 showed the height of a single cortex and M3 implied the distance between the two inner borders of the opposite cortices.

After 2 weeks μCT data showed in all five study groups no significant difference both in BV/TV and trabecular thickness in the cortical area and the medullary area. Increased callus formation was detected in the group fibrin + BMP-2 ($p < 0.05$) after

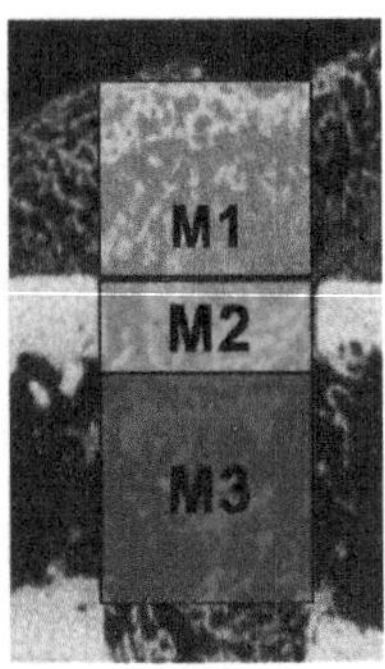

Fig. 12.3 Regions of interest: *M1* periosteal callus area; *M2* cortical area; *M3* medullary area (From Keibl et al. [42] with permission)

2 and 4 weeks compared to all other groups. It cannot be excluded that BMP-2 caused the periosteal callus formation due to heterotopic ossification.

In the medullary area the BV/TV significantly decreased at the 4 weeks time point ($p<0.05$). Group fibrin+BMP-2 showed lower BV/TV compared to control and fibrin+ASC at 4 weeks ($p<0.05$) indicating "high levels of remodelling" in the fibrin+BMP-2 group (Fig. 12.4a). The cortical BV/TV showed significant increases in all groups after 4 weeks compared to 2 weeks ($p<0.05$) (Fig. 12.4b).

No differences in the trabecular thickness (tb.th.) in the medullary area were shown between the groups after 2 weeks. A difference between the 2 and 4 week time points existed in the control, fibrin and fibrin+ASC group. After 4 weeks the trabecular thickness was significantly more pronounced in the group fibrin+ASC in comparison with fibrin+ASC+BMP-2 ($p<0.05$) (Fig. 12.5).

There was significantly increased callus in the group fibrin+BMP-2 after 2 weeks compared to all other groups ($p<0.05$). This callus amount decreased significantly after 4 weeks. However, it was still significantly higher than the other groups ($p<0.05$). In the group fibrin+ASC+BMP-2 a small but significant increase from the 2 to 4 week time point could be observed ($p<0.05$) (Fig. 12.6).

μCT longitudinal sections of specimens of both time points (2 and 4 weeks) are shown in Fig. 12.7.

12.7.2.3 Descriptive Histological Analysis and Results

The qualitative and morphological aspects of bone regeneration were analysed by means of histological preparation as previously described [45]. In brief, 20 femora in extracts were dehydrated in ascendending grades of ethanol (from 40% to 100%) and embedded in Technovit 7200 (Heraeus Kulzer GmbH, Wehrheim, Germany). With a thickness of about 30 μm undecalcified thin-ground sections were prepared and stained according to Levai-Laczko [46]. All the 20 histological specimens were photographed with a digital camera (Nikon DXM 1200, Nikon, Tokyo, Japan) mounted on a microscope (Nikon Mikrophot-FXA, Nikon, Tokyo, Japan).

The stage of bone regeneration after 2 and 4 weeks was the aim of the histological analysis to determine, which type of bone tissue (whether woven bone and/or

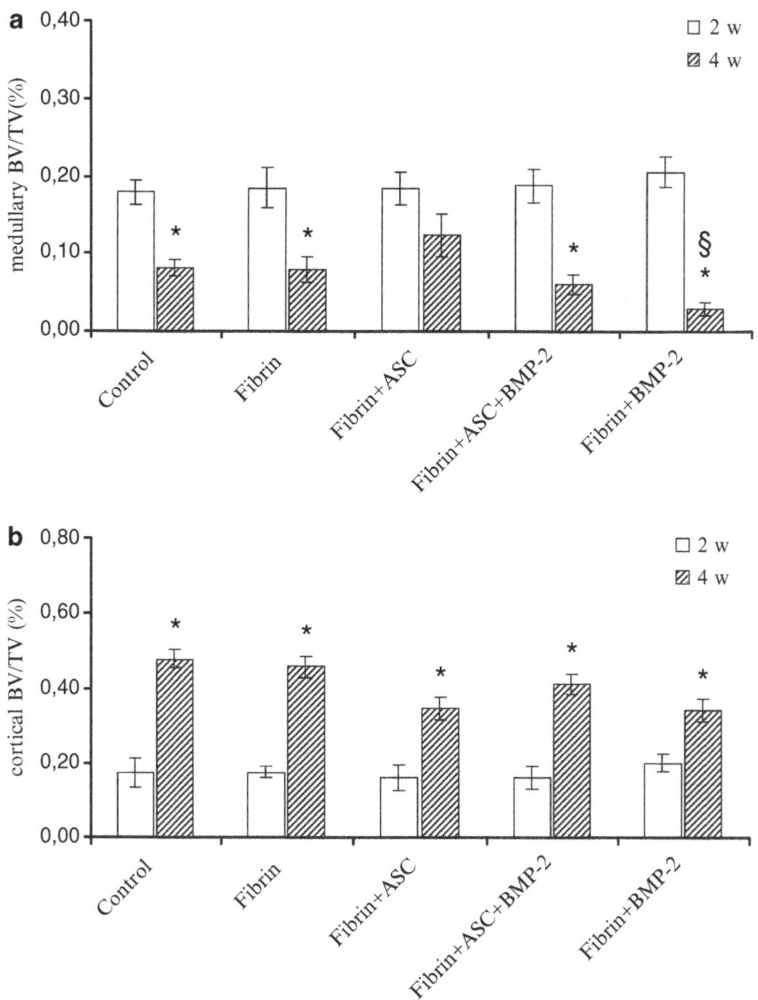

Fig. 12.4 Comparison of the μCT data after 2 versus 4 weeks: mean value (*white bar*: 2 week period, *spriped bar*: 4 week period) ± SEM. (**a**) medullary BV/TV (%): * $p<0.05$ 2 vs. 4 weeks; § $p<0.05$ Fibrin + BMP-2 vs. control and Fibrin + ASC at 4 weeks. (**b**) cortical BV/TV (%): * $p<0.05$ 2 vs. 4 weeks (Modified from Keibl et al. [42])

lamellar bone) was present, if a typical callus had developed, and if new bone tissue had been formed in the marrow space.

After 2 weeks the descriptive histological evaluation presented similar results in all five groups. Newly formed bone tissue were composed of a network of woven bone compacted with lamellar bone (i.e. plexiform bone) (Fig. 12.8). More bone tissue had developed in the marrow space in the group Fibrin + ASC + BMP-2 and in group Fibrin + BMP-2 compared to all other groups. In the control group there was

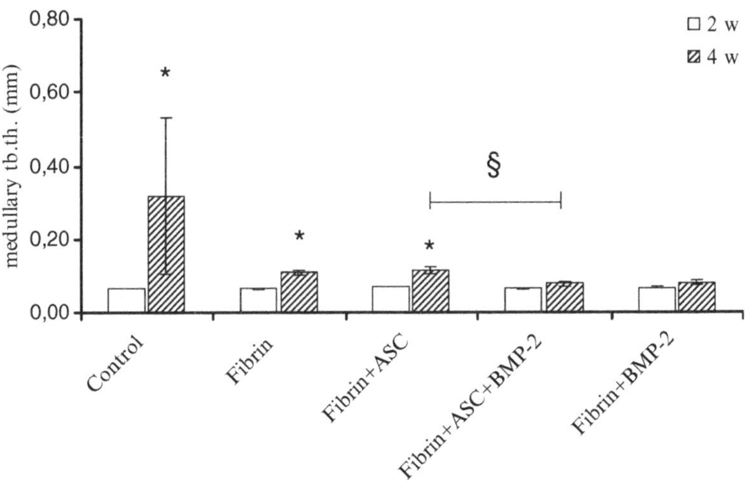

Fig. 12.5 Comparison of the μCT data after 2 versus 4 weeks: mean value (*white bar*: 2 week period, *spriped bar*: 4 week period) ± SEM medullary trabecular thickness (tb.th.) (mm): * $p<0.05$ 2 vs. 4 weeks; [§] $p<0.05$ Fibrin + ASC vs. F + ASC + BMP-2 at 4 weeks (Modified from Keibl et al. [42])

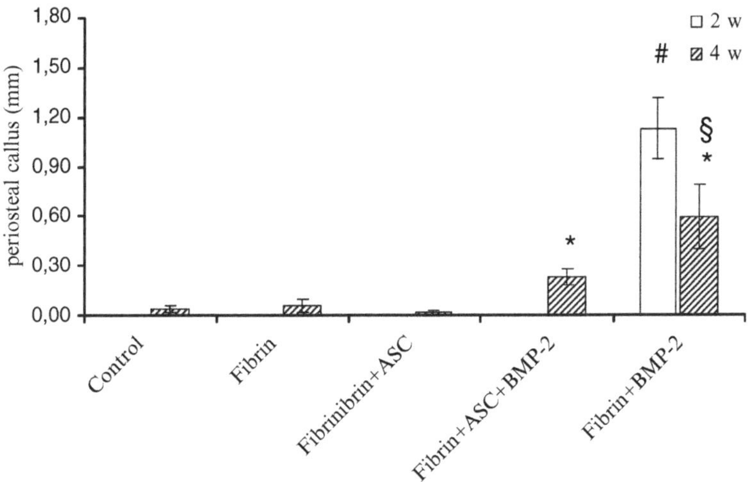

Fig. 12.6 Comparison of the μCT data after 2 versus 4 weeks: mean value (*white bar*: 2 week period, *spriped bar*: 4 week period) ± SEM periosteal callus (mm): * $p<0.05$ 2 vs. 4 weeks; [§] $p<0.05$ Fibrin + BMP-2 vs. control, Fibrin and Fibrin + ASC at 4 weeks; [#] $p<0.05$ Fibrin + BMP-2 vs. all other groups at 2 weeks (Modified from Keibl et al. [42])

only little periosteal callus formation whereas in the Fibrin + BMP-2 group this process was much more evident.

After 4 weeks, there was more lamellar bone developed, but the amount of woven bone in the cortical region remained almost unchanged compared to the 2 week time point. In all groups the bone tissue in the marrow space had mostly disappeared as in

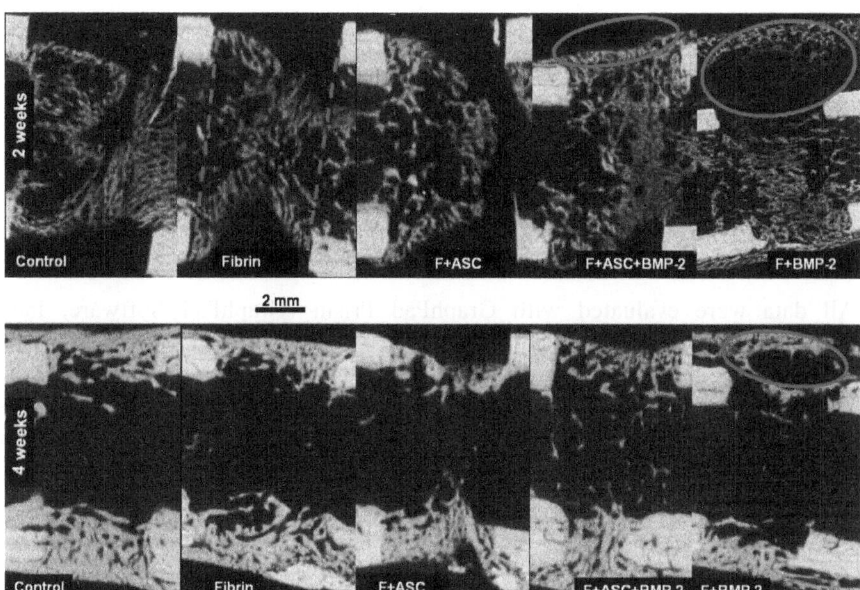

Fig. 12.7 μCT longitudinal sections of all groups after 2 and 4 weeks: 2 mm drilling channel is marked; all groups are shown at 2 weeks: massive callus formation is visible in the group Fibrin+BMP-2, whereas less or no callus formation is seen in group Fibrin+BMP-2+ASC. All groups are shown at 4 weeks: callus formation is still visible in group Fibrin+BMP-2 (From Keibl et al. [42] with permission)

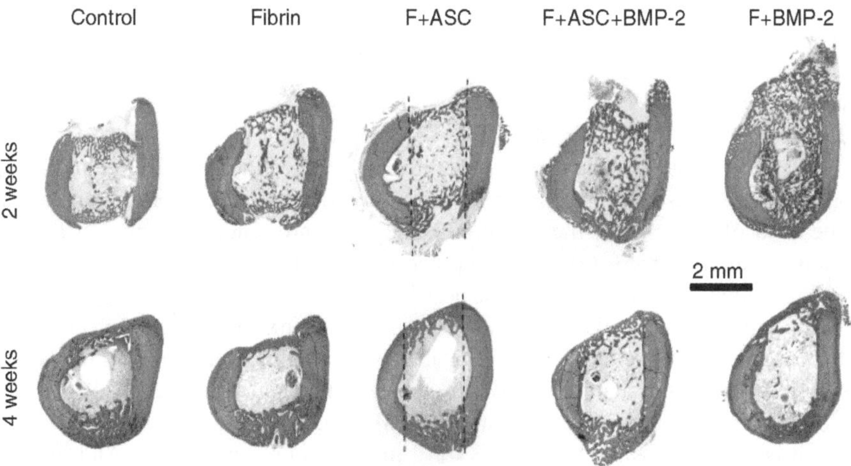

Fig. 12.8 Histological images of all groups after 2 and 4 weeks: 2 mm drilling channel is marked; all groups at 2 weeks are shown: massive callus formation is visible in the group Fibrin+BMP-2, whereas less or no callus formation is seen in group Fibrin+BMP-2+ASC. All groups are also shown at 4 weeks: callus formation is still visible in group Fibrin+BMP-2. Levai-Laczko staining; 2× magnification (From Keibl et al. [42] with permission)

approximately all of the periosteal callus tissue, except in the group Fibrin + BMP-2. The histological and mechanical integrity of the cortical bone had been widely restored and most of the bone tissue in the marrow space and the periosteal region, initially formed in excess after the trauma had been resorbed again after 4 weeks.

12.8 Statistical Evaluation

All data were evaluated with GraphPad Prism (GraphPad Software, Inc., San Diego, CA). Descriptive data are presented as means ± SEM. A Kruskal-Wallis analysis for global significant differences was used for differentiation between the groups. In case of significant results ($p < 0.05$), a Dunn`s multiple comparison test as post-test was used. Differences between groups with p values <0.05 were considered as significant.

12.9 Discussion

Bone defects resulting from orthopaedics or trauma surgery are caused posttraumatically, when a tumor grows, after an infection or in case of aseptic osteolysis. Transplantations of autogenous and allogenous bone are repeatedly carried out and bear several risks. Frequent complications may be intra- and postoperative problems, local complications (infection, haematoma), higher costs, re-operation, prolonged duration of surgery, long hospital stays and rehabilitation periods and a limited supply of autologous spongiosa as well as non-healing of the implant [3, 4].

Current medical research is mainly focused on the application of stimulating substances which accelerate the healing process (such as adult human adipose derived stem cells (ASC) or BMP-2) as used in our studies.

Literature describes the use of a diameter of 2 mm in a rat drill hole model as promising, because it represents an ideal size in order to analyse bone regeneration. The kind of movement, the age-group, the size and gender are of outstanding scientific significance for the healing process [3, 47, 48].

Our studies demonstrated that it is possible to get experimental bone regeneration results without osteosynthesis within a short period of time (after 2 and 4 weeks). Our defect model does not represent a critical size one, but more a screening model for bone regeneration. Usually BMP-2 is used on a collagen matrix [15, 38, 39, 49–51]. We have checked a possible variation by injecting BMP-2. This method proved that it is simply possible to apply BMP-2 in fibrin intra-operatively. And what is more, the application of our non-critical size model may lead to an underestimation of the results described. Studies in a critical size model may further illustrate the mechanism.

In former studies we have seen that the transcortical drill hole had closed after 6 weeks and could not be used for further evaluations. Due to the small dimension of the femur of the rat (length: about 5 cm) an accurate analysis using conventional radiological techniques is not adequate. Final results could be found by means of μCT and descriptive histology. The μCT and the descriptive histology data which we collected corresponded adequately.

The difficulty to place the drill channel perpendicular to the shaft axis was due to the high variability of bone volume. An obvious variability of the curvature of the femur also affected the medullary trabecular thickness measurement.

With the application of BMP-2 in fibrin there was a significant reaction of the callus after a period of 2 weeks. However, in the group Fibrin + ASC + BMP-2 there was no reaction concerning callus development. In the 4 weeks period the callus formation in the group Fibrin + BMP-2 was significantly smaller than in the 2 weeks period group. In the group Fibrin + ASC + BMP-2 there was a smaller callus in general. In group Fibrin + BMP-2 after 2 and 4 weeks a quick and strong effect of BMP-2 could be found as often described in literature [2]. There was a remarkable reduction of the callus surplus by means of the combination of ASC with BMP-2, but no acceleration in the healing process especially in the 2 weeks group. The mechanism that explains why the adipose derived stem cells had an effect in the BMP-2 still have to be researched in detail in vivo and in vitro. As a matter of fact the concentration of the BMP-2 was in all groups (with and without ASCs) the same and there was no dilution by adding the ASCs since they were mixed in the fibrinogen as a pellet. Possibly the number of applied ASCs in our model was too small in order to develop increased bone regeneration in the drill hole.

Tortelli et al. have stated that the presence of circulating MSCs was still in discussion in 2010. By means of immunohistochemical analysis they demonstrated in a murine model of ectopic bone formation the absence of marked MSCs after 30 days in comparison to the presence of bone tissue after 60 days [52]. Adding the MSCs however, showed a positive effect. This may be due to local, short acting cellular interactions or local, humoural effects. Tasso et al. stated in their mice model of ectopic bone formation that the interaction between graft and host has an enormous importance in tissue engineering and that basic scientific findings of the nature and the cross talk between stem cells in future models are of long-term concern [53].

After 4 weeks the amount of woven bone in the cortical region seemed rather unchanged in comparison to the results after 2 weeks, when there was more lamellar bone present. This can be explained by the continuous deposition of lamellar bone on the surface of the woven bone network. It could also be possible that BMP-2 caused a periosteal callus formation due to heterotopic ossification and that ASCs only had an effect on this process.

Neither after 2 nor after 4 weeks there was no reaction to the ASCs alone on the bone regeneration, corresponding to similar results with bone marrow stem cells mentioned in literature [54, 55]. The ASCs used in our studies have demonstrated an immune-suppressive effect in another in vitro research [43]. Presumably, in our pre-clinical model this may not have a particular effect, especially because the cells were transplanted in a xenogeneic setting. The application of adipose-derived stem cells of rats or human ASCs with systemic immune-suppression would have possibly achieved another perhaps better result [56]. However, this doesn't correspond to the clinical routine. The fate of the ASCs is not known in our model and for this reason further studies are necessary to investigate the role of the ASCs whether they remain in the defect or not [52].

The same amount of bone volume/tissue volume in the medullary area was measured after a period of 2 weeks. After 4 weeks the bone volume/tissue volume in the medullary area had significantly increased in the group Fibrin + BMP-2. For this reason we assume that BMP-2 delays the bone dissimilation in the marrow.

In all groups in which BMP-2 was applied the descriptive histological evaluation showed the accelerated remodelling process, especially in the 2 weeks groups. As a result a high amount of bone had developed in the medullary space.

Several studies illustrate the importance of mesenchymal stem cells, especially bone marrow derived stem cells in treating critical size bone defects and their bone repairing capacity. There are only a few in vivo studies comparing mesenchymal stem cells from bone marrow with those derived from adipose tissue. Niemeyer et al. described a critical size defect of sheep tibia which is transferable to the human patients and to a real clinical situation. They demonstrated that undifferentiated adipose stem cells are inferior to bone marrow stem cells with regards to their osteogenic potential [57, 58]. Runyan et al. stated positive results in their study on revitalization of a large-volume bone allograft using autologous adipose-derived stem cells stimulated by BMP-2 and porcine hemimandible allograft construct placed as thoracic and abdominal implants in a pig model [59]. Cowan et al. described the positive mineralizing effect of adipose derived cells using a novel osteoconductive apatite coated scaffold which healed critical size mouse calvarial defects [60]. Yoon et al. evaluated the successful osteogenic potential of human adipose-derived stem cells implanted also in an osteoconductive scaffold for bone regeneration in a nude rat critical size calvarial defects [61]. Shoji et al. also demonstrated in their study also the therapeutic potential consequence of adipose derived cells with a bioabsorbable scaffold in a rat model of non-union femoral fracture [62].

Few clinical bone regeneration studies have been carried out besides these preclinical reports. Quarto et al. described the successful treatment of long bone defects with ex vivo expanded mesenchymal stem cells (MSCs) on a hydroxy-apatite collagen scaffold and pointed out if these results were due to the presence of the MSCs or not [63]. Horwitz et al. reported that MSCs derived from the bone marrow can serve as long term precursors for bone regeneration, but may require extensive in vitro expansion [64]. Various studies applying cell based therapies proved positive effects of myocardial infarct, multiple sclerosis, graft versus host disease and osteogenesis imperfecta with human adult mesenchymal stem cells [65, 66].

Calori et al. have already described that it is significant to emphasize different responses to the BMP stimulus among different species [67, 68]. As a consequence the amount of ASC as well as BMP-2 should be titrated and analyzed in further projects on a bigger scale e.g.: a femur non union model in rats. A promising improvement in our studies could be the dosage of the applied BMP-2 [24, 69, 70].

The development of reliable clinical methods for orthopaedic and trauma surgery, veterinary treatment and regenerative human medicine in general seem to be a potential signpost for the future.

ASCs in combination with bone morphogenetic protein embedded in an optimal matrix constitute a promising aspect of bone regeneration which was also discussed by Lane et al. in 2005 and undoubtedly a rewarding subject of further scientific clinical research [21, 55, 71].

Acknowledgement We thank our colleagues Karin Hahn, Christoph Castellani, Asmita Banerjee, Daniela Dopler, Tatjana Morton, Martina Moritz, Susanne Wolbank for their support during the completion of the project as well as Gerald Zanoni and Stefan Tangl.

We also want to express our gratefulness to Monika Großauer and Mohammad Jafamadar for their excellent help during the preparations of the book chapter, Ilse Jung for statistically analysing our data.

This work was carried out under the scope of the European NoE EXPERTISSUES (NMP3-CT-2004-500283).

References

1. Axelrad TW, Kakar S, Einhorn TA (2007) New technologies for the enhancement of skeletal repair. Injury 38(Suppl 1):S49–S62
2. Gautschi OP, Frey SP, Zellweger R (2007) Bone morphogenetic proteins in clinical applications. ANZ J Surg 77:626–631
3. Stutzle H, Hallfeldt K, Mandelkow H, Kessler S et al (1998) Knochenneubildung durch Knochenersatzmaterial. Orthopade 27:118–125
4. Rueger JM (1998) Knochenersatzmittel—heutiger Stand und Ausblick. Orthopade 27:72–79
5. Schiebler TH (2000) Histologie. Springer, Heidelberg
6. Landry PS, Marino AA, Sadasivan KK, Albright JA (1996) Bone injury response. An animal model for testing theories of regulation. Clin Orthop Relat Res 332:260–273
7. Tscherne H, Oestern HJ (1982) Die Klassifizierung des Weichteilschadens bei offenen und geschlossenen Frakturen. Unfallheilkunde 85:111–115
8. Rüter A, Trentz O, Wagner M (1995) Unfallchirurgie. Urban & Schwarzenberg, München/ Wien/Baltimore
9. Boehler L (1944) Die Technik der Knochenbruchbehandlung im Frieden und im Kriege. III.Band. 5. – 8. Auflage. Verlag Wilhelm Maudrich, Wien
10. Einhorn TA (1998) The cell and molecular biology of fracture healing. Clin Orthop Relat Res 355:S7–S21
11. Einhorn TA (1995) Enhancement of fracture-healing. J Bone Joint Surg Am 77:940–956
12. Minuth WW, Strehl R, Schumacher K (2003) Zukunftstechnologie Tissue Engineering. Von der Zellbiologie zum künstlichen Gewebe. Wiley-VCH GmbH & Co. KGaA, Weinheim
13. Marzi I, Mutschler W (1999) Pathophysiologie des Traumas. In: Mutschler W, Haas N (eds) Praxis der Unfallchirurgie. Georg Thieme, Stuttgart/New York, pp 18–55
14. Schieker M, Seitz S, Gulkan H, Nentwich M et al (2004) Tissue Engineering von Knochen. Integration und Migration von humann mesenchymalen Stammzellen in besiedelten Konstrukten im Mausmodell. Orthopade 33:1354–1360
15. Minuth WW, Strehl R, Schumacher K (2002) Von der Zellkultur zum Tissue engineering. Pabst Science, Lengerich
16. Pittenger MF, Mackay AM, Beck SC, Jaiswal RK et al (1999) Multilineage potential of adult human mesenchymal stem cells. Science 284:143–147
17. Thomas ED (1999) Bone marrow transplantation: a review. Semin Hematol 36:95–103
18. Wikipedia (2009) Geschichte der Stammzellforschung. http://www nabelschnurblut-wiki de//index php?title=Geschichte_der_Stammzellforschung. Last update: 15 Apr 2009. Accessed 03 Dec 2009

19. Gomillion CT, Burg KJ (2006) Stem cells and adipose tissue engineering. Biomaterials 27:6052–6063
20. Dicker A, Le Blanc K, Astrom G, van Harmelen V et al (2005) Functional studies of mesenchymal stem cells derived from adult human adipose tissue. Exp Cell Res 308:283–290
21. Patterson TE, Kumagai K, Griffith L, Muschler GF (2008) Cellular strategies for enhancement of fracture repair. J Bone Joint Surg Am 90(Suppl 1):111–119
22. Strem BM, Hicok KC, Zhu M, Wulur I et al (2005) Multipotential differentiation of adipose tissue-derived stem cells. Keio J Med 54:132–141
23. Waese EY, Kandel RA, Stanford WL (2008) Application of stem cells in bone repair. Skeletal Radiol 37:601–608
24. Lendeckel S, Jodicke A, Christophis P, Heidinger K et al (2004) Autologous stem cells (adipose) and fibrin glue used to treat widespread traumatic calvarial defects: case report. J Craniomaxillofac Surg 32:370–373
25. Bergel S (1907) Über Wirkungen des Fibrins. Deut Med Wochenschr 9:663–665
26. Redl H (2004) History of tissue adhesives. In: Saltz R, Toriumi DM (eds) Tissue glues in cosmetic surgery. Quality Medical Publishing, St. Louis, pp 1–27
27. Le Guehennec L, Layrolle P, Daculsi G (2004) A review of bioceramics and fibrin sealant. Eur Cell Mater 8:1–11
28. Seelich T, Redl H (1980) Fibrinklebung 1. Theoretischer und experimenteller Teil. In: Kl S (ed) Fibrinogen, Fibrin und Fibrinkleber. Schattauer, Stuttgard/New York
29. Kaeser A, Dum N (1994) Wirkprinzip der Fibrinklebung. In: Manegold BC, Lange V, Salm R (eds) Technik der Fibrinklebung in der endoskopischen Chirurgie. Springer, Berlin/Heidelberg/New York/London/Paris/Tokyo/Hong Kong/Barcelona/Budapest, pp 3–11
30. Oehler G (1992) Grundprinzip der Fibrinklebung – Anforderungen an die Qualität und Sicherheit. In: Freigang B, Weerda H (eds) Fibrinklebung in der Otorhinolaryngologie. Springer, Berlin/Lübeck, pp 3–6
31. Gebhardt Ch (1992) Fibrinklebung in der Allgemein- und Unfallchirurgie, Orthopädie. Kinder- und Thoraxchirurgie. Springer, Nürnberg
32. Bösch P, Nowotny Ch, Schwägerl W, Leber H (1980) Über die Wirkung des Fibrinklebesystems bei orthopädischen Operationen an Hämophilen und bei anderen Blutgerinnungsstörungen. In: Schimpf K (ed) Fibrinogen, Fibrin und Fibrinkleber. Schattauer, Stuttgart/New York, pp 274–283
33. Urist MR (1965) Bone: formation by autoinduction. Science 150:893–899
34. Li X, Cao X (2006) BMP signaling and skeletogenesis. Ann N Y Acad Sci 1068:26–40
35. Reddi AH (1998) Role of morphogenetic proteins in skeletal tissue engineering and regeneration. Nat Biotechnol 16:247–252
36. Reddi AH (1998) Initiation of fracture repair by bone morphogenetic proteins. Clin Orthop Relat Res 355:S66–S72
37. Rosen V (2006) BMP and BMP inhibitors in bone. Ann N Y Acad Sci 1068:19–25
38. Dragoo JL, Choi JY, Lieberman JR, Huang J et al (2003) Bone induction by BMP-2 transduced stem cells derived from human fat. J Orthop Res 21:622–629
39. Dragoo JL, Lieberman JR, Lee RS, Deugarte DA et al (2005) Tissue-engineered bone from BMP-2-transduced stem cells derived from human fat. Plast Reconstr Surg 115:1665–1673
40. Einhorn TA (2003) Clinical applications of recombinant human BMPs: early experience and future development. J Bone Joint Surg Am 85-A(Suppl 3):82–88
41. Termaat MF, Den Boer FC, Bakker FC, Patka P et al (2005) Bone morphogenetic proteins. Development and clinical efficacy in the treatment of fractures and bone defects. J Bone Joint Surg Am 87:1367–1378
42. Keibl C, Fügl A, Zanoni G, Tangl S et al (2011) Human adipose derived stem cells reduce callus volume upon BMP-2 administration in bone regeneration. Injury 42:814–820
43. Wolbank S, Peterbauer A, Fahrner M, Hennerbichler S et al (2007) Dose-dependent immunomodulatory effect of human stem cells from amniotic membrane: a comparison with human mesenchymal stem cells from adipose tissue. Tissue Eng 13:1173–1183

44. Kalpakcioglu BB, Morshed S, Engelke K, Genant HK (2008) Advanced imaging of bone macrostructure and microstructure in bone fragility and fracture repair. J Bone Joint Surg Am 90(Suppl 1):68–78
45. Donath K (1988) Die Trenn-Dünnschliff-Technik zur Herstellung histologischer Präparate von nicht schneidbaren Geweben und Materialien. Der Präparator 34:197–206
46. Laczko J, Levai G (1975) A simple differential staining method for semi-thin sections of ossifying cartilage and bone tissues embedded in epoxy resin. Mikroskopie 31:1–4
47. Pereira AC, Fernandes RG, Carvalho YR, Balducci I et al (2007) Bone healing in drill hole defects in spontaneously hypertensive male and female rats' femurs. A histological and histometric study. Arq Bras Cardiol 88:104–109
48. Schoch T (1994) Knochenregeneration mit frischen und sterilisierten Auto- und Allografts – Experimentelle Untersuchungen mit morphometrischer Methode. Dissertation, TU München
49. Bishop GB, Einhorn TA (2007) Current and future clinical applications of bone morphogenetic proteins in orthopaedic trauma surgery. Int Orthop 31:721–727
50. Pountos I, Corscadden D, Emery P, Giannoudis PV (2007) Mesenchymal stem cell tissue engineering: techniques for isolation, expansion and application. Injury 38(Suppl 4):S23–S33
51. Pountos I, Giannoudis PV (2005) Biology of mesenchymal stem cells. Injury 36(suppl 3): S8–S12
52. Tortelli F, Tasso R, Loiacono F, Cancedda R (2010) The development of tissue-engineered bone of different origin through endochondral and intramembranous ossification following the implantation of mesenchymal stem cells and osteoblasts in a murine model. Biomaterials 31:242–249
53. Tasso R, Fais F, Reverberi D, Tortelli F et al (2010) The recruitment of two consecutive and different waves of host stem/progenitor cells during the development of tissue-engineered bone in a murine model. Biomaterials 31:2121–2129
54. Li H, Dai K, Tang T, Zhang X et al (2007) Bone regeneration by implantation of adipose-derived stromal cells expressing BMP-2. Biochem Biophys Res Commun 356:836–842
55. Peterson B, Zhang J, Iglesias R, Kabo M et al (2005) Healing of critically sized femoral defects, using genetically modified mesenchymal stem cells from human adipose tissue. Tissue Eng 11:120–129
56. McIntosh KR, Lopez MJ, Borneman JN, Spencer ND et al (2009) Immunogenicity of allogeneic adipose-derived stem cells in a rat spinal fusion model. Tissue Eng Part A 15:2677–2686
57. Niemeyer P, Fechner K, Milz S, Richter W et al (2010) Comparison of mesenchymal stem cell from bone marrow and adipose tissue for bone regeneration in a critical size defect of the sheep tibia and the influence of platelet-rich plasma. Biomaterials 31:3572–3579
58. Cancedda R, Bianchi G, Derubeis A, Quarto R (2003) Cell therapy for bone disease: a review of current status. Stem Cells 21:610–619
59. Runyan CM, Jones DC, Bove KE, Maercks RA et al (2010) Porcine allograft mandible revitalization using autologous adipose-derived stem cells, bone morphogenetic protein-2, and periosteum. Plast Reconstr Surg 125:1372–1382
60. Cowan CM, Shi YY, Aalami OO, Chou YF et al (2004) Adipose-derived adult stromal cells heal critical-size mouse calvarial defects. Nat Biotechnol 22:560–567
61. Yoon E, Dhar S, Chun DE, Gharibjanian NA et al (2007) In vivo osteogenic potential of human adipose-derived stem cells/poly lactideco-glycolic acid constructs for bone regeneration in a rat critical-wized calvarial defect model. Tissue Eng 13:619–627
62. Shoji T, Ii M, Mifune Y, Matsumoto T et al (2010) Local transplantation of human multipotent adipose-derived stem cells accelerates fracture healing via enhanced osteogenesis and angiogenesis. Lab Invest 90:637–649
63. Quarto R, Mastrogiacomo MCR, Kutepov SM, Mukhachev V et al (2001) Repair of large bone defects with the use of autologous bone marrow stromal cells. N Engl J Med 344:385–386
64. Horwitz EM, Gordon PL, Koo WK, Marx JC et al (2002) Isolated allogenic bone marrow-derived mesenchymal cells engraft and stimulate growth in children with osteogenesis imperfacta: implications for cell therapy of bone. Proc Natl Acad Sci USA 99:8932–8937

65. Mohyeddin Bonab M, Yazdanbakhsh S, Lotfi J, Alimoghaddom K et al (2007) Does mesenchymal stem cell therapy help multiple sclerosis patients? Report of a pilot study. Iran J Immunol 4:50–57
66. Mohyeddin Bonab M, Mohamad-Hassani MR, Alimoghaddam K, Sanatkar M et al (2007) Autologous in vitro expanded mesenchymal stem cell therapy for human old myocardial infarction. Arch Iran Med 10:467–473
67. Carlori GM, Donati D, DiBella C, Tagliabue L (2009) Bonemorphogenetic proteins and tissue engineering: future directions. Injury 40(Suppl 3):S67–S76
68. Westerhuis RJ, van Bezooijen RL, Kloen P (2005) Use of bone morphogenetic proteins in traumatology. Injury 36:1405–1412
69. Hasharoni A, Zilberman Y, Turgeman G, Helm GA et al (2005) Murine spinal fusion induced by engineered mesenchymal stem cells that conditionally express bone morphogenetic protein-2. J Neurosurg Spine 3:47–52
70. Schmoekel HG, Weber FE, Schense JC, Gratz KW et al (2005) Bone repair with a form of BMP-2 engineered for incorporation into fibrin cell ingrowth matrices. Biotechnol Bioeng 89:253–262
71. Lane JM (2005) Bone morphogenic protein science and studies. J Orthop Trauma 19:S17–S22

Chapter 13
Stem Cells and Leukemia

Vincenzo Giambra and Christopher R. Jenkins

Contents

Abstract In the last few years, the concepts governing our understanding of cancer have changed. In particular, the point of view that many leukemias are developmentally well-defined and, like in normal hematopoiesis, driven by a relatively small subset of cells called leukemia stem cells (LSCs) has become well-established. Recent studies suggest that defined subsets of LSCs within a tumor are capable of recreating the entire tumor and thus are responsible for relapse/recurrence and metastasis. This subset of "cancer stem cells" has been postulated to possess certain properties akin to those characterized in hematopoietic stem cells such as the capacity to (1) self-renew and to (2) give rise to non-self-renewing or "differentiated" progeny cells that make up the bulk of a tumor. Among the hematopoietic malignancies, acute myeloid leukemia (AML) is the best characterized thus far with respect to "leukemia stem cells" and much data support that the above two properties exist within a relatively rare subpopulation. Related studies have also demonstrated that leukemia stem

V. Giambra (✉) • C.R. Jenkins
Terry Fox Laboratory, BC Cancer Agency, Vancouver, BC V5Z 1 L3, Canada
e-mail: vgiambra@bccrc.ca; cjenkins@bccrc.ca

R.K. Srivastava and S. Shankar (eds.), *Stem Cells and Human Diseases*,
DOI 10.1007/978-94-007-2801-1_13, © Springer Science+Business Media B.V. 2012

cells are functionally distinct from bulk cells. These subsets are relatively quiescent or slowly cycling, whereas clonogenic progenitors ("differentiated" progeny which cannot self-renew) proliferate rapidly. Current antiproliferative chemotherapy usually affects these dividing progenitors and induces disease relapses as defined by decreased bulk tumor burden. Relapse is not uncommon however, suggesting the quiescent leukemia stem cells are not effectively removed from circulation by existing therapies. Therefore, new therapies that specifically target leukemia stem cells hold great promise for achieving the elusive cure for cancer.

Keywords Leukemia • Leukemia stem cells (LSCs) • Hematopoiesis • Self-renewal • Chemotherapy • Bone marrow niche

Abbreviations

ALL	Acute lymphoblastic leukemia
AML	Acute myeloid leukemia
APL	Acute promyelocytic leukemia
B-ALL	B cell acute lymphoblastic leukemia
CLL	Chronic lymphocytic leukemia
CML	Chronic myeloid leukemia
CMP	Common myeloid progenitor
CSC	Cancer stem cell
GMP	Granulocyte-macrophage progenitors
HH	Hedgehog
HIF	Hypoxia-inducible factor
HSC	Hematopoietic stem cell
ICN	Intracellular Notch
LSC	Leukemia stem cells
MDS	Myelodysplastic syndromes
MM	Multiple myeloma
PcG	Polycomb group
ROS	Reactive oxygen species
T-ALL	T cell acute lymphoblastic leukemia

13.1 Introduction: The Leukemia Stem Cell Theory

For many decades, cancer, whether solid or leukemic, has been characterized as a disease of anomalous, uncontrolled proliferation. Historically, individual cancers have been considered functionally homogeneous. Many drugs and therapies were developed with the hope to kill several orders of magnitude of cells, maintain a low level of toxicity for normal tissues and have a broad cytotoxic effect on all of the

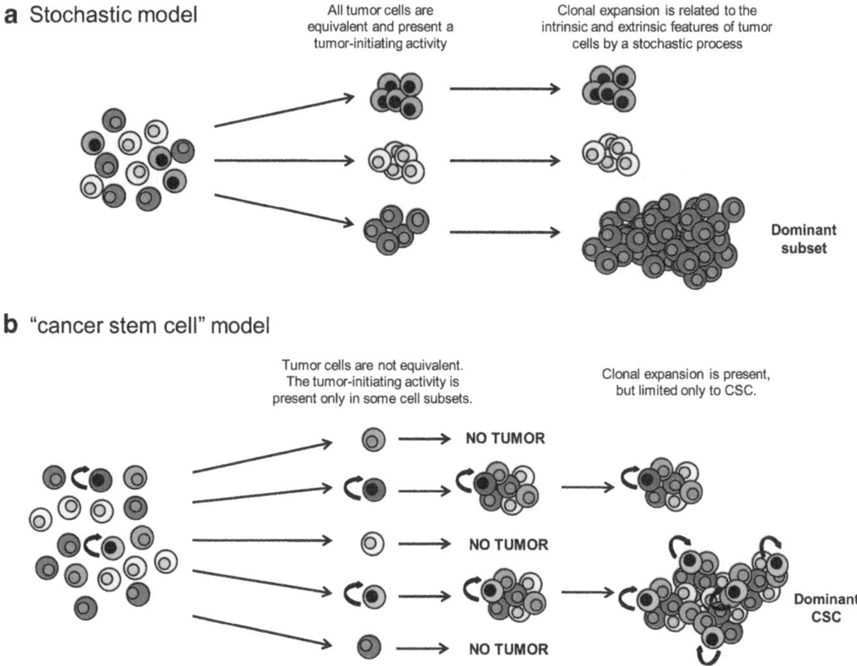

Fig. 13.1 Models of heterogeneity in cancer. (**a**) According to the stochastic model, all cancer cells are equal and have the same probability to proliferate and form new tumors, even if they can behave differently between each others, due to intrinsic and extrinsic factors. By clonal evolution, all subsets can become a dominant population and be expanded with a growth advantage. (**b**) For the cancer stem cells model, a limited subset of cells can only initiate new tumor. These CSCs are distinct from the bulk cells, present self-renewal activity and generate the heterogeneous non tumorigenic subsets. This model predicts that there are intrinsic features that distinguish the tumor-initiating cells from the rest of bulk cells. A clonal evolution of tumor is possible, but limited to CSCs

main cancer subsets [2]. However, the inability of traditional therapies to generate successful cures has caused many cancer researchers to alter their position and consider tumors to be decisively far from unvaried, but significantly heterogeneous. Leukemia, as in solid tumors, presents great heterogeneity in terms of surface markers, response to therapy, morphology, genetic mutation and cell proliferation kinetics. For a number of years, this high variability has been explained by the clonal evolution model as a stochastic response to extrinsic or intrinsic influences [3]. This theory postulates that differentially mutated tumor cells are selected for and expand in a manner in which the most "fit" clone is able to create a dominant population. From this point of view, all cancer cells have a similar potential for supporting/regenerating tumor growth (Fig. 13.1a) [4]. Within the last 15–20 years, new data have emerged to support an alternative explanation for the great heterogeneity of cancer cells, and suggests the retention of a hierarchical organization within the tumor. In this regard, like in normal hematopoiesis, many leukemias are driven by relatively small subsets of cells termed leukemia stem cells (LSCs) which form the pinnacle of the hierarchy and are capable of self-renewal and differentiation to regenerate the tumor's heterogeneity (Fig. 13.1b).

The presence of LSCs was proposed over 40 years ago [5], but the first experimental proof of their existence came from the revolutionary studies in acute myeloid leukemia (AML) achieved in John Dick's laboratory [1, 6, 7]. John Dick and his collaborators identified a rare subset of cells in human AML (around 0.01–1% of the bulk population) that had the potential to induce leukemia when transplanted into immunodeficient NOD-SCID mice. This cell fraction, analogous to hematopoietic stem cells, re-established the phenotypic heterogeneity present in the primary tumor and exhibited a self-renewing capacity upon serial transplantation.

It is important to remind oneself that the LSC model is not exclusive to the clonal evolution model. The stochastic mutational variation of LSCs has been well characterized [8, 9]. In this regard, LSCs themselves have the capacity to undergo clonal evolution and acquire mutations that confer more aggressive properties in terms of self-renewal or growth. It has been reported that serial transplantation can potentiate the generation of more aggressive tumors as a result of the *in vivo* selection of cells [10].

LSCs, as is the case with other "cancer stem cells" (CSCs), are conceptually distinct from the "cell of origin". These represent the cell type that receives the first oncogenic hit in the step-wise progression of leukemogenesis [11]. LSCs are not always the result of the transformation of normal HSCs, but can originate from progenitors or more differentiated cells that have acquired stem activity. In chronic myeloid leukemia (CML), mutations in the β-catenin gene are able to confer self-renewal properties on granulocyte-macrophage progenitors (GMP), transforming them into LSCs [12]. In another mouse model of leukemia, the expression of an MLL-AF9 fusion protein confers stem-like properties on committed progenitor cells [13, 14]. However, it seems that only GMPs with a high dosage of MLL-AF9 delivered by retrovirus are efficiently transformed. Progenitors in which the oncogene is regulated by an endogenous promoter exhibit poor transformative potential, indicating the importance of gene expression level in specific cell contexts [15].

The concepts defining CSCs are still under constant refinement. The technical issues used to identify CSCs present confounding interpretations. In the case of xenotransplantation, it is reasonable to posit that incomplete immunosuppression or discrepancies between species in cytokines or other factors of growth can result in confounding selection of different cell subsets. Even in syngeneic models, the normal niche of healthy recipient mice does not accurately reflect the cancer environment itself. It has also been shown that in some cancers, such as melanoma, modifications to xenotransplantation assays can alter the frequency of tumorigenic cells [16]. In melanoma in fact, all cancer cells appear to have unlimited tumorigenic capacity on serial transplantation without clear hierarchical organization [17], suggesting that some tumors may not obey the CSC model. Nevertheless, in recent years a high volume of data has confidently proven the existence of LSCs in different human leukemias and in some mouse models of these human hematopoietic malignancies.

These reports have also shown that the frequency of LSCs can vary dramatically. In a mouse model of CML originating from the MOZ-TIF2 translocation [18] or in Pten-deficient mice [19], only a small subset of cells have LSC activity. However, in other mouse models, such as Eμ-myc pre-B/B lymphomas, Eμ-N-RAS thymic lymphomas or PU.1−/− AML, more than 10% of cells have leukemia-forming capacity after transplantation into syngeneic C57BL/6 (Ly5.1+) recipient mice [20]. Moreover, at least 25% of granulocyte–myeloid progenitors or myeloid-lineage cells harboring the *Mll–Af9* oncogene, are leukemogenic when injected in recipient mice [13, 14]. In all of these models, the dominant cell populations have tumor regenerating capacity, suggesting that clarification is required to determine whether these tumors may have LSC subsets with high frequency or alternatively, these cells may not conform to the CSC model at all and instead obey a stochastic model of tumorigenesis.

In the next sections of this chapter, the current evidence for the leukemia stem cell model will be described and summarized, emphasizing those studies that have isolated LSC subsets from primary human cells and mouse models of leukemia by markers of cell surface immunophenotype. Finally, the primary pathways that modulate LSCs activity will be discussed with evidence supporting their clinical relevance and their potential impact in treatments of leukemic patients in the future.

13.2 Types of Leukemia Stem Cells

As paralleled by the heterogeneity described within normal development, hematopoietic tumor diversity is reflective of the various cell populations that are permissive to transformation in the hematopoietic system. For as many stem, progenitor and differentiated progeny populations that have been defined, it is believed that there are as many unique leukemias capable of developing with distinct pathologies. Primarily, leukemias are classified by the lineage from which they originate or phenotypically appear to originate, into myeloid and lymphoid compartments. These diseases can be further classified by the temporal progression to medical crisis. Most leukemias are roughly divided into chronic and acute disease. The malignancy is deemed chronic if it takes a relatively long time for a clonal expansion of cells to progress to frank leukemia with relatively mild numbers of blasts in the bone marrow and periphery. Acute leukemias are diagnosed in patients exhibiting symptoms requiring immediate intervention and typically involve an aggressive expansion of blasts in the bone marrow and periphery. The nature of these leukemias and their LSC activity also relates to their cell of origin. It is still unclear for many leukemias, if the malignant LSC activity occurs in a cell with endogenous self-renewal (multipotent hematopoietic stem cells) or if self-renewal is imparted on a committed progenitor/transit-amplifying cell. A key consideration in the identification of LSCs in each leukemia type is whether the immunophenotype is a good representation of functional heterogeneity. It is therefore likely that future refinement of stem immunophenotype will rely more on functional classifiers (e.g., ROS) and this is also likely to better inform us of the cell of origin. We focus our attention in this chapter on the most commonly diagnosed leukemias and discuss the LSC populations that have been identified for each, to date.

13.2.1 Acute Myeloid Leukemia (AML)

Acute Myeloid Leukemia (AML) is a malignancy of developmentally arrested myeloid cells that are found in increased number in the bone marrow and lead to incomplete systemic hematopoietic output. A medical emergency requiring immediate intervention, the disease and its symptoms are characterized by dramatically reduced numbers of terminally differentiated myeloid lineage cells and include fever and increased infections (granulocytopenia), bruising and bleeding (thrombocytopenia), along with pallor, fatigue and difficulty with breathing (anemia). The disease is known to be molecularly heterogeneous with numerous, generally well-characterized, chromosomal translocations believed to drive oncogenesis, some of which define specific subgroups that often have different prognoses. The genetic and epigenetic changes in AML pathogenesis are thought to result in changes in differentiation potential as well as survival and proliferation.

Experiments by John Dick and colleagues were the first to show that human samples of AML could be fractionated by cell surface immunophenotype and xenotransplanted into mice. They utilized a strain of mice displaying a highly immunodeficient phenotype (NOD-SCID) which was capable of engrafting human tissues. Cells representing all of the seven subtypes of the French-American-British classification system were fractionated into $CD34^+CD38^+$ and $CD34^+CD38^-$ populations and remarkably they found that only the $CD34^+CD38^-$ fraction was capable of engrafting into primary and secondary recipients [6]. The leukemic cells isolated from the secondary recipients closely retained the phenotype of the original sample demonstrating for the first time the prospective sorting and serial-transplantation of LSC populations that could recreate the original tumor heterogeneity. The experimental strategy employed by John Dick has been used for the identification of all subsequent LSCs in other malignancies and in AML has since been further refined to the cell surface immunophenotype: $CD34^+CD38^-CD90^-IL\text{-}3R^+CD71^-HLA^-DR^-CD117^-$ [21–24]. There is evidence that this human AML leukemic stem cell is heterogeneous in self-renewal capacity as seen in normal HSC biology [1, 25]. This functional heterogeneity is underscored by recent debate as to the immunophenotypic identity of the AML LSC which for years was believed to be $CD38^-$ (Table 13.1). The use of anti-CD38 antibodies typically used to identify putative cell subsets by flow cytometry was shown to inhibit engraftment of both leukemic and normal cord blood cells in immunodeficient mice, casting doubt onto the identity of the human LSC as being truly CD38 negative since many of the $CD38^+$ cells were capable of robust engraftment [26]. Another subtype of AML was recently discovered to contain leukemia-initiating cells in the $CD34^-$ fraction in which nucleophosmin has been mutated, a lesion that has been correlated with low CD34 expression [27]. These researchers

Table 13.1 Acute myeloid leukemia (AML)

Cancer type	CSC surface markers	Minimal tumorigenic dose [cell numbers]	Technique of transplant	Strain	Reference
Human AML	$CD34^{+}CD38^{-}$	200,000	IV	NSG	[6, 7]
MLL–AF9 mouse model	$IL\text{-}7R^{-}$ Lin^{-} Sca-1- $c\text{-}Kit^{+}$ $CD34^{+}$ $FcgRII/III^{+}$ [13]	4	IV	C57Bl/6J	[13, 14]
Acute Promyelocytic Leukemia (APL) mouse model	$CD34^{+}$, $c\text{-}Kit^{+}$, $Fc\gamma RIII/II^{+}$, $Gr1^{int}$	3,000	IV/ retro-orbital	C57Bl/6J	[34]
DEK/CAN mouse model	**LIC:** $Sca1+/c\text{-}Kit+/lin^{-}/Flk2^{-}$ (LT-HSC) **LMC:** $Sca1^{-}/c\text{-}Kit+/lin^{lo}$ (CMP/GMP)	LIC: 200,000 LMC: 2,000	Retro-orbital	C57Bl/6J	[32]

IV intravenous injection, *IF* intrafemoral injection

found that leukemia-initiating activity resided in the $CD34^-$ fraction in half of the patients studied and was heterogeneously distributed between $CD34^+$ and $CD34^-$ fractions in the remaining half of cases. Very recently, John Dick's group was able to see LSC activity in various populations (fractionated by CD34 and CD38 expression), however the majority of LSC activity was found to lie in the $CD34^+$ fraction [28]. Most interestingly, gene expression profiling of all fractions capable of engraftment in immunodeficient mice revealed an LSC-specific gene signature that was capable of predicting both overall- and event-free survival, suggesting that these leukemia stem cells are of clinical importance. A number of confounding factors could explain the inconsistency of these findings, including the use of primary human samples that are genetically heterogeneous in the overall genetic background of the patient, the choice of immunodeficient mouse strain, and the mutations that drive the leukemia as well as the dependency of LSC definition on cell surface immunophenotype. For these reasons, numerous groups have sought to model AML and their LSC compartments using mouse models involving transgenic expression of oncogenes known to cause specific subtypes of human disease (e.g., MLL-AF9) and many are beginning to study functional differences that may affect LSC activity (e.g., Oxygen sensing) as opposed to differential cell surface antigen expression[29].

In a mouse model of AML utilizing a retroviral version of the human chromosomal translocation (t9;11) that results in the fusion transcription factor MLL-AF9, the LSC activity is found to lie in cells not phenotypically resembling hematopoietic stem cells. Scott Armstrong and colleagues were able to show that self-renewal could be imparted on committed progenitors, granulocyte-macrophage progenitors (GMP), and that their prospective sorting enriched for transplantable activity [13]. Furthermore, they were able to expression profile these populations and show that these cells exhibited a "stem signature" that may represent, in part, the self-renewal programme imparted on these cells. In a set of follow-up experiments, it was found that Wnt/β-catenin signaling was necessary for transformation of hematopoietic stem cells (HSC) and GMPs [30].

The translocation event t(6;9) resulting in the DEK/CAN fusion protein is thought to define a particularly poor prognostic subset of AML and typically occurs in a much younger group than is typical for the disease with a median onset of 25–30 years of age [31]. In a retroviral mouse model of this disease, there was heterogeneity in the population of cells capable of initiating and maintaining the malignancy. Specifically, cells resembling the long-term hematopoietic stem cell (LT-HSC) compartment were found to be more permissive to transformation while those resembling a common myeloid progenitor (CMP) or granulocyte-macrophage progenitor (GMP) were capable of transferring disease to secondary recipients [32].

Another subtype of AML is Acute Promyelocytic Leukemia (APL), a hematologic malignancy best known for the t(15;17) translocation that generates a PML-RARα fusion. This subtype responds relatively poorly to conventional chemotherapy; however the cells are sensitive to treatment with all-*trans* retinoic acid (ATRA), a derivative of vitamin A, which acts as a differentiating agent to force the rapidly proliferating immature progenitors to terminally differentiate, often inducing remission. Relapse is common in APL, which has prompted researchers to identify the leukemic stem cell activity using mouse models of pathogenesis. Using a transgenic model of APL, the Gilliland lab was able to show that the PML-RARα gene was capable of inducing self-renewal in committed progenitors, promyelocytes [33]. Concurrently, another group found the initiating cell to lie in a committed myeloid progenitor ($CD34^{+}c\text{-}kit^{+}Fc\gamma RIII/II^{+}Gr1^{int}$) [34]. These studies seem to suggest that self-renewal acquisition in committed progenitors may be a critical step in APL pathogenesis, while the initiating cell population is still in need of further characterization. This disparity could be a result of differences in experimental design (differences in transgenic mice or retrovirus) or possibly reflective of plasticity in initiating cell identity.

There have been reports that have challenged the leukemia stem cell hypothesis in AML and other hematological malignancies and should temper even the staunchest proponent from assuming the leukemia stem cell is well-defined in this disease. The Strasser lab reported a startling finding in a mouse model of AML in which the *PU.1* gene has been knocked-out[20]. They found that in congenic murine recipients that are non-irradiaded, as few as ten cells were able to engraft and cause leukemia. Many have pondered if these reports are enough to abandon the leukemia stem cell hypothesis completely as one of the principle tenets was their low frequency. These reports have pointed out that rarity for some models of great aggressivity and homogeneity is not necessary for these stem cells and that not all leukemia stem cells may be rare.

There are important functional aspects of AML leukemia stem cells that have been taken advantage of in the development of new treatments. The quiescent nature of stem cells in general, and of leukemia stem cells, is thought to be responsible in part for the chemo- and radio-resistant nature of these cells. In an effort to sensitize these cells to therapy, there have been efforts to force these cells to enter the cell cycle. One such effort involves the use of G-CSF, a cytokine that forces the leukemic stem cells to mildly proliferate and sensitizes them to chemotherapy treatment resulting in programmed cell death [35].

13.2.2 Chronic Myeloid Leukemia (CML)

Chronic Myeloid Leukemia (CML) is a well-studied malignancy in which the hematopoietic stem cell is believed to once again, as in AML, to be the cell of origin. The defining invariable genetic lesion in CML is the Philadelphia chromosome, a chromosomal translocation between the breakpoint cluster region (BCR) and the Abelson tyrosine kinase (ABL), t(9;22)(q34;q11), resulting in a fusion protein-kinase known as BCR-ABL. BCR-ABL is a constitutively active kinase and generates a myeloproliferative disorder that occurs in three phases: chronic phase, accelerated phase and blast crisis [36]. The chronic phase is the most common stage of diagnosis with patients typically asymptomatic or complaining of fatigue or other mild symptoms. Accelerated phase generally involves progressive genetic evolution of the clone and increased numbers of blasts in the bone marrow. Blast crisis is the stage at which the malignant clone has accumulated the right number and type of genetic and epigenetic alterations to effectively evolve into frank leukemia, closely resembling Acute Myeloid Leukemia. At this stage, prognosis is poor and survival is rapidly shortened. The unopposed BCR-ABL kinase is thought to primarily influence proliferation and survival of transformed stem and downstream progenitors. In recent years, the development of small molecule tyrosine kinase inhibitors has been a positive development in disease treatment and Imatinib mesylate was developed to impair the driving force of this malignancy. Unfortunately, Imatinib resistance develops frequently, altering our view of the use of such inhibitors and bringing about novel treatment plans in which the drug is combined with traditional chemotherapeutics for better efficacy. The development of second-generation derivatives of Imatinib that respond to the most commonly mutated types of BCR-ABL have also been effective at reducing mortality.

Based on the observation that chemotherapeutics targeting cycling cells were ineffective in eradicating a population leading to disease relapse, Allen Eaves' group identified the $CD34^+$ fraction of human CML cells to be relatively quiescent by Hoescht 33342 and Pyronin Y staining and engraftable in immunodeficient mice to generate shorter-term $CD34^+CD38^+$ and more perpetual $CD34^+CD38^-$ subsets [37]. Subsequent data has suggested that the leukemia stem cell population for CML is more plastic than was previously appreciated (Table 13.2). In the chronic phase of leukemogenesis, the leukemia stem cell fraction appears to immunophenotypically lie in the hematopoietic stem compartment within a $CD34^+CD38^-Lin^-$ population which has self-renewal potential [38]. There are suggestions however, that $CD34^-Lin^-$ cells can also function as leukemia stem cells in CML as this population has been found to engraft in immunodeficient mice and show resistance to Imatinib treatment [39]. Data from Irving Weissman's lab suggests that the blast crisis leukemia stem cell is markedly different from that seen in earlier stages. They found that *in vitro* self-renewal was observed in the GMP population, implicating this progenitor population as a potential leukemia stem cell compartment [12]. This result suggests that as the disease progresses, self-renewal signals derived from Wnt/β-catenin signaling may be acquired by committed $CD34^+CD38^+$ progenitors, expanding the leukemia stem cell pool. These findings are similar to those of the Armstrong lab, which noted that GMPs served as leukemia stem cells in a mouse model of MLL-AF9 AML [13]. At about the same time as the Weissman group's findings, researchers utilized a mouse retroviral bone marrow transplantation model to transduce committed progenitors with one of two oncogenes, MOZ-TIF2 or BCR-ABL [18]. Common myeloid progenitors (CMP) and granulocyte-macrophage progenitors (GMP) were transformable by MOZ-TIF2 as evaluated by serial replating and *in vivo* generation of Acute Myeloid Leukemia, whereas BCR-ABL was incapable of these activities[18]. These results suggested that BCR-ABL is unable to provide an environment permissive for secondary mutations in progenitors and that the acquisition of self-renewal may be a very rare event in the context of this oncogene. In an E2A-knockout mouse model of CML, GMPs served as leukemia stem cells suggesting that in myeloid leukemias of an "acute" phase, either AML or blast crisis CML, there may be biologically significant ramifications of the acquisition of self-renewal in genetic collaboration in aggressiveness of disease [40].

Table 13.2 Chronic myeloid leukemia (CML)

Cancer type	CSC surface markers	Minimal tumorigenic dose [cell numbers]	Technique of transplant	Strain	Reference
Human samples	$CD34^{+}CD38^{-}Lin^{-}$	20,000	IV	NOD/SCID	[66]
MOZ-TIF2 mouse model	**CMP:** IL-7Rα–Thy1.1–Lin–Sca-1–c-Kit+FcγR^{lo} CD34+	CMP: 100,000	IV	C57Bl/6 J	[18]
	GMP: IL-7Rα–Thy1.1–Lin–Sca-1–c-Kit+ FcγR^{hi} CD34+ [231]	GMP: 10,000			
Inducible BCR-ABL mouse model	$Sca\text{-}1^{+}c\text{-}kit^{+}Lin^{-}$	10,000	IV	NOD/SCID	[42]
E2A−/− mouse model	Lin-IL-7R- Sca-1- c-Kit+ CD34+ FcγR+	50	IV	C57Bl/6J	[40]

IV intravenous injection

As in most myeloid leukemias, there have been great efforts to fractionate leukemia stem cells from normal hematopoietic stem and progenitor cells. As a method for the discovery of CML leukemia stem cell biomarkers, a group of researchers expression profiled LSC subsets of patient samples and found that IL1RAP was overexpressed on CML LSCs and cord blood $CD34^+$ cells transduced with BCR-ABL in reference to normal hematopoietic stem cells[41]. In the same study, it was particularly interesting that an antibody specific to ILRAP specifically targeted $CD34^+CD38^-$ Ph+cells while normal equivalents were unaffected[41]. Mouse models of CML have been crucial in our understanding of BCR-ABL pathogenesis with models involving retroviral bone marrow transplantation and inducible transgenic mice both contributing a wealth of knowledge with regards LSC biology. In the most effective inducible transgenic model, a Sca-1^+c-kit^+Lin^- population was found to enrich for transplantable activity into syngeneic recipient mice [42].

Highly complex and subject to interpretation, it has become clear that while much has been learned in the last decade about CML leukemia stem cell biology, there is a great deal of work to be done to delineate whether these populations are fixed or exhibit plasticity that has been observed in other diseases. Currently the only curative treatment for high-risk CML is allogeneic stem cell transplantation, a method that has numerous treatment-associated side effects that make the search for leukemia stem cell-specific treatments a worthwhile enterprise.

13.2.3 Myelodysplastic Syndromes (MDS)

Myelodysplastic syndromes (MDS) are a group of heterogeneous disorders of hematopoietic stem cells that present as defective hematopoiesis and dysplasia of multiple blood lineages. Mechanistically, our understanding of MDS is limited, with its high incidence and a high frequency of progression of Acute Myeloid Leukemia suggestive that further research is necessary. MDS frequently leads to bone marrow failure as patients develop peripheral cytopenias due to increased levels of apoptosis [43]. The best-characterized subset of Myelodysplastic syndromes is 5q- syndrome, defined by the interstitial deletion of the long arm of chromosome 5 (5q). Pathologically, dysplastic megakaryocytes, elevated platelet counts, refractory anemia and in some instances neutropenia are common hallmarks of this disease. Efforts in recent years to find deregulated genes in the commonly deleted region (CDR) have yielded mild benefits but have yet to represent all of the symptoms of this disease [44, 45]. Recent work by the Karsan lab identified two microRNAs, miR-145 and mIR-146a, that were found to modulate effects of innate immunity. Their results suggest that non-coding regions of 5q were responsible, in part for the phenotype of the syndrome [46].

Originally, MDS was suggested to have a myeloid-restricted progenitor as the cell of origin due to a lack of lymphoid involvement. However, roughly a decade ago it was shown that cells resembling hematopoietic stem cells with a phenotype: $CD34^+CD38^-CD90^+$ could transplant into NOD-SCID mice and upon isolation were

Table 13.3 Myelodysplastic Syndromes (MDS)

Cancer type	CSC surface markers	Minimal tumorigenic dose [cell numbers]	Technique of transplant	Strain	Reference
Human samples (5q-)	$CD34^+CD38^-$ $CD90^+$	700,000	IV	NOD/SCID	[47]

IV intravenous injection

found to be 5q deleted (Table 13.3) [47]. Interestingly, most cells had difficulty engrafting suggesting that they were defective in their ability to reconstitute the mouse with a human hematopoietic system. The $CD34^+CD38^-CD90^+$ phenotype is also representative of another subtype of MDS, trisomy 8, a secondary event to 5q-, strongly implicating the hematopoietic stem cells as the cell of origin which has effectively blocked any suggestion that autologous stem cell transplantation may be an option for curing MDS. Other studies have sought to refine the MDS LSC cell surface immunophenotype in order to better distinguish the malignant clone from normal HSCs and found that $CD34^+CD38^-$ MDS stem cells express higher levels of CD133 and CD13 specifically. Complete clinical and cytogenetic remission has been achieved through the use of the thalidomide analog, lenalidomide in 5q-syndrome, however 50% of patients relapse suggesting that the MDS clone is not being effectively eradicated. A recent study suggests that lenalidomide is effective at reducing the $CD34^+CD38^+$ progenitor fraction but has no efficacy against the $CD34^+CD38^-CD90^+$ HSC population and in some cases of recurrent MDS this population expanded following treatment [48]. The engraftment of MDS stem cell clones in immunodeficient mice has been technically quite challenging with even the most immunodeficient strains failing to recapitulate the original disease as is effectively achieved in most leukemia models [49–51]. The high frequency of bone marrow failure and secondary AML in MDS has spurred the discovery of the MDS LSC, but the generation of better mouse models may be necessary to gain insight due to the difficulty in the engraftment of human MDS clones in even the most immunodeficient of strains.

13.2.4 Acute Lymphoblastic Leukemia (ALL)

Acute Lymphoblastic Leukemia (ALL) is a broad group of malignancies that vary greatly in pathogenesis with many gross morphological and molecularly heterogeneous subtypes. The disease is united by the common accumulation of primitive lymphoid blasts in the bone marrow, secondary lymphoid organs with frequent infiltration of kidneys, liver and the central nervous system. Similarly to Acute Myeloid Leukemias, ALL is a medical emergency requiring a quick diagnostic response and treatment. Although developmentally and pathologically distinct entities, ALLs of both the T-lineage and B-lineage are effectively identical in treatment approaches. Many cases of ALL involve chromosomal translocations that may result in fusion proteins of both kinases and transcription factors that result in increased expression or altered activity, as well as the increased expression of proto-oncogenes and loss of tumor suppressors. As opposed to AML, ALL is primarily a disease of the young with a median onset of 13 years of age in the United States [232]. The age of the patient often determines the aggressiveness of treatment with younger patients recently receiving increasingly higher doses of standard chemotherapy, resulting in greatly reduced mortality. Unfortunately, treatment associated side effects are harsh in a developing child and relapse is still frequent, prompting researchers to find and eliminate any possible leukemia stem cells that are theorized to support the malignancy.

13.2.4.1 T Cell Acute Lymphoblastic Leukemia (T-ALL)

T cell Acute Lymphoblastic Leukemia is a disease primarily believed to be driven by deregulated T cell transcription factor expression/activity leading to increased survival, proliferation, stunted differentiation and increased self-renewal. In recent years, it was discovered that over 50% of human T-ALL cases harbored activating Notch1 mutations resulting in overactive signaling of the Notch pathway[53]. Phenocopying mutations in the E3 ubiquitin ligase FBW7 result in prolonged Notch signaling and increasing evidence suggests that overactive Notch signaling is a defining feature of the majority of T-ALL [54–56]. Deregulation of transcription factors such as LMO2, LYL1, TAL1 among others are believed to provide collaborating signals which may block differentiation and enhance self-renewal [57, 58]. Mutations and other forms of deregulation of RAS/ERK and P13K/AKT signaling are thought to be driving proliferative alterations in T-ALL. Several reports have established T-ALL to contain rare LSCs, however the identity of the leukemia stem cell immunophenotype remains contentious with more work needed to elucidate the transplantable activity of this disease (Table 13.4).

In the first studies to identify the T-ALL LSC population, Allison Blair's group found that at least 5×10^5 bulk cells were necessary for engraftment in NOD/SCID mice while $CD34^+CD7^-$ or $CD34^+CD4^-$ populations were proliferative *in vitro* and were capable of engrafting in NOD/SCID mice and self-renewing as monitored by serial transplantation [52]. In contrast to these findings, John Dick's group recently found that a $CD7^+CD1a^-$ population was best able to proliferate long-term via the well-established OP9-DL1 co-culture system and showed enhanced engraftment via the more sensitive NOD/SCID $IL2R\gamma^{-/-}$ mouse [59]. Most interestingly is that the $CD1a^-$ subset exhibited higher resistance to the glucocorticoid dexamethasone *in vivo* suggesting that it may be functionally relevant in modeling disease relapse following chemotherapy. These results suggest that further work in refining the human T-ALL leukemia stem cell activity is necessary and that the use of multiple types of immunodeficient mice may be necessary to best quantify engraftment potential of prospectively sorted subsets.

As in other malignancies, T-ALL leukemia stem cell activity has begun to be investigated using mouse models of pathogenesis. PTEN mutations are relatively frequent in T-ALL (5–10%) and as such, a mouse model in which PTEN is conditionally

Table 13.4 T cell acute lymphoblastic leukemia (T-ALL)

Cancer type	CSC surface markers	Minimal tumorigenic dose [cell numbers]	Technique of transplant	Strain	Reference
Human samples	CD34+CD4-CD7-	4000	IV	NSG	[52]
Human samples	CD7+CD1a-	1000	IF	NSG	[59]
Pten−/− mouse model	c-Kit^{mid}CD3+Lin-	100	IV	NSG	[60]
SCL^{tg}LMO1^{tg} mouse model	CD4-CD8-CD44-CD25+ (DN3 stage); CD4-CD8-CD44-CD25- (DN4 stage)	1000	IV	NSG	[62]
TG-B T-lymphoma mouse model	Sca-1^{+}c-Kit^{+}	100	IV/IP	B10.BR (H-2k)	[29]

IV intravenous injection, *IF* intrafemoral injection, *IP* intraperitoneal injection deleted in hematopoietic stem cells was found to lead to myeloproliferative disease followed by T cell acute lymphoblastic leukemia [60]. Selective pressure for further mutation in cells lacking PTEN is thought to drive leukemogenesis with a recurrent t(14;15) translocation leading to c-myc overexpression. Interestingly, the authors found that active (unphosphorylated) β-catenin was increased in a leukemia stem cell-enriched c-Kit^{mid}CD3^{+}Lin^{-} population and that conditional ablation of β-catenin led to delays in disease progression [60]. An interesting report recently described the imparting of self-renewal on committed T cells in the thymus by the transcription LMO2 [58, 61]. An expanded pool of CD4^{-}CD8^{-} T cells became permissive to further mutations, including Notch1 mutations, suggesting that this gene and the pathways it regulates are important regulators of T-ALL stem cell activity. In a similar set of experiments, transgenic TAL1/LMO1 mice were generated and were observed to have an expanded ETP/DN1 (DN refers to CD4^{-}CD8^{-}) pool of thymocytes, defined by a Lin^{-}Sca-1^{+}c-kit^{+}CD44^{+}CD25^{-} immunophenotype which was permissive to the accumulation of Notch1 mutations and the generation of frank leukemia [62]. These leukemias were found to have an enrichment of leukemia stem cell activity in CD4^{-}CD8^{-}CD44^{-}CD25^{+} (DN3) and CD4^{-}CD8^{-}CD44^{-}CD25^{-} (DN4) subsets. Immature thymic progenitors appear to be the leukemia cell of origin in this model, creating a permissive environment for further mutational selection and generating leukemia stem cells of various transplantable activities suggesting that stem activity may be somewhat plastic in T-ALL with different oncogenic hits generating different types of leukemia stem cells. These findings will most certainly spur the refinement of the T-ALL stem cell identity in humans.

13.2.4.2 B Cell Acute Lymphoblastic Leukemia (B-ALL)

B cell Acute Lymphoblastic Leukemias are a molecularly heterogeneous group of clonal immature pre-B cell blasts that impinge on normal hematopoiesis in the bone marrow and eventually infiltrate the periphery. They constitute the majority of ALL, and although various subtypes are being thought of as distinct entities, they are united by their phenotypically immature B cell phenotype. One of the most frequent genetic lesions in B-ALL is the chromosomal translocation that defines CML, the Philadelphia chromosome (Ph), which produces the constitutively active BCR-ABL tyrosine kinase. This mutation is found in 5% of childhood and 20–40% of adult disease and is thought to arise in lymphoid lineage progenitor cells as opposed to hematopoietic stem cells as is the case in CML [63–65]. In the progression of CML to "blast crisis", it is possible for this clonal disease to evolve into a B cell acute lymphoblastic leukemia, a subtype which is molecularly and prognostically unique. The development of small molecule tyrosine kinase inhibitors has been a positive development in CML treatment and Imatinib mesylate has been used in the treatment of Ph+B-ALL as well to some success. Unfortunately as in CML, mutations in the catalytic domain of the ABL kinase can lead to Imatinib resistance, bringing about the synthesis of second-generation inhibitors which are effective against mutated BCR-ABL. Interestingly, the same immunophenotype of the AML LSC,

$CD34^+CD38^-$, is also representative of the stem cell activity of BCR-ABL-positive B cell acute lymphoblastic leukemia [66]. For other subtypes of B-ALL, leukemia stem cell activity is believed to reside within a $CD34^+CD10^-CD19^-$ population [67]. This is again, not without contention however since transplantable activity has been found within the $CD34^+CD19^+$ fraction and within a $CD133^+CD19^-$ population but this may be due in part to the use of an intrafemoral route of injection, suggesting that the homing of these cells may be an important confounder [68, 69]. Whether or not these studies are again reflective of underlying genetic and/or epigenetic diversity will be important points for future consideration but much of the differences again are likely attributable to differences in the assays employed to measure the stem cell activity. It is interesting however that for TEL-AML1 B-ALL, that transplantable activity resides within the $CD34^+CD38^{low/-}CD19^+$ fraction in NOD-SCID transplantation and that this phenotype is seen in human umbilical cord blood cells transduced with lentiviral TEL-AML1 [70]. TEL-AML1 is thought to act in a dominant-negative manner, interfering with normal AML1 (RUNX1) signaling, enhancing self-renewal of lymphoblastic leukemia cells [71]. This study was also crucial in pointing to an *in utero* developmental origin of this subtype of B-ALL and more critically, it established a role for clonal evolution in the development of the leukemic stem cell. Using a pair of monochorionic twins, one in a preleukemic state and one with frank leukemia, they were able to show a common ancestral clone which had gained accessory lesions in order to generate frank leukemia [70]. Interestingly, both twins shared this $CD34^+CD38^{low/-}CD19^+$ leukemic stem cell compartment yet the frequency and genetic lesions in NOD/SCID transplantation differed between them suggesting that the stem cell itself had mutated, becoming more frequent/aggressive in the frank leukemia [70]. In another twin study, recurrent copy number alterations were found to be involved in driving leukemogenesis and may be involved in driving leukemic stem cell diversity [72]. The same group has further explored this by monitoring sub-clonal diversity of TEL-AML1 leukemia using fluorescence in situ hybridization directed towards several common copy number alterations [73]. The dynamics of these alterations show that constant selective pressure contextualized via relapse and immunodeficient mouse transplantation cause genetic variegation that likely mirrors leukemia stem cell diversity (Table 13.5) [73].

Researchers have sought to relate the poor prognostic subtypes of B-ALL that harbour the BCR-ABL, t(9;22), and MLL-AF4, t(4;11), translocations to their leukemic stem cell activity. Researchers found that a majority of $CD34^+CD19^-$ cells carried these translocations suggesting a more primitive cell of origin than a committed lymphoid progenitor [74]. Using fluorescence in situ hybridization, they were able to show that stem/progenitors harbouring the translocations were unable to differentiate into myeloerythroid colonies *in vitro*. Providing further support for certain subtypes of B-ALL to originate *in utero*, it was shown in an elegant set of experiments that bone marrow mesenchymal cells from the developing embryo/fetus, which share a developmental ancestor with hematopoietic stem cells, harboured the MLL-AF4 oncogene, suggesting that a common mesenchymal stem ancestor is the cell of origin for MLL-AF4+ B-ALL [75].

Table 13.5 B cell acute lymphoblastic leukemia (B-ALL)

Cancer type	CSC surface markers	Minimal tumorigenic dose [cell numbers]	Technique of transplant	Strain	Reference
Ph+ Human samples	CD34+CD38-	10,000	IV	NOD/SCID	[66]
Human samples	CD34+CD10-	70,000	IV	NOD/SCID	[67]
	CD34+CD19-	50,000			
Human samples	CD34+CD19-	15,000	IF	NOD/SCID	[68]
	CD34-CD19+	30,000			
	CD34+CD19+	40,000			
Human samples	CD133+CD19+	5,400	IV	NOD/SCID	[69]
TEL-AML1 human samples	CD34+CD38$^{-/low}$ CD19+	50,000	Intra-tibial	NOD/SCID	[70]

IV intravenous injection, *IF* intrafemoral injection

13.3 Regulation of Leukemia Stem Cell Activity

The molecular mechanisms involved in the regulation of LSC activity have yet to be fully elucidated; however it seems that normal and malignant stem cells share some degree of overlap in the pathways they are reliant upon. These regulatory mechanisms can be divided in three categories:

(1) Control of self renewal; including genes such as Notch [76–81], Wnt/β-catenin [30, 82–85], Hedgehog [86–88], Bmi1 (also known as *PCGF4)* and the other Polycomb Group (PcG) proteins [89, 90] and the HOX transcription factors [91–93];
(2) Control of metabolism; implying modulators of reactive oxygen species (ROS) levels [94, 95], such as the forkhead O (FoxO) family of transcription factors [96, 97] and IDH1/2[98–100];
(3) Control of interactions with cell niches and bone marrow microenvironment [101, 102]; including proteins related to hypoxia (e.g. Hif1α) [29, 103, 104], chemokine receptors (e.g. CXCR4) [105, 106] and adhesion molecules (e.g. CD44) [107, 108].

13.3.1 Self-renewal Pathways in Leukemia Stem Cells

LSCs are distinguished from the other cancer cells due to their inherent self-renewal activity. Self renewal is defined as the capacity of a stem cell to renew itself indefinitely. Stem cells must produce differentiated progeny to carry out specialized functions but retain copies of themselves for future use. To accomplish this task, asymmetric divisions are necessary to create two different daughter cells, one with stem cell characteristics identical to the mother cell, and another destined to produce committed cells with limited replicative potential (Fig. 13.2).

It is still unclear which pathways modulate self-renewal activity in LSCs. However, it is known that common pathways can regulate self-renewal activity of different types of stem cells and can induce cancer when they are deregulated. It has been suggested that networks of proto-oncogenes and tumor suppressors regulate self-renewal and age-related changes in stem cells and are related to the generation and proliferation of cancer cells in the same tissues as well [109]. The most well-known mechanisms involved in these self-renewal programs include Notch, Wnt/β-catenin, Hedgehog (Hh) and BMI1 signaling pathways.

13.3.1.1 Notch Signaling Pathway

Notch family members are essential during the development of embryonic hematopoietic stem cells (HSC), are involved in the commitment of myeloid and lymphoid cells [110] and potentiate the self-renewal activity of HSCs [76, 77] and

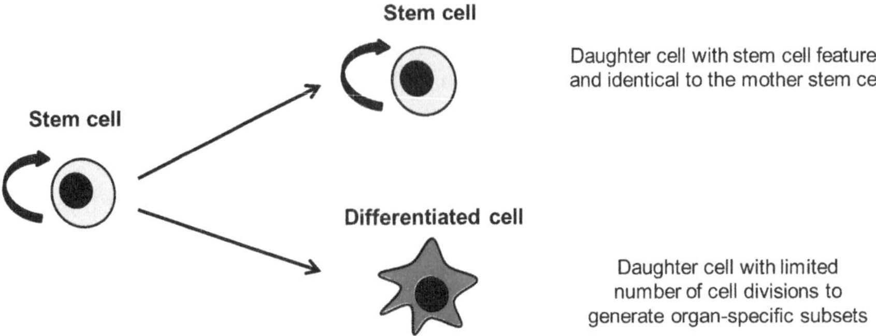

Fig. 13.2 Example of asymmetric division. One main feature of LSCs is the presence of self-renewal activity. It means the ability of a stem cell to have symmetric divisions and, so, to produce to different daughter cells, one, identical to mother cell, with stem properties and another one with limited divisions and committed to generate differentiated subsets

neural stem cells[111, 112]. Recent data have also shown that Notch has a key role in the activation of human T-cell acute leukemia (T-ALL) and in the maintenance of leukemia initiating cells. In fact, LSC activity of human T-ALL samples is abolished in *in vivo* and *in vitro* functional assays after treatment to block Notch signaling by γ-secretase inhibitor [78]. In mammals there are four homologous Notch family members. Each gene encodes a transmembrane receptor that recognizes cognate Jagged or Delta-like ligands. The binding of receptor with ligand activates a series of metalloprotease cleavage events in the extracellular portion of the receptor (e.g. γ-secretase), resulting in the release of intracellular Notch (ICN). ICN translocates to the nucleus where it activates the transcription of target genes by forming a multimeric complex with co-activators of the mastermind-like (MAML) family and the DNA-binding transcription factor CSL.

The function of Notch in the maintenance of HSCs is based on numerous gain-of-function studies. The number of HSCs and/or their self-renewal activity increase after the overexpression of ICN or the Notch target gene Hes1 in bone marrow progenitors[80, 113]. Notch signaling may also modulate HSC homeostasis via the Notch ligand Jagged1. Mice, genetically modified to express osteoblast-specific and activated PTH/PTHrP receptors (PPRs) show an expansion in the number of osteoblasts expressing Jagged1 and HSCs with an activated Notch pathway [114]. Of note, it is not yet clear how Notch contributes to the self-renewal activity of HSCs. In mice, the conditional inactivation of Notch1 or Jagged1 or both does not impair HSC maintenance. In addition, HSCs lacking Notch are able to reconstitute the hematopoietic compartment after transplantation [115]. It is possible that homologous Notch receptors or ligands might compensate for Notch1 and Jagged1, but this line of experimental evidence suggests that canonical Notch signaling is dispensable for HSC maintenance in bone marrow [79].

In T-ALL, over 50% of cases are associated with mutations of Notch1 gene [53], while about 15% have mutations in another gene in the Notch signaling pathway, FBW7/Sel10 [55]. In both cases, these mutations lead to hyperactivation of Notch signaling. The main genetic alterations of the Notch1 gene in T-ALL affect its heterodimerization domain (HD), leading to ligand-independent cleavage and subsequently increased and uncontrolled production of ICN [116], and the PEST domain, stabilizing ICN from degradation [117]. In this disease, Notch is believed to be involved in the self-renewal activity of LSCs. Coculture of primary human T-ALL with mouse stromal cell lines expressing Notch1-ligand delta-like1 improved LSC maintenance and long-term growth of leukemic cells. In addition, inhibition of Notch signaling by γ-secretase inhibitors inhuman T-ALL samples impairs serial transplantation of cultured cells into NOD-SCID mice [78]. These studies strongly suggest that sustained Notch activation is required for the maintenance of LSCs *in vitro* and *in vivo*.

Recent data has also shown that hypoxia-inducible factor 1α (HIF1α) and Notch are connected in an essential interaction for LSC activity in a mouse model of T cell lymphoma. HIF1α activity maintains self-renewal of LSCs by repressing a negative feedback loop in the Notch pathway. In fact, HIF1α potentiates the expression of Hes1 mediated by Notch. Hes1 is a Notch target gene, critical for stem and progenitor cells [118, 119]. HIF1α does not directly interact with Notch, but prevents Hes1 from binding to the N-box of Hes1 promoter, blocking a negative feedback regulation of Hes1 [29].

In general, the role of Notch in the maintenance of LSCs and tissue stem cells is still not clear and is not necessarily universal [120]. However, the most recent findings suggest an important functional conservation between LSCs and tissue stem cells where Notch signaling seems to have a predominant role. For this reason, the Notch pathway is considered an attractive therapeutic target to treat different types of tumors. New drugs, such as γ-secretase inhibitors or neutralizing antibodies against Notch1, have already been developed and are currently being tested in clinical trials.

13.3.1.2 Wnt/β-catenin Signaling Pathway

The Wnt/β-catenin signaling axis is a highly conserved pathway that controls different developmental processes, including the homeostasis and self renewal activity of stem cells in many tissues. Its deregulation has been associated with several types of tumors, most famously colon cancer. In the canonical Wnt pathway, when Wnt ligands are absent and the Frizzled/LRP receptors are not engaged, β-catenin forms a complex with the tumor suppressors, adenomatous polyposis coli (APC) and axin, known as the "β-catenin destruction complex". At this stage, β-catenin is phosphorylated in an evolutionarily conserved region of its amino terminus by CKI and GSK3, two kinases residing in the complex, and subsequently targeted for proteosomal degradation. When Wnt ligands are present, the stimulated receptor recruits axin which inactivates the destruction complex, leading to the stabilization of β-catenin. As a consequence of its increased stability, β-catenin accumulates and translocates to the nucleus where it interacts with DNA-binding proteins of Tcf/Lef family forming a transcriptionally active complex by displacing Groucho-family repressors [121].

Aberrant Wnt/β-catenin signaling induces proliferation of crypt progenitors and generation of benign polyps [122] and can transform intestinal stem cells into CSCs [123]. A recent study has also reported that in colon cancer cells, stromal myofibroblasts induce Wnt/β-catenin-dependent transcription by secreting stromal factors such as HGF[124]. After stimulation with these stromal factors, more differentiated tumor cells acquire self-renewal activity and acquire a CSC phenotype, suggesting a functional role for the microenvironment in controlling colon CSC activity, mediated by Wnt signaling. In mouse epidermal tumors, CSCs show related features to normal stem cells where β-catenin is preferentially located in the nucleus with respect to other non-stem cells [125]. Finally, the Wnt/β-catenin pathway has been shown to protect mouse mammary CSCs of p53 null mice from irradiation, reducing their DNA damage response [126].

In the hematopoietic system, HSCs themselves and the microenvironment produce Wnt proteins that are the ligands of Wnt/β-catenin pathway. Furthermore, Wnt3a or activated β-catenin directly stimulates HSCs *in vivo* increasing their self-renewal activity [83]. The persistence of these signals exhausts and, consequently, depletes the long-term stem cell pool [127]. Human HSCs are also affected by Wnt signaling. In fact Wnt5A-conditionated medium increases the transplantability of human HSCs in NOD-SCID mice and the proliferation of undifferentiated progenitors *in vitro* [128]. In addition, using mice engineered to overexpress the Wnt inhibitor Dkk1, under the control of an osteoblast specific promoter, it has been demonstrated that the Wnt pathway maintains the quiescent state of HSCs, preserving their self-renewing capability [129].

In leukemia, Wnt/β-catenin signaling has a critical role in the establishment and maintenance of LSCs [30, 85]. In AML mouse models generated by co-expression of the Hoxa9 and Meis1a oncogenes or the fusion oncoprotein MLL-AF9, Wang and collaborators have shown that leukemic transformation and the maintenance of established leukemia depends on activation of the β-catenin pathway. These results show that β-catenin is necessary for the self-renewal activity of LSCs generated from HSCs or more differentiated granulocyte macrophage progenitors (GMP). In normal GMPs, activation of β-catenin signaling is crucial and, thus, required, to transform committed progenitor cells [30]. In another mouse model of AML generated by fusion proteins encoded by *Mixed Lineage Leukemia* (MLL), β-catenin activation is involved in the development of LSCs. Conditional deletion of β-catenin modulates a phenotypic switch of LSCs to pre-LSC-like stage and induces drug sensitivity in MLL transformed cells [85]. These findings are very promising because they suggest new treatments for AML. In fact, β -catenin is frequently activated in AML but is dispensable for the self-renewal of adult HSCs [130]. Targeting the Wnt/β-catenin pathway could represent a new potential therapeutic approach for the selective eradication of AML stem cells.

13.3.1.3 Hedgehog Signaling Pathway

The Hedgehog (Hh) genes were first identified in the fruit fly *Drosophila melanogaster* by C. Nüsslein-Volhard and E. F. Wieschaus in the late 1970s. Today it is known that the Hh genes are highly conserved evolutionarily and are involved in embryonic development by regulation of processes like tissue patterning, cell proliferation and differentiation [131]. In mammals, this pathway has a functional role in the maintenance of stem cells and in tissue repair/regeneration [132, 133]. Importantly however, recent data has shown that Hh signaling is dispensable for adult murine HSC activity and definitive hematopoiesis [87, 134]. Alterations resulting in abberant Hh signaling can lead to basal cell carcinoma and medulloblastoma, with further evidence for roles in leukemias such as CML [88] and in solid tumors of the pancreas, prostate, lung, and breast[131].

The mature Hh proteins are ligands for the membrane receptors Patched1 and Patched2 (both abbreviated Ptch). In absence of Hh ligands, Ptch receptors inhibit the formation of active cell-membrane proteins termed *smoothened C* (SmoC), and consequently induce the phosphorylation and ubiquitination of Gli2/3 transcription factors for proteosomal degradation. In presence of binding with Hh ligands, Ptch receptors are internalized and degraded by lysosomes, generating active SmoC. This protein stabilizes the Gli2/3 transcription factors that translocate into the nucleus and activate the transcription of Hh target genes.

With respect to aberrations of the Hh signaling pathway, we can distinguish three different types of cancers. In the first type, Hh pathway is aberrantly over-activated due to mutations in one or more of its components. One example of this category is Gorlin syndrome, in which a high incidence of basal cell carcinoma, medulloblastoma, and rhabdomyosarcoma is observed [135]. In a second type of alteration, tumor cells produce and accept their own Hh ligands; a process known as autocrine signaling [136]. In the third class of abnormal pathway activation known as paracrine signaling, Hh ligands are secreted by tumor cells and stimulate stromal cell subsets. In the stroma, Hh signaling can modulate factors, such as insulin-like growth factor receptor (IGFR) and components of Wnt signaling pathway known to promote a favorable growth environment for the tumor and to support CSCs and tumor angiogenesis [137, 138]. Via a paracrine mechanism, tumors receive indirect sustenance from the stromal cells, mediated by Hh proteins. A variant of this type of mechanism is called "reverse paracrine" because Hh ligands are produced and secreted from stromal cells and directly affect cancer cells.

Hh signaling pathway has a key role in the maintenance and self renewal of CSCs in multiple myeloma (MM) [139], B-cell acute lymphocytic leukemia (B-ALL) [140] and in chronic myeloid leukemia (CML) [88, 141]. In MM, a plasma cell malignancy of the bone marrow and a non-Gorlin cancer, a marked asymmetry in the expression levels of Hh pathway components and Hh reporter activity between CSCs and differentiated tumor cells has been observed. In this malignancy, Hh ligands are required for the expansion of MM stem cells and to maintain those subsets in an undifferentiated state [139]. In human B-ALL, inhibition of the Hh pathway by cyclopamine and the SMO inhibitor IPI-926, primarily affects the highly clonogenic B-ALL cells expressing aldehyde dehydrogenase (ALDH) and limit their self-renewal activity *in vitro* and *in vivo* [140]. In a mouse model of CML generated by expression of *Bcr-Abl1* oncogene, the loss of SMO results in a depletion of LSCs. In addition, pharmacological inhibition of Hh signaling reduces the proliferation of imatinib-resistant mouse and human CML cells, increasing the time until relapse following therapy [88, 141]. These findings indicate that Hh antagonists may be useful in the design of new therapies to target this important stem cell pathway and to eradicate LSCs in potentially Hh-dependent malignancies such as MM, B-ALL and CML.

13.3.1.4 BMI1/Polycomb Group Genes

Polycomb Group (PcG) genes were identified for the first time in *Drosophila* and recognized as repressors of Hox genes and modulators of anterior-posterior body patterning [142]. The PcG family consists of many proteins involved in the regulation of global epigenetic changes of the genome. They are assembled into chromatin-associated complexes and are variable in their components in a menner related to the state of cells [143]. In mammals we distinguish two main types of PcG chromatin-modifying complexes, named Polycomb Repressive Complexes 1 and 2 (PRC1, PRC2). The core of PRC1 complex is constituted by one subunit of PCGF, CBX, PHC, SCML, and RING1 paralog factors [144, 145]. PRC2 complexes includes SUZ12 and EZH1 or EZH2 which are histone methyltransferases involved in trimethylation of lysine 27 on histone H3 (H3K27me3), a typical epigenetic mark of silencing [146–149]. The specific function of different PcG complexes is defined by the activities of distinct components. For example, the presence of the PRC1 protein Bmi1 is essential for the maintenance and cell proliferation in both normal and leukemia stem cells [89]. On the contrary, PCGF2/MEL18, a paralog PcG gene, is a tumor suppressor and blocks cell proliferation [150, 151], underscoring the complexity and structural diversification of these complexes.

In normal hematopoiesis and hematological malignancies, PcG genes play a major role in the regulation of many cell processes such as the modulation of cell-cycle checkpoints, DNA damage repair, cell differentiation, senescence, and apoptosis. In Bmi1-deficient mice, HSCs have reduced self-renewal activity and bone marrow progenitors show restricted proliferative capacities [89, 90, 152]. Moreover, enforced expression of Bmi1 promotes properties associated with "stemness" through cell divisions [153], while MEL18 appears to be more involved in cell differentiation [154]. In HSCs, Bmi1 functions are also related to the regulation of the *Ink4a/Arf* locus where Bmi1 acts as repressor. This tumor suppressor locus encodes two proteins that act as cell cycle regulators; $p16^{INK4A}$ acts to bind to and inhibit Cdk4 and Cdk6 while $p14^{ARF}$ binds to and inhibits Mdm2 resulting in p53-mediated cell cycle arrest. Interestingly in Bmi1-deficient HSCs, the expression levels of $p16^{INK4A}$ and $p14^{ARF}$ are higher than in corresponding wild type cells [90, 155]. Deletion of both genes partially rescues the proliferative defects of Bmi1−/− HSCs [156], suggesting that Bmi1 has a protective effect on the HSC compartment by inhibition of the *Ink4a/Arf* locus.

The identification of Bmi1 as a collaborator of c-Myc-induced lymphomagenesis was one of the first lines of evidence pointing to the importance of PcG proteins in cancer [157–159]. PcG proteins are essential in the maintenance of CSCs during tumorigenesis due to their capacity to regulate genes related to stemness and differentiation [160–163]. In hepatocellular carcinoma, lentiviral knockdown of Bmi1 drastically reduces the number of side population (SP) cells that are known to have CSC properties in this malignancy, and impair the self-renewal activity of this subset [164]. In a mouse model of AML involving the co-expression of the HoxA9 and Meis1 oncogenes, Bmi1 is required in the maintenance of the proliferative activity of LSCs. In fact leukemic stem and progenitors cells of this model generated from Bmi1-deficient fetal liver cells present signs of differentiation and apoptosis, undergoing a proliferative arrest and a resulting failure of transplantation into secondary recipients [89]. Moreover, the repression of Bmi1 by a lentiviral RNA interference approach in CD34+ cells of human AML reduces the long-term expansion of this subset *in vitro,* impairing its self-renewal capacity [165]. Finally, recent data have demonstrated that Bmi1 cooperates with the *Mll-Af9* fusion oncogene for leukemic transformation of murine myeloid progenitor cells [166] and with the *Bcr-abl* oncogene in inducing a CML-like leukemia in immunodeficient mice using human CD34+ cells [167]. In both cases, Bmi1 is essential for the proliferative capacity and self-renewal properties of LSCs. All these data are consistent with reports that show a correlation between Bmi1 expression in CD34+ cell subsets with prognosis and development of many leukemias such as AML, CML and myelodysplastic syndrome [168–170]. The role of PcG members in cancer underlies the relevance of deregulated chromatin machinery resulting in global epigenetic changes during tumorigenesis, especially in hematologic malignancies. These alterations, which are prognostic for some diseases, may be frequently involved in mechanisms of stemness and/or differentiation in LSCs. This has important implications for the generation of novel therapy and explains why PcG proteins are becoming more and more tractable to new therapeutic approaches.

13.3.2 Metabolic Pathways in Leukemia Stem Cells

In 2000, D. Hanahan and R.A. Weinberg rationalized the complexities of neoplastic diseases in six distinct hallmarks [171]. These highlighted principles including sustained proliferative signaling, tissue invasion and metastasis, enhanced replicative immortality, induced angiogenesis, resistance to cell death and evasion of growth suppressors. In the last few years however, the acquisition of further knowledge to our understanding of cancer biology suggest that two additional hallmarks should be added. The first is defined as the capability of tumor cells to evade immune destruction, mediated by macrophages, natural killer cells and T and B lymphocytes. The second involves the reprogramming of energy metabolism. In this last hallmark, cancer cells are able to alter their cellular metabolism to obtain the most efficient support to satisfy their energetic requests [172].

In terms of metabolism, proliferating tumor cells have to satisfy three basic needs: rapid and intense ATP production, forceful biosynthesis of macromolecules and tight control of cellular redox status [173]. Similar modifications have been described in proliferating cells and non-tumorigenic cells in response to physiological growth signals [174]. In cancer, these alterations are adaptations to varied and dynamic microenvironments where nutrients and oxygen are spatially and temporally heterogeneous. The best characterized metabolic switch in tumor cells is known as the Warburg effect. It describes a shift in the mechanism of ATP production from oxidative phosphorylation to glycolysis, despite an abundance of oxygen [175]. This metabolic state has been alternately described as "aerobic glycolysis" and involves the conversion of glucose to lactate as opposed to mitochondrial metabolism via the Krebs cycle [173].

Aerobic metabolism is less efficient in terms of energy production, generating less ATP per unit of glucose utilized and increasing the levels of reactive oxygen species (ROS) which has wide-ranging effects on cellular components. The damaging effects of ROS accumulation in cells have been well characterized during aging and tumorigenesis [176]. However recent data has demonstrated that redox status is essential in the maintenance and self-renewal activity of normal and cancer stem cells [95, 177, 178]. ROS subtypes includes all oxygen species more reactive than free oxygen such as superoxide (O_2^-), hydrogen peroxide (H_2O_2), hydroxyl radical (•OH), singlet oxygen ($^1O_2$) and nitric oxide (NO). Under physiological conditions, ROS are produced in mitochondria [179] and through an active ROS-generating system, a result of NADPH oxidation by NOX family members (NOX 1–5, Duox1, and 2) [180].

In HSCs, some regulatory mechanisms that control redox status have been characterized using Atm deficient mice. In fact, in this mouse model, HSCs present high levels of ROS related to a decreased capacity for self-renewal and activation of p38MAPK. Additionally, in serial transplantation assays, the lifespan of wild-type HSCs is improved by protracted treatments with the antioxidant N-acetyl L-cysteine (NAC) or an inhibitor of p38 MAPK [94]. Other reports have also demonstrated that FoxO proteins such as FoxO1, FoxO3 and FoxO4 play an essential role in the control of physiological oxidative stress in HSCs [96, 181, 182]. The deletion of FoxO1, FoxO3a and FoxO4 genes in mouse impairs the long-term repopulating capacity of HSCs and affects lymphoid development, inducing a myeloid lineage expansion. Furthermore, FoxO-deficient HSCs show increased cell cycling activity and apoptosis, correlating with changes in ROS levels and in the expression of genes involved in ROS regulation [96]. The expression of FoxO transcription factors is negatively modulated by the phosphoinositide 3-kinase (PI3K)/AKT pathway which is suppressed in HSCs and activated in hematopoietic progenitors [182]. Finally, FoxO3 and Atm directly interact to inhibit ROS accumulation and to potentiate the DNA damage response [183, 184].

Most recent data has also identified Lkb1 (*liver kinase B1*) as another regulator of HSC metabolism and stem cell homeostasis [185–187]. Lkb1 is a kinase that activates AMPK by phosphorylation in response to high AMP/ATP ratios. The phenotype of Lkb1 deficient mice is characterized by increased numbers of hematopoietic progenitors and defects in the maintenance of HSCs which escape from their quiescent state, defining a central role for Lkb1 in metabolic control of stem cells.

Although the redox status of cancer stem cells is still clearly not well-defined, it has been demonstrated that, like normal stem cells, some CSCs maintain low ROS levels in spite of non-stem tumor cells which present high ROS concentrations[177, 178]. It has been suggested that in terms of redox control, CSCs may share common metabolic pathways with normal stem cells. One example is provided by $CD24^{-/low}$/$CD44^{+}$ breast CSCs. In breast cancer, CSCs are more resistant to radiation-induced DNA damage with respect to their non-tumorigenic offspring and show significantly low levels of both basal and radiation-induced ROS. The high expression of ROS scavenging molecules in breast CSCs seems to participate in this phenotype and contribute to the radio-resistance of stem cells [177]. Recently it has also been reported in human gastrointestinal CSCs than an interaction between a variant of CD44 transmembrane receptor (CD44v) and a cysteine-glutamate exchange transporter (xCT) controls the cysteine transport for glutathione (GSH) synthesis and defense against toxic levels of ROS [188].

In hematopoietic malignancies, ROS levels are critical for the maintenance of different types of LSCs [97, 189, 190]. Data has recently suggested that the sesquiterpene lactone parthenolide (PTL) induces apoptosis of LSCs in AML and blast crisis CML without affecting non-leukemic cells. One observed molecular mechanism that has been associated with PTL activity in AML is the inhibition of NF-κB signaling with subsequent increases in ROS concentration. Furthermore, the redox status of LSCs appears to be critical for the efficacy of PTL treatment. In the presence of NAC antioxidant agent, AML cells were found to be more resistant to PTL-mediated apoptosis [189]. Recent reports have suggested that PTL preferentially targets LSCs with high levels of myeloperoxidase (MPO) in AML cells [191].

In CML, quiescent LSCs are sensitive to pharmacological inhibition of promyelocytic leukaemia (PML) protein tumour suppressor, mediated by As_2O_3 [190]. The mechanism of arsenic cytotoxicity involves the degradation of the PML-retinoic acid receptor-alpha (PML-RARalpha) fusion protein and concerted increases in ROS production by either NADPH oxidase or collapse of the mitochondrial transmembrane potential [192]. The effect of PML inhibition by As_2O_3 has been described in human primary cells and in a mouse model of CML, generated by the expression of the BCR-ABL fusion protein. Ito et al., have demonstrated how the arsenic treatment impairs the maintenance of LSCs and improves the efficiency of therapies by sensitizing LSCs to pro-apoptotic stimuli [190]. The self-renewal activity of CML stem cells is also potentially modulated by the TGF-β–FoxO pathway. The TGF-β pathway inhibits Akt activation and consequentially induces the nuclear localization of FoxO3 in LSCs. Additionally, treatment with a TGF-β inhibitor and Imatinib mesylate, a BCR-ABL inhibitor, is more efficient in leukemic FoxO3a-deficient cells than the correspondent wild type cells in depleting CML LSCs. This suggests that in a BCR-ABL mouse model of CML, FoxO3 is involved in the acquisition of LSC resistance to imatinib therapy [97]. Nevertheless, FoxO proteins seem to have distinct roles in different types of leukemia. Deletion of FoxO3 promotes a myeloproliferative syndrome, characterized by increased levels of hematopoietic progenitors, hypersensitivity to cytokines and uncharacteristically high ROS levels [193].

The role of the redox system in CSCs is becoming an emerging area of cancer research. Taken together, these recent studies highlight the importance of metabolic pathways in tumorigenesis and maintenance of CSCs activity in which ROS levels control cell fate [173]. Low ROS concentrations appear to support cell proliferation and survival pathways. On the contrary, an excessive increase of ROS levels can reach a critical point and lead to cell death. A better understanding of these pathways will be critical in the design of new strategies to slow tumor progression and result in a positive clinical outcome.

13.3.3 Leukemia Stem Cells and Microenvironment

The overall behavior of tumor cells is the result of the combined effects of intrinsic genetic mutations and signaling through extrinsic factors mediated by the microenvironment. If oncogenes and tumor suppressors control growth and survival in a cell autonomous way, abnormal microenvironmental conditions such as hypoxia and low pH are essential in shaping the cell's ability to evade death and acquire resistance to therapy [102]. It is only beginning to emerge that through these microenvironmental signals and stresses, the cell niche induces drug resistance to CSCs and promote the selection of more pharmacologically resistant clones through the acquisition of secondary genetic mutations [194].

In bone marrow, we can distinguish two distinct microenvironments in the growth and development of normal and leukemia stem cells: osteoblastic and vascular [195]. The osteoblastic niche is constituted by osteoblasts, octeoclasts and stromal cells and is confined to the inner side of the bone cavity. It is here that long-term quiescent HSCs are localized [114, 196]. The vascular niche is formed by sinusoidal endothelial cells and $CD146^+$ mesenchymal progenitors surrounding blood vessels that support homing, growth, transendothelial movements and commitment of normal HSCs and aberrant LSCs [197]. Stromal cells in both osteoblastic and vascular niches play an essential role in the survival, growth and commitment of LSCs [198–200]. In this context, leukemic populations move towards CXCL12 positive vascular niches in the bone marrow and create an abnormal microenvironment altering the normal niche for HSCs. Additionally, normal human $CD34^+$ cells injected in leukemic, immunocompromised mice move through the cancer niche responding to stem-cell factor (SCF) produced by leukemic subsets [101].

The migration or homing of LSCs to tumor niches is the first step in leukemic-stromal interactions in the bone marrow. This process involves the stromal cell–derived factor-1 alpha (CXCL12) and its receptor CXCR4 on leukemic cells. In AML, CXCR4 expression is associated with poor prognosis [201] and is significantly higher in Flt3/internal tandem duplication (ITD) AML compared with FLT3/wild-type AML [202], suggesting that the relation between Flt3 and CXCR4 is crucial in the trafficking of leukemic cells.

The adhesion to bone marrow niches is the second essential step in the interaction of LSCs with the microenvironment and requires adhesion molecules such as the glycoprotein CD44. This glycoprotein is a transmembrane receptor that binds hyaluronan (HA) and is expressed as variant isoforms (CD44v) in many cell types [203]. In HSCs and LSCs, CD44 binding to HA activates multiple signaling pathways which promote migration, adhesion, apoptotic resistance and the expression of additional adhesion molecules (e.g., integrins) [204, 205]. CD44 also contributes to matrix assembly of endosteal region and stem cell niches where HA is particularly abundant [206–208]. In AML, CD44 is a primary regulator of homing and stem cell activity. *In vivo* administration of anti-CD44 antibody to leukemic mice impairs the maintenance of LSCs and induces differentiation, pointing to the relevance of interactions between AML LSCs and the niche in the preservation of stem cell properties [107]. In a mouse model of CML, BCR-ABL-expressing CD44 deficient LSCs show reduced homing and engraftment in the bone marrow with respect to CD44 wild-type LSCs and normal HSCs [108]. This suggests that in CML, CD44 can be considered a promising therapeutic target for eradication of LSCs without altering HSCs activity.

Other transmembrane proteins involved in LSC maintenance are CD123 and CD47 [209, 210]. Both surface markers control the growth of human AML LSCs. It has been reported that blocking CD123 activity by specific antibodies inhibits the homing of leukemic cells and activates an innate immune response [209]. In contrast, CD47 is constitutively expressed on mouse and human leukemias and protects the cells from phagocytosis, mediated by macrophages of the immune system [210, 211].

Recent evidence also suggests that a hypoxic bone marrow microenvironment represents a conditional niche in which *hypoxia-inducible factors* (HIFs) induce hypoxic stability and regulate the tumorigenic capacity of LSCs [29]. In the hematopoietic system, the oxygen level gradually decreases from the vascular to the osteoblastic niches to create hypoxic microenvironments [102]. HIF transcription factors mediate the response of cells to hypoxia and modulate the expression of genes, involved in processes such as angiogenesis, motility, survival, membrane integrity and cell metabolism [212]. HIF proteins are consist of two subunits: α (HIF-α) and β (HIF-β or ARNT). In the presence of high oxygen levels (>8–10%), HIF-α is rapidly degraded after hydroxylation, catalyzed by specific proline hydroxylase (PHD) enzymes. The proteosomal destruction of hydroxylated HIFα is also mediated by by an E3 ubiquitin ligase, the von Hippel-Lindau protein (pVHL) complex. Under conditions of low oxygen (<8%), the HIF- α subunit is stabilized, binding to the HIF-β protein and translocates to the nucleus to constitute an active transcriptional complex with other coactivators such as CBP/p300 [213]. HIF-α genes include HIF-1α and HIF-2α members that control most HIF target genes such as the vascular endothelial growth factor (VEGF), glycolytic enzymes (PGK, ALDA), glucose transporters (GLUT1), and proteins regulating motility (lysl oxidase), metastasis (CXCR4, E-cadherin) and extracellular matrix remodeling (LOX, MMP1) [214].

In the normal hypoxic endosteal niche, HSCs are quiescent through the precise regulation of HIF-1α levels and adapt their metabolism to the oxygen concentrations of their local environment [104, 215]. In cancer, hypoxic regions arise by intense cell proliferation and aberrant patterns in blood vessel conformation. In these hypoxic tumor regions, the HIF activity may contribute to maintain the undifferentiated state in CSCs as well as in normal stem and progenitors cells [212, 216]. It has been demonstrated that hypoxia and HIFs induce the expression of genes that promote self-renewal in normal stem cells and CSCs, including Sox2 [217], Notch [218, 219] and β-catenin [103]. In glioblastoma, HIF2α is specifically expressed in a CD133$^+$ subset enriched for CSCs, while HIF1α is present in both tumorigenic and non-tumorigenic cells. Furthermore, the deletion of HIF impairs the survival and tumor-initiating capacity of glioma stem cells *in vitro* and *in vivo*, suggesting that the response to hypoxia in some CSCs might involve different HIF patterns [220]. In hematological malignancies, recent evidence shows that the interaction between HIF1α and the Notch signaling pathway is essential for the maintenance of LSCs under normal oxygen levels in AML and in a mouse model of lymphoma, generated by insertional mutation of *Epm2a* gene[29]. In these tumors, the reduction of HIF activity by both RNA interference and the HIF-inhibitor echinomycin decreases the self-renewal capacity of LSCs *in vitro* and *in vivo* as monitored by the transplantability of human AML cells in immunocompromised mice. Wang et al. have also demonstrated that HIF1α reinforces the expression of Hes1, an important Notch target gene involved in stem and progenitor cell functions [118, 119, 221], antagonizing the Hes1 binding to the N-boxes in autoregulative process of Hes1 promoter [29].

In the last decades, cumulative evidence has shown that the tumor microenvironment plays an essential role in tumorigenesis. Leukemic niches control the maintenance of LSC in many ways, such as (1) promoting self-renewal and pro-survival pathways mediated by stromal cells, (2) facilitating the homing and adhesion of leukemic cells and (3) maintaining the hypoxic and metabolic conditions that stabilize the activity of specific stem populations. This interaction between LSCs and tumor microenvironments results in an increase of resistance to anti-leukemic treatments and to a subsequently poor outcome. Therefore, targeting leukemic niches by using new and more specific approaches might affect leukemic progression and effectively eradicate LSCs, leading to therapies that are more successful in achieving cure with reduced toxicity for the patient.

13.4 Conclusion

Following the CSCs hypothesis, LSCs constitute a minority subset of self-renewing leukemic cells that sustain tumor growth and may be responsible for tumor relapse, staying silently hidden in patients following conventional therapies [222]. In this chapter, we have reported the successful identification and functional characterization of LSCs in some hematopoietic malignancies. Not all leukemias appear to obey the CSC model with a representative hierarchical organization [20]. Although it is still controversial and contentious that the majority of cancers are the result of highly self-renewing cancer subsets, it has been proposed that CSCs might not always be rare in every tumor and may represent a significant fraction of tumor burden[223].

The CSC theory also predicts that an effective cure will require the complete eradication of LSCs without affecting the normal stem cells. Because LSCs and HSCs share many similar properties, the identification of efficient drug targets, unique to cancer stem cells will be the first step in the design of novel therapies. This challenging goal will require further understanding of normal and cancer stem cell biology as well as the processes involved in the regulation of tumor microenvironments. In fact, the mechanisms that cause LSCs to be resistant to current therapies are unclear but likely involve some intrinsic and extrinsic properties of LSCs such as relative quiescence, localization in protective and hypoxic niches and expression of multi-drug transporters and/or more effective DNA repair processes.

In the past few years, a limited number of efficient agents against LSCs have been identified. Parthenolide (PTL) was one of the first. This molecule targets the Nf-κB pathway that is active in AML LSCs and not in HSCs [224–226]. Using a different approach, researchers found that LSC homing to bone marrow niches was impaired by blocking CXCR4 activity with pharmacologic inhibitors. In preclinical models of CLL [227], ALL [228] and AML [229, 230] the CXCR4 inhibition sensitizes leukemic cells to chemotherapy, impairing their displacement to more protective microenvironments. Other strategies to eradicate LSCs have attempted to use antibodies against specific transmembrane receptors such as CD44 [107], CD47 [210, 211] and CD123 [209] to prevent LSC homing and/or to activate an immune response. Finally, new data has emerged showing that targeting HIF1 activity by the HIF inhibitor echinomycin may represent a selective and more effective way to deplete LSCs in AML, reducing the side effects of cancer therapy [29]. Taken together, these novel experimental agents and contemporary developments that have evolved our understanding of LSCs mechanisms raise the encouraging possibility that effective and less toxic cures against hematological malignances could be designed and applied in the near future.

Acknowledgements We would like to thank Dr. Andrew P. Weng for supporting us, Melissa Howard for numerous discussions and help in the editing of manuscript, Sonya Lam and Olena Shevchuk for their collaboration and assistance in the lab.

References

1. Hope KJ et al (2004) Acute myeloid leukemia originates from a hierarchy of leukemic stem cell classes that differ in self-renewal capacity. Nat Immunol 5:738–743
2. Blagosklonny MV (2005) Carcinogenesis, cancer therapy and chemoprevention. Cell Death Differ 12:592–602
3. Dick JE (2008) Stem cell concepts renew cancer research. Blood 112:4793–4807
4. Nowell PC (1976) The clonal evolution of tumor cell populations. Science 194:23–28
5. Bruce WR, Van Der Gaag H (1963) A quantitative assay for the number of murine lymphoma cells capable of proliferation in vivo. Nature 199:79–80
6. Lapidot T et al (1994) A cell initiating human acute myeloid leukaemia after transplantation into SCID mice. Nature 367:645–648
7. Bonnet D, Dick JE (1997) Human acute myeloid leukemia is organized as a hierarchy that originates from a primitive hematopoietic cell. Nat Med 3:730–737

8. Barabe F et al (2007) Modeling the initiation and progression of human acute leukemia in mice. Science 316:600–604
9. Notta F et al (2011) Evolution of human BCR-ABL1 lymphoblastic leukaemia-initiating cells. Nature 469:362–367
10. Clark EA et al (2000) Genomic analysis of metastasis reveals an essential role for RhoC. Nature 406:532–535
11. Visvader JE, Lindeman GJ (2008) Cancer stem cells in solid tumours: accumulating evidence and unresolved questions. Nat Rev Cancer 8:755–768
12. Jamieson CH et al (2004) Granulocyte-macrophage progenitors as candidate leukemic stem cells in blast-crisis CML. N Engl J Med 351:657–667
13. Krivtsov AV et al (2006) Transformation from committed progenitor to leukaemia stem cell initiated by MLL-AF9. Nature 442:818–822
14. Somervaille TC, Cleary ML (2006) Identification and characterization of leukemia stem cells in murine MLL-AF9 acute myeloid leukemia. Cancer Cell 10:257–268
15. Chen W et al (2008) Malignant transformation initiated by Mll-AF9: gene dosage and critical target cells. Cancer Cell 13:432–440
16. Quintana E et al (2008) Efficient tumour formation by single human melanoma cells. Nature 456:593–598
17. Quintana E et al (2010) Phenotypic heterogeneity among tumorigenic melanoma cells from patients that is reversible and not hierarchically organized. Cancer Cell 18:510–523
18. Huntly BJ et al (2004) MOZ-TIF2, but not BCR-ABL, confers properties of leukemic stem cells to committed murine hematopoietic progenitors. Cancer Cell 6:587–596
19. Yilmaz OH et al (2006) Pten dependence distinguishes haematopoietic stem cells from leukaemia-initiating cells. Nature 441:475–482
20. Kelly PN et al (2007) Tumor growth need not be driven by rare cancer stem cells. Science 317:337
21. Jordan CT et al (2000) The interleukin-3 receptor alpha chain is a unique marker for human acute myelogenous leukemia stem cells. Leukemia 14:1777–1784
22. Blair A et al (1998) Most acute myeloid leukemia progenitor cells with long-term proliferative ability in vitro and in vivo have the phenotype CD34(+)/CD71(−)/HLA-DR. Blood 92:4325–4335
23. Blair A et al (1997) Lack of expression of Thy-1 (CD90) on acute myeloid leukemia cells with long-term proliferative ability in vitro and in vivo. Blood 89:3104–3112
24. Blair A, Sutherland HJ (2000) Primitive acute myeloid leukemia cells with long-term proliferative ability in vitro and in vivo lack surface expression of c-kit (CD117). Exp Hematol 28:660–671
25. Dykstra B et al (2007) Long-term propagation of distinct hematopoietic differentiation programs in vivo. Cell Stem Cell 1:218–229
26. Taussig DC et al (2008) Anti-CD38 antibody-mediated clearance of human repopulating cells masks the heterogeneity of leukemia-initiating cells. Blood 112:568–575
27. Taussig DC et al (2010) Leukemia-initiating cells from some acute myeloid leukemia patients with mutated nucleophosmin reside in the CD34(−) fraction. Blood 115:1976–1984
28. Eppert K et al (2011) Stem cell gene expression programs influence clinical outcome in human leukemia. Nat Med 17(9):1086–1093
29. Wang Y et al (2011) Targeting HIF1alpha eliminates cancer stem cells in hematological malignancies. Cell Stem Cell 8:399–411
30. Wang Y et al (2010) The Wnt/beta-catenin pathway is required for the development of leukemia stem cells in AML. Science 327:1650–1653
31. Chi Y et al (2008) Acute myelogenous leukemia with t(6;9)(p23;q34) and marrow basophilia: an overview. Arch Pathol Lab Med 132:1835–1837
32. Oancea C et al (2010) The t(6;9) associated DEK/CAN fusion protein targets a population of long-term repopulating hematopoietic stem cells for leukemogenic transformation. Leukemia 24:1910–1919
33. Wojiski S et al (2009) PML-RARalpha initiates leukemia by conferring properties of self-renewal to committed promyelocytic progenitors. Leukemia 23:1462–1471

34. Guibal FC et al (2009) Identification of a myeloid committed progenitor as the cancer-initiating cell in acute promyelocytic leukemia. Blood 114:5415–5425
35. Saito Y et al (2010) Induction of cell cycle entry eliminates human leukemia stem cells in a mouse model of AML. Nat Biotechnol 28:275–280
36. Vardiman JW et al (2002) The World Health Organization (WHO) classification of the myeloid neoplasms. Blood 100:2292–2302
37. Holyoake T et al (1999) Isolation of a highly quiescent subpopulation of primitive leukemic cells in chronic myeloid leukemia. Blood 94:2056–2064
38. Jorgensen HG, Holyoake TL (2007) Characterization of cancer stem cells in chronic myeloid leukaemia. Biochem Soc Trans 35:1347–1351
39. Lemoli RM et al (2009) Molecular and functional analysis of the stem cell compartment of chronic myelogenous leukemia reveals the presence of a CD34- cell population with intrinsic resistance to imatinib. Blood 114:5191–5200
40. Minami Y et al (2008) BCR-ABL-transformed GMP as myeloid leukemic stem cells. Proc Natl Acad Sci USA 105:17967–17972
41. Jaras M et al (2010) Isolation and killing of candidate chronic myeloid leukemia stem cells by antibody targeting of IL-1 receptor accessory protein. Proc Natl Acad Sci USA 107:16280–16285
42. Koschmieder S et al (2005) Inducible chronic phase of myeloid leukemia with expansion of hematopoietic stem cells in a transgenic model of BCR-ABL leukemogenesis. Blood 105:324–334
43. Li X et al (2004) Simultaneous demonstration of clonal chromosome abnormalities and apoptosis in individual marrow cells in myelodysplastic syndrome. Int J Hematol 80:140–145
44. Giagounidis AA et al (2006) Biological and prognostic significance of chromosome 5q deletions in myeloid malignancies. Clin Cancer Res 12:5–10
45. Ebert BL et al (2008) Identification of RPS14 as a 5q- syndrome gene by RNA interference screen. Nature 451:335–339
46. Starczynowski DT et al (2010) Identification of miR-145 and miR-146a as mediators of the 5q- syndrome phenotype. Nat Med 16:49–58
47. Nilsson L et al (2000) Isolation and characterization of hematopoietic progenitor/stem cells in 5q-deleted myelodysplastic syndromes: evidence for involvement at the hematopoietic stem cell level. Blood 96:2012–2021
48. Tehranchi R et al (2010) Persistent malignant stem cells in del(5q) myelodysplasia in remission. N Engl J Med 363:1025–1037
49. Kerbauy DM et al (2004) Engraftment of distinct clonal MDS-derived hematopoietic precursors in NOD/SCID-beta2-microglobulin-deficient mice after intramedullary transplantation of hematopoietic and stromal cells. Blood 104:2202–2203
50. Martin MG et al (2010) Limited engraftment of low-risk myelodysplastic syndrome cells in NOD/SCID gamma-C chain knockout mice. Leukemia 24:1662–1664
51. Thanopoulou E et al (2004) Engraftment of NOD/SCID-beta2 microglobulin null mice with multilineage neoplastic cells from patients with myelodysplastic syndrome. Blood 103:4285–4293
52. Cox CV et al (2007) Characterization of a progenitor cell population in childhood T-cell acute lymphoblastic leukemia. Blood 109:674–682
53. Weng AP et al (2004) Activating mutations of NOTCH1 in human T cell acute lymphoblastic leukemia. Science 306:269–271
54. Malyukova A et al (2007) The tumor suppressor gene hCDC4 is frequently mutated in human T-cell acute lymphoblastic leukemia with functional consequences for Notch signaling. Cancer Res 67:5611–5616
55. O'Neil J et al (2007) FBW7 mutations in leukemic cells mediate NOTCH pathway activation and resistance to gamma-secretase inhibitors. J Exp Med 204:1813–1824
56. Thompson BJ et al (2007) The SCFFBW7 ubiquitin ligase complex as a tumor suppressor in T cell leukemia. J Exp Med 204:1825–1835
57. Teitell MA, Pandolfi PP (2009) Molecular genetics of acute lymphoblastic leukemia. Annu Rev Pathol 4:175–198

58. McCormack MP et al (2010) The Lmo2 oncogene initiates leukemia in mice by inducing thymocyte self-renewal. Science 327:879–883
59. Chiu PP et al (2010) Leukemia-initiating cells in human T-lymphoblastic leukemia exhibit glucocorticoid resistance. Blood 116:5268–5279
60. Guo W et al (2008) Multi-genetic events collaboratively contribute to Pten-null leukaemia stem-cell formation. Nature 453:529–533
61. McCormack MP, Curtis DJ (2010) The thymus under siege: Lmo2 induces precancerous stem cells in a mouse model of T-ALL. Cell Cycle 9:2267–2268
62. Tremblay M et al (2010) Modeling T-cell acute lymphoblastic leukemia induced by the SCL and LMO1 oncogenes. Genes Dev 24:1093–1105
63. Gleissner B et al (2002) Leading prognostic relevance of the BCR-ABL translocation in adult acute B-lineage lymphoblastic leukemia: a prospective study of the German Multicenter Trial Group and confirmed polymerase chain reaction analysis. Blood 99:1536–1543
64. Ribeiro RC et al (1987) Clinical and biologic hallmarks of the Philadelphia chromosome in childhood acute lymphoblastic leukemia. Blood 70:948–953
65. Chan LC et al (1987) A novel abl protein expressed in Philadelphia chromosome positive acute lymphoblastic leukaemia. Nature 325:635–637
66. Cobaleda C et al (2000) A primitive hematopoietic cell is the target for the leukemic transformation in human philadelphia-positive acute lymphoblastic leukemia. Blood 95:1007–1013
67. Cox CV et al (2004) Characterization of acute lymphoblastic leukemia progenitor cells. Blood 104:2919–2925
68. le Viseur C et al (2008) In childhood acute lymphoblastic leukemia, blasts at different stages of immunophenotypic maturation have stem cell properties. Cancer Cell 14:47–58
69. Cox CV et al (2009) Expression of CD133 on leukemia-initiating cells in childhood ALL. Blood 113:3287–3296
70. Hong D et al (2008) Initiating and cancer-propagating cells in TEL-AML1-associated childhood leukemia. Science 319:336–339
71. Morrow M et al (2004) TEL-AML1 promotes development of specific hematopoietic lineages consistent with preleukemic activity. Blood 103:3890–3896
72. Bateman CM et al (2010) Acquisition of genome-wide copy number alterations in monozygotic twins with acute lymphoblastic leukemia. Blood 115:3553–3558
73. Anderson K et al (2011) Genetic variegation of clonal architecture and propagating cells in leukaemia. Nature 469:356–361
74. Hotfilder M et al (2005) Leukemic stem cells in childhood high-risk ALL/t(9;22) and t(4;11) are present in primitive lymphoid-restricted CD34+CD19- cells. Cancer Res 65:1442–1449
75. Menendez P et al (2009) Bone marrow mesenchymal stem cells from infants with MLL-AF4+ acute leukemia harbor and express the MLL-AF4 fusion gene. J Exp Med 206:3131–3141
76. Varnum-Finney B et al (2000) Pluripotent, cytokine-dependent, hematopoietic stem cells are immortalized by constitutive Notch1 signaling. Nat Med 6:1278–1281
77. Karanu FN et al (2000) The notch ligand jagged-1 represents a novel growth factor of human hematopoietic stem cells. J Exp Med 192:1365–1372
78. Armstrong F et al (2009) NOTCH is a key regulator of human T-cell acute leukemia initiating cell activity. Blood 113:1730–1740
79. Maillard I et al (2008) Canonical notch signaling is dispensable for the maintenance of adult hematopoietic stem cells. Cell Stem Cell 2:356–366
80. Stier S et al (2002) Notch1 activation increases hematopoietic stem cell self-renewal in vivo and favors lymphoid over myeloid lineage outcome. Blood 99:2369–2378
81. Duncan AW et al (2005) Integration of Notch and Wnt signaling in hematopoietic stem cell maintenance. Nat Immunol 6:314–322
82. Zhao C et al (2007) Loss of beta-catenin impairs the renewal of normal and CML stem cells in vivo. Cancer Cell 12:528–541
83. Reya T et al (2003) A role for Wnt signalling in self-renewal of haematopoietic stem cells. Nature 423:409–414

84. Zheng X et al (2004) Gamma-catenin contributes to leukemogenesis induced by AML-associated translocation products by increasing the self-renewal of very primitive progenitor cells. Blood 103:3535–3543
85. Yeung J et al (2010) Beta-catenin mediates the establishment and drug resistance of MLL leukemic stem cells. Cancer Cell 18:606–618
86. Zhang Y, Kalderon D (2001) Hedgehog acts as a somatic stem cell factor in the Drosophila ovary. Nature 410:599–604
87. Gao J et al (2009) Hedgehog signaling is dispensable for adult hematopoietic stem cell function. Cell Stem Cell 4:548–558
88. Zhao C et al (2009) Hedgehog signalling is essential for maintenance of cancer stem cells in myeloid leukaemia. Nature 458:776–779
89. Lessard J, Sauvageau G (2003) Bmi-1 determines the proliferative capacity of normal and leukaemic stem cells. Nature 423:255–260
90. Park IK et al (2003) Bmi-1 is required for maintenance of adult self-renewing haematopoietic stem cells. Nature 423:302–305
91. Sauvageau G et al (1995) Overexpression of HOXB4 in hematopoietic cells causes the selective expansion of more primitive populations in vitro and in vivo. Genes Dev 9:1753–1765
92. Antonchuk J et al (2002) HOXB4-induced expansion of adult hematopoietic stem cells ex vivo. Cell 109:39–45
93. Heuser M et al (2009) Modeling the functional heterogeneity of leukemia stem cells: role of STAT5 in leukemia stem cell self-renewal. Blood 114:3983–3993
94. Ito K et al (2006) Reactive oxygen species act through p38 MAPK to limit the lifespan of hematopoietic stem cells. Nat Med 12:446–451
95. Owusu-Ansah E, Banerjee U (2009) Reactive oxygen species prime Drosophila haematopoietic progenitors for differentiation. Nature 461:537–541
96. Tothova Z et al (2007) FoxOs are critical mediators of hematopoietic stem cell resistance to physiologic oxidative stress. Cell 128:325–339
97. Naka K et al (2010) TGF-beta-FOXO signalling maintains leukaemia-initiating cells in chronic myeloid leukaemia. Nature 463:676–680
98. Gross S et al (2010) Cancer-associated metabolite 2-hydroxyglutarate accumulates in acute myelogenous leukemia with isocitrate dehydrogenase 1 and 2 mutations. J Exp Med 207:339–344
99. Ward PS et al (2010) The common feature of leukemia-associated IDH1 and IDH2 mutations is a neomorphic enzyme activity converting alpha-ketoglutarate to 2-hydroxyglutarate. Cancer Cell 17:225–234
100. Figueroa ME et al (2010) Leukemic IDH1 and IDH2 mutations result in a hypermethylation phenotype, disrupt TET2 function, and impair hematopoietic differentiation. Cancer Cell 18:553–567
101. Colmone A et al (2008) Leukemic cells create bone marrow niches that disrupt the behavior of normal hematopoietic progenitor cells. Science 322:1861–1865
102. Konopleva MY, Jordan CT (2011) Leukemia stem cells and microenvironment: biology and therapeutic targeting. J Clin Oncol 29:591–599
103. Mazumdar J et al (2010) O2 regulates stem cells through Wnt/beta-catenin signalling. Nat Cell Biol 12:1007–1013
104. Takubo K et al (2010) Regulation of the HIF-1alpha level is essential for hematopoietic stem cells. Cell Stem Cell 7:391–402
105. Nagasawa T et al (1996) Defects of B-cell lymphopoiesis and bone-marrow myelopoiesis in mice lacking the CXC chemokine PBSF/SDF-1. Nature 382:635–638
106. Tavor S et al (2004) CXCR4 regulates migration and development of human acute myelogenous leukemia stem cells in transplanted NOD/SCID mice. Cancer Res 64:2817–2824
107. Jin L et al (2006) Targeting of CD44 eradicates human acute myeloid leukemic stem cells. Nat Med 12:1167–1174
108. Krause DS et al (2006) Requirement for CD44 in homing and engraftment of BCR-ABL-expressing leukemic stem cells. Nat Med 12:1175–1180

109. Pardal R et al (2005) Stem cell self-renewal and cancer cell proliferation are regulated by common networks that balance the activation of proto-oncogenes and tumor suppressors. Cold Spring Harb Symp Quant Biol 70:177–185
110. Radtke F et al (2010) Notch signaling in the immune system. Immunity 32:14–27
111. Shen Q et al (2004) Endothelial cells stimulate self-renewal and expand neurogenesis of neural stem cells. Science 304:1338–1340
112. Hitoshi S et al (2002) Notch pathway molecules are essential for the maintenance, but not the generation, of mammalian neural stem cells. Genes Dev 16:846–858
113. Kunisato A et al (2003) HES-1 preserves purified hematopoietic stem cells ex vivo and accumulates side population cells in vivo. Blood 101:1777–1783
114. Calvi LM et al (2003) Osteoblastic cells regulate the haematopoietic stem cell niche. Nature 425:841–846
115. Mancini SJ et al (2005) Jagged1-dependent Notch signaling is dispensable for hematopoietic stem cell self-renewal and differentiation. Blood 105:2340–2342
116. Malecki MJ et al (2006) Leukemia-associated mutations within the NOTCH1 heterodimerization domain fall into at least two distinct mechanistic classes. Mol Cell Biol 26:4642–4651
117. Chiang MY et al (2006) Identification of a conserved negative regulatory sequence that influences the leukemogenic activity of NOTCH1. Mol Cell Biol 26:6261–6271
118. Tomita K et al (1999) The bHLH gene Hes1 is essential for expansion of early T cell precursors. Genes Dev 13:1203–1210
119. Yu X et al (2006) HES1 inhibits cycling of hematopoietic progenitor cells via DNA binding. Stem Cells 24:876–888
120. Wilson A, Radtke F (2006) Multiple functions of Notch signaling in self-renewing organs and cancer. FEBS Lett 580:2860–2868
121. Wend P et al (2010) Wnt signaling in stem and cancer stem cells. Semin Cell Dev Biol 21:855–863
122. Reya T, Clevers H (2005) Wnt signalling in stem cells and cancer. Nature 434:843–850
123. Zhu L et al (2009) Prominin 1 marks intestinal stem cells that are susceptible to neoplastic transformation. Nature 457:603–607
124. Vermeulen L et al (2010) Wnt activity defines colon cancer stem cells and is regulated by the microenvironment. Nat Cell Biol 12:468–476
125. Malanchi I et al (2008) Cutaneous cancer stem cell maintenance is dependent on beta-catenin signalling. Nature 452:650–653
126. Zhang M et al (2010) Selective targeting of radiation-resistant tumor-initiating cells. Proc Natl Acad Sci USA 107:3522–3527
127. Scheller M et al (2006) Hematopoietic stem cell and multilineage defects generated by constitutive beta-catenin activation. Nat Immunol 7:1037–1047
128. Murdoch B et al (2003) Wnt-5A augments repopulating capacity and primitive hematopoietic development of human blood stem cells in vivo. Proc Natl Acad Sci USA 100:3422–3427
129. Fleming HE et al (2008) Wnt signaling in the niche enforces hematopoietic stem cell quiescence and is necessary to preserve self-renewal in vivo. Cell Stem Cell 2:274–283
130. Koch U et al (2008) Simultaneous loss of beta- and gamma-catenin does not perturb hematopoiesis or lymphopoiesis. Blood 111:160–164
131. Heretsch P et al (2010) Modulators of the hedgehog signaling pathway. Bioorg Med Chem 18:6613–6624
132. Trowbridge JJ et al (2006) Hedgehog modulates cell cycle regulators in stem cells to control hematopoietic regeneration. Proc Natl Acad Sci USA 103:14134–14139
133. Bhardwaj G et al (2001) Sonic hedgehog induces the proliferation of primitive human hematopoietic cells via BMP regulation. Nat Immunol 2:172–180
134. Dohle E et al (2010) Sonic hedgehog promotes angiogenesis and osteogenesis in a coculture system consisting of primary osteoblasts and outgrowth endothelial cells. Tissue Eng Part A 16:1235–1237

135. Teglund S, Toftgard R (2010) Hedgehog beyond medulloblastoma and basal cell carcinoma. Biochim Biophys Acta 1805:181–208
136. Varnat F et al (2009) Human colon cancer epithelial cells harbour active HEDGEHOG-GLI signalling that is essential for tumour growth, recurrence, metastasis and stem cell survival and expansion. EMBO Mol Med 1:338–351
137. Yauch RL et al (2008) A paracrine requirement for hedgehog signalling in cancer. Nature 455:406–410
138. Hsieh A et al (2011) Hedgehog/GLI1 regulates IGF dependent malignant behaviors in glioma stem cells. J Cell Physiol 226:1118–1127
139. Peacock CD et al (2007) Hedgehog signaling maintains a tumor stem cell compartment in multiple myeloma. Proc Natl Acad Sci USA 104:4048–4053
140. Lin TL et al (2010) Self-renewal of acute lymphocytic leukemia cells is limited by the Hedgehog pathway inhibitors cyclopamine and IPI-926. PLoS One 5:e15262
141. Dierks C et al (2008) Expansion of Bcr-Abl-positive leukemic stem cells is dependent on Hedgehog pathway activation. Cancer Cell 14:238–249
142. Sauvageau M, Sauvageau G (2010) Polycomb group proteins: multi-faceted regulators of somatic stem cells and cancer. Cell Stem Cell 7:299–313
143. Schwartz YB, Pirrotta V (2007) Polycomb silencing mechanisms and the management of genomic programmes. Nat Rev Genet 8:9–22
144. Levine SS et al (2002) The core of the polycomb repressive complex is compositionally and functionally conserved in flies and humans. Mol Cell Biol 22:6070–6078
145. Valk-Lingbeek ME et al (2004) Stem cells and cancer; the polycomb connection. Cell 118:409–418
146. Cao R et al (2002) Role of histone H3 lysine 27 methylation in Polycomb-group silencing. Science 298:1039–1043
147. Czermin B et al (2002) Drosophila enhancer of Zeste/ESC complexes have a histone H3 methyltransferase activity that marks chromosomal Polycomb sites. Cell 111:185–196
148. Kirmizis A et al (2004) Silencing of human polycomb target genes is associated with methylation of histone H3 Lys 27. Genes Dev 18:1592–1605
149. Kuzmichev A et al (2002) Histone methyltransferase activity associated with a human multiprotein complex containing the enhancer of zeste protein. Genes Dev 16:2893–2905
150. Guo WJ et al (2007) Mel-18 acts as a tumor suppressor by repressing Bmi-1 expression and down-regulating Akt activity in breast cancer cells. Cancer Res 67:5083–5089
151. Tetsu O et al (1998) Mel-18 negatively regulates cell cycle progression upon B cell antigen receptor stimulation through a cascade leading to c-myc/cdc25. Immunity 9:439–448
152. van der Lugt NM et al (1994) Posterior transformation, neurological abnormalities, and severe hematopoietic defects in mice with a targeted deletion of the bmi-1 proto-oncogene. Genes Dev 8:757–769
153. Iwama A et al (2004) Enhanced self-renewal of hematopoietic stem cells mediated by the polycomb gene product Bmi-1. Immunity 21:843–851
154. Kajiume T et al (2009) Reciprocal expression of Bmi1 and Mel-18 is associated with functioning of primitive hematopoietic cells. Exp Hematol 37:857–866 e2
155. Jacobs JJ et al (1999) The oncogene and Polycomb-group gene bmi-1 regulates cell proliferation and senescence through the ink4a locus. Nature 397:164–168
156. Oguro H et al (2006) Differential impact of Ink4a and Arf on hematopoietic stem cells and their bone marrow microenvironment in Bmi1-deficient mice. J Exp Med 203:2247–2253
157. Jacobs JJ et al (1999) Bmi-1 collaborates with c-Myc in tumorigenesis by inhibiting c-Myc-induced apoptosis via INK4a/ARF. Genes Dev 13:2678–2690
158. van Lohuizen M et al (1991) Identification of cooperating oncogenes in E mu-myc transgenic mice by provirus tagging. Cell 65:737–752
159. Haupt Y et al (1991) Novel zinc finger gene implicated as myc collaborator by retrovirally accelerated lymphomagenesis in E mu-myc transgenic mice. Cell 65:753–763
160. Martin-Perez D et al (2010) Polycomb proteins in hematologic malignancies. Blood 116:5465–5475

161. Lee TI et al (2006) Control of developmental regulators by Polycomb in human embryonic stem cells. Cell 125:301–313
162. Boyer LA et al (2006) Polycomb complexes repress developmental regulators in murine embryonic stem cells. Nature 441:349–353
163. Kim JY et al (2004) Defective long-term repopulating ability in hematopoietic stem cells lacking the Polycomb-group gene rae28. Eur J Haematol 73:75–84
164. Chiba T et al (2008) The polycomb gene product BMI1 contributes to the maintenance of tumor-initiating side population cells in hepatocellular carcinoma. Cancer Res 68:7742–7749
165. Rizo A et al (2009) Repression of BMI1 in normal and leukemic human CD34(+) cells impairs self-renewal and induces apoptosis. Blood 114:1498–1505
166. Yuan J et al (2011) Bmi1 is essential for leukemic reprogramming of myeloid progenitor cells. Leukemia 25(8):1335–1343
167. Rizo A et al (2010) BMI1 collaborates with BCR-ABL in leukemic transformation of human CD34+ cells. Blood 116:4621–4630
168. Mihara K et al (2006) Bmi-1 is useful as a novel molecular marker for predicting progression of myelodysplastic syndrome and patient prognosis. Blood 107:305–308
169. Chowdhury M et al (2007) Expression of Polycomb-group (PcG) protein BMI-1 predicts prognosis in patients with acute myeloid leukemia. Leukemia 21:1116–1122
170. Mohty M et al (2007) The polycomb group BMI1 gene is a molecular marker for predicting prognosis of chronic myeloid leukemia. Blood 110:380–383
171. Hanahan D, Weinberg RA (2000) The hallmarks of cancer. Cell 100:57–70
172. Hanahan D, Weinberg RA (2011) Hallmarks of cancer: the next generation. Cell 144:646–674
173. Cairns RA et al (2011) Regulation of cancer cell metabolism. Nat Rev Cancer 11:85–95
174. Vander Heiden MG et al (2009) Understanding the Warburg effect: the metabolic requirements of cell proliferation. Science 324:1029–1033
175. Warburg O (1956) On the origin of cancer cells. Science 123:309–314
176. Finkel T, Holbrook NJ (2000) Oxidants, oxidative stress and the biology of ageing. Nature 408:239–247
177. Diehn M et al (2009) Association of reactive oxygen species levels and radioresistance in cancer stem cells. Nature 458:780–783
178. Kobayashi CI, Suda T (2011) Regulation of reactive oxygen species in stem cells and cancer stem cells. J Cell Physiol. doi:10.1002/jcp. 22764
179. Balaban RS et al (2005) Mitochondria, oxidants, and aging. Cell 120:483–495
180. Katsuyama M (2010) NOX/NADPH oxidase, the superoxide-generating enzyme: its transcriptional regulation and physiological roles. J Pharmacol Sci 114:134–146
181. Storz P (2011) Forkhead homeobox type O transcription factors in the responses to oxidative stress. Antioxid Redox Signal 14:593–605
182. Miyamoto K et al (2007) Foxo3a is essential for maintenance of the hematopoietic stem cell pool. Cell Stem Cell 1:101–112
183. Tsai WB et al (2008) Functional interaction between FOXO3a and ATM regulates DNA damage response. Nat Cell Biol 10:460–467
184. Yalcin S et al (2008) Foxo3 is essential for the regulation of ataxia telangiectasia mutated and oxidative stress-mediated homeostasis of hematopoietic stem cells. J Biol Chem 283:25692–25705
185. Gan B et al (2010) Lkb1 regulates quiescence and metabolic homeostasis of haematopoietic stem cells. Nature 468:701–704
186. Gurumurthy S et al (2010) The Lkb1 metabolic sensor maintains haematopoietic stem cell survival. Nature 468:659–663
187. Nakada D et al (2010) Lkb1 regulates cell cycle and energy metabolism in haematopoietic stem cells. Nature 468:653–658
188. Ishimoto T et al (2011) CD44 variant regulates redox status in cancer cells by stabilizing the xCT subunit of system xc(−) and thereby promotes tumor growth. Cancer Cell 19:387–400

189. Guzman ML et al (2005) The sesquiterpene lactone parthenolide induces apoptosis of human acute myelogenous leukemia stem and progenitor cells. Blood 105:4163–4169
190. Ito K et al (2008) PML targeting eradicates quiescent leukaemia-initiating cells. Nature 453:1072–1078
191. Kim YR et al (2010) Myeloperoxidase expression as a potential determinant of parthenolide-induced apoptosis in leukemia bulk and leukemia stem cells. J Pharmacol Exp Ther 335:389–400
192. Chou WC, Dang CV (2005) Acute promyelocytic leukemia: recent advances in therapy and molecular basis of response to arsenic therapies. Curr Opin Hematol 12:1–6
193. Yalcin S et al (2010) ROS-mediated amplification of AKT/mTOR signalling pathway leads to myeloproliferative syndrome in Foxo3(−/−) mice. EMBO J 29:4118–4131
194. Meads MB et al (2009) Environment-mediated drug resistance: a major contributor to minimal residual disease. Nat Rev Cancer 9:665–674
195. Perry JM, Li L (2007) Disrupting the stem cell niche: good seeds in bad soil. Cell 129:1045–1047
196. Zhang J et al (2003) Identification of the haematopoietic stem cell niche and control of the niche size. Nature 425:836–841
197. Kopp HG et al (2005) The bone marrow vascular niche: home of HSC differentiation and mobilization. Physiology (Bethesda) 20:349–356
198. Wei J et al (2008) Microenvironment determines lineage fate in a human model of MLL-AF9 leukemia. Cancer Cell 13:483–495
199. Nilsson SK et al (2005) Osteopontin, a key component of the hematopoietic stem cell niche and regulator of primitive hematopoietic progenitor cells. Blood 106:1232–1239
200. Naveiras O, Daley GQ (2006) Stem cells and their niche: a matter of fate. Cell Mol Life Sci 63:760–766
201. Konoplev S et al (2007) Overexpression of CXCR4 predicts adverse overall and event-free survival in patients with unmutated FLT3 acute myeloid leukemia with normal karyotype. Cancer 109:1152–1156
202. Rombouts EJ et al (2004) Relation between CXCR-4 expression, Flt3 mutations, and unfavorable prognosis of adult acute myeloid leukemia. Blood 104:550–557
203. Zoller M (2011) CD44: can a cancer-initiating cell profit from an abundantly expressed molecule? Nat Rev Cancer 11:254–267
204. Lapidot T et al (2005) How do stem cells find their way home? Blood 106:1901–1910
205. Lundell BI et al (1997) Activation of beta1 integrins on CML progenitors reveals cooperation between beta1 integrins and CD44 in the regulation of adhesion and proliferation. Leukemia 11:822–829
206. Stern R (2008) Association between cancer and "acid mucopolysaccharides": an old concept comes of age, finally. Semin Cancer Biol 18:238–243
207. Girish KS, Kemparaju K (2007) The magic glue hyaluronan and its eraser hyaluronidase: a biological overview. Life Sci 80:1921–1943
208. Avigdor A et al (2004) CD44 and hyaluronic acid cooperate with SDF-1 in the trafficking of human CD34+ stem/progenitor cells to bone marrow. Blood 103:2981–2989
209. Jin L et al (2009) Monoclonal antibody-mediated targeting of CD123, IL-3 receptor alpha chain, eliminates human acute myeloid leukemic stem cells. Cell Stem Cell 5:31–42
210. Majeti R et al (2009) CD47 is an adverse prognostic factor and therapeutic antibody target on human acute myeloid leukemia stem cells. Cell 138:286–299
211. Jaiswal S et al (2009) CD47 is upregulated on circulating hematopoietic stem cells and leukemia cells to avoid phagocytosis. Cell 138:271–285
212. Keith B, Simon MC (2007) Hypoxia-inducible factors, stem cells, and cancer. Cell 129:465–472
213. Pouyssegur J et al (2006) Hypoxia signalling in cancer and approaches to enforce tumour regression. Nature 441:437–443
214. Majmundar AJ et al (2010) Hypoxia-inducible factors and the response to hypoxic stress. Mol Cell 40:294–309

215. Simsek T et al (2010) The distinct metabolic profile of hematopoietic stem cells reflects their location in a hypoxic niche. Cell Stem Cell 7:380–390
216. Yoshida Y et al (2009) Hypoxia enhances the generation of induced pluripotent stem cells. Cell Stem Cell 5:237–241
217. McCord AM et al (2009) Physiologic oxygen concentration enhances the stem-like properties of CD133+ human glioblastoma cells in vitro. Mol Cancer Res 7:489–497
218. Gustafsson MV et al (2005) Hypoxia requires notch signaling to maintain the undifferentiated cell state. Dev Cell 9:617–628
219. Chen Y et al (2007) Oxygen concentration determines the biological effects of NOTCH-1 signaling in adenocarcinoma of the lung. Cancer Res 67:7954–7959
220. Li Z et al (2009) Hypoxia-inducible factors regulate tumorigenic capacity of glioma stem cells. Cancer Cell 15:501–513
221. Janzen V et al (2006) Stem-cell ageing modified by the cyclin-dependent kinase inhibitor p16INK4a. Nature 443:421–426
222. Clarke MF et al (2006) Cancer stem cells–perspectives on current status and future directions: AACR workshop on cancer stem cells. Cancer Res 66:9339–9344
223. Trumpp A, Wiestler OD (2008) Mechanisms of disease: cancer stem cells–targeting the evil twin. Nat Clin Pract Oncol 5:337–347
224. Guzman ML et al (2001) Nuclear factor-kappaB is constitutively activated in primitive human acute myelogenous leukemia cells. Blood 98:2301–2307
225. Guzman ML et al (2007) An orally bioavailable parthenolide analog selectively eradicates acute myelogenous leukemia stem and progenitor cells. Blood 110:4427–4435
226. Guzman ML et al (2007) Rapid and selective death of leukemia stem and progenitor cells induced by the compound 4-benzyl, 2-methyl, 1,2,4-thiadiazolidine, 3,5 dione (TDZD-8). Blood 110:4436–4444
227. Burger M et al (2005) Small peptide inhibitors of the CXCR4 chemokine receptor (CD184) antagonize the activation, migration, and antiapoptotic responses of CXCL12 in chronic lymphocytic leukemia B cells. Blood 106:1824–1830
228. Juarez J et al (2003) Effects of inhibitors of the chemokine receptor CXCR4 on acute lymphoblastic leukemia cells in vitro. Leukemia 17:1294–1300
229. Nervi B et al (2009) Chemosensitization of acute myeloid leukemia (AML) following mobilization by the CXCR4 antagonist AMD3100. Blood 113:6206–6214
230. Zeng Z et al (2009) Targeting the leukemia microenvironment by CXCR4 inhibition overcomes resistance to kinase inhibitors and chemotherapy in AML. Blood 113:6215–6224
231. Na Nakorn T et al (2002) Myeloerythroid-restricted progenitors are sufficient to confer radioprotection and provide the majority of day 8 CFU-S. J Clin Invest 109:1579–1585
232. National Cancer Institute (2008) Surveillance, Epidemiology, and End Results (SEER) Program. http://www.seer.cancer.gov/statfacts/html/alyl.html. Accessed 20 June 2008

Chapter 14
The Role of Adult Bone Marrow Derived Mesenchymal Stem Cells, Growth Factors and Scaffolds in the Repair of Cartilage and Bone

Antal Salamon and Erzsébet Toldy

Contents

Abstract Mesenchymal stem cells are to be multipotent cells that have the potential to differentiate into different lineages of mesenchymal tissues including cartilage and bone. These cells have an excellent regeneration potential for tissue repair. Pluripotent mesenchymal progenitor cells are denoted as stromal or mesenchymal stem cells. These cells are relative easy to isolate from small aspirates of bone marrow and can be expand in culture. They are able to differentiate into lineage. This process is assisted by application of bioactive molecules, specific growth factors and biomaterials (scaffolds). Articular cartilage injury has a poor prognosis for repair. Mesenchymal stem cells, when exposed to growth factors can differentiate into cells which become chondroblasts and these are able to form cartilage. The formation of bone after injury requires mesenchymal stem cells which are capable to differentiate into osteoblasts. Together with growth factors, first of all with bone morphologic

---

A. Salamon M.D., Ph.D., D.Sc. (✉)
Department of Traumatology, Markusovszky Teaching Hospital of County Vas, Király utca 11, Szombathely 9700, Hungary
e-mail: salamon.antal@chello.hu

E. Toldy Pharm.D., Ph.D.
Central Laboratory of Markusovszky Teaching Hospital of County Vas, Szombathely, Hungary

Institute of Diagnostics University of Pécs, Szombathely, Hungary

R.K. Srivastava and S. Shankar (eds.), *Stem Cells and Human Diseases*,
DOI 10.1007/978-94-007-2801-1_14, © Springer Science+Business Media B.V. 2012

proteins excellent reparative process can be achieved. The authors deal in this chapter with the experimental investigations and clinical applications of the adult bone marrow derived mesenchymal stem cells, bioactive molecules and carriers.

Keywords Bone marrow • Mesenchymal stem cells • Growth factors • Carriers • Genes • Tissue repair

Abbreviations

APC	Adenomatous polyposis coli
Beta catenin	An integral component in the Wnt signaling pathway
BMP	Bone morphogenetic protein
BMPs	Bone morphologic proteins
DBM	Demineralized bone matrix
DMEM	Dulbecos's modified Eagle's medium
Dvl	Dishevelled
FGF(a,b)	Fibroblast growth factor (acid, basic)
Fz	Frizzled
GSK-3b	Glycogen synthase kinase 3b
IGF	Insuline like growth factor
LEF	Lymphoid enhancer factor
LRP	Low-density lipoprotein receptor related protein
MSC	Mesenchymal stem cell
PDGF	Platelet derived growth factor
PGA	Polyglycolic acid
PLA	Polylactic acid
PLLA	Poly (L-lactic acid)
rh BMP	Recombinant bone morphologic protein
Scaffold	Degradable carrier
SMAD	Merging the terms Sma and Mad genes
TCF	T cell factor
TGF beta	Transforming growth factor beta
Wnt	Combination of Wg (wingless) and Int genes

14.1 Mesenchymal Stem Cells, Growth Factors, Carriers

Adult stem cells exist in different tissues (periosteum, muscle, adipose tissue, blood, ligament) but the major enriched source is the bone marrow, which contains in humans arrays of spicules of osteotrabecular bone. Mesenchymal stem cells (MSC-s) which reside in adult bone marrow have excellent regeneration potential for tissue repair. They are able to differentiate in culture into cartilage,

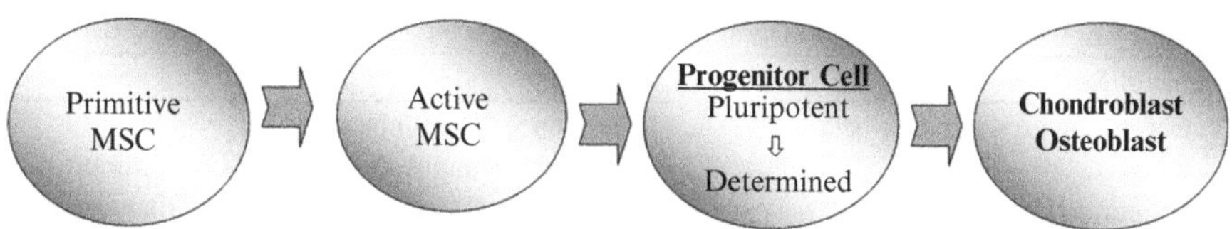

Fig. 14.1 Differentiation of stem cells

bone and into other tissues. Bone marrow contains two cell types: hemapoetic cells (precursors of blood lineage) and MSC-s. Human bone marrow MSC is an identical phenotype with marrow stromal cells which are an early differentiated progeny of MSC. The so called "relaxed primitive MSC" pass over different stages to activate MSC to pluripotent progenitor cell and is transformed into determined progenitor cell (Fig. 14.1).

Some groups used the term "marrow stromal cell", others mesenchymal stem cell. Therefore the term "multipotent stromal cells" has been proposed. Stem cells are undifferentiated multipotent precursor cells that share two characteristic properties: unlimited or prolonged self-renewal and potency for differentiation. There are a more specialized kinds of stem cells called precursor or progenitor cells which can produce specialized differentiated cells. Large number of genes can be expressed between the samples within the same group [1–6].

The success of the reparative process depends on understanding of molecular signals that control MSC differentiation, expansion during the early phase. Applying the culture condition differentiation may take place in application of bioactive molecules. The induction of chondrogenesis, osteogenesis depends on the activity of many factors, -cell density, cell adhesion, dexamethasone, collagen type hydrogels, aggrecan.

Bioactive molecules were used in different cell cultures. Chondrocytic differentiation was seen using pelleted micromass, the cells were controlled with TGF beta or BMP 4 growth factors [7, 8]. Others used ascorbic acid, beta glycerinphosphate and dexamethasone and BMP-2 [9–11]. Bosnakovski [12] used alginate and collagen type I and II hydrogels supplemented with TGF beta and treated with dexamethasone. Some authors applied the following method: bone marrow aspirates were suspended, centrifugated and resuspended in Dulbeco's modified Eagle's medium (DMEM) supplemented with fetal calb serum and antibiotics (penicillin 100 U, streptomycin 100 μ/g, amphotericin 0.25 μ/g) [13–16]. In these culture conditions MSC differentiation can be reached by the application of bioactive molecules, growth factors. It must be remarked that the culture conditions, inductive molecules can alter the behavior of bone marrow stromal cells and the microenvironment is critical for proper in vivo delivery.

Growth factors are proteins (polypeptides) that serve as signaling agents for cell enhancing, cellular function. Differentiation of bone marrow MSC requires the presence of growth factors, which have an important role for the regenerative process (Table 14.1).

Table 14.1 Cartilage or bone formation inducing growth factors

Name of factors	Functions
Transforming growth factor beta (TGF beta)	Influences of cellular activities Stimulates undifferentiated MSC proliferation
Bone morphogenetic proteins (BMP 2,4,7,9)	Important role in cell growth, bone formation. Ability to stimulate differentiation of MSCs into chondrocytes, osteoblasts, promotes osteoprogenitors into osteoblasts
Platelet derived growth factor (PDGF)	Mitogen for MSCs and osteoblasts. Extracellular matrix deposition
Insulin-like growth factor I-II (IGF I-II)	Promotes proliferation, differentiation of osteoprogenitor cells
Fibroblast growth factor basic (bFGF) Fibroblast growth factor acid (aFGF)	Promotes growth and differentiation of chondrocytes, osteoblasts

The transforming growth factor beta (TGF beta), the insulin like growth factor (IGF-I-II) and the fibroblast growth factors (FGF) stimulates MSC proliferation, but the most important role is played by the members of TGF-beta superfamily, the bone morphogenetic proteins (BMPs).

Cartilage and bone regeneration is a complex regenerative process, which remains to a great extent an unknown cascade of complex biological events, particularly at the interactions between various intracellular and extracellular molecular signaling pathways. Many signaling pathways are involved in bone regeneration. The BMP/Smad pathway and Wnt/b-catenin pathway are the most studed in this regards. Due to BMP's utmost importance in stimulating bone formation, the cross-talk between Wnt and BMP pathway during chondrogenic or osteogenic differentiation has received increasing attention.

The growth factors effect on the target cells by binding cell surface receptors before initiating signaling cascade, that results biologic activity of MSC. This binding is known as a ligand receptor interaction. These receptors have cellular domains that bind to and activate the signal transduction system. The signals are sent via specific proteins (transcriptions factors) to the nucleus resulting expression of genes that lead to the synthesis of macromolecules involved in cartilage and bone formation and MSC become chondroblast, osteoblast. BMP receptors are two types: I and II and are serin/threonine protein kinases. These kinases are enzymes that phosphorylate the intracellular proteines (SMADs) which participate in transcriptional regulation of expression of genes. The activated transcription factor and SMAD proteins are transfer to the nucleus and bind to DNA and induce expression of a new gene via specific protein [4, 17, 18]. The Wnt pathway is the most extensively studied biochemical process, which controls gene expression by stabilizing b-catenin in regulating a diverse array of biological processes. In the presence of an appropriate Wnt ligand, Wnt bind to the receptor Fz and co-receptor LRP-5/6, and this binding leads to the activation of the intracellular protein, Dvl, an intracellular mediator that plays a central role in transducing the signal from the receptor complex. The activation of Dvl leads to the inhibition of GSK-3b, results in the disassociation of the multi-protein complex and the intracellular accumulation of b-catenin. Hence, b-catenin

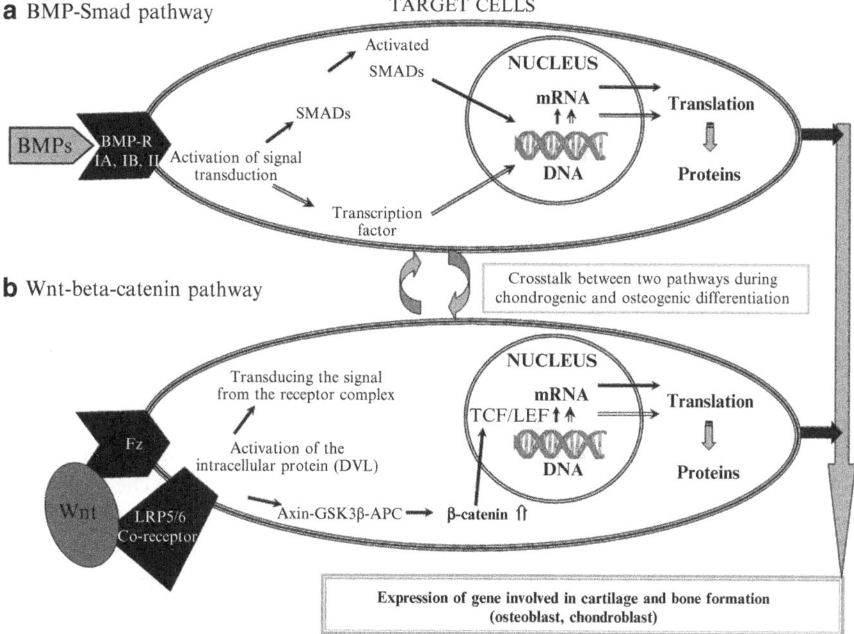

Fig. 14.2 An overview of BMP/Smad and Wnt-beta-catenin pathways. **The effects of growth factors-receptor interaction** (**a**): BMPs effect on target cells by binding to specific cell surface receptors, which activates the signal-transductions system. This signaling mechanism results activation of signal transduction and activates SMAD proteins, which migrates to the nucleus resulting expression of genes. **The Wnt/beta-catenin signaling pathway** (**b**): Wnt bind to the receptor Fz and co-receptor LRP-5/6, and this binding leads to the activation of the intracellular protein, Dvl, which is an important mediator for transducing the signal from the receptor complex. The activation of Dvl leads to the inhibition of GSK-3b, results the intracellular accumulation of beta-catenin. The b-catenin accumulates and translocates to the nucleus, where in concert with members of the TCF/LEF family, activates the transcription of a wide range of genes

cannot be targeted for degradation and it accumulates and translocates to the nucleus, where in concert with members of the TCF/LEF family, activates the transcription of a wide range of genes, including c-myc and cyclin D1. In the absence of Wnt ligand, b-catenin is targeted for phosphorylation and degradation by a multi-protein complex comprising GSK-3b, APC, and axin [19] (Fig. 14.2a, b).

Recently some authors provide evidences that Wnt pathway is crucial to bone regenerative process. Wnt/ beta catenin signaling is required for both BMP-2 induced ectopic enchondral ossification and fracture repair [20, 21].

MSC-s provide important tool for the tissue repair when combined with carriers (scaffolds). To the success of growth factor therapy is critical and the role of delivery system and of the environment is also important. Scaffolds provide a defined three dimensional structure for tissue development and biodegrade at a controlled rate. For selecting an appropriate carrier certain conditions must exist: (1) the delivery of the growth factor at the appropriate dose, (2) the presence of suitable

Table 14.2 Frequently used substances for carriers (scaffolds)

Name of substance
Type I collagen
Fibrin
Hyaluronic acid
Demineralized bone matrix (DEBM)
Hydroxyapatite
Bioglass
Synthetic polymers (PLA, PGA, PLLA)

substratum, (3) the presence of suitable space for cell migration, (4) the ability of delivery system to biodegrade [18].

Some **carriers** are popular because of their favorable degradation characteristics and carrying for bioactive molecules (Table 14.2).

The type I collagen is an abundant protein in the extracellular matrix. This protein is an attractive carrier because of its fibrillar structure [22–25]. The hyaluronic acid is an important constituent of articular cartilage and soft tissues. MSC embedded in hyaluron acid-based material may support differentiation and be useful for osteochondral defect repair [26–28]. The demineralised bone matrix (DEBM) is attributed to proteins, growth factors, it is also effective as a potential of carriers having osteoinductive potential [18, 29–31]. The hydroxyapatite is biocompatible and osteoconductive scaffold and can enhance the bone repair together with growth factors [1, 32, 33]. Bioglass was also used [34]. Polymers (PLA, PGA) are usual scaffolds because of their biocompatibility and ability to bind protein, to deliver peptide molecules [35–37]. In the presence of TGF beta -1 cartilage tissue developed on PLLA scaffolds with the cartilage specific gene expressed GAG and type II collagen accumulated [38]. Biochemical environment also directs cell functions in vitro mechanical loading can affect MSC proliferation and improves osteogenic, chondrogenic phenotype of MSCs. These effects are other parameters such as substrate nature or soluble environments [39].

An important approach utilising MSC- is gene therapy. With gene therapy, genes encoding for therapeutic growth factors can be expressed on a high level [40]. A single dose of exogenous protein does not induce always an adequate biological response, therefore an other useful strategy for protein delivery can be gene therapy involving genetic information to cells. In the expression of IGF-1 and TGF beta 1, BMP-2, higher levels of GAG synthesis, stronger staining of proteoglycan and type II collagen, greater expression of cartilage specific genes was seen. Gene induced chondrogenesis of MSC using multiple genes may reduce viral dosis and can be advantages for collagen repair [41]. MSC can be used as gene vehicle for gene therapy of trauma care, it is capable to deliver genes to improve regenerative process. Gene therapy performs the transfer of genetic information to target cells and these cells synthesise the protein encoded by the gene. The protein synthesis after gene therapy depends on the technique used to deliver gene to a specific anatomic site using in vivo technique or via ex vivo approach. Adenovirus vectors are increasingly popular gene transfer vehicles. Vectors are agent that enhances the expression of DNA in the target cell. They can be viral or non viral origin. The process from the virus into the target cell is the so called transduction. The transduced cell can produce and secrete growth factor in high concentration [18, 42, 43]. MSC-s or marrow stromal cells are effective not only in regenerative medicine, but have immunomodulatory functions they exhibit a powerful immunosuppressive activity. They are able to modulate the function of all major immune cell population. The mechanism of their function is today still not thoroughly clear. They represent important candidates for tissue regeneration and for immune response in graft rejection [44–47].

14.2 Chondrogenesis

Articular cartilage injury has a poor prognosis for repair after damage by injury. This is due first of all to their avascular nature. The cartilage is a simple structure, because of its single cell type and its extracellular matrix having three molecules: water, type II collagen and aggrecan (large aggregating proteoglycan). The main structural feature of the extracellular matrix is hyaline cartilage, therefore the cartilage tissue repair strategies have been on regenerating hyaline tissue. Damage of either the type II collagen or aggrecan may lead to loss of cartilage function. If cartilage is to be repaired in vivo, the balance between collagen and proteoglycan must be restored for the function. The focus of the cartilage tissue engineering is the reconstruction of regenerating hyaline tissue [48, 49]. Because of its limited potential for healing there is a great challenge to scientists over the last years. Many experimental investigations were performed to repair full thickness cartilage defects, but no methods produced entirely successful regenerated cartilage. MSCs have the capacity to differentiate into chondroblasts, chondrocytes with the synthesis of cartilage. These cells can expand in culture and they are able to form cartilage. [16, 50]. In order to improve the quality of regeneration during chondrogenesis, adult MSCs are regarded as a promising alternative. Chondrogenic differentiation of MSC induces expression of multiple cartilage specific molecules inducing collagen type II and aggrecan and result chondrocyte like phenotype. This process is combined with signals of growth factors that regulate the chondrogenesis. Recently many new experimental data have come to light dealing with the inducing markers of chondrogenesis. FGF-2 treatment of bone marrow MSC monolayers enhanced chondrogenic differentiation and is important for tissue engineering strategies on MSC expansion for cartilage repair [51]. MSCs can differentiate into chondroblasts when cultured with TGF beta [52]. The TGF beta superfamily of proteins and their members such as (BMPs) are important regulatory proteins in chondrogenesis. SMAD proteins are important substrates for BMP receptors of TGF/BMP signaling. BMPs have been shown to induce the differentiation of MSC-s into chondrocytes. According to Guo et al. [53] TGF beta 1 is a multifunctional molecule and has a central role in promotion of cartilage repair and inhibition of inflammatory and reactive immune response. Other authors used TGF beta 1 and dexamethasone in the present of collagen type II for expression of the chondrogenic phenotype. Their opinion is that collagen type II has the potential to induce and maintain MSC chondrogenesis and TGF beta 1 enhances the differentiation [12, 38, 54]. Lee et al. [55] tried to define the role of specific TGF-dependent signaling pathways involved in regulation of chondrogenesis from human MSCs. They found that chondrogenesis induced by TGF beta 3 in alginate bed system was confirmed by examining cartilage. Other authors find also that human MSCs have been shown to differentiate into chondrocytes when cultured with TGF beta 3 [56]. The BMP-2-/TGF beta-3 combination is the best culture condition to induce chondrocyte phenotype in pellet cultures of BMP [57]. In other experiments the authors investigated also the role of BMPs. Their opinion is that in the absence of BMP-4, MSC didn't exhibit chondrogenic differentiation. In the presence of BMP-4 the rate and extent of chondrogenesis increased. Their study tests the hypothesis that BMP is instructive to chondrogenesis due to specific cellular signal transducers [8]. Mijkovits et al. [58] emphasises also the important role of this growth factor. According to their opinion, BMP-4 is an important stimulator of chondrogenesis and can improve the healing process of a cartilage defect. Use of appropriate carrier for BMP-4 is crutial for successful repair of cartilage defects. BMP-2 and 9 were also used in MSC culture and in hydrogel alginate induced markers of chondrogenesis [59]. BMP-7 induced MSC secrete type II collagen and GAG [60].

14.3 Osteogenesis

The formation and regeneration of bone is a complex process involving the interaction of cellular elements and local regulators. These include growth factors, hormones, extracellular matrix components which have effect on bone. Bone marrow contains MSC-s which can produce osteoblastic differentiation in osteogenic medium with formation of bone matrix. The repair of the injured bone is a complex process.

Growth factors are agents to enhance the repair of bone. FGF-1 and FGF- 2 were identified during the early stage of fracture healing which promote growth and differentiation of variety of cells including osteoblasts. The recombinant human (rh) FGF 2 accelerate fracture healing. IGF 1 and IGF 2 are involved in healing fracture in rats. But the results varied and it is difficult to determine the potential role of these growth factors of fracture healing. Platelet derived growth factor (PDGF) was mitotic for osteoblasts, but the beneficial histological findings were analized only on a small number of animals [18, 61].

BMP 2 induces a significant increase in bone repair. The use of rhBMP 2 opened a new frontier in the treatment of bone defects, nonunions involving differentiation of MSC progenitor cells into osteoblasts. According to histological data from animal and human studies BMP 2 undergoes remodelling and integration with surrounding bone. BMP 2 is the first growth factor in recombinant form which is able to initiate osteoinductive activity [24, 62, 63]. According to YASKO et al. [25] implantation of rhBMP 2 with guanidine hydrochloride extracted DBM showed significant bone formation. Others demonstrated bone healing with histological and mechanical evidence after implantation of rhBMP 2 delivered in collagen sponge [64]. BMP 2 accelerated the healing of fractures in rabbit osteotomy model [65]. According to Cook et al. [66] investigations with rhBMP 2,4 and 7 showed osteogenetic activity in animals. CHENG et al. [67] studied the osteogenetic effect of BMPs and found that BMP 2,6 and 9 may play a role in the reparative process Adenoviral gene transfer into human bone marrow derived osteoprogenitor cells using carriers promote the bone cell differentiation. Recombinant adenoviral vectors expressing several BMP genes promote osteogenesis [66, 68–70].

14.4 Clinical Relevance

After injury articular cartilage has limited potential for repair and presently has a different outcome of current treatment. Injuries of the articular cartilage that do not penetrate the subchondral bone don't heal successfully and there is a progress to the degeneration of the articular surface. Injuries that penetrate the subchondral bone undergo repair with fibrous, fibrocartilaginous tissues. The exact repair of cartilage injuries is impeded by the avascular nature of cartilage. This is a great challenge to scientists and clinicians and this is widely researched area in orthopedic surgery. Traditional treatments are abrasions arthroplasty microfracture technics, spongiosa plastic, chondrocyte transplantation, but there are not achieved ideal results following current treatment. Lately many experimental investigations have been performed to repair full thickness cartilage defects but no methods have resulted successfully regenerated hyaline cartilage. Adult MSC-s isolated from bone marrow are attractive candidates as progenitor cells for cartilage repair because of their documented chondrogenic potential for tissue repair and may be differentiated in culture. Multipotential MSCs undergo a process of cell fate determination to become genitor cells and can differentiate into chondrocytes. This process is determined by combination of signals of growth factors that regulate the chondrogenesis. Despite advanced preclinical experimental investigations recently still relatively few clinical data can be found using this new method. In recent years, the first clinical trials on the utility of MSCs, growth factors and scaffolds have been initiated. New techniques of cell therapy and molecular medicine have been currently applied in human clinical trials.

The reconstruction of human articular cartilage-polymer construction can be effectively used in the treatment of human articular cartilage defects [71]. Heidrich et al. [72] studied the regenerating effect of hMSC and human chondrocytes. After 6 months they did not find any difference between the both cells -chondrocytes and MSCs have shown the same chondrogenic differentiation in patients. Wakitani et al. [73, 74] implanted culture expanded bone marrow derived MSCs into full thickness of articular defects of knee joint of three patients. Their clinical symptoms improve 6 months after transplantation. They believed that the procedure can be clinically useful method for the treatment of large cartilage defects. Subsequently 41 patients received MSC transportations. Neither tumors nor infections were observed. Their opinion is: autologous MSC transplantation will be widely used around the world.

There are more clinical data on the area of the treatment bone injuries, bone defects. In most instances autologous bone graft is used because of its osteogenetic properties. Osteoconduction supports the ingrowth of capillaries, perivascular tissues and osteoprogenitor cells into the structure of graft. Osteoinduction is a process attributed to BMPs that supports the proliferation of undifferentiated MSC-s, formation of osteoprogenitor cells with the capacity to form bone [30, 75]. Autologous cancellanous bone is harvested from the iliac crest. The problem is that the time of operation is longer and sometimes associated with postoperative pain or haematome and sometimes with the potential of infection. Other source of autologous bone material is to use MSCs derived from the bone marrow. According to the experimental investigations the bone marrow contains MSCs which are capable of giving rise to multiple lineages in cell culture. This material is osteogenetic and potentially osteoinductive through growth factors secreting by the transplanted cells. The osteoinductive capacity of bone morphologic proteins (BMPs) was discussed in preclinical animal models.

The efficacy of BMPs for treatment of orthopaedic patients is now being evaluated in clinical trials. According to Lieberman [18] there is a great interest in the development of clinical application for growth factors in the promotion and acceleration of fracture healing, in the treatment of nonunions, bone defects. Traditionally autologous bone grafts are used, but only limited amounts of autologous spongiosa can be found and the harvesting procedures come sometimes in the prominence. Therefore an alternative method can play an important role in the treatment. The biologically based method using MSCs, growth factors and carriers, sometime gene therapy promote of fracture healing. The foregoing clinical trials document the efficacy of recombinant protein in the healing process. First of all the application of BMP 2 and 7 can be a new osteoinductive therapy. According to Govender et al. [63] the mechanism of action of rh BMP-2 involves osteoinductive signaling and regulation, gene expression inducing differentiation of mesenchymal progenitor cells into osteoblasts. The rhBMP-2 on an absorbable collagen sponge matrix is efficacy for the treatment of fractures. Their group operated 450 patients with an open tibial fracture; either group received tibial fixation with intramedullary nail, the other group has got tibial fixation and medullary nail and an implant containing 0.75 or 1.50 mg/ml rh BMP-2 with absorbable collagen sponge. The patients who received rhBMP-2 had significantly less infection, fewer hardware failures and faster wound healing, accelerated fracture healing. In another study rhBMP 2 and collagen carrier used to induce spinal fusion. After 6 months all patients had evidence of fusion on radiographs and computer tomografs. The rhBMP 2 with absorbable collagen was also used for anterior interbody fusion of the lumbar spine. The patient had degenerative lumbar disc disease. The investigational group (143 patients) received rh BMP-2 on absorbable collagen sponge, the control group (136 patients) received autogenous iliac-crest bone graft. The fusion after 2 years was higher for the BMP treated group [75]. In another clinical trial the results of 124 tibial nonunions using intramedullary nail and autologous bone graft were compared to a control group where an intramedullary nail and implantation of Op 1 (BMP-7) and type I collagen carrier was used. After 9 months the healing rate was similar in two groups [76]. The efficacy of rhBMP-7 with type in collagen carrier was studied in 24 patients in the treatment of fibular defects. Healing of the defects was observed. [77]. Zimmermann et al. [22] reported also their experiences 21 patients were operated having nonunion of long bone fractures. The rhBMP -7 induced the formation of new bone differentiation of MSC. There were no postoperative complications. The bony healing was successful in all cases. Nonunions of long bone fractures are therapeutic and economic problems. If basic surgical methods (nailing, plate fixation with compression, autologous cancellous bone grafting) fail to work, alternative treatment options are needed.

Due to their easy isolation and differentiation potential MSC-s are being introduced into clinical medicine. Purified MSCs retain their osteogenic potential after growth in culture and cryopreservation could be useful for clinical application. MSCs are easy to isolate from aspirates of bone marrow. The most accessible source is the crista iliaca. Osteogenetic factors may be introduced in an injectable manner. The requirement of an injectable carrier is maintanance of suitable concentrations of the osteogenetic factors. This is important to allow to use large quantities of bone forming precursor cells. Different scaffolds can be used for fracture repair, they must be biocompatible to minimize inflammation and suitable porous to make cellular and vascular invasion. These carriers must be biodegradable [78]. Many scaffolds have these criteria, among others Type I collagen, fibrin, hyaluronate, alginate, DBM, Polymers (PLA, PGA). The delivery systems for the rhBMPs must adapt to the anatomic location and the vitality of the surrounding tissue is also important.

The use of MSCs, BMPs and carriers have opened a new frontier in the treatment of bone nonunions, bone defects and provide an alternative to autogenous and allogen bone grafts. Allogenic bone grafts can also be used but their osteogenic potential is lower, the resorption rate is higher and the revascularisation is inferior compared with an autologous graft. The host immunological response may affect the success of an allograft. That diminishes its effectiveness [66, 76]. What is the future of the successful treatment of cartilage and bone injuries?

The preclinical animal experiments have already shown the important role of MSCs in the healing of cartilage and bone injuries, defects. Application of growth factors and carriers may improve the healing process. The osteoinductive capacity of BMPs has been demonstrated in these experiments and the further development of the new carriers will help the orthopaedic surgeons to achieve successful results in the near future, but over the next years still new clinical trials should be completed for the efficacy of the new therapy.

There is an abundance of data from developmental and cellular models which suggests that Wnt pathway interacts with several other pathways. Obtaining a further understanding of how these pathways affect Wnt signaling during bone regeneration may provide knowledge for combined therapeutic interventions. In general, based on recent discoveries, despite the fact that not all underlying mechanisms are well understood yet, Wnt signaling pathway indeed plays a crucial role in bone regenerative process, such as ectopic bone formation and fracture repair. Modulation of Wnt pathway using pharmacological agents (e.g., lithium) may provide a promising therapeutic approach to improve bone regeneration.

14.5 Conclusions

According to the detailed preclinical animal investigations, MSC-s derived from bone marrow are excellent progenitor cells for regenerative cartilage and bone tissues. Some growth factors, first of all the BMPs effect as signaling agents for these cells and play an important role in cartilage and bone formation. Using carriers is important in the reparative process. The delivery system must degrade without generating an immune or inflammatory response. This new technology will obviously develop progress in the future and can be applied besides the treatment of cartilage defects, open fractures, nonunions, defects of long bones, fusion of spine. Application of minimally invasive surgical technique can reduce the complications. The current possibilities, options for the orthopaedic surgeons are to use autologous or allogenic cancellous or cortical bone, autologous bone marrow, but in the future the osteogenic proteins will come into prominence. The present preclinical trials give promising results and we hope that this new technology can improve the results in the clinical practice in the treatment of cartilage and bone injuries, defects.

So far our knowledge concerning how Wnt pathway regulates bone regeneration has been still at its infancy, and there are many issues remaining to be addressed. Finally we cite the opinion of Chen et al. [19] who asked the following questions: "what is the role of Wnt pathway in bone regeneration? Which of the Wnts are the critical ligands to mediate Wnt pathway in bone regeneration? These questions are topics of investigation that are being actively studied to understand cartilage and bone formation better. In addition, bone regeneration is a complex process involving multiple molecular signaling pathways, thus another attention should focus on how these pathways interact with Wnt signal".

References

1. Prockop JD (1997) Marrow stromal cells as stem cells for nonhemapoetic tissues. Science 276(4):71–74
2. Tuan RS, Boland G, Tuli R (2003) Adult mesenchymal stem cells and cell based tissue engineering. Arthritis Res Ther 3(1):532–544
3. Bianco P, Kuznetsov SA, Riminucci M, Gehron Robey P et al (2006) Postnatal skeletal stem cells. Methods Enzymol 419:117–148
4. Salamon A, Toldy E (2009) The role of adult bone marrow derived mesenchymal stem cells, growth factors and carriers in the treatment of cartilage and bone defects. J Stem Cells 4(1):71–80
5. Salamon A, Toldy E (2009) Adult bone marrow derived mesenchymal stem cells and stem cells for tissue repair. Clin Exp Med J 3(3):369–379

6. Xiao Y, Mareddy S, Crawford R (2010) Clonal characterisation of bone marrow derived stem cells and their application. Int Oral Sci 2(3):127–135
7. Pittinger MF, Mackay M, Beck SC et al (1999) Multilineage potential of adult human mesenchymal stem cells. Science 284(2):143–147
8. Hatekayama Y, Nagujen J, Wang X et al (2003) Smad signalling in mesenchymal and chondroprogenitor cells. J Bone Joint Surg 85-A(Suppl 3):13–18
9. Maegawa N, Kavamura K, Hirose M, Yajima H et al (2007) Enhancement of osteoblastic differentiation of mesenchymal stem cells cultured by selective combination of bone morphogenetic protein-2 (BMP-2) and fibroblast growth factor (FGF-2). Tissue Eng Regen Med 1(4):306–313
10. Niemeyer P, Krause V, Punzel M, Fellenberg J et al (2003) Mesenchymale Stammzellen zum Tissue Engineering von Knochen: dreidimensionale osteogene Differezierung auf mineralisierten Kollagen. Z Orthop 141(6):712–717
11. Song Su, Jin OJ, Seok Y, Keum H et al (2007) Effect of culture conditions osteogenic differentiation in human mesenchymal stem cells. J Microbiol Biotechnol 17(7):1113–1119
12. Bosnakovski D, Mizuno M, Kim G, Takagi S et al (2006) Chondrogenic differentiation of bovine bone marrow mesenchymal stem cells (MSCs) in different hydrogels: influence of collagen II type extracellular matrix MSC chondrogenesis. Biotechnol Bioeng 93(6):1152–1163
13. Cheng H, Jiang W, Philips FM, Haydon RC et al (2003) Osteogenetic activity of the forteen types of human bone morphogenetic proteins (BMPs). J Bone Joint Surg 85-A(8):1544–1552
14. Titorencu I, Jinga VV, Constantinescu E, Gefencu AV et al (2006) Proliferation, differentiation and characterisation of osteoblasts for human BM mesenchymal cells. Cytotherapy 9(7):682–696
15. Kim H, Lee JH, Suh H (2003) Interaction of mesenchymal stem cells and osteoblasts for in vitro osteogenesis. Yonsei Med 44(2):187–197
16. Nauman A, Dennis J, Staudenmayer R, Rotter N et al (2002) Mesenchymale Stammzellen, neue möglichkeiten der Gewebezuchtung für die plastisch-rekonstruktive Chirurgie. Laringo Rhino-Othologie 81(7):521–527
17. Reddi AH (2001) Bone morphogenetic proteins: from basic science to clinical applications. J Bone Joint Surg 83-A(Suppl 1):451–456
18. Lieberman JR, Daluiski A, Einhorn TA (2002) The role of growth factors in the repair of bone. J Bone Joint Surg 84-A(6):1032–1044
19. Chen Y, Alman BA (2009) Wnt pathway, an essential role in bone regeneration. J Cell Biochem 106(3):353–362
20. Zou L, Zou X, Myging T, Zeng Y et al (2006) Molecular mechanism of osteochondroprogenitor fate determination during bone formation. Adv Exp Med Biol 585:431–441
21. Silkstone D, Hong H, Alman BA (2008) Beta catenin in the race to fracture repair: in it to Wnt. Nat Clin Pract Rheumatol 4(8):413–419
22. Zimmermann G, Moghaddam A, Wagner C, Vock B et al (2006) Klinische Erfahrungen mit Bone Morphogenetic Protein 7 (BMP 7) bei Pseudoarthrosen langer Röhrenknochen. Der Unfallchirurg 109(7):528–537
23. Cook S, Baffes G, Wolfe MW, Sampath K et al (1994) The effect of recombinant human osteogenic protein-1 on healing of large segmental bone defects. J Bone Joint Surg 76-A(6):827–838
24. Bouxsein ML, Turek TJ, Blake CA, D'Augusta BS et al (2001) Recombinant bone morphogenetic protein-2 accerelates healing in a rabbit ulnar osteotomy. J Bone Joint Surg 83(8):219–1229
25. Yasko AW, Lane JM, Fellinger EJ, Rosen V et al (1992) The healing of segmental bone defects, induced by recombinant human bone morphogenetic protein (rhBMP-2). J Bone Joint Surg 74-A(5):659–670
26. Radomsky ML, Aufdemorte TB, Swain LD, Fox T (1999) Novel formulation of fibroblast growth factor in hyaluron gel accerelates fracture healing in nonhuman primates. J Orthop Res 17(4):607–614
27. Solchaga LA, Dennis JE, Goldberg VM, Caplan AI (1999) Hyaluronic acid-based polymers as cell carriers for tissue engineered repair of bone and cartilage. J Orthop Res 17(2):205–213

28. Kayakabe M, Tsutsumi S, Watanabe H, Kato Y et al (2006) Transplantation of autologous rabbit BM-derived stromal cells embedded in hyaluronic acid gel sponge into osteochondral defects of the knee. Cytotherapy 8(4):343–353
29. Johnson EE, Urist MR, Finerman GA (1992) Resistant nonunions and partial or complete segmental defects of long bones. Treatment with implants of a composite of human bone morphogenetic protein (BMP) and autolysed antigen extracted allogeneic bone. Clin Orthop 277:229–237
30. Finkemeier Ch (2002) Bone grafting and bone-graft substitutes. J Bone Joint Surg 84-A(3):454–464
31. Tiedeman JJ, Garvin KL, Klle TA, Conolly JF (1995) The role of a composite demiralized bone matrix and bone marrow in the treatment of osseous defects. Orthopedics 18(12):1153–1158
32. den Boer FC, Wippermann BW, Blokhuis TJ, Patka P et al (2006) Healing of segmental bone defects with granular porous hydroxyapatite augmented with recombinant human osteogenic protein-I or autologous bone marrow. J Orthop Res 21(3):521–528
33. Wolfe SW, Pike L, Slade JF, Katz LD (1999) Augmentation of distal radius fracture fixation with coralline hydroxyapatite bone graft substitute. J Hand Surg (Am) 24(4):816–827
34. Wheeler DL, Stokes KE, Park HM, Hollinger JO (1997) Evaluation of particulate bioglass in a rabbit radius osteotomy model. J Biomed Mater Res 35(2):249–254
35. Hollinger JO, Leong K (1996) Poly(alpha-hydroxy) acids: carriers for bone morphogenetic proteins. Biomaterials 17(2):187–194
36. Partridge K, Yang X, Clarke NMP, Okubo Y et al (2002) Adenoviral BMP-2 gene transfer in mesenchymal stem cells: in vitro and in vivo bone formation on biodegradable polymer scaffolds. Biochem Biophys Res Commun 292(1):144–152
37. El-Amin SF, Attawia M, Lu H, Shah A et al (2002) Integrin expression by human osteoblast cultured on degradable polymeric materials applicable for tissue engineered bone. J Orthop Res 20(1):20–28
38. Hu J, Liu X, Ma PX (2009) Chondrogenic and osteogenic differentiation of human bone marrow derived mesenchymal stem cells on a nanofibrous scaffold with designed pore network. Biomaterials 30(28):5061–5067
39. Potier E, Noailly J, Ito K (2010) Directing bone marrow-derived cell function with mechanics. J Biomed 43(5):807–817
40. Pelinkovic D, Horas U, Engelhard M, Lee JY et al (2002) Gentherapie von Knorpelgewebe. Z Orthop 140(2):153–159
41. Steinert AF, Palmer GD, Pilapil C, Nöth U et al (2009) Enhanced in vitro chondrogenesis of primary mesenchymal stem cells by combined gene transfer. Tissue Eng Part A 15(5):1127–1139
42. Evans Ch, Robbins PD (1995) Possible orthopaedic applications of gene therapy. J Bone Joint Surg 77-A(7):1103–11014
43. Scaduto AA, Lieberman JR (1999) Gene therapy for osteoinduction. Orthop Clin North Am 30(4):625–633
44. Spitkovsky D, Hescheler J (2008) Adult mesenchymal stromal cells for therapeutic applications. Minim Invasive Ther Allied Technol 17(2):79–90
45. Sotiropoulou PA, Papamichail M (2007) Immune properties of mesenchymal stem cells. Methods Mol Biol 407:225–243
46. Dazzi F, Horwood NJ (2007) Potential of mesenchymal stem cells therapy. Curr Opin Oncol 19(6):650–655
47. Brooke G, Cook M, Blair C, Han R et al (2007) Stromal cells. Semin Cell Dev Biol 18(6):845–858
48. Hollander AP, Heathfield TF, Webber C et al (1994) Increased damage to type II collagen in ostreoarthric articular cartilage detected by a new immunoassay. J Clin Invest 93(4):1722–1732
49. Hollander AP, Dickinson SC, Kafienah W (2010) Stem cells and cartilage development: complexities of a simple tissue. Stem Cells 28(11):1992–1996
50. Cook SD, Patron LP, Salkeld SL, Rueger D (2003) Repair of articular cartilage defects with osteogenic protein-1 (BMP-7) in dogs. J Bone Joint Surg 85-A(Suppl 3):116–123

51. Stewart AA, Byron CR, Pondenis H, Stewart MC (2007) Effect of fibroblast growth factor-2 on equine mesenchymal stem cell monolayer expansion and chondrogenesis. Am J Vet Res 68(9):941–945
52. Hunziker EB, Driesang IR, Morris EA (2001) Chondrogenesis in cartilage repair is induced by members of transforming growth factor beta superfamily. Clin Orthop 391(Suppl S):171–181
53. Guo X, Zheng Q, Kulbaski I, Yuan Q et al (2006) Bone regeneration with active angiogenesis by basic fibroblast growth factor gene transfered mesenchymal stem cells seeded on porous beta TPC ceramic scaffolds. Biomed Mater 1(3):93–99
54. Diao H, Wang J, Xia S, Dong L et al (2009) Improved cartilage regeneration utilizing mesenchymal stem cells in TGF-beta-1 gene activated scaffolds. Tissue Eng Part A 15(9):2687–2698
55. Lee JW, Kim Y, Kim SH, Han SH et al (2004) Chondrogenic differentiation of mesenchymal stem cells and its clinical application. Yonsei Med J 30(45 Suppl):41–47
56. Moiloi EK, Mao JJ (2006) Chondrogenesis of mesenchymal stem cells by controlled delivery of transforming growth factor beta-3. Conf Proc IEEE Eng Med Biol Soc 1:2647–2650
57. Ronziere MC, Perrier E, Mallein-Gerin F, Freyiria AM (2010) Chondrogenic potential of bone marrow and adipose tissue- derived adult human mesenchymal stem cells. Biomed Mater Eng 1(3):145–158
58. Miljkovic ND, Cooper GM, Marra KG (2008) Chondrogenesis, bone morphogenetic protein-4 and mesenchymal stem cells. Osteoarthritis Cartilage 16(10):1121–1130
59. Majundar MK, Wang E, Morris EA (2001) BMP-2 and BMP-9 promotes chondrogenic differentiation of human multipotential mesenchymal cells and overcomes the inhibitory effect of Il-1. Cell Physiol 189(3):275–284
60. Bai X, Zhao C, Duan H, Qu F (2011) BMP-7 induces the differentiation of bone marrow-derived mesenchymal cells into chondrocytes. Med Biol Eng Comput 49(6):687–692
61. Pacifici L, Casella F, Ripari M (2002) The principles of tissue engineering: role of growth factors in the bone regeneration. Minerva Stomatol 51(9):351–359
62. Kessler S, Mayr-Wohlfart U, Ignatius A, Puhl W et al (2003) Der Einfluss von Bone Morphogenetic Protein 2 (BMP-2), Vascular Endothelial Growth Factor (VEGF) und basischem Fibroblasten Wachstumfactor (bFGF) und Osteointegration. Degradation und biochemische Eigenschaften eines synthetischern Knochen Ersatzstoffes. Z Orthop 141(4):472–480
63. Govender S, Csimma C, Genant HK, Valentin-Opran A, BESTT (2002) Recombinant human bone morphogenetic protein-2 for treatment of open tibial fractures. J Bone Joint Surg 84-A(12):2123–2134
64. Sciadini MF, Johnson KD (2000) Evaluation of recombinant human bone morphogenetic protein-2 as a bone graft substitute in a canine segmental defect model. Orthop Res 18(2):289–302
65. Bostrom MP, Camacho NR (1998) Potential role of bone morphogenetic proteins in fracture healing. Clin Orthop 355(Suppl S):274–282
66. Cook SD, Baffes GC, Wolfe MW, Sampath K et al (1994) Recombinant human bone morphogenetic protein-7 induces healing in a canine long-bone segmental defect model. Clin Orthop 301(4):302–312
67. Cheng H, Wei Jiang BA, Phillips FM, Haydon RC et al (2003) Osteogenetic activity of the forteen type of human bone morphogenetic proteins (BMPs). J Bone Joint Surg 85/A(8):1544–1552
68. Koch H, Jadloviec JH, Whalen JD, Robbins P et al (2005) Osteoblastare Differenzierung von humanen adulten mesenchymalen Stammzellen durch transgenes BMP-2 in Abwesenheit von Dexamethasone. Z Orthop 143(6):684–690
69. Riew KD, Wright NM, Cheng S, Avioli LV et al (1998) Induction of bone formation using a recombinant adenoviral vector carrying the human BMP-2 gene in a rabbit spinal fusion model. Calcif Tissue Int 63(4):357–360
70. Lou J, Xu F, Merkel K, Manske P (1999) Gene therapy: adenovirus mediated human bone morphogenetic protein-2 gene transfer induces mesenchymal progenitor cell proliferation and differentiation in vivo. J Orthop Res 17(1):43–50
71. Berner A, Hendrich C, Battmann A, Schütze N et al (2003) Rekonstruktion von Gelenkknorpeldefekten mit Knorpel Polymer-Konstrukten hergestellt durch Beschichtung mit mesenchymalen Stammzellen. Z Orthop 141(S1):176

72. Hendrich C, Weber M, Battmann A, Steinert A et al (2003) Chondrogene Differenzierung von humanen Chondrozyten und mesenchymalen Stammzellen in einem in vivo Modell. Z Orthop 141:S1–S176
73. Wakitani S, Nawata M, Tensho K, Okabe T et al (2007) Repair of cartilage defects in patellofemoral joint with autologous bone marrow mesenchymal cell transplantation: three cases report involving nine defects in five knees. Tissue Eng Regen Med 1(1):74–79
74. Wakitani S, Okabe T, Horibe S, Mitsuoka T et al (2011) Safety of autologous bone marrow-derived mesenchymal stem cell transplantation for cartilage repair in 41 patients with 45 joints followed for up to 11 years and 5 months. J Tissue Eng Regen Med 5(2):146–150
75. Einhorn TA (2003) Clinical applications of recombinant human BMPs: early experience and future development. J Bone Joint Surg 2003/A(Suppl 3):82–88
76. Friedlaender GE, Perry CR, Cole JD, Cook SD et al (2001) Osteogenic protein −1 (bone morphogenetic protein-7) in the treatment of tibial nonunions. J Bone Joint Surg 83-A(Suppl 1):151–158
77. Geesink RG, Hoefnagels NH, Bulstra SK (1999) Osteogenic activity of OP-1 bone morphogenetic protein (BMP-7) in human fibular defect. J Bone Joint Surg Br 81(4):710–718
78. Seeherman H, Li R, Wozney J (2003) A review of preclinical program development for evaluating injectable carriers for osteogenetic factors. J Bone Joint Surg 85/A(Suppl 3):96–108

Chapter 15
Worth the Weight: Adipose Stem Cells in Human Disease

Saleh Heneidi and Gregorio Chazenbalk

Contents

Abstract Adipose Stem Cells (ASCs) are at the forefront of reconstructive and regenerative medicine, currently offering perhaps the most significant utilization of stem cells in the treatment of injured and dysfunctional connective or mesenchymal tissue; be it in bone, cartilage, fat, muscle, or heart tissues. However, these cells are also at the heart of adipogenesis and adipose endocrine function, playing key roles in numerous diseases and disorders relating to obesity, lipidystrophies, diabetes, and

---

S. Heneidi
Department of Obstetrics/Gynecology, Medical College of Georgia at Georgia Health Sciences University, Augusta, GA, USA

G. Chazenbalk, Ph.D. (✉)
Department of Obstetrics/Gynecology, The David Geffen School of Medicine at UCLA, 10833 Le Conte Avenue, Box 951740, Los Angeles, CA 90095-1740, USA
e-mail: gchazenbalk@mednet.ucla.edu

R.K. Srivastava and S. Shankar (eds.), *Stem Cells and Human Diseases*, DOI 10.1007/978-94-007-2801-1_15, © Springer Science+Business Media B.V. 2012

metabolic dysfunction. Adipose is an extremely heterogeneous tissue, with research indicating that cross-talk between numerous cells types creates a dynamic niche where cell populations are continually shifting to direct endogenous stem/progenitor cell activity. The study of these dynamics may elucidate additional aspects of stem cell influence over adipose health and dysfunction, as well as other mechanisms of cross-talk and cell differentiation which are mirrored in major tissues such as the liver, heart, and brain, thus allowing for new sources of adult stem/progenitor cells.

Keywords Adipose-derived Stem Cell • ASC • Adipose • Inflammation • Tissue regeneration • Adipocyte • Chronic wound therapy • Adipose tissue macrophage • Plasticity

15.1 Adipose-Derived Stem Cells

Adipose tissue is the source of adipose stem cells (ASCs) with documented multi-lineage differentiation potentials [1–6], which have opened new avenues for potential therapeutic strategies [7, 8]. Adult stem cells from bone marrow, stroma, and mesenchymal stem cells (MSCs), have been shown to differentiate into adipocytes, chondrocytes, osteoblasts, and myoblasts *in vitro* and to undergo differentiation *in vivo* [9–11]. Limitations to the use of bone marrow-derived stem cells (BMSC) as transplant materials are in the quantity of cells that can be collected from the patient and associated donor-site pain and morbidity. In addition, cells obtained from bone marrow must often be cultured and expanded for several weeks to generate sufficient numbers for therapeutic use; an expensive and heavily regulated procedure which fundamentally changes the biology of the cells [12]. Adipose-derived stem cells (ASCs) are similar in many ways to bone marrow-derived mesenchymal stem cells, but they have several significant advantages. Many more cells can be harvested from adipose tissue, as the number of ASCs present in adipose tissue is 100–1,000 times higher per centimeter than that of mesenchymal stem cells from circulation or in bone marrow [13, 14], and the harvest procedure is less invasive than harvesting bone marrow. ASCs can be maintained in culture for extended periods of time and can be induced *in vitro* to differentiate down adipogenic, chondrogenic, myogenic, and osteogenic pathways in the presence of lineage-specific induction factors [8, 15]. Moreover, ASCs can be safely and efficiently transplanted to autologous hosts, can be manufactured in accordance with current good manufacturing practice (GMP) guidelines, and can be used for a variety of therapeutic regenerative medicine therapies.

15.2 Components of Adipose Tissue

Adipose tissue is a key player in metabolic homeostasis through its role as both an energy depot and endocrine organ [16, 17]. In obesity, excess adiposity is associated with impaired lipid and glucose homeostasis, leading to hyperlipidemia,

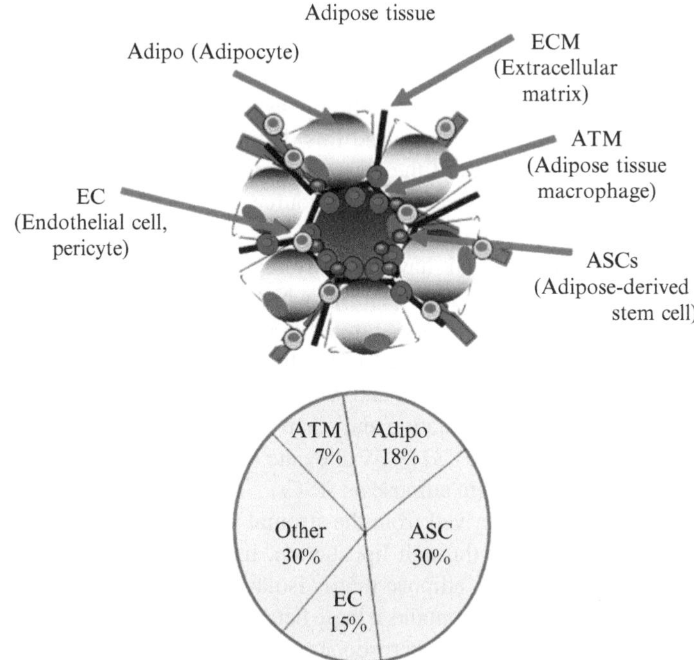

Fig. 15.1 Population of cell types present in adipose. The observed plasticity of ASCs may be attributable to the heterogeneity of adipose and its diverse microenvironment. Adipose is comprised of an ever-changing heterogeneous mixture of cells including adipocytes, adipose tissue macrophages (ATMs), endothelial cells (ECs) of the microvasculature, multiple lineages of adipose stem/progenitor cells (ASCs), and other cells that include myeloid dendritic cells, nerve tissue, stromal cells, and fibroblasts

insulin resistance, and type 2 diabetes [18]. Similar metabolic disorders are now known to be associated with adipose tissue deficiency [16, 19–21], which may result from altered paracrine and endocrine activity [22]. Adipose tissue secretes hormones and adipokines that modulate metabolism in tissues such as brain, skeletal muscle, liver, and pancreas. Known adipokine actions include regulation of food intake and satiety (leptin), energy balance and insulin sensitivity (adiponectin, leptin, resistin, etc.), and mediation of inflammation and the immune response (tumor necrosis factor-α (TNF-α), interleukins (IL)-1, -6, -10, plasminogen activator inhibitor 1, etc.) [17, 23–25].

Adipose is comprised of an ever-changing heterogeneous mixture of cells including adipocytes, adipose tissue macrophages (ATMs), endothelial cells (ECs) of the microvasculature including pericytes, myeloid dendritic cells, nerve tissue, cells present in the stroma and extracellular matrix (ECM) including fibroblasts, and multiple lineages of adipose stem/progenitor cells (ASCs) (Fig. 15.1). [26]. This diversity of cells can lead to confusion when addressing one cell type, as the stem/progenitor population consistently exists in between defined cell types, displaying an undefined and changing array of markers and morphologies. In an attempt to unify terminology the 2006 International Society for Cell Therapy sought to define

Multipotent Mesenchymal Stromal Cells as (1) having the potential to differentiate down adipogenic, osteogenic, and chondrogenic pathways; (2) being positive for CD73, CD90, and CD105; and (3) being negative for CD34 and CD31 [27]. However, even this definition needs redefining, as multiple protocols produce these cells from a wide variety of tissues, and this definition also does not address the vascular component that much of the research is now fixated on [28, 29]. For now one must take care to define the cell type by source, marker expression, differentiation potential, and morphology.

Interestingly, Daquinag et al. have recently reported finding the first truly adipose stem/progenitor marker, delta-decorin, which if truly ASC specific, will add great homogeneity to the field of ASC research [30]. As it stands, there are multiple adipose stem progenitor cells included in the literature, with a variety of lineages from numerous embryologic origins, and while cell types may appear to be homogeneous at times, different stimulation and reactivity can identify multiple subpopulations within each cell type [31]. To alleviate confusion, we will refer to all multi-potent cells derived from adipose as ASCs.

ASCs are most readily derived from the stromal vascular fraction (SVF) of adipose removed surgically or through liposuction, usually from the abdomen. The SVF is the component of the adipose that is isolated though enzymatic digestion, usually with collagenase, and contains a large heterogeneous mixture of cells. When observing these cells under the microscope one would see very large lipid filled adipocytes, flat and large cells correlating to mature preadipocytes and macrophages, and small round cells that include monocytes, macrophages, and a lineage of stem/progenitor cells. There are also long, thin, spindle-shaped cells that are composed of preadipocytes (with lipid droplets), fibroblasts, fibrocytes, pericytes, and other slightly different populations of ASCs.

Another available source of ASCs in the SVF can be found in sphere clusters that can be observed free floating in culture [21, 32]. Although the precise lineage of the spheres is unknown, these clusters are believed to be separate ASCs that interact with preadipocytes, and are generated by small round progenitors [21, 32, 33]. Interestingly, both De Francesco et al. and our own group have observed that cells of the sphere clusters are not necessarily clonally derived, as CD34+/CD90+ cells formed sphere clusters when placed in non-adherent growth conditions [21, 33]. Additionally, we have observed that the sphere formation develops as tiny progenitor cells home in on a given source, which they then cluster around before they differentiate (unpublished data). While additional studies must be done to characterize these stem/progenitor cells, they appear to express a unique marker profile with an especially high expression of S100A4, as well as CD34 expression that increases as the cells form clusters, before dissipating after differentiation [21]. De Francesco et al. were able to differentiate spheres into multiple cell populations [33]. When characterized, spheres expressed MSC markers including CD34, CD90,CD29, CD44, CD105, and CD117, as well as endothelial-progenitor-cell markers including CD34, CD90, CD44, and CD54 [33]. De Francesco also observed CD34+/CD90+

spheres quickly became adipocytes with adipogenic medium and differentiated into endothelial cells (CD31+/VEGF+/Flk-1+) when placed in methylcellulose [33]. Howson et al. isolated ASC spheres and found them to be CD34+, Tie-2+, NG2+, nestin+, PDGFR-α/β+, and negative for aSMA, and the endothelial CD31 and eNOS, suggesting an immature phenotype [32]. The addition of bFGF allowed the sphere formation to adhere to the dish and would differentiate to mural cells with αSMA in response to serum [32]. Finally, Howson made the surprising discovery that the spheres could differentiate to pericytes when co-cultured with aorta angiogenic outgrowths [32].

Many of the positive effects of ASC therapy on vascular health are attributed to a seemingly close relationship to pericytes, which reside in the vasculature, giving support to the endothelium and guiding vasculogenesis and angiogenesis [34]. Pericytes can be from both mesodermal and ectodermal, as pericytes in the brain vasculature can be induced down a neuronal or glial path [35, 36]. A population of ASCs is primarily located in the walls of adipose microvasculature, and these cells possess many characteristics of pericytes [37]. Further, these cells also possess pericyte-like ability to stabilize endothelial networks *in vitro* and participate *in vivo* with endothelial cells in the formation and stabilization of new vessels that connect with host vasculature [38]. Pericytes can differentiate to fibroblasts, myofibroblasts, preadipocytes, and can also exhibit macrophage-like qualities [39]. There has been confusion as to the ideal markers for pericytes identification. While neuronal-related markers NG2 and CD271 are commonly used [28], the walls of vasculature have been found to contain endothelial progenitors that were isolated in the blood and umbilical cord [40, 41] and can differentiate to macrophages and MSCs [28]. The walls of adipose microvasculature have also been found to contain resident CD34+, CD45+ hematopoietic progenitor cells, which may make up as much as 3.5% of the SVF [42, 43], with 0.75% of CD34+/45+ also expressing CD117 (c-kit). Finally, the SVF may also contain progenitor cells that consist of myeloid intermediates, which migrate to the adipose, and as such, CD14+ monocytes (with a fibroblast morphology) have been described as a source of multi-potent cells [44–46].

15.3 ASC Differentiation and Proliferation and Niche Plasticity

Adipose tissue is an endocrinologically active organ composed of adipocytes, macrophages, and vascular tissue, as well as ASCs [2, 47]. Both macrophages and adipocytes secrete adipokines that function as cytokines, chemokines, growth factors, and neurally active hormones. Obesity is understood to be a chronic low-grade inflammatory state. Increasing numbers of macrophages accumulate in the adipose tissue [48], which may result from increased secretion of pro-inflammatory cytokines. *In vitro* co-culture of differentiated 3 T3-L1 adipocytes and RAW 264 macrophages has been shown to result in significant upregulation of pro-inflammatory and downregulation of anti-inflammatory cytokines [49, 50]. Macrophages in turn influence adipocyte growth and proliferation, metabolism and secretory activity, through the production of cytokines and chemokines [51, 52] leading to increasing obesity and insulin resistance. Adipocytes may also modulate macrophage function, as macrophages express receptors to both leptin and adiponectin [53, 54] illustrating the importance of both paracrine and autocrine signaling in adipose tissue growth and function. As mature adipocytes do not undergo mitosis, an increase in adipocytes reflects differentiation of adipocyte precursors such as preadipocytes. Some studies suggest adipocytes can also dedifferentiate to preadipocytes [55], and adipocyte precursors and preadipocytes have also been recently observed to rapidly and efficiently differentiate into typical macrophages [56–58] demonstrating the considerable plasticity of these cells. On the other hand, macrophages and monocytes can differentiate to CD68(+), S100(+), CD14(−) dendritic cells, and monocytes in some studies demonstrate characteristics of circulating stem/progenitor cells [59]. Macrophages and adipocytes also both express adipsin (complement factor D), which is an adipocyte differentiation-dependent serine protease gene [60]. Additionally, macrophages in co-culture can accumulate cytoplasmic lipid vacuoles and, in direct co-culture with adipocytes, peritoneal macrophages can develop an elongated morphology with long cellular extensions [61]. Microarray studies comparing gene expression profiles of macrophages and progenitor/stem cells indicate that both cell types demonstrate the capacity for endocytosis, vesicle trafficking, and actin remodeling [62]. Monocytes/macrophages cooperate with progenitor cells during neovascularization and tissue repair [63]. There are several lines of evidence that suggest monocytes/macrophages have the ability to differentiate to endothelial cells *in vivo* and *in vitro* [45, 46, 63, 64].

It has been recently demonstrated that cell-to-cell co-culture between adipocytes and a 'macrophage fraction' containing adipose tissue macrophages (ATMS) and ASCs results in the robust proliferation of preadipocytes [21]. Chazenbalk et al. demonstrated that ASCs not only have the capacity to differentiate to preadipocytes/adipocytes, but also adipose tissue macrophages of hematopoietic cell origin can be converted to preadipocytes of a mesenchymal cell origin during cell-cell interaction between adipocytes and the 'macrophage fraction', underscoring the high degree of cell plasticity present in adipose tissue (Fig. 15.2). Cell-assisted lipotransfer (CAL) technology clearly indicates that interaction of ASCs and ASC-poor lipoaspirated material dramatically increased the number of ASCs in the aspirated fat [65, 66]. Taken together with the results of our co-culture experiments, these results suggest that ASCs delivered into an injured or diseased tissue not only secrete cytokines and growth factors that stimulate recovery in a paracrine fashion [67], but also utilize cell-cell interaction that may be critical for maximizing tissue regeneration. Another line of evidence as to the high degree of plasticity of adipose tissue is indicated by the capacity of adipocytes to lose fat/lipids and dedifferentiate to a new cell population termed dedifferentiated fat (DFAT) cells. Additionally, DFAT cells are a multi-potent cell population, differentiating to cardiomyocytes, smooth muscle, and osteoblast [68–71] (Fig. 15.3).

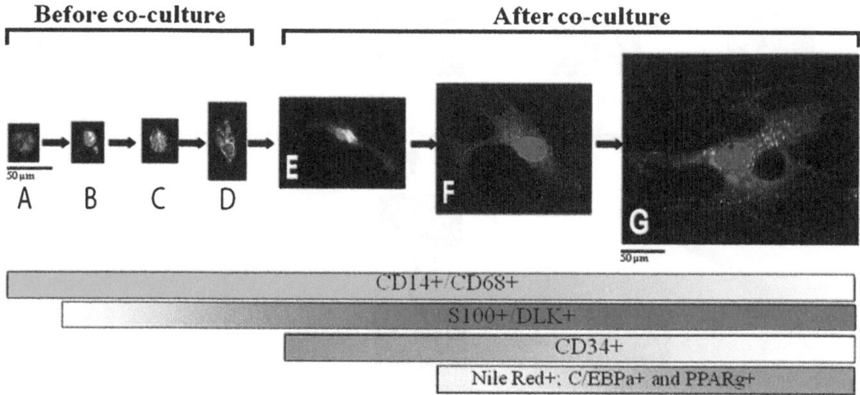

Fig. 15.2 Morphological changes exhibited during ATM differentiation to preadipocytes/adipocytes. Expression of markers CD14 (monocytes/macrophages) (*green*), S-100 (preadipocytes/adipocytes) (*red*), and CD34 (ASCs) (*dark green*), Nile Red (preadipocytes/adipocytes) (*orange*). **Before co-culture:** (**a**) most of the ATMs are CD14 (+); (**b–d**) very few ATMs are CD14 (+)/S-100 (+); **After co-culture:** (**e**) ATMs enlarged as they transform to preadipocytes (begin to express DLK/S-100 while maintaining CD14 expression); (**f**) Preadipocytes are S-100 (+) and DLK (+) and start losing CD14 and CD34 expression; (**g**) As preadipocytes start differentiating to adipocytes, there is an increase cell size, lipid accumulation (Nile Red (+) cells) and C/EBPα and PPARγgene expression. The gradient of the color inside the lower bars indicates the changes in cell expression markers in correlation with the morphological changes exhibited during ATMs differentiation to preadipocytes

15.4 Use of Adipose Tissue in Tissue Engineering and Regenerative Medicine

Tissue engineering and regenerative medicine combine the use of growth factors, biomaterials, artificial and synthetic scaffolds and matrixes, and stem cells to restore natural appearance or function. A critical constraint in tissue regeneration is neovascularization of the construct, as significant vasculogenesis may take up to 5 days, during which time graft cell death and tissue necrosis can occur [72]. Although autologous fat transplantation has been used since 1893, the clinical longevity of the graft is highly variable and the volume of large grafts decreases significantly over time [73]. Histologically, progressive loss of transplanted adipocytes is noted along with conversion of the graft to fibrous tissue and often, cyst formation, apparently the result of insufficient vascularization leading to hypoxia and cell death. However, some of the transplanted graft remains, adding to soft-tissue volume. It is thought that the recipient's native blood supply nourishes the portion of the graft adjacent to it and sustains it until vascularization occurs, while more central areas of the graft die. This theory has led to a decrease in fat graft sizes in the hope that smaller grafts will mean that more fat is adjacent to viable recipient tissue, resulting in improved availability and diffusion of cellular nutrition until neovascularization occurs [74, 75]. Using grafts rich in pluripotent adult adipose

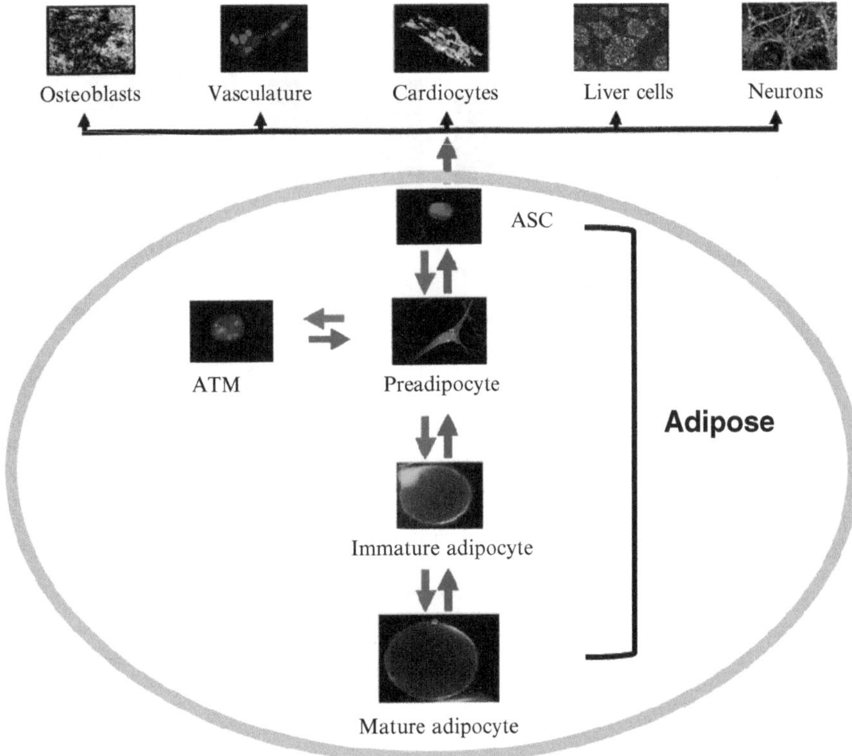

Fig. 15.3 Components of adipose tissue and cell plasticity: Adipose is composed of mature and immature adipocytes, preadipocytes, adipose tissue macrophages (ATMs), and Adipose-derived stem cells (ASCs). There is significant plasticity within this population, as indicated by *red* (differentiation) and *blue* (dedifferentiation) *arrows*. Additionally, ASCs have been shown to differentiate to: osteoblasts and chondrocytes; vasculature, both endothelial and smooth muscle components; cardiomyocytes, hepatocytes, and neuronal cells

stem cells represents a potential means of overcoming these limitations [76–78], which would have tremendous therapeutic implications for plastic, reconstructive, and regenerative medicine.

15.5 Wound Healing Pathogenesis

The most tangible and immediate use for ASCs in plastic and regenerative medicine is in the treatment of wounds. A multitude of treatment modalities have been utilized to treat a variety of wound pathologies. Ideally, physicians could achieve maximal effects by taking care to tailor ASC treatment to complement the patient's own healing processes, and enhance those processes that are dysfunctional. Additionally, basic pathophysiology of wound healing is observed, in some degree, in almost all disease states, and thus, serves as an invaluable guide in anticipating morbidities and dysfunction and in the determination ASC therapy for a given disease.

Determining the optimal therapy relies heavily on the current phase of healing at the area of injury allows one to anticipate dysfunction and choose. Cell migration, proliferation, extra cellular matrix (ECM) deposition, angiogenesis, and remodeling each take place during overlapping phases and have specific cells and cytokines associated with each step [78]. All healing consists of four key phases: hemostasis, inflammation, proliferation, and remodeling. Wound healing, regardless of tissue type, involves basic cellular mechanisms that one can target, to varying degrees with directed ASC therapy.

Inflammation begins after initial hemostasis, initiating with leukocyte migration to the site of injury. The first endogenous responders are often fibroblasts, spindle shaped mesenchymal cells that can be derived out of almost any connective tissue. During normal and healthy healing, inflammation can begin almost immediately after injury and lasts for days, often peaking between at about 2 days. Depending on the area and tissue type, MCP-1secretion draws endogenous macrophages or circulating monocytes to the site of injury where they secrete inflammatory cytokines such as TNF-α, IL-6, and NFκ-B. As the wound is bathed in inflammatory signals and cytokines a subset of fibroblasts differentiate into myofibroblasts, large contractile fibroblasts responsible for wound closure and tissue remodeling at the site of injury. While serving as anchors, these large flat cells secrete a multitude of cytokines and strongly influence the wound healing microenvironment during the first 2 weeks of injury. These cells actively contribute to ECM deposition before undergoing apoptosis, or more rarely, dedifferentiation to a fibroblast phenotype [79]. By this time angiogenesis has been initiated and pericytes guide endothelial invasion of microvasculature. After angiogenesis, the area of the wound may continually undergo remodeling and alteration of the ECM for up to 2 years after injury [80, 81].

If ASCs are injected directly into a wound site, a pool of fresh cells is offered to treat the site of injury and also take advantage of immunomodulatory effects. In this case ASCs would quickly be influenced by the surrounding paracrine or inflammatory signals and differentiate in response to endogenous cell signaling [82, 83]. With this approach the ASCs serve to support and enhance whichever phase of wound healing they have been transplanted into. For instance, if the wound is a recent injury undergoing initial cell migration, and these cells are actively secreting growth factors such as TGF-β1, ASC differentiation is driven to a fibroblast/ spindle-shaped preadipocyte type cell [78]. If the wound is undergoing angiogenesis with niche cells secreting factors such as VEGF and FGF-2, the ASCs will serve to enhance microvasculature as they differentiate towards a pericyte phenotype that offers support and guidance for endothelial cells [84], while an additional subset of ASCs undergoing mesenchymal to endothelial transformation [38, 64].

The second approach to ASC treatment of wounds would be to utilize the cells along with other biomaterials, synthetic or bio-matrix scaffolds, and a customized cocktail of growth factors and hormones to drive wound healing down a specific desired path [85–88]. This approach can be illustrated clearly when used to treat chronic wounds, where tissue dysfunction or wound pathogenesis is such that intervention is required. Chronic wounds present a host of difficulties due to a variety of factors and etiologies. For the most effective treatments ASCs are utilized in combination with therapies that take into account the particular type of injury and specific dysfunction in healing.

15.6 Inflammation and Immune Response in Determination of ASCs

Treatments of wounds are most effective when used in combinations that not only account for the presence of ischemia, senescence, and bacteria colonization, but also take into account the particular phase of healing; hemostasis, inflammation, proliferation, or remodeling. In regard to fibrotic tissue development in particular, one must take care that cells are not driven to blindly differentiate in an area of uncontrolled inflammation and fibrosis, especially in patient populations that are prone to forms of systemic fibrosis [89, 90]. Under optimal healing conditions wounds will often have scaring and fibrosis. Understanding the underlying cellular mechanisms of fibrosis can help direct ASC treatment with combinations of therapies that promote epithelialization or vascularization, as well as allow for the prediction of ASC action once being introduced to a microenvironment of active fibrosis.

Standard wound treatment includes methods of necrotic tissue debridement, draining and removal of edematous fluid, bacteria reduction, and moisture correction. Advances in tissue engineering and regenerative medicine have yielded novel tools that allow for targeted wound treatments. These tools include foams, gels, impregnated gauze, dressings, and new classes of self-assembling nanofibers that can accelerate hemostasis and serve as scaffolding for weeks after treatment [91–93]. New FDA approved therapies combine numerous proliferative growth factors, along with artificial scaffolding and even provide therapeutic fibroblasts [29]. These treatments can offer added healing efficiency by utilizing the synergistic effects of multiple "like-minded" growth factors, such as combining PDGF and FGF-2, a potent stimulator of mesenchymal proliferation and angiogenesis [94–96]. Additionally, new attention is being given to examine and improve common treatment modalities, such as pressure therapies including hyperbaric oxygen and negative pressure/vacuum therapy. Pressure treatments are utilized in a variety of wound types including chronic and necrotizing wounds, skin grafts, thermal and radiation burns, and infection [97, 98]. Characterized treatment indicates that benefits are the result of added access that allows for washing and draining of wounds, enhancing the formation of granulation tissue and reducing bacteria colonization [99]. However, recent evidence indicates pressure therapies can also promote low levels of the reactive oxygen species nitric oxide that activates and recruits endogenous stem progenitor cells of mesenchymal origin [100–103]. This in turn would modulate inflammation and immune response.

When hyper-activated, the inflammatory niche may cause monocytes, macrophages, fibroblasts, and myofiboblasts to produce inflammatory cytokines that can stimulate a positive feedback loop of constitutively activated cells that continually produce inflammatory signals, contractile stress fibers and add to ECM deposition [104]. Additionally, this feedback loop pushes mesenchymal cell proliferation, driving more fibroblast to differentiate and forego apoptosis, eventually forming fibrotic tissue. This inflammatory pathway is exacerbated in subsets of individuals that are susceptible to systemic inflammation and fibrosis [89, 104, 105]. In superficial wounds this process most commonly results in the formation of a hypertrophic scar or keloid. In other tissues and organs this deleterious process leads to fibrosis and eventually loss of function. Finally, dysfunctional matrix remodeling in fibrosis is characterized by continuous accumulation of ECM, which results in calcification [103, 106, 107]. With this in mind, care must be taken not to implant ASCs into an area of active fibrosis, as endogenous signals could easily activate the fibrotic process in the transplanted cells.

ASCs have demonstrated the ability to overcome chronic inflammation and fibrotic positive feedback loops through strong anti-inflammatory and immune modulation effects [62, 108, 109]. The anti-oxidation effects serve to increase blood vessel dilation and increase blood supply, thereby providing additional circulating stem cells to the area of injury [101, 108]. Numerous animal models have utilized ASCs to down regulate inflammatory cytokines [80, 110–112], or to stimulate the body's own immunoregulatory mechanisms, effectively altering the niche of the target tissue [113, 114]. Intravenous human MSCs were found to improve myocardial infarction in mice even after almost 97% of cells went to the lungs, liver, or spleen [115]. The protective effect was found to be a result of emboli formed in the lungs, where over 80% of the cells engrafted, and stimulated expression of TSG-6 was associated with reduced inflammatory response and reduced infarct size [115]. Immunomodulation along with reduced inflammation has been credited in additional studies that utilized MSCs and ASCs to effectively treat acute and chronic myocardial infarction, lung injury, diabetes, kidney disease, Graft versus Host, and neurodegenerative disorders [116–121]. Currently, a Phase 3 clinical trial is using intravenous infusion of MSCs have been effective in treating inflammatory (and perhaps auto-immune) conditions such as Crohn's disease, and while the vast majority of cells don't home in on damaged tissue but are believed to have immunosuppressive functions through paracrine secretions [122–124] [ClinicalTrials.gov: NCT00482092].

15.7 Chronic Wounds and Soft-Tissue Regeneration with ASCs

In therapeutic use, ASCs can be most readily and efficaciously be used in the treatment of superficial wounds and soft-tissue engraftment and regeneration [125]. When utilized in cosmetic or reconstructive surgery these cells greatly enhance recovery time while minimizing scar formation; effects attributed to both enhanced proliferation, angiogenesis, and immunomodulation [34, 125, 126]. These attributes are essential in ASC treatment of non-healing or chronic wounds.

The vast majority of chronic wounds consist of venous, pressure, and diabetic ulcers. Each year, approximately 500,000 people undergo treatment for Venous Ulcers alone [127]. In normal healthy wound healing, initial injury leads to hypoxia of surrounding tissue and cells causing the release of growth factors, breakdown of ECM, and believed activation of endogenous ASCs and MSCs to drive angiogenesis to enhance the formation of granulation tissue [111]. In contrast, chronic non-healing wounds are both complex and multi-factorial in etiology, the vast majority of which are characterized by the common causative factors of ischemia, senescence, and bacteria colonization.

Ischemia in chronic wounds is often compounded with repeated ischemia-reperfusion injury which is distinguished by both ischemic and oxidant stress. ASCs are uniquely adept at improving ischemia injuries, as they have a proven benefit to vascular health, promoting vascular growth and significantly increasing the expression of VEGF, angiopoietin-1, and FGF-2 [111, 128, 129]. Additionally, recent *in vivo* experiments found ASCs to protect kidneys against ischemia-reperfusion injury through suppression of inflammatory responses and oxidative stress [108]. Examination of anti-oxidative capacity of ASC treated mice displayed demonstrated increased NADPH quinone oxidoreductase-1, HO-1, glutathione peroxidase, and glutathione reductase activities [108].

Senescence includes the cellular and systemic changes observed with both aging tissues and patients. Chronic wound pathogenesis, along with associated morbidities such as obesity, insulin resistance, and diabetes increasingly manifest with age [130]. Additionally, aging patient populations have a reduced pool of cells that actively participate in healing, and while all stem and progenitor populations are diminished in the elderly and aging, ASCs represent the largest available source of adult stem cells [128, 131]. Interestingly, recent studies into human female ASC donor age and growth kinetics has shown that adipogenic capacity is not significantly altered with patient age, however, osteogenic proliferative potential was greatly reduced [132].

The final common factor observed in chronic wounds, bacteria colonization usually takes place with a dysfunctional host immune response, often concomitant with another pathologic condition. This may be amplified by a non-concordant inflammatory response, as bacterial invasion through closure failure of chronic wounds can be correlated directly with increases in metalloproteinases, intergrins, growth factors, and increased cytokine secretion [110, 125]. Moreover, wound closure failure rate is significantly higher if concentrations of bacteria (of any kind) exceed $10^6/mm^3$ [133]. ASCs can be utilized to reduce wound closure time through both cell driven closure and immunosuppressive paracrine effects, thereby reducing the window of bacterial invasion into a wound. However, the most beneficial effects of ASCs are synergistic and observed when combined with other remedial treatments [134].

In order to study the ability of ASCs to provide clinical soft-tissue support, Strem et al. developed an *in vivo* model for characterizing and optimizing stem cell-based methods for soft-tissue augmentation [4]. In the rodent, the subdermal scalp space represents a highly vascular and accessible yet demanding subcutaneous space for testing the implantation of soft-tissue constructs. Using this model, they found that freshly isolated ASCs, when mixed with minced donor fat, can improve the longevity and volume of the graft. Rosa ASCs or cultured Rosa ASCs were mixed with minced adipose tissue from B6;129 sF1/J mice and injected over the skulls of experimental athymic nude mice. Six months after transplantation, the grafted fat that was mixed with freshly isolated stem cells weighed 2.5 times more than the fat from the graft-only group. Further, the grafted fat that was mixed with ASCs maintained its adipocyte-rich appearance, whereas the grafted fat from the graft-only group, which lacked cell supplementation, had a more fibrous tissue appearance [135]. This suggests that ASCs may improve neovascularization in the graft, which is of key clinical importance.

By implementing techniques that can be less invasive, longer lasting, and almost scar free, the use of ASCs in soft-tissue regeneration stands to revolutionize the fields of cosmetic and reconstructive surgery. One novel method, the lipostructure technique was developed to preserve the fragile nature of the donor adipose tissue and gently transfer these small fat particles to locations near vascularized structures and associated nutrition [136]. Implementation of this technique has led to improved fat grafting outcomes for many surgeons. Combining stem cell therapy with lipostructure could potentially enhance tissue survivability, as stem cells have been shown to enhance angiogenesis [137, 138] and to minimize inflammatory responses [109, 139, 140] soft-tissue volume restoration in a patient with soft-tissue involution who had previously failed autologous fat grafting alone This case provided evidence that fat itself can act as a matrix for stem cells. Further, Yoshimura et al. have reported successful ASC-supplemented fat transplantation for soft-tissue fill in treating patients who either had demonstrable soft-tissue defects, or were undergoing breast augmentation [65, 141]. This technology, referred to as cell-assisted lipotransfer (CAL), has been successfully used for cosmetic breast augmentation and to treat facial lipoatrophy in hundreds of patients in Japan. This approach has been replicated to treat another soft-tissue disorder, Lipodystrophy, in a Phase 1 clinical trial that is active but not recruiting [ClinicalTrials.gov: NCT00715546].

15.8 ASCs in Musculoskeletal Regeneration

ASCs are easily coaxed into osteogenic, chondrogenic, and myogenic differentiation *in vitro* [1, 142–147]. Current therapies for cartilage, tendon, and muscle loss do not restore original function, [145] and disease and injury to these tissues is often progressive, debilitating, and extremely painful.

Studies of myocyte regeneration have demonstrated increased muscle mass and improved function [142, 148–151]. Interestingly, there are benefits beyond direct myocyte proliferation, as experiments utilizing ASCs to study peripheral nerve damage in a rat sciatic nerve model have found the added benefit of increased muscle mass in the hind limp after as little as 2 weeks, presumably due to enhanced nerve repair preventing muscle atrophy [152].

The original *in vivo* experiments that studied ASC differentiation to bone and cartilage derived the cells from human infrapatellar fat pads that were embedded into fibrin glue nodules, and could easily drive the cells to hyaline cartilage by culturing in chondrogenic media [153, 154]. Further, by transfecting ASCs to drive BMP-2 expression cells differentiated to bone and even established a marrow cavity [153, 155]. Follow up studies have since put into question the role of BMP2 in ASC osteogenic differentiation [156, 157]. Recent examination of canonical BMP2 signaling pathway showed no consistent effect on ECM mineralization, osteogenic markers, and no change in Smad1/5/8 phosphorylation, which lead the authors to call the utility of BMP2 induced in ASCs into question [156].

In a patient with major bone defect, Mesimaki et al. performed microvascular surgery in maxillary reconstruction using implanted ectopic bone produced by ASCs [158], and observed significant bone and vessel engraftment and growth. In this study, a micro-vascular flap was composed of cultured ASCs seeded onto a β-tricalcium phosphate scaffold with BMP-2. The flap was then implanted into the left rectus abdominis muscle for 8 months before being transplanted to the maxillofacial defect. While this procedure utilized an rh-BMP-2 containing scaffold, Mesimaki noted that there was likely no attributable change with the inclusion of BMP-2 [158].

The first major use of ASCs in human craniofacial defects was to stimulate bone regeneration in a calvarial defect [159]. Here, fibrin glue was engrafted with ASCs to promote significant osteogensis with bone formation three after months. These works have encouraged clinical trials to develop bone grafts using ASCs on various scaffolds, to repair large osseous defects by pre-engineering large synthetic bone and observing the vascularization *in vivo* [ClinicalTrials.gov: NCT01218945].

15.9 ASCs in Cardiovascular Therapies and Regeneration

Congestive heart failure and ischemic disease are the largest cause of morbidity and mortality in the Western world [160]. Prior to stem cell therapy, no treatment for acute or chronic ischemic heart disease could effectively replace or repair damaged and fibrotic cardiomyocytes and cardiofibroblast [161]. The ideal therapy would enhance cardiomyogenesis and vasculogenesis, while decreasing apoptosis and inflammation [162]. Treatment with ASCs provides a renewable source of functional cardiomyocytes as well as develops the vasculature needed to support new and recovering tissues in the ischemic myocardium [163, 164]. The strong angiogenic features of ASC therapy give it a distinct advantage BMSCs [165–167], as *in vitro* and *in vivo* models have shown hypoxic conditions to show greater ability in adipose stromal derived cells to stimulate blood vessel growth. Numerous studies have used cultured and freshly isolated ASCs to improve cardiac function in induced heart injury [167–172]. Additionally, cell transplantation of freshly-isolated ASCs to treat myocardial infarction provided a more beneficial effect than pre-differentiated ASC treatment, improving cell engraftment, differentiation, angiogenesis and fibrosis [166, 173].

While each study displayed a marked improvement with ASCs, several studies have shown conflicting results with *in vivo* treatments, with reports of cardiac marker expression [4] and more significantly, differentiation to a smooth muscle phenotype [168]. It is very likely that these variations are due to method of delivery and level of inflammation, as smooth muscle formation would most likely be do cardiofibroblast-to-cardiac myofibroblasts differentiation, which are activated by the injury and theoretically, may even show a response to the needle that delivers the cells [174, 175].

Initial clinical trials of adult stem cell therapy resulted in some improved function, but were greatly constrained by the number MSCs which could be isolated [176, 177]. Recently, Sanz-Ruiz et al. reported Phase 1 clinical trials using ASCs in patients with acute myocardial infarction (AMI), which is an on-going prospective, randomized, double-blind trial that has recruited 48 patients [162] [ClinicalTrials.gov: NCT00442806]. In this trial, ASCs are delivered through intracoronary infusion in patients with AMI and left ventricular (LV) ejection fraction impairment. Additionally, ASCs are being utilized for treatment of non-revascularizable ischemic myocardium [ClinicalTrials.gov: NCT00426868] in a Phase 1 clinical trial that has recruited 36 patients [162]. ASCs are isolated from lipoaspirate and injected transendocardial. While there have been advances, current knowledge and use of ASCs is still far behind that of BMSCs. Furthermore, a number of issues still persist, such as to other cell types, optimal cell dose, timing and mode of delivery, suitable patient cohorts, and demonstration of long-term safety and efficacy outcomes for more consistent clinical results [178–180].

15.10 ASCs in Nervous Tissue Regeneration

Adipose is dispersed throughout the body, with various depots displaying separate origins [31, 50, 181, 182]. While most adipose originates from the mesoderm, lineage studies have determined that some adipose, including the adipose of the skull, derives from primitive pericytes of neural crest origins [31]. Some ASCs and pericytes isolated from the adipose SVF are positive for both Neuron-glia Antigen 2 (NG2) and CD271 (L-NGFR). Perivascular derived pericytes are positive for NG2, a disialoganglioside found in the nervous system and neural crest cells, and these are believed to give rise to ectodermal MSCs and subsequent adipose [28, 140, 183]. CD271 or Low-Affinity Nerve Growth Factor Receptor (L-NGFR) is expressed in primitive stem cells from bone marrow and adipose origin, and is commonly expressed by neurons, Schwann cells, oligodendrocytes, astrocytes, dendritic cells, and neural crest cells [184, 185].

In vitro studies have demonstrated differentiation of ASCs into neural lineage cells and Schwann cells [186, 187]. ASC has been shown to enhance peripheral nerve healing, as well as potentially offer a suitable replacement for Schwann cells in therapeutic grafts [188]. Interestingly, when compared to bone marrow-derived MSCs, ASCs have significantly enhanced expression of the neural progenitor marker, nestin [189].

Positive effects of ASC therapy are attributed to differentiation, cytokine signaling, and anti-inflammatory effects [190–192]. Recent work by Lopatina et al. demonstrates that ASCs can stimulate the regeneration of nerves in innervated limbs of mice and also induce axon growth in subcutaneous matrigel implants [193]. Additionally, ASCs enhanced production of brain-derived neurotrophic factor (BDNF) as well as nerve fiber growth [193]. Treatments of spinal cord injury with ASC derived neurospheres resulted in less than 1% engraftment, and these cells differentiated to oligodendrocytes while stimulating endogenous oligodendrocyte proliferation [194]. Demylenating conditions, which can be induced by autoimmune reactions [195], may benefit from this neuroprotective effect. Diseases that result in injury to the oligodendroglial cells include multiple sclerosis [196] and leukodystrophies, and cerebral palsy [197].

Numerous *in vivo* animal models have demonstrated improvement of central nervous system injuries, documenting ASC differentiation to both astrocytes and Schwann cells [180, 194, 198, 199], although function was not gained in all studies [71]. Transplanted ASCs have also been demonstrated to slow both cellular degeneration and behavioral deterioration in quinolinic acid-induced mouse models of Huntington disease; improving limb clasping, attenuating striatal neuron death, and increasing overall survival [192]. Recent studies in the twitcher mouse model of Krabbe's disease, a neurodegenerative lysosomal storage disorder, demonstrated that intracerebroventricular administration of ASCs improved body weight and motor control [191]. These effects were presumably due in large part to strong anti-inflammatory regulation, with ASCs significantly reducing mRNA expression of TNF-α, MCP-1, and iNOS [191].

To this end, numerous clinical trials are currently being conducted to test the safety and efficiency of MSCs from bone marrow and adipose on a number of nervous system disorders including spinal cord injury, multiple sclerosis, ALS, stroke, Parkinson's, and neuroblastoma [29].

15.11 ASCs and Cancer: Cause or Cure?

A key concern in the application of ASCs, or any stem cell therapies for that matter, is the danger that implanted or infused cells will contribute to the progression of a cancer, either by negatively influencing the microenvironment or directly differentiating into cancerous stromal cells [200, 201]. In fact, because ASCs have the ability to regulate and differentiate into fibroblasts, myofibroblasts, fibrocytes, preadipocytes, macrophages, and pericytes, ASCs are in a position to greatly influence stromal niche promotion, especially in regards to regulating epithelial-to-mesenchymal transition (EMT) [200, 202]. This plasticity has also been demonstrated in carcinoma-associated MSCs, which promoted cancer growth by directly increasing cell number [11]. Dissecting the mechanism of this growth, cancer-associated MSCs were found to have higher expression of BMP-2, -4, and -6, and *in vivo* inhibition of BMP signaling was effective in abrogating MSC-associated tumor growth [11].

There has long been a known association between cancer and obesity [31, 203, 204]. Aside from having significantly larger pool of adipose and ASC to draw from, obese individuals often have significant low grade inflammation., often attributed to increased hypoxia and adipose death, but may also be partially due to the sympathetic state that is induced by obesity [204].

As obesity is a risk factor for prostate cancer, Lin et al. studied the influence of transplanted ASCs into transplanted PC3 prostate cancer cells and found ASCs to engraft as to tumors while doubling capillary density, correlating to increased FGF2 expression [205]. Additional studies confirmed that ASCs promoted tumor progression when co-injected with prostate cancer cells [206]. These results would implicate ASC angiogenic properties in the promotion of cancer. However, recent studies have indicated that ASC promotes breast cancer growth when co-injected with active cells, but not dormant tumor cells, indicating regenerative therapy with ASCs could be safely performed once there was no evidence of disease [207].

The majority of data has shown a positive relationship between cancers and ASCs, either through engraftment, cytokine signaling, or angiogenic effects [203, 205, 208, 209]. However, Cousin et al. demonstrated that ASCs can strongly inhibit the growth of pancreatic cancer cells *in vitro*, using co-cultures or conditioned medium, and *in vivo* [210]. In this study, pancreatic cell death was induced following G1-phase arrest without apoptosis. Additionally, ASC inhibitory effect was seen in a variety of epithelial cancers, including breast, liver, lung, and prostate [210, 211].

Interestingly, Altanerova et al. have recently reported a taking advantage of ASC seeming ability to ability to track and engraft into tumors and micrometastases to develop a novel cancer suicide gene therapy that utilizes ASCs that express a prodrug-converting enzyme to target and kill glioblastomas, human colon adenocarcinomas, and bone metastases of a prostate carcinoma [212–214]. These treatments hold the potential to target inoperable and aggressive cancers and could open up a new path of ASC therapy.

15.12 Conclusions

ASCs are now beginning to be effectively applied in cosmetic and reconstructive surgeries. However, much must now be done to clarify and develop techniques that allow precise and effective isolation, culturing, and engrafting/infusion of ASCs, in addition to studies aimed at determining ideal growth factor and biomaterial combinations for use in tissue engineering and regenerative medicine.

There also exists a large gap in knowledge as to ASC interaction with niche cells and the microenvironments of various target tissues. Our studies of adipose and its components have found modulation of the niche to generate ASCs, with co-cultures of macrophages and adipocytes driving macrophage to preadipocyte differentiation. This finding may not be unique, as macrophages are known to differentiate into macrophage-like cells that are key players in nearly every tissue and organ. Further elucidation of this cross-talk may open to understanding various disease pathologies, as well as open the door to more directed and efficient differentiation of ASCs.

While diverse, ASCs are perhaps the most attainable of all stem/progenitor cells. These cells demonstrate significant regenerative and therapeutic applications, increasing proliferation of damaged cells, while decreasing inflammation through immunomodulatory affects. Most significantly, as pericytes can be isolated from the adipose SVF, angiogenic properties of ASCs render these cells as perhaps the best option for ischemic-reperfusion injury and vascular related diseases. Many diseases of connective tissues lie at the intersection of defects in inflammation, auto-immunity, and ischemia. As such ASCs may be the ideal therapy hard to treat and non-healing conditions such as vasculitis, non-healing fractures, relapsing poly-chondritis, and immune associated conditions such as psoriasis and scleroderma.

Finally, more research is needed to determine the effects of ASCs on cancer, and to define the exact conditions and mechanisms that occur when and if ASCs engraft and promote tumor growth. On the other side of treating cancer, scientist are now utilizing ASCs ability to home into injured and cancerous tissues as they develop novel treatments for glioblastomas and prostate carcinomas, in the hopes of treating inoperable and chemo-resistant cancers.

References

1. Zuk PA, Zhu M, Mizuno H, Huang J, Futrell JW, Katz AJ et al (2001) Multilineage cells from human adipose tissue: implications for cell-based therapies. Tissue Eng 7(2):211–228
2. Zuk PA, Zhu M, Ashjian P, De Ugarte DA, Huang JI, Mizuno H et al (2002) Human adipose tissue is a source of multipotent stem cells. Mol Biol Cell 13(12):4279–4295, PMCID: 138633
3. Rodriguez AM, Pisani D, Dechesne CA, Turc-Carel C, Kurzenne JY, Wdziekonski B et al (2005) Transplantation of a multipotent cell population from human adipose tissue induces dystrophin expression in the immunocompetent mdx mouse. J Exp Med 201(9):1397–1405, PMCID: 2213197
4. Strem BM, Zhu M, Alfonso Z, Daniels EJ, Schreiber R, Beygui R et al (2005) Expression of cardiomyocytic markers on adipose tissue-derived cells in a murine model of acute myocardial injury. Cytotherapy 7(3):282–291
5. Tholpady SS, Aojanepong C, Llull R, Jeong JH, Mason AC, Futrell JW et al (2005) The cellular plasticity of human adipocytes. Ann Plast Surg 54(6):651–656
6. Katz AJ, Tholpady A, Tholpady SS, Shang H, Ogle RC (2005) Cell surface and transcriptional characterization of human adipose-derived adherent stromal (hADAS) cells. Stem Cells 23(3):412–423
7. Caplan AI (2007) Adult mesenchymal stem cells for tissue engineering versus regenerative medicine. J Cell Physiol 213(2):341–347
8. Gimble JM, Katz AJ, Bunnell BA (2007) Adipose-derived stem cells for regenerative medicine. Circ Res 100(9):1249–1260
9. D'Ippolito G, Schiller PC, Ricordi C, Roos BA, Howard GA (1999) Age-related osteogenic potential of mesenchymal stromal stem cells from human vertebral bone marrow. J Bone Miner Res 14(7):1115–1122
10. Savitz SI, Dinsmore JH, Wechsler LR, Rosenbaum DM, Caplan LR (2004) Cell therapy for stroke. NeuroRx 1(4):406–414, PMCID: 534949

11. McLean K, Gong Y, Choi Y, Deng N, Yang K, Bai S et al (2011) Human ovarian carcinoma-associated mesenchymal stem cells regulate cancer stem cells and tumorigenesis via altered BMP production. J Clin Invest 121(8):3206–3219
12. Vacanti V, Kong E, Suzuki G, Sato K, Canty JM, Lee T (2005) Phenotypic changes of adult porcine mesenchymal stem cells induced by prolonged passaging in culture. J Cell Physiol 205(2):194–201
13. Aust L, Devlin B, Foster SJ, Halvorsen YD, Hicok K, du Laney T et al (2004) Yield of human adipose-derived adult stem cells from liposuction aspirates. Cytotherapy 6(1):7–14
14. Pittenger MF, Mackay AM, Beck SC, Jaiswal RK, Douglas R, Mosca JD et al (1999) Multilineage potential of adult human mesenchymal stem cells. Science 284(5411):143–147
15. Miyazaki M, Zuk PA, Zou J, Yoon SH, Wei F, Morishita Y et al (2008) Comparison of human mesenchymal stem cells derived from adipose tissue and bone marrow for ex vivo gene therapy in rat spinal fusion model. Spine (Phila Pa 1976) 33(8):863–869
16. Garg A (2004) Regional adiposity and insulin resistance. J Clin Endocrinol Metab 89(9):4206–4210
17. Scherer PE (2006) Adipose tissue: from lipid storage compartment to endocrine organ. Diabetes 55(6):1537–1545
18. Dixit VD (2008) Adipose-immune interactions during obesity and caloric restriction: reciprocal mechanisms regulating immunity and health span. J Leukoc Biol 84(4):882–892, PMCID: 2638733
19. Hegele RA, Leff T (2004) Unbuckling lipodystrophy from insulin resistance and hypertension. J Clin Invest 114(2):163–165, PMCID: 449754
20. Rudich A, Ben-Romano R, Etzion S, Bashan N (2005) Cellular mechanisms of insulin resistance, lipodystrophy and atherosclerosis induced by HIV protease inhibitors. Acta Physiol Scand 183(1):75–88
21. Chazenbalk G, Bertolotto C, Heneidi S, Jumabay M, Trivax B, Aronowitz J et al (2011) Novel pathway of adipogenesis through cross-talk between adipose tissue macrophages, adipose stem cells and adipocytes: evidence of cell plasticity. PLoS One 6(3):e17834, PMCID: 3069035
22. Chazenbalk G, Trivax BS, Yildiz BO, Bertolotto C, Mathur R, Heneidi S et al (2010) Regulation of adiponectin secretion by adipocytes in the polycystic ovary syndrome: role of tumor necrosis factor-{alpha}. J Clin Endocrinol Metab 95(2):935–942, PMCID: 2840865
23. Lago F, Dieguez C, Gomez-Reino J, Gualillo O (2007) Adipokines as emerging mediators of immune response and inflammation. Nat Clin Pract Rheumatol 3(12):716–724
24. Steppan CM, Bailey ST, Bhat S, Brown EJ, Banerjee RR, Wright CM et al (2001) The hormone resistin links obesity to diabetes. Nature 409(6818):307–312
25. Ahima RS, Lazar MA (2008) Adipokines and the peripheral and neural control of energy balance. Mol Endocrinol 22(5):1023–1031, PMCID: 2366188
26. Gesta S, Tseng YH, Kahn CR (2007) Developmental origin of fat: tracking obesity to its source. Cell 131(2):242–256
27. Dominici M, Le Blanc K, Mueller I, Slaper-Cortenbach I, Marini F, Krause D et al (2006) Minimal criteria for defining multipotent mesenchymal stromal cells. The International Society for Cellular Therapy position statement. Cytotherapy 8(4):315–317
28. Tallone T, Realini C, Bohmler A, Kornfeld C, Vassalli G, Moccetti T et al (2011) Adult human adipose tissue contains several types of multipotent cells. J Cardiovasc Transl Res 4(2):200–210
29. Ankrum J, Karp JM (2010) Mesenchymal stem cell therapy: two steps forward, one step back. Trends Mol Med 16(5):203–209, PMCID: 2881950
30. Daquinag AC, Zhang Y, Amaya-Manzanares F, Simmons PJ, Kolonin MG (2011) An isoform of decorin is a resistin receptor on the surface of adipose progenitor cells. Cell Stem Cell 9(1):74–86
31. Billon N, Monteiro MC, Dani C (2008) Developmental origin of adipocytes: new insights into a pending question. Biol Cell 100(10):563–575
32. Howson KM, Aplin AC, Gelati M, Alessandri G, Parati EA, Nicosia RF (2005) The postnatal rat aorta contains pericyte progenitor cells that form spheroidal colonies in suspension culture. Am J Physiol Cell Physiol 289(6):C1396–C1407

33. De Francesco F, Tirino V, Desiderio V, Ferraro G, D'Andrea F, Giuliano M et al (2009) Human CD34/CD90 ASCs are capable of growing as sphere clusters, producing high levels of VEGF and forming capillaries. PLoS One 4(8):e6537, PMCID: 2717331
34. Hong SJ, Traktuev DO, March KL (2010) Therapeutic potential of adipose-derived stem cells in vascular growth and tissue repair. Curr Opin Organ Transplant 15(1):86–91
35. Korn J, Christ B, Kurz H (2002) Neuroectodermal origin of brain pericytes and vascular smooth muscle cells. J Comp Neurol 442(1):78–88
36. Dore-Duffy P, Katychev A, Wang X, Van Buren E (2006) CNS microvascular pericytes exhibit multipotential stem cell activity. J Cereb Blood Flow Metab 26(5):613–624
37. Traktuev DO, Merfeld-Clauss S, Li J, Kolonin M, Arap W, Pasqualini R et al (2008) A population of multipotent CD34-positive adipose stromal cells share pericyte and mesenchymal surface markers, reside in a periendothelial location, and stabilize endothelial networks. Circ Res 102(1):77–85
38. Traktuev DO, Prater DN, Merfeld-Clauss S, Sanjeevaiah AR, Saadatzadeh MR, Murphy M et al (2009) Robust functional vascular network formation in vivo by cooperation of adipose progenitor and endothelial cells. Circ Res 104(12):1410–1420
39. Diaz-Flores L, Gutierrez R, Madrid JF, Varela H, Valladares F, Acosta E et al (2009) Pericytes. Morphofunction, interactions and pathology in a quiescent and activated mesenchymal cell niche. Histol Histopathol 24(7):909–969
40. Ingram DA, Mead LE, Moore DB, Woodard W, Fenoglio A, Yoder MC (2005) Vessel wall-derived endothelial cells rapidly proliferate because they contain a complete hierarchy of endothelial progenitor cells. Blood 105(7):2783–2786
41. Ingram DA, Mead LE, Tanaka H, Meade V, Fenoglio A, Mortell K et al (2004) Identification of a novel hierarchy of endothelial progenitor cells using human peripheral and umbilical cord blood. Blood 104(9):2752–2760
42. Cousin B, Andre M, Arnaud E, Penicaud L, Casteilla L (2003) Reconstitution of lethally irradiated mice by cells isolated from adipose tissue. Biochem Biophys Res Commun 301(4):1016–1022
43. Varma MJ, Breuls RG, Schouten TE, Jurgens WJ, Bontkes HJ, Schuurhuis GJ et al (2007) Phenotypical and functional characterization of freshly isolated adipose tissue-derived stem cells. Stem Cells Dev 16(1):91–104
44. Majka SM, Fox KE, Psilas JC, Helm KM, Childs CR, Acosta AS et al (2010) De novo generation of white adipocytes from the myeloid lineage via mesenchymal intermediates is age, adipose depot, and gender specific. Proc Natl Acad Sci USA 107(33):14781–14786, PMCID: 2930432
45. Kuwana M, Okazaki Y, Kodama H, Izumi K, Yasuoka H, Ogawa Y et al (2003) Human circulating CD14+ monocytes as a source of progenitors that exhibit mesenchymal cell differentiation. J Leukoc Biol 74(5):833–845
46. Zhao Y, Glesne D, Huberman E (2003) A human peripheral blood monocyte-derived subset acts as pluripotent stem cells. Proc Natl Acad Sci USA 100(5):2426–2431, PMCID: 151357
47. Fraser JK, Wulur I, Alfonso Z, Hedrick MH (2006) Fat tissue: an underappreciated source of stem cells for biotechnology. Trends Biotechnol 24(4):150–154
48. Bouloumie A, Curat CA, Sengenes C, Lolmede K, Miranville A, Busse R (2005) Role of macrophage tissue infiltration in metabolic diseases. Curr Opin Clin Nutr Metab Care 8(4):347–354
49. Chen JS, Chen YL, Greenberg AS, Chen YJ, Wang SM (2005) Magnolol stimulates lipolysis in lipid-laden RAW 264.7 macrophages. J Cell Biochem 94(5):1028–1037
50. Greenberg AS, Obin MS (2006) Obesity and the role of adipose tissue in inflammation and metabolism. Am J Clin Nutr 83(2):461S–465S
51. Xu H, Barnes GT, Yang Q, Tan G, Yang D, Chou CJ et al (2003) Chronic inflammation in fat plays a crucial role in the development of obesity-related insulin resistance. J Clin Invest 112(12):1821–1830, PMCID: 296998
52. Suganami T, Nishida J, Ogawa Y (2005) A paracrine loop between adipocytes and macrophages aggravates inflammatory changes: role of free fatty acids and tumor necrosis factor alpha. Arterioscler Thromb Vasc Biol 25(10):2062–2068

53. Gainsford T, Willson TA, Metcalf D, Handman E, McFarlane C, Ng A et al (1996) Leptin can induce proliferation, differentiation, and functional activation of hemopoietic cells. Proc Natl Acad Sci USA 93(25):14564–14568, PMCID: 26173
54. Weisberg SP, McCann D, Desai M, Rosenbaum M, Leibel RL, Ferrante AW Jr (2003) Obesity is associated with macrophage accumulation in adipose tissue. J Clin Invest 112(12):1796–1808, PMCID: 296995
55. Yagi K, Kondo D, Okazaki Y, Kano K (2004) A novel preadipocyte cell line established from mouse adult mature adipocytes. Biochem Biophys Res Commun 321(4):967–974
56. Cousin B, Munoz O, Andre M, Fontanilles AM, Dani C, Cousin JL et al (1999) A role for preadipocytes as macrophage-like cells. FASEB J 13(2):305–312
57. Charriere G, Cousin B, Arnaud E, Andre M, Bacou F, Penicaud L et al (2003) Preadipocyte conversion to macrophage. Evidence of plasticity. J Biol Chem 278(11):9850–9855
58. Prunet-Marcassus B, Cousin B, Caton D, Andre M, Penicaud L, Casteilla L (2006) From heterogeneity to plasticity in adipose tissues: site-specific differences. Exp Cell Res 312(6):727–736
59. Seta N, Kuwana M (2007) Human circulating monocytes as multipotential progenitors. Keio J Med 56(2):41–47
60. Festy F, Hoareau L, Bes-Houtmann S, Pequin AM, Gonthier MP, Munstun A et al (2005) Surface protein expression between human adipose tissue-derived stromal cells and mature adipocytes. Histochem Cell Biol 124(2):113–121
61. Lumeng CN, Deyoung SM, Saltiel AR (2007) Macrophages block insulin action in adipocytes by altering expression of signaling and glucose transport proteins. Am J Physiol Endocrinol Metab 292(1):E166–E174
62. Charriere GM, Cousin B, Arnaud E, Saillan-Barreau C, Andre M, Massoudi A et al (2006) Macrophage characteristics of stem cells revealed by transcriptome profiling. Exp Cell Res 312(17):3205–3214
63. Anghelina M, Moldovan L, Zabuawala T, Ostrowski MC, Moldovan NI (2006) A subpopulation of peritoneal macrophages form capillarylike lumens and branching patterns in vitro. J Cell Mol Med 10(3):708–715
64. Rehman J, Traktuev D, Li J, Merfeld-Clauss S, Temm-Grove CJ, Bovenkerk JE et al (2004) Secretion of angiogenic and antiapoptotic factors by human adipose stromal cells. Circulation 109(10):1292–1298
65. Yoshimura K, Sato K, Aoi N, Kurita M, Hirohi T, Harii K (2008) Cell-assisted lipotransfer for cosmetic breast augmentation: supportive use of adipose-derived stem/stromal cells. Aesthetic Plast Surg 32(1):48–55, discussion 6-7. PMCID: 2175019
66. Yoshimura K, Sato K, Aoi N, Kurita M, Inoue K, Suga H et al (2008) Cell-assisted lipotransfer for facial lipoatrophy: efficacy of clinical use of adipose-derived stem cells. Dermatol Surg 34(9):1178–1185
67. Zannettino AC, Paton S, Arthur A, Khor F, Itescu S, Gimble JM et al (2008) Multipotential human adipose-derived stromal stem cells exhibit a perivascular phenotype in vitro and in vivo. J Cell Physiol 214(2):413–421
68. Oki Y, Watanabe S, Endo T, Kano K (2008) Mature adipocyte-derived dedifferentiated fat cells can trans-differentiate into osteoblasts in vitro and in vivo only by all-trans retinoic acid. Cell Struct Funct 33(2):211–222
69. Jumabay M, Matsumoto T, Yokoyama S, Kano K, Kusumi Y, Masuko T et al (2009) Dedifferentiated fat cells convert to cardiomyocyte phenotype and repair infarcted cardiac tissue in rats. J Mol Cell Cardiol 47(5):565–575
70. Jumabay M, Zhang R, Yao Y, Goldhaber JI, Bostrom KI (2010) Spontaneously beating cardiomyocytes derived from white mature adipocytes. Cardiovasc Res 85(1):17–27, PMCID: 2791054
71. Ohta Y, Takenaga M, Tokura Y, Hamaguchi A, Matsumoto T, Kano K et al (2008) Mature adipocyte-derived cells, dedifferentiated fat cells (DFAT), promoted functional recovery from spinal cord injury-induced motor dysfunction in rats. Cell Transplant 17(8):877–886
72. Borges J, Mueller MC, Padron NT, Tegtmeier F, Lang EM, Stark GB (2003) Engineered adipose tissue supplied by functional microvessels. Tissue Eng 9(6):1263–1270

73. Ersek RA, Chang P, Salisbury MA (1998) Lipo layering of autologous fat: an improved technique with promising results. Plast Reconstr Surg 101(3):820–826
74. Yamaguchi M, Matsumoto F, Bujo H, Shibasaki M, Takahashi K, Yoshimoto S et al (2005) Revascularization determines volume retention and gene expression by fat grafts in mice. Exp Biol Med (Maywood) 230(10):742–748
75. Carpaneda CA, Ribeiro MT (1994) Percentage of graft viability versus injected volume in adipose autotransplants. Aesthetic Plast Surg 18(1):17–19
76. Chang EI, Bonillas RG, El-ftesi S, Ceradini DJ, Vial IN, Chan DA et al (2009) Tissue engineering using autologous microcirculatory beds as vascularized bioscaffolds. FASEB J 23(3):906–915, PMCID: 2653982
77. Gurtner GC, Chang E (2008) "Priming" endothelial progenitor cells: a new strategy to improve cell based therapeutics. Arterioscler Thromb Vasc Biol 28(6):1034–1035
78. Gurtner GC, Werner S, Barrandon Y, Longaker MT (2008) Wound repair and regeneration. Nature 453(7193):314–321
79. Frangogiannis NG, Michael LH, Entman ML (2000) Myofibroblasts in reperfused myocardial infarcts express the embryonic form of smooth muscle myosin heavy chain (SMemb). Cardiovasc Res 48(1):89–100
80. Badillo AT, Redden RA, Zhang L, Doolin EJ, Liechty KW (2007) Treatment of diabetic wounds with fetal murine mesenchymal stromal cells enhances wound closure. Cell Tissue Res 329(2):301–311
81. Medina A, Scott PG, Ghahary A, Tredget EE (2005) Pathophysiology of chronic nonhealing wounds. J Burn Care Rehabil 26(4):306–319
82. Bai X, Alt E (2010) Myocardial regeneration potential of adipose tissue-derived stem cells. Biochem Biophys Res Commun 401(3):321–326
83. Kim WS, Park BS, Sung JH, Yang JM, Park SB, Kwak SJ et al (2007) Wound healing effect of adipose-derived stem cells: a critical role of secretory factors on human dermal fibroblasts. J Dermatol Sci 48(1):15–24
84. Blanton MW, Hadad I, Johnstone BH, Mund JA, Rogers PI, Eppley BL et al (2009) Adipose stromal cells and platelet-rich plasma therapies synergistically increase revascularization during wound healing. Plast Reconstr Surg 123(2 Suppl):56S–64S
85. Altman AM, Abdul Khalek FJ, Seidensticker M, Pinilla S, Yan Y, Coleman M et al (2010) Human tissue-resident stem cells combined with hyaluronic acid gel provide fibrovascular-integrated soft-tissue augmentation in a murine photoaged skin model. Plast Reconstr Surg 125(1):63–73
86. Nakamura S, Kishimoto S, Nambu M, Fujita M, Tanaka Y, Mori Y et al (2010) Fragmin/protamine microparticles as cell carriers to enhance viability of adipose-derived stromal cells and their subsequent effect on in vivo neovascularization. J Biomed Mater Res A 92(4):1614–1622
87. Park A, Hogan MV, Kesturu GS, James R, Balian G, Chhabra AB (2010) Adipose-derived mesenchymal stem cells treated with growth differentiation factor-5 express tendon-specific markers. Tissue Eng Part A 16(9):2941–2951, PMCID: 2928041
88. Takikawa M, Nakamura S, Nambu M, Ishihara M, Fujita M, Kishimoto S et al (2011) Enhancement of vascularization and granulation tissue formation by growth factors in human platelet-rich plasma-containing fragmin/protamine microparticles. J Biomed Mater Res B Appl Biomater 97(2):373–380
89. Edward M, Quinn JA, Mukherjee S, Jensen MB, Jardine AG, Mark PB et al (2008) Gadodiamide contrast agent 'activates' fibroblasts: a possible cause of nephrogenic systemic fibrosis. J Pathol 214(5):584–593
90. Bellini A, Mattoli S (2007) The role of the fibrocyte, a bone marrow-derived mesenchymal progenitor, in reactive and reparative fibroses. Lab Invest 87(9):858–870
91. Luo Z, Wang S, Zhang S (2011) Fabrication of self-assembling D-form peptide nanofiber scaffold d-EAK16 for rapid hemostasis. Biomaterials 32(8):2013–2020
92. Ruan L, Zhang H, Luo H, Liu J, Tang F, Shi YK et al (2009) Designed amphiphilic peptide forms stable nanoweb, slowly releases encapsulated hydrophobic drug, and accelerates animal hemostasis. Proc Natl Acad Sci USA 106(13):5105–5110, PMCID: 2663994

93. Murray MM, Spindler KP, Ballard P, Welch TP, Zurakowski D, Nanney LB (2007) Enhanced histologic repair in a central wound in the anterior cruciate ligament with a collagen-platelet-rich plasma scaffold. J Orthop Res 25(8):1007–1017
94. Thomopoulos S, Das R, Sakiyama-Elbert S, Silva MJ, Charlton N, Gelberman RH (2010) bFGF and PDGF-BB for tendon repair: controlled release and biologic activity by tendon fibroblasts in vitro. Ann Biomed Eng 38(2):225–234, PMCID: 2843401
95. Nihsen ES, Zopf DA, Ernst DM, Janis AD, Hiles MC, Johnson C (2007) Absorption of bioactive molecules into OASIS wound matrix. Adv Skin Wound Care 20(10):541–548
96. Chang Y, Ceacareanu B, Zhuang D, Zhang C, Pu Q, Ceacareanu AC et al (2006) Counter-regulatory function of protein tyrosine phosphatase 1B in platelet-derived growth factor- or fibroblast growth factor-induced motility and proliferation of cultured smooth muscle cells and in neointima formation. Arterioscler Thromb Vasc Biol 26(3):501–507
97. Impellizzeri P, Dardik H, Shah HJ, Brotman-O'Neil A, Ibrahim IM (2011) Vacuum-assisted closure therapy with omental transposition for salvage of infected prosthetic femoral-distal bypass involving the femoral anastomosis. J Vasc Surg 54(4):1154–1156
98. Nather A, Chionh SB, Han AY, Chan PP, Nambiar A (2010) Effectiveness of vacuum-assisted closure (VAC) therapy in the healing of chronic diabetic foot ulcers. Ann Acad Med Singapore 39(5):353–358
99. Tamhankar AP, Ravi K, Everitt NJ (2009) Vacuum assisted closure therapy in the treatment of mesh infection after hernia repair. Surgeon 7(5):316–318
100. Childress B, Stechmiller JK, Schultz GS (2008) Arginine metabolites in wound fluids from pressure ulcers: a pilot study. Biol Res Nurs 10(2):87–92
101. Guzik TJ, Korbut R, Adamek-Guzik T (2003) Nitric oxide and superoxide in inflammation and immune regulation. J Physiol Pharmacol 54(4):469–487
102. Feldmeier J, Carl U, Hartmann K, Sminia P (2003) Hyperbaric oxygen: Does it promote growth or recurrence of malignancy? Undersea Hyperb Med 30(1):1–18
103. Jorgensen TB, Sorensen AM, Jansen EC (2008) Iatrogenic systemic air embolism treated with hyperbaric oxygen therapy. Acta Anaesthesiol Scand 52(4):566–568
104. Desmouliere A, Chaponnier C, Gabbiani G (2005) Tissue repair, contraction, and the myofibroblast. Wound Repair Regen 13(1):7–12
105. Quaggin SE, Kapus A (2011) Scar wars: mapping the fate of epithelial-mesenchymal-myofibroblast transition. Kidney Int 80(1):41–50
106. Kim L, Kim do K, Yang WI, Shin DH, Jung IM, Park HK et al (2008) Overexpression of transforming growth factor-beta 1 in the valvular fibrosis of chronic rheumatic heart disease. J Korean Med Sci 23(1):41–48, PMCID: 2526480
107. Yoon YS, Park JS, Tkebuchava T, Luedeman C, Losordo DW (2004) Unexpected severe calcification after transplantation of bone marrow cells in acute myocardial infarction. Circulation 109(25):3154–3157
108. Chen YT, Sun CK, Lin YC, Chang LT, Chen YL, Tsai TH et al (2011) Adipose-derived mesenchymal stem cell protects kidneys against ischemia-reperfusion injury through suppressing oxidative stress and inflammatory reaction. J Transl Med 9:51, PMCID: 3112438
109. Puissant B, Barreau C, Bourin P, Clavel C, Corre J, Bousquet C et al (2005) Immunomodulatory effect of human adipose tissue-derived adult stem cells: comparison with bone marrow mesenchymal stem cells. Br J Haematol 129(1):118–129
110. Ebrahimian TG, Pouzoulet F, Squiban C, Buard V, Andre M, Cousin B et al (2009) Cell therapy based on adipose tissue-derived stromal cells promotes physiological and pathological wound healing. Arterioscler Thromb Vasc Biol 29(4):503–510
111. Eto H, Suga H, Inoue K, Aoi N, Kato H, Araki J et al (2011) Adipose injury-associated factors mitigate hypoxia in ischemic tissues through activation of adipose-derived stem/progenitor/stromal cells and induction of angiogenesis. Am J Pathol 178(5):2322–2332, PMCID: 3081200
112. Kubo N, Narumi S, Kijima H, Mizukami H, Yagihashi S, Hakamada K et al (2011) Efficacy of adipose tissue-derived mesenchymal stem cells for fulminant hepatitis in mice induced by concanavalin A. J Gastroenterol Hepatol doi: 10.111/j. 1440–1746.2011.06798.x

113. Lin G, Garcia M, Ning H, Banie L, Guo YL, Lue TF et al (2008) Defining stem and progenitor cells within adipose tissue. Stem Cells Dev 17(6):1053–1063, PMCID: 2865901
114. Siegel G, Schafer R, Dazzi F (2009) The immunosuppressive properties of mesenchymal stem cells. Transplantation 87(9 Suppl):S45–S49
115. Lee RH, Pulin AA, Seo MJ, Kota DJ, Ylostalo J, Larson BL et al (2009) Intravenous hMSCs improve myocardial infarction in mice because cells embolized in lung are activated to secrete the anti-inflammatory protein TSG-6. Cell Stem Cell 5(1):54–63
116. Ortiz LA, Dutreil M, Fattman C, Pandey AC, Torres G, Go K et al (2007) Interleukin 1 receptor antagonist mediates the antiinflammatory and antifibrotic effect of mesenchymal stem cells during lung injury. Proc Natl Acad Sci USA 104(26):11002–11007, PMCID: 1891813
117. Ortiz LA, Gambelli F, McBride C, Gaupp D, Baddoo M, Kaminski N et al (2003) Mesenchymal stem cell engraftment in lung is enhanced in response to bleomycin exposure and ameliorates its fibrotic effects. Proc Natl Acad Sci USA 100(14):8407–8411, PMCID: 166242
118. Kunter U, Rong S, Djuric Z, Boor P, Muller-Newen G, Yu D et al (2006) Transplanted mesenchymal stem cells accelerate glomerular healing in experimental glomerulonephritis. J Am Soc Nephrol 17(8):2202–2212
119. Lee RH, Seo MJ, Reger RL, Spees JL, Pulin AA, Olson SD et al (2006) Multipotent stromal cells from human marrow home to and promote repair of pancreatic islets and renal glomeruli in diabetic NOD/scid mice. Proc Natl Acad Sci USA 103(46):17438–17443, PMCID: 1634835
120. Ringden O, Uzunel M, Rasmusson I, Remberger M, Sundberg B, Lonnies H et al (2006) Mesenchymal stem cells for treatment of therapy-resistant graft-versus-host disease. Transplantation 81(10):1390–1397
121. Phinney DG, Isakova I (2005) Plasticity and therapeutic potential of mesenchymal stem cells in the nervous system. Curr Pharm Des 11(10):1255–1265
122. Erhayiem B, Dhingsa R, Hawkey CJ, Subramanian V (2011) Ratio of visceral to subcutaneous fat area is a biomarker of complicated Crohn's disease. Clin Gastroenterol Hepatol 9(8):684.e1–687.e1
123. Garcia-Olmo D, Herreros D, De-La-Quintana P, Guadalajara H, Trebol J, Georgiev-Hristov T et al (2010) Adipose-derived stem cells in Crohn's rectovaginal fistula. Case Rep Med 2010:961758, PMCID: 2833320
124. Olivier I, Theodorou V, Valet P, Castan-Laurell I, Guillou H, Bertrand-Michel J et al (2011) Is Crohn's creeping fat an adipose tissue? Inflamm Bowel Dis 17(3):747–757
125. Cherubino M, Rubin JP, Miljkovic N, Kelmendi-Doko A, Marra KG (2011) Adipose-derived stem cells for wound healing applications. Ann Plast Surg 66(2):210–215
126. Brem H, Kodra A, Golinko MS, Entero H, Stojadinovic O, Wang VM et al (2009) Mechanism of sustained release of vascular endothelial growth factor in accelerating experimental diabetic healing. J Invest Dermatol 129(9):2275–2287
127. Simka M, Majewski E (2003) The social and economic burden of venous leg ulcers: focus on the role of micronized purified flavonoid fraction adjuvant therapy. Am J Clin Dermatol 4(8):573–581
128. Kim JH, Jung M, Kim HS, Kim YM, Choi EH (2011) Adipose-derived stem cells as a new therapeutic modality for ageing skin. Exp Dermatol 20(5):383–387
129. Tse KH, Kingham PJ, Novikov LN, Wiberg M (2011) Adipose tissue and bone marrow-derived stem cells react similarly in an ischaemia-like microenvironment. J Tissue Eng Regen Med. doi: 10.1002/term.452
130. Guo W, Pirtskhalava T, Tchkonia T, Xie W, Thomou T, Han J et al (2007) Aging results in paradoxical susceptibility of fat cell progenitors to lipotoxicity. Am J Physiol Endocrinol Metab 292(4):E1041–E1051
131. Madonna R, Renna FV, Cellini C, Cotellese R, Picardi N, Francomano F et al (2011) Age-dependent impairment of number and angiogenic potential of adipose tissue-derived progenitor cells. Eur J Clin Invest 41(2):126–133

132. Zhu M, Kohan E, Bradley J, Hedrick M, Benhaim P, Zuk P (2009) The effect of age on osteogenic, adipogenic and proliferative potential of female adipose-derived stem cells. J Tissue Eng Regen Med 3(4):290–301
133. Robson MC (1997) Wound infection. A failure of wound healing caused by an imbalance of bacteria. Surg Clin North Am 77(3):637–650
134. Madonna R, De Caterina R (2011) Stem cells and growth factor delivery systems for cardiovascular disease. J Biotechnol 154(4):291–297
135. Moseley TA, Zhu M, Hedrick MH (2006) Adipose-derived stem and progenitor cells as fillers in plastic and reconstructive surgery. Plast Reconstr Surg 118(3 Suppl):121S–128S
136. Coleman SR (2006) Structural fat grafting: more than a permanent filler. Plast Reconstr Surg 118(3 Suppl):108S–120S
137. Hausman DB, Lu J, Ryan DH, Flatt WP, Harris RB (2004) Compensatory growth of adipose tissue after partial lipectomy: involvement of serum factors. Exp Biol Med (Maywood) 229(6):512–520
138. Planat-Benard V, Menard C, Andre M, Puceat M, Perez A, Garcia-Verdugo JM et al (2004) Spontaneous cardiomyocyte differentiation from adipose tissue stroma cells. Circ Res 94(2):223–229
139. Ryden M, Dicker A, Gotherstrom C, Astrom G, Tammik C, Arner P et al (2003) Functional characterization of human mesenchymal stem cell-derived adipocytes. Biochem Biophys Res Commun 311(2):391–397
140. Tang W, Zeve D, Suh JM, Bosnakovski D, Kyba M, Hammer RE et al (2008) White fat progenitor cells reside in the adipose vasculature. Science 322(5901):583–586, PMCID: 2597101
141. Yoshimura K, Asano Y, Aoi N, Kurita M, Oshima Y, Sato K et al (2010) Progenitor-enriched adipose tissue transplantation as rescue for breast implant complications. Breast J 16(2):169–175
142. Mizuno H, Zuk PA, Zhu M, Lorenz HP, Benhaim P, Hedrick MH (2002) Myogenic differentiation by human processed lipoaspirate cells. Plast Reconstr Surg 109(1):199–209, discussion 10-11
143. Hao W, Pang L, Jiang M, Lv R, Xiong Z, Hu YY (2010) Skeletal repair in rabbits using a novel biomimetic composite based on adipose-derived stem cells encapsulated in collagen I gel with PLGA-beta-TCP scaffold. J Orthop Res 28(2):252–257
144. Hattori H, Sato M, Masuoka K, Ishihara M, Kikuchi T, Matsui T et al (2004) Osteogenic potential of human adipose tissue-derived stromal cells as an alternative stem cell source. Cells Tissues Organs 178(1):2–12
145. Oakes BW (2004) Orthopaedic tissue engineering: from laboratory to the clinic. Med J Aust 180(5 Suppl):S35–S38
146. Obaid H, Connell D (2010) Cell therapy in tendon disorders: what is the current evidence? Am J Sports Med 38(10):2123–2132
147. Peterson B, Zhang J, Iglesias R, Kabo M, Hedrick M, Benhaim P et al (2005) Healing of critically sized femoral defects, using genetically modified mesenchymal stem cells from human adipose tissue. Tissue Eng 11(1–2):120–129
148. Di Rocco G, Iachininoto MG, Tritarelli A, Straino S, Zacheo A, Germani A et al (2006) Myogenic potential of adipose-tissue-derived cells. J Cell Sci 119(Pt 14):2945–2952
149. Gimble JM, Grayson W, Guilak F, Lopez MJ, Vunjak-Novakovic G (2011) Adipose tissue as a stem cell source for musculoskeletal regeneration. Front Biosci (Schol Ed) 3:69–81
150. Jeon ES, Moon HJ, Lee MJ, Song HY, Kim YM, Bae YC et al (2006) Sphingosylphosphorylcholine induces differentiation of human mesenchymal stem cells into smooth-muscle-like cells through a TGF-beta-dependent mechanism. J Cell Sci 119(Pt 23):4994–5005
151. Kim M, Choi YS, Yang SH, Hong HN, Cho SW, Cha SM et al (2006) Muscle regeneration by adipose tissue-derived adult stem cells attached to injectable PLGA spheres. Biochem Biophys Res Commun 348(2):386–392

152. Scholz T, Sumarto A, Krichevsky A, Evans GR (2011) Neuronal differentiation of human adipose tissue-derived stem cells for peripheral nerve regeneration in vivo. Arch Surg 146(6):666–674
153. Dragoo JL, Choi JY, Lieberman JR, Huang J, Zuk PA, Zhang J et al (2003) Bone induction by BMP-2 transduced stem cells derived from human fat. J Orthop Res 21(4):622–629
154. Dragoo JL, Carlson G, McCormick F, Khan-Farooqi H, Zhu M, Zuk PA et al (2007) Healing full-thickness cartilage defects using adipose-derived stem cells. Tissue Eng 13(7):1615–1621
155. Dragoo JL, Samimi B, Zhu M, Hame SL, Thomas BJ, Lieberman JR et al (2003) Tissue-engineered cartilage and bone using stem cells from human infrapatellar fat pads. J Bone Jt Surg Br 85(5):740–747
156. Zuk P, Chou YF, Mussano F, Benhaim P, Wu BM (2011) Adipose-derived stem cells and BMP2: part 2. BMP2 may not influence the osteogenic fate of human adipose-derived stem cells. Connect Tissue Res 52(2):119–132
157. Chou YF, Zuk PA, Chang TL, Benhaim P, Wu BM (2011) Adipose-derived stem cells and BMP2: part 1. BMP2-treated adipose-derived stem cells do not improve repair of segmental femoral defects. Connect Tissue Res 52(2):109–118
158. Mesimaki K, Lindroos B, Tornwall J, Mauno J, Lindqvist C, Kontio R et al (2009) Novel maxillary reconstruction with ectopic bone formation by GMP adipose stem cells. Int J Oral Maxillofac Surg 38(3):201–209
159. Lendeckel S, Jodicke A, Christophis P, Heidinger K, Wolff J, Fraser JK et al (2004) Autologous stem cells (adipose) and fibrin glue used to treat widespread traumatic calvarial defects: case report. J Craniomaxillofac Surg 32(6):370–373
160. Heron M, Hoyert DL, Murphy SL, Xu J, Kochanek KD, Tejada-Vera B (2009) Deaths: final data for 2006. Natl Vital Stat Rep 57(14):1–134
161. Psaltis PJ, Zannettino AC, Worthley SG, Gronthos S (2008) Concise review: mesenchymal stromal cells: potential for cardiovascular repair. Stem Cells 26(9):2201–2210
162. Sanz-Ruiz R, Gutierrez Ibanes E, Arranz AV, Fernandez Santos ME, Fernandez PL, Fernandez-Aviles F (2010) Phases I-III clinical trials using adult stem cells. Stem Cells Int 2010:579142, PMCID: 2975079
163. Laflamme MA, Murry CE (2005) Regenerating the heart. Nat Biotechnol 23(7):845–856
164. Madonna R, De Caterina R (2010) Adipose tissue: a new source for cardiovascular repair. J Cardiovasc Med (Hagerstown) 11(2):71–80
165. Efimenko A, Starostina EE, Rubina KA, Kalinina NI, Parfenova EV (2010) Viability and angiogenic activity of mesenchymal stromal cells from adipose tissue and bone marrow in hypoxia and inflammation in vitro. Tsitologiia 52(2):144–154
166. van der Bogt KE, Schrepfer S, Yu J, Sheikh AY, Hoyt G, Govaert JA et al (2009) Comparison of transplantation of adipose tissue- and bone marrow-derived mesenchymal stem cells in the infarcted heart. Transplantation 87(5):642–652, PMCID: 2866004
167. Wang L, Deng J, Tian W, Xiang B, Yang T, Li G et al (2009) Adipose-derived stem cells are an effective cell candidate for treatment of heart failure: an MR imaging study of rat hearts. Am J Physiol Heart Circ Physiol 297(3):H1020–H1031
168. Cai L, Johnstone BH, Cook TG, Tan J, Fishbein MC, Chen PS et al (2009) IFATS collection: human adipose tissue-derived stem cells induce angiogenesis and nerve sprouting following myocardial infarction, in conjunction with potent preservation of cardiac function. Stem Cells 27(1):230–237, PMCID: 2936459
169. Schenke-Layland K, Strem BM, Jordan MC, Deemedio MT, Hedrick MH, Roos KP et al (2009) Adipose tissue-derived cells improve cardiac function following myocardial infarction. J Surg Res 153(2):217–223, PMCID: 2700056
170. van der Bogt KE, Sheikh AY, Schrepfer S, Hoyt G, Cao F, Ransohoff KJ et al (2008) Comparison of different adult stem cell types for treatment of myocardial ischemia. Circulation 118(14 Suppl):S121–S129
171. Miyahara Y, Nagaya N, Kataoka M, Yanagawa B, Tanaka K, Hao H et al (2006) Monolayered mesenchymal stem cells repair scarred myocardium after myocardial infarction. Nat Med 12(4):459–465

172. Danoviz ME, Nakamuta JS, Marques FL, dos Santos L, Alvarenga EC, dos Santos AA et al (2010) Rat adipose tissue-derived stem cells transplantation attenuates cardiac dysfunction post infarction and biopolymers enhance cell retention. PLoS One 5(8):e12077, PMCID: 2919414
173. Mazo M, Planat-Benard V, Abizanda G, Pelacho B, Leobon B, Gavira JJ et al (2008) Transplantation of adipose derived stromal cells is associated with functional improvement in a rat model of chronic myocardial infarction. Eur J Heart Fail 10(5):454–462
174. Thompson SA, Copeland CR, Reich DH, Tung L (2011) Mechanical coupling between myofibroblasts and cardiomyocytes slows electric conduction in fibrotic cell monolayers. Circulation 123(19):2083–2093
175. Freyman T, Polin G, Osman H, Crary J, Lu M, Cheng L et al (2006) A quantitative, randomized study evaluating three methods of mesenchymal stem cell delivery following myocardial infarction. Eur Heart J 27(9):1114–1122
176. Piao H, Youn TJ, Kwon JS, Kim YH, Bae JW, Bora S et al (2005) Effects of bone marrow derived mesenchymal stem cells transplantation in acutely infarcting myocardium. Eur J Heart Fail 7(5):730–738
177. Zimmet JM, Hare JM (2005) Emerging role for bone marrow derived mesenchymal stem cells in myocardial regenerative therapy. Basic Res Cardiol 100(6):471–481
178. Payne GA, Kohr MC, Tune JD (2011) Epicardial perivascular adipose tissue as a therapeutic target in obesity-related coronary artery disease. Br J Pharmacol; doi: 10.1111/j.1476–5381.2011.01370
179. Mazo M, Gavira JJ, Pelacho B, Prosper F (2011) Adipose-derived stem cells for myocardial infarction. J Cardiovasc Transl Res 4(2):145–153
180. Tobita M, Orbay H, Mizuno H (2011) Adipose-derived stem cells: current findings and future perspectives. Discov Med 11(57):160–170
181. Casteilla L, Planat-Benard V, Laharrague P, Cousin B (2011) Adipose-derived stromal cells: their identity and uses in clinical trials, an update. World J Stem Cells 3(4):25–33, PMCID: 3097937
182. Sepe A, Tchkonia T, Thomou T, Zamboni M, Kirkland JL (2011) Aging and regional differences in fat cell progenitors - a mini-review. Gerontology 57(1):66–75, PMCID: 3031153
183. Takashima Y, Era T, Nakao K, Kondo S, Kasuga M, Smith AG et al (2007) Neuroepithelial cells supply an initial transient wave of MSC differentiation. Cell 129(7):1377–1388
184. Quirici N, Scavullo C, de Girolamo L, Lopa S, Arrigoni E, Deliliers GL et al (2010) Anti-L-NGFR and -CD34 monoclonal antibodies identify multipotent mesenchymal stem cells in human adipose tissue. Stem Cells Dev 19(6):915–925
185. Ishimura D, Yamamoto N, Tajima K, Ohno A, Yamamoto Y, Washimi O et al (2008) Differentiation of adipose-derived stromal vascular fraction culture cells into chondrocytes using the method of cell sorting with a mesenchymal stem cell marker. Tohoku J Exp Med 216(2):149–156
186. Fujimura J, Ogawa R, Mizuno H, Fukunaga Y, Suzuki H (2005) Neural differentiation of adipose-derived stem cells isolated from GFP transgenic mice. Biochem Biophys Res Commun 333(1):116–121
187. Kingham PJ, Kalbermatten DF, Mahay D, Armstrong SJ, Wiberg M, Terenghi G (2007) Adipose-derived stem cells differentiate into a Schwann cell phenotype and promote neurite outgrowth in vitro. Exp Neurol 207(2):267–274
188. di Summa PG, Kingham PJ, Raffoul W, Wiberg M, Terenghi G, Kalbermatten DF (2010) Adipose-derived stem cells enhance peripheral nerve regeneration. J Plast Reconstr Aesthet Surg 63(9):1544–1552
189. Safford KM, Hicok KC, Safford SD, Halvorsen YD, Wilkison WO, Gimble JM et al (2002) Neurogenic differentiation of murine and human adipose-derived stromal cells. Biochem Biophys Res Commun 294(2):371–379
190. Zavan B, Vindigni V, Gardin C, D'Avella D, Della Puppa A, Abatangelo G et al (2010) Neural potential of adipose stem cells. Discov Med 10(50):37–43
191. Ripoll CB, Flaat M, Klopf-Eiermann J, Fisher-Perkins JM, Trygg CB, Scruggs BA et al (2011) Mesenchymal lineage stem cells have pronounced anti-inflammatory effects in the twitcher mouse model of Krabbe's disease. Stem Cells 29(1):67–77

192. Lee ST, Chu K, Jung KH, Im WS, Park JE, Lim HC et al (2009) Slowed progression in models of Huntington disease by adipose stem cell transplantation. Ann Neurol 66(5):671–681
193. Lopatina T, Kalinina N, Karagyaur M, Stambolsky D, Rubina K, Revischin A et al (2011) Adipose-derived stem cells stimulate regeneration of peripheral nerves: BDNF secreted by these cells promotes nerve healing and axon growth de novo. PLoS One 6(3):e17899, PMCID: 3056777
194. Zhang HT, Cheng HY, Cai YQ, Ma X, Liu WP, Yan ZJ et al (2009) Comparison of adult neurospheres derived from different origins for treatment of rat spinal cord injury. Neurosci Lett 458(3):116–121
195. Ben-Hur T, Einstein O, Mizrachi-Kol R, Ben-Menachem O, Reinhartz E, Karussis D et al (2003) Transplanted multipotential neural precursor cells migrate into the inflamed white matter in response to experimental autoimmune encephalomyelitis. Glia 41(1):73–80
196. Kuhlmann T, Miron V, Cui Q, Wegner C, Antel J, Bruck W (2008) Differentiation block of oligodendroglial progenitor cells as a cause for remyelination failure in chronic multiple sclerosis. Brain 131(Pt 7):1749–1758
197. Girolamo F, Ferrara G, Strippoli M, Rizzi M, Errede M, Trojano M et al (2011) Cerebral cortex demyelination and oligodendrocyte precursor response to experimental autoimmune encephalomyelitis. Neurobiol Dis 43(3):678–689
198. Ryu HH, Lim JH, Byeon YE, Park JR, Seo MS, Lee YW et al (2009) Functional recovery and neural differentiation after transplantation of allogenic adipose-derived stem cells in a canine model of acute spinal cord injury. J Vet Sci 10(4):273–284, PMCID: 2807262
199. Kang SK, Shin MJ, Jung JS, Kim YG, Kim CH (2006) Autologous adipose tissue-derived stromal cells for treatment of spinal cord injury. Stem Cells Dev 15(4):583–594
200. Bissell MJ, Radisky DC, Rizki A, Weaver VM, Petersen OW (2002) The organizing principle: microenvironmental influences in the normal and malignant breast. Differentiation 70(9-10):537–546, PMCID: 2933198
201. Donnenberg VS, Zimmerlin L, Rubin JP, Donnenberg AD (2010) Regenerative therapy after cancer: what are the risks? Tissue Eng Part B Rev 16(6):567–575, PMCID: 3011999
202. Spaeth EL, Dembinski JL, Sasser AK, Watson K, Klopp A, Hall B et al (2009) Mesenchymal stem cell transition to tumor-associated fibroblasts contributes to fibrovascular network expansion and tumor progression. PLoS One 4(4):e4992, PMCID: 2661372
203. Iyengar P, Combs TP, Shah SJ, Gouon-Evans V, Pollard JW, Albanese C et al (2003) Adipocyte-secreted factors synergistically promote mammary tumorigenesis through induction of anti-apoptotic transcriptional programs and proto-oncogene stabilization. Oncogene 22(41):6408–6423
204. Zhang Y, Bellows CF, Kolonin MG (2010) Adipose tissue-derived progenitor cells and cancer. World J Stem Cells 2(5):103–113, PMCID: 3097931
205. Lin G, Yang R, Banie L, Wang G, Ning H, Li LC et al (2010) Effects of transplantation of adipose tissue-derived stem cells on prostate tumor. Prostate 70(10):1066–1073, PMCID: 2877148
206. Prantl L, Muehlberg F, Navone NM, Song YH, Vykoukal J, Logothetis CJ et al (2010) Adipose tissue-derived stem cells promote prostate tumor growth. Prostate 70(15):1709–1715
207. Zimmerlin L, Donnenberg AD, Rubin JP, Basse P, Landreneau RJ, Donnenberg VS (2011) Regenerative therapy and cancer: in vitro and in vivo studies of the interaction between adipose-derived stem cells and breast cancer cells from clinical isolates. Tissue Eng Part A 17(1-2):93–106, PMCID: 3011910
208. Muehlberg FL, Song YH, Krohn A, Pinilla SP, Droll LH, Leng X et al (2009) Tissue-resident stem cells promote breast cancer growth and metastasis. Carcinogenesis 30(4):589–597
209. Pinilla S, Alt E, Abdul Khalek FJ, Jotzu C, Muehlberg F, Beckmann C et al (2009) Tissue resident stem cells produce CCL5 under the influence of cancer cells and thereby promote breast cancer cell invasion. Cancer Lett 284(1):80–85

210. Cousin B, Ravet E, Poglio S, De Toni F, Bertuzzi M, Lulka H et al (2009) Adult stromal cells derived from human adipose tissue provoke pancreatic cancer cell death both in vitro and in vivo. PLoS One 4(7):e6278, PMCID: 2707007
211. Sun B, Roh KH, Park JR, Lee SR, Park SB, Jung JW et al (2009) Therapeutic potential of mesenchymal stromal cells in a mouse breast cancer metastasis model. Cytotherapy 11(3):289–298, 1 p following 98
212. Altanerova V, Horvathova E, Matuskova M, Kucerova L, Altaner C (2009) Genotoxic damage of human adipose-tissue derived mesenchymal stem cells triggers their terminal differentiation. Neoplasma 56(6):542–547
213. Altanerova V, Cihova M, Babic M, Rychly B, Ondicova K, Mravec B et al (2011) Human adipose tissue-derived mesenchymal stem cells expressing yeast cytosinedeaminase::uracil phosphoribosyltransferase inhibit intracerebral rat glioblastoma. Int J Cancer; doi: 10.1002/ijc.26278
214. Cavarretta IT, Altanerova V, Matuskova M, Kucerova L, Culig Z, Altaner C (2010) Adipose tissue-derived mesenchymal stem cells expressing prodrug-converting enzyme inhibit human prostate tumor growth. Mol Ther 18(1):223–231, PMCID: 2839205

Chapter 16
Neural Crest and Hirschsprung's Disease

Kim Hei-Man Chow, Paul Kwong-Hang Tam, and Elly Sau-Wai Ngan

Contents

K.H.-M. Chow
Department of Surgery, Li Ka Shing Faculty of Medicine, University of Hong Kong, Pokfulam, Faculty of Medicine Building, 21 Sassoon Road, Hong Kong, SAR, China

Stem Cell and Regenerative Medicine Consortium, Li Ka Shing Faculty of Medicine, Li Ka Shing Faculty of Medicine, University of Hong Kong, Pokfulam, Hong Kong, China

P.K.-H. Tam
Department of Surgery, Li Ka Shing Faculty of Medicine, University of Hong Kong, Pokfulam, Faculty of Medicine Building, 21 Sassoon Road, Hong Kong, SAR, China

Centre for Reproduction, Development and Growth, Li Ka Shing Faculty of Medicine, University of Hong Kong, Pokfulam, Hong Kong, China

E.S.-W. Ngan (✉)
Department of Surgery, Li Ka Shing Faculty of Medicine, University of Hong Kong, Pokfulam, Faculty of Medicine Building, 21 Sassoon Road, Hong Kong, SAR, China

Centre for Reproduction, Development and Growth, Li Ka Shing Faculty of Medicine, University of Hong Kong, Pokfulam, Hong Kong, China

Stem Cell and Regenerative Medicine Consortium, Li Ka Shing Faculty of Medicine, Li Ka Shing Faculty of Medicine, University of Hong Kong, Pokfulam, Hong Kong, China
e-mail: engan@hku.hk

R.K. Srivastava and S. Shankar (eds.), *Stem Cells and Human Diseases*, DOI 10.1007/978-94-007-2801-1_16, © Springer Science+Business Media B.V. 2012

Abstract Neural crest cells are a transient population of stem cells in vertebrates that give rise to the entire peripheral nervous system (PNS) as well as various non-neural progenies. A peculiar control and coordination of proliferation, migration and differentiation is required for neural crest cells to generate a full diversity of progenies, navigate different organs and establish functional domains in their target organs. Defects in such developmental process may lead to a board spectrum of congenital disorders, and in some cases, also cancer. In this review, we will focus on one specific neurocristopathy in the PNS: the Hirschsprung's disease (colonic aganglionosis), to emphasize how unraveling the molecular mechanisms underlying the neural crest cell fate determination and progression may facilitate our understanding of the disease etiologies and future development of therapies.

Keywords Hirschsprung's disease • Epithelial-mesenchymal transition • RET receptor tyrosine kinase • Neural stem cells • Hedgehog • Notch

16.1 Introduction

Neural crest cells are a population of multipotent and highly migratory progenitor cells originated from the borders of the neural fold in vertebrate embryo. After undergoing epithelial-mesenchymal transition (EMT), neural crest cells delaminate from the neuroepithelium, migrate to different organs, and give rise to various cell types, including the sensory and autonomic ganglia of the peripheral nervous system (PNS), melanocytes in skin, endocrine cells in the thyroid and adrenal glands, smooth muscle cells in the heart, craniofacial cartilage and bone, adipocytes and connective tissues. Abnormalities of neural crest development in human yield a large array of diseases, such as multiple neoplasia, skeletal syndromes, Hirschsprung's disease (aganglionic megacolon), pigment defects and conotruncal heart malformations.

In this review, we will first give an overview of our current understanding on the multipotency nature of neural crest cells. Then we will use Hirschsprung's disease (HSCR) as a model to summarize the recent insights into the disease etiology gained by revealing the molecular basis of neural crest development. Lastly, we will also highlight the current state-of the art studies on stem cell biology which may facilitate a systematic analysis of disease pathogenesis and development of cell-based therapies for these diseases.

16.2 Neural Crest Cells in Vertebrate Development

In vertebrates, neural crest is formed as a transient structure upon the fusion of neural folds, become mesenchymal in nature and delaminated from the neuroepithelium [1]. The delaminated neural crest cells undergo a phase of extensive

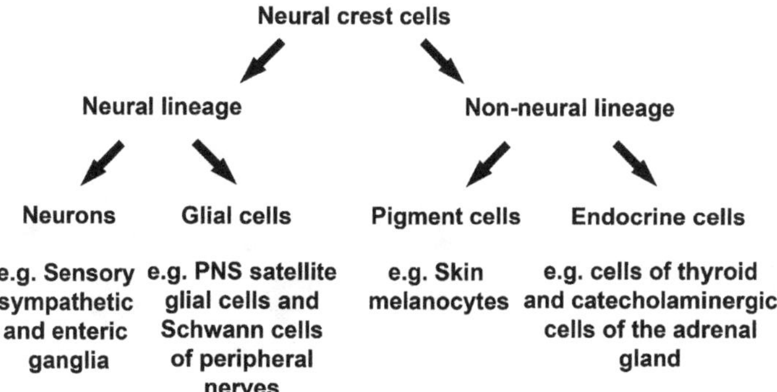

Fig. 16.1 The main cell types derived from the neural crest

migration and become widely distributed within the embryo. The induction process is regulated by a blander of signals originated from the surrounding mesoderm and non-neural ectoderm, including those of bone morphological protein (BMP), Notch, Wingless (Wnt), and fibroblast growth factor (FGF) signaling pathways [2, 3]. Neural crest cells are multipotent in nature and can give rise to a hierarchy of cell progenies in adulthood. In general, the progenies can be sub-classified into neural and non-neural lineages (Fig. 16.1). Neural cell lineages include the sensory, sympathetic and enteric neurons, satellite glial cells, and Schwann cells making up the peripheral nervous system (PNS); while non-neural cells are the pigment cells (skin melanocytes), endocrine cells of the thyroid and adrenal glands (chromaffin cells), and mesenchymal cells in head and neck, which can eventually be further differentiated into connective tissue cells, tendons, cartilage, bone, vascular smooth muscles, and adipocytes.

Lineage commitment of neural crest cells is found to either be stochastic or environmentally determined by extracellular signals which are not mutually exclusive [4, 5]. These include cell-to-cell interactions, exposure to different soluble cues and cell autonomous genetic programs. For example during neural lineage commitment, bone morphogenetic proteins (BMP) are found to promote the differentiation of neurons but prevent the maturation of glial cells [6]. Neuregulin (NRG) promotes the commitment towards Schwann cell lineage [7] while brain-derived neurotrophic factor (BDNF) directs the differentiation towards primary sensory neuron lineage [8] and neurotrophin-3 (NTP-3) promotes the neural and glial cell commitment [9]. Extracellular matrix favors the formation of catecholaminergic neurons. Culture of quail trunk neural crest with a gel of reconstituted basement membrane (RBM) components induces the formation of catecholaminergic-positive cells [5]. In addition, nerve growth factor (NGF) is known to be a mitogen favoring sympathetic and sensory neuron maturation and survival. Long-term culture of chromaffin cells with NGF can induce the conversion of adrenal chromaffin cells into neurons [10]; whereas glucocorticoid is required for the maintenance of the endocrine phenotype of the chromaffin cells [10]. Acidic (aFGF) and basic FGF (bFGF) are similar to

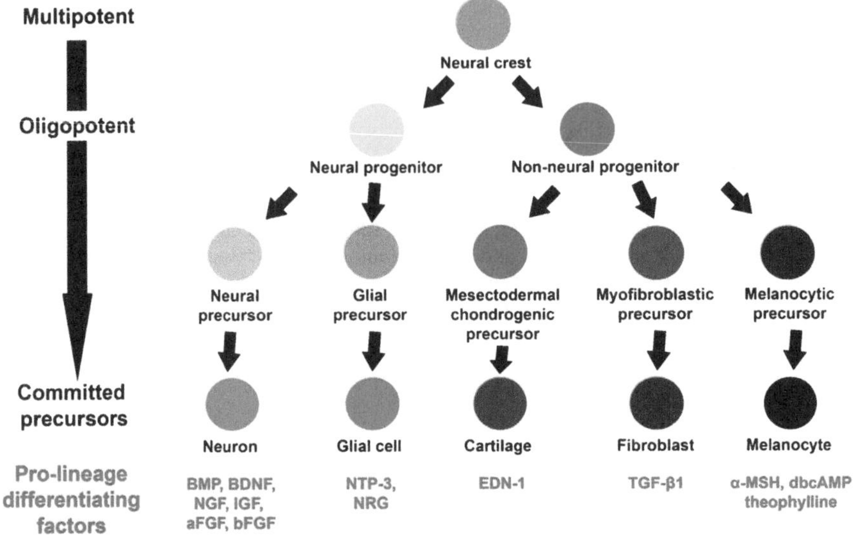

Fig. 16.2 A hierarchical model for neural crest lineage diversification

NGF, which can act locally to stimulate mitotic expansion and initial axon growth of developing ganglia [10, 11]. Such effect of both FGFs and NGF can be further enhanced by addition of insulin-like growth factors (IGFs) which significantly increases the responsiveness of chromaffin cell to these factors [12]. For non-neuronal lineage commitment, sonic hedgehog (Shh) is found to contribute to the tissue patterning in addition to promote cell survival and proliferation of neural crest cells [13]. Other factors like α-melanocyte stimulating hormone (α-MSH) or dibutyryl cyclic AMP (dbcAMP) plus theophylline promotes the neural crest commitment towards melanogenic sub-lineage [14]. Treatment of quail neural crest cells in the presence of these factors results in a faster pigmentation and larger pigmented colonies formation than the untreated cells, suggesting α-MSH not only accelerates melanogenic differentiation but also affects the fate of neural crest towards melanogenic differentiation *in vitro* [13]. The endothelin (EDN) signaling pathways have discrete roles during embryonic development by influencing the fate of different neural crest cell populations. EDNRA receptor/EDN-1 signaling primarily acts on the cephalic neural crest supporting their invasion to the pharyngeal arches and the formation of bones and cartilage of the facial skeleton, dermis, and smooth muscle of the great arteries. EDNRB receptor, on the other hand, is expressed in the vagal and truncal neural crest, through which to promote their differentiation into sympathetic and sensory neurons, and melanocytes [15]. Transforming growth factor beta-1 (TGF-β1) is a factor that activates calcineurin signaling in neural crest cells and eventually leads to their differentiation into smooth muscle cells (Fig. 16.2) [16].

The "stemness" nature of neural crest cells has been demonstrated in many different ways. First, neural crest cells are remarkably multipotent, giving rise to a variety of cell types of different organs as summarized above. Indeed, the plasticity of neural crest cells was first observed in a culture of pre-migratory neural crest cells isolated from quail embryos, in which clones of pigmented and non-pigmented cells were found [17]. Phenotypic analysis then proves these cells with diverse developmental potencies and they are able to differentiate into neurons, non-neuronal cells, melanocytes, and cells with intermediate developmental potencies [18]. Lineage analysis on quail neural crest cells subsequently illustrates that these cells can generate a diverse types of progenies consisted of sensory neurons, presumptive pigment cells, ganglionic supportive cells, and adrenomedullary cells [19]. Much more work has been done in mouse, and more recently, even with neural crest cells isolated from human tissues (see below). Secondly, neural crest cells possess self-renewal property. Early studies shows the presence of undifferentiated neural crest cells in colonies of varied types of differentiated cells, which supports the idea of their self-renewal capability [17, 18, 20–22]. This idea has been further reinforced recently following the increasing evidence suggesting the presence of undifferentiated neural crest cells in the peripheral systems of adults (such as skin [23–32], bone marrow [33–38] and postnatal bowel [39–46]). Thirdly, it is believed that neural crest cells also follow the hierarchical stem cell model for their lineage diversification. Highly multipotent neural crest cells generate heterogeneous progenies consisting of all oligopotent and bipotent progenitors, as well as the more restricted unipotent neural crest derivatives (Fig. 16.2). Lastly, the global expression profile analysis has also revealed that human neural crest cells share a high similarity to pluripotent embryonic stem cells, in particular, on the expression profile of the cellular mediators responsible for cell proliferation and growth [47]. In addition, the human neural crest cells are also found to express the pluripotency markers such as NANOG, POU5F1 and SOX2, which may endow them with great differentiation plasticity.

16.3 Enteric Nervous System Development and Hirschsprung's Disease (HSCR)

Commitment of neural crest to neural lineage mainly contributes to the formation of PNS. PNS is a neural network which makes up the sensory ganglia of the cranial, Rachidian nerves, and the autonomic nervous system (ANS) that innervates the viscera, heart, and smooth muscles. A distinct part of the PNS is the enteric nervous system (ENS), which is originated from the vagal and sacral crests, and innervates the gastrointestinal tract [48, 49]. HSCR is resulted from the absence of ganglion cells in the colon, representing one of the most common neurocristopathies in the PNS. In the following sections, we will focus on this disease and summarize how we can reveal key insights into the disease etiology by defining the mechanisms of neural crest development.

16.3.1 *Neural Crest and Enteric Nervous System (ENS) Development*

The ENS is considered as the "second brain" of a body which consists of some hundred million neurons and more than that in the spinal cord [50]. Varied classes of neurons are clustered together with glia (the supporting cells) to form two major plexuses along the gut. The myenteric plexus is located between the circular and longitudinal muscle layers; and the submucosal plexus is ramifying in the intestinal submucosa. These two plexuses mediate motility, regulate blood flow within the gut wall and control water and electrolyte transport across the mucosal epithelium.

The ENS is mainly derived from the vagal neural crest cells which enter the foregut at embryonic day (E) 9–9.5 in mouse or after 4 weeks of gestation in humans [51]. These enteric neural crest cells (ENCCs) then migrate rostral-caudally along the gut wall mesenchyme and sequentially colonize the foregut, midgut and down to the rectum of the hindgut. The whole process is completed by E15 in mouse or after 7 weeks gestation in humans [52, 53]. Indeed, neural crest cells from sacral region also contribute up to 17% of ENS, they penetrate the wall of hindgut and colonize the hindgut by distal-proximal migration [54, 55]. Therefore, ganglion cells of both the myenteric and submucosal plexuses of the pre-umbilical gut are originated from the vagal region of the entire neural crest, whereas ganglia of the post-umbilical region are contributed by the cells from both the vagal and sacral neural crests.

During the migration process, the enteric neural crest cells actively divide, increase cell population for the migration and invasion along the entire gut. Time-lapse live imaging of ENCCs that were genetically-labeled with fluorescent protein showed that apart from few isolated cells at the migration wave front, the majority of the ENCCs were inter-connected and migrated as chains of cells extending and moving caudally along and within the strands, suggesting the presence of intercellular interactions among these migrating neural crest cells [56–58]. Neural guidance molecules are likely from the gut mesenchyme by which the local microenvironment may guide or support the migration of ENCCs. In particular, the unique tissue microenvironment in caecum can induce a conspicuous pause in migration of ENCCs and accumulation of isolated cells in the migratory wave front, leading to a transient fragmentation of migrating cell chains [52, 58–60]. After passing through the caecum, the migrating cell strands will be reformed and cells will continue moving caudally until they fully colonize the rectum.

Along with migration, another aspect in fulfilling the development of ENS is cell proliferation. Within the small intestine and colon, there are over a million neurons and a similar number of glial cells, which are derived from only several thousands of neural crest cells from vagal crest at the time when they enter the foregut. In achieving such expansion in number, these ENCCs must undergo extensive proliferation during the colonization of the gut [61, 62]. Importantly, cell number and population density of the ENCCs can influence the migratory activities. As supported by both mathematical and experimental results, proliferation of ENCCs at the wave front is believed to be the major driving force for the gut invasion process [63].

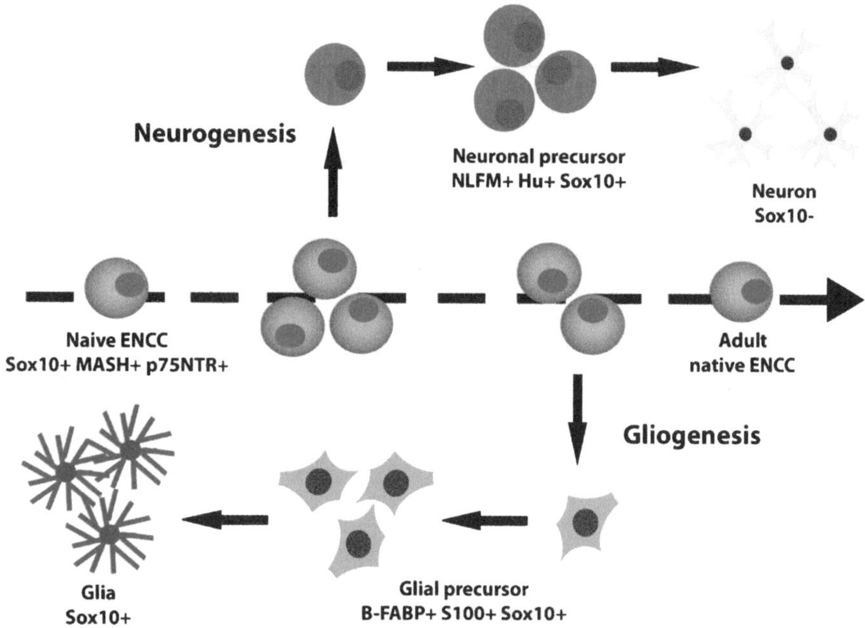

Fig. 16.3 A schematic diagram showing the homeostasis between maintenance and differentiation of ENCCs into neurons and glia in functional ENS

For the proper ENS development, ENCCs need to be able to differentiate into millions of neuron and glia which are organized into the network to coordinate the complex behaviors of the gut. Gradual increases in the number of neuronal and glial precursors within the population of neural crest-derived cells are normally observed during ENS development [62, 64, 65]. This developmental process involves sequential waves of neurogenesis and gliogenesis, and requires an appropriate balance between the proliferation and differentiation of ENCCs and their progenies (Fig. 16.3). Intriguingly, unlike in the central nervous systems, the enteric neural crest-derived neuronal and glial precursor cells continue to proliferate while progressively differentiating. Nevertheless, how such a complex regulatory mechanism coordinating these seemingly diverse cellular processes is achieved is still not fully understood.

16.3.2 Hirschsprung's Disease (HSCR)

HSCR was first reported by Harald Hirschsprung in 1966. It is also known as aganglionic megacolon, a malformation disorder characterized by the absence of the enteric ganglia along the intestine at variable lengths, resulting in intestinal obstruction and massive distension of the bowel. It is the most common identifiable

Table 16.1 Twelve known Hirschsprung's disease genes

Gene name	Full name	Location	Inheritance	Penetrance
RET	Rearranged during transfection	10q11.2	Dominant	50–72%
GDNF	Glial cell line-derived neurotrophic factor	5q13.1	Non-Mendelian	Unknown
NRTN	Neurturin	19p13.3	Non-Mendelian	Very rare
EDNRB	Endothelin receptor type B	13q22	Dominant/recessive	30–85%
EDN3	Endothelin-3	20q13	Dominant	Incomplete
ECE1	Endothelin converting enzyme-1	1p36	Dominant	Unknown
SOX10	Sex determining region Y box 10	22q13	Dominant/recessive	>80%
PHOX2B	Paired-like homeobox 2b	4p12	Dominant	Unknown
KIAA1279	KIF-binding protein	10q22.1	Recessive	Unknown
NRG1	Neuregulin-1	8p12	Dominant	Incomplete
ZEB2	Zinc finger E-box homeobox-2	2q22	Sporadic	Unknown
NTN	Neurturin	19p13	Unknown	Unknown

neurocristopathy, with an incidence about 1:4,500 live births and appears more often in males (male: female = 4:1) [66]. Dependent on the extent of aganglionosis, HSCR can be classified as short-segment (S-HSCR) or long-segment (L-HSCR) [66]. In majority, cases identified are S-HSCR and are sporadic in nature (>80%); whereas the L-HSCR and total aganglionosis are less common, and represent the inherited form of the disease [67–69]. Genetic causes of the disease are known to be multigenic which can be resulted from different genetic mutations and/or accumulation of genetic changes in multiple genes [69]. To date, around 14 known HSCR genes have been identified (Table 16.1). They are mainly implicated in the maintenance, migration and differentiation of ENCCs. Although mutations in these genes only accounts for 50% of the cases [70–72], significant insights on understanding the etiology of HSCR have been made by revealing their roles in ENS development.

16.3.3 Molecular Mechanisms Underlying ENS Development

16.3.3.1 RET Receptor Tyrosine Kinase/GDNF Family Receptor-Alpha (GFRα1)/Glial Cell Line-Derived Neurotrophic Factor (GDNF)

RET-GDNF-GFRα1 constitutes the key signaling pathway for ENS development. The rearranged during transfection (*RET*) proto-oncogene encodes a tyrosine kinase receptor that is expressed in ENCCs [73]. Glial cell line-derived neurotrophic factor (GDNF) family members including GDNF, neurturin, artemin, and persephin are the ligands for RET receptor and are secreted from the gut mesenchyme. Activation of RET by these ligands is mediated by their initial interaction with one of the four different glycosyl phosphatidylinositol anchored co-receptors, the GDNF family receptor-alpha (GFRα) 1–4 [74]. Each ligand has a preferential co-receptor, for instance GDNF has high affinity towards GFRα1, neuturin, artemin and persephin to GFRα2, GFRα3 and GFRα4, respectively [73, 75–77]. GDNF-GFRα complex subsequently dimerizes with RET and results in stimulation of a number of signal transduction pathways including mitogen-activated protein kinase (MAPK), phosphoinositide 3-kinase (PI3K)/protein kinase B (Akt), secretory (Sec), phospholipase C-gamma (PLC-γ) pathways, in turn, they mediate the migration, proliferation and differentiation of ENCCs [78].

In *Ret*, *Gdnf* or *Gfrα1* null mice, the ENCCs fail to colonize the gut beyond the esophagus and the knockout mice exhibit complete intestinal aganglionosis, due to a combination of cell death and failure of migration of ENCCs [79–83]. Conditional ablation of *Ret* or *Gfrα1* in post-migratory ENCCs also resulted in impaired migration of the precursor cells and massive death of enteric neurons, leading to colon aganglionosis [84, 85], implied that Ret signaling is also implicated in late ENS development in mice. Notably, such cell death in the colon is not associated with caspase signaling, suggesting it is distinct from that of apoptosis, but the mechanisms behind remain unknown. Furthermore, the Ret-Gdnf pathway may also indirectly interfere with glial differentiation. When enteric progenitor cells are grafted into aganglionic *Ret* null gut, these cells do not undergo glial cell differentiation. Therefore, it is believed that the diseased gut lacks critical diffusible or cell-cell contact signals, probably derived from the progenitors of enteric neurons that are reduced or absent in *Ret*-deficient embryos, to support glial cell differentiation [86, 87]. In the *in vitro* cell and gut explant cultures, Gdnf is also required for maintenance of ENCCs survival, proliferation, differentiation as well as migration [88, 89].

Mutations in *RET* receptor tyrosine kinase are most commonly found in HSCR. Currently, over 100 mutations have been identified, ranging from large deletions encompassing the *RET* gene, to micro-deletions, insertions, nonsense, missense, and splicing mutations, and all these mutations are found to be heterozygous in humans [90–93]. Mutations in the coding sequences of the *RET* gene account for up to 50% of familial HSCR patients and 15–35% of sporadic cases. Missense, frameshift and complex mutations in *RET* coding-region have been detected in patients with both long-and short-segment HSCR and in syndromic cases associated with maternal deafness, talipes and mairotation of the gut. Some rare mutations, such as multiple endocrine neoplasia type 2A (*MEN2A*) mutations (C609Y and C620R) have also been identified in HSCR patient presenting with medullary thyroid carcinoma (MTC). A comprehensive structural and functional analysis of a panel of *RET* mutations has been recently performed and demonstrated that most of the HSCR mutations residing in the extracellular domain of RET indeed interferes the protein maturation of RET, leading to reduced RET expression [94]. On the other hand, non-coding mutations and single nucleotide polymorphisms (SNP) spanning throughout the *RET* gene are found to be more common in sporadic cases and suggested to impart susceptibility in the non-familial HSCR cases. Case-control and transmission disequilibrium test in several ethnic backgrounds have found that a frequent c135G>A SNP (rs_1800858) lying in exon 2 and resulted to a silent change is over represented and transmitted in HSCR patients [95–97]. Comparative genomics study subsequently identified another SNP (rs_2435357) lying within intron 1 is also highly associated to HSCR susceptibility, contributing a 20-fold higher of risk than coding sequence mutations [98]. This T>C SNP lies within an enhancer region and the presence of T allele is found to reduce the enhancer activity in the *in vitro* reporter assay [98]. Given that the penetrance of the T allele for HSCR trait is both dose-dependent and has a sexual preference towards males [98], it is believed that SNP may directly contribute to the disease by reducing *RET* expression level in the diseased bowel and account for the higher risk in males. In addition, another two HSCR-associated *RET* promoter SNPs -5G>A (rs_10900296) and -1A>C (rs_10900297) are found to explain for the higher incidence of HSCR in Chinese [99–104]. These SNPs are residing at the *RET* promoter region and directly interfere the thyroid transcription factor-1 (TITF-1) binding to the *RET* promoter and reduce the *RET* transcription [99]. Subsequent expression profiling study using the patients' gut biopsies further demonstrated that these genetic variants in *RET* promoter confer a significant reduction of *RET* expression in the ganglion cells of HSCR patients [103]. All these studies indicated that both coding- and non-coding mutations may lead to aberrant RET expression, suggesting that sufficient amount of functional RET in ENCCs is critical for the proper ENS development. Therefore, it is believed that haploinsufficiency of the RET may represent a cause of HSCR. Intriguingly, heterozygous deletion or mutations of Ret are not sufficient to cause HSCR-like phenotype in mice. This substantial difference from the human disease may limit the use of mouse model for revealing the implication of *RET* variants in HSCR development.

Despite *Gfnf* and *Gfrα1* homozygous knockout mice are phenotypically resembling the *Ret* knockout mice, mutations of *GDNF* and *GFRα1* are less common (<5%) in HSCR [105–107]. So far only six HSCR patients were reported for carrying mutations in *GDNF*, among them, four were complicated with other contributing factors, including *RET* mutation and trisomy 21 [105, 106]. It is suggested these *GDNF* mutations may not be sufficient to cause HSCR. In addition, human mutation in *GFRα1* has never been identified in patients except a deletion at the locus with incomplete penetrance was reported in one family [82, 83, 108]. Similarly, mutation of neurtirin, another family member of the *GDNF* family, is also rare and probably not sufficient for inducing HSCR. Such mutation has only been identified in one family, in conjunction with *RET* mutation [109].

16.3.3.2 Endothelin-3 (EDN3)/Endothelin Receptor B (EDNRB)

Endothelin receptor (EDNRB) is a G-protein coupled seven-transmembrane receptor that can be activated by binding to one of the three isoforms of endothelin 1–3 (EDN1 to EDN3). Among these isoforms, only EDN3 is essential for the formation of ENS [110]. Activation of EDNRB is known to stimulate a variety of downstream signaling, including phosphatidylinositol phosphate turnover, intracellular Ca2+ accumulation, activation of MAPK, release of arachidonic acid via phospholipase A2, and inhibition of cAMP production [110], but current knowledge is still limited on which of them are essential for neural crest migration and fate determination.

In mouse, Edn3 is predominantly expressed in mesenchyme of the developing gut while Ednrb is expressed by both migrating neural crest cells and some mesenchymal cells [111–113]. *Edn3* and *Ednrb* null-mice display distal colon aganglionosis due to delayed neural crest migration and subsequent failure of colonization of the distal gut [114–116]. Nevertheless, the ENCC survival and proliferation are not affected in these mice [112, 113, 117–120]. The role of Edn3/Ednrb signaling in enteric neural crest development has been directly illustrated in the conditional knock-out mice. Deletion of *Ednrb* in neural crest cells, resembled the phenotypes observed in the conventional knockout mice, which characterized with colonic aganglionosis associated with pigmentation defect in mouse and the mutants die within 5 week due to megacolon [52, 114, 121]. In addition, the presence of cross-talk between the Edn3/Ednrb and Ret/Gfrα1/Gdnf signaling has been demonstrated directly in the *in vitro* studies. Treatment of ENCCs with Edn3 and Gdnf synergistically promoted their proliferation, and provided the critical force for their migration into the distal bowel of rodent gut [63]. Consistently, mice with hypomorphic *Edn3* mutant allele mixed with a *Ret*-heterozygous background display a higher incidence of intestinal aganglionosis than those carrying two copies of *Ret* alleles [122].

In human, *EDNRB* is mapped within the locus 13q22, a susceptibility locus for HSCR [123–125]. Interstitial 13q22 deletions are frequently found in HSCR patients suggesting that *EDNRB* haploinsufficiency may represent a disease mechanism for HSCR. Both *EDN3* and *EDNRB* mutations are found to be associated with HSCR and mutations of the *EDNRB* have been identified in 5% of HSCR patients [66, 126–129]. Moreover, mutation of *ECE1* gene encoding the enzyme that activates EDN3, has been reported in a syndromic HSCR patient associated with craniofacial, and cardiac defect [130]. Consistent with the findings in mice, presence of genetic interactions between *EDNRB* and *RET* has been demonstrated [122, 131]. Missense mutations of *EDNRB* and non-coding mutations of *RET* alleles are preferentially being co-transmitted in affected individuals [122]. All these clinical findings support the implication of EDN3/EDNRB signaling in HSCR disease.

16.3.3.3 Bone Morphogenetic Protein (BMP) Signaling

Bone morphogenetic proteins comprise a subgroup of TGF-β family of secreted signaling molecules, including BMP2, BMP4 and BMP7. These molecules binds to a type II (BMPRII) receptor dimer, which phosphorylates and recruits a type I receptor dimer, BMPR1A or BMPR1B (or in the case of BMP7, activin receptor-like-2 (Alk-2)), forming a hetero-tetrameric complex with the ligand. This would lead to phosphorylation and activation of SMAD signaling cascade, and subsequently the transcription of target genes [120]. BMP signaling is important in early neural crest cell induction and migration [52, 121]. In chicks, BMP2, BMP4, BMPRII and phosphorylated-Smad are strongly expressed in both the submucosal and myenteric plexuses of the developing gut [123, 124]. *In ovo* inhibition of BMP in the presumptive gut mesoderm with noggin caused impaired migration of ENCCs by interfering with their responsiveness to GDNF [124]. However, BMP alone is insufficient to promote migration of neural crest cells. Interaction between GDNF and BMP4 in ENS is required for gut colonization and neuronal differentiation [123]. Ectopic expression of noggin in ENS progenitors in mouse results in an increase in neuronal numbers in both enteric plexuses and smooth muscle as well as a reduction of TrkC expression neurons, implying that BMP signaling is also implicated in limiting the progenitor pool size and formation of TrkC expressing neurons. In human, mutations in *ZFHX1B*, the gene that encodes Smad-interacting protein-1 (SIP1), have been identified in several patients with Mowat-Wilson syndrome, a dominant form of Hirschsprung's disease with mental retardation syndrome. *Zfhx1b*-knockout mice exhibits defects in development of vagal neural crest cells and displayed a delamination arrest of cranial neural crest cells. These suggested that Sip1 is essential for the development of vagal neural crest precursors and the migratory behavior of cranial neural crest cells in the mouse [128, 129].

16.3.3.4 Hedgehog (Hh) Signaling

Hedgehog gene (Hh) family encodes the secreted signaling molecules, Sonic hedgehog (Shh), Indian hedgehog (Ihh) and Desert hedgehog (Dhh), that are known to be essential for tissue patterning and development of varied systems [132] and implicated in various diseases [133]. Hh signal transduction is mediated by two transmembrane proteins: Smoothened (Smo) the obligatory signal transducer and the Hh-binding receptor Patched (Ptch), which constitutively blocks Smo activity in the absence of Hh. Hh binding to Ptch unleashes Smo activity, and activates the downstream cascade by blocking the formation of the repressor form of Gli family of zinc-finger transcription factor, Gli^R, and promoting the formation of Gli activator Gli^A [87, 134]. Both graded activity and stepwise regulation of Hh signaling are important for its actions. Importantly, Shh signaling is important for spatial development of neural plexuses and mice lacking the Shh secreted proteins showed ectopic ganglia formation in the bowel [135]. Data obtained from the *in vitro* studies using neurosphere and gut explants cultures further suggest that Shh not only promotes proliferation of the ENCCs, but also modulates their responsiveness towards GDNF, which in turn inhibits the neuronal differentiation [136]. *Ihh-/-* mice, on the other hand, exhibited dilated colon with abnormally thin wall accompanied by partial intestinal aganglionsis [135]. All these studies highlighted the biological relevance of Hh signaling in ENS development. Nevertheless, so far, none of the genes involved in HH pathways are found to be associated with HSCR, neither in a genome-wide association study (GWAS), nor mutation screening conducted on HSCR patients [137].

16.3.3.5 Notch Signaling

Notch is a transmembrane receptor with four known isoforms (NOTCH 1-4) that can be activated by juxtacrine interactions with its specific ligands like Delta-like ligand-1, -3, and -4 (Dll1, 3, and 4) or Jagged-1 and -2 (Jag-1, -2) [138]. Upon activation, Notch receptors undergo a series of proteolytic cleavages to release the Notch intracellular domain (NICD) [139], which subsequently translocates into the nucleus and binds with the transcription factor recombination signal binding protein 1 for J-Kappa (Rbp-jκ) to generate a transcription complex. This initiates the transcription of target genes such as hairy and enhancer of split-1 *(Hes1)* which in turn, represses the expression of pro-neural proteins (*e.g.* Mash1) to switch differentiation from neurogenesis to gliogenesis [140]. For Notch signaling, there are multiple layers of regulatory mechanism for signal refinement [141]. One of which is mediated by the protein O-fucosyltransferase-1 (Pofut1) which mainly modulates the Notch receptor trafficking [142–147].

Notch receptors, Notch 1 and Notch 2, and their ligands, Dll1 and Dll3, are the main forms expressed in the ENCCs [148] and they play a key role in various aspects of the ENS development. Activation of Notch signaling is required not only for the maintenance of these cells [148], but also for switching of these cells from a program of neurogenesis to gliogenesis *in vitro* and *in vivo* [149]. Perturbation of Notch signaling by knocking out *Pouft1* specifically in neural crest cells results in a significant reduction in the ENCCs. Also, it was found that Notch signaling suppresses *Sox10* expression in the mouse ENCCs through up-regulation of *Mash-1*, resulting in premature neurogenesis and reduction of glia in the bowel [149].

Increasing evidence shows that interaction between the Shh and Notch signaling pathways exists in the switching mechanism from neurogenesis to gliogenesis by coordinating some common targets [150–153]. More recently, pathway-based epistasis analysis of data generated by a GWAS on HSCR disease indicates that specific genotype constellations of *PTCH1* (rs357552A/G, rs10512248A/C) and *DLL3* (rs2354225A/G) SNPs confer higher risk to HSCR (OR = 2.582). Importantly, deletion of *Ptch1* in mouse ENCCs induces a robust *Dll1* expression and activation of the Notch pathway, leading to premature gliogenesis and reduction of enteric progenitors in mutant bowels. Dll integrates Hedgehog and Notch pathways to coordinate neuronal and glial cell differentiation during ENS development. In addition, HH-mediated gliogenesis is highly conserved such that HH is consistently able to promote gliogenesis of human neural crest-related precursors. Collectively, *PTCH1* and *DLL3* are the additional HSCR susceptibility genes and HH-NOTCH-induced premature gliogenesis represent a new disease mechanism for HSCR [154].

16.3.3.6 SRY-Related Homeobox (HMG)-Box Transcription Factors (SOX)

Sox family transcription factors process an HMG type of DNA-binding motif. More than 30 of them have been identified in vertebrates and particularly, Sox10 and Sox2 are known to be expressed in neural crest and its derivatives, and implicated in ENS development [155, 156].

Sox10 is a member of subgroup E of Sox genes, which is expressed in neural crest cells for their survival and maintenance prior to lineage segregation [60, 157, 158]. Given that Sox10 is specifically expressed in ENCCs of the early developing gut, it is considered as a marker of enteric progenitors [60]. However, upon neural differentiation, Sox10 expression becomes restricted to glial cells but not in neurons [159, 160]. Functionally, Sox10 is highly conserved among rat [161], mouse [162, 163] and human [163, 164]. Mice carrying a dominant mutation in *Sox10* (Sox10$^{Dom/+}$) exhibits pigmentation and ENS defects accompanied with the loss of Ednrb expression [162]. Deletion of *Sox10* in mouse by knocking in *lacZ* gene to replace the complete open reading frame of *Sox10*, also leads to similar outcomes [165]. Homozygous *Sox10* mutant mice shows a severe degeneration of sensory and motor neurons due to the absence of peripheral glial cells, while haploinsufficiency of *Sox10* results in pigmentation defect and HSCR like phenotypes [165]. In zebrafish, knocking down of *Sox10* also causes an ENS defect and down-regulation of the *colorless* gene [166, 167]. As illustrated in the *in vitro* assay with neural crest cells isolated from rat embryos, Sox10 was found to maintain multipotency of neural crest by inhibiting the neurogenesis [157]. Consistently, constitutive expression of Sox10 in mouse ENCCs abolishes their neuronal and glial differentiation, which further supports the role of Sox10 in the maintenance of ENS progenitor pool [168]. An increasing number of genetic studies suggest that functional interactions between Sox10 and other pathways are crucial for proper ENS development. Single nucleotide polymorphism-based genome scan in Sox10$^{Dom/+}$ F1 intercross progeny has identified multiple modifier loci on mouse chromosomes 3, 5, 8, 11 and 14 from which these modifiers may exert distinct effect on penetrance and severity of aganglionosis. Importantly, different loci associations also account for the phenotypes of Sox10$^{Dom/+}$ ranging over a continuum from severe aganglionosis to no detectable phenotype in the mutant gut [169]. In particular, paired box-3 (*Pax3), Sox8* and *Ednrb* genes have been found to be the modifier genes, and contribute to an increased penetrance and more severe aganglionosis in Sox10$^{Dom/+}$ mutants [170–172]. In human, mutations of *SOX10* are also implicated in the development of Waardenburg-Hirschsprung's disease and Mowat-Wilson Syndrome. In Mowat-Wilson Syndrome, genetic interaction between *SOX10* and *ZFH1XB* has been demonstrated and the functional coordination between these two transcription factors in maintaining the proliferation capacities of ENCCs are crucial for proper ENS development [173].

Another family member, Sox2, which belongs to the B1-type of HMG box family transcription factors, is co-expressed with Sox10 in neural crest [174–176]. During ENS development, Sox2 expression is highly overlapped with that of Sox10 and both are expressed in the ENS progenitors and glial cells of developing gut [177]. Moreover, Sox2 expressing cells, like Sox10 expressing cells, are able to migrate and differentiate upon transplantation into the mouse gut [177], suggesting the stemness nature of these cells. Sox2 is also playing differential roles in neural induction as well as varied stages during ENS development. In avian embryo, overexpression of Sox2 inhibits neural crest formation and the subsequent EMT; while in later stage of development, it inhibits Notch induced glial differentiation [178]. More recently, with a human embryonic stem cell model, *SOX2* is also found to be down-regulated in migratory neural crest progenitors during EMT; but its expression is turned on again when the neural crest started to differentiate. SOX2 promotes the formation of dorsal root ganglion-like clusters through up-regulating neurogenin-1 (*NGN1*) and *MASH1* expressions[179]. All these findings highlight the importance of Sox2 in neural crest development and its roles in peripheral neurogenesis are highly conserved among vertebrates.

16.3.3.7 Retinoic Acid (RA) and Other Transcription Factors

Retinoic acid (RA), a vitamin A-derived morphogen, is implicated in various aspects of embryogenesis. Excess levels of endogenous RA signaling during fetal development leads to a delayed colonization of the distal bowel by ENS precursors, and accompanied with abnormal gut mesenchyme differentiation, and other abnormalities of gut looping, rotation and morphogenesis [180]. Retinaldehyde dehydrogenase 2 (*Raldh2/Aldh1a2*)-deficient mice lacking the enzyme for production of RA display intestinal aganglionosis, defects in gut mesenchyme, and in other neural crest derivatives. These phenotypes can be partially "rescued" by treatment of the pregnant mothers with RA [181]. In addition, subsequent studies using isolated ENCC culture directly illustrate that RA regulates ENS precursor proliferation, enhances their neuronal differentiation and the GDNF-induced migration by reducing Pten accumulation [182, 183]. Other factors such as prokineticins (Prok-1 and Prok-2), and their receptors (PK-R1 and PK-R2) are expressed in both the ENS precursors and mature enteric neurons, and this signaling pathway is found to be crucial in ENS development as well as its postnatal function [184, 185]. Serotonin (5-HT), on the other hand, mainly mediates neurogenesis of mature ENS, as mice lacking 5-HT4 receptors shows improper postnatal ENS growth and maintenance [186].

Other transcription factors such as paired-like homeobox 2b (Phox2b), paired box-3 (Pax3), homeobox protein-B5 (Hoxb5) and thyroid transcription factor-1 (Titf-1) are also implicated in ENS development and HSCR disease. For instance, *Phox2b* deficient mice lack enteric neurons throughout the entire gastrointestinal tract [162, 187, 188] and *PHOX2B* mutations have also been identified in patients with Ondine-HSCR disease. Pax3 is expressed during early embryogenesis in the dorsal neural tube before neural tube closure and the emigrating neural crest cells. *PAX3* mutations are known to cause Waardenburg syndrome (WS)-HSCR disease in humans [189] and mouse Splotch mutant which carries *Pax3* mutation also display similar neural crest-derived defects [190]. Transgenic mice expressing dominant negative *Hoxb5* in ENCCs exhibit intestinal hypoganglionosis [191]. It is noteworthy that all these transcription factors may contribute to HSCR disease and ENS development primarily through modulating *RET* gene expression [100, 188, 192–194].

16.3.4 Advances in Stem Cell Research

The usual treatment for HSCR is "pull-through" surgery where the portion of the colon that does have nerve cells is pulled through and sewn over the part that lacks nerve cells. Nevertheless, the functional outcome of surgery is variable and a significant number of patients still suffer from life-long complications ranging from intractable constipation, incontinence, enterocolitis, to devastating short bowel syndrome. Recent advances in identification and isolation of lineage specific progenitor cells of high neuronal plasticity and diversity pave a way for the establishment of an effective cell-based therapy for HSCR. In this following section, we will summarize the achievements on the establishment of transplantation models for HSCR, and highlight the use of multipotent neural stem cells and ENCCs to reconstitute/replenish the absent ganglia in HSCR bowel. Moreover, alternative source of somatic stem cells and the potential applications of the induced pluripotent stem cells (iPSC) derived from fetal and adult tissue for cell-based therapies and disease modeling for HSCR will also be discussed.

16.3.4.1 Neural Stem Cells (NSCs)

Neural stem cells are predominantly localized in the hippocampus and subventricular zone (SVZ) of the brain. They are present in fetal brain and persist into adulthood to support CNS development, as well as neuron self-renewal in the adult brain [195]. NSCs have been successfully isolated from both fetal and adult mammalian brain and propagated in culture supplemented with mitogens such as epidermal growth factor (EGF) or fibroblast growth factor (FGF). These cells are capable to self-renew as illustrated in neural colony-forming cell assay, in which a single cell could give rise to a neurosphere and further propagate to generate secondary and tertiary neurospheres in culture. In addition, NSCs are multi-potent and possess a high degree of neural plasticity to generate neurons, oligodendrocytes and astrocytes [195, 196].

The use of NSCs for neural repairing has been demonstrated in various CNS disease- and spinal cord injury-animal models. For example, transplantation of NSC-derived dopamine neurons to a Parkinson's disease rat model results in a significant functional improvement [197, 198]. Similarly, restoration of behavioral performance in ischemic rats has also been observed after intra-hippocampal implantation of multipotent neuroepithelium cells [199]. Moreover, with spinal cord injury model, transplantation of *in vitro* expanded NSCs isolated from rat embryonic spinal cord into adult rat spinal cord at site with contusion injury also results in mitotic neurogenesis, formation of synaptic structures, which subsequently leads to significant functional recovery. All these indicate that *in vitro* expanded neural stem cells are potential source for transplantable materials.

Recent studies have also shown that these NSCs can generate enteric neurons upon transplantation into the gastrointestinal tract. Transplantation into the gastric wall of adult mice results in differentiation into nitric oxide synthase (nNOS) expressing neurons, a major enteric neurons [200, 201]. In addition, transplantation of these NSCs into the pyloric wall of an animal model of gastroparesis (nNOS knockout mice) results in neurons and glia formation including nNOS expressing neurons. Importantly, significant functional improvements in the rate of gastric emptying and in electrical field stimulation-induced relations have been observed [201].

More recently, transplantation of rat fetal cerebral cortex-derived NSCs, as well as those from mid-embryonic rat neural tube into the rectum of adult rats where enteric neurons had been destroyed chemically, also promotes regeneration of neural and glial cells with a concomitant increase in both the expressions of nNOS and choline acetyltransferase (ChAT) that eventually restores the rectoanal inhibitory reflex [40, 202]. Taken these findings together, both fetal and adult NSCs are capable to form mature, terminally differentiated enteric neurons and glia that can eventually integrate into neural network of the recipients [203].

16.3.4.2 Enteric Neural Crest Cells (ENCCs)

ENCCs are multipotent cells harbored within the gastrointestinal tract that can give rises to enteric neurons and glia. These cells have been identified in the ENS of neonates as well as adults, suggesting that they may represent a source of stem cells for ENS regeneration [204]. ENCCs are first isolated from mouse embryonic gut, they grow as neurospheres in *in vitro* culture and can subsequently differentiate into neurons and glial cells [44]. These ENCCs are able to colonize both the wild-type and *Ret*-deficient aganglionic gut wall and give rise to neurons and glia when they are microinjected into mouse bowel [39]. ENCCs isolated from the neonatal rat also possess similar neural plasticity, they are able to differentiate into neurons and glia after transplantation into the muscular distal denervated colon of rats whose neural plexuses are chemically eliminated with benzalkonium chloride [205]. ENCCs also persist in adult gut of rats, and their neurogenic potential has also been illustrated in both *in vitro* and *in vivo* data. However, in contrast to those isolated at younger ages, these adult progenitors preferentially generate glia but not neurons upon transplantation into developing peripheral nerve, suggesting the presence of perinatal changes in their responsiveness to lineage determination factors [204]. More importantly, human ENCC cultures have also been established using human fetal tissues (the 9th week of gestation) and gut biopsies at all ages (from 9 month to 17 years) [206–208]. Neurospheres cultivated from myenteric plexus of human fetus and children bowel samples have been maintained for a long period of time in culture while retaining their differentiation capability [206]. In other studies, culturing single cell suspensions of ENCCs have also been achieved with human neonatal colon/ileum samples, followed by formation of neurospheres. Subsequent transplantation into aganglionic embryonic mouse hindgut also indicates that these cells can proliferate extensively, migrate and differentiate [209, 210]. Interestingly, apart from the intestinal plexuses, multipotent ENCCs are also present in the gut mucosa. Neurospheres derived from ENCCs isolated from postnatal human gut mucosal tissues have been transplanted into the cultured chick and human aganglionic gut, and the formation of ganglia-like structures, enteric neurons and glia is observed [211]. All these promising results suggest that both NSCs and ENCCs may provide potential therapeutic tools to reconstitute/replenish absent ganglia in HSCR bowel. Nevertheless, how to obtain stem cells in sufficient numbers and of full neural plasticity are still the most formidable challenges to overcome. Therefore, identification of a useful source of stem cells of high neural plasticity and diversity will pave a way for the establishment of an effective cell-based therapy for HSCR.

16.3.4.3 Skin-Derived Progenitors (SKP)

Over the past decade, it has become apparent that somatic tissue stem cells are present in most adult tissues for tissue homeostasis. Mammalian skin is a sophisticated organ that dynamically regenerates over the entire life of an animal. The epidermal skin-derived stem cells, keratinocytes, have been located in the bulge of the skin. On one hand, they contribute to the epidermal homeostasis in the skin through maintaining compartments of hair follicles, sebaceous gland, and the interfollicular epidermis [212]; while on the other, they provide support during wound healing and regeneration [213, 214]. More recently, a distinct type of stem cells has been identified in dermis, they are called skin-derived progenitors (SKPs). SKPs are multipoint dermal stem cells residing within a hair follicle niche that arise during embryogenesis and persist in adulthoods. They exhibit high neural plasticity, but show distinct transcription profiles when compared to NSCs in the CNS [215]. Several lines of evidences suggest these SKPs are neural crest-related cells which share properties with embryonic neural crest precursors, such as the expressions of nestin, $p75^{NTR}$, fibronectin, and Sox10 [216–218]. Several independent groups have been successfully isolate SKPs from both embryonic and adult porcine skin. These cells express pluripotency related (Oct3/4; Sox2; Stat3 and Nanog) and neural crest ($p75^{NTR}$, Sox10, Snail) markers [217, 219, 220]. Similar properties are also observed in cell isolated from human [31, 221–226] and rodent skins [218, 227, 228].

Due to the fact that skin is an easily accessible reservoir of adult progenitor cells and SKPs endow high neural plasticity, SKPs is of a high therapeutic value to be used in cell therapy for neurodegenerative diseases. SKPs have been isolated from rodents' skin at all developmental stages which are capable to differentiate into peripheral catecholamergic neurons and Schwann cells [24]. The same has also been achieved in human by culturing adult dermis in pro-neural medium, and a large number of nestin-positive neural precursors have been obtained [28]. Subsequent exposure of these precursors to hippocampal-astrocyte conditional medium further drives these cells towards to neural differentiation, resulting in neuronal morphology, stable expression of neuronal differentiation markers (neurofilament, β-tubulin); and the presence of voltage-dependent calcium transients [28]. The use of neonatal SKPs to generate myelinating Schwann cells for the injured and dysmyelinated nervous system has also been demonstrated. SKP-derived Schwann cells can robustly associate with and myelinate regenerating axon, generating compact myelin that is indistinguishable from that made by resident Schwann cells in the nerve. In addition, the *in vivo* environments of the regenerating peripheral nerve and the neonatal brain are sufficient to direct naïve SKPs to differentiate into Schwann cells that myelinate axons in both the peripheral and central nervous systems [224]. Taken all these together, SKPs represent a potential source of neural precursors that can be used to regenerate the damaged/diseased CNS and PNS, particularly, neurons and glia cells derived from these progenitors may be used to replenish the missing ENS in the future.

16.3.4.4 Stem Cells in Bone Marrow

Bone marrow-derived mesenchymal stem cells (BM-MSCs) are multipotent cells that are capable to give rise to a variety of cells from all three germ layers, including marrow stromal cells, adipocytes, osteoblastic cells, chondrocytes, tendinocytes, and myocytes [229–238]. They exist in various locations including adipose tissue, cord blood, embryos and bone marrow. Among which, BM-MSCs are recently suggested as an alternative source for neural regeneration, due to their ability to give rise to neurons and glia as supported by both *in vitro* and *in vivo* experiments. In mice, despite the overall proportion of differentiated cells expressing neural markers derived from BM-MSCs are slightly less than those derived from NSCs, nestin-expression neural progenitors and differentiated neurons derived from both origins shows highly similar electrophysiological and functional characteristics [239]. Intriguingly, BM-MSCs has been found to differentiate into neurons and astrocytes, migrate throughout forebrain and cerebellum when transplanted into the brain [33, 240, 241] or peritoneum [242]. BM-MSCs can also generate neurons expressing enteric neural markers such as protein gene product 9.5 (PGP9.5), neural nitric oxide synthase (nNOS), and enteric neural transmitter vasoactive intestinal polypeptide (VIP), NGF, and GDNF *in vitro* [243]. More recently, neural crest progenitors have also been identified in the bone marrow (BM-NCSC) which express the neural crest markers and are able to generate all neural crest derivatives including neurons, glia and myofibroblast [244]. Using double-transgenic reporter mouse lines, *Wnt1* and *P0-Cre/Floxed-EGFP* mice, and certain population of BM-MSCs has been shown to be derived from these multipotent BM-NCSCs. Given that neural crest progenitors are detected in aorta-gonad-mesonephros region, circulating blood, and liver at the embryonic stage, it is believed that these neural crest progenitors originated from these organs join the migration pathway of hematopoietic cells on the way and get to the bone marrow from the embryonic period through adulthood. Although BM-NCSCs are highly accessible source of stem cells in adult, significant tissue-source-dependent differences have been revealed, in particular, on their responses to lineage-determination factors. Therefore, careful consideration of these differences will be necessary if these cells are to be used for cell-based therapy.

16.3.4.5 Induced Pluripotent Stem Cells (iPSCs)

The recent breakthrough in reprogramming of somatic cells has opened up a window for delineation of molecular mechanisms that underlie various aspects of human development, ageing and different diseases, simply by deriving pluripotent cells directly from somatic cells of affected individuals. Similar to human embryonic stem cells, iPSCs retain the ability to self-renew and differentiate into all three germ layers which may provide chances to obtain renewable source of healthy cells to treat various diseases, including neurological disorders. It has been demonstrated that dopaminergic neurons derived from iPSCs are capable to alleviate some of the locomotor abnormalities in a rat model of Parkinson's disease [245, 246]. Moreover, recent iPSCs disease models have also led to identification of small-molecule candidates for various treatments. For instance, continuous kinetin treatment, a plant hormone, significantly increases the percentage of differentiating neurons and expression of key peripheral neuron markers in neural crest cells derived from iPSCs of patients with familial dysautonomia. Treatments with insulin-like growth factor 1 (IGF1) and gentamicin also effectively enhance glutamatergic synapse formation and methyl CpG binding protein 2 (MeCP2) expression in Rett Syndrome patient-iPSCs derived neurons, respectively [247]. In another case with spinal muscular atrophy patient-iPSCs, treatment with valproic acid or tobramycin also induces survival motor neuron-1 (SMN) protein expression and improves neuronal functions [248]. It is of great potential that these cell based screenings could eventually be applied to the development of monoclonal antibodies and other protein therapeutics, as well as to identify drugs that are specifically effective against diseased neural crest cells and to reduce the cost of therapeutic testing in the future [249].

16.3.4.6 Limitations and Implications

Despite results obtained from various *in vivo* and *in vitro* studies using different model systems are very encouraging, the idea of using these various primitive progenitors as a source to replenish or treatment of HSCR disease remains unfeasible at this stage due to a number of reasons. One important hurdle is to establish an effective cell delivery strategy. It has been shown that many cells do not survive beyond a brief period after transplantation nor be able to migrate far from the transplantation site to efficiently regenerate the lost tissues. More importantly, how the newly transplanted cells generate the specific subtypes of neurons and glia, and how can they connect with the existing nervous system of the recipient to establish a functional neuronal networks in the gut, still are the major challenges. Another issue is regarding to the generation of sufficient progenitors for therapy in humans. It likely requires repeated harvesting of human tissues or long-term culture and expansion of progenitors, which is neither ethical nor practical [250]. Moreover, as cell-based therapy will be mostly carried out post-natally, the remarkably different environment harbored by the gut at that age when compared to that of embryonic gut has proven to be a factor hindering the efficient colonization and survival of implanted cells [250, 251].

Although the advances in iPSC technology may circumvent the problem of tissue re-harvesting due to its unlimited proliferation capability, the use of genetically corrected patient specific iPSC in cell-based therapy remains controversial due to several concerns. Up to this stage, most of human iPSCs that display disease specific phenotypes of patients are reprogrammed with virus-mediated methods. From which reprogramming factors might remain integrated in these cells, thus it cannot be excluded that residue vector expression may contribute to the observed phenotype. An example is the use of c-Myc as a reprogramming factor that might lead to high incidence of tumors observed mouse chimeras generated from mouse iPSCs [252]. As a realization of this, several strategies are emerged to improve the reprogramming technology in order to generate genetically unmodified-factor free human iPSCs, however no optimized protocols have been found yet and they share different strengths and disadvantages. The use of Cre-recombinase excisable viruses suggests that integrated vectors can be excised upon expression of the enzyme after reprogramming. However, regions excised do not include the flanking LoxP sites and the residual viral elements remains as a concern [253]. Similar to this strategy, the PiggyBac transposition method also aims to remove any exogenous reprogramming factors by transient expression of the transposase enzyme [253–255]. However, subsequent processes involved in the identification of iPSC clones with minimal copy vector inserts, mapping of integration sites; excision of reprogramming cassettes and validation of factor free clones are known to be laborious and time consuming [254, 255]. Other non-integrating methods including the use of episomes [256], adenoviral transformation [257], RNA viruses [258], and protein transfection [259, 260] have also been developed. Although these approaches eliminate the risks of transgene integration into the host genome accompanied with the use of expression vector, the reprogramming efficiencies are low and protein production is tedious. In addition to those modeling considerations, it is also important to monitor cell karyotype and detect if there are potential chromosomal abnormalities arose from prolonged cell culturing [261]. These are factors that may potentially lead to tumorigenicity and teratoma formation after transplantation into patients.

Despite the use of cell-based therapies remains immature and limited at this stage, the knowledge derived from various models does lead to new understanding of the molecular mechanisms behind the various aspects of the life cycle of the enteric neurons. The use of iPSCs as a platform to systematically delineate the molecular signaling mechanisms behind specific human diseases is also relatively feasible, especially for understanding various neurodegenerative diseases (Table 16.2). Our current understanding of these disorders is primarily based on the findings using human tumor cell lines and animal models. Nevertheless, the substantial differences between human and animals may hamper the use of these models to fully reveal the molecular pathways underlying human diseases; or for the discovery of new therapeutic agents [262]. Recently, generation of specific subtypes of neurons from iPSCs using defined culture conditions has been achieved. Thus, the emergence of patient specific iPSCs offers an unlimited source of materials to understand how defects in development and differentiation of these primitive cells to neural cells may cause diseases. To date, iPSCs have already been derived from individuals carrying inherited defects or neurodegenerative disorders, such as amyotrophic lateral sclerosis (ALS) [263], familial dysautonomia [264], spinal muscular atrophy [248], Parkinson's

Table 16.2 Human induced pluripotent stem cell model for neural disorders

Disease	Brief description	Reference
Amyotrophic lateral sclerosis	Degeneration of neurons located in ventral horn of spinal cord and cortical neurons	[263]
Parkinson's disease	Degeneration of CNS	[265–267]
Spinal muscular atrophy	Neuromuscular disease with degeneration of motor neurons	[248]
Rett syndrome	Neuro-developmental disorder of the grey matter of the brain	[247]
Familial dysautonomia	Suboptimal development and survival of neurons of autonomous nervous system	[264]
Angelman syndrome	Neural disorder characterized by intellectual and development delay	[268]
Friedreich's ataxia	Degeneration of the nervous system	[269]
Huntington's disease	Neurodegenerative disorder that affect muscle coordination and leads to cognitive decline and dementia	[270, 271]

disease [265–267], and Rett Syndrome [247], which offer invaluable opportunities for gaining insights into their disease mechanisms. In addition, human iPSCs and differentiated progenies can also be used as a platform for high-throughput screening for identification of the therapeutic targets, and eventually contribute to the development of new treatments for the cure and prevention of these diseases.

16.4 Chapter Summary

Neural crest cells contribute to the development of various organs and are implicated in a board spectrum of human diseases. In this chapter, we have summarized the current knowledge regarding the some of the most studied factors that regulate the developmental process of neural crest; such findings may provide foundation to meet our long-term goal of improving treatments not only for HSCR, but also other neurodegenerative diseases. Significant advances have been made in unveiling the roles of various pathways in neural crest induction and differentiation, while in contrast it is only at the beginning stage to understand their roles in different neural crest-related disorders. Thus taking the advantage of the establishment of patient specific iPSCs, continued attention to this area should unveil the knowledge of the disease etiology as well as that for regenerative medicine and therapies in the future.

Acknowledgement This work was supported by a seed funding grant for basic research from the University of Hong Kong and by research grant HKU775710 from the Hong Kong Research Grants Council to E.S.W.N.

References

1. Etchevers HC, Amiel J, Lyonnet S (2006) Molecular bases of human neurocristopathies. Adv Exp Med Biol 589:213–234
2. Meulemans D, Bronner-Fraser M (2004) Gene-regulatory interactions in neural crest evolution and development. Dev Cell 7(3):291–299
3. Monsoro-Burq AH, Wang E, Harland R (2005) Msx1 and Pax3 cooperate to mediate FGF8 and WNT signals during Xenopus neural crest induction. Dev Cell 8(2):167–178
4. Morrison SJ, Shah NM, Anderson DJ (1997) Regulatory mechanisms in stem cell biology. Cell 88(3):287–298
5. Maxwell GD, Forbes ME (1987) Exogenous basement-membrane-like matrix stimulates adrenergic development in avian neural crest cultures. Development 101(4):767–776
6. Dore JJ, Crotty KL, Birren SJ (2005) Inhibition of glial maturation by bone morphogenetic protein 2 in a neural crest-derived cell line. Dev Neurosci 27(1):37–48
7. Topilko P, Murphy P, Charnay P (1996) Embryonic development of Schwann cells: multiple roles for neuregulins along the pathway. Mol Cell Neurosci 8(2–3):71–75
8. Sieber-Blum M (1991) Role of the neurotrophic factors BDNF and NGF in the commitment of pluripotent neural crest cells. Neuron 6(6):949–955
9. Chalazonitis A (2004) Neurotrophin-3 in the development of the enteric nervous system. Prog Brain Res 146:243–263
10. Anderson DJ, Axel R (1986) A bipotential neuroendocrine precursor whose choice of cell fate is determined by NGF and glucocorticoids. Cell 47(6):1079–1090
11. Stemple DL, Mahanthappa NK, Anderson DJ (1988) Basic FGF induces neuronal differentiation, cell division, and NGF dependence in chromaffin cells: a sequence of events in sympathetic development. Neuron 1(6):517–525
12. Frodin M, Gammeltoft S (1994) Insulin-like growth factors act synergistically with basic fibroblast growth factor and nerve growth factor to promote chromaffin cell proliferation. Proc Natl Acad Sci USA 91(5):1771–1775
13. Jeong J et al (2004) Hedgehog signaling in the neural crest cells regulates the patterning and growth of facial primordia. Genes Dev 18(8):937–951
14. Satoh M, Ide H (1987) Melanocyte-stimulating hormone affects melanogenic differentiation of quail neural crest cells in vitro. Dev Biol 119(2):579–586
15. Sommer L (2011) Generation of melanocytes from neural crest cells. Pigment Cell Melanoma Res 24(3):411–421
16. Mann KM et al (2004) Calcineurin initiates smooth muscle differentiation in neural crest stem cells. J Cell Biol 165(4):483–491
17. Sieber-Blum M, Cohen AM (1980) Clonal analysis of quail neural crest cells: they are pluripotent and differentiate in vitro in the absence of noncrest cells. Dev Biol 80(1):96–106
18. Baroffio A, Dupin E, Le Douarin NM (1988) Clone-forming ability and differentiation potential of migratory neural crest cells. Proc Natl Acad Sci USA 85(14):5325–5329
19. Bronner-Fraser M, Fraser SE (1988) Cell lineage analysis reveals multipotency of some avian neural crest cells. Nature 335(6186):161–164
20. Vincent M, Thiery JP (1984) A cell surface marker for neural crest and placodal cells: further evolution in peripheral and central nervous system. Dev Biol 103(2):468–481
21. Rickmann M, Fawcett JW, Keynes RJ (1985) The migration of neural crest cells and the growth of motor axons through the rostral half of the chick somite. J Embryol Exp Morphol 90:437–455
22. Bronner-Fraser M (1986) Analysis of the early stages of trunk neural crest migration in avian embryos using monoclonal antibody HNK-1. Dev Biol 115(1):44–55
23. Toma JG et al (2001) Isolation of multipotent adult stem cells from the dermis of mammalian skin. Nat Cell Biol 3(9):778–884
24. Fernandes KJ et al (2004) A dermal niche for multipotent adult skin-derived precursor cells. Nat Cell Biol 6(11):1082–1093

25. Shi C et al (2006) Stem cells and their applications in skin-cell therapy. Trends Biotechnol 24(1):48–52
26. Nurse CA, Macintyre L, Diamond J (1984) Reinnervation of the rat touch dome restores the Merkel cell population reduced after denervation. Neuroscience 13(2):563–571
27. Reynolds BA, Tetzlaff W, Weiss S (1992) A multipotent EGF-responsive striatal embryonic progenitor cell produces neurons and astrocytes. J Neurosci 12(11):4565–4574
28. Joannides A et al (2004) Efficient generation of neural precursors from adult human skin: astrocytes promote neurogenesis from skin-derived stem cells. Lancet 364(9429):172–178
29. Reynolds BA, Weiss S (1992) Generation of neurons and astrocytes from isolated cells of the adult mammalian central nervous system. Science 255(5052):1707–1710
30. Song H, Stevens CF, Gage FH (2002) Astroglia induce neurogenesis from adult neural stem cells. Nature 417(6884):39–44
31. Belicchi M et al (2004) Human skin-derived stem cells migrate throughout forebrain and differentiate into astrocytes after injection into adult mouse brain. J Neurosci Res 77(4):475–486
32. Uchida N et al (2000) Direct isolation of human central nervous system stem cells. Proc Natl Acad Sci USA 97(26):14720–14725
33. Kopen GC, Prockop DJ, Phinney DG (1999) Marrow stromal cells migrate throughout forebrain and cerebellum, and they differentiate into astrocytes after injection into neonatal mouse brains. Proc Natl Acad Sci USA 96(19):10711–10716
34. Woodbury D et al (2000) Adult rat and human bone marrow stromal cells differentiate into neurons. J Neurosci Res 61(4):364–370
35. Munoz JR et al (2005) Human stem/progenitor cells from bone marrow promote neurogenesis of endogenous neural stem cells in the hippocampus of mice. Proc Natl Acad Sci USA 102(50):18171–18176
36. Yoo SW et al (2008) Mesenchymal stem cells promote proliferation of endogenous neural stem cells and survival of newborn cells in a rat stroke model. Exp Mol Med 40(4):387–397
37. Rubio D et al (2005) Spontaneous human adult stem cell transformation. Cancer Res 65(8):3035–3039
38. Serakinci N et al (2004) Adult human mesenchymal stem cell as a target for neoplastic transformation. Oncogene 23(29):5095–5098
39. Natarajan D et al (1999) Multipotential progenitors of the mammalian enteric nervous system capable of colonising aganglionic bowel in organ culture. Development 126(1):157–168
40. Liu W et al (2007) Neuroepithelial stem cells differentiate into neuronal phenotypes and improve intestinal motility recovery after transplantation in the aganglionic colon of the rat. Neurogastroenterol Motil 19(12):1001–1009
41. Martucciello G et al (2007) Neural crest neuroblasts can colonise aganglionic and ganglionic gut in vivo. Eur J Pediatr Surg 17(1):34–40
42. Rauch U et al (2006) Expression of intermediate filament proteins and neuronal markers in the human fetal gut. J Histochem Cytochem 54(1):39–46
43. Blaugrund E et al (1996) Distinct subpopulations of enteric neuronal progenitors defined by time of development, sympathoadrenal lineage markers and Mash-1-dependence. Development 122(1):309–320
44. Schafer KH, Hagl CI, Rauch U (2003) Differentiation of neurospheres from the enteric nervous system. Pediatr Surg Int 19(5):340–344
45. Gritti A et al (1996) Multipotential stem cells from the adult mouse brain proliferate and self-renew in response to basic fibroblast growth factor. J Neurosci 16(3):1091–1100
46. Micci MA et al (2005) Caspase inhibition increases survival of neural stem cells in the gastrointestinal tract. Neurogastroenterol Motil 17(4):557–564
47. Thomas SK et al (2004) Nestin is a potential mediator of malignancy in human neuroblastoma cells. J Biol Chem 279(27):27994–27999
48. Yamashita T et al (1984) Autonomic nervous system in human palatine tonsil. Acta Otolaryngol Suppl 416:63–71

49. Gershon MD (1981) The enteric nervous system. Annu Rev Neurosci 4:227–272
50. Karaosmanoglu T et al (1996) Regional differences in the number of neurons in the myenteric plexus of the guinea pig small intestine and colon: an evaluation of markers used to count neurons. Anat Rec 244(4):470–480
51. Newgreen D, Young HM (2002) Enteric nervous system: development and developmental disturbances–part 2. Pediatr Dev Pathol: the official journal of the Society for Pediatric Pathology and the Paediatric Pathology Society 5(4):329–349
52. Druckenbrod NR et al (2008) Targeting of endothelin receptor-B to the neural crest. Genesis 46(8):396–400
53. Fu M et al (2003) HOXB5 expression is spatially and temporarily regulated in human embryonic gut during neural crest cell colonization and differentiation of enteric neuroblasts. Dev Dyn: an official publication of the American Association of Anatomists 228(1):1–10
54. Burns AJ, Douarin NM (1998) The sacral neural crest contributes neurons and glia to the post-umbilical gut: spatiotemporal analysis of the development of the enteric nervous system. Development 125(21):4335–4337
55. Serbedzija GN et al (1991) Vital dye labelling demonstrates a sacral neural crest contribution to the enteric nervous system of chick and mouse embryos. Development 111(4):857–866
56. Young HM et al (1998) A single rostrocaudal colonization of the rodent intestine by enteric neuron precursors is revealed by the expression of Phox2b, Ret, and p75 and by explants grown under the kidney capsule or in organ culture. Dev Biol 202(1):67–84
57. Conner PJ et al (2003) Appearance of neurons and glia with respect to the wavefront during colonization of the avian gut by neural crest cells. Dev Dyn 226(1):91–98
58. Young HM et al (2004) Dynamics of neural crest-derived cell migration in the embryonic mouse gut. Dev Biol 270(2):455–473
59. Druckenbrod NR, Epstein ML (2005) The pattern of neural crest advance in the cecum and colon. Dev Biol 287(1):125–133
60. Heanue TA, Pachnis V (2007) Enteric nervous system development and Hirschsprung's disease: advances in genetic and stem cell studies. Nat Rev Neurosci 8(6):466–479
61. Young HM, Newgreen D (2001) Enteric neural crest-derived cells: origin, identification, migration, and differentiation. Anat Rec 262(1):1–15
62. Baetge G, Gershon MD (1989) Transient catecholaminergic (TC) cells in the vagus nerves and bowel of fetal mice: relationship to the development of enteric neurons. Dev Biol 132(1):189–211
63. Landman KA, Simpson MJ, Newgreen DF (2007) Mathematical and experimental insights into the development of the enteric nervous system and Hirschsprung's disease. Dev Growth Differ 49(4):277–286
64. Wallace AS, Burns AJ (2005) Development of the enteric nervous system, smooth muscle and interstitial cells of Cajal in the human gastrointestinal tract. Cell Tissue Res 319(3):367–382
65. Noakes PG, Hornbruch A, Wolpert L (1993) The relationship between migrating neural crest cells and growing limb nerves in the developing chick forelimb. Prog Clin Biol Res 383A:381–390
66. Amiel J, Lyonnet S (2001) Hirschsprung disease, associated syndromes, and genetics: a review. J Med Genet 38(11):729–739
67. Passarge E (1967) The genetics of Hirschsprung's disease. Evidence for heterogeneous etiology and a study of sixty-three families. N Engl J Med 276(3):138–143
68. Spouge D, Baird PA (1985) Hirschsprung disease in a large birth cohort. Teratology 32(2):171–177
69. Badner JA et al (1990) A genetic study of Hirschsprung disease. Am J Hum Genet 46(3):568–580
70. Kapur RP (2005) Multiple endocrine neoplasia type 2B and Hirschsprung's disease. Clin Gastroenterol Hepatol 3(5):423–431
71. Parisi MA, Kapur RP (2000) Genetics of Hirschsprung disease. Curr Opin Pediatr 12(6):610–617

72. Gariepy CE (2004) Developmental disorders of the enteric nervous system: genetic and molecular bases. J Pediatr Gastroenterol Nutr 39(1):5–11
73. Manie S et al (2001) The RET receptor: function in development and dysfunction in congenital malformation. Trends Genet 17(10):580–589
74. Airaksinen MS, Titievsky A, Saarma M (1999) GDNF family neurotrophic factor signaling: four masters, one servant? Mol Cell Neurosci 13(5):313–325
75. Sanicola M et al (1997) Glial cell line-derived neurotrophic factor-dependent RET activation can be mediated by two different cell-surface accessory proteins. Proc Natl Acad Sci USA 94(12):6238–6243
76. Trupp M et al (1998) Multiple GPI-anchored receptors control GDNF-dependent and independent activation of the c-Ret receptor tyrosine kinase. Mol Cell Neurosci 11(1–2):47–63
77. Cik M et al (2000) Binding of GDNF and neurturin to human GDNF family receptor alpha 1 and 2. Influence of cRET and cooperative interactions. J Biol Chem 275(36):27505–27512
78. Heuckeroth RO (2003) Finding your way to the end: a tale of GDNF and endothelin-3. Neuron 40(5):871–873
79. Durbec PL et al (1996) Common origin and developmental dependence on c-ret of subsets of enteric and sympathetic neuroblasts. Development 122(1):349–358
80. Schuchardt A et al (1994) Defects in the kidney and enteric nervous system of mice lacking the tyrosine kinase receptor Ret. Nature 367(6461):380–383
81. Pichel JG et al (1996) Defects in enteric innervation and kidney development in mice lacking GDNF. Nature 382(6586):73–76
82. Cacalano G et al (1998) GFRalpha1 is an essential receptor component for GDNF in the developing nervous system and kidney. Neuron 21(1):53–62
83. Enomoto H et al (1998) GFR alpha1-deficient mice have deficits in the enteric nervous system and kidneys. Neuron 21(2):317–324
84. Uesaka T et al (2007) Conditional ablation of GFRalpha1 in postmigratory enteric neurons triggers unconventional neuronal death in the colon and causes a Hirschsprung's disease phenotype. Development 134(11):2171–2181
85. Uesaka T et al (2008) Diminished Ret expression compromises neuronal survival in the colon and causes intestinal aganglionosis in mice. J Clin Invest 118(5):1890–1898
86. Bondurand N et al (2003) Neuron and glia generating progenitors of the mammalian enteric nervous system isolated from foetal and postnatal gut cultures. Development 130(25):6387–6400
87. Jiang J, Hui CC (2008) Hedgehog signaling in development and cancer. Dev Cell 15(6):801–812
88. Young HM et al (2001) GDNF is a chemoattractant for enteric neural cells. Dev Biol 229(2):503–516
89. Pachnis V et al (1998) *III.* Role Of the RET signal transduction pathway in development of the mammalian enteric nervous system. Am J Physiol 275(2 Pt 1):G183–G186
90. Hofstra RM et al (2000) RET and GDNF gene scanning in Hirschsprung patients using two dual denaturing gel systems. Hum Mutat 15(5):418–429
91. Angrist M et al (1995) Mutation analysis of the RET receptor tyrosine kinase in Hirschsprung disease. Hum Mol Genet 4(5):821–830
92. Attie T et al (1995) Diversity of RET proto-oncogene mutations in familial and sporadic Hirschsprung disease. Hum Mol Genet 4(8):1381–1386
93. Seri M et al (1997) Frequency of RET mutations in long- and short-segment Hirschsprung disease. Hum Mutat 9(3):243–249
94. Kjaer S et al (2010) Mammal-restricted elements predispose human RET to folding impairment by HSCR mutations. Nat Struct Mol Biol 17(6):726–731
95. Fitze G et al (1999) Association of RET protooncogene codon 45 polymorphism with Hirschsprung disease. Am J Hum Genet 65(5):1469–1473
96. Borrego S et al (1999) Specific polymorphisms in the RET proto-oncogene are over-represented in patients with Hirschsprung disease and may represent loci modifying phenotypic expression. J Med Genet 36(10):771–774

97. Borrego S et al (2000) RET genotypes comprising specific haplotypes of polymorphic variants predispose to isolated Hirschsprung disease. J Med Genet 37(8):572–578
98. Emison ES et al (2005) A common sex-dependent mutation in a RET enhancer underlies Hirschsprung disease risk. Nature 434(7035):857–863
99. Garcia-Barcelo M et al (2005) TTF-1 and RET promoter SNPs: regulation of RET transcription in Hirschsprung's disease. Hum Mol Genet 14(2):191–204
100. Garcia-Barcelo M et al (2004) Highly recurrent RET mutations and novel mutations in genes of the receptor tyrosine kinase and endothelin receptor B pathways in Chinese patients with sporadic Hirschsprung disease. Clin Chem 50(1):93–100
101. Garcia-Barcelo MM et al (2003) Chinese patients with sporadic Hirschsprung's disease are predominantly represented by a single RET haplotype. J Med Genet 40(11):e122
102. Lui VC et al (2005) Novel RET mutation produces a truncated RET receptor lacking the intracellular signaling domain in a 3-generation family with Hirschsprung disease. Clin Chem 51(8):1552–1554
103. Miao X et al (2007) Role of RET and PHOX2B gene polymorphisms in risk of Hirschsprung's disease in Chinese population. Gut 56(5):736
104. Miao X et al (2010) Reduced RET expression in gut tissue of individuals carrying risk alleles of Hirschsprung's disease. Hum Mol Genet 19(8):1461–1467
105. Angrist M et al (1996) Germline mutations in glial cell line-derived neurotrophic factor (GDNF) and RET in a Hirschsprung disease patient. Nat Genet 14(3):341–344
106. Salomon R et al (1996) Germline mutations of the RET ligand GDNF are not sufficient to cause Hirschsprung disease. Nat Genet 14(3):345–347
107. Ivanchuk SM et al (1996) De novo mutation of GDNF, ligand for the RET/GDNFR-alpha receptor complex, in Hirschsprung disease. Hum Mol Genet 5(12):2023–2026
108. Ferlin A et al (2003) The human Y chromosome's azoospermia factor b (AZFb) region: sequence, structure, and deletion analysis in infertile men. J Med Genet 40(1):18–24
109. Doray B et al (1998) Mutation of the RET ligand, neurturin, supports multigenic inheritance in Hirschsprung disease. Hum Mol Genet 7(9):1449–1452
110. Heuckeroth RO et al (1998) Neurturin and GDNF promote proliferation and survival of enteric neuron and glial progenitors in vitro. Dev Biol 200(1):116–129
111. Leibl MA et al (1999) Expression of endothelin 3 by mesenchymal cells of embryonic mouse caecum. Gut 44(2):246–252
112. Barlow A, de Graaff E, Pachnis V (2003) Enteric nervous system progenitors are coordinately controlled by the G protein-coupled receptor EDNRB and the receptor tyrosine kinase RET. Neuron 40(5):905–916
113. Lee HO, Levorse JM, Shin MK (2003) The endothelin receptor-B is required for the migration of neural crest-derived melanocyte and enteric neuron precursors. Dev Biol 259(1):162–175
114. Baynash AG et al (1994) Interaction of endothelin-3 with endothelin-B receptor is essential for development of epidermal melanocytes and enteric neurons. Cell 79(7):1277–1285
115. Hosoda K et al (1994) Targeted and natural (piebald-lethal) mutations of endothelin-B receptor gene produce megacolon associated with spotted coat color in mice. Cell 79(7):1267–1276
116. Yanagisawa H et al (1998) Dual genetic pathways of endothelin-mediated intercellular signaling revealed by targeted disruption of endothelin converting enzyme-1 gene. Development 125(5):825–836
117. Wu JJ et al (1999) Inhibition of in vitro enteric neuronal development by endothelin-3: mediation by endothelin B receptors. Development 126(6):1161–1173
118. Kruger GM et al (2003) Temporally distinct requirements for endothelin receptor B in the generation and migration of gut neural crest stem cells. Neuron 40(5):917–929
119. Hearn CJ, Murphy M, Newgreen D (1998) GDNF and ET-3 differentially modulate the numbers of avian enteric neural crest cells and enteric neurons in vitro. Dev Biol 197(1):93–105
120. Woodward MN et al (2003) Analysis of the effects of endothelin-3 on the development of neural crest cells in the embryonic mouse gut. J Pediatr Surg 38(9):1322–1328

121. Puffenberger EG et al (1994) A missense mutation of the endothelin-B receptor gene in multigenic Hirschsprung's disease. Cell 79(7):1257–1266
122. Carrasquillo MM et al (2002) Genome-wide association study and mouse model identify interaction between RET and EDNRB pathways in Hirschsprung disease. Nat Genet 32(2):237–244
123. Cohen IT, Gadd MA (1982) Hirschsprung's disease in a kindred: a possible clue to the genetics of the disease. J Pediatr Surg 17(5):632–634
124. Puffenberger EG et al (1994) Identity-by-descent and association mapping of a recessive gene for Hirschsprung disease on human chromosome 13q22. Hum Mol Genet 3(8):1217–1225
125. Van Camp G et al (1995) Chromosome 13q deletion with Waardenburg syndrome: further evidence for a gene involved in neural crest function on 13q. J Med Genet 32(7):531–536
126. Chakravarti A (1996) Endothelin receptor-mediated signaling in hirschsprung disease. Hum Mol Genet 5(3):303–307
127. Amiel J et al (1996) Heterozygous endothelin receptor B (EDNRB) mutations in isolated Hirschsprung disease. Hum Mol Genet 5(3):355–357
128. Kusafuka T, Wang Y, Puri P (1996) Novel mutations of the endothelin-B receptor gene in isolated patients with Hirschsprung's disease. Hum Mol Genet 5(3):347–349
129. Auricchio A et al (1996) Endothelin-B receptor mutations in patients with isolated Hirschsprung disease from a non-inbred population. Hum Mol Genet 5(3):351–354
130. Hofstra RM et al (1999) A loss-of-function mutation in the endothelin-converting enzyme 1 (ECE-1) associated with Hirschsprung disease, cardiac defects, and autonomic dysfunction. Am J Hum Genet 64(1):304–308
131. McCallion AS et al (2003) Phenotype variation in two-locus mouse models of Hirschsprung disease: tissue-specific interaction between Ret and Ednrb. Proc Natl Acad Sci USA 100(4):1826–1831
132. Murone M, Rosenthal A, de Sauvage FJ (1999) Hedgehog signal transduction: from flies to vertebrates. Exp Cell Res 253(1):25–33
133. Nieuwenhuis E, Hui CC (2005) Hedgehog signaling and congenital malformations. Clin Genet 67(3):193–208
134. Ingham PW, McMahon AP (2001) Hedgehog signaling in animal development: paradigms and principles. Genes Dev 15(23):3059–3087
135. Ramalho-Santos M, Melton DA, McMahon AP (2000) Hedgehog signals regulate multiple aspects of gastrointestinal development. Development 127(12):2763–2772
136. Fu M et al (2004) Sonic hedgehog regulates the proliferation, differentiation, and migration of enteric neural crest cells in gut. J Cell Biol 166(5):673–684
137. Garcia-Barcelo MM et al (2009) Genome-wide association study identifies NRG1 as a susceptibility locus for Hirschsprung's disease. Proc Natl Acad Sci USA 106(8):2694–2699
138. Artavanis-Tsakonas S, Rand MD, Lake RJ (1999) Notch signaling: cell fate control and signal integration in development. Science 284(5415):770–776
139. Mumm JS et al (2000) A ligand-induced extracellular cleavage regulates gamma-secretase-like proteolytic activation of Notch1. Mol Cell 5(2):197–206
140. Kageyama R, Ohtsuka T, Tomita K (2000) The bHLH gene Hes1 regulates differentiation of multiple cell types. Mol Cells 10(1):1–7
141. Bray SJ (2006) Notch signalling: a simple pathway becomes complex. Nat Rev Mol Cell Biol 7(9):678–689
142. Okajima T, Irvine KD (2002) Regulation of notch signaling by o-linked fucose. Cell 111(6):893–904
143. Okajima T, Xu A, Irvine KD (2003) Modulation of notch-ligand binding by protein O-fucosyltransferase 1 and fringe. J Biol Chem 278(43):42340–42345
144. Okajima T et al (2005) Chaperone activity of protein O-fucosyltransferase 1 promotes notch receptor folding. Science 307(5715):1599–1603
145. Okamura Y, Saga Y (2008) Pofut1 is required for the proper localization of the Notch receptor during mouse development. Mech Dev 125(8):663–673

146. Sasamura T et al (2007) The O-fucosyltransferase O-fut1 is an extracellular component that is essential for the constitutive endocytic trafficking of Notch in Drosophila. Development 134(7):1347–1356
147. Shi S, Stanley P (2003) Protein O-fucosyltransferase 1 is an essential component of Notch signaling pathways. Proc Natl Acad Sci USA 100(9):5234–5239
148. Okamura Y, Saga Y (2008) Notch signaling is required for the maintenance of enteric neural crest progenitors. Development 135(21):3555–3565
149. Wakamatsu Y, Maynard TM, Weston JA (2000) Fate determination of neural crest cells by NOTCH-mediated lateral inhibition and asymmetrical cell division during gangliogenesis. Development 127(13):2811–2821
150. Androutsellis-Theotokis A et al (2006) Notch signalling regulates stem cell numbers in vitro and in vivo. Nature 442(7104):823–826
151. Elkabetz Y et al (2008) Human ES cell-derived neural rosettes reveal a functionally distinct early neural stem cell stage. Genes Dev 22(2):152–165
152. Solecki DJ et al (2001) Activated Notch2 signaling inhibits differentiation of cerebellar granule neuron precursors by maintaining proliferation. Neuron 31(4):557–568
153. Ingram WJ et al (2008) Sonic Hedgehog regulates Hes1 through a novel mechanism that is independent of canonical Notch pathway signalling. Oncogene 27(10):1489–1500
154. Ngan ESW, MG B, Yip BHK, Yip BHK, Poon H-C, Lau ST, Kwok CKM, Sat T, Sham MH, Wong KKY, Wainwright BJ, Cherny SS, Hui CC, Sham PC, Lui VCH, Tam PKH (2011) Hedgehog/Notch-induced premature gliogenesis represents a new disease mechanism for Hirschsprung's disease in mice and humans. J Clin Invest 121(9):3467–3478
155. Pevny LH, Lovell-Badge R (1997) Sox genes find their feet. Curr Opin Genet Dev 7(3):338–344
156. Schepers GE, Teasdale RD, Koopman P (2002) Twenty pairs of sox: extent, homology, and nomenclature of the mouse and human sox transcription factor gene families. Dev Cell 3(2):167–170
157. Kim J et al (2003) SOX10 maintains multipotency and inhibits neuronal differentiation of neural crest stem cells. Neuron 38(1):17–31
158. Honore SM, Aybar MJ, Mayor R (2003) Sox10 is required for the early development of the prospective neural crest in Xenopus embryos. Dev Biol 260(1):79–96
159. Bondurand N et al (1998) Expression of the SOX10 gene during human development. FEBS Lett 432(3):168–172
160. Cheng Y et al (2000) Chick sox10, a transcription factor expressed in both early neural crest cells and central nervous system. Brain Res Dev Brain Res 121(2):233–241
161. Kuhlbrodt K et al (1998) Sox10, a novel transcriptional modulator in glial cells. J Neurosci: the official journal of the Society for Neuroscience 18(1):237–250
162. Southard-Smith EM, Kos L, Pavan WJ (1998) Sox10 mutation disrupts neural crest development in Dom Hirschsprung mouse model. Nat Genet 18(1):60–64
163. Herbarth B et al (1998) Mutation of the Sry-related Sox10 gene in Dominant megacolon, a mouse model for human Hirschsprung disease. Proc Natl Acad Sci USA 95(9):5161–5165
164. Pingault V et al (1998) SOX10 mutations in patients with Waardenburg-Hirschsprung disease. Nat Genet 18(2):171–173
165. Britsch S et al (2001) The transcription factor Sox10 is a key regulator of peripheral glial development. Genes Dev 15(1):66–78
166. Kelsh RN, Eisen JS (2000) The zebrafish colourless gene regulates development of non-ectomesenchymal neural crest derivatives. Development 127(3):515–525
167. Dutton KA et al (2001) Zebrafish colourless encodes sox10 and specifies non-ectomesenchymal neural crest fates. Development 128(21):4113–4125
168. Bondurand N et al (2006) Maintenance of mammalian enteric nervous system progenitors by SOX10 and endothelin 3 signalling. Development 133(10):2075–2086
169. Owens SE et al (2005) Genome-wide linkage identifies novel modifier loci of aganglionosis in the Sox10Dom model of Hirschsprung disease. Hum Mol Genet 14(11):1549–1558

170. Maka M, Stolt CC, Wegner M (2005) Identification of Sox8 as a modifier gene in a mouse model of Hirschsprung disease reveals underlying molecular defect. Dev Biol 277(1):155–169
171. Cantrell VA et al (2004) Interactions between Sox10 and EdnrB modulate penetrance and severity of aganglionosis in the Sox10Dom mouse model of Hirschsprung disease. Hum Mol Genet 13(19):2289–2301
172. Lang D, Epstein JA (2003) Sox10 and Pax3 physically interact to mediate activation of a conserved c-RET enhancer. Hum Mol Genet 12(8):937–945
173. Stanchina L et al (2010) Genetic interaction between Sox10 and Zfhx1b during enteric nervous system development. Dev Biol 341(2):416–428
174. Pevny LH, Nicolis SK (2010) Sox2 roles in neural stem cells. Int J Biochem Cell Biol 42(3):421–424
175. Wegner M, Stolt CC (2005) From stem cells to neurons and glia: a Soxist's view of neural development. Trends Neurosci 28(11):583–588
176. Heanue TA, Pachnis V (2006) Expression profiling the developing mammalian enteric nervous system identifies marker and candidate Hirschsprung disease genes. Proc Natl Acad Sci USA 103(18):6919–6924
177. Heanue TA, Pachnis V (2011) Prospective identification and isolation of enteric nervous system progenitors using Sox2. Stem Cells 29(1):128–140
178. Wakamatsu Y et al (2004) Multiple roles of Sox2, an HMG-box transcription factor in avian neural crest development. Dev Dyn: an official publication of the American Association of Anatomists 229(1):74–86
179. Cimadamore F et al (2011) Human ESC-derived neural crest model reveals a key role for SOX2 in sensory neurogenesis. Cell Stem Cell 8(5):538–551
180. Pitera JE et al (2001) Embryonic gut anomalies in a mouse model of retinoic Acid-induced caudal regression syndrome: delayed gut looping, rudimentary cecum, and anorectal anomalies. Am J Pathol 159(6):2321–2329
181. Niederreither K et al (2003) The regional pattern of retinoic acid synthesis by RALDH2 is essential for the development of posterior pharyngeal arches and the enteric nervous system. Development 130(11):2525–2534
182. Sato Y, Heuckeroth RO (2008) Retinoic acid regulates murine enteric nervous system precursor proliferation, enhances neuronal precursor differentiation, and reduces neurite growth in vitro. Dev Biol 320(1):185–198
183. Fu M et al (2010) Vitamin A facilitates enteric nervous system precursor migration by reducing Pten accumulation. Development 137(4):631–640
184. Ngan ES et al (2007) Prokineticin-1 modulates proliferation and differentiation of enteric neural crest cells. Biochim Biophys Acta 1773(4):536–545
185. Ngan ES et al (2008) Prokineticin-1 (Prok-1) works coordinately with glial cell line-derived neurotrophic factor (GDNF) to mediate proliferation and differentiation of enteric neural crest cells. Biochim Biophys Acta 1783(3):467–478
186. Liu MT et al (2009) 5-HT4 receptor-mediated neuroprotection and neurogenesis in the enteric nervous system of adult mice. J Neurosci: the official journal of the Society for Neuroscience 29(31):9683–9699
187. Kapur RP (1999) Early death of neural crest cells is responsible for total enteric aganglionosis in Sox10(Dom)/Sox10(Dom) mouse embryos. Pediatr Dev Pathol 2(6):559–569
188. Pattyn A et al (1999) The homeobox gene Phox2b is essential for the development of autonomic neural crest derivatives. Nature 399(6734):366–370
189. Corry GN, Hendzel MJ, Underhill DA (2008) Subnuclear localization and mobility are key indicators of PAX3 dysfunction in Waardenburg syndrome. Hum Mol Genet 17(12):1825–1837
190. Schubert FR et al (2001) Early mesodermal phenotypes in splotch suggest a role for Pax3 in the formation of epithelial somites. Dev Dyn: an official publication of the American Association of Anatomists 222(3):506–521

191. Lui VC et al (2008) Perturbation of hoxb5 signaling in vagal neural crests down-regulates ret leading to intestinal hypoganglionosis in mice. Gastroenterology 134(4):1104–1115
192. Lang D et al (2000) Pax3 is required for enteric ganglia formation and functions with Sox10 to modulate expression of c-ret. J Clin Invest 106(8):963–971
193. Guillemot F et al (1993) Mammalian achaete-scute homolog 1 is required for the early development of olfactory and autonomic neurons. Cell 75(3):463–476
194. Leon TY et al (2009) Transcriptional regulation of RET by Nkx2-1, Phox2b, Sox10, and Pax3. J Pediatr Surg 44(10):1904–1912
195. Johansson CB et al (1999) Neural stem cells in the adult human brain. Exp Cell Res 253(2):733–736
196. Qu Q, Shi Y (2009) Neural stem cells in the developing and adult brains. J Cell Physiol 221(1):5–9
197. Carvey PM et al (2001) A clonal line of mesencephalic progenitor cells converted to dopamine neurons by hematopoietic cytokines: a source of cells for transplantation in Parkinson's disease. Exp Neurol 171(1):98–108
198. Studer L, Tabar V, McKay RD (1998) Transplantation of expanded mesencephalic precursors leads to recovery in parkinsonian rats. Nat Neurosci 1(4):290–295
199. Sinden JD et al (1997) Recovery of spatial learning by grafts of a conditionally immortalized hippocampal neuroepithelial cell line into the ischaemia-lesioned hippocampus. Neuroscience 81(3):599–608
200. Micci MA et al (2005) Neural stem cell transplantation in the stomach rescues gastric function in neuronal nitric oxide synthase-deficient mice. Gastroenterology 129(6):1817–1824
201. Micci MA et al (2001) Neural stem cells express RET, produce nitric oxide, and survive transplantation in the gastrointestinal tract. Gastroenterology 121(4):757–766
202. Dong YL et al (2008) Neural stem cell transplantation rescues rectum function in the aganglionic rat. Transplant Proc 40(10):3646–3652
203. Bliss TM, Andres RH, Steinberg GK (2010) Optimizing the success of cell transplantation therapy for stroke. Neurobiol Dis 37(2):275–283
204. Kruger GM et al (2002) Neural crest stem cells persist in the adult gut but undergo changes in self-renewal, neuronal subtype potential, and factor responsiveness. Neuron 35(4):657–669
205. Pan WK et al (2011) Transplantation of neonatal gut neural crest progenitors reconstructs ganglionic function in benzalkonium chloride-treated homogenic rat colon. J Surg Res 167(2):e221–e230
206. Rauch U et al (2006) Isolation and cultivation of neuronal precursor cells from the developing human enteric nervous system as a tool for cell therapy in dysganglionosis. Int J Color Dis 21(6):554–559
207. Metzger M et al (2009) Expansion and differentiation of neural progenitors derived from the human adult enteric nervous system. Gastroenterology 137(6):2063–2073, e4
208. Metzger M (2010) Neurogenesis in the enteric nervous system. Arch Ital Biol 148(2):73–83
209. Almond S et al (2007) Characterisation and transplantation of enteric nervous system progenitor cells. Gut 56(4):489–496
210. Lindley RM et al (2008) Human and mouse enteric nervous system neurosphere transplants regulate the function of aganglionic embryonic distal colon. Gastroenterology 135(1):205–216, e6
211. Metzger M et al (2009) Enteric nervous system stem cells derived from human gut mucosa for the treatment of aganglionic gut disorders. Gastroenterology 136(7):2214–2225, e1-3
212. Blanpain C, Fuchs E (2009) Epidermal homeostasis: a balancing act of stem cells in the skin. Nat Rev Mol Cell Biol 10(3):207–217
213. Biernaskie J et al (2009) SKPs derive from hair follicle precursors and exhibit properties of adult dermal stem cells. Cell Stem Cell 5(6):610–623
214. Li L et al (2010) Human dermal stem cells differentiate into functional epidermal melanocytes. J Cell Sci 123(Pt 6):853–860

215. Zhao MT et al (2010) Deciphering the mesodermal potency of porcine skin derived progenitors (SKP) by microarray analysis. Cell Reprogram 12(2):161–173
216. Wong CE et al (2006) Neural crest-derived cells with stem cell features can be traced back to multiple lineages in the adult skin. J Cell Biol 175(6):1005–1015
217. Zhao M et al (2009) Tracing the stemness of porcine skin-derived progenitors (pSKP) back to specific marker gene expression. Cloning Stem Cells 11(1):111–122
218. Fernandes KJ, Toma JG, Miller FD (2008) Multipotent skin-derived precursors: adult neural crest-related precursors with therapeutic potential. Philos Trans R Soc Lond B Biol Sci 363(1489):185–198
219. Dyce PW et al (2004) Stem cells with multilineage potential derived from porcine skin. Biochem Biophys Res Commun 316(3):651–658
220. Lermen D et al (2010) Neuro-muscular differentiation of adult porcine skin derived stem cell-like cells. PLoS One 5(1):e8968
221. Gago N et al (2009) Age-dependent depletion of human skin-derived progenitor cells. Stem Cells 27(5):1164–1172
222. Biernaskie J et al (2007) Skin-derived precursors generate myelinating Schwann cells that promote remyelination and functional recovery after contusion spinal cord injury. J Neurosci: the official journal of the Society for Neuroscience 27(36):9545–9559
223. Gorio A et al (2004) Fate of autologous dermal stem cells transplanted into the spinal cord after traumatic injury (TSCI). Neuroscience 125(1):179–189
224. Kyung KS, Ho CW, Kwan CB (2007) Potential therapeutic clue of skin-derived progenitor cells following cytokine-mediated signal overexpressed in injured spinal cord. Tissue Eng 13(6):1247–1258
225. Sieber-Blum M et al (2006) Characterization of epidermal neural crest stem cell (EPI-NCSC) grafts in the lesioned spinal cord. Mol Cell Neurosci 32(1–2):67–81
226. Tunici P et al (2006) Brain engraftment and therapeutic potential of stem/progenitor cells derived from mouse skin. J Gene Med 8(4):506–1513
227. Qiu Z et al (2010) Skeletal myogenic potential of mouse skin-derived precursors. Stem Cells Dev 19(2):259–268
228. Fernandes KJ, Miller FD (2009) Isolation, expansion, and differentiation of mouse skin-derived precursors. Methods Mol Biol 482:159–170
229. Pittenger MF et al (1999) Multilineage potential of adult human mesenchymal stem cells. Science 284(5411):143–147
230. Tremain N et al (2001) MicroSAGE analysis of 2,353 expressed genes in a single cell-derived colony of undifferentiated human mesenchymal stem cells reveals mRNAs of multiple cell lineages. Stem Cells 19(5):408–418
231. Le Blanc K, Pittenger M (2005) Mesenchymal stem cells: progress toward promise. Cytotherapy 7(1):36–45
232. Lee RH et al (2004) Characterization and expression analysis of mesenchymal stem cells from human bone marrow and adipose tissue. Cell Physiol Biochem: international journal of experimental cellular physiology, biochemistry, and pharmacology 14(4–6):311–324
233. Wang G et al (2005) Adult stem cells from bone marrow stroma differentiate into airway epithelial cells: potential therapy for cystic fibrosis. Proc Natl Acad Sci USA 102(1):186–191
234. Brazelton TR et al (2000) From marrow to brain: expression of neuronal phenotypes in adult mice. Science 290(5497):1775–1779
235. Zuk PA et al (2001) Multilineage cells from human adipose tissue: implications for cell-based therapies. Tissue Eng 7(2):211–228
236. Katz AJ et al (2005) Cell surface and transcriptional characterization of human adipose-derived adherent stromal (hADAS) cells. Stem Cells 23(3):412–423
237. Panepucci RA et al (2004) Comparison of gene expression of umbilical cord vein and bone marrow-derived mesenchymal stem cells. Stem Cells 22(7):1263–1278
238. In 't Anker PS et al (2004) Isolation of mesenchymal stem cells of fetal or maternal origin from human placenta. Stem Cells 22(7):1338–1345

239. Song S et al (2007) Comparison of neuron-like cells derived from bone marrow stem cells to those differentiated from adult brain neural stem cells. Stem Cells Dev 16(5):747–756
240. Triaca V, Aloe L (2005) Neuronal markers expression of NGF-primed bone marrow cells (BMCs) transplanted in the brain of 6-hydroxydopamine and ibotenic acid lesioned littermate mice. Neurosci Lett 384(1–2):82–86
241. Azizi SA et al (1998) Engraftment and migration of human bone marrow stromal cells implanted in the brains of albino rats–similarities to astrocyte grafts. Proc Natl Acad Sci USA 95(7):3908–3913
242. Mezey E et al (2000) Turning blood into brain: cells bearing neuronal antigens generated in vivo from bone marrow. Science 290(5497):1779–1782
243. Gao YJ et al (2006) Differentiation potential of bone marrow stromal cells to enteric neurons in vitro. Chin J Dig Dis 7(3):156–163
244. Nagoshi N et al (2008) Ontogeny and multipotency of neural crest-derived stem cells in mouse bone marrow, dorsal root ganglia, and whisker pad. Cell Stem Cell 2(4):392–403
245. Wernig M et al (2008) Neurons derived from reprogrammed fibroblasts functionally integrate into the fetal brain and improve symptoms of rats with Parkinson's disease. Proc Natl Acad Sci USA 105(15):5856–5861
246. Rodriguez-Gomez JA et al (2007) Persistent dopamine functions of neurons derived from embryonic stem cells in a rodent model of Parkinson disease. Stem Cells 25(4):918–928
247. Marchetto MC et al (2010) A model for neural development and treatment of Rett syndrome using human induced pluripotent stem cells. Cell 143(4):527–539
248. Ebert AD et al (2009) Induced pluripotent stem cells from a spinal muscular atrophy patient. Nature 457(7227):277–280
249. Rubin LL (2008) Stem cells and drug discovery: the beginning of a new era? Cell 132(4):549–552
250. Hotta R et al (2010) Effects of tissue age, presence of neurones and endothelin-3 on the ability of enteric neurone precursors to colonize recipient gut: implications for cell-based therapies. Neurogastroenterol Motil: the official journal of the European Gastrointestinal Motility Society 22(3):331–e86
251. Druckenbrod NR, Epstein ML (2009) Age-dependent changes in the gut environment restrict the invasion of the hindgut by enteric neural progenitors. Development 136(18):3195–3203
252. Nakagawa M et al (2008) Generation of induced pluripotent stem cells without Myc from mouse and human fibroblasts. Nat Biotechnol 26(1):101–106
253. Saha K, Jaenisch R (2009) Technical challenges in using human induced pluripotent stem cells to model disease. Cell Stem Cell 5(6):584–595
254. Kaji K et al (2009) Virus-free induction of pluripotency and subsequent excision of reprogramming factors. Nature 458(7239):771–775
255. Woltjen K et al (2009) piggyBac transposition reprograms fibroblasts to induced pluripotent stem cells. Nature 458(7239):766–770
256. Yu J et al (2009) Human induced pluripotent stem cells free of vector and transgene sequences. Science 324(5928):797–801
257. Stadtfeld M et al (2008) Induced pluripotent stem cells generated without viral integration. Science 322(5903):945–949
258. Fusaki N et al (2009) Efficient induction of transgene-free human pluripotent stem cells using a vector based on Sendai virus, an RNA virus that does not integrate into the host genome. Proc Jpn Acad Ser B Phys Biol Sci 85(8):348–362
259. Kim D et al (2009) Generation of human induced pluripotent stem cells by direct delivery of reprogramming proteins. Cell Stem Cell 4(6):472–476
260. Zhou H et al (2009) Generation of induced pluripotent stem cells using recombinant proteins. Cell Stem Cell 4(5):381–384
261. Spits C et al (2008) Recurrent chromosomal abnormalities in human embryonic stem cells. Nat Biotechnol 26(12):1361–1363
262. Dragunow M (2008) The adult human brain in preclinical drug development. Nat Rev Drug Discov 7(8):659–666

263. Dimos JT et al (2008) Induced pluripotent stem cells generated from patients with ALS can be differentiated into motor neurons. Science 321(5893):1218–1221
264. Lee G et al (2009) Modelling pathogenesis and treatment of familial dysautonomia using patient-specific iPSCs. Nature 461(7262):402–406
265. Soldner F et al (2009) Parkinson' disease patient-derived induced pluripotent stem cells free of viral reprogramming factors. Cell 136(5):964–977
266. Hargus G et al (2010) Differentiated Parkinson patient-derived induced pluripotent stem cells grow in the adult rodent brain and reduce motor asymmetry in Parkinsonian rats. Proc Natl Acad Sci USA 107(36):15921–15926
267. Cooper O et al (2010) Differentiation of human ES and Parkinson's disease iPS cells into ventral midbrain dopaminergic neurons requires a high activity form of SHH, FGF8a and specific regionalization by retinoic acid. Mol Cell Neurosci 45(3):258–266
268. Chamberlain SJ, Li XJ, Lalande M (2008) Induced pluripotent stem (iPS) cells as in vitro models of human neurogenetic disorders. Neurogenetics 9(4):227–235
269. Ku S et al (2010) Friedreich's ataxia induced pluripotent stem cells model intergenerational GAATTC triplet repeat instability. Cell Stem Cell 7(5):631–637
270. Park IH et al (2008) Disease-specific induced pluripotent stem cells. Cell 134(5):877–886
271. Zhang N et al (2010) Characterization of human Huntington's disease cell model from induced pluripotent stem cells. PLoS Curr 2:RRN1193

Chapter 17
Common Denominators of Self-renewal and Malignancy in Neural Stem Cells and Glioma

Grzegorz Wicher, Karin Holmqvist, and Karin Forsberg-Nilsson

Contents

G. Wicher
Department of Immunology, Genetics and Pathology, Science for Life Laboratory, Uppsala University, 751 85 Uppsala, Sweden

K. Holmqvist
MSD, Solbakken 1, Postboks 458, Brakeroya N-3002 Drammen, Norway

K. Forsberg-Nilsson (✉)
Department of Immunology, Genetics and Pathology, Science for Life Laboratory, Uppsala University, 751 85 Uppsala, Sweden
e-mail: karin.nilsson@igp.uu.se

R.K. Srivastava and S. Shankar (eds.), *Stem Cells and Human Diseases*,
DOI 10.1007/978-94-007-2801-1_17, © Springer Science+Business Media B.V. 2012

Abstract Regulation of neural stem cell number needs to be tightly controlled. Mutations affecting stem cells or progenitor cells may result in uncontrolled proliferation and ultimately cancer. A growing body of evidence suggests that this mechanism underlies the genesis of several brain tumors, e.g. gliomas. The most malignant form is called glioblastoma multiforme and unfortunately its prognosis remains poor. The concept of tumor-causing stem cell-like cells, also called cancer stem cells, in solid tumors has attracted a lot of interest over the last 5–10 years. A glioma-initiating cell bearing stem cell characteristics has been proposed as the origin of glioma, with the ability to seed new tumors through the capacity to evade chemotherapy and irradiation. This would be a unique feature for glioma-initiating cells, not shared by the bulk of tumor cells. Neural progenitors and glioma-initiating cells have several common traits, such as sustained proliferation and a highly efficient migratory capacity in the brain. There are similarities between then neurogenic niche where adult neural stem cells reside, and the tumorigenic niche. These include interactions with the extracellular matrix, and many of the matrix components are deregulated in glioma. The signaling pathways that are mutated in glioma are in general important neural stem cell pathways that regulate cell proliferation/self renewal, differentiation, migration and survival. Molecular changes in these pathways due to mutations are associated with brain tumor development and so present therapy targets. Novel molecular classification of glioblastoma gives hope for more stratified treatment, and we are hopefully on the threshold to patient-specific treatments which may finally change the outcome in this devastating disease.

Keywords Cancer stem cell • Glioblastoma • Glioma initiating cell • Neural progenitor • Receptor tyrosine kinase

Abbreviations

BMP	Bone morphogenetic protein
BTSC	Brain tumor stem cell
CSC	Cancer stem cells
CNS	Central nervous system
DG	Dentate gyrus
EGF	Epidermal growth factor
EGFR	Epidermal growth factor receptor
ECM	Extracellular matix
FGF-2	Fibroblast growth factor-2
GBM	Glioblastoma multiforme
Gli	Glioma-associated oncogene homolog
GIC	Glioma-initiating cell
HA	Hyaluronan
MMP	Matrix metalloprotease
NSC	Neural stem cell

Ptch	Patched
PTEN	Phosphatase and tensin homolog
PDGF	Platelet derived growth factor
RTK	Receptor tyrosine kinase
RB1	Retinoblastoma 1
Shh	Sonic hedgehog
SCF	Stem cell factor
SVZ	Sub ventricular zone
SGZ	Subgranular zone
TGF-β	Transforming growth factor β
TP53	Tumor protein 53
Wnt	Wingless

17.1 Neural Stem Cells

17.1.1 Embryonic Neural Stem Cells

Neural stem cells (NSCs) are the origin of neurons, astrocytes and oligodendrocytes (Fig. 17.1). An immense body of studies over the past two decades has characterized these immature, highly proliferative cells with self-renewal capacity that are found in large numbers during central nervous system (CNS) development [1, 2]. In addition, NSCs, albeit in smaller numbers, persist into adulthood and can generate neurons even in old age [3, 4]. Early in development, the future brain and spinal

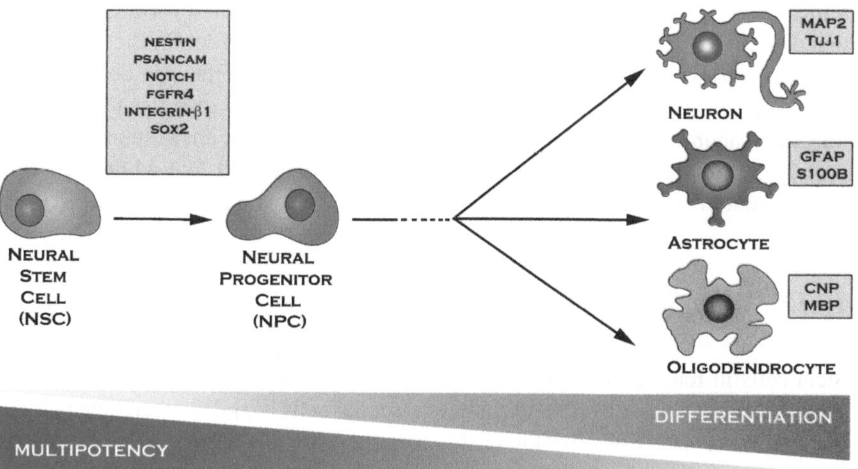

Fig. 17.1 Neural stem cell differentiation. Neural stem cells give rise to neural progenitor cells that differentiate to neurons, astrocytes and oligodendrocytes. As differentiation proceeds, multipotency decreases

cord consists of a single cell layer of stem cells, the neuroepithelium. As the neural tube matures and neurogenesis begins, it develops a pseudostratified architecture with proliferating NSCs closest to the lumen and intermediate progenitor cells migrating towards the periphery. The proportion of stem cells to more restricted offspring is controlled by symmetric and asymmetric division, which ensures the maintenance of a stem cell pool [5–7]. A distinct, transient type of progenitor during development is the radial glial cell [8]. Radial glia serve both as self-renewing stem cells, and constitute a scaffold through their extended morphology, which stretches from the lumen to the pial surface and on which nascent neurons climb to their final destinations [9, 10]. Radial glia can thus be considered a continuum of the neural stem cell compartment [11, 12].

17.1.2 Adult Neural Stem Cells

The majority of NSCs will differentiate and form mature lineages with time but two germinal zones remain in the brain throughout adulthood. These developmentally different neural stem niches are the subgranular zone (SGZ) of the dentate gyrus (DG) in the hippocampus and the sub ventricular zone (SVZ) of the lateral ventricular [4, 13–15]. The SVZ is the major source of proliferating cells in the adult brain, making this region interesting from of regenerative and tumor perspectives. The SVZ stem cells, located beneath the ependymal layer of the wall of the lateral ventricle are in contact with nutrients and other factors of the cerebrospinal fluid through their primary cilium [16]. Both the rodent and the human SVZ composition include three distinct cell types. Type-A cells are neuroblasts, originating from type-B cells, which are the neural stem cells, expressing GFAP. The type-C cells, also called transit amplifying cells and are generated by type-B cells, and form type-A cells that end up in the olfactory bulb [17]. Newly born neurons from SVZ cells migrate along a pathway called the rostral migratory stream to the olfactory bulb [18, 19]. Neurogenesis also occurs in the DG of the hippocampus and give rise to functional integrated neurons [20] and this process is under influence of environmental factors such as stress and exercise [21]. It is now well established that adult SGZ cells contribute to learning and memory where they reside in the hippocampus [22].

17.1.3 The Stem Cell Niche and Markers of Neural Stem Cells

Stem cells in renewable adult tissues are usually located in specialized niches that offer protection and nourishment to these essential cells [23]. Whether stem cell fate is quiescence, apoptosis, division or differentiation it is under the control of interactions between the stem cell and its microenvironment in the so-called neurogenic niche [24, 25]. Enclosed in this specialized microenvironment NCSs give rise to numerous neuronal and non-neuronal cells during development and in the adult brain. Influence on NSC fate in the niche comes from several sources. Local astrocytes provide both structural support and stimulatory factors for neurogenesis [26, 27] and endothelial cells were identified as critical components of stem cell self-renewal in the niche [28, 29]. When the stem cell exits its compartment, the factors that maintain its 'stemness' are no longer available, and the cell is likely to enter the differentiation pathway as a result of new local environmental influences [23].

The interest in neural stem cells and their promise for clinical application has triggered a thorough search for specific markers to enable their identification and isolation. To date, there is no single neural stem cell marker. Instead, stem cell identification relies on a combination of several immature markers, and the exclusion of more mature markers for committed cell lineages (Fig. 17.1). The first described marker for NSCs was the intermediate filament nestin [30], which is still widely used albeit not exclusive for NSCs. Markers used to define the anatomical location of NSCs have often been surface markers, PSA-NCAM [31], PDGFR-alfa [32, 33], Notch [34], FGFR4 [35] and CD133 [36], to name a few. With regard to cell adhesion and binding to extracellular proteins of the neurogenic niche, the integrin family should be mentioned. Inactivation of integrin β1 severely affects proliferation, survival and migration of the neural stem cell lineage [37]. The involvement of integrins in NSC regulation has led researchers to search for integrin signatures that may enrich NSCs [38]. Endothelial cells of the niche are a rich source of laminin and NSCs express the laminin receptor α6β1, which is important for maintenance of the NSC lineage [39].

Among intermediate filament proteins used to identify NSCs are GFAP, a marker both for adult neural stem cells and astrocytes [17], nestin as mentioned above and vimentin [40]. Other identifiers of the neural stem cell lineage are the transcription factors Sox2 [41] and Sox9 [42]. With the increasing number of conditions that have been reported to influence NSC differentiation, it is important to reassess the linage specificity of genes that are commonly used as marker genes. Still, more than 20 years after the identification of NSCs the best alternative for a safe assessment of neural stem cell number is the clonogenic neurosphere assay [43].

17.1.4 Self-renewal, Symmetric and Asymmetric Cell Division

During embryogenesis stem cells of the nervous system are self-renewing at a rapid rate to fulfill the need to build the entire nervous system. Whether a NSC possesses the capacity to go through an unlimited number of cell divisions without losing its multipotency is not clear. However, as a result of the NSC self-renewal program a stable and continuous reconstitution of undifferentiated cells in neurogenic niche is maintained (Fig. 17.2). The germinal zones of the developing mammalian nervous system change their composition as gestation proceeds, from neuroepithelial stem cells that only generates neurons to radial glial cells, that give rise to both neurons and glia. In mice, this shift takes place generally between E10 and E12 [44, 45].

At any given time, the key to NSC regulation in the adult brain is the balance between quiescence and proliferation. Throughout the life of mammalians, stem cells transition between the quiescent state and cell division, which generates new neurons [46]. Their division could either be symmetric which generates daughter

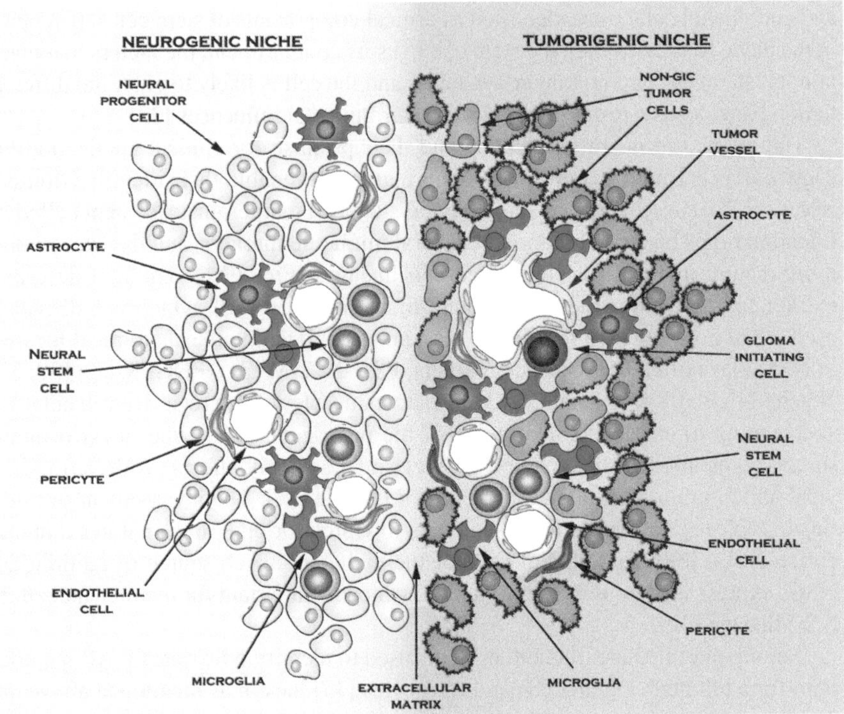

Fig. 17.2 The neurogenic and the tumorigenic niche. The neurogenic niche is built from NSC, neural progenitor cells, endothelial cells, pericytes, astrocytes and microglia. The microenvironment of the tumorigenic niche is composed of glioma-initiating cells and non-GIC tumor cells, as well as cell types also present in the normal niche

cells with similar fates, or asymmetric, which means that one of the daughter cells will be differentiating [5, 6]. A single, unipotent neural progenitor arising from asymmetric division is capable of producing numerous differentiated progeny, while symmetric division expands the NSCs pool [47]. The proper balance between symmetrical and asymmetrical division is crucial to maintain the neurogenic niche as well as cell type and diversity in CNS (Fig. 17.2). NSCs self-renew to replenish the stem cell pool and produce transient amplifying cells as progenitors for various glia and neurons, which leave the stem cell niche, proliferate a limited number of times and become post-mitotic as they move into a terminal differentiation pathway [48].

17.2 Neural Stem Cell Culture

In rodents, NSCs can be isolated from different regions of the embryonic nervous system as well as the adult hippocampus and SVZ, and be propagated either as neurosphere cultures in spheroid, free floating aggregates [49–51] or on matrices in

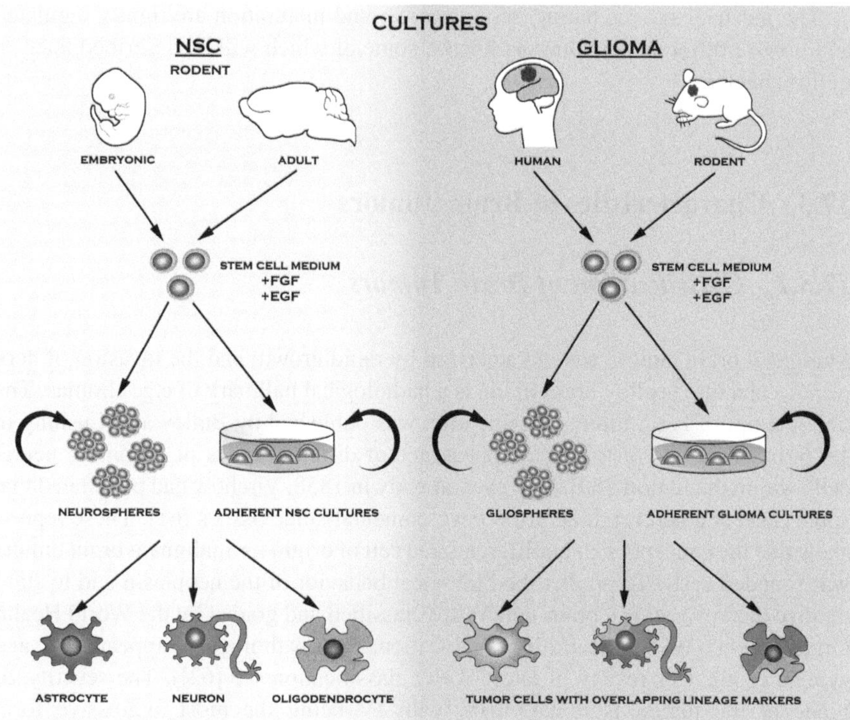

Fig. 17.3 NCS and glioma cell culture. NSCs from the embryonic or adult rodent brain, as well as glioma cells isolated from human or rodent brains can be propagated in stem cell medium (containing FGF and EGF) either as neurosphere/gliosphere cultures or as adherent monolayer culture. While NSC differentiate to mature cell lineages, glioma cultures give rise to tumor cells with overlaps in cell lineage markers that are normally not expressed by the same cell

monolayer culture [52] (Fig. 17.3). By adding fibroblast growth factor-2 (FGF-2) and/or epidermal growth factor (EGF), stem cell self-renewal capacity and multipotency is maintained in defined, serum-free culture [1, 49]. The neurosphere culture, which is the most frequent method to propagate neural stem cells, is efficient but quite heterogeneous in nature [53] containing neural progenitors with varying degree of potency. Approximately 2.4% *bona fide* neural stem cells are found in the neurospheres, and upon re-plating only around 10% of these will survive [54] further illustrating the heterogeneity of the system. It has been demonstrated that neurospheres are highly motile and prone to fuse even under stringent clonal culture conditions [55] and that estimating neural stem cell number in vitro may need to employ assays, which exclude progenitor cells [43]. The adherent primary cultures from embryonic neural tissue are of short-term use because of the limited number of passages, which they can be propagated with maintained undifferentiated properties [56]. In terms of its continuous expansion and stability of marker expression, in vitro-differentiated embryoynic stem cells have shown promising capacity as a neural stem cell source [57–59].

The mechanisms regulating NSC turnover and maturation are tightly regulated and involve different signaling pathways, some of which will be described later on in this chapter.

17.3 Characteristics of Brain Tumors

17.3.1 Classification of Brain Tumors

Malignant brain tumors are characterized by rapid growth and the invasion of neoplastic cells into healthy brain tissue is a pathological hallmark of e.g. gliomas. The first attempt of brain tumor classification was published by Bailey and Cushing in 1926, based on the histological appearance of different types of immature neural cells within the tumors [60]. However, already in 1858, Virchow had postulated that some cancers could originate from developmental-stage tissues [61]. These reports show that the concept of an undifferentiated cell of origin for malignant brain tumors was founded early. To predict the biological behavior of the neoplasm and to standardize therapy regimes brain tumors are classified and graded by the World Health Organization (WHO) according to location, histopathological appearance, and genetic profile (for review of latest WHO classification see [62]). The severity, or grade of the tumor, is ranked I-IV, with IV being the most aggressive form (Table 17.1). Of these, neuroepithelial tumors, medulloblastomas and glioblastomas are the most aggressive forms with a WHO grade of IV, while ependymomas and astrocytomas depending on subtype vary between grade I and III in severity [62]. Grade I tumors are benign. The typical life expectancy for a patient diagnosed with a WHO grade II tumor exceeds 5 years, a type III tumor 2–3 years but survival for patients with type IV tumors largely depend on the presence of effective treatment for that particular tumor [62]. For most brain tumors, surgical removal and radiation therapy is recommended, and chemotherapies may further prolong the survival. These measures however, are not efficient for high-grade gliomas.

17.3.2 Glioblastoma Multiforme

Gliomas can be of grade I-IV, where grade III and IV have the poorest outcome. Grade IV glioma is also called glioblastoma multiforme (GBM) and can be of primary or secondary type. The primary GBM, which constitutes 90% of the cases develop rapidly without any previous symptoms while the remaining 10% are of secondary type, originating from less malignant neoplasms. Both grade II and grade III glioma have a tendency to progress to GBM. Despite combined surgery, radiation and new drugs such as temozolamide, GBM is a devastating disease with a median survival of 14–15 months [63]. It is also the most common primary brain

Table 17.1 WHO grades of CNS glioma

Tumor	Grade
Astrocytic tumours	
Subependymal giant cell astrocytoma	I
Pilocytic astrocytoma	I
Pilomyxoid astrocytoma	II
Diffuse astrocytoma	II
Pleomorphic xanthoastrocytoma	II
Anaplastic astrocytoma	III
Glioblastoma (multiforme)	IV
Giant cell glioblastoma	IV
Gliosarcoma	IV
Oligodendroglial tumours	
Oligodendroglioma	II
Anaplastic oligodendroglioma	III
Oligoastrocytic tumours	
Oligoastrocytoma	II
Anaplastic oligoastrocytoma	III

tumor in adults. GBM have a high proliferative rate, large degree of heterogeneity of cell morphology and they are highly invasive [64]. Because GBM effectively infiltrates the normal brain tissue, surgical removal is almost impossible and remaining tumor cells give rise to rapid recurrence of symptoms.

Because of the poor outcome for GBM patients, the WHO classification, based on morphological criteria needs to be complemented and concerted efforts to identify subgroups of GBM based on genetic and epigenetic events have been made (Table 17.2). Genetic loss affects almost all chromosomes in GBM, but loss of heterozygosity (LOH) for chromosome 10 is most frequent [65]. Amplification of genetic material also occurs, the most common being epidermal growth factor receptor (EGFR) amplification [66]. The molecular pathogenesis of GBM has been investigated in great detail. Mutations in the tumor suppressor genes PTEN, TP53 and RB1, affecting the central growth regulatory pathways have been found in virtually all GBM [67, 68]. Perturbed signaling pathways include tyrosine kinase receptor pathways and TGFβ signaling, affecting the PI3K/AKT axis and the RAS/MAPK pathway [69]. Thus, all major means to control proliferation and survival are affected in GBM. Analysis of DNA methylation, microRNA and proteomic changes in glioma patients contributes to build an integrated knowledge base for future therapies [70].

The stratification of GBM based on signal transduction pathway activation and/or on mutations in pathway member genes may be particularly valuable for future development of targeted therapies. Brennan et al. recently reported signature events that can be used to a divide GBM into three classes: EGF receptor activation, PDGF pathway activation and loss of NF1 expression [71]. The Cancer Genome Atlas network was established to generate a large-scale catalog of cancer abnormalities, including glioma [67]. Using the TCGA database, Verhaak et al. [72] recently defined four subtypes of GBM by proteome and transcriptome analyses: Proneural,

Table 17.2 GBM subtype classification and frequently mutated genes

	GBM subtypes			
	Proneural	Neural	Classical	Mesenchymal
Frequently mutated genes	TP53 (54%) PIK3R1 (19%) IDH1 (30%) PTEN (16%) EGFR (16%) PDGFRA (11%)	EGFR (26%) TP53 (21%) PTEN (21%) NF1 (16%) ERBB2 (16%) PIK3R1 (11%)	EGFR (32%) PTEN (23%) EGFRvIII (23%)	NF1 (37%) TP53 (32%) PTEN (32%) RB1 (13%)

Modified from Verhaak et al. [72]

Neural, Classical and Mesenchymal. The Proneural group is strongly related to PDGF signaling, the Classical group to EGFR activation and the Mesenchymal group to NF1 inactivation. Thus, the classification by Verhaak et al. largely matches that of Brennan et al. Designation of the Neural subtype was based on expression of neural markers. Intriguingly, enrichment of genes expressed in oligodendrocytes was associated with the Proneural type, the Mesenchymal type was associated with a genetic signature of cultured astrocytes while the neural and classical types were enriched for genes expressed in astrocytes [72]. While the above data cannot be viewed as evidence for the cell of origin in GBM, it is interesting to note the relationship between GBM subtypes and cellular lineages.

17.3.3 Glioma Microenvironment

Glioma are very heterogenous with regard to the composition of tumor cells, and in addition the high grade glioma cells are mixed with various specialized cell types that support the tumor [73]. The general concept that the tumor microenvironment (Fig. 17.2), where a multitude of normal cells add to the complexity of a cancer, is a determinant for tumor development has increasingly been recognized, as reviewed by Hanahan and Weinberg [74]. Thus, tumor cells together with extracellular matrix (ECM) components and tumor blood vessels enable communication between the tumor cells and the surroundings. This microenvironment influences the process of tumor progression from of the first steps of transformation to a malignant tumor development. Brain parenchymal cell populations involved in the tumor microenvironment include endothelial cells, pericytes, vascular smooth muscle cells and astrocytes that form the neurovascular unit [75]. In addition to providing support, tumor associated astrocytes may also mediate invasion through secretion of matrix degrading enzymes [76] and provide trophic factors for the tumor [77]. Abundant non-transformed cells in the tumor environment are microglia/macrophages that can either come from resident microglia in the brain or originate from the periphery [78, 79]. As for other immune cells, similar to other cancer types; GBM can avoid the immune system by secreting immunomodulatory factors and by decreased expression of MHC class I, thus exhibiting only weak immunogenicity [80].

17.3.4 *Gliomas Are Highly Invasive*

Migration is a key event in CNS development and for the rapid spread of tumor cells in the brain. It is a particularly important aspect of glioma where the grade of malignancy increases with invasion of surrounding tissue [81]. This characteristic may reflect a reminiscence of the immature cell from which the tumor arose. Migration during CNS development is essential for forming and stabilizing the spatial organization of tissues and cell types. GBM successfully invades the normal brain, but rarely metastasizes outside the brain. Therapies to target GBM invasion are highly warranted since it efficiently migrates e.g. along white matter tracts [82]. To be able to establish new tumors distant from the original tumor, cancer cells must first detach and move across the microenvironment, including breakdown of extracellular matrix molecules. Glioma cell expression of cell adhesion molecules to specific ECM proteins plays a crucial part in the invasion [83]. The matrix degradation, by a variety of enzymatic processes [64] must continue as the tumor cells migrate. This ECM remodeling is accompanied by rapid formation by the tumor cell of its own de novo ECM, made primarily of molecules that support migration [84]. The movement of cells over large distances towards a chemotactic gradient is one mechanism by which migration is controlled. Several classes of molecules possess chemoattractive properties, including growth factors and chemokines. EGF [85] and PDGF [86] are examples of chemoattractants for glioma cells whose signaling pathways, as mentioned above, are involved in gliomagenesis. Interactions between the chemokine CXCL12 and its receptor CXCR4 also mediate glioma migration [87].

17.4 Glioblastoma Cell Culture

Studies of human glioma cells in vitro has until recently heavily relied on patient-derived cell lines established already during the 1960s and 1970s [88]. A large number of malignant glioma cell lines (e.g. U373MG, U251MG, U87MG, T968G, LN229, SF763) were established from patients and have been used in a multitude of studies over the years. Similarly, cell lines from rats and mice in which tumors were chemically induced exhibit morphological phenotypes of gliomas (RG2, C6, 9L, F98, GL261). All the patient-derived lines and the rodent cell lines have in common that they were established and maintained in medium containing fetal calf serum, and thus exposed to a variety of factors of undefined concentrations that have likely influenced their characteristics.

Naturally, primary human GBM cells are also used, but for practical reasons and to obtain reproducibility despite inter-patient variability, there is a need for cell lines that better represent the patient's tumor than the serum-derived cell lines [89]. Since the finding that stem cell-like cells can be isolated and cultured from brain tumors [90, 91] the number of new lines generated in several laboratories, including our own, is growing. These are established using defined serum-free conditions that are identical to those used for culture of normal neural stem cells [92] (Fig. 17.3). Fresh tumor samples from GBM patients are obtained directly from neurosurgery, dissociated and propagated either as spheres (so called gliospheres) or adherent monolayer culture [93]. Care must be taken to cryopreserve low passage cells for proper characterization of genotype and phenotype. Cell cultures obtained from GBM patients do not always form lines, and the percentage of tumors giving rise to cell lines varies. However, a large fraction of the GBM lines, either as spheres or adherent cultures can be easily propagated and may be expected to maintain at least partly the properties related to the original tumor.

17.5 Common Traits of Neural Stem Cells and Brain Tumor Cells

17.5.1 Glioma Cell of Origin and Its Relation to Neural Stem Cells

Because malignant brain tumors, such as GBM and medulloblastoma are known since long to contain undifferentiated cells, scientists have searched for stem cells or stem cell-like cells in these tumors with the aim to identify the cell of origin for the tumors. The concept of cancer stem cells (CSC) originates from hematopoietic tumors where they have been subject of study since the seminal findings by Dick [94]. Expression of the neural stem cell marker nestin in various neuroepithelial tumors was presented in 1990s [95, 96]. Some 10 years later, several groups identified a population of cells, which is referred to as CSC or brain tumor stem cells (BTSCs) within different types of CNS tumors such as epenymomas, medulloblastomas, astrocytomas, and gliomas [34, 36, 90, 97, 98]. The concept of CSC has been widely investigated, and the identification and in vitro propagation of stem cell-like cells from brain tumor patients as well as from animal models of brain tumors have given support to this view. Tumor cells with stem cell properties can be grown in stem cell medium, passaged by limiting dilution to show self renewal, induced to change cell lineage marker expression reminiscent of neural stem cell differentiation, and can recapitulate tumors in mice that are indistinguishable from the original tumor when transplanted at very low cell number [99–101].

When applying the stem cell terminology to cells of origin for tumors one must bear in mind, however, that CSC can only be identified experimentally through their capacity to form tumors that can be serially passaged in mice. A CSC is not synonymous with a stem cell, and there are several arguments for a note of caution of the CSC concept. First, stem cell-like glioma cultures where the mitogen is withdrawn, i.e. the classical stem cell differentiation procedure, co-express markers of neurons and glia, which doesn't occur in normal differentiated offspring of neural stem cells. Second, early CSC markers, such as CD133 [91] have later shown not to be a fully reliable identifier of cancer initiating cells from glioma [102, 103]. Third, the reason that only a small fraction of tumor cells can recapitulate tumors in mice could have alternative explanations. The microenvironment has a major influence on tumor development, and the difference between human and mouse, as well as the absence of an immune system must be taken into account when performing xenografts of human cancer cells [104, 105].

In addition to neural stem cells and neural progenitor cells, more differentiated glial cells have been suggested as tumor-initiating cells. The concept of de-differentiation, namely that lineage-specified progenitors, astrocytes or oligodendrocyte would take on a phenotype similar to their progenitors and proliferate in an uncontrolled manner, must also be taken into account [106] and could provide important keys to understand brain tumor biology. Because the exact nature of the GBM initiating cell(s) is not yet fully identified we will use the term glioma-initiating cell (GIC) throughout this review when referring to the possible GBM founding cells. Successful treatment of GBM calls for the identification of the GIC and if different GBM could arise from different cell types, as indicated by novel GBM subtype stratification [72] it would explain the high phenotypic heterogeneity of these tumors. This could be an alternative explanation to the CSC theory according to which the ability of a multipotent CSC is believed to form various progeny as an explanation for the cellular diversity within GBM tumors. All key GBM pathways, as mentioned previously in this paper, are neural stem and progenitor cell pathways [71] and the transcriptional profile of NSC and experimental glioma is largely overlapping [107]. The novel categorization of GBM into four subtypes based on transcriptional and proteomic data opens new possibilities to design stratified therapeutic approaches based on GBM subtype.

17.5.2 Neural Stem Cells as GBM Therapy?

Another aspect of NSC in GBM is the quite unexpected and efficient homing of grafted NSC to experimental brain tumors found by Aboody et al. who showed the ability of NSC to track even single scattered tumor cells [108]. Furthermore, Glass et al. observed migration of normal NSC from the subventricular zone of the hippocampus to the tumor [109]. A tumor suppressing activity by cultured NSC has been demonstrated and co-transplantation of NSC and GBM suppress tumor formation in mice [109, 110]. GBM preferentially affects adults and the peak of incidence is between 45 and 70 years of age, while it is a very rare disease in children. Since 80% of the patients are older than 50 years and neurogenesis is known to decrease in the aging brain [111] it can be speculated that the neural stem and progenitor cell population of the young brain counteracts glioma formation. However, it is not known if NSC in the human brain has any ability to migrate to the tumor or suppress GBM development. Whether this has any relevance for potential therapies remains to be investigated.

17.5.3 The Neurogenic and the Tumorigenic Niche

The tumor perivascular niche (Fig. 17.2) is believed to harbor specific brain tumor-initiating subpopulations, i.e. GIC similarly to how the vasculature provides a niche for NSCs in the normal brain. GICs, possessing stem cell characteristics have been shown to be resistant to conventional treatment [112, 113]. This could explain GBM recurrence since both chemotherapy and radiation requires cell division to be efficient, and the tumor stem cell would be in a quiescent state. A key to treat GBM more efficiently will likely be through inhibiting invasion, and for that purpose the ECM of the tumor niche must be well characterized. Many components of the ECM are deregulated in malignant glioma such as hyaluronan (HA). vitronectin, collagen I and IV, osteopontin and tenacin-C.

Hyaluronan is one of the most abundant ECM components in the normal brain is. It is part of the neurogenic niche and is important for NSC maintenance [114]. HA is enriched in primary brain tumors [115] and in the white matter tracts [116] which are regions commonly invaded by brain tumors [117]. Gliomas often migrate extensively via white matter tracts, perivascular spaces and the subependyma [82, 84].

Notably, normal neural stem/progenitor cells that are transplanted into the intact or injured rodent brain also prefer white tracts for widespread migration [118, 119]. The two receptors for HA, CD44 and RHAMM (Receptor for Hyaluronic Acid Mediated Motility) mediate primary tumor cell invasion and migration [120, 121]. Interestingly, CD44 has also been suggested to have HA- independent roles in cell adhesion/migration since CD44 can act as a cell-surface anchor for the ECM degrading enzyme [122] (MMP-9).

In the normal rodent brain, the large glycoprotein tenacin-C is present in the stem cell niches during development [123, 124] and in the adult brain after injury [124–126]. In glioma, tenacin-C expression correlates inversely with prognosis and survival [127]. Sarkar et al. showed that tenacin-C up-regulation of glioma invasiveness was inhibited by blocking by MMP-12 siRNA [128], illustrating that tenacin-C is a permissive substrate for glioma invasiveness.

MMPs are proteolytic enzymes belonging to a family of 25 members that are critical for tumor cell invasion. MMP-2 and MMP-9 are highly expressed in malignant glioma [129, 130]. The increase of MMP-2 and MMP-9 expression correlates with increased malignancy as measured in primary gliomas [131]. Down-regulation of MMP2 in an intracranial glioma model inhibited tumor growth via the p16 pathway [132]. Inhibition of MMP has been attempted for glioma, but serious side effects and poor bioavailability has hampered its use [133].

Integrins are instrumental by playing a role in cell survival, proliferation, migration and differentiation. They provide a physical link between the cell and the ECM molecules that is important for adhesion, and also transmits signals from the extracellular space into the cell. In the normal adult brain the most abundant integrins are: $\alpha 1$, $\alpha 3$, $\alpha 5$, αv, and $\beta 1$. In GBM, the $\alpha 2\beta 1$, $\alpha 5\beta 1$, $\alpha 6\beta 1$ integrins are often up-regulated in glioma [134, 135]. As mentioned previously, the infiltration of tumor cells also often occur along blood vessels, and $\alpha v\beta 3$ and $\alpha v\beta 5$ expression is associated with the tumor vasculature in high grade glioma, but not for low grade astrocytomas or normal brain tissue [136, 137]. The $\alpha v\beta 3$ expression co-localize with MMP2 expression in tumor cells of the leading front [137, 138] further illustrating the importance of integrins and MMPs in tumor infiltration.

17.6 Common Traits in Signaling Pathways

17.6.1 Receptor Tyrosine Kinases

Receptor tyrosine kinases (RTKs) compose one of the largest groups of structurally conserved membrane bound transmembrane receptors. Ligand binding results in a conformational change in the extracellular domain, which enables receptor dimerization and kinase activation. The resulting autophosphorylation of tyrosine residues on the carboxy-terminal part of the receptor triggers activation of down-stream intra cellular signal cascades regulating diverse cellular functions such as survival, differentiation, proliferation, and migration (Fig. 17.4) [139]. The RTKs can be divided into subtypes based on their ligand binding affinity and structure. Below we will focus on two RTK classes because of their importance for neural stem cells and GBM: the RTK type I (EGF receptor) and RTK type III (PDGF receptor, colony stimulating factor-1 receptor c-fms, and stem cell factor (SCF)/c-KIT receptor).

17.6.1.1 Class I RTK: Epidermal Growth Factor Receptor

The Epidermal growth factor receptor (EGFR) is activated by ligand dimerization and binding, whereupon receptor autophophorylation triggers an intracellular signaling cascade. This pathway regulates NSCs number and self-renewal both during early and late embryonic CNS development. Specific mitotic response to EGF was associated with the appearance of late neural progenitor cells that express high level of EGFR [140]. Therefore, EGFR expression and stimulation of NSCs and late progenitor cells tend to facilitate postnatal gliogenesis [140, 141]. Aberrant EGFR activation plays a significant role in tumor development especially for GBM. Amplification, over-expression or rearrangement of EGFR leads to cell transformation and development of brain tumors [142, 143]. The altered EFGR signaling allows for an autocrine loop, which induces dysregulation of additional genes such as TGF-alfa, FOSL1, IL8 or EMP1, thus amplifying the process of oncogenesis [142]. Indeed, a mutant form of EGFR fuels glioma establishment and is also required for maintenance. Inhibition of mutated EGFR signaling has naturally been in focus for attempts to cure GBM. Small molecule clinical trials have included erlotinib (Tarceva®) and gefitinib (Iressa®). However, none of these drugs have yet given any real breakthrough for GBM therapy as reviewed by Krakstad and Chekenya [144]. Among several plausible explanations for this failure is a need for

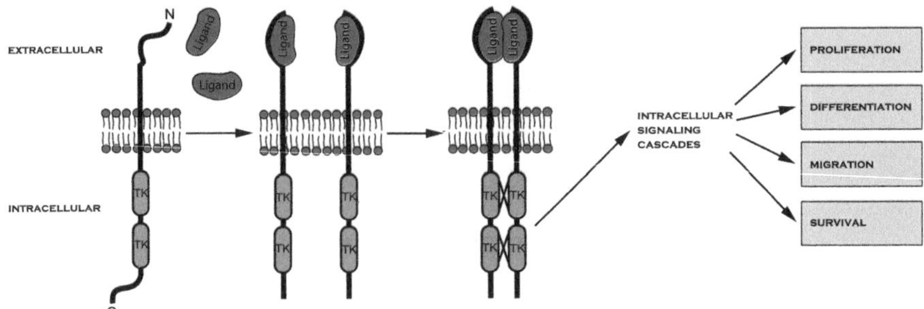

Fig. 17.4 Receptor tyrosine kinase (RTK) activation. RTKs are activated by ligand binding (here illustrated by dimeric ligands such in the case of EGFR and PDGFR) and subsequent receptor phosphorylation leads to intracellular signal transduction affecting one or several cellular fates

combined treatment of GBM proliferation, invasion and apoptosis. Another reason being that GBM can preserve aggressiveness even if the constitutive EGFR signaling is lost due to selective pressure by compensatory pathways [145]. EGFR however remains a strong candidate for tumor therapy due to its profound involvement in gliomagenesis.

17.6.1.2 Class III RTK

Several signaling pathways with great importance for neural stem cell biology and GBM belong to this class of RTK. Small molecule inhibitors to class III RTK have been developed, the most important one so far being imatinib mesylate (Imatinib®). In addition to the transmembrane receptors PDGFR, c-fms and c-KIT, imatinib is an efficient blocker of ABL, which is a non-receptor tyrosine kinase. Treatment with imatinib has dramatically changed the outcome for patients with chronic myeloid leukemia, which is characterized by a constitutive BCR-ABL activation [146]. For GBM the prospects for patients treated with imatinib has not been very promising. Progression free survival was not, or only minimally increased [147].

Platelet-Derived Growth Factor Receptor

Platelet-derived growth factor (PDGF) receptors -α and -β bind different combinations of the PDGF homo or heterodimeric ligands, with the PDGF $\alpha\alpha$ receptor binding PDGF -AA, -BB, -CC, and -AB, the PDGF $\beta\beta$ receptor binding PDGF -BB and -DD, the heterodimer of the α and β receptor binds PDGF -AB and -BB. PDGF is a necessary for normal development and has a multitude of functions in embryogenesis [148]. In the brain, PDGFRα expression is first detected at embryonic day 9 [149] and at E13.5 it is found mainly in oligodendrocyte progenitors [150]. Numerous studies, including knockout mice have shown that PDGF-A signaling is crucial for formation of the oligodendrocyte lineage [148]. Abnormal activity of PDGF contributes to a variety of pathological conditions, including brain tumors, as discussed above [151]. PDGF expression by viral vector insertion, often in combination with tumor suppressor inactivation, has shown to be a powerful inducer of different types of brain tumors in embryonic, newborn and adult mice [152–154]. However, it is not clear to what extent PDGF paracrine signaling, by itself, can be instrumental in glioma formation [153, 155]. Transgenic overexpression of PDGF ligands at various stages of CNS development has not resulted in brain tumors [156–158] unless a second genetic alteration is present [159].

The function of PDGF in neural progenitors was first thought to be of neurogenic mode, increasing the number of neurons formed from neural stem/progenitor cells [52, 160]. However, PDGF is not an instructive differentiation factor, but rather acts to expand the progenitor pool [32]. Furthermore, neural progenitors have an endogenous production of PDGF-BB, which in an autocrine/paracrine fashion slows down its own cell maturation process [161]. PDGF signaling in the adult neural stem cell compartment was proposed by Jackson et al. [33]. They showed that GFAP positive NSC in the rodent and human adult brain express PDGFα. By infusing PDGF-A into the lateral ventricle, the pool of GFAP positive type B cells was increased, and hyperplasia resembling glioma were formed. These, however disappeared when infusion was discontinued [33]. The above conclusions have been challenged by Chojnacki et al. [162] who suggest that instead of NSC, the PDGF-responsive cell is an oligodendrocyte precursor, which proliferates close to the ventricular wall in response to PDGF administration. The recent GBM stratification [72] where the proneural subtype has been shown to share gene expression pattern with the oligodendrocytic lineage and has perturbations in the PDGF signaling pathway as it's main characteristics would speak in favor of oligodendrocyte precursor as the GIC for these GBM. However, this doesn't rule out that other glioma subtypes could be derived from adult SVZ NSC. Regardless of the GIC identity and the rather dismal results of imatinib therapy so far, PDGFR remains a promising target, especially for the proneural class of GBM.

Colony Stimulating Factor-1 Receptor

Colony stimulating factor-1 (CSF-1) is expressed in the developing brain [163], and its receptor c-fms is exclusively found in microglial cells of the CNS [164]. In addition, CSF-1 expression is upregulated in animal models of Alzheimer's disease [165, 166] and in patient material from Alzheimer's patients [167]suggesting important roles in CNS disease inflammation. By over-expressing the c-fms receptor in microglia, the inflammation response is increased [168]. In a model for Parkinson's disease, the survival of human neural progenitors transplanted into the 6-hydroxydopamine-lesioned rat brain was improved when cells were grafted near blood vessels or cerebrospinal fluid compartments [169], suggesting that CSF-1 positive microglia have supportive effects on grafted stem cells. The role for CSF-1 as a neuroprotective agent has since been confirmed [170] and it promotes differentiation of neural stem cells [171]. The CSF-1/c-fms complex is also expressed in normal human astrocytes, glioma cell lines, and up regulated in gliomas [172]. Komohara suggested that CSF-1 released from glioma cells act to recruit infiltrating microglia/macrophages, which aggravates the disease [173]. Furthermore, insertional mutatgeneis of the CSF1 locus [174] promoted formation of high-grade astrocytoma. Whether beneficial effects of CSF-1 targeting in GBM could be achieved through suppression of tumor-associated macrophages remains to be seen.

Stem Cell Factor Receptor

Stem cell factor (SCF) and its receptor c-KIT act as a mitogen and a chemoattractant for many cells of the hematopoetic system (reviewed in [175]). The SCF/c-KIT complex is important in regulating gametogenesis, hematopoiesis, and melanogenesis [176–179]. The expression pattern for SCF and c-KIT has been studied during the mouse nervous system development, and expression is localized to the neural tube as early as E11.5 [180]. Both cultured neural stem/progenitor cells [181] and neurons [182] in culture express SCF/c-KIT receptor complex in vitro. Although SCF acts as a survival factor and a chemoattractant for embryonic neural progenitors, it does no alter the differentiation potential [181]. In the mammalian ciliary epithelium of the eye, c-KIT receptors are expressed by adult neural stem cells and SCF increased the proliferation and neurosphere formation capacity while inhibiting the differentiation in vitro [183]. Expression of c-KIT is a diagnostic tool for gastrointestinal tumors and it is involved in several other tumor types [184]. The role for c-KIT in human glioma is less clear, and different studies give somewhat different results. Sun et al. reported that SCF is highly expressed in human glioma cell lines, and that in primary human glioma the level of SCF increases with increasing WHO grade [185]. Down regulation of SCF inhibited tumor growth and glioma angiogenesis in mouse glioma models [185]. On the other hand, Went et al. found c-KIT expression only in a small number of the investigated patients [184]. The value of blocking c-KIT as a way to repress tumor angiogenesis in glioma is therefore unclear.

17.6.2 Notch Signaling

The Notch transmembrane-spanning receptor regulates many important developmental processes such as patterning of the neural plate, stem cell self-renewal capacity, and proliferation during early neurogenesis [186, 187]. It is also involved in glioma development and progression [188]. Activated by its ligands Delta and Jagged, that are both over-expressed in glioma cells [189], the intracellular domain of Notch is cleaved and translocates to the nucleus where it acts as a transcription factor, activating down-stream target genes. The adaptor protein, interacting directly with cytoplasmic domain of Notch is called Numb, and was originally identified as an inhibitor of Notch signaling and play pivotal role in NSCs differentiation and maturation [190]. During early neurogenesis and development of glioblastoma, Notch stimulates self-renewal of NSCs and CD133-posotive GICs via activation of common target genes e.g Hes-1 [191]. Notch promotes invasive glioma formation and the expression level is directly correlated with glioma grade [192]. Low-grade gliomas are characterized by inactive Notch and high-grade gliomas by active Notch signaling. Interestingly, Notch activation promotes cell proliferation and neural stem cell-like (neurosphere) formation from human glioma cells [193]. Additionally, Numb is expressed by all grades of glioblastomas and seems to participate in tumor growth and progression [194]. Downregulation of Notch ligands or Notch induces apoptosis and inhibits proliferation of glioma cells [195]. Moreover, Notch signaling is instrumental in the cellular response to hypoxia in glioblastomas, were endothelial cell activation plays a crucial role [196, 197]. Bao and colleagues showed that the fraction of CD133 positive GICs in gliomas are enriched after radiation treatment compared to the overall proportion of cells both in culture and after transplantation into the brains of immuno-compromised mice, possibly explanation the radiation resistance observed in gliomas [112]. Recently it was reported that Notch inhibition using a gamma-secretase inhibitor enhanced temozolamide treatment in a mouse model of glioma [198].

17.6.3 The Sonic Hedgehog Pathway

Sonic hedgehog (Shh) signaling is essential during morphogenesis and pattering of the nervous system. Shh is necessary for oligodendrocyte linage development and drives proliferation in CNS precursor cells in adult brain [199]. The receptor for Shh is the transmembrane protein Patched (Ptch). In the absence of Shh, Ptch represses the Shh/Ptch pathway by acting on the transmembrane receptor Smoothened (Smo). In gliomas, the Shh/Pych pathway is inappropriately active during tumor cell proliferation and migration. Shh expressing cells were found in the microenvironment of the perivascular niche of gliomas, correlating to angiogenesis [200]. Additionally, the Shh/Ptch pathway display tumor-grade specificity and was shown to be activated in grade II and III gliomas, by not grade IV [201]. Smo activates the canonical Shh pathway through glioma-associated oncogene homolog (Gli) –dependent pathway, including activation N-myc, Sox2, cyclin D, Gli-1 or Gli-2 [202]. Gli, was first described as a gene amplified in all grades of gliomas and is a crucial component of the Shh/Ptch pathway [200]. Due to the vital role of the Shh pathway in neural stem cells it has been suggested that GICs may also be susceptible to Shh pathway manipulation. Over-expression of Gli-1 induced brain tumor [202] while decreased levels of Gli-1 or Smo induced a significant decrease in tumor volume, as well as the extent of migration and invasion of CD133-positive GICs [203–205]. This may give the opportunity to use the transcription factor, Gli-1 in anti-invasive therapeutic intervention for human gliomas.

17.6.4 TGFβ/BMP Signaling

The transforming growth factor β (TGF-β) superfamily consists of ligands of the TGFβ, BMP, GDF, Activin and AMH pathways [206]. TGF-β family members play crucial roles in neural stem cell biology during normal brain development e.g. by apoptotic selection, as well as in the pathobiology of gliomas [207, 208]. Binding to type I and type II serine/threonine kinase receptors, TGFβ activate Smad signaling which influences glioma cell proliferation and migration. Recent studies showed that TGF-β1 increased migration and invasiveness in glioma cells while TGF-β2 has been identified as a key factor in the progression of malignant gliomas [209, 210]. By inducing expression of self-renewal genes such as Sox4, Sox2 or LIF, TGF-β directly supports preservation of stemness by GICs [210, 211]. As a result of Sox2 knockdown attenuation of tumorigenic activity of GIC was obtained [210]. Moreover, combined immunization/gene delivery of TGF-β signaling component antisense oligonucleotides undergo intensive testing, and may be a promising approach for glioma therapy [212, 213].

Bone morphogenetic proteins (BMP), as is known for TGFβ, may have either tumor suppressing activity, inducing apoptosis, or tumor promoting functions depending on cell type and stage of differentiation [214]. It was demonstrated that BMP-7 induces cell cycle arrest and therefore reduces proliferation of human gliom [215, 216]. Moreover, BMP-4, involved in normal nervous system development promotes differentiation and reduces the number of glioma-initiating cells [217].

17.6.5 Wingless (Wnt)/β-Catenin

As the signaling pathways described above, Wingless (Wnt) signaling is a developmentally conserved pathway that regulates neural progenitor proliferation, migration and fate determination [218–220]. In the canonical pathway initiated by binding of Wnt to the Frizzled –LRP5/6 receptor complex, activation of β-catenin/T cell factor signaling leads to activation of proneural transcription factor activation [221, 222] β-catenin exhibit bidirectional functions and from one side promotes neural precursor cell proliferation when it binds to LEF/TCF transcription factors, and on the other side its contributes to the maintenance of neural precursor cells in an undifferentiated state through activation N1IC [222]. In addition to their role during development, the alteration of components of Wnt/ β-catenin pathway has been implicated in tumor formation and progression. The activity of the Wnt pathway is dependent on the cytosolic concentration of free β-catenin and accumulation has frequently been associated in various brain tumors. The level of Wnt2 and β-catenin expression is significantly higher in glioblastomas and positively correlates with higher tumor grade [223]. Additionally, β-catenin (CTNNB1) gene mutations, influencing protein phosphorylation, is frequently found in brain tumors [224].

17.7 Concluding Remarks

Many similarities support the notion that glioblastoma arises from transformed neural stem or progenitor cells, but the exact nature of the glioma-initiating cell remains elusive. The very poor prognosis for glioblastoma patients is one of the major challenges in cancer research, not least because it is the most common brain tumor with a median survival of 1 year. The high degree of heterogeneity in glioblastoma, together with its invasive growth pattern and resistance to treatment contribute to the failure so far to obtain efficient treatment. In order to achieve major breakthroughs in glioblastoma therapy individual targeted combination therapies aiming at cancer cell proliferation/survival, invasion and angiogenesis at the same time will most likely be needed. Recent knowledge about the motility of glioblastoma cells provide clues to new anti-invasion therapy. These are based on characterization of the normal neurogenic niche and the tumorigenic niche. Stem cell-like cell lines from glioblastoma patients can be established and maintained using protocols that were developed for neural stem cells. These cell lines present promising new in vitro models for glioma biology and for screening of new drugs. Furthermore, molecular subclassification of glioblastoma presents a novel opportunity for more individualized treatment. This is especially promising because the patterns of expression not only define perturbed signal transduction pathways, but they may also link glioblastoma subtypes to specific tumor-initiating cells. The most efficient way to combine the above advancement in glioblastoma research to beneficial outcome for the patient is now the focus for both scientists and clinicians.

References

1. McKay R (1997) Stem cells in the central nervous system. Science 276:66–71
2. Gage F (2000) Mammalian neural stem cells. Science 287:1433–1438
3. Gage F (1998) Stem cells in the central nervous system. Curr Opin Neurobiol 8:671–676
4. Eriksson P, Perfilieva E, Björk-Eriksson T, Alborn A-M, Nordborg C, Peterson D, Gage F (1998) Neurogenesis in the adult human hippocampus. Nat Med 4:1313–1317
5. Temple S (2001) The development of neural stem cells. Nature 414:112–117
6. Alvarez-Buylla A, Garcia-Verdugo JM, Tramontin AD (2001) A unified hypothesis on the lineage of neural stem cells. Nat Rev Neurosci 2:287–293
7. Guillemot F (2005) Cellular and molecular control of neurogenesis in the mammalian telencephalon. Curr Opin Cell Biol 17:639–647
8. Merkle FT, Tramontin AD, Garcia-Verdugo JM, Alvarez-Buylla A (2004) Radial glia give rise to adult neural stem cells in the subventricular zone. Proc Natl Acad Sci USA 101:17528–17532
9. Rakic P (1990) Principles of neural cell migration. Experientia 46:880–891
10. Rakic P (1972) Mode of cell migration of the superficial layers of fetal monkey neocortex. J Comp Neurol 145:61–84
11. Tramontin AD, Garcia-Verdugo JM, Lim DA, Alvarez-Buylla A (2003) Postnatal development of radial glia and the ventricular zone (VZ): a continuum of the neural stem cell compartment. Cereb Cortex 13:580–587

12. McMahon SS, McDermott KW (2007) Developmental potential of radial glia investigated by transplantation into the developing rat ventricular system in utero. Exp Neurol 203:128–136
13. Sanai N, Tramontin AD, Quinones-Hinojosa A, Barbaro NM, Gupta N, Kunwar S, Lawton MT, McDermott MW, Parsa AT, Manuel-Garcia Verdugo J, Berger MS, Alvarez-Buylla A (2004) Unique astrocyte ribbon in adult human brain contains neural stem cells but lacks chain migration. Nature 427:740–744
14. Lois C, Alvarez-Buylla A (1993) Proliferating subventricular zone cells in the adult mammalian forebrain can differentiate into neurons and glia. Proc Natl Acad Sci USA 90:2074–2077
15. Doetsch F, Alvarez-Buylla A (1996) Network of tangential pathways for neuronal migration in adult mammalian brain. Proc Natl Acad Sci USA 93:14895–14900
16. Han YG, Spassky N, Romaguera-Ros M, Garcia-Verdugo JM, Aguilar A, Schneider-Maunoury S, Alvarez-Buylla A (2008) Hedgehog signaling and primary cilia are required for the formation of adult neural stem cells. Nat Neurosci 11:277–284
17. Doetsch F, Caille I, Lim DA, Garcia-Verdugo JM, Alvarez-Buylla A (1999) Subventricular zone astrocytes are neural stem cells in the adult mammalian brain. Cell 97:703–716
18. Lois C, Alvarez-Buylla A (1994) Long-distance neuronal migration in the adult mammalian brain. Science 264:1145–1148
19. Curtis MA, Kam M, Nannmark U, Anderson MF, Axell MZ, Wikkelso C, Holtas S, van Roon-Mom WM, Bjork-Eriksson T, Nordborg C, Frisen J, Dragunow M, Faull RL, Eriksson PS (2007) Human neuroblasts migrate to the olfactory bulb via a lateral ventricular extension. Science 315:1243–1249
20. van Praag H, Schinder AF, Christie BR, Toni N, Palmer TD, Gage FH (2002) Functional neurogenesis in the adult hippocampus. Nature 415:1030–1034
21. Ma DK, Kim WR, Ming GL, Song H (2009) Activity-dependent extrinsic regulation of adult olfactory bulb and hippocampal neurogenesis. Ann N Y Acad Sci 1170:664–673
22. Vukovic J, Blackmore DG, Jhaveri D, Bartlett PF (2011) Activation of neural precursors in the adult neurogenic niches. Neurochem Int 59(3):341–346
23. Fuchs E, Tumbar T, Guasch G (2004) Socializing with the neighbors: stem cells and their niche. Cell 116:769–778
24. Palmer TD, Willhoite AR, Gage FH (2000) Vascular niche for adult hippocampal neurogenesis. J Comp Neurol 425:479–494
25. Lim DA, Tramontin AD, Trevejo JM, Herrera DG, Garcia-Verdugo JM, Alvarez-Buylla A (2000) Noggin antagonizes BMP signaling to create a niche for adult neurogenesis. Neuron 28:713–726
26. Lim DA, Alvarez-Buylla A (1999) Interaction between astrocytes and adult subventricular zone precursors stimulates neurogenesis. Proc Natl Acad Sci USA 96:7526–7531
27. Seri B, Garcia-Verdugo JM, Collado-Morente L, McEwen BS, Alvarez-Buylla A (2004) Cell types, lineage, and architecture of the germinal zone in the adult dentate gyrus. J Comp Neurol 478:359–378
28. Shen Q, Goderie SK, Jin L, Karanth N, Sun Y, Abramova N, Vincent P, Pumiglia K, Temple S (2004) Endothelial cells stimulate self-renewal and expand neurogenesis of neural stem cells. Science 304:1338–1340
29. Ward NL, Lamanna JC (2004) The neurovascular unit and its growth factors: coordinated response in the vascular and nervous systems. Neurol Res 26:870–883
30. Lendahl U, Zimmerman L, McKay RDG (1990) CNS stem cells express a new class of intermediate filament protein. Cell 60:585–595
31. Nguyen L, Rigo JM, Malgrange B, Moonen G, Belachew S (2003) Untangling the functional potential of PSA-NCAM-expressing cells in CNS development and brain repair strategies. Curr Med Chem 10:2185–2196
32. Erlandsson A, Enarsson M, Forsberg-Nilsson K (2001) Immature neurons from CNS stem cells proliferate in response to PDGF. J Neurosci 21:3483–3491
33. Jackson EL, Garcia-Verdugo JM, Gil-Perotin S, Roy M, Quinones-Hinojosa A, VandenBerg S, Alvarez-Buylla A (2006) PDGFR alpha-positive B cells are neural stem cells in the adult

SVZ that form glioma-like growths in response to increased PDGF signaling. Neuron 51:187–199

34. Yoon K, Gaiano N (2005) Notch signaling in the mammalian central nervous system: insights from mouse mutants. Nat Neurosci 8:709–715
35. Cai J, Wu Y, Mirua T, Pierce JL, Lucero MT, Albertine KH, Spangrude GJ, Rao MS (2002) Properties of a fetal multipotent neural stem cell (NEP cell). Dev Biol 251:221–240
36. Singh SK, Clarke ID, Hide T, Dirks PB (2004) Cancer stem cells in nervous system tumors. Oncogene 23:7267–7273
37. Leone DP, Relvas JB, Campos LS, Hemmi S, Brakebusch C, Fassler R, Ffrench-Constant C, Suter U (2005) Regulation of neural progenitor proliferation and survival by beta1 integrins. J Cell Sci 118:2589–2599
38. Campos LS (2005) Beta1 integrins and neural stem cells: making sense of the extracellular environment. Bioessays 27:698–707
39. Shen Q, Wang Y, Kokovay E, Lin G, Chuang SM, Goderie SK, Roysam B, Temple S (2008) Adult SVZ stem cells lie in a vascular niche: a quantitative analysis of niche cell-cell interactions. Cell Stem Cell 3:289–300
40. Alonso G (2001) Proliferation of progenitor cells in the adult rat brain correlates with the presence of vimentin-expressing astrocytes. Glia 34:253–266
41. Miyagi S, Kato H, Okuda A (2009) Role of SoxB1 transcription factors in development. Cell Mol Life Sci 66:3675–3684
42. Scott CE, Wynn SL, Sesay A, Cruz C, Cheung M, Gomez Gaviro MV, Booth S, Gao B, Cheah KS, Lovell-Badge R, Briscoe J (2010) SOX9 induces and maintains neural stem cells. Nat Neurosci 13:1181–1189
43. Louis SA, Rietze RL, Deleyrolle L, Wagey RE, Thomas TE, Eaves AC, Reynolds BA (2008) Enumeration of neural stem and progenitor cells in the neural colony-forming cell assay. Stem Cells (Dayton, Ohio) 26:988–996
44. Hartfuss E, Galli R, Heins N, Gotz M (2001) Characterization of CNS precursor subtypes and radial glia. Dev Biol 229:15–30
45. Noctor SC, Flint AC, Weissman TA, Wong WS, Clinton BK, Kriegstein AR (2002) Dividing precursor cells of the embryonic cortical ventricular zone have morphological and molecular characteristics of radial glia. J Neurosci 22:3161–3173
46. Ma DK, Bonaguidi MA, Ming GL, Song H (2009) Adult neural stem cells in the mammalian central nervous system. Cell Res 19:672–682
47. Egger B, Gold KS, Brand AH (2010) Notch regulates the switch from symmetric to asymmetric neural stem cell division in the Drosophila optic lobe. Development (Cambridge, England) 137:2981–2987
48. Kageyama R, Ohtsuka T, Shimojo H, Imayoshi I (2009) Dynamic regulation of Notch signaling in neural progenitor cells. Curr Opin Cell Biol 21:733–740
49. Reynolds BA, Tetzlaff W, Weiss S (1992) A multipotent EGF-responsive striatal embryonic progenitor cell produces neurons and astrocytes. J Neurosci 12:4565–4574
50. Reynolds BA, Weiss S (1996) Clonal and population analyses demonstrate that an EGF-responsive mammalian embryonic CNS precursor is a stem cell. Dev Biol 175:1–13
51. Gritti A, Parati EA, Cova L, Frolichsthal P, Galli R, Wanke E, Faravelli L, Morassutti DJ, Roisen F, Nickel DD, Vescovi AL (1996) Multipotential stem cells from the adult mouse brain proliferate and self-renew in response to basic fibroblast growth factor. J Neurosci 16:1091–1100
52. Johe KK, Hazel TG, Muller T, Dugich Djordjevic MM, McKay RD (1996) Single factors direct the differentiation of stem cells from the fetal and adult central nervous system. Genes Dev 10:3129–3140
53. Suslov ON, Kukekov VG, Ignatova TN, Steindler DA (2002) Neural stem cell heterogeneity demonstrated by molecular phenotyping of clonal neurospheres. Proc Natl Acad Sci USA 99:14506–14511
54. Reynolds BA, Rietze RL (2005) Neural stem cells and neurospheres–re-evaluating the relationship. Nat Methods 2:333–336

55. Singec I, Knoth R, Meyer RP, Maciaczyk J, Volk B, Nikkhah G, Frotscher M, Snyder EY (2006) Defining the actual sensitivity and specificity of the neurosphere assay in stem cell biology. Nat Methods 3:801–806
56. Pardo B, Honegger P (2000) Differentiation of rat striatal embryonic stem cells in vitro: monolayer culture vs. three-dimensional coculture with differentiated brain cells. J Neurosci Res 59:504–512
57. Okabe S, Forsberg-Nilsson K, Spiro AC, Segal M, McKay RD (1996) Development of neuronal precursor cells and functional postmitotic neurons from embryonic stem cells in vitro. Mech Dev 59:89–102
58. Andang M, Moliner A, Doege CA, Ibanez CF, Ernfors P (2008) Optimized mouse ES cell culture system by suspension growth in a fully defined medium. Nat Protoc 3:1013–1017
59. Conti L, Pollard SM, Gorba T, Reitano E, Toselli M, Biella G, Sun Y, Sanzone S, Ying QL, Cattaneo E, Smith A (2005) Niche-independent symmetrical self-renewal of a mammalian tissue stem cell. PLoS Biol 3:e283
60. Bailey P, Cushing H (1926) A classification of tumors of the glioma group on a histogenic basis. J. Lippincott, Philadelphia
61. Virchow R (1858) Cellular pathology. Berlin
62. Louis DN, Ohgaki H, Wiestler OD, Cavenee WK, Burger PC, Jouvet A, Scheithauer BW, Kleihues P (2007) The 2007 WHO classification of tumours of the central nervous system. Acta Neuropathol 114:97–109
63. Stupp R, Mason WP, van den Bent MJ, Weller M, Fisher B, Taphoorn MJ, Belanger K, Brandes AA, Marosi C, Bogdahn U, Curschmann J, Janzer RC, Ludwin SK, Gorlia T, Allgeier A, Lacombe D, Cairncross JG, Eisenhauer E, Mirimanoff RO (2005) Radiotherapy plus concomitant and adjuvant temozolomide for glioblastoma. N Engl J Med 352:987–996
64. Nakada M, Nakada S, Demuth T, Tran NL, Hoelzinger DB, Berens ME (2007) Molecular targets of glioma invasion. Cell Mol Life Sci 64:458–478
65. Rasheed BK, McLendon RE, Friedman HS, Friedman AH, Fuchs HE, Bigner DD, Bigner SH (1995) Chromosome 10 deletion mapping in human gliomas: a common deletion region in 10q25. Oncogene 10:2243–2246
66. Wong AJ, Bigner SH, Bigner DD, Kinzler KW, Hamilton SR, Vogelstein B (1987) Increased expression of the epidermal growth factor receptor gene in malignant gliomas is invariably associated with gene amplification. Proc Natl Acad Sci USA 84:6899–6903
67. Cancer Genome Atlas Research Network (2008) Comprehensive genomic characterization defines human glioblastoma genes and core pathways. Nature 455:1061–1068
68. Parsons DW, Jones S, Zhang X, Lin JC, Leary RJ, Angenendt P, Mankoo P, Carter H, Siu IM, Gallia GL, Olivi A, McLendon R, Rasheed BA, Keir S, Nikolskaya T, Nikolsky Y, Busam DA, Tekleab H, Diaz LA Jr, Hartigan J, Smith DR, Strausberg RL, Marie SK, Shinjo SM, Yan H, Riggins GJ, Bigner DD, Karchin R, Papadopoulos N, Parmigiani G, Vogelstein B, Velculescu VE, Kinzler KW (2008) An integrated genomic analysis of human glioblastoma multiforme. Science 321:1807–1812
69. Rich JN, Bigner DD (2004) Development of novel targeted therapies in the treatment of malignant glioma. Nat Rev Drug Discov 3:430–446
70. Riddick G, Fine H (2011) Integration and analysis of genome-scale data from gliomas. Nat Rev Neurol 7:439–450
71. Brennan C, Momota H, Hambardzumyan D, Ozawa T, Tandon A, Pedraza A, Holland E (2009) Glioblastoma subclasses can be defined by activity among signal transduction pathways and associated genomic alterations. PLoS One 4:e7752
72. Verhaak RG, Hoadley KA, Purdom E, Wang V, Qi Y, Wilkerson MD, Miller CR, Ding L, Golub T, Mesirov JP, Alexe G, Lawrence M, O'Kelly M, Tamayo P, Weir BA, Gabriel S, Winckler W, Gupta S, Jakkula L, Feiler HS, Hodgson JG, James CD, Sarkaria JN, Brennan C, Kahn A, Spellman PT, Wilson RK, Speed TP, Gray JW, Meyerson M, Getz G, Perou CM, Hayes DN (2010) Integrated genomic analysis identifies clinically relevant subtypes of glioblastoma characterized by abnormalities in PDGFRA, IDH1, EGFR, and NF1. Cancer Cell 17:98–110

73. Charles NA, Holland EC, Gilbertson R, Glass R, Kettenmann H (2011) The brain tumor microenvironment. Glia 59:1169–1180
74. Hanahan D, Weinberg RA (2011) Hallmarks of cancer: the next generation. Cell 144:646–674
75. Wen PY, Kesari S (2008) Malignant gliomas in adults. N Engl J Med 359:492–507
76. Le DM, Besson A, Fogg DK, Choi KS, Waisman DM, Goodyer CG, Rewcastle B, Yong VW (2003) Exploitation of astrocytes by glioma cells to facilitate invasiveness: a mechanism involving matrix metalloproteinase-2 and the urokinase-type plasminogen activator-plasmin cascade. J Neurosci 23:4034–4043
77. Hoelzinger DB, Demuth T, Berens ME (2007) Autocrine factors that sustain glioma invasion and paracrine biology in the brain microenvironment. J Natl Cancer Inst 99:1583–1593
78. Badie B, Schartner JM (2000) Flow cytometric characterization of tumor-associated macrophages in experimental gliomas. Neurosurgery 46:957–961, Discussion 961–952
79. Parney IF, Waldron JS, Parsa AT (2009) Flow cytometry and in vitro analysis of human glioma-associated macrophages. Laboratory investigation. J Neurosurg 110:572–582
80. Pollack IF, Okada H, Chambers WH (2000) Exploitation of immune mechanisms in the treatment of central nervous system cancer. Semin Pediatr Neurol 7:131–143
81. Packer R (1999) childhood medulloblastoma: progress and future challenges. Brain Dev 21:75–81
82. Giese A, Westphal M (1996) Glioma invasion in the central nervous system. Neurosurgery 39:235–250, Discussion 250–232
83. Giese A, Rief MD, Loo MA, Berens ME (1994) Determinants of human astrocytoma migration. Cancer Res 54:3897–3904
84. Bellail AC, Hunter SB, Brat DJ, Tan C, Van Meir EG (2004) Microregional extracellular matrix heterogeneity in brain modulates glioma cell invasion. Int J Biochem Cell Biol 36:1046–1069
85. Lund-Johansen M, Bjerkvig R, Humphrey PA, Bigner SH, Bigner DD, Laerum OD (1990) Effect of epidermal growth factor on glioma cell growth, migration, and invasion in vitro. Cancer Res 50:6039–6044
86. Lund-Johansen M, Forsberg K, Bjerkvig R, Laerum OD (1992) Effects of growth factors on a human glioma cell line during invasion into rat brain aggregates in culture. Acta Neuropathol 84:190–197
87. Hong X, Jiang F, Kalkanis SN, Zhang ZG, Zhang XP, DeCarvalho AC, Katakowski M, Bobbitt K, Mikkelsen T, Chopp M (2006) SDF-1 and CXCR4 are up-regulated by VEGF and contribute to glioma cell invasion. Cancer Lett 236:39–45
88. Ponten J, Westermark B (1978) Properties of human malignant glioma cells in vitro. Med Biol 56:184–193
89. Lee J, Kotliarova S, Kotliarov Y, Li A, Su Q, Donin NM, Pastorino S, Purow BW, Christopher N, Zhang W, Park JK, Fine HA (2006) Tumor stem cells derived from glioblastomas cultured in bFGF and EGF more closely mirror the phenotype and genotype of primary tumors than do serum-cultured cell lines. Cancer Cell 9:391–403
90. Hemmati HD, Nakano I, Lazareff JA, Masterman-Smith M, Geschwind DH, Bronner-Fraser M, Kornblum HI (2003) Cancerous stem cells can arise from pediatric brain tumors. Proc Natl Acad Sci USA 100:15178–15183
91. Singh SK, Clarke ID, Terasaki M, Bonn VE, Hawkins C, Squire J, Dirks PB (2003) Identification of a cancer stem cell in human brain tumors. Cancer Res 63:5821–5828
92. Pollard SM, Yoshikawa K, Clarke ID, Danovi D, Stricker S, Russell R, Bayani J, Head R, Lee M, Bernstein M, Squire JA, Smith A, Dirks P (2009) Glioma stem cell lines expanded in adherent culture have tumor-specific phenotypes and are suitable for chemical and genetic screens. Cell Stem Cell 4:568–580
93. Kelly JJ, Stechishin O, Chojnacki A, Lun X, Sun B, Senger DL, Forsyth P, Auer RN, Dunn JF, Cairncross JG, Parney IF, Weiss S (2009) Proliferation of human glioblastoma stem cells occurs independently of exogenous mitogens. Stem Cells (Dayton, Ohio) 27:1722–1733

94. Lapidot T, Sirard C, Vormoor J, Murdoch B, Hoang T, Caceres-Cortes J, Minden M, Paterson B, Caligiuri MA, Dick JE (1994) A cell initiating human acute myeloid leukaemia after transplantation into SCID mice. Nature 367:645–648
95. Tohyama T, Lee V, Rorke L, Marvin M, McKay R, Trojanowsky J (1992) Nestin expression ion embryonic human neuroepithelim and in human neuroepithelial tumor cells. Lab Invest 66:303–313
96. Dahlstrand J, Lardelli M, Lendahl U (1995) Nestin mRNA expression correlates with the central nervous system progenitor cell state in many, but not all, regions of developing central nervous system. Brain Res Dev Brain Res 84:109–129
97. Galli R, Binda E, Orfanelli U, Cipelletti B, Gritti A, De Vitis S, Fiocco R, Foroni C, Dimeco F, Vescovi A (2004) Isolation and characterization of tumorigenic, stem-like neural precursors from human glioblastoma. Cancer Res 64:7011–7021
98. Ignatova TN, Kukekov VG, Laywell ED, Suslov ON, Vrionis FD, Steindler DA (2002) Human cortical glial tumors contain neural stem-like cells expressing astroglial and neuronal markers in vitro. Glia 39:193–206
99. Dirks PB (2010) Brain tumor stem cells: the cancer stem cell hypothesis writ large. Mol Oncol 4:420–430
100. Park DM, Rich JN (2009) Biology of glioma cancer stem cells. Mol Cells 28:7–12
101. Huse JT, Holland EC (2010) Targeting brain cancer: advances in the molecular pathology of malignant glioma and medulloblastoma. Nat Rev 10:319–331
102. Ogden AT, Waziri AE, Lochhead RA, Fusco D, Lopez K, Ellis JA, Kang J, Assanah M, McKhann GM, Sisti MB, McCormick PC, Canoll P, Bruce JN (2008) Identification of A2B5 + CD133- tumor-initiating cells in adult human gliomas. Neurosurgery 62:505–514, Discussion 514–505
103. Wang J, Sakariassen PO, Tsinkalovsky O, Immervoll H, Boe SO, Svendsen A, Prestegarden L, Rosland G, Thorsen F, Stuhr L, Molven A, Bjerkvig R, Enger PO (2008) CD133 negative glioma cells form tumors in nude rats and give rise to CD133 positive cells. Int J Cancer 122:761–768
104. Kelly PN, Dakic A, Adams JM, Nutt SL, Strasser A (2007) Tumor growth need not be driven by rare cancer stem cells. Science 317:337
105. Rosen JM, Jordan CT (2009) The increasing complexity of the cancer stem cell paradigm. Science 324:1670–1673
106. Dai C, Celestino J, Okada Y, Louis D, Fuller G, Holland E (2001) PDGF autocrine stimulation dedifferentiates cultured astrocyets and induces oligodendrogliomas and oligoastrocytomas from neural progenitors and astroyctes in vivo. Genes Dev 15:1913–1925
107. Demoulin J-B, Enarsson M, Larsson J, Essaghir A, Heldin C-H, Forsberg-Nilsson K (2006) The gene expression profile of PDGF-treated neural stem cells corresponds to partially differentiated neurons and glia. Growth Factors 24:184–196
108. Aboody KS, Brown A, Rainov NG, Bower KA, Liu S, Yang W, Small JE, Herrlinger U, Ourednik V, Black PM, Breakefield XO, Snyder EY (2000) Neural stem cells display extensive tropism for pathology in adult brain: evidence from intracranial gliomas. Proc Natl Acad Sci USA 97:12846–12851
109. Glass R, Synowitz M, Kronenberg G, Walzlein JH, Markovic DS, Wang LP, Gast D, Kiwit J, Kempermann G, Kettenmann H (2005) Glioblastoma-induced attraction of endogenous neural precursor cells is associated with improved survival. J Neurosci 25:2637–2646
110. Staflin K, Honeth G, Kalliomaki S, Kjellman C, Edvardsen K, Lindvall M (2004) Neural progenitor cell lines inhibit rat tumor growth in vivo. Cancer Res 64:5347–5354
111. Tropepe V, Craig CG, Morshead CM, van der Kooy D (1997) Transforming growth factor-alpha null and senescent mice show decreased neural progenitor cell proliferation in the forebrain subependyma. J Neurosci 17:7850–7859
112. Bao S, Wu Q, McLendon RE, Hao Y, Shi Q, Hjelmeland AB, Dewhirst MW, Bigner DD, Rich JN (2006) Glioma stem cells promote radioresistance by preferential activation of the DNA damage response. Nature 444:756–760

113. Calabrese C, Poppleton H, Kocak M, Hogg TL, Fuller C, Hamner B, Oh EY, Gaber MW, Finklestein D, Allen M, Frank A, Bayazitov IT, Zakharenko SS, Gajjar A, Davidoff A, Gilbertson RJ (2007) A perivascular niche for brain tumor stem cells. Cancer Cell 11:69–82
114. Preston M, Sherman LS (2011) Neural stem cell niches: roles for the hyaluronan-based extracellular matrix. Front Biosci (Schol Ed) 3:1165–1179
115. Delpech B, Maingonnat C, Girard N, Chauzy C, Maunoury R, Olivier A, Tayot J, Creissard P (1993) Hyaluronan and hyaluronectin in the extracellular matrix of human brain tumour stroma. Eur J Cancer 29A:1012–1017
116. Bignami A, Perides G, Asher R, Dahl D (1992) The astrocyte–extracellular matrix complex in CNS myelinated tracts: a comparative study on the distribution of hyaluronate in rat, goldfish and lamprey. J Neurocytol 21:604–613
117. De Clerck YA, Shimada H, Gonzalez-Gomez I, Raffel C (1994) Tumoral invasion in the central nervous system. J Neurooncol 18:111–121
118. Englund U, Bjorklund A, Wictorin K, Lindvall O, Kokaia M (2002) Grafted neural stem cells develop into functional pyramidal neurons and integrate into host cortical circuitry. Proc Natl Acad Sci USA 99:17089–17094
119. Tabar V, Panagiotakos G, Greenberg ED, Chan BK, Sadelain M, Gutin PH, Studer L (2005) Migration and differentiation of neural precursors derived from human embryonic stem cells in the rat brain. Nat Biotechnol 23:601–606
120. Akiyama Y, Jung S, Salhia B, Lee S, Hubbard S, Taylor M, Mainprize T, Akaishi K, van Furth W, Rutka JT (2001) Hyaluronate receptors mediating glioma cell migration and proliferation. J Neurooncol 53:115–127
121. Koochekpour S, Pilkington GJ, Merzak A (1995) Hyaluronic acid/CD44H interaction induces cell detachment and stimulates migration and invasion of human glioma cells in vitro. Int J Cancer 63:450–454
122. Yu Q, Stamenkovic I (1999) Localization of matrix metalloproteinase 9 to the cell surface provides a mechanism for CD44-mediated tumor invasion. Genes Dev 13:35–48
123. Garcion E, Halilagic A, Faissner A, ffrench-Constant C (2004) Generation of an environmental niche for neural stem cell development by the extracellular matrix molecule tenascin C. Development (Cambridge, England) 131:3423–3432
124. Steindler DA, Settles D, Erickson HP, Laywell ED, Yoshiki A, Faissner A, Kusakabe M (1995) Tenascin knockout mice: barrels, boundary molecules, and glial scars. J Neurosci 15:1971–1983
125. Gates MA, Fillmore H, Steindler DA (1996) Chondroitin sulfate proteoglycan and tenascin in the wounded adult mouse neostriatum in vitro: dopamine neuron attachment and process outgrowth. J Neurosci 16:8005–8018
126. Nishio T, Kawaguchi S, Yamamoto M, Iseda T, Kawasaki T, Hase T (2005) Tenascin-C regulates proliferation and migration of cultured astrocytes in a scratch wound assay. Neuroscience 132:87–102
127. Maris C, Rorive S, Sandras F, D'Haene N, Sadeghi N, Bieche I, Vidaud M, Decaestecker C, Salmon I (2008) Tenascin-C expression relates to clinicopathological features in pilocytic and diffuse astrocytomas. Neuropathol Appl Neurobiol 34:316–329
128. Sarkar S, Nuttall RK, Liu S, Edwards DR, Yong VW (2006) Tenascin-C stimulates glioma cell invasion through matrix metalloproteinase-12. Cancer Res 66:11771–11780
129. Yong VW (1999) The potential use of MMP inhibitors to treat CNS diseases. Expert Opin Investig Drugs 8:255s–268s
130. Rao VH, Lees GE, Kashtan CE, Nemori R, Singh RK, Meehan DT, Rodgers K, Berridge BR, Bhattacharya G, Cosgrove D (2003) Increased expression of MMP-2, MMP-9 (type IV collagenases/gelatinases), and MT1-MMP in canine X-linked Alport syndrome (XLAS). Kidney Int 63:1736–1748
131. Wang M, Wang T, Liu S, Yoshida D, Teramoto A (2003) The expression of matrix metalloproteinase-2 and -9 in human gliomas of different pathological grades. Brain Tumor Pathol 20:65–72

132. Rao JS, Bhoopathi P, Chetty C, Gujrati M, Lakka SS (2007) MMP-9 short interfering RNA induced senescence resulting in inhibition of medulloblastoma growth via p16(INK4a) and mitogen-activated protein kinase pathway. Cancer Res 67:4956–4964
133. Gabelloni P, Da Pozzo E, Bendinelli S, Costa B, Nuti E, Casalini F, Orlandini E, Da Settimo F, Rossello A, Martini C (2010) Inhibition of metalloproteinases derived from tumours: new insights in the treatment of human glioblastoma. Neuroscience 168:514–522
134. Gingras MC, Roussel E, Roth JA, Moser RP (1995) Little expression of cytokine mRNA by fresh tumour-infiltrating mononuclear leukocytes from glioma and lung adenocarcinoma. Cytokine 7:580–588
135. Mahesparan R, Read TA, Lund-Johansen M, Skaftnesmo KO, Bjerkvig R, Engebraaten O (2003) Expression of extracellular matrix components in a highly infiltrative in vivo glioma model. Acta Neuropathol 105:49–57
136. Gladson CL (1996) Expression of integrin alpha v beta 3 in small blood vessels of glioblastoma tumors. J Neuropathol Exp Neurol 55:1143–1149
137. Bello L, Francolini M, Marthyn P, Zhang J, Carroll RS, Nikas DC, Strasser JF, Villani R, Cheresh DA, Black PM (2001) Alpha(v)beta3 and alpha(v)beta5 integrin expression in glioma periphery. Neurosurgery 49:380–389, Discussion 390
138. Brooks PC, Stromblad S, Sanders LC, von Schalscha TL, Aimes RT, Stetler-Stevenson WG, Quigley JP, Cheresh DA (1996) Localization of matrix metalloproteinase MMP-2 to the surface of invasive cells by interaction with integrin alpha v beta 3. Cell 85:683–693
139. Heldin CH (2001) Signal transduction: multiple pathways, multiple options for therapy. Stem Cells (Dayton, Ohio) 19:295–303
140. Burrows RC, Wancio D, Levitt P, Lillien L (1997) Response diversity and the timing of progenitor cell maturation are regulated by developmental changes in EGFR expression in the cortex. Neuron 19:251–267
141. Zhu G, Mehler MF, Mabie PC, Kessler JA (1999) Developmental changes in progenitor cell responsiveness to cytokines. J Neurosci Res 56:131–145
142. Ramnarain DB, Park S, Lee DY, Hatanpaa KJ, Scoggin SO, Otu H, Libermann TA, Raisanen JM, Ashfaq R, Wong ET, Wu J, Elliott R, Habib AA (2006) Differential gene expression analysis reveals generation of an autocrine loop by a mutant epidermal growth factor receptor in glioma cells. Cancer Res 66:867–874
143. Nishikawa R, Ji XD, Harmon RC, Lazar CS, Gill GN, Cavenee WK, Huang HJ (1994) A mutant epidermal growth factor receptor common in human glioma confers enhanced tumorigenicity. Proc Natl Acad Sci USA 91:7727–7731
144. Krakstad C, Chekenya M (2010) Survival signalling and apoptosis resistance in glioblastomas: opportunities for targeted therapeutics. Mol Cancer 9:135
145. Mukasa A, Wykosky J, Ligon KL, Chin L, Cavenee WK, Furnari F (2010) Mutant EGFR is required for maintenance of glioma growth in vivo, and its ablation leads to escape from receptor dependence. Proc Natl Acad Sci USA 107:2616–2621
146. Santos FP, Quintas-Cardama A (2011) New drugs for chronic myelogenous leukemia. Curr Hematol Malig Rep 6:96–103
147. Morris PG, Abrey LE (2010) Novel targeted agents for platelet-derived growth factor receptor and c-KIT in malignant gliomas. Target Oncol 5:193–200
148. Andrae J, Gallini R, Betsholtz C (2008) Role of platelet-derived growth factors in physiology and medicine. Genes Dev 22:1276–1312
149. Orr-Urtreger A, Lonai P (1992) Platelet-derived growth factor-A and its receptor are expressed in separate, but adjacent cell layers of the mouse embryo. Development (Cambridge, England) 115:1045–1058
150. Pringle N, Mudhar H, Collarini E, Richardson W (1992) PDGF receptors in the rat CNS: during late neurogenesis, PDGF alpha-receptor expression appears to be restricted to glial cells of the oligodendrocyte lineage. Development (Cambridge, England) 115:535–551
151. Heldin C-H, Westermark B (1999) Mechanism of action and in vivo role of platelet-derived growth factor. Physiol Rev 4:1283–1316

152. Uhrbom L, Hesselager G, Nister M, Westermark B (1998) Induction of brain tumors in mice using a recombinant platelet-derived growth factor B-chain retrovirus. Cancer Res 58:5275–5279
153. Appolloni I, Calzolari F, Tutucci E, Caviglia S, Terrile M, Corte G, Malatesta P (2009) PDGF-B induces a homogeneous class of oligodendrogliomas from embryonic neural progenitors. Int J Cancer 124:2251–2259
154. Lindberg N, Kastemar M, Olofsson T, Smits A, Uhrbom L (2009) Oligodendrocyte progenitor cells can act as cell of origin for experimental glioma. Oncogene 28:2266–2275
155. Assanah MC, Bruce JN, Suzuki SO, Chen A, Goldman JE, Canoll P (2009) PDGF stimulates the massive expansion of glial progenitors in the neonatal forebrain. Glia 57:1835–1847
156. Forsberg-Nilsson K, Erlandsson A, Zhang X-Q, Ueda H, Svensson K, Nister M, Trapp B, Peterson A, Westermark B (2003) Oligodendrocyte precursor hypercellularity and abnormal retina development in mice overexpressing PDGF-B in myelinating tracts. Glia 41:276–289
157. Fruttiger M, Karlsson L, Hall A, Abrahamsson A, Calver A, Boström H, Willets K, Bertold C-H, Heath J, Betsholtz C, Richardson W (1999) Defective oligodendrocyte development and severe hypomyelination in PDGF-A knockout mice. Development (Cambridge, England) 126:457–467
158. Niklasson M, Bergstrom T, Zhang XQ, Gustafsdottir SM, Sjogren M, Edqvist PH, Vennstrom B, Forsberg M, Forsberg-Nilsson K (2010) Enlarged lateral ventricles and aberrant behavior in mice overexpressing PDGF-B in embryonic neural stem cells. Exp Cell Res 316:2779–2789
159. Hede SM, Hansson I, Afink GB, Eriksson A, Nazarenko I, Andrae J, Genove G, Westermark B, Nister M (2008) GFAP promoter driven transgenic expression of PDGFB in the mouse brain leads to glioblastoma in a Trp53 null background. Glia 57(11)
160. Williams B, Park J, Alberta J, Muhlebach S, Hwang G, Roberts T, Stiles C (1997) A PDGF-regulated immediate early gene response initiates neuronal differentiation in ventricular zone progenitor cells. Neuron 18:553–562
161. Erlandsson A, Brannvall K, Gustafsdottir S, Westermark B, Forsberg-Nilsson K (2006) Autocrine/paracrine platelet-derived growth factor regulates proliferation of neural progenitor cells. Cancer Res 66:8042–8048
162. Chojnacki A, Mak G, Weiss S (2011) PDGFRα expression distinguishes GFAP-expressing neural stem cells from PDGF-responsive neural precursors in the adult periventricular area. J Neurosci 31:9503–9512
163. Thery C, Hetier E, Evrard C, Mallat M (1990) Expression of macrophage colony-stimulating factor gene in the mouse brain during development. J Neurosci Res 26:129–133
164. Raivich G, Gehrmann J, Kreutzberg GW (1991) Increase of macrophage colony-stimulating factor and granulocyte-macrophage colony-stimulating factor receptors in the regenerating rat facial nucleus. J Neurosci Res 30:682–686
165. Murphy GM Jr, Zhao F, Yang L, Cordell B (2000) Expression of macrophage colony-stimulating factor receptor is increased in the AbetaPP(V717F) transgenic mouse model of Alzheimer's disease. Am J Pathol 157:895–904
166. Lue LF, Rydel R, Brigham EF, Yang LB, Hampel H, Murphy GM Jr, Brachova L, Yan SD, Walker DG, Shen Y, Rogers J (2001) Inflammatory repertoire of Alzheimer's disease and nondemented elderly microglia in vitro. Glia 35:72–79
167. Du Yan S, Zhu H, Fu J, Yan SF, Roher A, Tourtellotte WW, Rajavashisth T, Chen X, Godman GC, Stern D, Schmidt AM (1997) Amyloid-beta peptide-receptor for advanced glycation endproduct interaction elicits neuronal expression of macrophage-colony stimulating factor: a proinflammatory pathway in Alzheimer disease. Proc Natl Acad Sci USA 94:5296–5301
168. Mitrasinovic OM, Perez GV, Zhao F, Lee YL, Poon C, Murphy GM Jr (2001) Overexpression of macrophage colony-stimulating factor receptor on microglial cells induces an inflammatory response. J Biol Chem 276:30142–30149
169. Yang M, Donaldson AE, Marshall CE, Shen J, Iacovitti L (2004) Studies on the differentiation of dopaminergic traits in human neural progenitor cells in vitro and in vivo. Cell Transplant 13:535–547

170. Schabitz WR, Kruger C, Pitzer C, Weber D, Laage R, Gassler N, Aronowski J, Mier W, Kirsch F, Dittgen T, Bach A, Sommer C, Schneider A (2008) A neuroprotective function for the hematopoietic protein granulocyte-macrophage colony stimulating factor (GM-CSF). J Cereb Blood Flow Metab 28:29–43
171. Kruger C, Laage R, Pitzer C, Schabitz WR, Schneider A (2007) The hematopoietic factor GM-CSF (granulocyte-macrophage colony-stimulating factor) promotes neuronal differentiation of adult neural stem cells in vitro. BMC Neurosci 8:88
172. Alterman RL, Stanley ER (1994) Colony stimulating factor-1 expression in human glioma. Mol Chem Neuropathol 21:177–188
173. Komohara Y, Ohnishi K, Kuratsu J, Takeya M (2008) Possible involvement of the M2 anti-inflammatory macrophage phenotype in growth of human gliomas. J Pathol 216:15–24
174. Bender AM, Collier LS, Rodriguez FJ, Tieu C, Larson JD, Halder C, Mahlum E, Kollmeyer TM, Akagi K, Sarkar G, Largaespada DA, Jenkins RB (2010) Sleeping beauty-mediated somatic mutagenesis implicates CSF1 in the formation of high-grade astrocytomas. Cancer Res 70:3557–3565
175. Glaspy J (1996) Clinical applications of stem cell factor. Curr Opin Hematol 3:223–229
176. Dolci S, Williams DE, Ernst MK, Resnick JL, Brannan CI, Lock LF, Lyman SD, Boswell HS, Donovan PJ (1991) Requirement for mast cell growth factor for primordial germ cell survival in culture. Nature 352:809–811
177. Lowry PA, Zsebo KM, Deacon DH, Eichman CE, Quesenberry PJ (1991) Effects of rrSCF on multiple cytokine responsive HPP-CFC generated from SCA + Lin- murine hematopoietic progenitors. Exp Hematol 19:994–996
178. Heinrich MC, Dooley DC, Freed AC, Band L, Hoatlin ME, Keeble WW, Peters ST, Silvey KV, Ey FS, Kabat D et al (1993) Constitutive expression of steel factor gene by human stromal cells. Blood 82:771–783
179. Migliaccio AR, Migliaccio G, Mancini G, Ratajczak M, Gewirtz AM, Adamson JW (1993) Induction of the murine "W phenotype" in long-term cultures of human cord blood cells by c-kit antisense oligomers. J Cell Physiol 157:158–163
180. Keshet E, Lyman SD, Williams DE, Anderson DM, Jenkins NA, Copeland NG, Parada LF (1991) Embryonic RNA expression patterns of the c-kit receptor and its cognate ligand suggest multiple functional roles in mouse development. EMBO J 10:2425–2435
181. Erlandsson A, Larsson J, Forsberg-Nilsson K (2004) Stem cell factor is a chemoattractant and a survival factor for CNS stem cells. Exp Cell Res 301:201–210
182. Zhang SC, Fedoroff S (1997) Cellular localization of stem cell factor and c-kit receptor in the mouse nervous system. J Neurosci Res 47:1–15
183. Das AV, James J, Zhao X, Rahnenfuhrer J, Ahmad I (2004) Identification of c-Kit receptor as a regulator of adult neural stem cells in the mammalian eye: interactions with Notch signaling. Dev Biol 273:87–105
184. Went PT, Dirnhofer S, Bundi M, Mirlacher M, Schraml P, Mangialaio S, Dimitrijevic S, Kononen J, Lugli A, Simon R, Sauter G (2004) Prevalence of KIT expression in human tumors. J Clin Oncol 22:4514–4522
185. Sun L, Hui AM, Su Q, Vortmeyer A, Kotliarov Y, Pastorino S, Passaniti A, Menon J, Walling J, Bailey R, Rosenblum M, Mikkelsen T, Fine HA (2006) Neuronal and glioma-derived stem cell factor induces angiogenesis within the brain. Cancer Cell 9:287–300
186. Aguirre A, Rubio ME, Gallo V (2010) Notch and EGFR pathway interaction regulates neural stem cell number and self-renewal. Nature 467:323–327
187. Wang L, Chopp M, Zhang RL, Zhang L, Letourneau Y, Feng YF, Jiang A, Morris DC, Zhang ZG (2009) The Notch pathway mediates expansion of a progenitor pool and neuronal differentiation in adult neural progenitor cells after stroke. Neuroscience 158:1356–1363
188. Ying M, Wang S, Sang Y, Sun P, Lal B, Goodwin CR, Guerrero-Cazares H, Quinones-Hinojosa A, Laterra J, Xia S (2011) Regulation of glioblastoma stem cells by retinoic acid: role for Notch pathway inhibition. Oncogene 30(31):3454–3467
189. Kanamori M, Kawaguchi T, Nigro JM, Feuerstein BG, Berger MS, Miele L, Pieper RO (2007) Contribution of Notch signaling activation to human glioblastoma multiforme. J Neurosurg 106:417–427

190. Petersen PH, Zou K, Krauss S, Zhong W (2004) Continuing role for mouse Numb and Numbl in maintaining progenitor cells during cortical neurogenesis. Nat Neurosci 7:803–811
191. Schreck KC, Taylor P, Marchionni L, Gopalakrishnan V, Bar EE, Gaiano N, Eberhart CG, Schreck KC, Taylor P, Marchionni L, Gopalakrishnan V, Bar EE, Gaiano N, Eberhart CG (2010) The Notch target Hes1 directly modulates Gli1 expression and Hedgehog signaling: a potential mechanism of therapeutic resistance. Clin Cancer Res 16:6060–6070
192. Pierfelice TJ, Schreck KC, Dang L, Asnaghi L, Gaiano N, Eberhart CG (2011) Notch3 activation promotes invasive glioma formation in a tissue site-specific manner. Cancer Res 71:1115–1125
193. Zhang XP, Zheng G, Zou L, Liu HL, Hou LH, Zhou P, Yin DD, Zheng QJ, Liang L, Zhang SZ, Feng L, Yao LB, Yang AG, Han H, Chen JY (2008) Notch activation promotes cell proliferation and the formation of neural stem cell-like colonies in human glioma cells. Mol Cell Biochem 307:101–108
194. Yan B, Omar FM, Das K, Ng WH, Lim C, Shiuan K, Yap CT, Salto-Tellez M (2008) Characterization of Numb expression in astrocytomas. Neuropathology 28:479–484
195. Purow BW, Haque RM, Noel MW, Su Q, Burdick MJ, Lee J, Sundaresan T, Pastorino S, Park JK, Mikolaenko I, Maric D, Eberhart CG, Fine HA (2005) Expression of Notch-1 and its ligands, Delta-like-1 and Jagged-1, is critical for glioma cell survival and proliferation. Cancer Res 65:2353–2363
196. Seidel S, Garvalov BK, Wirta V, von Stechow L, Schanzer A, Meletis K, Wolter M, Sommerlad D, Henze AT, Nister M, Reifenberger G, Lundeberg J, Frisen J, Acker T (2010) A hypoxic niche regulates glioblastoma stem cells through hypoxia inducible factor 2 alpha. Brain 133:983–995
197. Li JL, Sainson RC, Shi W, Leek R, Harrington LS, Preusser M, Biswas S, Turley H, Heikamp E, Hainfellner JA, Harris AL (2007) Delta-like 4 Notch ligand regulates tumor angiogenesis, improves tumor vascular function, and promotes tumor growth in vivo. Cancer Res 67:11244–11253
198. Gilbert CA, Daou MC, Moser RP, Ross AH (2010) Gamma-secretase inhibitors enhance temozolomide treatment of human gliomas by inhibiting neurosphere repopulation and xenograft recurrence. Cancer Res 70:6870–6879
199. Ahn S, Joyner AL (2005) In vivo analysis of quiescent adult neural stem cells responding to Sonic hedgehog. Nature 437:894–897
200. Becher OJ, Hambardzumyan D, Fomchenko EI, Momota H, Mainwaring L, Bleau AM, Katz AM, Edgar M, Kenney AM, Cordon-Cardo C, Blasberg RG, Holland EC (2008) Gli activity correlates with tumor grade in platelet-derived growth factor-induced gliomas. Cancer Res 68:2241–2249
201. Ehtesham M, Sarangi A, Valadez JG, Chanthaphaychith S, Becher MW, Abel TW, Thompson RC, Cooper MK (2007) Ligand-dependent activation of the hedgehog pathway in glioma progenitor cells. Oncogene 26:5752–5761
202. Dahmane N, Sanchez P, Gitton Y, Palma V, Sun T, Beyna M, Weiner H, Ruiz i Altaba A (2001) The Sonic Hedgehog-Gli pathway regulates dorsal brain growth and tumorigenesis. Development (Cambridge, England) 128:5201–5212
203. Uchida H, Arita K, Yunoue S, Yonezawa H, Shinsato Y, Kawano H, Hirano H, Hanaya R, Tokimura H (2011) Role of sonic hedgehog signaling in migration of cell lines established from CD133-positive malignant glioma cells. J Neurooncol 104(3):697–704
204. Wang K, Pan L, Che X, Cui D, Li C (2010) Sonic Hedgehog/GLI signaling pathway inhibition restricts cell migration and invasion in human gliomas. Neurol Res 32:975–980
205. Clement V, Sanchez P, de Tribolet N, Radovanovic I, Ruiz i Altaba A (2007) HEDGEHOG-GLI1 signaling regulates human glioma growth, cancer stem cell self-renewal, and tumorigenicity. Curr Biol 17:165–172
206. Moustakas A, Heldin CH (2009) The regulation of TGFbeta signal transduction. Development (Cambridge, England) 136:3699–3714
207. Bodmer S, Strommer K, Frei K, Siepl C, de Tribolet N, Heid I, Fontana A (1989) Immunosuppression and transforming growth factor-beta in glioblastoma. Preferential production of transforming growth factor-beta 2. J Immunol 143:3222–3229

208. Yamada N, Kato M, Yamashita H, Nister M, Miyazono K, Heldin CH, Funa K (1995) Enhanced expression of transforming growth factor-beta and its type-I and type-II receptors in human glioblastoma. Int J Cancer 62:386–392
209. Lu Y, Jiang F, Zheng X, Katakowski M, Buller B, To SS, Chopp M (2011) TGF-beta1 promotes motility and invasiveness of glioma cells through activation of ADAM17. Oncol Rep 25:1329–1335
210. Ikushima H, Todo T, Ino Y, Takahashi M, Miyazawa K, Miyazono K (2009) Autocrine TGF-beta signaling maintains tumorigenicity of glioma-initiating cells through Sry-related HMG-box factors. Cell Stem Cell 5:504–514
211. Penuelas S, Anido J, Prieto-Sanchez RM, Folch G, Barba I, Cuartas I, Garcia-Dorado D, Poca MA, Sahuquillo J, Baselga J, Seoane J (2009) TGF-beta increases glioma-initiating cell self-renewal through the induction of LIF in human glioblastoma. Cancer Cell 15:315–327
212. Bogdahn U, Hau P, Stockhammer G, Venkataramana NK, Mahapatra AK, Suri A, Balasubramaniam A, Nair S, Oliushine V, Parfenov V, Poverennova I, Zaaroor M, Jachimczak P, Ludwig S, Schmaus S, Heinrichs H, Schlingensiepen KH (2011) Targeted therapy for high-grade glioma with the TGF-beta2 inhibitor trabedersen: results of a randomized and controlled phase IIb study. Neuro Oncol 13:132–142
213. Schneider T, Becker A, Ringe K, Reinhold A, Firsching R, Sabel BA (2008) Brain tumor therapy by combined vaccination and antisense oligonucleotide delivery with nanoparticles. J Neuroimmunol 195:21–27
214. Yamada N, Kato M, ten Dijke P, Yamashita H, Sampath TK, Heldin CH, Miyazono K, Funa K (1996) Bone morphogenetic protein type IB receptor is progressively expressed in malignant glioma tumours. Br J Cancer 73:624–629
215. Klose A, Waerzeggers Y, Monfared P, Vukicevic S, Kaijzel EL, Winkeler A, Wickenhauser C, Lowik CW, Jacobs AH (2011) Imaging bone morphogenetic protein 7 induced cell cycle arrest in experimental gliomas. Neoplasia 13:276–285
216. Pistollato F, Chen HL, Rood BR, Zhang HZ, D'Avella D, Denaro L, Gardiman M, te Kronnie G, Schwartz PH, Favaro E, Indraccolo S, Basso G, Panchision DM (2009) Hypoxia and HIF1alpha repress the differentiative effects of BMPs in high-grade glioma. Stem Cells (Dayton, Ohio) 27:7–17
217. Piccirillo SG, Reynolds BA, Zanetti N, Lamorte G, Binda E, Broggi G, Brem H, Olivi A, Dimeco F, Vescovi AL (2006) Bone morphogenetic proteins inhibit the tumorigenic potential of human brain tumour-initiating cells. Nature 444:761–765
218. Henderson BR, Fagotto F (2002) The ins and outs of APC and beta-catenin nuclear transport. EMBO Rep 3:834–839
219. Munji RN, Choe Y, Li G, Siegenthaler JA, Pleasure SJ (2011) Wnt signaling regulates neuronal differentiation of cortical intermediate progenitors. J Neurosci 31:1676–1687
220. Kalderon D (2002) Similarities between the Hedgehog and Wnt signaling pathways. Trends Cell Biol 12:523–531
221. Hirabayashi Y, Itoh Y, Tabata H, Nakajima K, Akiyama T, Masuyama N, Gotoh Y (2004) The Wnt/beta-catenin pathway directs neuronal differentiation of cortical neural precursor cells. Development (Dayton, Ohio) 131:2791–2801
222. Shimizu T, Kagawa T, Inoue T, Nonaka A, Takada S, Aburatani H, Taga T (2008) Stabilized beta-catenin functions through TCF/LEF proteins and the Notch/RBP-Jkappa complex to promote proliferation and suppress differentiation of neural precursor cells. Mol Cell Biol 28:7427–7441
223. Pu P, Zhang Z, Kang C, Jiang R, Jia Z, Wang G, Jiang H (2009) Downregulation of Wnt2 and beta-catenin by siRNA suppresses malignant glioma cell growth. Cancer Gene Ther 16:351–361
224. Koch A, Waha A, Tonn JC, Sorensen N, Berthold F, Wolter M, Reifenberger J, Hartmann W, Friedl W, Reifenberger G, Wiestler OD, Pietsch T (2001) Somatic mutations of WNT/wingless signaling pathway components in primitive neuroectodermal tumors. Int J Cancer 93:445–449

Chapter 18
Stem-Like Cells from Brain Tumours or *Vice Versa*?

Sara G.M. Piccirillo

Contents

Abstract In the last years the idea that tumours arise from a sub-population of cells endowed with "stem cell" features completely revolutionized the field of Oncology. This hypothesis found its initial confirmation by studies on non-solid tumours and then has been extended also to solid tumours and in particular to the most aggressive brain cancer i.e. glioblastoma multiforme (GBM).

It is a common believe that the so called "cancer stem cells" (CSCs) concept could also provide new insights into the cellular and molecular mechanisms of tumour growth and in the future to the development of new therapeutic strategies for the treatment of incurable cancers.

The CSC theory has also contributed to change our idea of treating brain cancers with normal neural stem cells since it seems that cancer cells and stem cells share functional properties and regulatory mechanisms.

As a consequence a significant effort is currently underway to identify both CSC-specific markers and the molecular mechanisms that underpin the tumorigenic potential of these cells, for this will have a critical impact on our understanding of origin and growth of tumours.

S.G.M. Piccirillo (✉)
Department of Clinical Neurosciences, Cambridge Centre for Brain Repair, University of Cambridge, Cambridge CB2 0PY, UK
e-mail: sp577@cam.ac.uk

R.K. Srivastava and S. Shankar (eds.), *Stem Cells and Human Diseases*,
DOI 10.1007/978-94-007-2801-1_18,

Keywords Central nervous system • Neural stem cells • Cancer stem cells • Glioblastoma multiforme • Tumorigenicity

Abbreviations

BMPs	Bone morphogenetic proteins
CNS	Central nervous system
CSC	Cancer stem cells
GBM	Glioblastoma multiforme
GFAP	Glial fibrillary acidic protein
NSCs	Neural stem cells

18.1 Introduction

Cancers are thought to be a consequence of genetic and epigenetic alterations in a single cell or a group of cells. These alterations could lead to uncontrolled growth and spread of tumour cells. It has been suggested that this can occur according to two models:

- Stochastic: where all the tumour cells share the ability to proliferate and reconstitute the tumour [1],
- Hierarchical-stem cell driven: where on the contrary only a small subset of cells is endowed with this ability [1], similarly to the repopulating ability of stem cells in a tissue.

This latter model found a lot of consensus in the recent years when several groups demonstrated that non-solid and solid tumours are organized according to this model and reported the existence of the so called "cancer stem cells" at the top of the hierarchy [2]. This concept has been applied also to the brain and in particular to glioblastoma multiforme (GBM) which is the most aggressive form accounting for over 60% of brain tumours. Median life expectancy in optimally managed patients is only 17–62 weeks with only 25% surviving 2 years [3]. Although it is widely accepted that GBM derives from glial cells, this hypothesis has never been adequately demonstrated. The discovery that the adult brain contains discrete neurogenic areas with their resident neural stem cells (NSCs) (reviewed in [4, 5]) radically challenged this concept and it is now under investigation that NSCs could be candidates for transformation leading to brain tumours.

Together, these findings have completely revolutionized the field of Neuro-oncology at a time when neural stem cells were exploited as therapeutic carrier to deliver anti-tumour factors to brain cancer cells.

18.2 Stem Cells as the Origin of Tumours

The hypothesis that cancers have similar properties to stem cells is not new, but it is only in the last years that the prospective identification and purification of CSCs was possible. This concept was first described by Rudolf Virchow and Julius Conheim in the nineteenth century. Virchow's embryonal-rest hypothesis suggested that cancer arises from activation of dormant cells present in mature tissue that are remainders of embryonic cells [6]. The hypothesis was based on the histological similarities between developing foetus and certain types of cancer such as teratocarcinomas, and the observation that both tissues have an enormous capacity for both proliferation and differentiation, albeit aberrant differentiation in the case of tumours [6]. This theory was re-awakened in the mid twentieth century predominantly through colony forming assays demonstrating that a small proportion of cells were capable of forming colonies. It was demonstrated that in media conditioned by spleen cells *in vitro,* there was a variation in colony forming ability of different solid cancers and even different specimens of the same type of cancer. Adenocarcinoma of the ovary samples showed 1 in 625 to 1 in 5,000 colony forming cells and a neuroblastoma had 1 in 2,000 cells were capable of colony forming [7].

This suggested that the cancer tissue was organized hierarchically into clonally derived populations with different proliferative potential therefore only certain cells from a tumour would be tumour forming *in vivo*. It is attractive to think that somatic stem cells may be a model for putative CSCs as they are undifferentiated cells, have an extensive self renewal ability, long term replication potential and multi-lineage differentiation potential. The heterogeneity of the disease also suggests a multipotential cell of origin.

The first evidence of CSCs existence was demonstrated in acute myeloid leukaemia (AML). Bonnet and Dick demonstrated that only a small subset of human AML cells that were phenotypically similar to normal haematopoietic stem cells, $CD34^{+}CD38^{-}$, could transfer AML into immuno-deficient mice and only this cell fraction was able to differentiate *in vivo* into leukemic blasts [8]. These findings supported the similarities between cancer stem cells and normal stem cells and the need to selectivcly target CSCs from their normal counterpart to define new therapeutic approaches [2]. This has lead to a continued search for the "stem cell marker" in this and other cancers including breast cancer [9], colon [10, 11] and brain [12–17].

18.3 Cancer Stem Cells in Brain Tumours

Although brain tumours account for less than 2% of all primary tumours they are responsible for 7% of the years of life lost from cancer before age 70 (Office for National Statistics, 2006).

If the burden of disease is considered in terms of the average years of life lost per patient, brain tumours are one of the most lethal cancers with over 20 years of life lost [18]. The high rates of mortality make these rare cancers into the third leading cause of cancer-related death among economically active men between 15 and 54 years of age and the fourth leading cause of cancer-related death among economically active women between 15 and 34 years of age [19].

Tumours that derive from the glial cells and in particular from astrocytes are the commonest class of human brain cancers in adults and are called astrocytomas. The commonest astrocytoma is the highly malignant glioblastoma (GBM) which accounts for over 60% of brain tumours. Median life expectancy in optimally managed patients is only 17–62 weeks with only 25% surviving 2 years [3]. The current clinical management of patients diagnosed with a GBM involves a combination of surgery, radiotherapy and chemotherapy. Radiotherapy has been the principle therapeutic modality since the late seventies [20] and the latest survival trends for patients with malignancies of the Central Nervous System (CNS) have remained largely static with slight improvement in the last few years upon the introduction of alkylating chemotherapy with temozolomide [3, 21].

Historically it was suggested that GBM arise from differentiated glial cells. In this context, several genetic mouse models have been developed and expression of oncogenes in specific cell types has provided useful insights in this direction. In particular, several studies have shown that early cortical astrocytes can be targeted *in vitro* or *in vivo* with oncogenes or activated signal generating proteins to produce tumours in animal models with convincing glioma histology. One of these studies has demonstrated that overexpression of Ras under a GFAP promoter leads to over activation of the p21-ras signalling pathway and promotes the growth of multifocal malignant astrocytoma [22]. More recently, it was demonstrated that astrocytes and NSCs cultured from $p16^{Ink4a}/p19^{Arf}$ null mice could form xenograft tumours showing GBM cardinal features when they expressed the constitutively active from of EGFR suggesting that either NSCs or astrocytes could be the GBM cell of origin [23].

In the same years, by directly accessing human brain samples, transformed, stem-like neural progenitors have been found in brain tumour tissues, i.e. GBM and medulloblastoma, ependymoma and neurocytoma [24–31]. In particular, it has been possible to isolate *bona fide* CSCs from GBMs, which possess both the full complement of normal stem cells functional characteristics and the ability to produce tumours which closely resemble the main histological, cytological and architectural features of the GBM, even when challenged through serial transplantation [27] (Fig. 18.1). These long-term expanding CSCs represent an interesting source to reproduce human GBM-like systems in culture and in the mouse brain.

Similarities between CSCs in GBM and NSCs stretch to the organisational hierarchy, the proliferative potential and the pathways which are up-regulated. Both CSCs and NSCs are able to proliferate extensively in the presence of growth factors (EGF and FGF) and differentiate into neurons, astrocytes and oligodendrocytes upon the withdrawal of growth factors and the addition of serum [27]. Furthermore, microarray data suggests that CSCs have a transcriptional state more closely related to foetal neural stem cells than to adult brain tissue [32]. The expression of

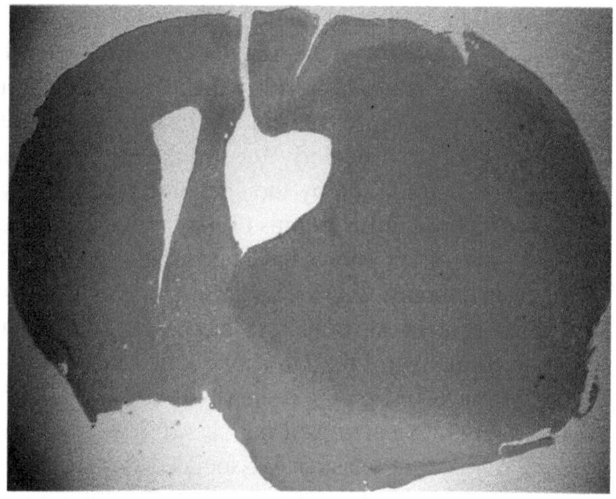

Fig.18.1 Haematoxylin and Eosin staining of a mouse brain (Nod/Scid animal) injected with 3×10^5 CSCs from human GBM. After 6 weeks, the tumour mass appears as a phenocopy of the original GBM and has infiltrated the right hemisphere and some cells are migrating to the contralateral hemisphere of the mouse brain (×16 objective)

telomerase, an enzyme which is able to maintain telomere length in spite of divisions, is a defining feature of stem cells and is also expressed in the majority of GBM [33]. Many of the stem cell pathways which are associated with normal stem cell development are associated with cancers such as Mbi-1, Notch, SHH and Wnt [1, 34]. Key regulatory factors like bone morphogenetic proteins (BMPs) regulate the fate of NSCs and CSCs [35].

Further evidence of a stem cell driven hierarchical model for GBM comes from the expression of markers and transcription factors also expressed in neural stem cells and in neural development. Sox-2 is one of the earliest known transcription factors expressed in the developing neural tube and is shown to be responsible for neural induction of the ectoderm. During embryonic development, self-renewing and multipotent neuroepithelial stem cells express Sox-2 [36]. In adult rats Sox-2 was found in the neurogenic regions both in the subventricular zone and sub granular zone of the dentate gyrus [37]. Sox-2 has been demonstrated to be expressed at higher levels in astrocytomas of increasing grades [38] and in CSCs and its targeting results in loss of tumorigenicity [39]. Nestin is a cytoplasmic intermediate filament protein, expressed by neural stem cell and their progenitors. It was originally known as Rat 401 when found to be expressed in the proliferative zone of the neural tube which is enriched in stem cells and was first cloned in 1990 [40]. It is also expressed in glioblastomas at increasing levels with increasing grade [38]. GFAP (glial fibrillary acidic protein), first described in 1971 [41], is a member of the cytoskeletal protein family and is widely expressed in astrocytes and in neural stem cells [42], this marker is also prevalent in astrocytomas.

There is also circumstantial evidence to suggest that GBM growth is hierarchical organized. Mutations may be more likely to accumulate in a slowly turning over population and stem cells may be resistant to therapy hence recurrence after treatment. It is thought that the transformation of a normal cell to a cancer cell is the result of a series of mutations to a defined number of pathways [43] and it is hypothesised that these mutations would be unlikely to occur in restricted progenitors and differentiated cells due to their short half life but would be far more likely to occur in a slowly turning over population of cells, which have the machinery for self-renewal already activated making the stem cell population a likely target for mutations that go on to cause cancer [1]. The cancer stem cell hypothesis is supported by clinical findings, particularly recurrence after chemotherapy. It is suggested that stem cells are not killed by the chemotherapy as they are slow proliferating, they may express drug efflux pumps or may not express the target protein of the chemotherapy. Drug efflux pumps have been shown to be present in glial tumours sections of untreated tumours. Both the lung resistance-related protein and multi drug resistance-associated protein were up regulated. The drug efflux pump P-glycoprotein was upregulated in 90% of the 27 primary glioblastoma cases investigated. However this study did not investigate the distribution of these proteins with reference to cells expressing putative stem cell markers. P-glycoprotein, lung resistance-related protein and multi drug resistance-associated protein were diffusely scattered throughout the tumour [44, 45]. The distribution of these efflux pumps does not suggest increased or sole expression on CSCs in GBM. It is important to be aware that germ cell cancers such as testicular cancer and chronic myelogenous leukaemia have a stem cell compartment yet do respond to chemotherapy [46] which weakens this argument for the cancer stem cells causing drug resistance.

Despite statements on the presence of CSCs in brain tumours and GBMs in many reviews [6, 46] there is not yet any definitive proof that they derive directly from NSCs. In support of this view, some substantial differences between CSCs and NSCs have been highlighted: in many cases the marker expression pattern of CSCs in GBM does not match the pattern of NSCs [27, 47], their propagation rate *in vitro* and *in vivo* is higher [33], they show a partial mitogen independence [48] and their tumorigenicity does not correlate with expression of specific markers [49, 50].

18.4 Identification of Glioblastoma Stem-Like Cells

In 2004, an intriguing study by Singh et al. [12] suggested that the CD133-positive (CD133+) cell fraction from GBM contained cells which were able to form new tumours after serial transplantation and, therefore, identified the CD133+ population as the one containing *bona fide* CSCs, whereas the CD133– cell pool did not. This antigen, that is a 120 kDa five transmembrane domain glycoprotein with unknown function, is expressed on primitive cell populations, i.e. CD34 bright hematopoietic stem and progenitor cells, neural [51] and endothelial stem cells, and other primitive cells, such as retina, retinoblastoma and developing epithelium [50]. Later, following the initial discovery of a subpopulation of CD133+ GBM cells with stem cell properties, the use of this antigen in other tumour types has been extensively investigated. In particular, *in vitro* proliferation assays and *in vivo* tumour-initiating experiments, carried out starting from both primary tumours and cancer cell lines with stem cell properties, provided evidence for the existence of CD133+ cells in ependymoma [30], prostate cancer [52], colon cancer [10, 11, 53] lung cancer [54], hepatocellular carcinoma [55–58], laryngeal carcinoma [59], melanoma [60], ovarian cancer [61] and pancreatic cancer [62, 63]. This notion is now being challenged, however. In fact, a growing number of reports demonstrated that the CD133+ cell fraction from freshly dissociated GBM tumours could be considered different from the negative one in terms of angiogenic and radiotherapy-resistance properties, but not always on the basis of tumour-initiating ability [64, 65], thereby indicating that distinct populations of CD133+ and – cells may contain cancer stem cells [14, 66–70]. Since then, however, conflicting reports hypothesized that CD133+ tumour vessels represent a perivascular niche for brain tumour stem cells [71] and identified CD133 marker as prognostic factor for adverse progression-free survival in high-grade glioma [72, 73] and oligodendroglial tumours [74].

This situation underlines the need to broaden and refine the studies in this area and, in particular, to search for more selective CSCs markers, that will lead to the unequivocal identification of these cells and of their molecular regulators. In this view, recent *in vitro* and *in vivo* findings have began to identify a series of specific proteins as key regulators of tumour and normal stem cells [35, 75, 76] and more recently, other cell surface markers have been proposed to enrich for stem-like cells in GBM, i.e. CD15/SSEA1, L1CAM, A2B5 and integrinalpha6 [13–17] although there is no general consensus on any of them. CSCs from brain tumours can be also enriched in the side population by dye exclusion [77, 78]. An alternative strategy that has been recently proposed by Clement and Deleyrolle is to evaluate autofluorescence or label-retaining properties of a quiescent population in GBM to identify with a functional assay the CSCs pool [79, 80]. This possibility is very intriguing although marker-driven targeting of CSCs is not possible.

More importantly, before a definitive identification of CSCs based on markers or functional properties will be possible, it is critical to demonstrate whether the CSC model mutually excludes the stochastic model in GBM or if the two models can be converted into each other during tumour evolution, as recently demonstrated for human leukemia [81, 82]. In this view, it will be also important to take into account that the concept of a rare CSC population has been recently challenged in melanoma [83, 84] and that the obsessive search for a definitive marker for GBM could be misleading.

18.5 Therapeutic Implications: Strategies and Challenges

The cancer stem cell model suggests that only a minority of cells is responsible for tumour growth and recurrence. The isolation of CSCs from the human brain has made possible the development of *in vivo* assay that can be used to assess the therapeutic efficacy of a drug. For the first time, by using CSCs from human GBM phenocopies of the original pathology of the patient can be retrieved in a mouse brain and this represents an unprecedented model to establish multiple copies of the same tumours starting from a limited amount of cells and to target these cells *in vivo*. The self-renewal ability of these cells also make possible to study them *ex vivo*, since it is possible to re-establish mouse models of the disease by serial transplantation.

First cancer stem cells targeting therapies for GBM have been developed in the last years and it has been demonstrated that they are successful in slowing down the disease in mouse models. In particular, it has been proposed that it is possible to inhibit the tumorigenic potential of CSCs by forcing differentiation with bone morphogenetic proteins (BMPs) [35] or by counteracting the pathway sustaining self-renewal (SHH, FGF and Notch) [85–87]. These approaches can be extremely helpful in targeting CSCs since conventional therapies kill only fast-cycling cells. Very recently, it has been proposed that actually CSCs are slow-cycling cells [80] supporting the idea to define new therapeutic strategies to complement conventional treatments. It will be extremely important to define CSC therapies which might also enhance radio- and chemo-sensitivity. In this view it has been recently demonstrated that in CSCs Notch promotes the survival to genotoxic stress, i.e. radiation [88]. An alternative therapeutic strategy to target CSCs come from the observation that in GBM they reside in proximity of vessels [89] and promote tumour growth by secreting VEGF [64]. By using the anti-VEGF antibody Avastin it has been demonstrated that tumour growth is inhibited in xenograft models [64]. This finding has been recently complemented by two studies demonstrating that CSCs can be converted both *in vitro* and *in vivo* in endothelial cells and therefore strategies aimed at targeting the tumour vasculatures should take into account the ability of CSCs to reconstitute vessels [90, 91]. Thus, the combination of novel, CSC-specific and conventional therapies may, in the future prove a formidable gun against incurable tumours.

Although the research on CSCs has the potential to revolutionize the way we look at and treat cancers, all the information about the identity and the function of these cells are still missing and therefore basic cellular and molecular studies are required to truly understand which is the origin of these cells and how we can exploit them to develop new therapeutic approaches for the benefit of cancer patients. In this perspective, it is critical to understand how complex are these cells at the stage of tumour resection from the brain of patients diagnosed with GBM: are they disseminated into the tumours? Does more than one population of CSCs exist in human GBM? Recently, some progress has been made in this direction. By combining a biological and genetic approach on tumour tissues and their corresponding CSCs it has been possible to demonstrate that the periphery of the tumour contains aberrant cells which partially share stem cells features but are less tumorigenic in comparison with CSCs originated from the tumour mass [92]. Whether these peripheral cells are responsible for the growth of recurrences in GBM patients, whether they can fulfil all the stem cell criteria when the recurrences occur and, most importantly, whether they represent a target for therapy remain to be determined.

18.6 Conclusions

The concept of CSCs in brain tumours provides a likely explanation for the failure of current therapies (surgery, radio- and chemotherapy), which remove the bulk of tumour tissue, but not the tumour stem cells, which are regarded as the root of tumour growth and seem to be particularly resistant to radio- and chemotherapy [65, 93]. Only very recently, experimental evidences for the concept of a hierarchical organisation of brain tumours with the tumour stem cells on top of the hierarchy was provided, which now opens up novel and promising alternative research and therapy avenues [94].

The failure of many molecule-targeted therapies developed in the field of oncology can likely be explained by the fact that they did not target the right cells (tumour stem cells) and that the targets were not selected based on a system understanding of tumour cells [49]. Several cellular features should be considered in the target discovery process, which can be summarized as the robust system of a tumour stem cell: its ability to maintain a stable state despite perturbations. To disrupt the glioma stem cell state, it is necessary (1) to understand their cellular state, which allows them to indefinitely self-renew and to escape apoptosis and senescence and (2) to achieve a basic understanding of all system components and interactions. This knowledge will allow to predict and then test perturbations that will disrupt the system. Investigations of such cell-intrinsic properties can only now be initiated, since strategies to isolate and functionally test glioma stem cells were unknown so far. From a therapeutic perspective, an ideal single drug or drug combination should specifically target components of central pathways that maintain the tumour stem cell state. Brain tumour stem cells share certain features with normal neural stem cell in the mature brain and might even be derived from them [95, 96]. Thus, key to developing a specific and curative therapy for glioma is to identify drug targets, which (a) mainly target glioma stem cells and (b) do not or only minimally affect normal neural stem cells and reduce long-term neurological side effects [97]. Strong support for such an approach is given by a study on hematopoietic and leukemic stem cells [98]. These authors demonstrated that it is possible to identify and to therapeutically target pathways that have different effects on normal and cancer stem cells within the same tissue.

What remains unclear, however, is which pathway components would need to be targeted to effectively block the described cell-intrinsic properties of these cells. At the moment, in fact, basic systems biological knowledge about these particular pathways and their interaction in brain tumour stem cells is missing: the overall pathway architecture including hierarchy structures, the exact nature of component interactions (in particular self-regulatory components) and the robustness of system sections. A systems understanding of these relevant pathways should allow predictions about the outcome of interfering at the level of single components, and would in this way significantly contribute to the selection of relevant molecular drug targets against CSCs in GBM [97]. The same pathways should ideally be investigated in parallel in NSCs to find out functional differences allowing for the selection of drug targets, which specifically affect the maintenance of tumour stem cells, but not normal neural stem cells in the mature brain.

In conclusion, the discovery that brain cancers contain CSCs deeply influenced the field of Neuro-oncology. As for all the newborn fields of investigation, we have only begun to understand how it will be possible to definitely interrogate these cells and develop new therapeutic approaches.

It is a common believe that this is an exciting time in cancer research and that the cancer stem cell field will reserve us a lot of new challenges and will surely impact our understanding of tumour biology. The challenge will be to identify ways of rapidly identifying and interrogating the stem cell population to uncover specific targets and screen libraries of small molecular therapeutic agents and collate this data into a tailor-made treatment for each patient.

References

1. Reya T et al (2001) Stem cells, cancer, and cancer stem cells. Nature 414(6859):105–111
2. Piccirillo SG et al (2009) Brain cancer stem cells. J Mol Med 87(11):1087–1095
3. Stupp R et al (2005) Radiotherapy plus concomitant and adjuvant temozolomide for glioblastoma. N Engl J Med 352(10):987–996
4. Gritti A, Vescovi AL, Galli R (2002) Adult neural stem cells: plasticity and developmental potential. J Physiol Paris 96(1–2):81–90
5. Lie DC et al (2004) Neurogenesis in the adult brain: new strategies for central nervous system diseases. Annu Rev Pharmacol Toxicol 44:399–421
6. Huntly BJ, Gilliland DG (2005) Leukaemia stem cells and the evolution of cancer-stem-cell research. Nat Rev Cancer 5(4):311–321
7. Hamburger AW, Salmon SE (1977) Primary bioassay of human tumor stem cells. Science 197(4302):461–463
8. Bonnet D, Dick JE (1997) Human acute myeloid leukemia is organized as a hierarchy that originates from a primitive hematopoietic cell. Nat Med 3(7):730–737
9. Al-Hajj M et al (2003) Prospective identification of tumorigenic breast cancer cells. Proc Natl Acad Sci USA 100(7):3983–3988
10. O'Brien CA et al (2007) A human colon cancer cell capable of initiating tumour growth in immunodeficient mice. Nature 445(7123):106–110
11. Ricci-Vitiani L et al (2007) Identification and expansion of human colon-cancer-initiating cells. Nature 445(7123):111–115
12. Singh SK et al (2004) Cancer stem cells in nervous system tumors. Oncogene 23(43):7267–7273
13. Bao S et al (2008) Targeting cancer stem cells through L1CAM suppresses glioma growth. Cancer Res 68(15):6043–6048
14. Ogden AT et al (2008) Identification of A2B5+CD133- tumor-initiating cells in adult human gliomas. Neurosurgery 62(2):505–514, discussion 514-5
15. Son MJ et al (2009) SSEA-1 is an enrichment marker for tumor-initiating cells in human glioblastoma. Cell Stem Cell 4(5):440–452
16. Anido J et al (2010) TGF-beta receptor inhibitors target the CD44(high)/Id1(high) glioma-initiating cell population in human glioblastoma. Cancer Cell 18(6):655–668
17. Lathia JD et al (2010) Integrin alpha 6 regulates glioblastoma stem cells. Cell Stem Cell 6(5):421–432
18. Burnet NG et al (2005) Years of life lost (YLL) from cancer is an important measure of population burden–and should be considered when allocating research funds. Br J Cancer 92(2):241–245
19. Kesari S et al (2008) Phase II study of temozolomide, thalidomide, and celecoxib for newly diagnosed glioblastoma in adults. Neuro Oncol 10(3):300–308
20. Walker MD et al (1978) Evaluation of BCNU and/or radiotherapy in the treatment of anaplastic gliomas. A cooperative clinical trial. J Neurosurg 49(3):333–343
21. Hegi ME et al (2005) MGMT gene silencing and benefit from temozolomide in glioblastoma. N Engl J Med 352(10):997–1003

22. Ding H et al (2001) Astrocyte-specific expression of activated p21-ras results in malignant astrocytoma formation in a transgenic mouse model of human gliomas. Cancer Res 61(9):3826–3836
23. Bachoo RM et al (2002) Epidermal growth factor receptor and Ink4a/Arf: convergent mechanisms governing terminal differentiation and transformation along the neural stem cell to astrocyte axis. Cancer Cell 1(3):269–277
24. Ignatova TN et al (2002) Human cortical glial tumors contain neural stem-like cells expressing astroglial and neuronal markers in vitro. Glia 39(3):193–206
25. Hemmati HD et al (2003) Cancerous stem cells can arise from pediatric brain tumors. Proc Natl Acad Sci USA 100(25):15178–15183
26. Singh SK et al (2003) Identification of a cancer stem cell in human brain tumors. Cancer Res 63(18):5821–5828
27. Galli R et al (2004) Isolation and characterization of tumorigenic, stem-like neural precursors from human glioblastoma. Cancer Res 64(19):7011–7021
28. Singh SK et al (2004) Identification of human brain tumour initiating cells. Nature 432(7015):396–401
29. Yuan X et al (2004) Isolation of cancer stem cells from adult glioblastoma multiforme. Oncogene 23(58):9392–9400
30. Taylor MD et al (2005) Radial glia cells are candidate stem cells of ependymoma. Cancer Cell 8(4):323–335
31. Sim FJ et al (2006) Neurocytoma is a tumor of adult neuronal progenitor cells. J Neurosci 26(48):12544–12555
32. Pollard SM et al (2009) Glioma stem cell lines expanded in adherent culture have tumor-specific phenotypes and are suitable for chemical and genetic screens. Cell Stem Cell 4(6):568–580
33. Varghese M et al (2008) A comparison between stem cells from the adult human brain and from brain tumors. Neurosurgery 63(6):1022–1033, discussion 1033-4
34. Pardal R, Clarke MF, Morrison SJ (2003) Applying the principles of stem-cell biology to cancer. Nat Rev Cancer 3(12):895–902
35. Piccirillo SG et al (2006) Bone morphogenetic proteins inhibit the tumorigenic potential of human brain tumour-initiating cells. Nature 444(7120):761–765
36. Ellis P et al (2004) SOX2, a persistent marker for multipotential neural stem cells derived from embryonic stem cells, the embryo or the adult. Dev Neurosci 26(2–4):148–165
37. Komitova M, Eriksson PS (2004) Sox-2 is expressed by neural progenitors and astroglia in the adult rat brain. Neurosci Lett 369(1):24–27
38. Ma YH et al (2008) Expression of stem cell markers in human astrocytomas of different WHO grades. J Neurooncol 86(1):31–45
39. Gangemi RM et al (2009) SOX2 silencing in glioblastoma tumor-initiating cells causes stop of proliferation and loss of tumorigenicity. Stem Cells 27(1):40–48
40. Lendahl U, Zimmerman LB, McKay RD (1990) CNS stem cells express a new class of intermediate filament protein. Cell 60(4):585–595
41. Eng LF et al (1971) An acidic protein isolated from fibrous astrocytes. Brain Res 28(2):351–354
42. Doetsch F et al (1999) Subventricular zone astrocytes are neural stem cells in the adult mammalian brain. Cell 97(6):703–716
43. Hanahan D, Weinberg RA (2000) The hallmarks of cancer. Cell 100(1):57–70
44. Tews DS et al (2000) Drug resistance-associated factors in primary and secondary glioblastomas and their precursor tumors. J Neurooncol 50(3):227–237
45. Dean M, Fojo T, Bates S (2005) Tumour stem cells and drug resistance. Nat Rev Cancer 5(4):275–284
46. Stiles CD, Rowitch DH (2008) Glioma stem cells: a midterm exam. Neuron 58(6):832–846
47. Zhang QB et al (2006) Differentiation profile of brain tumor stem cells: a comparative study with neural stem cells. Cell Res 16(12):909–915
48. Kelly JJ et al (2009) Proliferation of human glioblastoma stem cells occurs independently of exogenous mitogens. Stem Cells 27(8):1722–1733

49. Piccirillo SG, Vescovi AL (2007) Brain tumour stem cells: possibilities of new therapeutic strategies. Expert Opin Biol Ther 7(8):1129–1135
50. Bidlingmaier S, Zhu X, Liu B (2008) The utility and limitations of glycosylated human CD133 epitopes in defining cancer stem cells. J Mol Med 86(9):1025–1032
51. Uchida N et al (2000) Direct isolation of human central nervous system stem cells. Proc Natl Acad Sci USA 97(26):14720–14725
52. Collins AT et al (2005) Prospective identification of tumorigenic prostate cancer stem cells. Cancer Res 65(23):10946–10951
53. Ieta K et al (2008) Biological and genetic characteristics of tumor-initiating cells in colon cancer. Ann Surg Oncol 15(2):638–648
54. Eramo A et al (2008) Identification and expansion of the tumorigenic lung cancer stem cell population. Cell Death Differ 15(3):504–514
55. Suetsugu A et al (2006) Characterization of CD133+ hepatocellular carcinoma cells as cancer stem/progenitor cells. Biochem Biophys Res Commun 351(4):820–824
56. Ma S et al (2007) Identification and characterization of tumorigenic liver cancer stem/progenitor cells. Gastroenterology 132(7):2542–2556
57. Ma S et al (2008) CD133+ HCC cancer stem cells confer chemoresistance by preferential expression of the Akt/PKB survival pathway. Oncogene 27(12):1749–1758
58. Yin S et al (2007) CD133 positive hepatocellular carcinoma cells possess high capacity for tumorigenicity. Int J Cancer 120(7):1444–1450
59. Zhou L et al (2007) CD133, one of the markers of cancer stem cells in Hep-2 cell line. Laryngoscope 117(3):455–460
60. Monzani E et al (2007) Melanoma contains CD133 and ABCG2 positive cells with enhanced tumourigenic potential. Eur J Cancer 43(5):935–946
61. Ferrandina G et al (2008) Expression of CD133-1 and CD133-2 in ovarian cancer. Int J Gynecol Cancer 18(3):506–514
62. Hermann PC et al (2007) Distinct populations of cancer stem cells determine tumor growth and metastatic activity in human pancreatic cancer. Cell Stem Cell 1(3):313–323
63. Olempska M et al (2007) Detection of tumor stem cell markers in pancreatic carcinoma cell lines. Hepatobiliary Pancreat Dis Int 6(1):92–97
64. Bao S et al (2006) Stem cell-like glioma cells promote tumor angiogenesis through vascular endothelial growth factor. Cancer Res 66(16):7843–7848
65. Bao S et al (2006) Glioma stem cells promote radioresistance by preferential activation of the DNA damage response. Nature 444(7120):756–760
66. Beier D et al (2007) CD133(+) and CD133(-) glioblastoma-derived cancer stem cells show differential growth characteristics and molecular profiles. Cancer Res 67(9):4010–4015
67. Wang J et al (2008) CD133 negative glioma cells form tumors in nude rats and give rise to CD133 positive cells. Int J Cancer 122(4):761–768
68. Gunther HS et al (2008) Glioblastoma-derived stem cell-enriched cultures form distinct subgroups according to molecular and phenotypic criteria. Oncogene 27(20):2897–2909
69. Joo KM et al (2008) Clinical and biological implications of CD133-positive and CD133-negative cells in glioblastomas. Lab Invest 88(8):808–815
70. Yi JM et al (2008) Abnormal DNA methylation of CD133 in colorectal and glioblastoma tumors. Cancer Res 68(19):8094–8103
71. Christensen K, Schroder HD, Kristensen BW (2008) CD133 identifies perivascular niches in grade II-IV astrocytomas. J Neurooncol 90(2):157–170
72. Zeppernick F et al (2008) Stem cell marker CD133 affects clinical outcome in glioma patients. Clin Cancer Res 14(1):123–129
73. Rebetz J et al (2008) Glial progenitor-like phenotype in low-grade glioma and enhanced CD133-expression and neuronal lineage differentiation potential in high-grade glioma. PLoS One 3(4):e1936
74. Beier D et al (2008) CD133 expression and cancer stem cells predict prognosis in high-grade oligodendroglial tumors. Brain Pathol 18(3):370–377

75. Jackson EL et al (2006) PDGFR alpha-positive B cells are neural stem cells in the adult SVZ that form glioma-like growths in response to increased PDGF signaling. Neuron 51(2):187–199
76. Ligon KL et al (2007) Olig2-regulated lineage-restricted pathway controls replication competence in neural stem cells and malignant glioma. Neuron 53(4):503–517
77. Harris MA et al (2008) Cancer stem cells are enriched in the side population cells in a mouse model of glioma. Cancer Res 68(24):10051–10059
78. Bleau AM, Huse JT, Holland EC (2009) The ABCG2 resistance network of glioblastoma. Cell Cycle 8(18):2936–2944
79. Clement V et al (2010) Marker-independent identification of glioma-initiating cells. Nat Methods 7(3):224–228
80. Deleyrolle LP et al (2011) Evidence for label-retaining tumour-initiating cells in human glioblastoma. Brain 134(Pt 5):1331–1343
81. Notta F et al (2011) Evolution of human BCR-ABL1 lymphoblastic leukaemia-initiating cells. Nature 469(7330):362–367
82. Anderson K et al (2011) Genetic variegation of clonal architecture and propagating cells in leukaemia. Nature 469(7330):356–361
83. Quintana E et al (2008) Efficient tumour formation by single human melanoma cells. Nature 456(7222):593–598
84. Quintana E et al (2010) Phenotypic heterogeneity among tumorigenic melanoma cells from patients that is reversible and not hierarchically organized. Cancer Cell 18(5):510–523
85. Bar EE et al (2007) Cyclopamine-mediated hedgehog pathway inhibition depletes stem-like cancer cells in glioblastoma. Stem Cells 25(10):2524–2533
86. Loilome W et al (2009) Glioblastoma cell growth is suppressed by disruption of fibroblast growth factor pathway signaling. J Neurooncol 94(3):359–366
87. Fan X et al (2010) NOTCH pathway blockade depletes CD133-positive glioblastoma cells and inhibits growth of tumor neurospheres and xenografts. Stem Cells 28(1):5–16
88. Wang J et al (2010) Notch promotes radioresistance of glioma stem cells. Stem Cells 28(1):17–28
89. Calabrese C et al (2007) A perivascular niche for brain tumor stem cells. Cancer Cell 11(1):69–82
90. Ricci-Vitiani L et al (2010) Tumour vascularization via endothelial differentiation of glioblastoma stem-like cells. Nature 468(7325):824–828
91. Wang R et al (2010) Glioblastoma stem-like cells give rise to tumour endothelium. Nature 468(7325):829–833
92. Piccirillo SG et al (2009) Distinct pools of cancer stem-like cells coexist within human glioblastomas and display different tumorigenicity and independent genomic evolution. Oncogene 28(15):1807–1811
93. Liu G et al (2006) Analysis of gene expression and chemoresistance of CD133+ cancer stem cells in glioblastoma. Mol Cancer 5:67
94. Chen R et al (2010) A hierarchy of self-renewing tumor-initiating cell types in glioblastoma. Cancer Cell 17(4):362–375
95. Sanai N, Alvarez-Buylla A, Berger MS (2005) Neural stem cells and the origin of gliomas. N Engl J Med 353(8):811–822
96. Vescovi AL, Galli R, Reynolds BA (2006) Brain tumour stem cells. Nat Rev Cancer 6(6):425–436
97. Nicolis SK (2007) Cancer stem cells and "stemness" genes in neuro-oncology. Neurobiol Dis 25(2):217–229
98. Yilmaz OH et al (2006) Pten dependence distinguishes haematopoietic stem cells from leukaemia-initiating cells. Nature 441(7092):475–482

Chapter 19
Translating Mammary Stem Cell and Cancer Stem Cell Biology to the Clinics

Rajneesh Pathania, Vadivel Ganapathy, and Muthusamy Thangaraju

Contents

Abstract Breast cancer, one of the most deadly diseases in women, is a hierarchical entity comprising heterogeneous populations of cells with genetic or epigenetic alterations that allow them to grow as a tumor and subsequently cause metastasis. Since past 70 years, several classes of chemotherapeutic agents have been developed

R. Pathania • V. Ganapathy
Department of Biochemistry and Molecular Biology, Georgia Health Sciences University, Augusta, GA 30912, USA
e-mail: rpathania@georgiahealth.edu

M. Thangaraju, Ph.D. (✉)
Department of Biochemistry and Molecular Biology, Medical College Georgia, Georgia Health Sciences University, Augusta, GA 30912, USA
e-mail: mthangaraju@georgiahealth.edu

R.K. Srivastava and S. Shankar (eds.), *Stem Cells and Human Diseases*,
DOI 10.1007/978-94-007-2801-1_19, © Springer Science+Business Media B.V. 2012

which are used widely for treatment of breast cancer, and yet the breast cancer has not been eradicated. In the past two decades, stem cells have become the holy grail of biomedical research because the biology of these cells has potential to contribute to a better understanding of the molecular basis of not only cancer but also several other diseases as well as to foster new avenues for the design and development of novel classes of drugs for the treatment of these diseases. Further, stem cells can be used as a vector for gene therapy to treat diseases like cancer because stem cells can migrate relatively long distances, not only to the sites of injury and infection, but also to initial sites of tumor. Identification of mammary stem cells and cancer stem cells raises new hopes for the treatment of breast cancer. Previous studies have shown that cancer stem cells have similar property to the normal stem cells, but have the characteristic feature of increased self-renewal compared to normal stem cells. Stem cells also play an important role in carcinogenesis; thus understanding the role of stem cells in malignant transformation will have far-reaching implications in our understanding of the molecular mechanisms of cancer as well as for the discovery of new treatment modalities to completely obliterate several human malignancies.

Keywords Stem cells • Cancer stem cells • Self-renewal • Drug resistance

19.1 Introduction

Stem cells have potential to repair damaged tissues and hence in the treatment of degenerative diseases. Because of their tremendous therapeutic potential, stem cells have caught the attention of both basic and clinical cancer research communities [38, 62]. These cells are undifferentiated, long-lived, quiescent, and have limited ability for both self-renewal and differentiation [50, 54, 99]. The rate of their self-renewal varies during their lifetime and is also dependent on their tissue origin [2, 4, 73, 91, 94, 101]. The self-renewal of mammary gland stem cells increases dramatically during mid-pregnancy [4]. In recent years, impressive technical advancements have been made in the isolation and characterization of the mammary gland stem cells and cancer stem cells; however, the signaling pathways that regulate their self-renewal, proliferation, survival and differentiation remain almost unknown [4, 9, 98, 107, 112]. Cancer stem cells are resistant to currently available chemotherapeutic agents, radiation, and reactive oxygen species [1, 30]. Identification of specific drugs that have the ability to kill cancer stem cells is a difficult process because of the scarcity of these cells within tumors and their relative instability under culture conditions [44]. Studies aimed at the assessment of the efficacy of specific agents to kill cancer stem cells involve isolation of pure populations of these cells using cell sorters, determination of their number, in vitro techniques to monitor their ability to form tumorospheres, and in vivo approaches such as mouse xenograft assays to monitor their growth in live animals [23]. There is increasing evidence indicating that stem cells, due to their enhanced longevity, may place themselves at increased risk for mutations, consequently transforming into cancer stem cells [17, 66, 94, 101, 102]. The field of stem cell biology might shed light on the mechanisms of the pathogenesis of breast cancer and development of resistance to chemotherapy, thereby aiding in the development of better and more effective therapeutic strategies for the treatment of breast cancer [112].

Mammary tumors are composed of heterogeneous cells. Two models have been proposed to explain this heterogeneity: stochastic model and cancer stem cell model [28, 29]. The stochastic model is a non-hierarchical model that postulates that all cells are biologically similar and that they have equal potential to acquire mutations and initiate tumor [28, 29, 105]. In contrast, the hierarchical model, upon which the cancer stem cell (CSC) hypothesis is based, postulates that stem cells are heterogeneous with differing functional abilities and that only a few cells possess the ability of self-renewal and tumor initiation and can give rise to both tumorigenic and non-tumorigenic cell populations [28, 29, 113]. Recent studies show that tumor-initiating cells can be identified and purified from tumor tissues based on specific cell-surface markers and/or cell proliferation capabilities [3, 28, 86]. When tumor-initiating cells are xenografted in limited serial dilutions in immuno-compromised mice, these cells are able to produce tumors. In contrast, other cells do not have this ability to form tumors in xenograft mouse models. It is currently believed that the heterogeneous nature and dissimilar molecular characteristics of different types of solid tumors, including breast cancer, arise because of the differences in the cancer stem cell content of these tumors [28, 86].

Mammary tumors are heterogeneous (Luminal A, Luminal B, ErBB2, Basal and Normal-like), in terms of histopathology, gene expression, and molecular profiles; hence it is possible that tumors originate from different cell types, either from long-term stem cells, short-term stem cell, progenitor cells or differentiated cells [106]. Human breast tumors have been historically categorized into 18 subtypes according to histological subtypes and five subtypes according to microarray gene expression profiling [103, 104, 106].

19.2 Markers for Stem/Progenitor and Cancer Stem Cells

A variety of cell surfacc markers, aldehyde dehydrogenase activity, and PKH dye can be used to identify and purify stem/progenitor cells and cancer stem cells.

19.2.1 Cell-Surface Markers

Cancer stem cells or tumor-initiating cells were isolated from mammary tumor tissues by Al-Hajj and colleagues [3]. These investigators identified a subpopulation of cells in breast cancer that has tumor-initiating capacity. Their studies showed that as few as 100 cells carrying the $CD44^{+}CD24^{-/low}$ $Lineage^{-}$ phenotype, isolated through flow cytometry, were able to form tumors in NOD/SCID mice whereas cells with other phenotypes failed to form tumors. In addition, the tumors formed by these cancer stem cells were heterogeneous and possessed the capability of self-renewal [3]. This technical feasibility to isolate and purify tumorigenic and non-tumorigenic populations of cells based on specific cell-surface markers allows characterization of cancer stem cells at the molecular level and elucidation of the signaling pathways that are responsible for the abnormal self-renewal of cancer stem cells [3].

After successful isolation of cancer stem cells from mammary tumors, the central question was whether normal stem cells and progenitor cells exist in normal mammary gland. This question was answered within 3 years of the identification of cancer stem cells with successful isolation of normal stem cells/progenitor cells. These cells were isolated by mechanical and enzymatic dissociation of normal mouse mammary glands and subsequent removal of lineage positive populations such as the cells that are positive for Ter119, CD45 and CD31 (markers for red blood cells and endothelial cells). Stem cells sorted with the use of multiparameter cell sorting, labeled with $CD29^{hi}$ $CD24^{+}$ or $CD49f^{hi}$ $CD24^{+}$, were characterized with limiting dilution transplantation assay by two different groups [98, 107]. It was amazing that a single stem cell, when transplanted into mouse mammary fat pad, was able to regenerate the whole mammary gland [98, 107]. The identification of these normal stem cells is an important discovery because of their purported role in breast tumorigenesis [98]. These stem cells represent a rapidly cycling population of cells in the normal adult mammary gland. They exist mainly in G_0 and G_1 phase, and give rise to mammary progenitor cells (Lin^{-} $CD49f^{+}$ CD 24^{+}) that produce adherent colonies *in vitro*. The mammary stem cells and progenitor cells have molecular features indicative of a basal position and luminal position in the mammary epithelium [107]. Subsequent studies have shown that various cell-surface markers can be used for the isolation of these stem cells. These markers include $CD29^{hi}$, $CD49f^{hi}$, EpCAM, CD44 and CD24 [87]. Studies by Meyer group [75] show that, in human breast cancer, $CD44^{+}CD24^{+}$, $CD44^{+}CD24^{-}$ cells have tumor-initiating capacity. In addition to these markers, they have found that $CD44^{+}CD49f^{hi}CD133/2^{hi}$ subtypes have more tumorigenic capability than $CD44^{+}CD24^{+}$, $CD44^{+}CD24^{-}$ subtypes [75]. Interestingly, CD44v6 has also been reported to be involved in the genesis of mammary carcinomas [47]. But, this cell-surface epitope is also expressed in myoepithelial cells, which first appear during puberty and estrous cycle, decrease during lactation, and again increase during involution. It is also expressed in mammary normal epithelial cells. The expression of CD44v6 in normal mammary epithelial cells as well as cancer provides evidence for the stem cell origin of breast cancer [47].

19.2.2 ALDH 1

Aldehyde dehydrogenases (ALDH), consisting of several isoforms and mainly expressed in liver, are involved in detoxification of drugs and other xenobiotics as well as in oxidation of endogenous aldehydes [6, 19]. They are also involved in the conversion of retinol to retinoic acid, and function as a key regulator of stem cell differentiation [6, 19]. Ginestier and coworkers [39] showed increased aldehyde dehydrogenase activity (ALDH1) in both normal stem/progenitor cells and cancer stem cells in mammary gland. Their studies also showed that ALDH activity was increased in cancer stem cells and that the increased activity was associated with enhanced self-renewal and generation of heterogeneous populations of cells in the tumor [39]. Cancer stem cells that exhibit increased aldehyde dehydrogenase activity are referred to as Aldefluor^{+} cells because of the particular fluorescent technique used to identify these cells, and this Aldefluor^{+} population makes up approximately 15–20% of the population of the cells with the CD44^{+}/CD24^{-} phenotype [3, 6, 39]. Further, increased expression of ALDH1 in breast cancer tissues, detected by immunostaining, correlates with poor prognosis [39]. It has been shown in another study with African breast cancer patients that high prevalence of ALDH1 expression in cancer stem cells correlates with aggressiveness of cancer [79]. Recently, Marcato group [71] characterized the expression of all 19 ALDH isoforms in breast cancer patients as well as in breast cancer cell lines. Their studies have revealed that ALDH1A3 expression in breast cancer patients correlates significantly with tumor grade, metastasis, and cancer stage [71].

19.2.3 PKH Dye

Recently, Pece and coworkers [86] have shown that PKH26, a fluorescent dye that binds to cell membrane and segregates in daughter cells after each cell division, labels stem cells as well as cancer stem cells because of their quiescent nature. PKH26-positive cells possess all the characteristics of stem cells and cancer stem cells. Their studies also showed that poorly differentiated cancers displayed higher content of PKH26-positive cells than well-differentiated cancers [86].

Another study showed that in ErbB2 transgenic mice, a mouse mammary tumor model, the self-renewing divisions of cancer stem cells are more frequent and unlimited than their normal stem cells [20]. Loss of the tumor suppressor p53 gene in stem cells in the p53-null premalignant mammary gland leads to increased self-renewal. This study has also shown that p53 regulates polarity of cell division in stem cells and suggested that loss of p53 favors symmetric divisions of cancer stem cells, contributing to more self-renewal and tumor growth [20].

19.2.4 Fluorescent Dye Efflux

Multiple drug resistance is one of the major causes of failure of chemotherapy. ATP-binding cassette (ABC) transporters family is one of largest family of proteins (e.g., ABCG2/BCRP and ABCB1/MDR1) that play a major role in protecting stem cells and cancer stem cells from xenobiotics and anti-cancer drugs by actively removing them from the cells through efflux. Over-expression of these efflux transporters in cancer cells underlies the drug resistance [55, 96, 120]. These transporters are also expressed at high levels in stem cells. This provides a useful means to isolate stem cells. The DNA binding fluorescent dyes such as Rhodamine and Hoecht 33342 are substrates for these efflux transporters. Since the transporters are expressed at high levels in stem cells, the ability of the cells to mediate active efflux of these fluorescent dyes can be used as a marker for stem cells, aiding in their isolation/purification [42, 43, 115]. Published reports have shown that mice lacking ABC transporter such as ABCG2 have normal stem cells but are more responsive to chemotherapy [26, 97, 119, 121].

19.3 Self-renewal Pathway

Stem cells (SC) are defined by their ability to generate more SCs (by self-renewal process) and also to produce cells that differentiate. These two tasks can be achieved by a single self-renewing mitotic division called asymmetric self-renewing division, in which one progeny retains SC identity and the other (progenitor) undergoes multiple rounds of divisions before entering a post-mitotic fully differentiated state. These two cells, generated by asymmetric divisions, differ markedly in their proliferative potential. The SC remains quiescent or slowly proliferative whereas the progenitor cell divides actively. This ensures the production of large numbers of differentiated progeny while maintaining a relatively small pool of long-lived SCs [77]. However, SCs possess the ability to expand in number, as seen during development as well as in adult after tissue injuries, by a process called symmetric self-renewing division. Recent studies have demonstrated that asymmetric division functions as a mechanism of tumor suppression while symmetric division leads to proliferation, tissue growth and ultimately tumorigenesis [41]. In self-renewal, at least one of the daughter cells has the developmental potential identical to the mother cell [46]. Aberration of a few genes in the self-renewal pathway turns these cells into cancer stem cells [69]. Moreover, imbalance between proto-oncogenes, tumor suppressor genes and gatekeeper genes can enable the cells to increase the rate of self-renewal [46]. Selective targeting of self-renewal pathways may provide better outcome for breast cancer therapeutics [68]. The notch, hedgehog (HH), Her-2, and Wnt pathways are of prime interest because these pathways are altered in many types of cancers leading to more self-renewal of cancer stem cells [53].

19.3.1 Notch Signaling

Notch is a local signaling mechanism that plays an important role during self-renewal, proliferation, angiogenesis, epithelial-mesenchymal transition and differentiation [11, 92]. Studies by Bouras group [12] showed that knockdown of the Notch effector Cbf-1 leads to more self-renewal of mammary stem cells whereas constitutive activation leads to tumorigenesis. These findings support the notion that

Notch pathway plays an essential role in the maintenance of the stem cell pool by inhibiting excess proliferation [12]. This pathway is involved in the transformation of normal cells into cancer cells and also in the protection of cancer cells from drug-induced apoptosis. The Notch pathway has been implicated in several human cancers, such as leukemia, cervical cancer, lung carcinoma, and neuroblastoma [108]. Notch antibodies and a gamma secretase inhibitor block the self-renewal of cancer stem cells [34]. Treatment with tamoxifen in estrogen receptor positive breast cancer turns the Notch pathway on, leading to resistance to the anti-estrogen therapy [45]. Delta-like 4 ligand (DLL-4) is a component of the Notch pathway that plays an important role in self-renewal. Studies have shown that human DLL-4 antibody inhibits tumor growth by decreasing the expression of Notch target genes, and reduces the self-renewal and proliferation of cancer stem cells [49]. Further, it has been noted that modification of Notch signaling pathway converts the transformed phenotype into normal cell phenotype [108]. Therefore, the Notch pathway represents a good target for effective treatment of breast cancer.

19.3.2 Numb Signaling

Numb plays a critical role in the maintenance of the normal stem cell compartment and also serves as a tumor suppressor in the context of the cancer stem cell hypothesis [85]. In a recent study, it has been shown that loss of Numb expression correlates with aggressiveness of primary breast cancer and higher risk of metastasis [93].

19.3.3 Hedgehog Signaling

The Hedgehog signaling pathway also plays critical role in self-renewal of the cancer stem cells [110]. High expression of Sonic Hh and Gli2 has been observed in breast cancer [25, 78].

19.3.4 Wnt Signaling

Wnt signaling is one of major signal transduction pathways that is related to self-renewal of stem cells and induction of epithelial-mesenchymal transition [81]. Neth group [81] propose a model for induction of MT1MMPs (Membrane type1 matrix metalloproteinases) that involves a crosstalk between TGF-β and Wnt, leading to migration, invasion and metastasis of mesenchymal stem cells [81]. In addition, WNT genes are over-expressed in human breast cancer and also play a critical role in mouse mammary tumor formation [14]. Furthermore, over-expression of WNT 1 using the mammary specific promoter (MMTV) resulted in sevenfold increase in mammary stem cell population in mice [98]. In addition, knockdown of the WNT receptor LRP5 leads to reduction in proliferation rate as well dramatically diminished cancer stem cell invasiveness, emphasizing the importance of LRP5-mediated Wnt signaling [81].

19.3.5 Her-2/neu

Her2/neu is a potent oncogene and its expression at normal levels is necessary for mammary gland development [100, 117]. Amplification and/or over-expression of Her-2/neu is associated with increased aggressiveness of cancer, metastasis, and decreased survival of the patients [24]. HER2 activation signals through RAS-MAPK and PI3K/AKT/mTOR pathways that promote cell proliferation, survival and angiogenesis [24, 100]. In addition, Her-2 collaborates with the Notch signaling pathway leading to filling of the lumen of ductal carcinoma in situ with precursor cells, which evade apoptosis [90]. HER2 gene is amplified in 20–30% of breast cancer, and this amplification disrupts cell cycle regulation and p53-MDM2-ARF signaling. Over-expression of HER2 also impacts on stem cells, leading to mammary tumor formation and metastasis [57, 117].

Taken collectively, genetic and epigenetic alterations in proteins of the self-renewal pathway such as Notch, Hedgehog, Numb, Her2 and Wnt may allow stem cells to undergo uncontrolled self-renewal, resulting in the promotion of cancer stem cell formation, tumor initiation and growth, and ultimately metastasis [10, 46, 69].

19.4 Cancer Stem Cells in Metastasis

Metastatic breast cancer is a complex multi-step process involving spread of cancerous cells from the breast to distant organs and tissues, and is responsible for majority of deaths in breast cancer patients [21, 63]. Cancer stem cells have the capability to escape the primary tumor site and colonize in distant organs such as lung and bone [18]. Our current understanding of the metastatic capability of cancer stem cells is still at its infancy. Metastasis is a complex process involving multiple events [111]. Different models have been proposed in which cancer stem cells are the originators of metastatic cancer [89, 111, 114]. Cancer stem cells are multi-potent; they can give rise to endothelial cells that line the tumor vasculature and help in tumor growth and metastasis [8]. In a recent study, Liu group [67] have shown that bone marrow derived mesenchymal stem cells expand the breast cancer stem cell pool by positive feedback loop between IL-6 and CXCL7. In their studies, mesenchymal stem cells introduced at tibia traffic toward breast tumor and help in maintaining the bulk tumor and cancer stem cells pool [67].

19.5 Resistance to Breast Cancer Stem Cell Therapeutics

Breast cancer stem cells, like normal stem cells, are resistant to conventional chemotherapy; the self-renewal of cancer stem cells actually increases after chemotherapy [64]. In addition, most of the anti-cancer drugs principally target proliferating cells; therefore, the self-renewing, long-lived, unspecialized and relatively quiescent cancer stem cell population is more resistant to such drugs. The current cancer stem cell hypothesis postulates that recurrence of cancer after chemotherapy is due the proliferation of the cancer stem cells that survive the treatment with anti-cancer drugs. This concept is gaining traction in the field of cancer research with several studies reporting the aberrant increase in the number of cancer stem cells in tumors after conventional treatments [1, 82]. The observed resistance of cancer stem cells to anti-cancer drugs may be due to ABCG2, which is expressed at high levels in these cells. This transporter also plays an important role in the promotion of stem cell proliferation and the maintenance of the stem cell phenotype [32]. The epithelial-mesenchymal transition transcription factors such as snail, snug and Twist are associated with cell invasion, metastasis, increased expression of ABC transporters, and multi-drug resistance [95]. Recent studies have shown that the expression of ABCG2 is regulated microRNAs mir-519c/mir-328 [65].

Further, it has been shown that ALDH that is highly expressed in the stem cells and cancer stem cells in normal mammary gland and breast cancers, respectively, metabolizes anti-cancer drugs such as cyclophosphamide [15, 39]. Cancer stem cells are also resistant to radiation [31]. The increased resistance to radiation is a result of decreased reactive oxygen species production, followed by decreased double-strand break formation and activation of notch-1 gene leading to more number of cancer stem cell self- renewal [88]. In addition, Pajonk group [84] suggest that failure of radiation therapy is due to capability of cancer stem cells to repair DNA damage, redistribution, repopulation and reoxygenation of hypoxic area. Recent studies show that X-ray treatment induces radio-resistance in tumorospheres but not in adherent cells [116]. In conclusion, cancer stem cells are radio-resistant and have ability to survive even under unfavorable conditions.

Recent studies by Ishimoto group [51] have shown that CD44v controls the defense against reactive oxygen species by interacting with the plasma membrane transporter xCT (glutamate/cystine exchanger) and consequently stabilizing the transporter protein. This increases the entry of extracellular cystine into tumor cells, which results in increased synthesis of the antioxidant glutathione. Ablation of CD44v leads to reduction of proliferating cells, growth arrest, cell differentiation and senescence; all of these events are associated with decreased intracellular levels of glutathione, activation of p38/MAPK pathway, and upregulation of $p21^{CIP1/WAF1}$ [51].

19.6 Breast Cancer Stem Cell Therapeutics

Cancer stem cells are resistant to radio- and chemo-therapy; therefore it is important to design therapeutic strategies that selectively target these populations [23]. Many chemotherapeutic agents have been screened for their ability to target proliferating cancer cells. The primary end point assay in most of the clinical trials is on reduction in the tumor size. Unfortunately, the Response Evaluation Criteria in Solid Tumors (RECIST), which is used to judge the therapeutic efficacy on solid tumors, does not correlate with patient survival and clinical outcome [13, 23]. In addition tumors that develop after chemotherapy are resistant to multiple anti-cancer drugs resulting from alterations in enzymes and transporters that are involved in the transport and metabolism of drugs [26].

Another emerging and promising approach is selective targeting and elimination of cancer stem cells by inhibiting self-renewing pathway [58]. In addition to targeting proliferating population of cells in breast cancer, there is also a need to target the self-renewing population of cancer stem cells, which is the real culprit responsible for the initiation and progression of the tumor [83]. Identification and characterization of cancer stem cell antigens for antibody therapeutics are still in its early stage. In a recent study in glioblastoma, antibodies against integrin alpha 6 were effective in inhibiting the self-renewal of cancer stem cells and the formation of tumorospheres [61]. The major limitations of the therapeutic approach using antibodies for the treatment of cancer is that the markers for cancer stem cells often overlap with those of normal stem cells, thus seriously limiting the specificity of such antigens as selective targets for antibody-mediated killing of cancer stem cells [27]. For breast cancer, useful cancer stem cell markers are CD44, PKH dye and ALDH1, which are enriched in the tumorigenic population [3, 39, 86]. CD44 is the cell-surface extracellular matrix receptor and can participate in adhesion as well as signal transduction [72].

A recent study by Korkaya group [59] has shown that Her2 over-expression has influence on stem cells and progenitor cells, leading to tumorigenesis. Targeting of HER2 using the monoclonal antibody trastuzumab or lapatinib (epidermal growth factor inhibitor) has led to an increase in the survival rate of women with metastatic breast cancer [16, 33, 59, 70]. Lapatinib, which is in phase III trial, is a dual tyrosine kinase inhibitor of epidermal growth factor receptor (EGFR) and human-epidermal growth factor receptor type 2 (HER-2). The clinical trials involve women with HER-2-positive advanced or metastatic breast cancer [37, 82]. In trastuzumab-resistant or advanced metastatic patients, lapatinib and capecitabine are given in combination [37]. The observed decrease in the number of breast cancer stem cells and tumorospheres formation following treatment with lapatinib may be because of the selective toxicity of this drug for breast cancer stem cells [82].

Parthenolide inhibits tumorospheres proliferation through inhibition of NF-kB activity; in combination with docetaxel, it reduces metastasis and improves survival rate [82, 109, 118]. A recent study shows that combination of Cisplatin and TRAIL treatment enhances breast cancer stem cell death in vitro [22].

Recently, Merchant and coworkers [74] have proposed that the self- renewal genes such as Sonic Hedge hog may be required early in tumor formation or in the pre-malignant stage, but once tumor growth reaches a certain stage, these may not be required. Therefore, it may be better to simultaneously target the proliferating population as well as the self-renewal population by blocking the SHH signaling pathway [7]. In another study, it has been shown that cyclopamine, a natural plant alkaloid, reduces cancer stem cell population by reducing ABCG2 efflux [5].

Metformin, an oral hypoglycemic agent used clinically to treat diabetes, has recently been shown to have anti-cancer effects through induction of apoptosis and death in cancer stem cells [48, 60]. Clinical studies also suggest that Metformin treatment is associated with a decreased risk of developing breast cancer and with a better response to chemotherapy [36, 52, 60].

An oncolytic virus has been shown to specifically target, infect and lyse cancer cells that harbor activating mutations in H-ras leaving normal cells unaffected [82]. Oncolytic viruses have several advantages as suitable agents for eradication of cancer stem cells because the entry of such viruses into cancer stem cells is not a problem as the process occurs via infection. Further, this approach would not be undermined by development of drug resistance. Since the targeted cancer stem cells lyse releasing the viral particles in to circulation, cancer stem cells at distant metastatic sites can also be targeted. It has been shown that cancer stem cells infected with Adeno virus fail to form tumors in an orthotopic breast tumor model [35, 82]. The studies by Kondratyev group [56] show that gamma-secretase inhibitors, an antagonist of Notch signaling, target both tumor-initiating cells and bulk tumor cells in Her2-Neu mouse mammary tumors.

With robotic high-throughput screening of different chemotherapeutic compounds, it has been shown that Salinomycin has selective toxicity for the cancer stem cells and reduces mammary tumor formation in vivo as well reduces tumorospheres formation [44]. Salinomycin is a 751 Da monocarboxylic polyether ionophore that mainly blocks the transmembrane potassium potential and increases the efflux of K^+ ions from cytoplasm and mitochondria [80]. Salinomycin selectively kills malignant cells in comparison to normal mammary cells. The ability of this compound to effectively kill both cancer stem cells and apoptosis-resistant malignant cells may underscore the potential of this compound as a novel and valuable anti-cancer agent [80]. In a recent study, treatment with Salinomycin or Lapatinib reduced the cancer stem cells by tenfold, and increased the sensitivity of breast cancer cells to chemotherapy [40].

19.7 Clinical Implications

Isolation and characterization of the normal mammary stem cells, transient amplifying cells, progenitor cells, stromal cells and differentiated cells along with their cell surface markers are essential for providing a framework for determining the cellular targets of different oncogenic mutations and how these mutations promote tumor formation [106]. Failure of conventional cancer therapies and the emerging concept of cancer stem cells in solid tumors emphasize the need for the design of new cancer therapies that selectively target cancer stem cell populations. Because a small population of these cells remaining after therapy can lead to tumor formation, we must assess the efficacy of current therapies in terms of their ability to kill these cells. A better understanding of the self-renewal pathways that are responsible for survival of cancer stem cells and development of effective strategies to selectively target these pathways are absolutely essential for complete eradication of breast cancer [115].

Our current understanding of tumor origin and progression is poor and incomplete because of the limited knowledge that we have on the self-renewal of mammary stem cells and cancer stem cells. Exactly how and to what extent aberrant self-renewal plays a role in tumor recurrence remains a critical question. There is increasing evidence suggesting that self-renewal might be an important factor in tumor cell survival after chemotherapy and radiation therapy. Further, research must focus on targeting aberrant self-renewal of cancer stem cell population along with proliferating cell population [76]. Most of the currently available chemotherapeutic agents not only fail to eradicate cancer stem cells but also help in the expansion of cancer stem cell pool. Selective and effective targeting of cancer stem cells holds potential for successful eradication of breast cancer. A better understanding of the biology of cancer stem cells and their self-renewal pathways should make this goal achievable in near future.

References

1. Al-Ejeh F, Smart CE, Morrison BJ, Chenevix-Trench G, Lopez JA, Lakhani SR, Brown MP, Khanna KK (2011) Breast cancer stem cells: treatment resistance and therapeutic opportunities. Carcinogenesis 32:650–658
2. Al-Hajj M, Clarke MF (2004) Self-renewal and solid tumor stem cells. Oncogene 23:7274–7282
3. Al-Hajj M, Wicha MS, Benito-Hernandez A, Morrison SJ, Clarke MF (2003) Prospective identification of tumorigenic breast cancer cells. Proc Natl Acad Sci USA 100:3983–3988
4. Asselin-Labat ML, Vaillant F, Sheridan JM, Pal B, Wu D, Simpson ER, Yasuda H, Smyth GK, Martin TJ, Lindeman GJ, Visvader JE (2010) Control of mammary stem cell function by steroid hormone signalling. Nature 465:798–802
5. Balbuena J, Pachon G, Lopez-Torrents G, Aran JM, Castresana JS, Petriz J (2011) ABCG2 is required to control the Sonic Hedgehog pathway in side population cells with stem-like properties. Cytometry A. doi:10.1002/cyto.a.21103
6. Balicki D (2007) Moving forward in human mammary stem cell biology and breast cancer prognostication using ALDH1. Cell Stem Cell 1:485–487
7. Bar EE, Chaudhry A, Farah MH, Eberhart CG (2007) Hedgehog signaling promotes medulloblastoma survival via Bc/II. Am J Pathol 170:347–355
8. Bautch VL (2010) Cancer: tumour stem cells switch sides. Nature 468:770–771
9. Berry DA, Cronin KA, Plevritis SK, Fryback DG, Clarke L, Zelen M, Mandelblatt JS, Yakovlev AY, Habbema JD, Feuer EJ (2005) Effect of screening and adjuvant therapy on mortality from breast cancer. N Engl J Med 353:1784–1792
10. Bolos V, Blanco M, Medina V, Aparicio G, Diaz-Prado S, Grande E (2009) Notch signalling in cancer stem cells. Clin Transl Oncol 11:11–19

11. Bolos V, Grego-Bessa J, de la Pompa JL (2007) Notch signaling in development and cancer. Endocr Rev 28:339–363
12. Bouras T, Pal B, Vaillant F, Harburg G, Asselin-Labat ML, Oakes SR, Lindeman GJ, Visvader JE (2008) Notch signaling regulates mammary stem cell function and luminal cell-fate commitment. Cell Stem Cell 3:429–441
13. Brekelmans CT, Tilanus-Linthorst MM, Seynaeve C, vd Ouweland A, Menke-Pluymers MB, Bartels CC, Kriege M, van Geel AN, Burger CW, Eggermont AM, Meijers-Heijboer H, Klijn JG (2007) Tumour characteristics, survival and prognostic factors of hereditary breast cancer from BRCA2-, BRCA1- and non-BRCA1/2 families as compared to sporadic breast cancer cases. Eur J Cancer 43:867–876
14. Brown AM (2001) Wnt signaling in breast cancer: have we come full circle? Breast Cancer Res 3:351–355
15. Bunting KD, Lindahl R, Townsend AJ (1994) Oxazaphosphorine-specific resistance in human MCF-7 breast carcinoma cell lines expressing transfected rat class 3 aldehyde dehydrogenase. J Biol Chem 269:23197–23203
16. Cameron DA, Stein S (2008) Drug insight: intracellular inhibitors of HER2–clinical development of lapatinib in breast cancer. Nat Clin Pract Oncol 5:512–520
17. Chang CC, Sun W, Cruz A, Saitoh M, Tai MH, Trosko JE (2001) A human breast epithelial cell type with stem cell characteristics as target cells for carcinogenesis. Radiat Res 155:201–207
18. Chiang AC, Massague J (2008) Molecular basis of metastasis. N Engl J Med 359:2814–2823
19. Chute JP, Muramoto GG, Whitesides J, Colvin M, Safi R, Chao NJ, McDonnell DP (2006) Inhibition of aldehyde dehydrogenase and retinoid signaling induces the expansion of human hematopoietic stem cells. Proc Natl Acad Sci USA 103:11707–11712
20. Cicalese A, Bonizzi G, Pasi CE, Faretta M, Ronzoni S, Giulini B, Brisken C, Minucci S, Di Fiore PP, Pelicci PG (2009) The tumor suppressor p53 regulates polarity of self-renewing divisions in mammary stem cells. Cell 138:1083–1095
21. Comen E, Norton L, Massague J (2011) Clinical implications of cancer self-seeding. Nat Rev Clin Oncol 8:369–377
22. Cui Y, Parra I, Zhang M, Hilsenbeck SG, Tsimelzon A, Furukawa T, Horii A, Zhang ZY, Nicholson RI, Fuqua SA (2006) Elevated expression of mitogen-activated protein kinase phosphatase 3 in breast tumors: a mechanism of tamoxifen resistance. Cancer Res 66:5950–5959
23. Dave B, Chang J (2009) Treatment resistance in stem cells and breast cancer. J Mammary Gland Biol Neoplasia 14:79–82
24. Davies E, Hiscox S (2010) New therapeutic approaches in breast cancer. Maturitas 68:121–128
25. Daya-Grosjean L, Couve-Privat S (2005) Sonic hedgehog signaling in basal cell carcinomas. Cancer Lett 225:181–192
26. Dean M, Fojo T, Bates S (2005) Tumour stem cells and drug resistance. Nat Rev Cancer 5:275–284
27. Deonarain MP, Kousparou CA, Epenetos AA (2009) Antibodies targeting cancer stem cells: a new paradigm in immunotherapy? MAbs 1:12–25
28. Dick JE (2008) Stem cell concepts renew cancer research. Blood 112:4793–4807
29. Dick JE (2009) Looking ahead in cancer stem cell research. Nat Biotechnol 27:44–46
30. Diehn M, Cho RW, Lobo NA, Kalisky T, Dorie MJ, Kulp AN, Qian D, Lam JS, Ailles LE, Wong M, Joshua B, Kaplan MJ, Wapnir I, Dirbas FM, Somlo G, Garberoglio C, Paz B, Shen J, Lau SK, Quake SR, Brown JM, Weissman IL, Clarke MF (2009) Association of reactive oxygen species levels and radioresistance in cancer stem cells. Nature 458:780–783
31. Diehn M, Clarke MF (2006) Cancer stem cells and radiotherapy: new insights into tumor radioresistance. J Natl Cancer Inst 98:1755–1757
32. Ding XW, Wu JH, Jiang CP (2010) ABCG2: a potential marker of stem cells and novel target in stem cell and cancer therapy. Life Sci 86:631–637

33. Dinh P, de Azambuja E, Cardoso F, Piccart-Gebhart MJ (2008) Facts and controversies in the use of trastuzumab in the adjuvant setting. Nat Clin Pract Oncol 5:645–654
34. Dontu G, Jackson KW, McNicholas E, Kawamura MJ, Abdallah WM, Wicha MS (2004) Role of Notch signaling in cell-fate determination of human mammary stem/progenitor cells. Breast Cancer Res 6:R605–R615
35. Eriksson M, Guse K, Bauerschmitz G, Virkkunen P, Tarkkanen M, Tanner M, Hakkarainen T, Kanerva A, Desmond RA, Pesonen S, Hemminki A (2007) Oncolytic adenoviruses kill breast cancer initiating CD44+CD24-/low cells. Mol Ther 15:2088–2093
36. Evans JM, Donnelly LA, Emslie-Smith AM, Alessi DR, Morris AD (2005) Metformin and reduced risk of cancer in diabetic patients. BMJ 330:1304–1305
37. Frampton JE (2009) Lapatinib: a review of its use in the treatment of HER2-overexpressing, trastuzumab-refractory, advanced or metastatic breast cancer. Drugs 69:2125–2148
38. Gage FH, Verma IM (2003) Stem cells at the dawn of the 21st century. Proc Natl Acad Sci USA 100(Suppl 1):11817–11818
39. Ginestier C, Hur MH, Charafe-Jauffret E, Monville F, Dutcher J, Brown M, Jacquemier J, Viens P, Kleer CG, Liu S, Schott A, Hayes D, Birnbaum D, Wicha MS, Dontu G (2007) ALDH1 is a marker of normal and malignant human mammary stem cells and a predictor of poor clinical outcome. Cell Stem Cell 1:555–567
40. Gong C, Yao H, Liu Q, Chen J, Shi J, Su F, Song E (2010) Markers of tumor-initiating cells predict chemoresistance in breast cancer. PLoS One 5:e15630
41. Gonzalez C (2007) Spindle orientation, asymmetric division and tumour suppression in Drosophila stem cells. Nat Rev Genet 8:462–472
42. Goodell MA (2002) Multipotential stem cells and 'side population' cells. Cytotherapy 4:507–508
43. Goodell MA, Brose K, Paradis G, Conner AS, Mulligan RC (1996) Isolation and functional properties of murine hematopoietic stem cells that are replicating in vivo. J Exp Med 183:1797–1806
44. Gupta PB, Onder TT, Jiang G, Tao K, Kuperwasser C, Weinberg RA, Lander ES (2009) Identification of selective inhibitors of cancer stem cells by high-throughput screening. Cell 138:645–659
45. Hao L, Rizzo P, Osipo C, Pannuti A, Wyatt D, Cheung LW, Sonenshein G, Osborne BA, Miele L (2009) Notch-1 activates estrogen receptor-alpha-dependent transcription via IKKalpha in breast cancer cells. Oncogene 29:201–213
46. He S, Nakada D, Morrison SJ (2009) Mechanisms of stem cell self-renewal. Annu Rev Cell Dev Biol 25:377–406
47. Hebbard L, Steffen A, Zawadzki V, Fieber C, Howells N, Moll J, Ponta H, Hofmann M, Sleeman J (2000) CD44 expression and regulation during mammary gland development and function. J Cell Sci 113(Pt 14):2619–2630
48. Hirsch HA, Iliopoulos D, Tsichlis PN, Struhl K (2009) Metformin selectively targets cancer stem cells, and acts together with chemotherapy to block tumor growth and prolong remission. Cancer Res 69:7507–7511
49. Hoey T, Yen WC, Axelrod F, Basi J, Donigian L, Dylla S, Fitch-Bruhns M, Lazetic S, Park IK, Sato A, Satyal S, Wang X, Clarke MF, Lewicki J, Gurney A (2009) DLL4 blockade inhibits tumor growth and reduces tumor-initiating cell frequency. Cell Stem Cell 5:168–177
50. Hombach-Klonisch S, Panigrahi S, Rashedi I, Seifert A, Alberti E, Pocar P, Kurpisz M, Schulze-Osthoff K, Mackiewicz A, Los M (2008) Adult stem cells and their trans-differentiation potential–perspectives and therapeutic applications. J Mol Med (Berl) 86:1301–1314
51. Ishimoto T, Nagano O, Yae T, Tamada M, Motohara T, Oshima H, Oshima M, Ikeda T, Asaba R, Yagi H, Masuko T, Shimizu T, Ishikawa T, Kai K, Takahashi E, Imamura Y, Baba Y, Ohmura M, Suematsu M, Baba H, Saya H (2011) CD44 variant regulates redox status in cancer cells by stabilizing the xCT subunit of system xc(-) and thereby promotes tumor growth. Cancer Cell 19:387–400
52. Jiralerspong S, Palla SL, Giordano SH, Meric-Bernstam F, Liedtke C, Barnett CM, Hsu L, Hung MC, Hortobagyi GN, Gonzalez-Angulo AM (2009) Metformin and pathologic com-

plete responses to neoadjuvant chemotherapy in diabetic patients with breast cancer. J Clin Oncol 27:3297–3302

53. Kakarala M, Wicha MS (2008) Implications of the cancer stem-cell hypothesis for breast cancer prevention and therapy. J Clin Oncol 26:2813–2820
54. Kalirai H, Clarke RB (2006) Human breast epithelial stem cells and their regulation. J Pathol 208:7–16
55. Kim M, Turnquist H, Jackson J, Sgagias M, Yan Y, Gong M, Dean M, Sharp JG, Cowan K (2002) The multidrug resistance transporter ABCG2 (breast cancer resistance protein 1) effluxes Hoechst 33342 and is overexpressed in hematopoietic stem cells. Clin Cancer Res 8:22–28
56. Kondratyev M, Kreso A, Hallett RM, Girgis-Gabardo A, Barcelon ME, Ilieva D, Ware C, Majumder PK, Hassell JA (2011) Gamma-secretase inhibitors target tumor-initiating cells in a mouse model of ERBB2 breast cancer. Oncogene. doi:10.1038/onc.2011.212
57. Korkaya H, Paulson A, Iovino F, Wicha MS (2008) HER2 regulates the mammary stem/progenitor cell population driving tumorigenesis and invasion. Oncogene 27:6120–6130
58. Korkaya H, Wicha MS (2007) Selective targeting of cancer stem cells: a new concept in cancer therapeutics. BioDrugs 21:299–310
59. Korkaya H, Wicha MS (2009) HER-2, notch, and breast cancer stem cells: targeting an axis of evil. Clin Cancer Res 15:1845–1847
60. Kourelis TV, Siegel RD (2011) Metformin and cancer: new applications for an old drug. Med Oncol. doi:10.1007/s12032-011-9846-7
61. Lathia JD, Gallagher J, Heddleston JM, Wang J, Eyler CE, Macswords J, Wu Q, Vasanji A, McLendon RE, Hjelmeland AB, Rich JN (2010) Integrin alpha 6 regulates glioblastoma stem cells. Cell Stem Cell 6:421–432
62. Lawson DA, Xin L, Lukacs RU, Cheng D, Witte ON (2007) Isolation and functional characterization of murine prostate stem cells. Proc Natl Acad Sci USA 104:181–186
63. Li F, Tiede B, Massague J, Kang Y (2007) Beyond tumorigenesis: cancer stem cells in metastasis. Cell Res 17:3–14
64. Li X, Lewis MT, Huang J, Gutierrez C, Osborne CK, Wu MF, Hilsenbeck SG, Pavlick A, Zhang X, Chamness GC, Wong H, Rosen J, Chang JC (2008) Intrinsic resistance of tumorigenic breast cancer cells to chemotherapy. J Natl Cancer Inst 100:672–679
65. Li X, Pan YZ, Seigel GM, Hu ZH, Huang M, Yu AM (2011) Breast cancer resistance protein BCRP/ABCG2 regulatory microRNAs (hsa-miR-328, -519c and -520 h) and their differential expression in stem-like ABCG2+ cancer cells. Biochem Pharmacol 81:783–792
66. Li Y, Rosen JM (2005) Stem/progenitor cells in mouse mammary gland development and breast cancer. J Mammary Gland Biol Neoplasia 10:17–24
67. Liu S, Ginestier C, Ou SJ, Clouthier SG, Patel SH, Monville F, Korkaya H, Heath A, Dutcher J, Kleer CG, Jung Y, Dontu G, Taichman R, Wicha MS (2011) Breast cancer stem cells are regulated by mesenchymal stem cells through cytokine networks. Cancer Res 71:614–624
68. Liu S, Wicha MS (2010) Targeting breast cancer stem cells. J Clin Oncol 28:4006–4012
69. Lobo NA, Shimono Y, Qian D, Clarke MF (2007) The biology of cancer stem cells. Annu Rev Cell Dev Biol 23:675–699
70. Magnifico A, Albano L, Campaner S, Delia D, Castiglioni F, Gasparini P, Sozzi G, Fontanella E, Menard S, Tagliabue E (2009) Tumor-initiating cells of HER2-positive carcinoma cell lines express the highest oncoprotein levels and are sensitive to trastuzumab. Clin Cancer Res 15:2010–2021
71. Marcato P, Dean CA, Pan D, Araslanova R, Gillis M, Joshi M, Helyer L, Pan L, Leidal A, Gujar S, Giacomantonio CA, Lee PW (2011) Aldehyde dehydrogenase activity of breast cancer stem cells is primarily due to isoform ALDH1A3 and its expression is predictive of metastasis. Stem Cells 29:32–45
72. Marhaba R, Zoller M (2004) CD44 in cancer progression: adhesion, migration and growth regulation. J Mol Histol 35:211–231
73. Martin DR, Cox NR, Hathcock TL, Niemeyer GP, Baker HJ (2002) Isolation and characterization of multipotential mesenchymal stem cells from feline bone marrow. Exp Hematol 30:879–886

74. Merchant AA, Matsui W (2010) Targeting Hedgehog–a cancer stem cell pathway. Clin Cancer Res 16:3130–3140
75. Meyer MJ, Fleming JM, Lin AF, Hussnain SA, Ginsburg E, Vonderhaar BK (2010) CD44posCD49fhiCD133/2hi defines xenograft-initiating cells in estrogen receptor-negative breast cancer. Cancer Res 70:4624–4633
76. Moore N, Lyle S (2010) Quiescent, slow-cycling stem cell populations in cancer: a review of the evidence and discussion of significance. J Oncol 2011. doi:10.1155/2011/396076
77. Morrison SJ, Kimble J (2006) Asymmetric and symmetric stem-cell divisions in development and cancer. Nature 441:1068–1074
78. Mullor JL, Sanchez P, Ruiz i Altaba A (2002) Pathways and consequences: hedgehog signaling in human disease. Trends Cell Biol 12:562–569
79. Nalwoga H, Arnes JB, Wabinga H, Akslen LA (2009) Expression of aldehyde dehydrogenase 1 (ALDH1) is associated with basal-like markers and features of aggressive tumours in African breast cancer. Br J Cancer 102:369–375
80. Naujokat C, Fuchs D, Opelz G (2010) Salinomycin in cancer: a new mission for an old agent. Mol Med Rep 3:555–559
81. Neth P, Ries C, Karow M, Egea V, Ilmer M, Jochum M (2007) The Wnt signal transduction pathway in stem cells and cancer cells: influence on cellular invasion. Stem Cell Rev 3:18–29
82. Nguyen NP, Almeida FS, Chi A, Nguyen LM, Cohen D, Karlsson U, Vinh-Hung V (2010) Molecular biology of breast cancer stem cells: potential clinical applications. Cancer Treat Rev 36:485–491
83. O'Brien CS, Howell SJ, Farnie G, Clarke RB (2009) Resistance to endocrine therapy: are breast cancer stem cells the culprits? J Mammary Gland Biol Neoplasia 14:45–54
84. Pajonk F, Vlashi E, McBride WH (2010) Radiation resistance of cancer stem cells: the 4 R's of radiobiology revisited. Stem Cells 28:639–648
85. Pece S, Confalonieri S, Romano PR, Di Fiore PP (2010) NUMB-ing down cancer by more than just a NOTCH. Biochim Biophys Acta 1815:26–43
86. Pece S, Tosoni D, Confalonieri S, Mazzarol G, Vecchi M, Ronzoni S, Bernard L, Viale G, Pelicci PG, Di Fiore PP (2010) Biological and molecular heterogeneity of breast cancers correlates with their cancer stem cell content. Cell 140:62–73
87. Petersen OW, Polyak K (2010) Stem cells in the human breast. Cold Spring Harb Perspect Biol 2:a003160
88. Phillips TM, McBride WH, Pajonk F (2006) The response of CD24(-/low)/CD44+ breast cancer-initiating cells to radiation. J Natl Cancer Inst 98:1777–1785
89. Ponti D, Costa A, Zaffaroni N, Pratesi G, Petrangolini G, Coradini D, Pilotti S, Pierotti MA, Daidone MG (2005) Isolation and in vitro propagation of tumorigenic breast cancer cells with stem/progenitor cell properties. Cancer Res 65:5506–5511
90. Pradeep CR, Kostler WJ, Lauriola M, Granit RZ, Zhang F, Jacob-Hirsch J, Rechavi G, Nair HB, Hennessy BT, Gonzalez-Angulo AM, Tekmal RR, Ben-Porath I, Mills GB, Domany E, Yarden Y (2011) Modeling ductal carcinoma in situ: a HER2-Notch3 collaboration enables luminal filling. Oncogene. doi:10.1038/onc.2011.279
91. Presnell SC, Petersen B, Heidaran M (2002) Stem cells in adult tissues. Semin Cell Dev Biol 13:369–376
92. Raouf A, Zhao Y, To K, Stingl J, Delaney A, Barbara M, Iscove N, Jones S, McKinney S, Emerman J, Aparicio S, Marra M, Eaves C (2008) Transcriptome analysis of the normal human mammary cell commitment and differentiation process. Cell Stem Cell 3:109–118
93. Rennstam K, McMichael N, Berglund P, Honeth G, Hegardt C, Ryden L, Luts L, Bendahl PO, Hedenfalk I (2010) Numb protein expression correlates with a basal-like phenotype and cancer stem cell markers in primary breast cancer. Breast Cancer Res Treat 122:315–324
94. Reya T, Morrison SJ, Clarke MF, Weissman IL (2001) Stem cells, cancer, and cancer stem cells. Nature 414:105–111
95. Saxena M, Stephens MA, Pathak H, Rangarajan A (2011) Transcription factors that mediate epithelial-mesenchymal transition lead to multidrug resistance by upregulating ABC transporters. Cell Death Dis 2:e179

96. Scharenberg CW, Harkey MA, Torok-Storb B (2002) The ABCG2 transporter is an efficient Hoechst 33342 efflux pump and is preferentially expressed by immature human hematopoietic progenitors. Blood 99:507–512
97. Schinkel AH, Smit JJ, van Tellingen O, Beijnen JH, Wagenaar E, van Deemter L, Mol CA, van der Valk MA, Robanus-Maandag EC, te Riele HP et al (1994) Disruption of the mouse mdr1a P-glycoprotein gene leads to a deficiency in the blood-brain barrier and to increased sensitivity to drugs. Cell 77:491–502
98. Shackleton M, Vaillant F, Simpson KJ, Stingl J, Smyth GK, Asselin-Labat ML, Wu L, Lindeman GJ, Visvader JE (2006) Generation of a functional mammary gland from a single stem cell. Nature 439:84–88
99. Sieburg HB, Rezner BD, Muller-Sieburg CE (2011) Predicting clonal self-renewal and extinction of hematopoietic stem cells. Proc Natl Acad Sci USA 108:4370–4375
100. Slamon DJ, Clark GM, Wong SG, Levin WJ, Ullrich A, McGuire WL (1987) Human breast cancer: correlation of relapse and survival with amplification of the HER-2/neu oncogene. Science 235:177–182
101. Smalley M, Ashworth A (2003) Stem cells and breast cancer: a field in transit. Nat Rev Cancer 3:832–844
102. Smith GH (2002) Mammary cancer and epithelial stem cells: a problem or a solution? Breast Cancer Res 4:47–50
103. Sorlie T, Tibshirani R, Parker J, Hastie T, Marron JS, Nobel A, Deng S, Johnsen H, Pesich R, Geisler S, Demeter J, Perou CM, Lonning PE, Brown PO, Borresen-Dale AL, Botstein D (2003) Repeated observation of breast tumor subtypes in independent gene expression data sets. Proc Natl Acad Sci USA 100:8418–8423
104. Sotiriou C, Neo SY, McShane LM, Korn EL, Long PM, Jazaeri A, Martiat P, Fox SB, Harris AL, Liu ET (2003) Breast cancer classification and prognosis based on gene expression profiles from a population-based study. Proc Natl Acad Sci USA 100:10393–10398
105. Sottoriva A, Vermeulen L, Tavare S (2011) Modeling evolutionary dynamics of epigenetic mutations in hierarchically organized tumors. PLoS Comput Biol 7:e1001132
106. Stingl J, Caldas C (2007) Molecular heterogeneity of breast carcinomas and the cancer stem cell hypothesis. Nat Rev Cancer 7:791–799
107. Stingl J, Eirew P, Ricketson I, Shackleton M, Vaillant F, Choi D, Li HI, Eaves CJ (2006) Purification and unique properties of mammary epithelial stem cells. Nature 439:993–997
108. Stylianou S, Clarke RB, Brennan K (2006) Aberrant activation of notch signaling in human breast cancer. Cancer Res 66:1517–1525
109. Sweeney CJ, Mehrotra S, Sadaria MR, Kumar S, Shortle NH, Roman Y, Sheridan C, Campbell RA, Murry DJ, Badve S, Nakshatri H (2005) The sesquiterpene lactone parthenolide in combination with docetaxel reduces metastasis and improves survival in a xenograft model of breast cancer. Mol Cancer Ther 4:1004–1012
110. Tanaka H, Nakamura M, Kameda C, Kubo M, Sato N, Kuroki S, Tanaka M, Katano M (2009) The Hedgehog signaling pathway plays an essential role in maintaining the CD44+CD24-/low subpopulation and the side population of breast cancer cells. Anticancer Res 29:2147–2157
111. Visvader JE, Lindeman GJ (2008) Cancer stem cells in solid tumours: accumulating evidence and unresolved questions. Nat Rev Cancer 8:755–768
112. Wang RH (2006) The new portrait of mammary gland stem cells. Int J Biol Sci 2:186–187
113. Welte Y, Adjaye J, Lehrach HR, Regenbrecht CR (2010) Cancer stem cells in solid tumors: elusive or illusive? Cell Commun Signal 8:6
114. Wicha MS (2006) Cancer stem cells and metastasis: lethal seeds. Clin Cancer Res 12:5606–5607
115. Woodward WA, Chen MS, Behbod F, Rosen JM (2005) On mammary stem cells. J Cell Sci 118:3585–3594
116. Zhan JF, Wu LP, Chen LH, Yuan YW, Xie GZ, Sun AM, Liu Y, Chen ZX (2011) Pharmacological inhibition of AKT sensitizes MCF-7 human breast cancer-initiating cells to radiation. Cell Oncol (Doedr) 34:451–456

117. Zhou BP, Hung MC (2003) Dysregulation of cellular signaling by HER2/neu in breast cancer. Semin Oncol 30:38–48
118. Zhou J, Zhang H, Gu P, Bai J, Margolick JB, Zhang Y (2008) NF-kappaB pathway inhibitors preferentially inhibit breast cancer stem-like cells. Breast Cancer Res Treat 111:419–427
119. Zhou S, Morris JJ, Barnes Y, Lan L, Schuetz JD, Sorrentino BP (2002) Bcrp1 gene expression is required for normal numbers of side population stem cells in mice, and confers relative protection to mitoxantrone in hematopoietic cells in vivo. Proc Natl Acad Sci USA 99:12339–12344
120. Zhou S, Schuetz JD, Bunting KD, Colapietro AM, Sampath J, Morris JJ, Lagutina I, Grosveld GC, Osawa M, Nakauchi H, Sorrentino BP (2001) The ABC transporter Bcrp1/ABCG2 is expressed in a wide variety of stem cells and is a molecular determinant of the side-population phenotype. Nat Med 7:1028–1034
121. Zhou S, Zong Y, Lu T, Sorrentino BP (2003) Hematopoietic cells from mice that are deficient in both Bcrp1/Abcg2 and Mdr1a/1b develop normally but are sensitized to mitoxantrone. Biotechniques 35:1248–1252

Chapter 20
Breast Cancer Stem Cells

Shane R. Stecklein and Roy A. Jensen

Contents

S.R. Stecklein
Department of Pathology and Laboratory Medicine, University of Kansas Medical Center, 3901 Rainbow Boulevard, Mail Stop 1027, 66160 Kansas City, KS, USA

The University of Kansas Cancer Center, Kansas City, KS, USA

R.A. Jensen, M.D. (✉)
Department of Pathology and Laboratory Medicine, University of Kansas Medical Center, 3901 Rainbow Boulevard, Mail Stop 1027, 66160 Kansas City, KS, USA

The University of Kansas Cancer Center, Kansas City, KS, USA

Department of Anatomy and Cell Biology, University of Kansas Medical Center, 3901 Rainbow Boulevard, Mail Stop 1027, 66160 Kansas City, KS, USA

Department of Molecular Biosciences, University of Kansas, Lawrence, KS, USA
e-mail: rjensen@kumc.edu

R.K. Srivastava and S. Shankar (eds.), *Stem Cells and Human Diseases*,
DOI 10.1007/978-94-007-2801-1_20,

Abstract Breast cancer is a heterogeneous disease at both the histological and molecular levels. The current model of breast tumorigenesis suggests that the normal mammary stem cell and the various progenitors that arise thereof can be transformed and generate lineage-restricted tumor phenotypes. This model is supported by observations that the different subtypes of breast cancer share transcriptional signatures intrinsic to normal components of the mammary epithelium. This chapter aims to review seminal studies that allowed the isolation and characterization of both normal and cancer-associated breast cancer stem cells, to describe the link between cellular ancestry and subtype-specific cancer stem cells, and to discuss therapeutic challenges directly resulting from cellular plasticity and stem-cell like phenotypes in human breast cancer.

Keywords Breast cancer • Stem cells • Heterogeneity

20.1 Introduction

Breast cancer remains the second leading cause of cancer-related death worldwide. The World Health Organization (WHO) estimates that more than one million women are diagnosed with breast cancer annually, and more than 400,000 will die from the disease this year [8]. Though the global incidence of breast cancer appears to be increasing, the 5-year relative breast cancer survival rate has increased dramatically in developed countries over the last 50 years due to early detection and treatment of *in situ* and early stage disease and improvements in targeted therapies for specific subtypes of breast cancer.

At both the histological and molecular levels, human breast cancer is a heterogeneous group of diseases. Invasive ductal carcinoma (IDC), which comprises 80% of all breast cancers, can be divided into more than a dozen histological subtypes and at least six distinct molecular families [22, 38, 52, 54]. Though the precise causes of molecular and phenotypic diversity in human breast cancer remains poorly understood, lineage-specific breast 'cancer stem cells' (CSCs) (alternatively, "tumor-initiating cells" (TICs)) are thought to arise from transformation of discrete cellular lineages within the mammary gland. In this chapter, we describe the seminal works that identified breast CSCs, highlight the link between the CSC model and the observed molecular and phenotypic heterogeneity of breast cancer, and discuss the challenge of targeting the stem-like population in modern breast cancer therapy.

20.2 The Cancer Stem Cell Theory

Despite our growing knowledge of the molecular events that lead to malignancy, a number of fundamental questions remain concerning the cellular etiology of cancer. A growing body of evidence suggests that only a small number of cells within most human cancers truly possess tumorigenic properties. The first conclusive demonstration of a tumor-initiating cell was borne from studies in the laboratory of John Dick, where only a rare population of acute myeloid leukemia cells was able to recapitulate malignancy in an immunocompromised host [3, 27]. This pioneering study established that not call cancer cells are the same, and that, at least in some cancers, perpetuation of malignancy may be accomplished by an extreme minority of cells with special stem-like properties. Accumulating evidence now supports the existence of CSCs in a number of solid cancers, including those of the breast, brain, colon, and prostate [1, 7, 36, 41, 49].

Tissues which undergo regeneration, remodeling, and renewal during an organism's lifetime are thought to possess a small population of quiescent stem cells which respond to injury and microenvironmental cues in order to maintain tissue homeostasis. It is this long-lived population of cells, seated at the apex of a hierarchical differentiation pathway that establishes the various cellular components of adult tissues through asymmetric division. Given that most human epithelial cancers arise in tissues with relatively rapid cellular turnover rates, it can be implied that most of the cells within these tissues do not live long enough to accumulate the requisite number of mutations required for malignant transformation. The long-lived resident tissue stem cell, however, provides an elegant target for mutational transformation. By slowly amassing mutations throughout its lifetime (i.e., the lifetime of the organism), this cell may reach the threshold for tumorigenic conversion. Once transformed, this cell is thought to continue to undergo symmetric and asymmetric division and generate lineage-committed progeny, albeit the hierarchical nature of this differentiation is perturbed. The retention of self-renewal capacity only in the CSC is responsible for the limited tumorigenic potential of the cells which comprise the bulk of the tumor.

20.3 Cellular Hierarchy and Architecture of the Human Breast

The mammary gland is a tubuloalveolar epithelial structure embedded in a fat pad with supportive adipocytes, fibroblasts, endothelial cells and nervous tissue (Fig. 20.1). In response to hormonal cues associated with puberty, menstruation, pregnancy and lactation, the cells that make up the breast ductal and lobular epithelium exhibit dramatic cycles of growth, remodeling and apoptosis. Indeed, the remarkable capacity of this gland to replenish during and after these events suggested long ago the existence of a mammary stem cell.

The epithelial component of the human breast is histologically simple. The ducts and lobules are composed of two layers of cells, an outer layer of myoepithelial cells which express basal cytokeratins (e.g., CK5 and CK14), alpha-smooth muscle actin, and the mesenchymal intermediate filament vimentin, and an inner layer of luminal epithelial cells which express CK8 and CK18 and may or may not express ERα (Fig. 20.1). Though histologically simple, there is increasing evidence that a complex epithelial cellular hierarchy exists within the mammary gland. It is believed that the normal stem cell gives rise to one or more

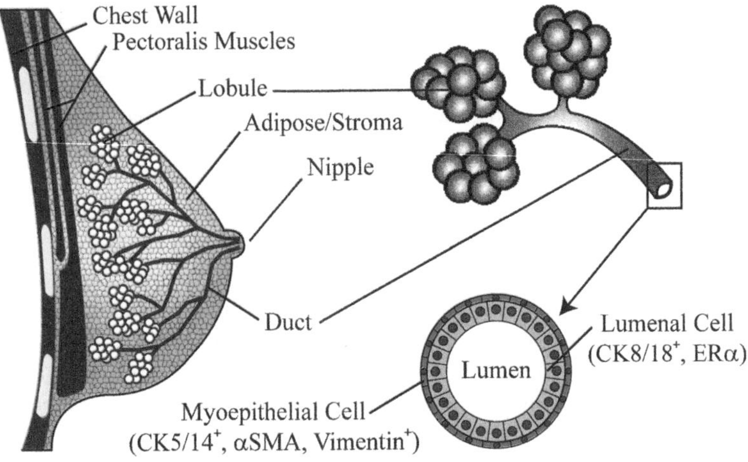

Fig. 20.1 Gross and microscopic anatomy of the human female breast

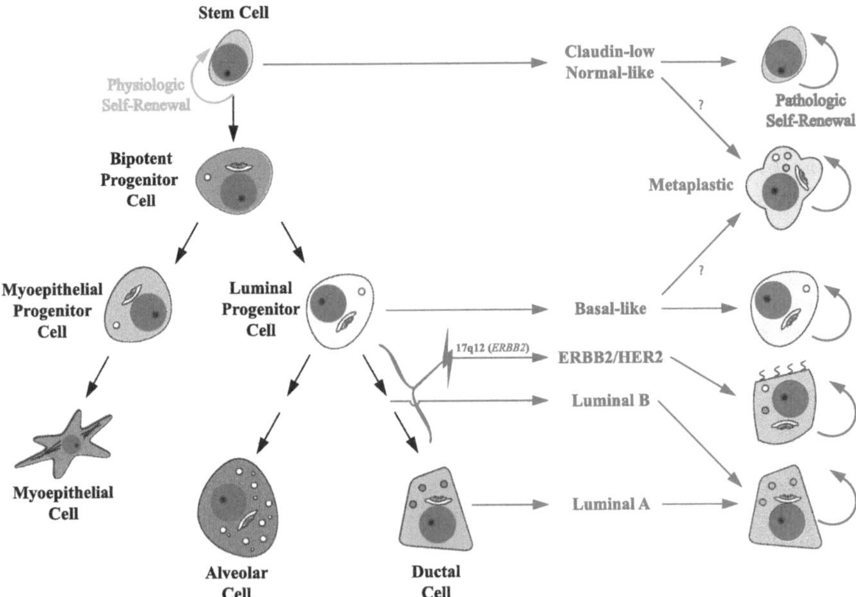

Fig. 20.2 Hierarcheal schema for breast epithelial precursors cells and corresponding breast cancer subtypes. Modified from visvader [60]

lineages of progenitors, a pool of cells poised for proliferation and subsequent differentiation into either myoepithelial- or luminal-restricted progenitor cells. These committed progenitor cells then undergo terminal differentiation to generate mature myoepithelial and ductal cells in a resting gland, or during lactation, the alveolar cells responsible for milk production (Fig. 20.2).

Approximately 80% of breast cancers arise from the luminal epithelial cells that make up the milk ducts, 5–10% arise from lobular epithelial cells, and the remaining fraction comprising a heterogeneous group of rarer subtypes. Histologically, these lesions are categorized into more than a dozen subtypes and grouped based on expression of estrogen receptor α (ER), progesterone receptor (PR) and the epidermal growth factor receptor 2 (ERBB2/HER2). Approximately 60% of all ductal and lobular lesions arise from the mature luminal epithelial cells, are ER^+ and are dependent on estrogen for growth. Roughly one quarter of all invasive breast cancers exhibit amplification of HER2, and the remaining 15% are referred to as "triple negative" (TN) meaning they lack expression of ER, PR and HER2. There is now strong evidence supporting discrete cellular ancestries for these different groups of breast cancers, and it is likely that unique cancer stem cell populations and even unique mechanisms of CSC self-renewal are at play in each.

20.4 Identification of Breast Cancer Stem Cells

A landmark paper in 1959 by DeOme and colleagues demonstrated that transplantation of isolated epithelial components of the mouse mammary gland into a syngeneic host could reconstitute a functional mammary gland [10]. There has since been immense interest in isolating and characterizing the properties of these stem cells. Such efforts have employed numerous techniques, including isolation of side population (SP) cells, generation of tumorospheres *in vitro*, vital assessment of enzymatic activity, and multiparameter flow cytometry.

SP cells are defined by their ability to efflux vital dyes, such as Hoechst 33342, via the action of membrane ATP-binding cassette transporters (e.g., BCRP1). A study by Welm et al. examined the frequency of dye effluxing cells in freshly dissected mouse mammary epithelium and reported that 2–3% of cells are SP cells. Moreover, the SP pool appeared to be significantly enriched in mammary stem cells as less than 4,000 SP cells were capable of regenerating a functional mammary gland upon transplantation into a cleared fatpad, while greater than 100,000 unsorted mammary epithelial cells are generally required to form a mammary outgrowth [61].

A cell culture technique developed by Max Wicha and colleagues termed mammosphere culture has enabled the propagation and characterization of putative normal and cancer-associated stem-like cells *in vitro*. This system employs suspension culture of mammary epithelial cells in serum-free defined growth medium. This technique is based on the fact that stem/progenitor cells are believed to be able to survive in serum-free suspension culture, while differentiated cells should undergo anoikis. The spherical structures generated from this system are enriched in early stem/progenitor cells and are capable of differentiating into all three mature epithelial cell lineages and forming complex functional structures in three dimensional culture systems [11].

Another study by Max Wicha's group identified that the detoxifying enzyme aldehyde dehydrogenase 1 (ALDH1) is highly expressed in a subpopulation of both normal and cancerous mammary epithelial cells that appear to have functional properties consistent with stem cells, including enhanced tumor xenograft formation in immunocompromised mice. This same study also demonstrated that expression of ALDH1 in human breast carcinomas is a strong predictor of poor prognosis [17].

In seminal papers from the laboratories of Jane Visvader and Connie Eaves, a functional murine mammary gland was successfully recapitulated by transplantation of a single cell with the $CD29^{hi}/CD24^{+}$ cell surface phenotype [45] or the $CD45^{-}/Ter119^{-}/CD31^{-}/Sca1^{low}/CD24^{med}/CD49f^{hi}$ phenotype [55]. Though the precise identity of the normal human mammary stem cell remains elusive, many of the same techniques that have been informative in characterization of the murine mammary gland epithelial hierarchy have provided insights into the biology of human breast cancers.

In 2003, researchers in the laboratory of Michael Clarke prospectively isolated a small population of cells from human breast cancers that possessed the $ESA^{+}/CD44^{+}/CD24^{-/low}$ surface phenotype and demonstrated that, when transplanted into immunocompromised mice, as few as 200 of these cells generate tumors which recapitulate the phenotypic heterogeneity observed in the primary tumor, while tens of thousands of cells with other surface profiles failed to incite malignancy [1]. This landmark study enabled tremendous enrichment of breast tumor initiating cells, and was thus fundamental in allowing the phenotypic and functional characterization of breast CSCs. Interestingly, the $CD44^{+}/CD24^{-/low}$ and ALDH1-derived CSCs appear to be overlapping but not identical cell populations [17].

20.5 Cell Fate and the Intrinsic Subtypes of Breast Cancer

First generation gene expression profiling of human breast cancers established at least six major types of invasive breast cancer: luminal type A, luminal type B, luminal type C, basal-like, ERBB2/HER2-overexpressing, and normal breast-like [22, 38, 52, 54]. Retrospective analysis of patient outcomes in these studies demonstrated that specific molecular taxonomies are strongly correlated with unfavorable clinical behavior and poor overall survival [52]. Refinements in profiling have since identified additional intrinsic subtypes of breast cancer. The morphological and molecular heterogeneity observed in human breast cancers likely stems from transformation of different cellular elements of the mammary epithelium (Fig. 20.2). This model is supported by the fact that each subtype of human breast cancer can be shown to share transcriptional, morphological and/or biochemical properties inherent to discrete cellular elements of the normal mammary gland [29].

20.5.1 Breast Tumors of Luminal Origin

The majority of invasive ductal and lobular carcinomas exhibit evidence of luminal differentiation. These tumors usually express ER and are thus amenable, to varying degrees, to therapies aimed at regulating ER signaling. Based on similarities in gene expression and morphology, it is believed that tumors comprising the luminal A and luminal B subtypes arise from transformation of cells in the terminal stages of luminal fate commitment [60]. Specifically, luminal A tumors exhibit robust expression of ER, PR, and other markers of mature luminal epithelial cells including the transcription factor GATA3 and luminal cytokeratins (CK8 and CK18), and likely arise from malignant transformation of the mature luminal ductal or lobular epithelial cell [38, 60]. Luminal B tumors commonly express ER, albeit at a lower level than luminal A tumors, and likely stem from transformation of a cell with an intermediate degree of terminal luminal commitment. Accordingly, these tumors usually exhibit lower expression of estrogen-related genes, higher mitotic indices and histological grade and a significantly poorer prognosis than luminal A malignancies [52, 56].

Several studies have confirmed the critical nature of the Notch morphogenetic pathway and the GATA3 transcription factor in the specification and maintenance of the luminal epithelium [2, 4, 6, 59]. GATA3 likely functions pleiotropically in breast tumorigenesis, simultaneously promoting terminal differentiation of ductal and alveolar epithelial cells while antagonizing the epithelial-to-mesenchymal transition and metastasis [2, 13, 62]. In agreement with experimental studies of GATA3 in the mammary gland, high expression of this transcription factor in human breast cancer correlates with lower grade, higher expression of ER and PR and improved survival [63]. Among the intrinsic subtypes of breast cancer, tumors with robust luminal phenotypes are associated with significantly better disease-free and overall survival than tumors with less differentiated phenotypes and those with exaggerated expression of ERBB2/HER2 [52]. Endocrine therapies aimed at regulating the synthesis and/or cellular responses to estrogen have led to significantly improved outcomes for women with hormone receptor positive breast cancer.

Given the critical role of estrogen in the growth and progression of these lesions, there has been much interest in defining the role(s) of sex hormones in the biology of breast CSCs. Though the $CD44^{Hi}/CD24^{-/Low}$ population is generally ER^- and is relatively infrequent in luminal type tumors, such cells may be indirectly maintained and perpetuated indirectly by estrogen and/or progesterone. Consistent with this notion, a recent report from Charlotte Kuperwasser's group demonstrated that estrogen can increase the CSC population in ER^+ MCF7 cells through the action of FGF [15].

20.5.2 ERBB2/HER2 Breast Tumors

Tumors overexpressing the ERBB2/HER2 receptor tyrosine kinase (RTK) appear to originate from a luminally-restricted cell, but have a significantly poorer prognosis than either the luminal A or luminal B subtypes without amplification or overexpression of this molecule. Overexpression of HER2 is observed in approximately 25–30% of human breast cancers, is usually caused by amplification of the 17q12 locus (containing the *ERBB2* gene), and results in exaggerated expression of wild-type HER2 RTK at the membrane [32, 48]. Though only a quarter of invasive malignancies exhibit amplification of HER2, this molecular abnormality is observed in nearly half of all ductal carcinoma *in situ* (DCIS) lesions, suggesting that *ERBB2* amplification is an early event in the pathogenesis of this subtype of breast cancer [30, 37] and represents an intrinsic subtype of breast cancer rather than an artifact of advanced disease. Overexpression of HER2 at the cell surface appears to promote dimerization-dependent signaling events that activate numerous signaling nodes and influence proliferation, differentiation and apoptosis [5]. ERBB2/HER2 cancers follow a more aggressive clinical course than do luminal tumors without amplification of this gene, are more resistant to chemotherapeutic agents and have an increased risk of distant metastasis [44, 52, 53]. The introduction of trastuzumab (Herceptin®) into the treatment paradigm of HER2-positive breast cancer dramatically improved survival for women with this subtype of disease [51]. Newer agents like the small-molecule tyrosine kinase inhibitor lapatinib (Tykerb®) can also inhibit HER2-associated signaling events [14, 35].

The dramatic improvement in clinical course, recurrence and prognosis in $HER2^{+}$ breast cancers following implementation of HER2-targeting agents suggested that, at least in part, the efficacy of these agents may be due to their ability to target CSCs. Indeed, one study demonstrated that forced expression of HER2 in mammary epithelial cells increased the expression of ALDH1, enhanced mammosphere forming potential and exaggerated tumorigenesis in immunocompromised mice [26]. Though cytotoxic chemotherapeutic agents have been demonstrated to select for CSC-like cells, treatment of $HER2^{+}$ cells with lapatinib has been shown to decrease the $CD44^{+}/CD24^{-/low}$ CSC population and abrogate mammosphere forming capacity following therapy in human breast cancers [28]. Taken together, these studies argue for a role of HER2 in maintaining the CSC population in $HER2^{+}$ breast cancers and suggest that the reduced tumor recurrence and improved survival attributed to HER2 targeting agents is due, at least in part, to targeting the CSC.

20.5.3 Basal-like Breast Cancers

Basal-like breast cancers (BLBCs) were so named because these neoplasms consistently express molecules normally confined to the basal/myoepithelial compartment of the ductal and lobular epithelium, including basal cytokeratins (CK5, CK6, CK14, CK17), α-smooth muscle actin and vimentin [19]. BLBCs account for approximately 15% of all invasive breast cancers and are typically of high histological grade, demonstrate high mitotic indices, mutations in the *TP53* tumor suppressor gene, dysfunction or silencing of breast cancer susceptibility gene 1 (BRCA1), and almost uniformly lack expression of ER, progesterone receptor PR and HER2 and are thus triple-negative [57, 58]. Due to the absence of these receptors, BLBCs are not amenable to the targeted anti-estrogen and anti-HER2 therapies that have dramatically improved survival of patients diagnosed with luminal-type or HER2-positive tumors. Because of the aggressive biological features inherent to these tumors as well lack of targeted therapies, the basal-like malignancies are associated with the most aggressive clinical behavior and poorest prognosis among all molecular classifications of breast cancer [52]. Interestingly, human breast cancer cell lines derived from basal-like malignancies show exaggerated self-renewal capacity *in vitro* and are almost uniformly composed of $CD44^{Hi}/CD24^{-/low}$ cells, suggesting they may be enriched for cells that possess stem/progenitor-like properties [16]. These similarities logically pointed to the mammary stem cell as the likely origin of BLBCs. Unexpectedly, comparison of the BLBC transcriptional profile with the profiles of normal mammary epithelial components revealed great similarity between the BLBC and the $CD49f^{+}/EpCAM^{+}$ luminal progenitor signature [29]. It has also since been demonstrated that deletion of *Brca1* in the luminal-progenitor population of the mouse mammary epithelium generates tumors which phenocopy human BLBCs at both the histological and molecular level, while the identical genetic change in the murine mammary stem cell generates adenomyoepitheliomas, an exceedingly rare form of human breast cancer [33].

20.5.4 Mammary Stem Cell-Derived Tumors

The claudin-low subtype of breast cancer was identified in 2007 by examining similarities between mouse and human mammary tumors [24]. This molecular subtype is characterized by low expression of components of the tight and adherens junctions, including claudins 3, 4, 7, and E-cadherin [24, 39]. When compared to all breast tumors, those classified as claudin-low also were enriched for expression of genes involved in immunological responses, cellular communication, extracellular matrix, migration and angiogenesis and showed recurrent copy-number amplification of the *KRAS2* locus [39]. Further studies revealed that claudin-low tumors display molecular features consistent with the mammary stem cell and exhibit transcriptional evidence of epithelial-to-mesenchymal transition, including high proportions of $CD44^{Hi}/CD24^{-/Low}$ cells, high expression of *TWIST1* and *SNAI3* and repression of E-cadherin [22]. Comparing the transcriptional profiles of claudin-low tumors with components of the normal mammary epithelium revealed that the gene expression patterns of these tumors closely mirrored those observed in the mammary stem cell-enriched population [29].

Metaplastic breast cancers (MBCs) are a morphologically diverse group of mostly TN malignancies that exhibit mesenchymal, sarcomatoid and/or squamous metaplasia [20, 21, 23, 40]. Transcriptional profiling of these tumors originally classified them as basal-like malignancies [40]. By refining the criteria used for classification and including the recently-identified claudin-low subtype, MBCs were shown to be molecularly heterogeneous and may cluster with the basal-like, claudin-low or normal breast-like subtypes [22, 39].

20.6 Epithelial-Mesenchymal Plasticity and Breast Cancer Stem Cells

The epithelial-mesenchymal transition (EMT) is a latent embryonic program that allows polarized epithelial cells to assume a mesenchymal phenotype and gain migratory and invasive capacity and resist apoptosis. This process is critical in early development, as embryo implantation, placentation, specification of the mesendoderm from the primitive streak and formation of neural crest cells from the neuroectoderm all require EMT. Moreover, in both physiologic wound healing as well as pathologic fibrosis, inflammation-associated signaling induces epithelial cell changes that facilitate wound closure and deposition of extracellular matrix components [25]. An intimate association has been described between EMT and the acquisition and maintenance of the stem cell phenotype in breast cancer. Clinically, EMT has been implicated in conferring two of the most insidious properties of cancer: the ability of epithelial cancer cells to invade and migrate away from the primary site, and the ability to resist cytotoxic chemotherapy and ionizing radiation. Several studies have demonstrated that forced induction of EMT in mammary epithelial cells by expression mesenchyme-associated transcription factors SNAI1 or TWIST1 or treatment with transforming growth factor beta (TGFβ) increased expression of stemness-associated genes, dramatic expansion of the $CD44^{Hi}/CD24^{-/Low}$ stem-like population, increased mammosphere-forming capacity and improved outgrowth potential in xenotransplant assays [31, 34]. Importantly, several studies have demonstrated that expression of EMT-associated genes, as well as those specifically associated with invasion, metastasis and angiogenesis were significantly enriched in the $CD44^{Hi}/CD24^{-/Low}$ stem-like population compared to differentiated cell populations [31, 46, 47]. Moreover, migrating breast cancer cells isolated from the peripheral circulation or bone marrow appear to be enriched in the $CD44^{Hi}/CD24^{-/Low}$ population [42, 43, 50]. Thus, aberrant induction of developmental transcriptional programs like EMT may allow cells without intrinsic self-renewal capacity to gain this property and alter the plasticity and developmental potential of even highly differentiated cells.

20.7 Breast Cancer Stem Cells and Cancer Therapy

The propensity for breast cancer recurrence following prolonged periods of remission suggests the existence of a small population of cells capable of maintaining quiescence during cytotoxic injury. Since most chemotherapeutic agents and radiation target actively dividing cells, slowly cycling quiescent CSCs would be expected to be intrinsically resistant to such insults. The efficacy of most modern therapeutic agents is evaluated based on tumor regression on radiographic evaluation (e.g., Response Evaluation Criteria in Solid Tumors (RECIST)). Since CSCs are thought to be a rare therapy-resistant population of cells within a solid tumor, gross regression of such lesions is likely due to death of rapidly dividing non-CSCs that comprise the bulk of the tumor. Traditional therapies may thus select for the CSC population that is responsible for local recurrence. A report by Jenny Chang's group has demonstrated that breast cancer cells isolated 12 weeks after treatment with neoadjuvant chemotherapy are enriched for the $CD44^{Hi}/CD24^{-/Low}$ population and exhibit increased mammosphere forming potential compared to paired specimens acquired before cytotoxic therapy [28]. In an extension of these studies, the same group reported that specimens acquired after endocrine or cytotoxic therapy are enriched for the mammary stem cell/claudin-low gene expression signature, including mesenchymal-associated genes [9].

The intrinsic resistance of CSCs to traditional therapies may explain why marked regression of gross tumor volume during therapy does not always correspond to improved survival. This highlights the necessity of developing agents that specifically target the CSC population. Since the defining characteristic of both normal and cancer-associated stem cells is their ability to undergo self- renewal, targeting morphogenetic pathways that are dysregulated in the CSCs, and in effect permit aberrant self-renewal, may provide novel strategies to combat cancer therapeutic resistance and tumor recurrence.

Several groups have demonstrated recurrent perturbations in the Notch, Wnt and Hedgehog pathways in both murine and human breast cancers. Hyperactivation of the Notch signaling pathway has been shown to increase self-renewal capacity of cells *in vitro* and is observed in a significant fraction of human breast cancers, especially those with luminal phenotypes [2, 4, 12]. Liberation of the Notch intracellular domain (NICD) is an obligate event for Notch signaling and requires the activity of γ-secretase. In an effort to prevent Notch-mediated self-renewal, phase I and II clinical trials are currently underway to evaluate the utility of γ-secretase inhibitors in combination breast cancer therapy. Further progress in understanding the role(s) of these and other morphogenetic pathways are certain to identify additional targets in CSCs. Using a screening approach, Weinberg and colleagues interrogated a library of approximately 16,000 compounds against an immortalized mammary epithelial cell line and this same cell line that had acquired stem-like properties by forced EMT. This screen identified the compound salinomycin, which was capable of selectively depleting the $CD44^{Hi}/CD24^{-/Low}$ population, decreasing mammosphere forming efficiency, slowing xenograft tumor growth and impeding metastasis *in vivo* [18].

20.8 Closing Remarks

The stem cell theory of cancer has revolutionized our understanding of the etiology of breast cancer and provides insights into the biological basis of breast cancer heterogeneity. Despite immense progress in the understanding of molecular perturbations that incite and elaborate malignancy, dramatic improvements in breast cancer therapy are unlikely unless we are able to develop and institute therapies that target

CSCs. Before such targeted approaches can become a reality, we must delineate the precise ancestry and histogenesis of each type of breast cancer and fundamentally understand if and to what extent a small population of self-renewing cells contributes to their perpetuation and progression. Because the same morphogenetic pathways that have been hijacked by cancer cells remain critical in resident normal tissue stem cells, we must identify unique alterations in CSCs that lend to therapeutic exploitation. Furthermore, because CSCs appear to be relatively resistant to the effects of conventional chemo- and radiotherapy, strategies which mitigate the selection of these quiescent cells and methods that enable detection of residual CSCs are sorely needed.

References

1. Al-Hajj M, Wicha MS, Benito-Hernandez A, Morrison SJ, Clarke MF (2003) Prospective identification of tumorigenic breast cancer cells. Proc Natl Acad Sci U S A 100:3983–3988
2. Asselin-Labat ML, Sutherland KD, Barker H, Thomas R, Shackleton M, Forrest NC, Hartley L, Robb L, Grosveld FG, van der Wees J, Lindeman GJ, Visvader JE (2007) Gata-3 is an essential regulator of mammary-gland morphogenesis and luminal-cell differentiation. Nat Cell Biol 9:201–209
3. Bonnet D, Dick JE (1997) Human acute myeloid leukemia is organized as a hierarchy that originates from a primitive hematopoietic cell. Nat Med 3:730–737
4. Bouras T, Pal B, Vaillant F, Harburg G, Asselin-Labat ML, Oakes SR, Lindeman GJ, Visvader JE (2008) Notch signaling regulates mammary stem cell function and luminal cell-fate commitment. Cell Stem Cell 3:429–441
5. Casalini P, Iorio MV, Galmozzi E, Menard S (2004) Role of HER receptors family in development and differentiation. J Cell Physiol 200:343–350
6. Chou J, Provot S, Werb Z (2010) GATA3 in development and cancer differentiation: cells GATA have it! J Cell Physiol 222:42–49
7. Collins AT, Berry PA, Hyde C, Stower MJ, Maitland NJ (2005) Prospective identification of tumorigenic prostate cancer stem cells. Cancer Res 65:10946–10951
8. Coughlin SS, Ekwueme DU (2009) Breast cancer as a global health concern. Cancer Epidemiol 33:315–318
9. Creighton CJ, Li X, Landis M, Dixon JM, Neumeister VM, Sjolund A, Rimm DL, Wong H, Rodriguez A, Herschkowitz JI, Fan C, Zhang X, He X, Pavlick A, Gutierrez MC, Renshaw L, Larionov AA, Faratian D, Hilsenbeck SG, Perou CM, Lewis MT, Rosen JM, Chang JC (2009) Residual breast cancers after conventional therapy display mesenchymal as well as tumor-initiating features. Proc Natl Acad Sci USA 106:13820–13825
10. Deome KB, Faulkin LJ Jr, Bern HA, Blair PB (1959) Development of mammary tumors from hyperplastic alveolar nodules transplanted into gland-free mammary fat pads of female C3H mice. Cancer Res 19:515–520
11. Dontu G, Abdallah WM, Foley JM, Jackson KW, Clarke MF, Kawamura MJ, Wicha MS (2003) In vitro propagation and transcriptional profiling of human mammary stem/progenitor cells. Genes Dev 17:1253–1270
12. Dontu G, Jackson KW, McNicholas E, Kawamura MJ, Abdallah WM, Wicha MS (2004) Role of Notch signaling in cell-fate determination of human mammary stem/progenitor cells. Breast Cancer Res 6:R605–R615
13. Dydensborg AB, Rose AA, Wilson BJ, Grote D, Paquet M, Giguere V, Siegel PM, Bouchard M (2009) GATA3 inhibits breast cancer growth and pulmonary breast cancer metastasis. Oncogene 28:2634–2642

14. Esteva FJ, Yu D, Hung MC, Hortobagyi GN (2010) Molecular predictors of response to trastuzumab and lapatinib in breast cancer. Nat Rev Clin Oncol 7:98–107
15. Fillmore CM, Gupta PB, Rudnick JA, Caballero S, Keller PJ, Lander ES, Kuperwasser C (2010) Estrogen expands breast cancer stem-like cells through paracrine FGF/Tbx3 signaling. Proc Natl Acad Sci USA 107:21737–21742
16. Fillmore CM, Kuperwasser C (2008) Human breast cancer cell lines contain stem-like cells that self-renew, give rise to phenotypically diverse progeny and survive chemotherapy. Breast Cancer Res 10:R25
17. Ginestier C, Hur MH, Charafe-Jauffret E, Monville F, Dutcher J, Brown M, Jacquemier J, Viens P, Kleer CG, Liu S, Schott A, Hayes D, Birnbaum D, Wicha MS, Dontu G (2007) ALDH1 is a marker of normal and malignant human mammary stem cells and a predictor of poor clinical outcome. Cell Stem Cell 1:555–567
18. Gupta PB, Onder TT, Jiang G, Tao K, Kuperwasser C, Weinberg RA, Lander ES (2009) Identification of selective inhibitors of cancer stem cells by high-throughput screening. Cell 138:645–659
19. Gusterson BA, Ross DT, Heath VJ, Stein T (2005) Basal cytokeratins and their relationship to the cellular origin and functional classification of breast cancer. Breast Cancer Res 7:143–148
20. Gutman H, Pollock RE, Janjan NA, Johnston DA (1995) Biologic distinctions and therapeutic implications of sarcomatoid metaplasia of epithelial carcinoma of the breast. J Am Coll Surg 180:193–199
21. Hennessy BT, Giordano S, Broglio K, Duan Z, Trent J, Buchholz TA, Babiera G, Hortobagyi GN, Valero V (2006) Biphasic metaplastic sarcomatoid carcinoma of the breast. Ann Oncol 17:605–613
22. Hennessy BT, Gonzalez-Angulo AM, Stemke-Hale K, Gilcrease MZ, Krishnamurthy S, Lee JS, Fridlyand J, Sahin A, Agarwal R, Joy C, Liu W, Stivers D, Baggerly K, Carey M, Lluch A, Monteagudo C, He X, Weigman V, Fan C, Palazzo J, Hortobagyi GN, Nolden LK, Wang NJ, Valero V, Gray JW, Perou CM, Mills GB (2009) Characterization of a naturally occurring breast cancer subset enriched in epithelial-to-mesenchymal transition and stem cell characteristics. Cancer Res 69:4116–4124
23. Hennessy BT, Krishnamurthy S, Giordano S, Buchholz TA, Kau SW, Duan Z, Valero V, Hortobagyi GN (2005) Squamous cell carcinoma of the breast. J Clin Oncol 23:7827–7835
24. Herschkowitz JI, Simin K, Weigman VJ, Mikaelian I, Usary J, Hu Z, Rasmussen KE, Jones LP, Assefnia S, Chandrasekharan S, Backlund MG, Yin Y, Khramtsov AI, Bastein R, Quackenbush J, Glazer RI, Brown PH, Green JE, Kopelovich L, Furth PA, Palazzo JP, Olopade OI, Bernard PS, Churchill GA, Van Dyke T, Perou CM (2007) Identification of conserved gene expression features between murine mammary carcinoma models and human breast tumors. Genome Biol 8:R76
25. Kalluri R, Weinberg RA (2009) The basics of epithelial-mesenchymal transition. J Clin Invest 119:1420–1428
26. Korkaya H, Paulson A, Iovino F, Wicha MS (2008) HER2 regulates the mammary stem/progenitor cell population driving tumorigenesis and invasion. Oncogene 27:6120–6130
27. Lapidot T, Sirard C, Vormoor J, Murdoch B, Hoang T, Caceres-Cortes J, Minden M, Paterson B, Caligiuri MA, Dick JE (1994) A cell initiating human acute myeloid leukaemia after transplantation into SCID mice. Nature 367:645–648
28. Li X, Lewis MT, Huang J, Gutierrez C, Osborne CK, Wu MF, Hilsenbeck SG, Pavlick A, Zhang X, Chamness GC, Wong H, Rosen J, Chang JC (2008) Intrinsic resistance of tumorigenic breast cancer cells to chemotherapy. J Natl Cancer Inst 100:672–679
29. Lim E, Vaillant F, Wu D, Forrest NC, Pal B, Hart AH, Asselin-Labat ML, Gyorki DE, Ward T, Partanen A, Feleppa F, Huschtscha LI, Thorne HJ, Fox SB, Yan M, French JD, Brown MA, Smyth GK, Visvader JE, Lindeman GJ (2009) Aberrant luminal progenitors as the candidate target population for basal tumor development in BRCA1 mutation carriers. Nat Med 15:907–913

30. Liu E, Thor A, He M, Barcos M, Ljung BM, Benz C (1992) The HER2 (c-erbB-2) oncogene is frequently amplified in in situ carcinomas of the breast. Oncogene 7:1027–1032
31. Mani SA, Guo W, Liao MJ, Eaton EN, Ayyanan A, Zhou AY, Brooks M, Reinhard F, Zhang CC, Shipitsin M, Campbell LL, Polyak K, Brisken C, Yang J, Weinberg RA (2008) The epithelial-mesenchymal transition generates cells with properties of stem cells. Cell 133:704–715
32. Moasser MM (2007) The oncogene HER2: its signaling and transforming functions and its role in human cancer pathogenesis. Oncogene 26:6469–6487
33. Molyneux G, Geyer FC, Magnay FA, McCarthy A, Kendrick H, Natrajan R, Mackay A, Grigoriadis A, Tutt A, Ashworth A, Reis-Filho JS, Smalley MJ (2010) BRCA1 basal-like breast cancers originate from luminal epithelial progenitors and not from basal stem cells. Cell Stem Cell 7:403–417
34. Morel AP, Lievre M, Thomas C, Hinkal G, Ansieau S, Puisieux A (2008) Generation of breast cancer stem cells through epithelial-mesenchymal transition. PLoS One 3:e2888
35. Murphy CG, Fornier M (2010) HER2-positive breast cancer: beyond trastuzumab. Oncology (Williston Park) 24:410–415
36. O'Brien CA, Pollett A, Gallinger S, Dick JE (2007) A human colon cancer cell capable of initiating tumour growth in immunodeficient mice. Nature 445:106–110
37. Park K, Han S, Kim HJ, Kim J, Shin E (2006) HER2 status in pure ductal carcinoma in situ and in the intraductal and invasive components of invasive ductal carcinoma determined by fluorescence in situ hybridization and immunohistochemistry. Histopathology 48:702–707
38. Perou CM, Sorlie T, Eisen MB, van de Rijn M, Jeffrey SS, Rees CA, Pollack JR, Ross DT, Johnsen H, Akslen LA, Fluge O, Pergamenschikov A, Williams C, Zhu SX, Lonning PE, Borresen-Dale AL, Brown PO, Botstein D (2000) Molecular portraits of human breast tumours. Nature 406:747–752
39. Prat A, Parker JS, Karginova O, Fan C, Livasy C, Herschkowitz JI, He X, Perou CM (2010) Phenotypic and molecular characterization of the claudin-low intrinsic subtype of breast cancer. Breast Cancer Res 12:R68
40. Reis-Filho JS, Milanezi F, Steele D, Savage K, Simpson PT, Nesland JM, Pereira EM, Lakhani SR, Schmitt FC (2006) Metaplastic breast carcinomas are basal-like tumours. Histopathology 49:10–21
41. Ricci-Vitiani L, Lombardi DG, Pilozzi E, Biffoni M, Todaro M, Peschle C, De Maria R (2007) Identification and expansion of human colon-cancer-initiating cells. Nature 445:111–115
42. Riethdorf S, Pantel K (2008) Disseminated tumor cells in bone marrow and circulating tumor cells in blood of breast cancer patients: current state of detection and characterization. Pathobiology 75:140–148
43. Riethdorf S, Wikman H, Pantel K (2008) Review: biological relevance of disseminated tumor cells in cancer patients. Int J Cancer 123:1991–2006
44. Rodriguez-Pinilla SM, Sarrio D, Honrado E, Hardisson D, Calero F, Benitez J, Palacios J (2006) Prognostic significance of basal-like phenotype and fascin expression in node-negative invasive breast carcinomas. Clin Cancer Res 12:1533–1539
45. Shackleton M, Vaillant F, Simpson KJ, Stingl J, Smyth GK, Asselin-Labat ML, Wu L, Lindeman GJ, Visvader JE (2006) Generation of a functional mammary gland from a single stem cell. Nature 439:84–88
46. Sheridan C, Kishimoto H, Fuchs RK, Mehrotra S, Bhat-Nakshatri P, Turner CH, Goulet R Jr, Badve S, Nakshatri H (2006) CD44+/CD24- breast cancer cells exhibit enhanced invasive properties: an early step necessary for metastasis. Breast Cancer Res 8:R59
47. Shipitsin M, Campbell LL, Argani P, Weremowicz S, Bloushtain-Qimron N, Yao J, Nikolskaya T, Serebryiskaya T, Beroukhim R, Hu M, Halushka MK, Sukumar S, Parker LM, Anderson KS, Harris LN, Garber JE, Richardson AL, Schnitt SJ, Nikolsky Y, Gelman RS, Polyak K (2007) Molecular definition of breast tumor heterogeneity. Cancer Cell 11:259–273
48. Shiu KK, Natrajan R, Geyer FC, Ashworth A, Reis-Filho JS (2010) DNA amplifications in breast cancer: genotypic-phenotypic correlations. Future Oncol 6:967–984

49. Singh SK, Hawkins C, Clarke ID, Squire JA, Bayani J, Hide T, Henkelman RM, Cusimano MD, Dirks PB (2004) Identification of human brain tumour initiating cells. Nature 432:396–401
50. Slade MJ, Payne R, Riethdorf S, Ward B, Zaidi SA, Stebbing J, Palmieri C, Sinnett HD, Kulinskaya E, Pitfield T, McCormack RT, Pantel K, Coombes RC (2009) Comparison of bone marrow, disseminated tumour cells and blood-circulating tumour cells in breast cancer patients after primary treatment. Br J Cancer 100:160–166
51. Slamon DJ, Leyland-Jones B, Shak S, Fuchs H, Paton V, Bajamonde A, Fleming T, Eiermann W, Wolter J, Pegram M, Baselga J, Norton L (2001) Use of chemotherapy plus a monoclonal antibody against HER2 for metastatic breast cancer that overexpresses HER2. N Engl J Med 344:783–792
52. Sorlie T, Perou CM, Tibshirani R, Aas T, Geisler S, Johnsen H, Hastie T, Eisen MB, van de Rijn M, Jeffrey SS, Thorsen T, Quist H, Matese JC, Brown PO, Botstein D, Eystein Lonning P, Borresen-Dale AL (2001) Gene expression patterns of breast carcinomas distinguish tumor subclasses with clinical implications. Proc Natl Acad Sci USA 98:10869–10874
53. Sorlie T, Tibshirani R, Parker J, Hastie T, Marron JS, Nobel A, Deng S, Johnsen H, Pesich R, Geisler S, Demeter J, Perou CM, Lonning PE, Brown PO, Borresen-Dale AL, Botstein D (2003) Repeated observation of breast tumor subtypes in independent gene expression data sets. Proc Natl Acad Sci USA 100:8418–8423
54. Sotiriou C, Neo SY, McShane LM, Korn EL, Long PM, Jazaeri A, Martiat P, Fox SB, Harris AL, Liu ET (2003) Breast cancer classification and prognosis based on gene expression profiles from a population-based study. Proc Natl Acad Sci USA 100:10393–10398
55. Stingl J, Eirew P, Ricketson I, Shackleton M, Vaillant F, Choi D, Li HI, Eaves CJ (2006) Purification and unique properties of mammary epithelial stem cells. Nature 439:993–997
56. Tamimi RM, Baer HJ, Marotti J, Galan M, Galaburda L, Fu Y, Deitz AC, Connolly JL, Schnitt SJ, Colditz GA, Collins LC (2008) Comparison of molecular phenotypes of ductal carcinoma in situ and invasive breast cancer. Breast Cancer Res 10:R67
57. Turner NC, Reis-Filho JS (2006) Basal-like breast cancer and the BRCA1 phenotype. Oncogene 25:5846–5853
58. Turner NC, Reis-Filho JS, Russell AM, Springall RJ, Ryder K, Steele D, Savage K, Gillett CE, Schmitt FC, Ashworth A, Tutt AN (2007) BRCA1 dysfunction in sporadic basal-like breast cancer. Oncogene 26:2126–2132
59. Usary J, Llaca V, Karaca G, Presswala S, Karaca M, He X, Langerod A, Karesen R, Oh DS, Dressler LG, Lonning PE, Strausberg RL, Chanock S, Borresen-Dale AL, Perou CM (2004) Mutation of GATA3 in human breast tumors. Oncogene 23:7669–7678
60. Visvader JE (2009) Keeping abreast of the mammary epithelial hierarchy and breast tumorigenesis. Genes Dev 23:2563–2577
61. Welm BE, Tepera SB, Venezia T, Graubert TA, Rosen JM, Goodell MA (2002) Sca-1(pos) cells in the mouse mammary gland represent an enriched progenitor cell population. Dev Biol 245:42–56
62. Yan W, Cao QJ, Arenas RB, Bentley B, Shao R (2010) GATA3 inhibits breast cancer metastasis through the reversal of epithelial-mesenchymal transition. J Biol Chem 285:14042–14051
63. Yoon NK, Maresh EL, Shen D, Elshimali Y, Apple S, Horvath S, Mah V, Bose S, Chia D, Chang HR, Goodglick L (2010) Higher levels of GATA3 predict better survival in women with breast cancer. Hum Pathol 41:1794–1801

Chapter 21
Translin/TRAX Deficiency Affects Mesenchymal Differentiation Programs and Induces Bone Marrow Failure

Reiko Ishida, Katsunori Aoki, Kazuhiko Nakahara, Yuko Fukuda, Momoko Ohori, Yumi Saito, Kimihiko Kano, Junichiro Matsuda, Shigetaka Asano, Richard T. Maziarz, and Masataka Kasai

Contents

R. Ishida • Y. Fukuda
Department of Immunology, National Institute of Infectious Diseases, 1-23-1 Toyama, Shinjuku-ku, Tokyo 162-8640, Japan

K. Aoki
Department of Immunology, National Institute of Infectious Diseases, 1-23-1 Toyama, Shinjuku-ku, Tokyo 162-8640, Japan

Department of Hematology (Internal Medicine), The University of Tokyo, Tokyo, Japan

K. Nakahara
National Institution for Academic Degrees and University Evaluation, Tokyo, Japan

M. Ohori • Y. Saito • K. Kano • S. Asano
Faculty of Science and Engineering, Waseda University, Tokyo, Japan

J. Matsuda
Laboratory of Animal Models for Human Diseases, National Institute of Biomedical Innovation, Osaka, Japan

R.T. Maziarz
Bone Marrow Transplantation Program, Center for Hematological Malignancies, Oregon Health and Science University, Portland, OR, USA

M. Kasai (✉)
Department of Immunology, National Institute of Infectious Diseases, 1-23-1 Toyama, Shinjuku-ku, Tokyo 162-8640, Japan

Center for Stem Cell and Regenerative Medicine, Institute of Medical Science, The University of Tokyo, Tokyo, Japan
e-mail: masataka@nih.go.jp

R.K. Srivastava and S. Shankar (eds.), *Stem Cells and Human Diseases*,
DOI 10.1007/978-94-007-2801-1_21,

Abstract The decision regarding self-renewal versus differentiation of hematopoietic stem cells (HSCs) is a crucial issue in bone marrow hematopoiesis. We have generated mice homozygous for an inactivating mutation of the whole Translin gene (Translin$^{-/-}$) and investigated their hematopoietic status during early and later in life. Here we show that Translin deficiency affects mesenchymal differentiation and results in perturbation of self-renewal HSCs. Young Translin$^{-/-}$ mice, especially around 3 weeks of age, displayed markedly reduced lymphocyte counts in the peripheral blood, attributable to developmental arrest of B-lymphocytes in the earliest progenitor stage. With aging, progressive bone marrow failure was displayed, with developmental arrest of myeloid cells and B lymphocytes in a stroma-dependent manner, and eventually ectopic osteogenesis, vasculogenesis and adipogenesis resulted. Despite apparent hematopoietic aplasia, however, the frequency of HSCs in the bone marrow of mutant mice was remarkably increased. Furthermore, knockdown of Translin and its binding partner protein, TRAX, up-regulated genes associated with mesenchymal differentiation in a mesenchymal stem cell line. Taken together, these findings suggest that the Translin and TRAX complex influences both self-renewal and multilineage differentiation of HSCs by targeting mesenchymal stem/progenitor cells.

Keywords Translin • TRAX • MSCs • HSCs • Bone marrow failure

21.1 Introduction

Hematopoietic progenitors differentiate from hematopoietic stem cells (HSCs), whose total number is kept constant by a process termed self-renewal [1]. Although this homeostasis is considered to be controlled by both intrinsic pathways and extrinsic signals, the precise mechanisms regulating stem cell fate are poorly understood. This is at least partly due to the lack of specific HSC markers and the structural complexity of bone. Studies in model organisms such as *Caenorhabditis elegans* and *Drosophila melanogaster* have provided evidence of two main mechanisms governing asymmetric cell division, one responsible for partitioning intrinsic regulators and the other involving the cell machinery responding to extrinsic signals [2]. Although HSCs seem to divide asymmetrically under steady-state conditions, it is conceivable that the balance between asymmetric and symmetric division of bone marrow HSCs is defective in some disease states.

A great deal of progress has been made in characterization of the specialized microenvironments in the bone marrow, known as the HSC 'niche' [3]. First, it was demonstrated that mutant mice with conditional inactivation of the bone morphogenetic protein (BMP) receptor type 1A (BMPR1A) may exhibit an increase in the number of osteoblasts and HSCs [4]. Furthermore, spindle-shaped osteoblasts lining the bone surface, which are termed SNO cells expressing a high level of N-cadherin, were found to constitute a component of the niche supporting HSCs (endosteal niche). Subsequently, it was reported that ligand-dependent activation of the PTH/PTHrP receptor (PPR) with parathyroid hormone (PTH) increased the number of osteoblasts and HSCs *in vivo* and *in vitro* respectively, leading to the conclusion that osteoblasts function as a niche for HSCs [5]. However, it should be noted that not all HSCs are closely associated with an endosteal niche [6]. Several studies have indicated that sinusoidal bone marrow endothelial cells, rather than osteoblasts, create a niche which supports the proliferation and differentiation of HSC/progenitor cells (vascular niche) [7]. More recently, Méndez-Ferrer et al. [8] identified Nestin^{+} mesenchymal cells (MSCs) closely associated with HSCs and reached a conclusion that they constitute a unique niche for HSCs.

In this chapter, we focus on the link between Translin/TRAX proteins and hematopoietic diseases such as progressive bone marrow failure. Translin was originally found as a DNA binding protein interacting with breakpoint junctions of chromosomal translocations in many cases of lymphoid neoplasms [9–14]. The human Translin gene encodes a protein of 228 amino acids with a predicted molecular size of 27 kDa. EM (electron microscopy) and crystallographic investigations to determine its three-dimensional character indicated a ring-shaped structure with an assembly of eight subunits [15, 16]. Amino acid homologies with human Translin are 99% and 86% for mouse and chicken forms, respectively, and point mutation analysis supported the view that the DNA binding domain of Translin is formed in the evolutionarily conserved ring-shaped structure in combination with the basic region (amino acids 86–97) polypeptides. Subsequent molecular analysis revealed that Translin participates in a variety of cellular activities central to nucleic metabolism, either of ssDNA or ss/dsRNA [17, 18]. In order to provide further insights into Translin function, we examined whether it might interact with other proteins using a yeast two-hybrid system and identified an associated 33 kDa protein partner, TRAX (Translin-associated factor X) showing extensive homology with Translin (38% identity of the C-terminal amino acid residues) [19, 20]. TRAX contains bipartite nuclear targeting sequences in its N-terminal region, suggesting a possible role in the selective nuclear transport of Translin, which itself lacks any nuclear targeting motifs. More recently, an asymmetric octameric assembly of Translin/TRAX subunits was shown to be a key activator of RISC (RNA-induced silencing complex) by degrading Ago2 (Argonautes 2)-nicked passenger strands [21–24], indicating that the association of these protein subunits plays a central role in eukaryotic cell biology.

To address the functional significance of Translin in the hematopoietic system, mice homozygous for an inactivating mutation of the whole Translin gene were generated. The knockout mutant (Translin$^{-/-}$) mice demonstrated delayed hematopoietic colony formation in the spleen after sublethal irradiation exposure, suggesting a significant role of the protein in hematopoietic regeneration [25] In the present study, young Translin$^{-/-}$ mice displayed markedly reduced lymphocyte counts in the peripheral blood, attributable to a developmental arrest of B-lymphocytes in the earliest progenitor stage. With aging, progressive bone marrow failure became apparent, with a dramatic increase of trabecular bones, vascular endothelial cells and brown adipocytes. Reciprocal bone marrow transplantation studies indicated these drastic changes to be stroma-dependent rather than stem cell autonomous. However, despite the apparent aplasia, the frequency of HSCs in the bone marrow was remarkably increased. Furthermore, in a population of mutant mice examined at approximately 18 months of age, hematopoietic tumors associated with malignant histiocytosis (MH) were identified. Finally, small interfering RNA (siRNA)-mediated knockdown of Translin and TRAX in mesenchymal progenitor cells resulted in up-regulation of several genes associated with mesenchymal differentiation programs. All these findings suggest that the Translin and TRAX complex affects bone marrow hematopoiesis by targeting mesenchymal stem/progenitor cells.

21.2 Results

21.2.1 Peripheral Cytopenia in Young Translin$^{-/-}$ Mice

Translin$^{-/-}$ mice were significantly smaller in body size than their wild-type littermate controls at birth, and growth retardation was most pronounced at 4–5-weeks of age. In addition, their mortality during this period was higher than that of wild-type or heterozygous littermates. As shown in Fig. 21.1a, the total number of white blood cells (WBCs) was markedly reduced in young Translin$^{-/-}$ mice, especially in those around 3 weeks of age. Further evaluation of peripheral blood revealed the lymphocyte counts in the mutant mice to be distinctly reduced relative to wild-type littermate controls, while there were no significant changes in the levels of granulocytes, monocytes and platelets. Since two-thirds of the WBCs in peripheral blood at this stage consist of B lymphocytes, it is reasonable to assume that the decrease in lymphocyte count may have been due to developmental failure of B lymphocytes. To examine this question, cells from peripheral blood and spleen were analyzed by flow cytometry using antibodies for lineage- and stage-specific cell-surface markers. On the whole, the development of B lymphocytes indeed appeared to be preferentially impaired in young Translin$^{-/-}$ mice and staining with antibodies to B220 and IgM revealed the percentage of B220^{+}/IgM^{+} cells (mature B cells) to be markedly reduced, whereas the percentage of B220^{-}/IgM^{-} cells was significantly increased (Fig. 21.1b). On the other hand, there were no significant changes in B220^{+}/IgM^{+} and B220^{-}/IgM^{-} cells in spleen (Fig. 21.1c). Reduction of mature B lymphocytes in peripheral blood was further confirmed by three-color flow cytometry analysis using PE-anti- IL-7R, PerCP-anti-B220 and FITC-anti-CD43 antibodies. The results clearly showed increased percentages of B220^{-} CD43^{+} IL-7R^{+} cells in peripheral blood of Translin$^{-/-}$ mice (Fig. 21.1d), indicating B-lymphocyte development to be arrested at the earliest progenitor stage.

Thus, our results demonstrated that young Translin$^{-/-}$ mice, especially those around 3 weeks of age, display peripheral cytopenia attributable to a developmental arrest of B lymphocytes. Impaired B lymphocyte development has been reported in mice deficient in various proteins required for recombination or repair [26, 27] and recombination-activating gene (RAG)1 and RAG2-deficient mice exhibit a complete block at the progenitor stage (pro B cells). Mice deficient in the DNA binding subunits of the DNA-dependent protein kinase (DNA-PK) [28, 29], Ku70 and Ku80 [30, 31] also exhibit arrest of lymphocyte development at the pro-B stage, as do large catalytic subunit (DNA-PKcs) deficient (SCID) mice [32]. In all of these cases, lymphocyte development is arrested at the pro-B cell stage, at which IgH chain gene

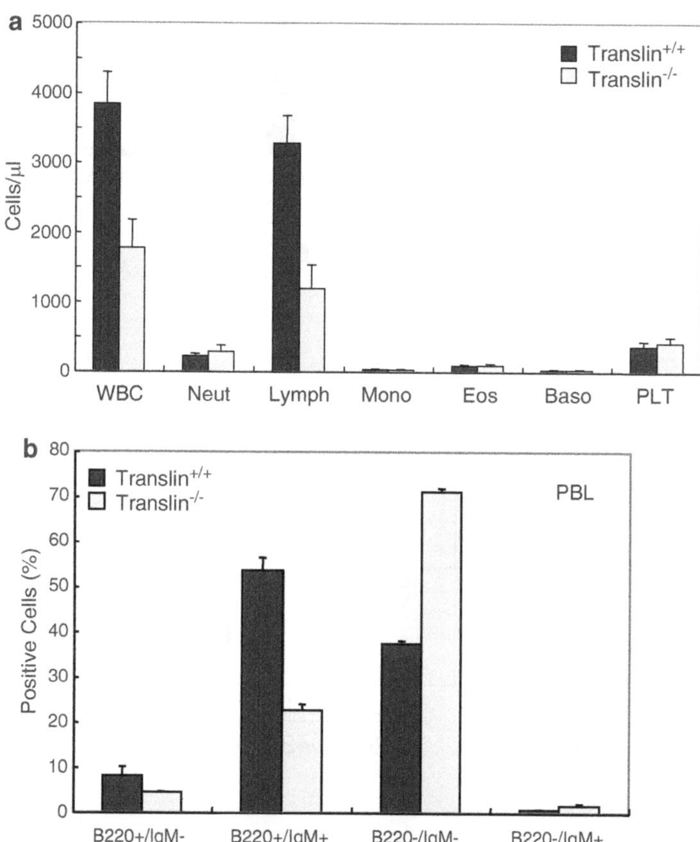

Fig. 21.1 Young Translin$^{-/-}$ mice exhibit developmental failure of peripheral lymphoid cells. (**a**) Peripheral blood analysis of Translin$^{+/+}$ and Translin$^{-/-}$ mice. Peripheral blood counts of Translin$^{+/+}$ and Translin$^{-/-}$ littermates at 3 weeks of age were determined by ADVIA120 (Bayer). The results shown are the averages for five mice of each genotype and represent the mean ± SD for whole blood cell (*WBC*), neutrophil (*Neut*), lymphocyte (*Lymph*), monocyte (*Mono*), eosinocyte (*Eos*), basophil (*Baso*) and platelet (*PLT*) values. (**b**, **c**) Reduction of mature B lymphocytes in the peripheral blood of Translin$^{-/-}$ mice. Peripheral blood lymphocytes (*PBL*) and spleen cells (*Spl*) of Translin$^{+/+}$ and Translin$^{-/-}$ littermates at 3 weeks of age were analyzed by flow cytometry using fluorescein (FITC)-labeled anti-B220 and phycoerythrin (PE)-labeled anti-IgM. (**d**) Developmental arrest of B lymphocytes in the peripheral blood of Translin$^{-/-}$ mice. PBL, Spl and bone marrow (*BM*) cells of Translin$^{+/+}$ and Translin$^{-/-}$ littermates at 3 weeks of age were analyzed by three-color flow cytometry using PerCP-anti-B220, FITC-anti-CD43 and PE-anti-IL-7R

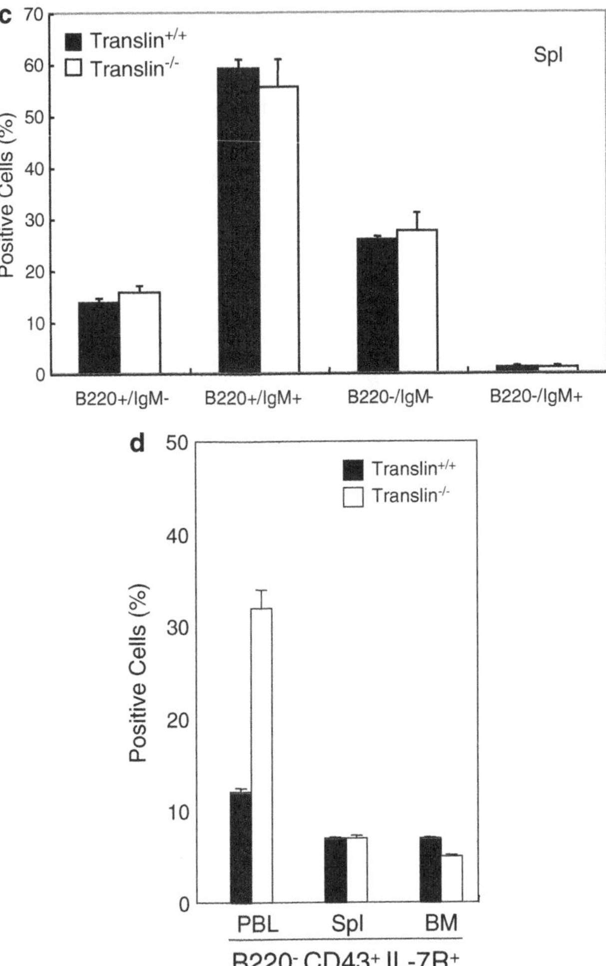

Fig. 21.1 (continued)

rearrangement is initiated. In Translin$^{-/-}$ mice, however, lymphocyte development was arrested at the stage of (B220^{-} CD43^{+} IL-7R^{+}) cells, which may be judged as uncommitted lymphohematopoietic progenitors but could in fact committed to the B lymphocyte lineage because we only observed marked reduction of mature B lymphocytes.

The general belief is that lymphocyte development in mammals occurs in the liver during embryogenesis and that hematopoiesis after birth takes place in the bone marrow. However, there has been substantial progress in elucidation of differences in processes of blood cell formation between fetal and adult life [33]. The appearance of the first HSCs in the developing mouse embryo is now known to occur in the aorta-gonad-mesonephros (AGM) region [34]. Subsequently, HSCs are maintained in the liver and bone marrow although many uncertainties remain as to how they are related in these two sites. Impaired lymphopoiesis in peripheral blood of young Translin$^{-/-}$ mice might support the idea that the adult pattern of lymphopoiesis occurring in bone marrow is not fully established shortly after birth [35], and that Translin plays a crucial role in peripheral lymphopoiesis during this period.

21.2.2 Stroma-Dependent Bone Marrow Failure in Aged Translin$^{-/-}$ Mice

The reduction in B lymphocytes in the peripheral blood of mutant mice was generally no longer evident at the age of 8–10 weeks. After around 8 months, however, some of the mutant mice began to exhibit progressive bone marrow failure. Histological sections of bone marrow demonstrated a reduction in myeloblasts and erythroblasts compared with Translin$^{+/+}$ littermate control (Fig. 21.2a, b). Eventually, as aging proceeded, the mutant mice were characterized by a decrease in the number of bone marrow cells and massive splenomegaly with extramedullary hematopoiesis. Simultaneously, an increase in the number of reticulocytes (over 10%) and circulating progenitors (metamyelocytes and orthochromatic erythroblasts) was noted in peripheral blood (Fig. 21.2c). Flow cytometric characterization using CD11b and Gr-1, markers of late myeloid differentiation, revealed a significant decrease of immature myeloid cells, CD11b^{low} Gr-1^{-} in the bone marrow from 12 months old Translin$^{-/-}$ mice (Fig. 21.3a). As shown in Fig. 21.3b, a reduction of sIgM^{+}B220^{+} B cells was observed in the bone marrow of aged Translin$^{-/-}$ mice with evaluation by flow cytometry using antibodies for lineage- and stage-specific cell-surface markers. Further analysis of B220 and CD43 expression on gated sIgM^{-} showed a decrease in CD43^{-} B220^{+} sIgM^{-} pre B cells and a significant increase in CD43^{+} B220^{-} sIgM^{-} immature cells including bi-potential precursors (Fig. 21.3c). Taken together, these results suggest that Translin deficiency results in developmental arrest of myeloid cells and B lymphocytes at an early progenitor stage. Concomitant developmental inhibition of myeloid and B lymphoid cells might be explained by the hypothesis that granulopoiesis and B lymphopoiesis are coupled in the bone marrow by development in a common niche [36].

We next asked whether the bone marrow failure in Translin$^{-/-}$ mice is stem cell autonomous or stroma-dependent. For this purpose, bone marrow cells from 8-week-old Translin$^{+/+}$ or Translin$^{-/-}$ (both with a Ly5.2 genotype) animals were transplanted into lethally irradiated Ly5.1 mice. We found that there was no difference in the number of donor-derived bone marrow cells (Fig. 21.3d, left panel). However, when bone marrow cells from Ly5.1 mice were transplanted into lethally irradiated animals, the number of donor-derived bone marrow cells in the Translin$^{-/-}$ mice was three times lower than in their Translin$^{+/+}$ littermate controls (Fig. 21.3d, right panel). Therefore, severe bone marrow failure in Translin$^{-/-}$ mice could be attributed to defects manifest within the hematopoietic microenvironment.

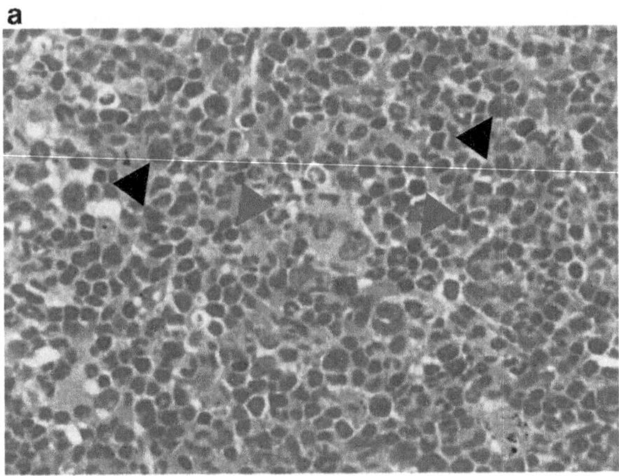

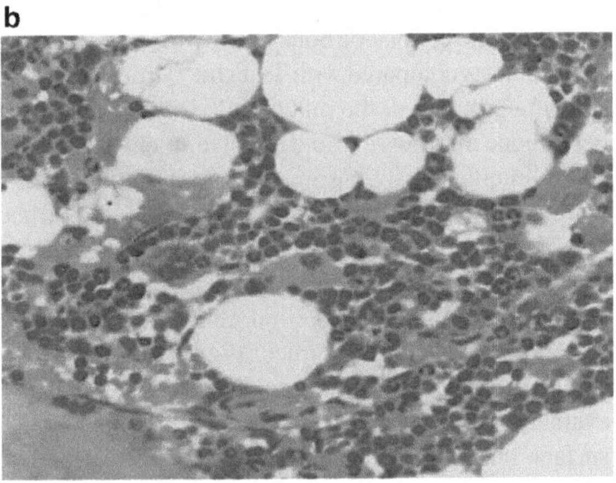

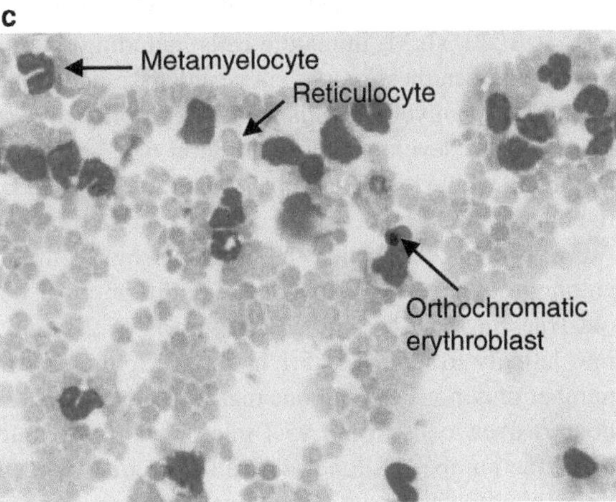

Fig. 21.2 Bone marrow failure and extramedullary hematopoiesis in aged Translin$^{-/-}$ mice. (**a**, **b**) Reduction of myeloblasts and erythroblasts in the bone marrow of aged Translin$^{-/-}$ mice (8 months old) as compared to Translin$^{+/+}$ littermates. Myeloblasts and erythroblasts in Translin$^{+/+}$ mice are shown by *black* and *red arrows* respectively. (**c**) Peripheral blood smears of aged Translin$^{-/-}$ mice (12 months old). Note increased number of reticulocytes, unusual appearance of metamyelocytes and orthochromatic erythroblasts (shown by *arrows*)

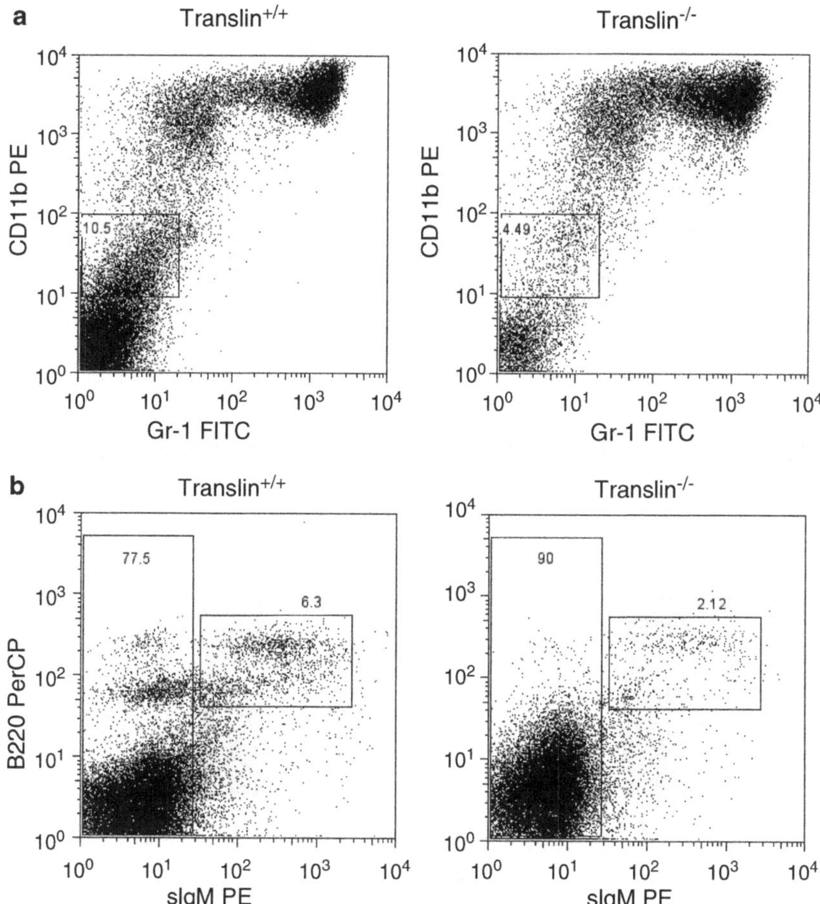

Fig. 21.3 Bone marrow failure in aged Translin$^{-/-}$ mice is stroma-dependent. (**a**) A significant decrease of immature myeloid cells, CD11b^{low} Gr-1^{-} in the bone marrow of aged Translin$^{-/-}$ mice as compared to Translin$^{+/+}$ mice (12 months old). (**b**) Reduction of B220^{+} sIgM^{+} B cells in the bone marrow of aged Translin$^{-/-}$ as compared to Translin$^{+/+}$ mice. (**c**) Reduction of CD43^{-} B220^{+} sIgM^{-} pre B cells and a significant increase of CD43^{+} B220^{-} sIgM^{-} immature cells in the bone marrow of aged Translin$^{-/-}$ as compared to Translin$^{+/+}$ mice. (**d**) Bone marrow failure of Translin$^{-/-}$ mice is stroma-dependent. Bone marrow cells (5×10^4) from 8 months old Translin$^{-/-}$ or littermate control mice (Ly5.2) were transplanted into irradiated 8 weeks old wild-type Ly5.1 mice (*left panel*). A reciprocal assay was also carried out by transplanting bone marrow cells derived from 8 weeks old wild-type donor mice (Ly5.1) into irradiated 8 months old Translin$^{-/-}$ or littermate control mice (*right panel*). Two months after transplantation, the bone marrow cells of the recipient mice were analyzed for engraftment of donor cells

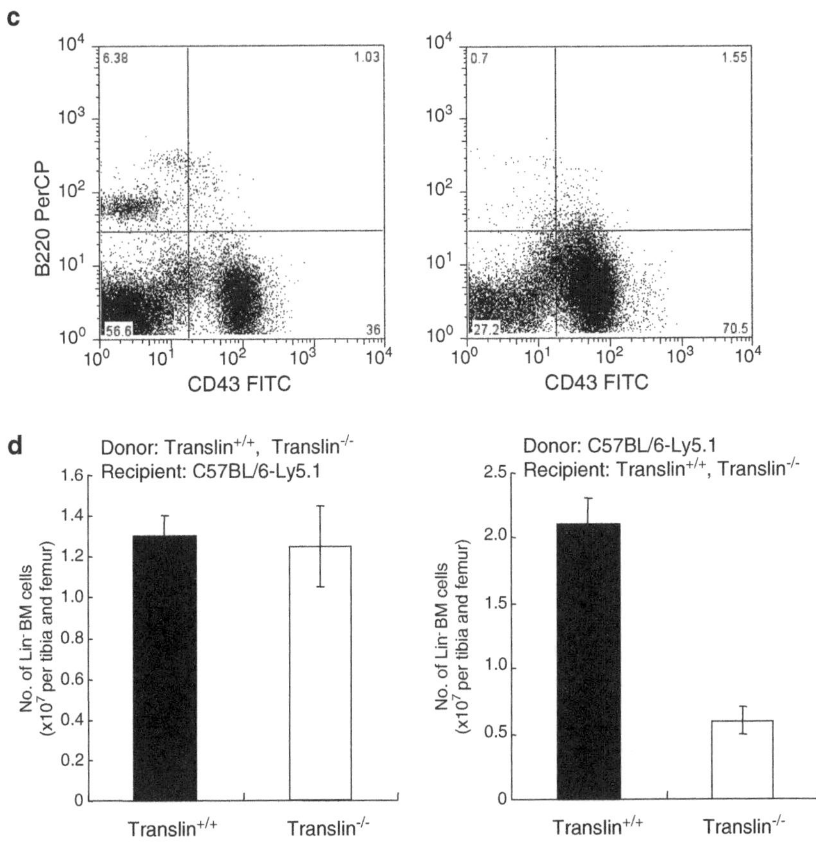

Fig. 21.3 (continued)

21.2.3 Increased Frequency of HSCs and Hematopoietic Tumors

To assess the role that Translin may play in the generation of stem cells, we first investigated the frequency of HSCs in bone marrow from aged Translin$^{-/-}$ mice exhibiting bone marrow failure by flow cytometry. Surprisingly, however, the frequency of HSCs (Lin^{-}, CD34^{-}, Sca1^{+}, c-Kit^{+},) was increased tenfold in the Translin$^{-/-}$ mice (Fig. 21.4a, right panel), compared with Translin$^{+/+}$ littermate controls (Fig. 21.4a, left panel). The total number of bone marrow cells per tibia and femur was reduced in the mutant mice, but the absolute number of HSCs was still increased fourfold. These results are in line with the previous observation that elimination of c-Myc activity in the bone marrow causes severe cytopenia and accumulation of HSCs, presumably through impaired interactions with niche elements [37]. Our preliminary analysis of the cell cycle status of HSCs revealed that there is a significant delay in the S/G2/M phase of bone marrow HSCs of Translin$^{-/-}$ mice as compared to the wild type case. Given the importance of the correct orientation and position

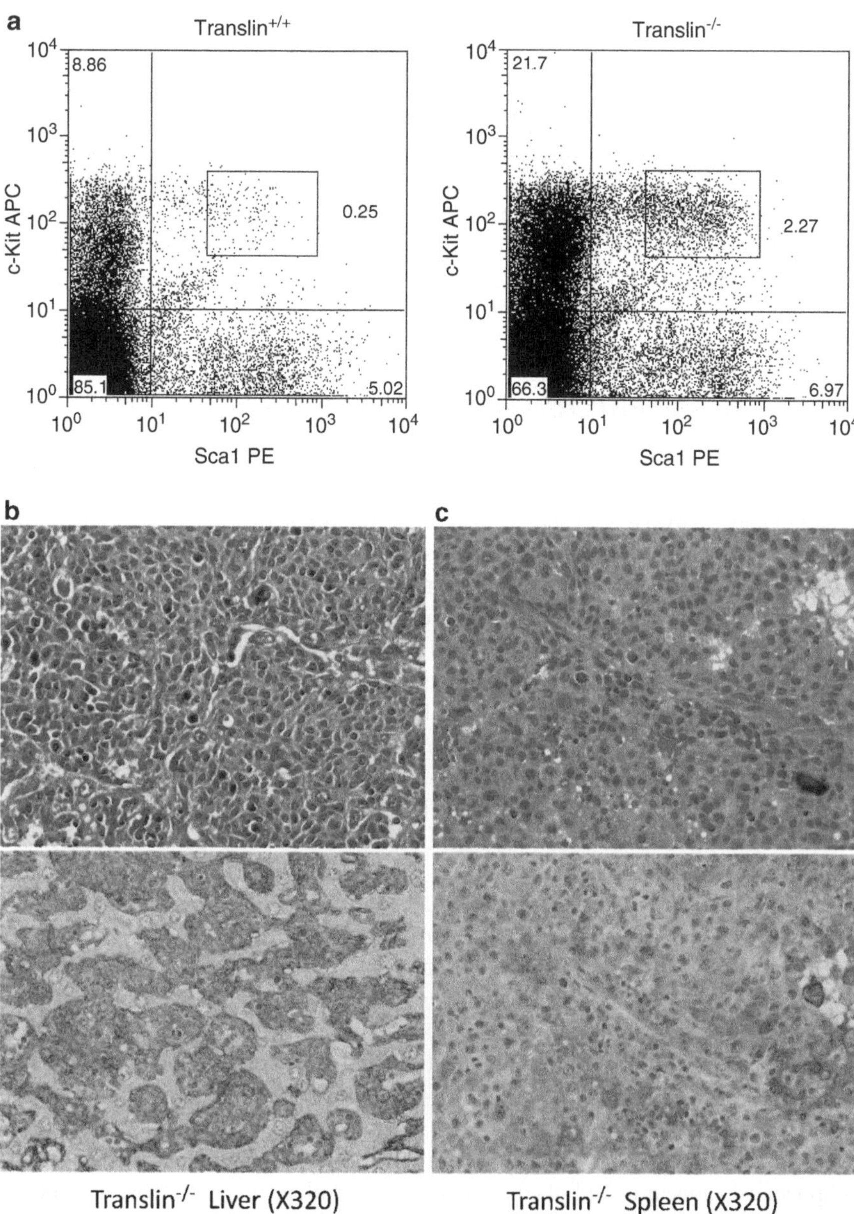

Fig. 21.4 Increased frequency of HSCs and hematopoietic tumors in aged Translin$^{-/-}$ mice. (**a**) The percentage of HSCs (Lin^{-}, CD34^{-}, Sca1^{+}, c-Kit^{+}) was increased tenfold in Translin$^{-/-}$ mice (*right panel*) compared with Translin$^{+/+}$ mice (12 months old) (*left panel*). (**b**) The liver specimens of 18 months old Translin$^{-/-}$ mice were subjected to staining with H&E (*upper panel*) and immunoperoxidase staining with anti-F4/80 (*lower panel*). (**c**) The spleen specimens of 18 months old Translin$^{-/-}$ mice were subjected to staining with H&E (*upper panel*) and immunoperoxidase staining with anti-F4/80 (*lower panel*)

of the mitotic spindle during asymmetric cell division [38], these results suggest that deficiency of Translin also can impact on the mitotic phase of the cell cycle which determines the fate of bone marrow stem cells.

Finally, we carried out a long term observation (18 months) of a group of Translin$^{-/-}$ mice. In this group of animals, ~5% developed tumors in the liver or spleen. Our histopathology studies showed highly invasive masses in the veins and sinusoids (Fig. 21.4b, c, upper panel), positive for F4/80 (Fig. 21.4b, c, lower panel), and most closely resembling hematopoietic tumors associated with malignant histiocytosis (MH), suggesting an origin from undifferentiated hematopoietic stem cells.

21.2.4 Accelerated Osteogenesis, Vasculogenesis and Adipogenesis

Histological analysis revealed obvious abnormalities in bone formation of the mutant mice (12 months old). In the Translin$^{+/+}$ mice, trabecular bone was restricted to the metaphyseal area (Fig. 21.5a), but dramatic increase in ectopically formed trabecular tissue was evident in the long-bone regions of the femurs from Translin$^{-/-}$ mice (Fig. 21.5b). To gain further insights into the role of Translin in the hematopoietic microenvironment, the femurs of aged mice (8 months old) were flushed with medium to remove marrow cells. When bone fractions were cut into small pieces and treated with collagenase and dispase, microarray analysis of total RNA from collected stromal cells demonstrated a 32 kDa highly o-glycosylated cell surface protein, CD99, expressed on leukocytes and endothelial junctions to be most differentially expressed among 45,000 genes of mouse gene chips from Affymetrix. Quantitative reverse transcription-PCR (qRT-PCR) revealed an approximately 30-fold increase in the mRNA level of CD99 in the stromal cells of Translin$^{-/-}$ mice compared with Translin$^{+/+}$ littermate control mice (Fig. 21.6a). Interestingly, immunoperoxidase staining for CD99 in paraffin-embedded bone tissues of aged Translin$^{-/-}$ mice revealed abundant vascular endothelial cells (black arrows) (Fig. 21.6b). Increased vascularization was further confirmed by staining with MECA-32 antibody specific for panendothelial cell antigen (data not shown). Furthermore, we found brown adipocyte-like cells (red arrows), and this was confirmed by immunoperoxidase staining with brown-fat-specific antibody (anti-UCP-1) [39] (Fig. 21.6c). Thus, in addition to osteogenesis, vasculogenesis and adipogenesis appear to be linked to bone marrow failure in aged Translin$^{-/-}$ mice.

21.2.5 Knockdown of Translin/TRAX Affects Gene Expression Profiling of MSCs

In contrast to the accelerated osteogenesis, vasculogenesis and adipogenesis, our histological analysis revealed delay in chondrocyte development during fetal and

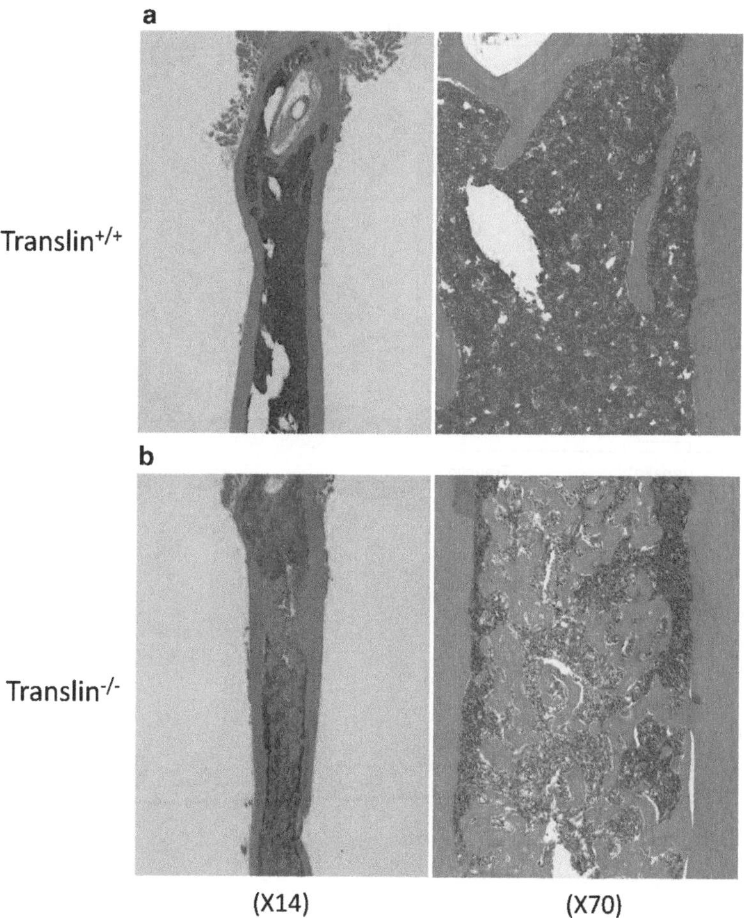

Fig. 21.5 Ectopically formed trabecular tissue in femurs of aged Translin$^{-/-}$ mice. (**a**, **b**) Histology of femurs with ectopically formed trabecular bone in 12 months old Translin$^{-/-}$ mice compared with Translin$^{+/+}$ mice. Magnifications, ×14, ×70

neonatal stages of Translin$^{-/-}$ mice (manuscript in preparation). Collectively, these results suggest that Translin is linked to differentiation of MSCs into various lineages of mesenchymal tissues. Since TRAX levels are known to be subjected to post-transcriptional regulation by Translin, we asked whether siRNA-mediated knockdown of Translin and TRAX affects gene expression in the C3H10T1/2 (10T1/2) stem cell line, originally isolated from mouse embryos and behaving like MSCs, ideal for studying the functional role of Translin/TRAX in mesenchymal differentiation. Translin/TRAX siRNA-mediated gene silencing followed by gene expression profiling revealed several genes associated with mesenchymal differentiation programs to be specifically up-regulated (Fig. 21.7). Given that collagen, type VIII, alpha1 (Col8a1) is known to be a product of microvascular endothelial cells, it is important to determine the potential role of endogenous fibromodulin

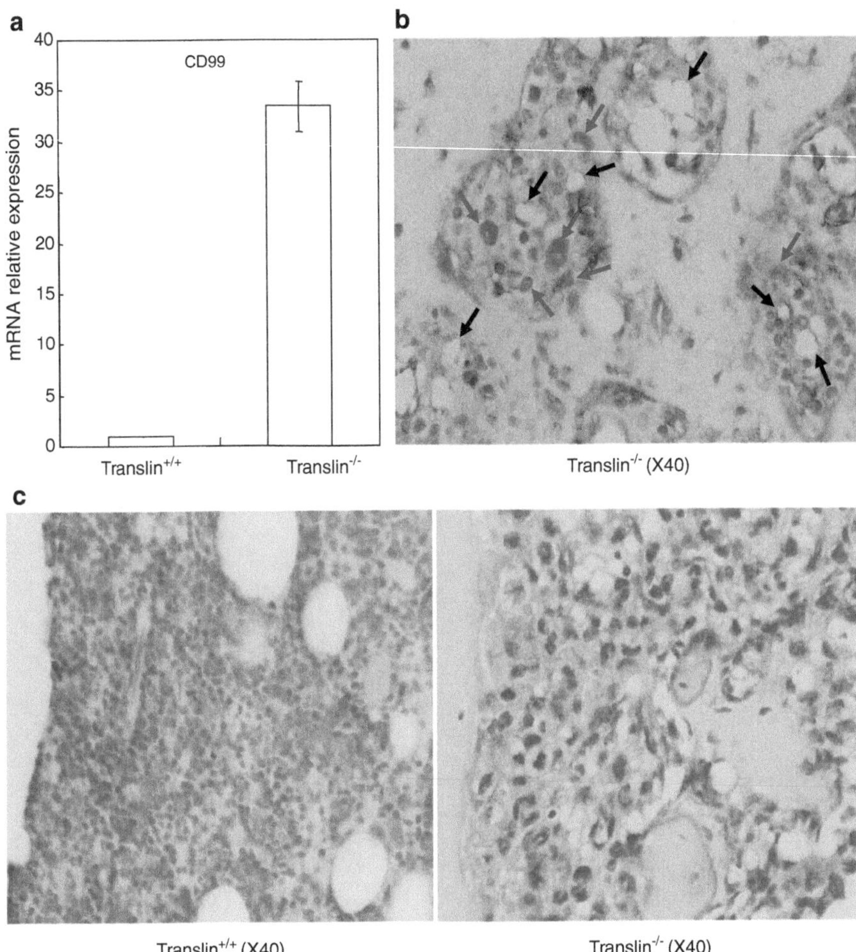

Fig. 21.6 Increased angiogenesis and adipogenesis in femurs of aged Translin$^{-/-}$ mice. (**a**) Increased mRNA for CD99 in the bone marrow stromal cells of 8 months old Translin$^{-/-}$ mice. The femurs of 8 months old *Translin*$^{-/-}$ or littermate control mice were separated into bone and bone marrow fractions by flushing with medium, the former being cut into small pieces and treated with collagenase and dispase. The released stromal cells were collected and cultured for 6 days. Total RNAs from the cultured cells were extracted and applied to microarray analysis. The most differentially expressed gene was a 32 kDa highly o-glycosylated cell surface protein, CD99. Quantitative RT-PCR for CD99 mRNA detection was performed in triplicate using Fast SYBR Green Master Mix on StepOnePlus Real-Time PCR Systems(Applied Biosystems)with GAPDH as an internal control for all reactions. (**b**) Increased frequencies of vascular endothelial cells and brown adipocyte-like cells in femurs of Translin$^{-/-}$ mice. The femurs of Translin$^{+/+}$ and Translin$^{-/-}$ mice (12 months old) were subjected to immunoperoxidase staining with anti-CD99 and counter-stained with hematoxylin. Abundant vascular endothelial cells and brown adipocyte-like cells were shown by *black* and *red arrows*, respectively. Magnifications, ×40. (**c**) Increased expression of the brown-fat-specific protein UCP-1 in femurs of Translin$^{-/-}$ mice. The femurs of Translin$^{+/+}$ and Translin$^{-/-}$ mice (12 months old) were subjected to immunoperoxidase staining with anti-UCP-1 and counterstained with hematoxylin. Magnifications, ×40

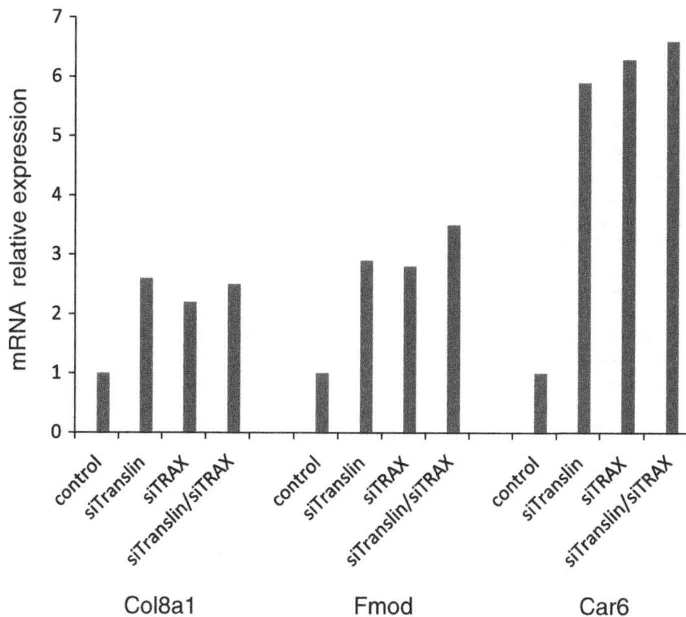

Fig. 21.7 Gene expression profiling of 10T1/2 cells after knockdown of Translin/TRAX. 10T1/2 cells were treated with siRNAs specific for Translin, TRAX, or both. Total RNAs from each sample were then extracted and applied to microarray analysis. The most differentially expressed genes were Col8a1, Fmod and Car6. Quantitative RT-PCR for each mRNA was performed in triplicate using Fast SYBR Green Master Mix on StepOnePlus Real-Time PCR Systems with GAPDH as an internal control for all reactions

(Fmod) and Col8a1 synthesis in MSC differentiation and stem cell maintenance. Carbonic anhydrase 6 (Car6) is one of the most inducible gene by hypoxia, via hypoxia inducible factor (HIF). Considering the essential role of hypoxia in the HSC niche, expression of Car6 at various stages of MSC differentiation is of major interest. Since activation of these proteins is also affected by either Translin or TRAX siRNA, it appears likely that assembly of Translin/TRAX subunits is involved in specific gene silencing through activation of RISC. Understanding under what physiological settings Translin subunits are assembled into homo-octamers or asymmetric hetero-octamers with TRAX (Fig. 21.8) is a new aspect of epigenetic regulation deserving of further exploration.

21.3 Conclusions

The present studies highlight a potentially important role for Translin in mesenchymal differentiation and bone marrow hematopoiesis. Given that TRAX levels are post-transcriptionally regulated by Translin, and both proteins have been shown to be key enhancers of the RNAi machinery, it seems likely that assembly of Translin/

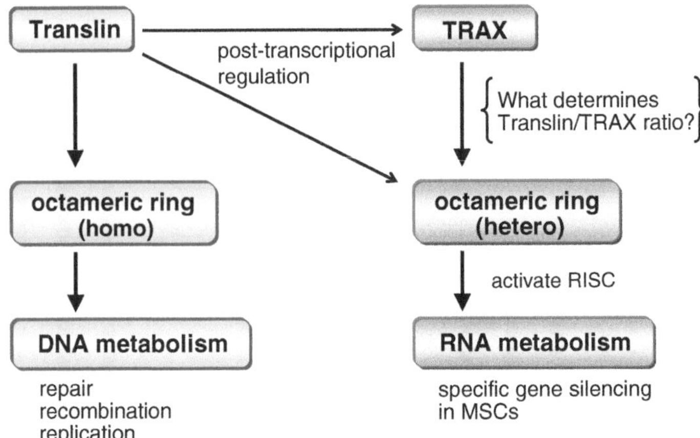

Fig. 21.8 Schematic model for the role of Translin and Translin/TRAX octamer subunits in DNA and RNA metabolism

TRAX subunits is responsible for gene silencing in MSCs through activation of RISC. Future studies should focus on physiological settings impacting on functions of Translin or Translin /TRAX octamer subunits in DNA and RNA metabolism.

Acknowledgements We are grateful to Drs. N. Kaneki, G.C. Bagby, J.C. Wang and J.L. Strominger for valuable suggestions. This work was supported by grant to M. K. from the Japan Health Sciences Foundation (JHSF) (SHC 4432).

References

1. Spangrude GJ, Heimfeld S, Weissman IL (1988) Purification and characterization of mouse hematopoietic stem cells. Science 241:58–62
2. Morrison SJ, Kimble J (2006) Asymmetric and symmetric stem-cell divisions in development and cancer. Nature 441:1068–74
3. Wilson A, Trumpp A (2006) Bone-marrow haematopoietic-stem-cell niches. Nat Rev Immunol 6:93–106
4. Zhang J, Niu C, Ye L, Huang H, He X, Tong WG, Ross J, Haug J, Johnson T, Feng JQ, Harris S, Wiedemann LM, Mishina Y, Li L (2003) Identification of the haematopoietic stem cell niche and control of the niche size. Nature 425:836–841
5. Calvi LM, Adams GB, Weibrecht KW, Weber JM, Olson DP, Knight MC, Martin RP, Schipani E, Divieti P, Bringhurst FR, Milner LA, Kronenberg HM, Scadden DT (2003) Osteoblastic cells regulate the haematopoietic stem cell niche. Nature 425:841–846
6. Kiel M, Yilmaz O, Iwashita T, Yilmaz O, Terhorst C, Morrison S (2005) SLAM family receptors distinguish hematopoietic stem and progenitor cells and reveal endothelial niches for stem cells. Cell 121:1109–1121
7. Avecilla S, Hattori K, Heissig B, Tejada R, Liao F, Shido K, Jin DK, Dias S, Zhang F, Hartman TE, Hackett NR, Crystal RG, Witte L, Hicklin DJ, Bohlen P, Eaton D, Lyden D, de Sauvage F,

Rafii S (2004) Chemokine-mediated interaction of hematopoietic progenitors with the bone marrow vascular niche is required for thrombopoiesis. Nat Med 10:64–71

8. Méndez-Ferrer S, Michurina TV, Ferraro F, Mazloom AR, Macarthur BD, Lira SA, Scadden DT, Ma'ayan A, Enikolopov GN, Frenette PS (2010) Mesenchymal and haematopoietic stem cells form a unique bone marrow niche. Nature 466:829–834
9. Kasai M, Maziarz R, Aoki K, Macintyre E, Strominger J (1992) Molecular involvement of the pvt-1 locus in a g/dT-cell leukemia bearing a variant t(8;14)(q24;q11) translocation. Mol Cell Biol 12:4751–4757
10. Kasai M, Aoki K, Matsuo Y, Minowada J, Maziarz R, Strominger J (1994) Recombination hotspot associated factors specifically recognize novel target sequences at the site of interchromosomal rearrangements in T-ALL patients with t(8;14)(q24;q11) and t(1;14)(p32;q11). Int Immunol 6:1017–1025
11. Aoki K, Nakahara K, Ikegawa C, Seto M, Takahashi T, Minowada J, Strominger JL, Maziarz RT, Kasai M (1994) Nuclear proteins binding to a novel target sequence within the recombination hotspot regions of Bcl-2 and the immunoglobulin DH gene family. Oncogene 9:1109–1115
12. Aoki K, Suzuki K, Sugano T, Tasaka T, Nakahara K, Kuge O, Omori A, Kasai M (1995) A novel gene, Translin, encodes a recombination hotspot binding protein associated with chromosomal translocations. Nat Genet 10:167–174
13. Kasai M, Matsuzaki T, Katayanagi K, Omori A, Maziarz RT, Strominger JL, Aoki K, Suzuki K (1997) The translin ring specifically recognizes DNA ends at recombination hot spots in the human genome. J Biol Chem 272:11402–11407
14. Aoki K, Inazawa J, Takahashi T, Nakahara K, Kasai M (1997) Genomic structure and chromosomal localization of the gene encoding translin, a recombination hotspot binding protein. Genomics 43:237–241
15. VanLoock MS, Yu X, Kasai M, Egelman EH (2001) Electron microscopic studies of the translin octameric ring. J Struct Biol 135:58–66
16. Sugiura I, Sasaki C, Hasegawa T, Kohno T, Sugio S, Moriyama H, Kasai M, Matsuzaki T (2004) Structure of human translin at 2.2 A resolution. Acta Crystallogr D Biol Crystallogr 60:674–679
17. Han J, Gu W, Hecht N (1995) Testis-brain RNA-binding protein, a testicular translational regulatory NA-binding protein, is present in the brain and binds to the 3' untranslated regions of transported brain mRNAs. Biol Reprod 53:707–717
18. Ishida R, Okado H, Sato H, Shionoiri C, Aoki K, Kasai M (2002) A role for the octameric ring protein, Translin, in mitotic cell division. FEBS Lett 525:105–110
19. Aoki K, Ishida R, Kasai M (1997) Isolation and characterization of a cDNA encoding a Translin-like protein, TRAX. FEBS Lett 401:109–112
20. Meng G, Aoki K, Tokura K, Nakahara K, Inazawa J, Kasai M (2000) Genomic structure and chromosomal localization of the gene encoding TRAX, a Translin-associated factor X. J Hum Genet 45:305–308
21. Liu Y, Ye X, Jiang F, Liang C, Chen D, Peng J, Kinch LN, Grishin NV, Liu Q (2009) C3PO, an endoribonuclease that promotes RNAi by facilitating RISC activation. Science 325:750–753
22. Jaendling A, McFarlane RJ (2010) Biological roles of translin and translin-associated factor-X: RNA metabolism comes to the fore. Biochem J 429:225–234
23. Ye X, Huang N, Liu Y, Paroo Z, Huerta C, Li P, Chen S, Liu Q, Zhang H (2011) Structure of C3PO and mechanism of human RISC activation. Nat Struct Mol Biol 18:650–657
24. Tian Y, Simanshu DK, Ascano M, Diaz-Avalos R, Park AY, Juranek SA, Rice WJ, Yin Q, Robinson CV, Tuschl T, Patel DJ (2011) Multimeric assembly and biochemical characterization of the Trax-translin endonuclease complex. Nat Struct Mol Biol 18:658–664
25. Fukuda Y, Ishida R, Aoki K, Nakahara K, Takashi T, Mochida K, Suzuki O, Matsuda J, Kasai M (2008) Contribution of Translin to hematopoietic regeneration after sublethal ionizing irradiation. Biol Pharm Bull 31:207–211

26. Mombaerts P, Iacomini J, Johnson RS, Herrup K, Tonegawa S, Papaioannou VE (1992) RAG-1-deficient mice have no mature B and T lymphocytes. Cell 68:869–877
27. Shinkai Y, Rathbun G, Lam KP, Oltz EM, Stewart V, Mendelsohn M, Charron J, Datta M, Young F, Stall AM, Alt FW (1992) RAG-2-deficient mice lack mature lymphocytes owing to inability to initiate V(D)J rearrangement. Cell 68:855–867
28. Gao Y, Chaudhuri J, Zhu C, Davidson L, Weaver DT, Alt FW (1998) A targeted DNA-PKcs-null mutation reveals DNA-PK-independent functions for KU in V(D)J recombination. Immunity 9:367–376
29. Taccioli GE, Amatucci AG, Beamish HJ, Gell D, Xiang XH, Torres Arzayus MI, Priestley A, Jackson SP, Rothstein AM, Jeggo PA, Herrera VL (1998) Targeted disruption of the catalytic subunit of the DNA-PK gene in mice confers severe combined immunodeficiency and radiosensitivity. Immunity 9:355–366
30. Nussenzweig A, Chen C, da Costa SV, Sanchez M, Sokol K, Nussenzweig MC, Li GC (1996) Requirement for Ku80 in growth and immunoglobulin V(D)J recombination. Nature 382:551–555
31. Gu Y, Seidl KJ, Rathbun GA, Zhu C, Manis JP, van der Stoep N, Davidson L, Cheng HL, Sekiguchi JM, Frank K, Stanhope-Baker P, Schlissel MS, Roth DB, Alt FW (1997) Growth retardation and leaky SCID phenotype of Ku70-deficient mice. Immunity 7:653–665
32. Bosma MJ, Carroll AM (1991) The SCID mouse mutant: definition, characterization, and potential uses. Annu Rev Immunol 9:323–350
33. Kincade PW, Owen JJ, Igarashi H, Kouro T, Yokota T, Rossi MI (2002) Nature or nurture? Steady-state lymphocyte formation in adults does not recapitulate ontogeny. Immunol Rev 187:116–125
34. Muller AM, Medvinsky A, Strouboulis J, Grosveld F, Dzierzak E (1994) Development of hematopoietic stem cell activity in the mouse embryo. Immunity 1:291–301
35. Dzierzak E (2002) Hematopoietic stem cells and their precursors: developmental diversity and lineage relationships. Immunol Rev 187:126–138
36. Ueda Y, Kondo M, Kelsoe G (2005) Inflammation and the reciprocal production of granulocytes and lymphocytes in bone marrow. J Exp Med 201:1771–1780
37. Wilson A, Murphy M, Oskarsson T, Kaloulis K, Bettess MD, Oser GM, Pasche AC, Knabenhans C, Macdonald HR, Trumpp A (2004) c-Myc controls the balance between hematopoietic stem cell self-renewal and differentiation. Genes Dev 18:2747–2763
38. Kaltschmidt J, Davidson C, Brown N, Brand A (2000) Rotation and asymmetry of the mitotic spindle direct asymmetric cell division in the developing central nervous system. Nat Cell Biol 2:7–12
39. Gesta S, Tseng YH, Kahn CR (2007) Developmental origin of fat: tracking obesity to its source. Cell 131:242–256

Chapter 22
Cancer Therapies and Stem Cells

Hiromichi Kimura

Contents

Abstract Current cancer treatment has improved in terms of quality of life, but adverse side effects due to chemotherapy still have a severe burden, despite the better treatment outcomes during and after the course of anticancer regimens. Molecular targeted therapies and so-called personalized medicine should reduce these problems. Organs and biological systems in a hematopoietic environment have tissue regeneration ability through utilization of stem cells. However, anticancer treatment is complicated due to the presence of cancer stem cells or cancer-initiating cells, which often share the characteristics of normal stem cells, despite the presence of genetic mutations. Many developmental signaling pathways are involved in proliferation, differentiation and self-renewal in stem cells and progenitors at critical steps that commit the direction of development. Here, we review recent findings for key signaling pathways that control stem cell and cancer properties, and that may be the target of therapeutic interventions based on small molecule inhibitors.

H. Kimura, Ph.D. (✉)
Drug Development Unit-Oncology, Takeda Pharmaceutical Company Limited,
26-1 Muraoka-Higashi 2-chome, Fujisawa, Kanagawa 251-8555, Japan
e-mail: Kimura_Hiromichi@takeda.co.jp

R.K. Srivastava and S. Shankar (eds.), *Stem Cells and Human Diseases*,
DOI 10.1007/978-94-007-2801-1_22,

Keywords Cancer-initiating cells • Cancer stem cells • Cancer treatment • Developmental signaling pathway • Small molecule inhibitor

22.1 Introduction

Developmental signaling pathways are often terminated or changed in mature cells, compared to progenitors, but are occasionally active in a small population of mature cells. These pathways are associated with stem cell functions including proliferation, differentiation and self-renewal, as well as with cancer development and cell death due to deregulated signaling. Currently, most cancer treatment is based on chemotherapy using cytotoxic agents. However, therapies have been developed against targets such as BCR-ABL, EGFR, ERBB2 over the last decade, and many groups in industry and academia have addressed the concept of targeted therapy to find better cancer treatment. Despite the number of targeted drugs used clinically, use of chemotherapy and radiation therapy is still widespread for cancer treatment, even though these therapies produce adverse side effects. For example, chemotherapy using paclitaxel and adriamycin induces persistent hair loss during therapy, indicating that normal stem cells are affected in this treatment. However, after the treatment is terminated, hair growth generally resumes. This suggests that stem cells maintain a dormant status during treatment or may be generated from surrounding tissues. The recovery of tissue function by normal stem cells suggests that many tissues at least partially possess plasticity. However, many cancer cells use stem cell signaling pathway activity as their proliferative driving forces. For improvement of cancer therapy, stem cell potency in normal cells may be used to kill malignant aggressive cancer cells that use stem cell factors for growth.

22.1.1 Targeting of Developmental Signaling in Cancer Treatment

Developmental signals control intracellular homeostasis and extracellular communication at particular times during cell growth in tissues. After organs mature, these signaling pathways are maintained and prepared for stimuli including physical, hormonal and internal responses as needed. Cancer cells also grow using developmental signaling pathways [1], while adult stem cells are widely present in the body to maintain homeostasis, such as that in skin and hair turnover.

22.1.2 Hedgehog Signaling

Activation of hedgehog (Hh) signaling has been found in many cancers, including basal cell carcinoma, medulloblastoma and pancreatic ductal adenocarcinoma. Three types of activation have been classified based on mutations of hedgehog signaling pathway genes. The first is based on a mutation of Patched1 (PTCH), a repressor for Smoothened (SMO), that leads to derepression of SMO activity, resulting in induction of activation of downstream Gli transcription factors in cancer cells. A constitutively active mutation of SMO causes sustained Gli transcriptional activation in cancer cells. This is an autonomous cell activation. The second class involves cancer cells expressing Hh protein. Hh is secreted into the extracellular environment and binds to PTCH, thus releasing SMO from suppression by PTCH in cancer cells. This type of activation does not require mutations in Hh signaling pathway genes. In the third class, Hh is also expressed in cancer cells, but the cancer cell is not activated; however, stromal cells respond to activation of Hh signaling and support cancer cell proliferation and survival via soluble proteins or extracellular attachments.

Many companies are developing SMO inhibitors to target autocrine and paracrine Hh signaling activation in cancer cell proliferation. Anticancer effects of SMO inhibitors have been shown in preclinical tests (http://clinicaltrials.gov), and clinical trials are ongoing for metastatic colorectal carcinoma, basal cell carcinoma, medulloblastoma, metastatic pancreatic cancer, advanced ovarian cancer, and prostate cancer. Hedgehog signaling controls hair growth, skeletal formation, neuronal development, and gut formation during normal development and maintenance of homeostasis in adults [2]. However, the exact balance of Hh signaling activity for treatment between normal cells and cancer cells that is required to minimize tissue damage after Hh inhibitor treatment is unknown. It is difficult to control all Hh signaling activity by small molecules, but it may be possible to reach a compromise drug concentration in targeted and non-targeted tissues. This concentration will depend on tissue penetration of the drug and tissue sensitivity to the inhibition of the responsible pathways. In preclinical experiments, juvenile mice treated with a Hh antagonist (HhAntag) showed dose-dependent defects in longitudinal bone development [3]. Generally, bone formation including fracture is regulated by remodeling homeostasis for maintenance and recovery. However, only 4 days of treatment of HhAntag caused permanent bone malformation with massive disruption of chondrocyte differentiation in epiphyseal plates. Interestingly, inhibition of Hh signaling by HhAntag in young mice led to enhanced terminal differentiation of chondrocytes after treatment for only 2 days. Furthermore, 2 days after cessation of treatment, accelerated ossification of the epiphyses occurred.

In skin, a small population of cells in hair follicles displays pluripotent characteristics [3]. The WNT, BMP, and FGF signaling pathways are involved in bone growth [4]. This implies that balanced signaling activity is important to maintain normal growth control and to start differentiation properly. Since bone defects are not restored through the bone remodeling system, adult mice after HhAntag treatment were shorter and had lower body weights compared to control mice, but no coat hair loss was observed during and after HhAntag treatment. Thus, chondrocyte progenitor cells in young mice may be sensitive to Hh signaling activity, unlike skin hair stem cells. The bulge is a niche for skin hair stem cells and the number of hair follicle stem cells changes little in humans between <40 and ≥70 years old [5]. Interestingly, expression of genes in the Hh signaling pathway in the young human skin bulge is similar to that in the aged skin bulge [5], whereas young mice at <14 weeks of age have a high level of Hh signaling activity, whereas adult mice show little Hh signaling [3]. With respect to tissue repair, the skin heals faster in younger individuals. This suggests that the Hh activity is balanced by surrounding stimuli including secreted proteins and cell-to-cell communications. Stem cells are typically dormant until they receive the proper level of signaling activity, but uni-committed progenitors may already be prepared for full functional activation in response to extracellular signals under conditions that maximize Hh signaling. Normal stem cells committed in the wrong direction can be removed during differentiation by programmed cell death such as apoptosis, autophagy and necrosis. Once the balanced Hh signaling is distorted by Hh inhibitors and genetic mutations in a cell, the differentiation process has progressed too far to be reversed in tissues in which multiple biological pathways, including WNT signaling, are active.

22.1.3 WNT Signaling

To date, 16 WNT secreted glycoproteins and 10 FZD seven-transmembrane protein receptors for WNT have been reported in humans, and LRP5/6 have also been described as alternative receptors (http://www.stanford.edu/group/nusselab/cgi-bin/wnt/). The WNT signaling pathway is involved in cancer. The WNT gene was first discovered in the integration site of the MMTV provirus in mouse chromosome 15 [6]. Subsequently, conventional gene targeting technology revealed the role of Wnt1 in brain development [7, 8], and WNT signaling is now understood to have a critical role in stem cell biology. WNT genes are also involved in some types of cancer, and are also associated with type II diabetes, skeletal dysplasia, polycystic kidney disease, and schizophrenia (http://www.stanford.edu/group/nusselab/cgi-bin/wnt/) [9]. During development, WNT plays pivotal roles in formation of the cardiovascular system, central nervous system, kidney, lung, bone, hair growth, and intestine crypts [10].

Cancer stem cells (CSCs) or cancer-initiating cells (CICs) are strongly associated with WNT signaling in some cancers. The majority of a tumor mass is composed of tumor cells and stromal cells, whereas CSCs or CICs are a relatively rare population. This raises the question of whether CSCs or CICs can be targeted without damaging normal stem cells and other normal cells? Key mutations in genes in the WNT signaling pathway are well known to cause aberrant cell proliferation. β-catenin mutations have been reported in digestive tract tissues [11], hepatocellular carcinoma [12], melanoma [13], medulloblastoma [14] and ovarian cancer [15], but not in triple negative/basal-like breast carcinoma [16]. Importantly, canonical WNT signaling activity was detected in triple negative/basal-like breast carcinoma, which is associated with a poor prognosis. Activation of the WNT signaling pathway is controlled by endogenous inhibitors in negative feedback loops, including AXIN2 [17] and DKK1 [18].

Several compounds have been developed for inhibition of canonical WNT signaling activity downstream of β-catenin. ICG-001 was discovered using the TOP/FOP FLASH reporter system [19]. ICG-001 specifically binds to CBP, which interferes with an interaction with β-catenin, resulting in downregulation of TCF transcriptional activity. CPG049090 was screened from a natural compound library in a high-throughput assay for inhibition of the protein-protein interaction between β-catenin and TCF [20]. NC043 was found in cell-based reporter TOP/FOP system screening [21]. siRNA screening in *Drosophila* cells identified three small molecules (iCRT3, 5, 14) that inhibit β-catenin/TCF target gene expression in human cancer cells. Taking advantage of removal of genetic redundancy in the screening system by using *Drosophila* cells also revealed another interesting aspect of these compounds. Disruption of the β-catenin/TCF interaction did not affect the interaction of β-catenin with E-cadherin or α-catenin, even though these complexes are thought to have overlapping interfaces. The structures of the β-catenin/TCF and β-catenin/iCRT complexes may provide more information on unique conformational changes associated with these interactions.

CSCs or CICs derived from glioblastoma showed radiation resistance with activation of DNA damage checkpoint genes [22]. In mammary gland progenitor cells, WNT signaling contributes to radiation resistance of stem cell antigen 1 (Sca1)-positive cells and stabilizes β-catenin expression, with a resultant increase in the $Sca1^+$ progenitor population [23]. With a certain radiation level (2 and 4 Gy, but not 6 Gy), $Sca1^+$ cells displayed enhanced growth activity compared with $Sca1^-$ cells [23]. CSCs or CICs may possess sensitivity to WNT/β-catenin signaling under conditions of genotoxic stress. In fact, stabilized β-catenin can potentiate neural progenitor expansion in the cerebral cortex during development in mice without changing the cell cycle rate and while sustaining cycling division phases [24]. β-catenin activity also directly contributes to development of mixed lineage leukemia (MLL) leukemic stem cells (LSCs) from pre-LSCs [25]. Downregulation of β-catenin in LSCs induced the reversion of transformation into pre-LSCs, resulting in prolonged disease-free survival by delaying the onset of leukemia. Yeung et al. showed that the LSC population was resistant to GSK3 inhibitor (LiCl) treatment, but that pre-LSCs did not show this behavior [25]. These results imply that CSCs or CICs have distinct phenotypes, even in the same lineage. Presumably, the balance between β-catenin signaling activity and others functions including developmental signaling and extracellular requirements can change to be more malignant, but remain reversible. Importantly, knockdown of β-catenin by shRNA delivery into MLL LSCs reversed resistance against treatment with a GSK3 inhibitor. The GSK3 inhibitor stabilizes β-catenin, and therefore full downregulation of β-catenin activity can be effective. This provides key information and a unique scenario in treatment of cancer refractory to chemotherapy.

22.1.4 NOTCH Signaling

In the early 1980s, the Notch gene was cloned from the notch locus in *Drosophila melanogaster* as a responsible gene that affects development [26]. Notch protein was subsequently found to be a transmembrane receptor and the cytoplasmic domain of Notch was shown to transduce signals [27]. The Notch1 gene is also essential during development in mice [28]. Notch protein is glycosylated and has specific functions. Single polypeptide Notch is initially cleaved by the ADAM/TACE protease family at an extracellular site, and then the rest of the Notch polypeptide associated with the extracellular domain is cleaved by the γ-secretase complex including presenilin and nicastrin, resulting in an intracellular Notch domain that can translocate into the nucleus and bind to CSL to form a transcriptionally active complex. Presenilin plays a role in catalysis and nicastrin promotes maturation of the complex [29]. Translocation of the Notch gene due to a chromosomal break occurs in T lymphoblastic neoplasms, resulting in expression of a truncated NOTCH1 protein [30].

Notch signaling has been suggested to have an important role in regulation of the determination of cell fate [31]. A mouse bone marrow, lineage-negative, $Sca1^{+}$ $c\text{-}kit^{+}$ population with constitutive expression of the Notch1 intracellular domain led to cellular immortalization with pluripotency and cytokine dependency [32]. Thus, NOTCH signaling pathway activity is not only oncogenic, but is a cell fate factor. Interestingly, conditional knockout of Notch1 caused development of a skin tumor [33] that showed sustained Gli2 expression and resembled basal cell carcinoma. Furthermore, β-catenin enhanced cell proliferation in the $Notch1^{-/-}$ epidermis. These findings show that canonical WNT signaling activity is suppressed by Notch1. Kolev et al. found that suppression of EGFR/ERK pathway activity by an EGFR or MEK inhibitor induces Notch1 gene expression in human primary keratinocytes [34], while the Notch2 protein level remained unchanged. These data provide insights into Notch1 tumor suppressor function and cell-type or tissue-specific function, including stem cell function.

Oncogenic mutations of NOTCH1 cause T-cell lymphoblastic leukemias and lymphomas (TLLs), but treatment of TLL lacking PTEN with a pharmacological Notch1 inhibitor failed to repress tumor cell propagation [35]. Thus, loss of PTEN leads to resistance in TLLs. These results suggest that the genetic background plays a critical role in response to targeted treatment. Phase I clinical trials of γ-secretase inhibitors (MK-0753, Merck; RO4929097, Roche; LY450139, Eli Lilly) have been performed in patients with hematopoietic cancer and solid tumors, including TLLs (http://clinicaltrials.gov/). Genetic status in tumor tissues may be required to distinguish responsive cases from treatment-insensitive cases. The effects of combination treatment with a Hh antagonist (GDC0449, Genentech) and a NOTCH inhibitor (RO492097, Roche) are also under investigation (http://clinicaltrials.gov/). The key point of this approach is to treat the self-renewal potential of a small population of tumor cells that are maintained by deregulated developmental signaling pathway activities. This may produce improved cancer therapy, especially in cases with tumor relapse.

22.1.5 FGF Signaling

Utilization of fibroblast growth factors (FGFs) to maintain embryonic stem cells and induce cellular functions and differentiation in the bone and skin is well established. FGFs are secreted proteins, with the human FGF family consisting of

22 members divided into 6 subfamilies [36]. FGFRs are receptor-type tyrosine kinases. Stem cell leukemia-lymphoma syndrome is caused by a reciprocal chromosomal translocation (8;13) [37], and is a distinct myeloproliferative disorder and TLL. The translocation generates a fusion product ZMYM2-FGFR1, which possesses a zinc finger domain in ZMYM2 and a tyrosine kinase domain in FGFR1. The zinc finger domain induces dimerization, which results in activation of FGFR1 and development of the disease [37]. Ren et al. reported that ZMYM2-FGFR1 in a mouse model of TLL induces constitutive Notch pathway activation [38]. Expression of Notch1 target genes, but not Notch1 protein, was inhibited by treatment with Notch inhibitors in ZNF112 cells carrying a ZMYM2-FGFR1 fusion gene [38]. A Notch inhibitor prolonged the survival-free period in mice with transplanted ZNF112 cells [38]. Although ZMYM2-FGFR1 fusion protein activity does not contribute to TLL tumorigenesis directly, other developmental Notch signaling activity is positively involved in TLL development. This indicates that a single genetic mutation related to a stem cell factor depends on other oncogenic support in stem cell signaling or via gene transcription.

22.1.6 *TGF-β/BMP Family Signaling*

Activation of TGF-β family receptors requires two steps of receptor assembly. After a ligand binds to the type I receptor, the type II receptor forms a tertiary complex. The type II receptor phosphorylates the type I receptor, resulting in activation of the receptor serine/threonine kinase. The activated kinase receptor type I then phosphorylates Smad proteins that transduce TGF-β signals to the cell nucleus. During transduction of signals from the receptor, Co-smad protein (Smad 4) helps Smad proteins to enter the nucleus and bind to DNA. BMPs are TGF-β family members that have their own receptors. BMP1, which is not related to the major BMP ligand family, is a protease that serves as an antagonist of secreted proteins by cleavage of procollagen fibrils and chordin. TGF-β/BMPs do not usually cause cell proliferation. In contrast, BMP receptor type IA inhibits proliferation of stem cells in the intestine and skin. Tissue-specific deletion of BMP receptor type IA in mice induces expansion of stem cells populations in hair follicles, intestine and bone marrow; and tumors in the hair follicles and polyposis in the intestines develop as a consequence of loss of BMP receptor type IA function [39–41]. These results suggest that BMP type IA protein or signaling is responsible for maintenance of tissue homeostasis as a negative regulator that maintains balance with proliferative signaling pathways.

In the cerebellum, granule neuron precursors (GNPs) are the origin of a type of medulloblastoma. GNPs proliferate by exposing Shh from Purkinje cells in the external granule layer in the cerebellum. Zhao et al. found that *in vitro* Shh-dependent GNP proliferation is blocked by BMP2 or BMP4 treatment because of induction of Math1 protein degradation (Zhao, G&D, 2008). BMP4 expression suppressed proliferation in medulloblastoma cells derived from ptch1$^{+/-}$, cdkn2c$^{-/-}$ mouse medulloblastoma [42]. Some human medulloblastoma cell lines exhibited retinoid-associated apoptosis through BMP2 expression [43]. Withdrawal of BMP signaling from tumor cells leads to uncontrolled cellular differentiation and proliferation by WNT signaling [44]. These results suggest that BMP signaling controls pluripotent stem cell function and self-renewal capability, in combination with other stem cell signaling pathways. Thus, BMP agonists may be useful for treatment of distinct cancers that utilize stem cell signaling pathways for tumor growth.

In contrast, TGF-β signaling has dual roles as a tumor suppressor and an oncogenic factor [45]. TGF-β contributes to the epithelial-mesenchymal transition (EMT) involved in Snail [46], Twist [47], and Slug [48] gene functions, stromal cell transformation [49, 50] and immunosuppression [51]. TGF-β also plays important roles in multiple steps of the metastasis process [52]. The TGF-β RIA kinase inhibitor SB431542 antagonizes BMP4-induced osteoblastic differentiation [53]. Furthermore, in a culture of glioma-initiating cells (GICs), Sox2 induction was downregulated by a TGF-β inhibitor, resulting in inhibition of GIC-derived tumor growth and sphere formation [54]. Conversely, TGF-β inhibitors could be used to replace Sox2 in reprogramming stem cells from mouse embryonic fibroblasts during induction of pluripotent stem cells (iPSC generation) [55]. This is a case in which the function of TGF-β signaling is fully distinguishable between tumor cells and non-tumor cells. Since TGF-β supports the metastasis transition process, a TGF-β inhibitor is likely to be useful for treatment of some aggressive cancers [52]. Treatment with doxorubicin, an agent that causes DNA damage, enhanced TGF-β1-mediated EMT in a breast cancer cell line; and MDA-MB-231 and doxorubicin in combination with a TGF-β inhibitor was more effective than either single treatment for suppressing tumor growth in *in vivo* xenograft models [56]. Many TGF-β receptor type IA inhibitors have been reported [45]. Collectively, the effects of antagonism of the TGF-β signaling pathway are complicated and require further investigation with an integrated analysis of genetic mutations and the activities of compensatory signaling pathways during treatment of disease.

22.2 Conclusions

Anticancer therapeutics are needed for management of tumors refractory to standard chemotherapy. CSCs or CICs are rare cells that form a treatment-resistant population in the mass of tumor tissues. Importantly, stem cells have been characterized to be in a long quiescent G0 phase, whereas there seem to be two phases, proliferative and quiescent, in CSCs or CICs. It is difficult to dissect the levels of activity of the many kinds of developmental signaling pathways that contribute to maintenance of balance between these phases. Some evidence has emerged from use of arsenite trioxide as a therapeutic agent for acute promyelocytic leukemia [57–59] via degradation of PML tumor suppressor protein [60]. In chronic myeloid leukemia (CML), arsenite trioxide treatment led to an exit from a quiescent state and an increase of proliferating stem cells in CSCs or CICs [61]. A number of small molecules are being developed to inhibit developmental signaling-driven cancer growth in CSCs or CICs, instead of chemotherapy. However, it is clear that further investigations of differences in genetic status, control of signaling regulation, internal and external stress responses, and mechanism of drug tolerance between normal cells and cancer cells are required for accurate diagnosis and development of new therapies.

References

1. Reya T, Morrison SJ, Clarke MF, Weissman IL (2001) Stem cells, cancer, and cancer stem cells. Nature 414(6859):105–111
2. Lum L, Beachy PA (2004) The Hedgehog response network: sensors, switches, and routers. Science 304(5678):1755–1759
3. Kimura H, Ng JM, Curran T (2008) Transient inhibition of the Hedgehog pathway in young mice causes permanent defects in bone structure. Cancer Cell 13(3):249–260
4. Baldridge D, Shchelochkov O, Kelley B, Lee B (2010) Signaling pathways in human skeletal dysplasias. Annu Rev Genomics Hum Genet 11:189–217
5. Rittie L, Stoll SW, Kang S, Voorhees JJ et al (2009) Hedgehog signaling maintains hair follicle stem cell phenotype in young and aged human skin. Aging Cell 8(6):738–751
6. Nusse R, van Ooyen A, Cox D, Fung YK et al (1984) Mode of proviral activation of a putative mammary oncogene (int-1) on mouse chromosome 15. Nature 307(5947):131–136
7. McMahon AP, Bradley A (1990) The Wnt-1 (int-1) proto-oncogene is required for development of a large region of the mouse brain. Cell 62(6):1073–1085
8. Thomas KR, Capecchi MR (1990) Targeted disruption of the murine int-1 proto-oncogene resulting in severe abnormalities in midbrain and cerebellar development. Nature 346(6287):847–850
9. Moon RT, Kohn AD, De Ferrari GV, Kaykas A (2004) WNT and beta-catenin signalling: diseases and therapies. Nat Rev Genet 5(9):691–701
10. Grigoryan T, Wend P, Klaus A, Birchmeier W (2008) Deciphering the function of canonical Wnt signals in development and disease: conditional loss- and gain-of-function mutations of beta-catenin in mice. Genes Dev 22(17):2308–2341
11. Polakis P (2000) Wnt signaling and cancer. Genes Dev 14(15):1837–1851
12. Wong CM, Fan ST, Ng IO (2001) Beta-Catenin mutation and overexpression in hepatocellular carcinoma: clinicopathologic and prognostic significance. Cancer 92(1):136–145
13. Omholt K, Platz A, Ringborg U, Hansson J (2001) Cytoplasmic and nuclear accumulation of beta-catenin is rarely caused by CTNNB1 exon 3 mutations in cutaneous malignant melanoma. Int J Cancer 92(6):839–842
14. Zurawel RH, Chiappa SA, Allen C, Raffel C (1998) Sporadic medulloblastomas contain oncogenic beta-catenin mutations. Cancer Res 58(5):896–899
15. Wright K, Wilson P, Morland S, Campbell I et al (1999) Beta-catenin mutation and expression analysis in ovarian cancer: exon 3 mutations and nuclear translocation in 16% of endometrioid tumours. Int J Cancer 82(5):625–629
16. Geyer FC, Lacroix-Triki M, Savage K, Arnedos M et al (2011) Beta-Catenin pathway activation in breast cancer is associated with triple-negative phenotype but not with CTNNB1 mutation. Mod Pathol 24(2):209–231
17. Jho EH, Zhang T, Domon C, Joo CK et al (2002) Wnt/beta-catenin/Tcf signaling induces the transcription of Axin2, a negative regulator of the signaling pathway. Mol Cell Biol 22(4):1172–1183
18. Niida A, Hiroko T, Kasai M, Furukawa Y et al (2004) DKK1, a negative regulator of Wnt signaling, is a target of the beta-catenin/TCF pathway. Oncogene 23(52):8520–8526
19. Emami KH, Nguyen C, Ma H, Kim DH et al (2004) A small molecule inhibitor of beta-catenin/CREB-binding protein transcription [corrected]. Proc Natl Acad Sci USA 101(34):12682–12687

20. Lepourcelet M, Chen YN, France DS, Wang H et al (2004) Small-molecule antagonists of the oncogenic Tcf/beta-catenin protein complex. Cancer Cell 5(1):91–102
21. Wang W, Liu H, Wang S, Hao X et al (2011) A diterpenoid derivative 15-oxospiramilactone inhibits Wnt/beta-catenin signaling and colon cancer cell tumorigenesis. Cell Res 21(5):730–740
22. Rich JN (2007) Cancer stem cells in radiation resistance. Cancer Res 67(19):8980–8984
23. Chen MS, Woodward WA, Behbod F, Peddibhotla S et al (2007) Wnt/beta-catenin mediates radiation resistance of Sca1+ progenitors in an immortalized mammary gland cell line. J Cell Sci 120(Pt 3):468–477
24. Chenn A, Walsh CA (2002) Regulation of cerebral cortical size by control of cell cycle exit in neural precursors. Science 297(5580):365–369
25. Yeung J, Esposito MT, Gandillet A, Zeisig BB et al (2010) Beta-Catenin mediates the establishment and drug resistance of MLL leukemic stem cells. Cancer Cell 18(6):606–618
26. Artavanis-Tsakonas S, Muskavitch MA, Yedvobnick B (1983) Molecular cloning of Notch, a locus affecting neurogenesis in Drosophila melanogaster. Proc Natl Acad Sci USA 80(7):1977–1981
27. Lyman D, Young MW (1993) Further evidence for function of the Drosophila Notch protein as a transmembrane receptor. Proc Natl Acad Sci USA 90(21):10395–10399
28. Swiatek PJ, Lindsell CE, del Amo FF, Weinmaster G et al (1994) Notch1 is essential for post-implantation development in mice. Genes Dev 8(6):707–719
29. Lai EC (2002) Notch cleavage: Nicastrin helps Presenilin make the final cut. Curr Biol 12(6):R200–R202
30. Ellisen LW, Bird J, West DC, Soreng AL et al (1991) TAN-1, the human homolog of the Drosophila notch gene, is broken by chromosomal translocations in T lymphoblastic neoplasms. Cell 66(4):649–661
31. Artavanis-Tsakonas S, Rand MD, Lake RJ (1999) Notch signaling: cell fate control and signal integration in development. Science 284(5415):770–776
32. Varnum-Finney B, Xu L, Brashem-Stein C, Nourigat C et al (2000) Pluripotent, cytokine-dependent, hematopoietic stem cells are immortalized by constitutive Notch1 signaling. Nat Med 6(11):1278–1281
33. Nicolas M, Wolfer A, Raj K, Kummer JA et al (2003) Notch1 functions as a tumor suppressor in mouse skin. Nat Genet 33(3):416–421
34. Kolev V, Mandinova A, Guinea-Viniegra J, Hu B et al (2008) EGFR signalling as a negative regulator of Notch1 gene transcription and function in proliferating keratinocytes and cancer. Nat Cell Biol 10(8):902–911
35. Palomero T, Sulis ML, Cortina M, Real PJ et al (2007) Mutational loss of PTEN induces resistance to NOTCH1 inhibition in T-cell leukemia. Nat Med 13(10):1203–1210
36. Itoh N (2007) The Fgf families in humans, mice, and zebrafish: their evolutional processes and roles in development, metabolism, and disease. Biol Pharm Bull 30(10):1819–1825
37. Roumiantsev S, Krause DS, Neumann CA, Dimitri CA et al (2004) Distinct stem cell myeloproliferative/T lymphoma syndromes induced by ZNF198-FGFR1 and BCR-FGFR1 fusion genes from 8p11 translocations. Cancer Cell 5(3):287–298
38. Ren M, Cowell JK (2011) Constitutive Notch pathway activation in murine ZMYM2-FGFR1-induced T-cell lymphomas associated with atypical myeloproliferative disease. Blood 117(25):6837–6847
39. Ming Kwan K, Li AG, Wang XJ, Wurst W et al (2004) Essential roles of BMPR-IA signaling in differentiation and growth of hair follicles and in skin tumorigenesis. Genesis 39(1):10–25
40. Howe JR, Bair JL, Sayed MG, Anderson ME et al (2001) Germline mutations of the gene encoding bone morphogenetic protein receptor 1A in juvenile polyposis. Nat Genet 28(2):184–187
41. Bleuming SA, He XC, Kodach LL, Hardwick JC et al (2007) Bone morphogenetic protein signaling suppresses tumorigenesis at gastric epithelial transition zones in mice. Cancer Res 67(17):8149–8155

42. Zhao H, Ayrault O, Zindy F, Kim JH et al (2008) Post-transcriptional down-regulation of Atoh1/Math1 by bone morphogenic proteins suppresses medulloblastoma development. Genes Dev 22(6):722–727
43. Hallahan AR, Pritchard JI, Chandraratna RA, Ellenbogen RG et al (2003) BMP-2 mediates retinoid-induced apoptosis in medulloblastoma cells through a paracrine effect. Nat Med 9(8):1033–1038
44. He XC, Zhang J, Tong WG, Tawfik O et al (2004) BMP signaling inhibits intestinal stem cell self-renewal through suppression of Wnt-beta-catenin signaling. Nat Genet 36(10):1117–1121
45. Yingling JM, Blanchard KL, Sawyer JS (2004) Development of TGF-beta signalling inhibitors for cancer therapy. Nat Rev Drug Discov 3(12):1011–1022
46. Peinado H, Quintanilla M, Cano A (2003) Transforming growth factor beta-1 induces snail transcription factor in epithelial cell lines: mechanisms for epithelial mesenchymal transitions. J Biol Chem 278(23):21113–21123
47. Yang J, Mani SA, Donaher JL, Ramaswamy S et al (2004) Twist, a master regulator of morphogenesis, plays an essential role in tumor metastasis. Cell 117(7):927–939
48. Romano LA, Runyan RB (2000) Slug is an essential target of TGFbeta2 signaling in the developing chicken heart. Dev Biol 223(1):91–102
49. Bhowmick NA, Chytil A, Plieth D, Gorska AE et al (2004) TGF-beta signaling in fibroblasts modulates the oncogenic potential of adjacent epithelia. Science 303(5659):848–851
50. Franco OE, Jiang M, Strand DW, Peacock J et al (2011) Altered TGF-beta signaling in a subpopulation of human stromal cells promotes prostatic carcinogenesis. Cancer Res 71(4):1272–1281
51. Du C, Wang Y (2011) The immunoregulatory mechanisms of carcinoma for its survival and development. J Exp Clin Cancer Res 30:12
52. Massague J (2008) TGFbeta in cancer. Cell 134(2):215–230
53. Maeda S, Hayashi M, Komiya S, Imamura T et al (2004) Endogenous TGF-beta signaling suppresses maturation of osteoblastic mesenchymal cells. EMBO J 23(3):552–563
54. Ikushima H, Todo T, Ino Y, Takahashi M et al (2009) Autocrine TGF-beta signaling maintains tumorigenicity of glioma-initiating cells through Sry-related HMG-box factors. Cell Stem Cell 5(5):504–514
55. Ichida JK, Blanchard J, Lam K, Son EY et al (2009) A small-molecule inhibitor of tgf-Beta signaling replaces sox2 in reprogramming by inducing nanog. Cell Stem Cell 5(5):491–503
56. Bandyopadhyay A, Wang L, Agyin J, Tang Y et al (2010) Doxorubicin in combination with a small TGFbeta inhibitor: a potential novel therapy for metastatic breast cancer in mouse models. PLoS One 5(4):e10365
57. Bradstock K, Matthews J, Benson E, Page F et al (1994) Prognostic value of immunophenotyping in acute myeloid leukemia. Australian Leukaemia Study Group. Blood 84(4):1220–1225
58. Soignet SL, Maslak P, Wang ZG, Jhanwar S et al (1998) Complete remission after treatment of acute promyelocytic leukemia with arsenic trioxide. N Engl J Med 339(19):1341–1348
59. Shen ZX, Chen GQ, Ni JH, Li XS et al (1997) Use of arsenic trioxide (As2O3) in the treatment of acute promyelocytic leukemia (APL): II. Clinical efficacy and pharmacokinetics in relapsed patients. Blood 89(9):3354–3360
60. Lallemand-Breitenbach V, Zhu J, Puvion F, Koken M et al (2001) Role of promyelocytic leukemia (PML) sumolation in nuclear body formation, 11 S proteasome recruitment, and As2O3-induced PML or PML/retinoic acid receptor alpha degradation. J Exp Med 193(12):1361–1371
61. Ito K, Bernardi R, Morotti A, Matsuoka S et al (2008) PML targeting eradicates quiescent leukaemia-initiating cells. Nature 453(7198):1072–1078

Chapter 23
Cancer Stem Cells and Head and Neck Squamous Cell Carcinomas

Shi-Long Lu

Contents

Abstract Recurrence and resistance to standard chemo- or radiotherapy in head and neck squamous cell carcinoma contribute significantly to the limited improvement of overall survival in head and neck cancer patients. The cancer stem cell hypothesis postulated that a subpopulation of cancer cells possess self-renewal ability, and produce differentiated cells to form the tumor mass. These cells may have distinct genotype and phenotype, and fail to respond to most of the current chemo- and radiotherapy, leading to tumor recurrence and resistance to therapy. Thus, identifying, characterizing, and understanding the cellular and molecular biology of putative cancer stem cells will provide a novel and possibly yield a more effective therapeutic approach. In this chapter, I will review recent advances on cancer stem cells in head and neck cancer and summarize them into four topics: (1) Putative cancer stem cell markers for head and neck cancer. (2) The PI3K/PTEN/AKT signaling as regulator

S.-L. Lu, MD, Ph.D. (✉)
Departments of Otolaryngology, Dermatology, Pathology, University of Colorado Anschutz Medical Campus, Research Complex 2, Room 7112, 12700 E 19th Avenue, Aurora, CO 80016, USA
e-mail: shi-long.lu@ucdenver.edu

R.K. Srivastava and S. Shankar (eds.), *Stem Cells and Human Diseases*,
DOI 10.1007/978-94-007-2801-1_23,

for cancer stem cells. (3) Metastatic cancer stem cells and epithelial-mesenchymal transition. (4) Targeting head and neck cancer stem cells for cancer therapy

Keywords Cancer stem cells • Head and neck cancer • PI3K/PTEN/AKT pathway • Epithelial mesenchymal transition • Metastasis • Targeted therapy

23.1 Introduction

The cancer stem cells (CSCs) theory, although controversial, reveals a novel hypothesis that explains tumor "regeneration", or tumor development at both primary and distant metastatic sites [1–3]. According to this theory, there is a specific subpopulation of cancer cells, called "CSCs", among the tumor mass, which have capacities to self-renew and generate differentiated cancer cells. In detail, these cells are tumorigenic when transplanted into immunocompromised mice, and can be serially transplanted through multiple generations, indicating the self-renewal ability. In addition, CSCs give rises to the heterogeneous lineages of cancer cells, indicating a cell hierarchy generated from certain "stemness" cells [1–3].

The CSCs hypothesis does not imply that CSCs are necessarily derived from normal tissue stem cells, but rather, it addresses the ability of self-renewal, and differentiation into non-self-renewing cell population. Therefore, the term "cancer stem cells" is not considered to be perfect, and alternative terms, such as "tumor-initiating cells" or "tumor propagating cells" have been proposed [1–3]. Being consistent with other chapters in this book, I will still use "cancer stem cell" throughout my section.

CSC concept emphasizes that tissue-resident stem or progenitor cells are the most common targets of transformation. This hypothesis has great impact on cancer therapy, since current therapies target proliferative cells rather than CSCs. This would explain why tumor recurred and are resistant to conventional chemo- or radiotherapies clinically [1–3]. Thus, identification, characterization, and design therapies directly targeting CSCs may achieve a true cure for cancer patients.

23.2 Putative Cancer Stem Cell Markers for Head and Neck Cancer

Head and neck squamous cell carcinomas (HNSCCs) refer to squamous cell carcinomas arising from oral cavity, tongue, pharyngeal and laryngeal regions, and is the sixth most common human cancer worldwide [4, 5]. There are about 600,000 new cases and 350,000 cancer deaths worldwide each year. HNSCCs usually occur in a relatively late age, and higher in male with well-known etiological factors of tobacco and/or alcohol [4, 5]. However, the incidence of HNSCC is increasing recently in women with relatively young age, and correlates with human papilloma virus (HPV) infection [4, 5]. Although there has been improvement in survival associated with HPV-associated oropharyngeal cancers, the survival rate has barely changed for HNSCC associated with tobacco and alcohol use in the past three decades. The mortality remains high mainly due to tumor recurrence, resistance to chemo- and radiotherapies, and metastasis [4, 5]. It is urgent to understand the biology of this disease to develop novel therapeutic approaches.

Stratified squamous epithelium, including head and neck epithelium, contains a basal layer of cells where normal stem cells reside. These normal stem cells are resources of epithelial replenishment and a reservoir for the normal homeostasis. Normal stem cells generate cells on cell division, with some remaining stem-ness (self-renew) and some remain more differentiated, constituting heterogeneity both morphologically and functionally. Cancer stem cells share several similarities with normal stem cells in the head and neck region, such as morphology, stem cell markers, and hierarchies [6–8].

23.2.1 CSCs Morphology

Morphology heterogeneity is a typical feature of malignant cell line. Dr. Mackenzie described three different morphology of cell growth in a culture dish [9, 10]. Holoclones refer to small tightly packed cells with high proliferative potential and are thought to contain stem cells. In addition, the growth of CSCs in a serum-free low attachment condition is prone to produce "spheroid" morphology [9, 10]. Meroclones are larger cells than holoclones. They are less proliferative, not be able to self-renew, and correlate to transit amplifying cells. Paraclones are large flattened cells and correspond to early differencing cells. Only cells from holoclone patterns are capable for self-renewal, and express higher level of CSCs markers, such as CD44 and keratin 15 [9, 10].

23.2.2 CD44

CD44 is the receptor for hyaluronan, and represents a group of transmembrane glycoproteins. CD44 has reported a useful CSCs marker for various cancer types, including breast, prostate, colon, and pancreas [11–14]. Using a similar approach, Prince et al. identified a subpopulation of CD44+ cells (usually less than 10%) in human HNSCC tumor specimen [8, 15]. The CD44+ tumor cells are more tumorigenic because as few as 2×10^3 of the CD44+ cells produced a tumor. Further analysis of the re-produced tumors by CD44+ cell population revealed the recapitulation of the primary tumor histology, and a mixture of both CD44+ and CD44- tumor cells, indicating that CD44+ cells give rise to CD44- progeny [8, 15]. The tumorigenicity of CD44+ cells through several series transplant also suggest that at least a portion of CD44+ cells are capable of self-renewal. CD44+ cell population also showed higher expression of another CSC marker, CD133, stem cell-related gene Bmi-1, and multiple drug resistance transporter gene ABCG2 [8, 15, 16]. CD44+ cells are also resistant to apoptosis-stimuli [17], and correlate with tumor aggressiveness in HNSCCs [18]. Thus, CD44 meets the general criteria as the first CSC marker in HNSCCs. However, recent report of CD44 splice variant added the complexity of using CD44 as a CSC marker [19]. Besides the standard form of CD44 (CD44s), which was shown to be CSCs markers in several types of cancer, there was a variant of CD44 (CD44v6), which regulates tumor progression, invasion and metastasis. In one study, either CD44s or CD44v6 were abundantly present in a majority of head and neck tissues and tumors, and none of these two CD44 isoforms distinguish normal from benign or malignant epithelia of head and neck [19]. Further validation of CD44 as CSC marker in head and neck cancer is still warranted.

23.2.3 Side Population

Side population (SP) cells are a subset of cells with elevated ability to efflux the vital DNA binding dye, Hoechst 33342 dye. This ability has been contributed to the increased expression of multiple drug resistance transporter proteins, such as ABCG2 [20]. SP cells were originally identified as a small population of bone marrow cells but recently have been found in various solid tumors that are clonogenic, both in vitro and in vivo. SP cells share many characteristics with stem cells. The SP has been used to enrich putative CSCs in several solid tumors, e.g. breast, prostate, liver, and skin [21–23].

SP cells have recently been utilized as a CSCs marker in HNSCC as well. Using HNSCC cell lines, SP cells were initially found in about 0.5% of HNSCC cells and increased to about 4.8% after the culture of sorted-SP+ cells [24, 25]. SP cells were also identified in primary HNSCC specimens, occupying about 0.5% of the total cell population [25]. In addition, SP cells express higher levels of Bmi-1, ABCG2, and are resistant to chemotherapy [24, 25]. Abnormal activation of Wnt/β-catenin signaling was also found in SP cells [24]. Interestingly, there is a report that EGFR regulates SP cells in HNSCCs [26], and imatinib-mediated inactivation of AKT reduced SP cells [27], providing another rationale for using EGFR targeted therapy in HNSCC patients. SP cells were also correlated to HNSCC invasion and metastasis [24, 28], but these findings need further validation.

23.2.4 CD133

CD133 (Prominin-1) is a 120-kDa cell surface glycoprotein with a potential role in the organization of plasma membrane topology. The AC133 antigen, which represents a hyper-glycosalyted version of CD133, has been used to enrich human hematopoietic stem cells, endothelial precursors, and fetal neural stem cells. CD133 was firstly used to identify CSCs in brain tumors, especially glioblastoma [29]. It is also broadly served to isolate CSC markers for lung, pancreatic, liver, colon, prostate cancers [30–33]. However, the validity of CD133 as CSC marker has been challenged by Shmelkov et al., who showed that both CD133+ and CD133- cells are tumorigenic in metastatic colon cancers [34].

Utilization of CD133 to identify putative CSCs in HNSCC has been described in a few reports [35, 36]. CD133+ cells were identified in a range of 0.2–3% of various HNSCC cell lines [35, 36]. In addition, an analysis of HNSCC primary tumor tissues also detected a subpopulation of cells that expressed CD133 [35, 36]. These CD133+ cells are highly tumorigenic, invasive, and resistant to standard chemotherapy compared to CD133- counterparts [35, 36]. Using a laryngeal SCC cell line, Hep-2, Zhou et al. found that less than 5% of Hep-2 cells expressed CD133, which possess self-renewal, and differentiation abilities [37]. Bmi-1, a critical gene for CSC proliferation, is overexpressed in CD133+ cells [38]. Injection of CD133+ cells increase the tumorigenicity in vivo compared to CD133- cells [39]. However, there was still a significant portion of CD133- cell possessing a tumor forming ability in vivo [39]. Therefore, the roles of CD133 as a CSCs marker in HNSCC need further investigation.

23.2.5 Aldehyde Dehydrogenase (ALDH1)

ALDH1 belongs to the ALDH family of enzymes which is responsible for oxidizing intracellular aldehydes and contributing to the oxidation of retinol to retinoic acid in early stem differentiation. High ALDH activity was initially noticed in leukemia CSCs, and has been subsequently used to isolate CSCs from multiple solid tumors [40–44]. In addition, the activity of the ALDH enzyme might be responsible for the resistance of CSCs to chemotherapy. ALDH activity can be measured by immunohistochemistry or western blotting methods. However, a newly developed method, which uses the substrate of the ALDH1 enzyme BODIPY to measure ALDH activity, called “Aldefluor”, has been widely used to characterize CSCs in many primary tumors including breast, colon, lung, prostate and pancreas [40–44]. In addition, a report illustrated that cells with high ALDH activity form distant metastases with strongly enhanced tumor progression at both orthotopic and metastatic sites in a prostate cancer model [43].

Using ALDH1 as a CSCs marker in HNSCC has recently initiated. Using primary HNSCC tumor tissues, Clay et al. found the $ALDH^{high}$ cells represented a small subpopulation of tumor cells ranging from 1% to 7%, and formed tumors from as low as 500 cells, as compared to $ALDH^{low}$ cells [45]. Chen et al. also showed ALDH1 is a specific marker for CSCs of HNSCC [46]. Furthermore, the $ALDH^{high}$ cells exhibit epithelial mesenchymal transition (EMT) and increase an endogenous EMT regulator, Snail1 [46]. $ALDH^{high}$ cells also express higher levels of Bmi-1. A knockdown of Bmi-1 in the $ALDH^{high}$ HNSCC cells inhibits the tumorigenicity and enhances radiochemosensitivity [47].

23.3 PI3K/PTEN/AKT Signaling as Regulator for Cancer Stem Cells

Although PI3K/PTEN/AKT is best known as the key regulator for cell survival and growth, recent studies suggest this pathway plays roles in the regulation and maintenance of stem cells as well [48]. The majority of the results came from extensive studies of PTEN. For example, PTEN dependence distinguishes hematopoietic stem cells from leukemia-initiating cells [49]. PTEN also maintains hematopoietic stem cells and acts in lineage choice and leukemia prevention [50]. Furthermore, PTEN deletion in bronchoalveolar epithelium increases the putative initiators of lung adenocarcinomas, bronchioalveolar stem cells, and increases sonic hedgehog signaling [51]. Also, PTEN deletion in prostate epithelium leads to the expansion of a prostatic stem/progenitor cell population and tumor initiation [52]. Similarly, deletion of PTEN in mammary epithelium enriches normal and malignant mammary stem/progenitor cells with AKT activation [53]. Further analysis showed the AKT-driven stem/progenitor cell enrichment is mediated by the activation of the Wnt/β-catenin pathway through phosphorylation of GSK-3β [53]. PTEN has also been involved in the side population regulation in esophageal SCC [54]. Overexpression of AKT in head and neck epithelium promote cancer progression and metastases, partly due to the enrichment of putative cancer stem cells in head and neck cancer, as revealed by increased numbers of CD44+ and CD133+ subpopulation in AKT transgenic mice [55]. These data suggest targeting PI3K/PTEN/AKT pathway may have more therapeutic potential because it may directly target putative cancer stem cells.

23.4 Metastatic Cancer Stem Cells and Epithelial-Mesenchymal Transition

Similar to the CSCs hypothesis, the concept of metastatic CSCs, or migrating CSCs (mCSCs), has also been proposed as one mechanism to explain tumor metastasis [56, 57]. The cellular and molecular similarities and differences between CSCs (also known as tumor initiating cells), and metastatic CSCs (mCSCs), (also known as metastasis initiating cells) is still unknown [56, 57]. However, two major factors seem to affect CSCs behaviors in tumor metastasis. They are epithelial-mesenchymal transition (EMT) and "niche", or the microenvironment of CSCs [56, 57]. EMT refers to a transition between epithelial and mesenchymal states, plays crucial roles in embryonic development, and has been closely connected with tumor invasion and metastasis. Recently, EMT has been found to confer stemness properties to differentiated epithelial cells, including expressing stem cell markers, forming mammospheres, and self-renewal traits. EMT has also been associated with tumor cell dissemination, the very early stage of tumor metastasis [58]. Whether these disseminated tumor cells are metastatic CSCs or not, or whether EMT is necessary for connecting tumor cell dissemination with mCSCs, are important questions to be answered in the future. Interestingly, a recent study showed evidence for EMT in CSCs of head and neck cancer. In this study, spheroid-derived cells, which enriched CSCs, expressed not only a higher level of stem cell markers, such as ALDH, CD44, and stemness transcriptional factor genes, such as Sox2, Nanog, and Oct3/4, but also higher level of Twist, and properties of EMT [59].

23.5 Targeting Head and Neck Cancer Stem Cells for Cancer Therapy

Multiple signaling pathways have been found to regulate the growth survival, and self-renewal of CSCs, providing opportunities for designing therapeutic approaches aiming directly at CSCs. This may prevent tumor recurrence, metastasis, and enhance sensitivities of chemo- radiotherapies, and will have a tremendous impact on improving a patient's survival. As detailed in Chap. 3, the PI3K/PTEN/AKT pathway has been shown as one of the critical pathways to maintain CSCs. This pathway has particular interest since many therapeutic designs targeting this pathway have been tested in clinical trials [60]. Preliminary studies have shown some promising results. For example, treatment of HNSCC cell lines by AKT kinase inhibitor, shRNA AKT, or tyrosine kinase inhibitor imatinib, significantly decrease the side population by 50–70% and decreases the amount of ABCG2 nuclear translocation [27]. Another promising therapeutic approach is targeting EMT, a process connecting CSCs and tumor metastasis as detailed in Chap. 4. Besides Twist and Snail1 molecules, a recent study on HNSCC showed S100A4 is an EMT mediator and maintains cancer-initiating cells in head and neck cancer [61]. While the S100A4 expression was positively correlated with clinical grading, stemness markers, and poorer patient survival, attenuation of endogenous S100A4 reduces stemness markers and tumorigenicity. Consistent with the biological findings, the PI3K/AKT pathway is down regulated along with an upregulation of PTEN upon a reduced level of S100A [61]. Further studies of CSCs regulators, particularly the recently discovered microRNAs [62], and targeting CSCs "niche" [63], will provide more therapeutic options.

23.6 Conclusion

The CSCs hypothesis, although promising, still has many questions that need to be answered. There are lacks of insight into the biology or characterization of CSCs. Current CSCs markers for most tumor types, particularly on head and neck cancer, are still controversial. However, mounting evidence does support the existence of subpopulation of self-renewing tumors cells in the bulk of tumor mass, including head and neck cancer. Further research results on the cellular and molecular mechanisms regulating and maintaining the putative CSCs, will be ultimately translated into a novel and possibly result in more effective therapeutic approaches.

References

1. Clevers H (2011) The cancer stem cell: premises, promises and challenges. Nat Med 17:313–319
2. Dalerba P, Cho RW, Clarke MF (2007) Cancer stem cells: models and concepts. Annu Rev Med 58:267–284
3. Marotta LL, Polyak K (2009) Cancer stem cells: a model in the making. Curr Opin Genet Dev 19:44–50
4. Argiris A, Karamouzis MV, Raben D, Ferris RL (2008) Head and neck cancer. Lancet 371:1695–1709
5. Leemans CR, Braakhuis BJ, Brakenhoff RH (2011) The molecular biology of head and neck cancer. Nat Rev Cancer 11:9–22
6. Mackenzie IC (2004) Growth of malignant oral epithelial stem cells after seeding into organotypical cultures of normal mucosa. J Oral Pathol Med 33:71–78
7. Locke M, Heywood M, Fawell S, Mackenzie IC (2005) Retention of intrinsic stem cell hierarchies in carcinoma-derived cell lines. Cancer Res 65:8944–8950
8. Prince ME, Ailles LE (2008) Cancer stem cells in head and neck squamous cell cancer. J Clin Oncol 26:2871–2875
9. Costea DE, Tsinkalovsky O, Vintermyr OK, Johannessen AC, Mackenzie IC (2006) Cancer stem cells – new and potentially important targets for the therapy of oral squamous cell carcinoma. Oral Dis 12:443–454
10. Harper LJ, Piper K, Common J, Fortune F, Mackenzie IC (2007) Stem cell patterns in cell lines derived from head and neck squamous cell carcinoma. J Oral Pathol Med 36:594–603
11. McDermott SP, Wicha MS (2010) Targeting breast cancer stem cells. Mol Oncol 4:404–419
12. Lukacs RU, Lawson DA, Xin L, Zong Y, Garraway I, Goldstein AS, Memarzadeh S, Witte ON (2008) Epithelial stem cells of the prostate and their role in cancer progression. Cold Spring Harb Symp Quant Biol 73:491–502
13. Lee CJ, Dosch J, Simeone DM (2008) Pancreatic cancer stem cells. J Clin Oncol 26:2806–2812
14. Vries RG, Huch M, Clevers H (2010) Stem cells and cancer of the stomach and intestine. Mol Oncol 4:373–384
15. Prince ME, Sivanandan R, Kaczorowski A, Wolf GT, Kaplan MJ, Dalerba P, Weissman IL, Clarke MF, Ailles LE (2007) Identification of a subpopulation of cells with cancer stem cell properties in head and neck squamous cell carcinoma. Proc Natl Acad Sci USA 104:973–978
16. Okamoto A, Chikamatsu K, Sakakura K, Hatsushika K, Takahashi G, Masuyama K (2009) Expansion and characterization of cancer stem-like cells in squamous cell carcinoma of the head and neck. Oral Oncol 45:633–639
17. Chikamatsu K, Ishii H, Takahashi G, Okamoto A, Moriyama M, Sakakura K, Masuyama K (2011) Resistance to apoptosis-inducing stimuli in CD44+ head and neck squamous cell carcinoma cells. Head Neck [Epub ahead of print]
18. Joshua B, Kaplan MJ, Doweck I, Pai R, Weissman IL, Prince ME, Ailles LE (2011) Frequency of cells expressing CD44, a Head and Neck cancer stem cell marker: Correlation with tumor aggressiveness. Head Neck [Epub ahead of print]
19. Mack B, Gires O (2008) CD44s and CD44v6 expression in head and neck epithelia. PLoS One 3:e3360
20. An Y, Ongkeko WM (2009) ABCG2: the key to chemoresistance in cancer stem cells? Expert Opin Drug Metab Toxicol 5:1529–1542
21. Patrawala L, Calhoun T, Schneider-Broussard R, Zhou J, Claypool K, Tang DG (2005) Side population is enriched in tumorigenic, stem-like cancer cells, whereas ABCG2+ and ABCG2- cancer cells are similarly tumorigenic. Cancer Res 65:6207–6219
22. Chiba T, Kita K, Zheng YW, Yokosuka O, Saisho H, Iwama A, Nakauchi H, Taniguchi H (2006) Side population purified from hepatocellular carcinoma cells harbors cancer stem cell-like properties. Hepatology 44:240–251

23. Loebinger MR, Giangreco A, Groot KR, Prichard L, Allen K, Simpson C, Bazley L, Navani N, Tibrewal S, Davies D et al (2008) Squamous cell cancers contain a side population of stem-like cells that are made chemosensitive by ABC transporter blockade. Br J Cancer 98:380–387
24. Song J, Chang I, Chen Z, Kang M, Wang CY (2010) Characterization of side populations in HNSCC: highly invasive, chemoresistant and abnormal Wnt signaling. PLoS One 5:e11456
25. Tabor MH, Clay MR, Owen JH, Bradford CR, Carey TE, Wolf GT, Prince ME (2011) Head and neck cancer stem cells: the side population. Laryngoscope 121:527–533
26. Chen JS, Pardo FS, Wang-Rodriguez J, Chu TS, Lopez JP, Aguilera J, Altuna X, Weisman RA, Ongkeko WM (2006) EGFR regulates the side population in head and neck squamous cell carcinoma. Laryngoscope 116:401–406
27. Chu TS, Chen JS, Lopez JP, Pardo FS, Aguilera J, Ongkeko WM (2008) Imatinib-mediated inactivation of Akt regulates ABCG2 function in head and neck squamous cell carcinoma. Arch Otolaryngol Head Neck Surg 134:979–984
28. Chen CY, Chiou SH, Huang CY, Jan CI, Lin SC, Tsai ML, Lo JF (2009) Distinct population of highly malignant cells in a head and neck squamous cell carcinoma cell line established by xenograft model. J Biomed Sci 16:100
29. Singh SK, Hawkins C, Clarke ID, Squire JA, Bayani J, Hide T, Henkelman RM, Cusimano MD, Dirks PB (2004) Identification of human brain tumour initiating cells. Nature 432:396–401
30. Chen YC, Hsu HS, Chen YW, Tsai TH, How CK, Wang CY, Hung SC, Chang YL, Tsai ML, Lee YY et al (2008) Oct-4 expression maintained cancer stem-like properties in lung cancer-derived CD133-positive cells. PLoS One 3:e2637
31. Hermann PC, Huber SL, Herrler T, Aicher A, Ellwart JW, Guba M, Bruns CJ, Heeschen C (2007) Distinct populations of cancer stem cells determine tumor growth and metastatic activity in human pancreatic cancer. Cell Stem Cell 1:313–323
32. Ma S, Chan KW, Lee TK, Tang KH, Wo JY, Zheng BJ, Guan XY (2008) Aldehyde dehydrogenase discriminates the CD133 liver cancer stem cell populations. Mol Cancer Res 6:1146–1153
33. O'Brien CA, Pollett A, Gallinger S, Dick JE (2007) A human colon cancer cell capable of initiating tumour growth in immunodeficient mice. Nature 445:106–110
34. Shmelkov SV, Butler JM, Hooper AT, Hormigo A, Kushner J, Milde T, St Clair R, Baljevic M, White I, Jin DK et al (2008) CD133 expression is not restricted to stem cells, and both CD133+ and CD133- metastatic colon cancer cells initiate tumors. J Clin Invest 118:2111–2120
35. Zhang Q, Shi S, Yen Y, Brown J, Ta JQ, Le AD (2010) A subpopulation of CD133(+) cancer stem-like cells characterized in human oral squamous cell carcinoma confer resistance to chemotherapy. Cancer Lett 289:151–160
36. Damek-Poprawa M, Volgina A, Korostoff J, Sollecito TP, Brose MS, O'Malley BW Jr, Akintoye SO, DiRienzo JM (2011) Targeted inhibition of CD133+ cells in oral cancer cell lines. J Dent Res 90:638–645
37. Zhou L, Wei X, Cheng L, Tian J, Jiang JJ (2007) CD133, one of the markers of cancer stem cells in Hep-2 cell line. Laryngoscope 117:455–460
38. Chen H, Zhou L, Dou T, Wan G, Tang H, Tian J (2011) BMI1'S maintenance of the proliferative capacity of laryngeal cancer stem cells. Head Neck [Epub ahead of print]
39. Wei XD, Zhou L, Cheng L, Tian J, Jiang JJ, Maccallum J (2009) In vivo investigation of CD133 as a putative marker of cancer stem cells in Hep-2 cell line. Head Neck 31:94–101
40. Ginestier C, Hur MH, Charafe-Jauffret E, Monville F, Dutcher J, Brown M, Jacquemier J, Viens P, Kleer CG, Liu S et al (2007) ALDH1 is a marker of normal and malignant human mammary stem cells and a predictor of poor clinical outcome. Cell Stem Cell 1:555–567
41. Huang EH, Hynes MJ, Zhang T, Ginestier C, Dontu G, Appelman H, Fields JZ, Wicha MS, Boman BM (2009) Aldehyde dehydrogenase 1 is a marker for normal and malignant human colonic stem cells (SC) and tracks SC overpopulation during colon tumorigenesis. Cancer Res 69:3382–3389
42. Jiang F, Qiu Q, Khanna A, Todd NW, Deepak J, Xing L, Wang H, Liu Z, Su Y, Stass SA et al (2009) Aldehyde dehydrogenase 1 is a tumor stem cell-associated marker in lung cancer. Mol Cancer Res 7:330–338

43. van den Hoogen C, van der Horst G, Cheung H, Buijs JT, Lippitt JM, Guzman-Ramirez N, Hamdy FC, Eaton CL, Thalmann GN, Cecchini MG et al (2010) High aldehyde dehydrogenase activity identifies tumor-initiating and metastasis-initiating cells in human prostate cancer. Cancer Res 70:5163–5173
44. Kim MP, Fleming JB, Wang H, Abbruzzese JL, Choi W, Kopetz S, McConkey DJ, Evans DB, Gallick GE (2011) ALDH activity selectively defines an enhanced tumor-initiating cell population relative to CD133 expression in human pancreatic adenocarcinoma. PLoS One 6:e20636
45. Clay MR, Tabor M, Owen JH, Carey TE, Bradford CR, Wolf GT, Wicha MS, Prince ME (2010) Single-marker identification of head and neck squamous cell carcinoma cancer stem cells with aldehyde dehydrogenase. Head Neck 32:1195–1201
46. Chen YC, Chen YW, Hsu HS, Tseng LM, Huang PI, Lu KH, Chen DT, Tai LK, Yung MC, Chang SC et al (2009) Aldehyde dehydrogenase 1 is a putative marker for cancer stem cells in head and neck squamous cancer. Biochem Biophys Res Commun 385:307–313
47. Chen YC, Chang CJ, Hsu HS, Chen YW, Tai LK, Tseng LM, Chiou GY, Chang SC, Kao SY, Chiou SH et al (2010) Inhibition of tumorigenicity and enhancement of radiochemosensitivity in head and neck squamous cell cancer-derived ALDH1-positive cells by knockdown of Bmi-1. Oral Oncol 46:158–165
48. Ito K, Bernardi R, Pandolfi PP (2009) A novel signaling network as a critical rheostat for the biology and maintenance of the normal stem cell and the cancer-initiating cell. Curr Opin Genet Dev 19:51–59
49. Yilmaz OH, Valdez R, Theisen BK, Guo W, Ferguson DO, Wu H, Morrison SJ (2006) Pten dependence distinguishes haematopoietic stem cells from leukaemia-initiating cells. Nature 441:475–482
50. Zhang J, Grindley JC, Yin T, Jayasinghe S, He XC, Ross JT, Haug JS, Rupp D, Porter-Westpfahl KS, Wiedemann LM et al (2006) PTEN maintains haematopoietic stem cells and acts in lineage choice and leukaemia prevention. Nature 441:518–522
51. Yanagi S, Kishimoto H, Kawahara K, Sasaki T, Sasaki M, Nishio M, Yajima N, Hamada K, Horie Y, Kubo H et al (2007) Pten controls lung morphogenesis, bronchioalveolar stem cells, and onset of lung adenocarcinomas in mice. J Clin Invest 117:2929–2940
52. Wang S, Garcia AJ, Wu M, Lawson DA, Witte ON, Wu H (2006) Pten deletion leads to the expansion of a prostatic stem/progenitor cell subpopulation and tumor initiation. Proc Natl Acad Sci USA 103:1480–1485
53. Korkaya H, Paulson A, Charafe-Jauffret E, Ginestier C, Brown M, Dutcher J, Clouthier SG, Wicha MS (2009) Regulation of mammary stem/progenitor cells by PTEN/Akt/beta-catenin signaling. PLoS Biol 7:e1000121
54. Li H, Gao Q, Guo L, Lu SH (2011) The PTEN/PI3K/Akt pathway regulates stem-like cells in primary esophageal carcinoma cells. Cancer Biol Ther 11:950–958
55. Moral M, Segrelles C, Lara MF, Martinez-Cruz AB, Lorz C, Santos M, Garcia-Escudero R, Lu J, Kiguchi K, Buitrago A et al (2009) Akt activation synergizes with Trp53 loss in oral epithelium to produce a novel mouse model for head and neck squamous cell carcinoma. Cancer Res 69:1099–1108
56. Hermann PC, Huber SL, Heeschen C (2008) Metastatic cancer stem cells: a new target for anti-cancer therapy? Cell Cycle 7:188–193
57. Brabletz T, Jung A, Spaderna S, Hlubek F, Kirchner T (2005) Opinion: migrating cancer stem cells – an integrated concept of malignant tumour progression. Nat Rev Cancer 5:744–749
58. Tomaskovic-Crook E, Thompson EW, Thiery JP (2009) Epithelial to mesenchymal transition and breast cancer. Breast Cancer Res 11:213
59. Chen C, Wei Y, Hummel M, Hoffmann TK, Gross M, Kaufmann AM, Albers AE (2011) Evidence for epithelial-mesenchymal transition in cancer stem cells of head and neck squamous cell carcinoma. PLoS One 6:e16466
60. Freudlsperger C, Burnett JR, Friedman JA, Kannabiran VR, Chen Z, Van Waes C (2011) EGFR-PI3K-AKT-mTOR signaling in head and neck squamous cell carcinomas: attractive targets for molecular-oriented therapy. Expert Opin Ther Targets 15:63–74

61. Lo JF, Yu CC, Chiou SH, Huang CY, Jan CI, Lin SC, Liu CJ, Hu WY, Yu YH (2011) The epithelial-mesenchymal transition mediator S100A4 maintains cancer-initiating cells in head and neck cancers. Cancer Res 71:1912–1923
62. Gangaraju VK, Lin H (2009) MicroRNAs: key regulators of stem cells. Nat Rev Mol Cell Biol 10:116–125
63. Cabarcas SM, Mathews LA, Farrar WL (2011) The cancer stem cell niche – there goes the neighborhood? Int J Cancer [Epub ahead of print]

Chapter 24
The Biology of Lung Cancer Stem Cells

Sandeep Singh and Srikumar P. Chellappan

Contents

Abstract Lung cancer is strongly correlated with tobacco smoking and causes the maximum number of cancer related deaths in the world. While there are many plausible models for the genesis and metastasis of cancers, the recently evolved cancer stem cell model offers new insights into the potential mechanisms underlying the development of cancer. According to this model, cancers can only be originated from a subset of the cells, which have the ability to self-renew and differentiate into the heterogeneous lineages of cancer cells. These cells are thought to have originated as a result of oncogenic transformation of the normal stem cells or its progenitor cells. The site of origin of various types of lung cancers coincides with specifically localized airway stem cell niches in murine models and therefore supports the stem cell origin of lung cancer. Various functional and cell surface markers have been used to identify and characterize lung cancer stem cells within the tumors; however, the clinical significance of these markers remains to be confirmed. Understanding

S. Singh • S.P. Chellappan (✉)
Department of Tumor Biology, H. Lee Moffitt Cancer Center and Research Institute, 12902 Magnolia Drive, Mail Code SRB3, Tampa, FL 33612, USA
e-mail: Srikumar.Chellappan@moffitt.org

R.K. Srivastava and S. Shankar (eds.), *Stem Cells and Human Diseases*,
DOI 10.1007/978-94-007-2801-1_24,

the molecular mechanisms governing the deregulated self-renewal of these cells may identify novel therapeutic targets to combat lung cancer. This chapter describes the current state of knowledge of the biology of putative lung cancer stem cells.

Keywords Lung cancer stem cells • Squamous cell carcinomas • Adenocarcinoma • Bronchioalveolar carcinoma

24.1 Lung Cancer and Cancer Stem Cells

Despite significant therapeutic advances, lung cancer causes the maximum number of cancer related death worldwide [111]. According to an estimate made by the World Health Organization (WHO), lung cancer will cause about 2.5 million deaths per year by the year 2030 [118]. Smoking is established as a causal factor for 90% of the lung cancers worldwide [118]. Although public awareness regarding the ill-effects of smoking is the primary strategy in the fight against lung cancer, the risk of lung cancer remains significantly high for long-term heavy smokers even after smoking cessation. Fifty percent of new lung cancer patients are former smokers, and many of them stopped smoking 5 years or more previously [59, 151]. Therefore, given the large number of current and former smokers, lung cancer will remain a major health issue for several decades. In the United States, approximately 85% of the patients diagnosed with lung cancer die of this disease within 5 years and this rate has not changed significantly since the 1970s [72, 73]. Moreover, there is a chance of developing a second primary lung cancer for every 1–3% patients per year even after their curative treatment of the first cancer. Further, treatment of the primary lesions rarely prevents the development of the distant metastases [73]. These facts highlight a need for better understanding the cellular and molecular events underlying the genesis of this disease for designing novel therapeutic strategies.

There are two models proposed for cancer initiation and progression. The prevailing theory supports the "clonal evolution model" [107]. This model suggests that cancers derive from the serial acquisition of genetic mutations by normal somatic cells resulting in their deregulated cell proliferation, inhibition of apoptosis and inhibition of differentiation. Each oncogenic mutation would further transform the cells resulting in their expansion and become a dominant population within the tumor. All these mutant tumor cells would have a similar proliferative capacity and potential for regenerating tumor growth [107, 156]. However, historically it has been observed that the proliferating cells of a primary malignant tumor are markedly heterogeneous in respect to phenotypic markers, sensitivity to the drugs, tumorigenic capacity upon transplantation, metastatic capacity as well as morphology and growth rate in tissue culture [163]. In recent years, the "cancer stem cell model" has been able to provide plausible explanations for tumor initiation, growth, metastasis as well as intratumoral heterogeneity [110, 156]. According to this model, cancers can only be originated through cancer stem cells. These are defined as the 'subset of the cells with the ability to self-renew and recapitulate the heterogeneous lineages

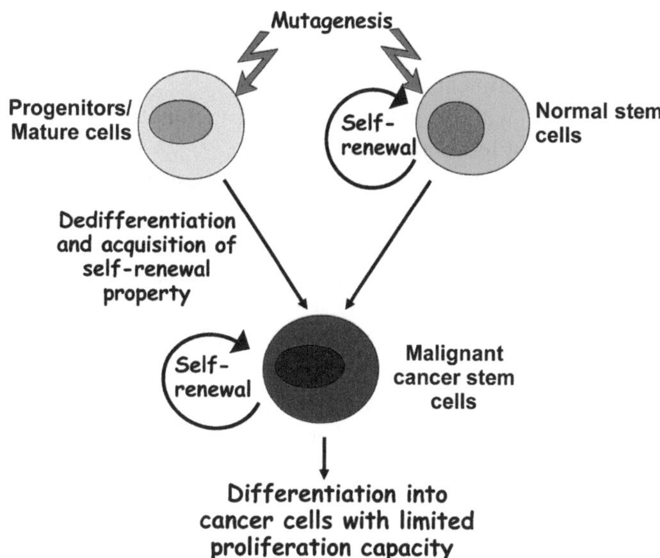

Fig. 24.1 Cancer stem cell model. Stem cells with intrinsic properties of self-renew and multi-lineage differentiation are present in adult tissues. These normal stem cells acquire the oncogenic mutations which results in its deregulated self-renewal and give rise to cancer stem cells. Additionally, mutations might also cause restricted progenitor cells to de-differentiate and acquire self renewal property and become cancer stem cells. Cancer stem cells self-renew themselves as well as differentiate to generate phenotypically diverse non-tumorigenic cancer cells, which constitute the bulk of the heterogeneous tumor

of cancer cells that comprise the tumor' [14, 28]. Self-renewal is the property of stem cells to maintain their numbers through symmetric or asymmetric mitotic cell division [104]. During asymmetric division, each stem cell generates one daughter cell with stem cell fate (self-renewal) and one daughter cell (progenitor cell) that is destined to differentiate [29]. However, upon injuries or when stem cell pool has to be developed during development, stem cells undergo symmetric cell division where all the divided cells have stem cell fate [104]. The fine balance between symmetric and asymmetric modes of division maintains the number of stem cells and its differentiated progeny depending on the developmental signals [104]. It is believed that oncogenic transformations of the normal stem cells or the progenitor cells that have recently been derived from normal stem cells results in deregulated self-renewal and give rise to cancer stem cells [46, 65, 92, 135]. Aberrant symmetric or asymmetric cell division maintains the number of cancer stem cells within the tumor, whereas the progenitor cells and its further differentiated progeny constitute the bulk of the heterogeneous tumor [29, 104, 110, 156]. These cells eventually become terminally differentiated and stop dividing, therefore are non-tumorigenic (Fig. 24.1). Moreover, it is also important to note that the clonal evolution model may co-exist with the cancer stem cell model. Cancer stem cells may themselves undergo "clonal evolution" by acquiring further mutations to undergo more aggressive self-renewal and growth as demonstrated for leukemia stem cells [112, 156].

One of the first experimental evidences for the existence of cancer stem cells came in the year 1997, with the identification of leukemia stem cells. Only a small subset of acute myeloid leukemia cells characterized as $CD34^{+}/CD38^{-}$ cells were able to transfer the disease from patients to the Non-obese diabetic severe combined immunodeficient (NOD-SCID) mice [14, 86]. Later, in the year 2003, the first evidence for hierarchical stem cell origin of solid tumor was experimentally demonstrated in breast cancer. The finding suggested that only a small subset of tumor cells, characterized as $CD44^{+}$, $CD24^{-/low}$ cells, were capable of recapitulating the original tumor heterogeneity in NOD-SCID mice [1]. Since the discovery of cancer stem cells in breast cancer, putative cancer stem cells have been identified for various solid tumors (Table 24.1) [2]. Comparatively less is known about the biology of lung cancer stem cells compared to other solid tumor stem cells. Identification of lung cancer stem cells has been challenging in part due to the complexity of the disease in terms of its phenotypical diversity and anatomically distinct sites of origin in pulmonary airways. Histologically, lung cancer can be subdivided into four major types which are small cell lung cancer (SCLC), and three types of non-small lung cancer (NSCLC), which include squamous cells carcinoma (SCC), adenocarcinoma and large cell carcinoma. SCLC and SCC occur in the proximal region of the respiratory tract whereas adenocarcinoma originates in the distal airway [50]. This regional diversity is hypothesized to be dependent upon the presence of a diverse pool of self-renewing lung epithelial stem cells present in different regions of the respiratory tract. In support of this hypothesis, originating sites of SCLC, SCC, and adeno-/broncheoalviolar carcinomas appear to coincide with recently identified airway stem cell niches in murine models [49, 120]. This article describes the current state of knowledge of the biology of putative lung cancer stem cells.

24.2 Normal Lung Epithelial Stem Cells

The epithelium of the adult airway consists of distinct cell types, arranged along the proximal to distal anatomical locations of the respiratory tract. Identification of resident stem cells among these cell types is challenging due to the slow turnover of the adult lung epithelial cells. The adult lung epithelium cells divide only every 30–50 days, therefore, the proliferative fraction has been estimated to be as small as 1.3% in the tracheal epithelium and 0.06% in bronchiolar epithelium [79, 120]. However, upon lung injuries, turnover of the specific cells increases in the specific regions of the lung, suggesting them as putative stem cells in their respective niches [49].

The proximal airway that is composed of trachea and main bronchi are mainly lined by ciliated columnar and mucus-secreting goblet cells. Cell turnover in this region is thought to be driven by specific cells known as basal cells. These cells are mostly restricted to the trachea in rodents, however they are ubiquitously found in proximal and distal airways in humans [12, 45]. Using mouse models of epithelial injury, various studies have confirmed the presence of tracheal Keratin 5 (K5)

Table 24.1 Surface markers of solid tumor stem cells

Cancer type	Markers	References
Breast carcinoma	$CD44^+CD24^-Lineage^-$	[1]
	$CD44^+CD24^-ALDH$	[52]
	ALDH	[23]
Bladder carcinoma	SP	[109]
Colorectal carcinoma	$CD133^+$	[126]
	$CD133^+$	[108]
	$CD133^+$ and $CD133^-$	[131]
	$CD44^+$	[39]
	$EpCAM^+CD44^+CD166^+$	[34]
	$CD44^+ALDH1$	[26]
	ALDH1	[69]
	$EpCAM^+ALDH1$	[22]
Endometrial carcinoma	$CD133^+$	[127]
	SP	[47]
Ewing sarcoma	$CD133^+$	[145]
Gastric carcinoma	$CD44^+$	[147]
Lung carcinoma (NSCLC)	$CD133^+$	[24]
	$CD133^+$	[41]
	SP	[89]
	SP	[143]
Lung carcinoma (SCLC)	SP	[37]
	$CD133^+ASCL^+ALDH1$	[76]
Medulloblastoma, glioma	$CD133^+$	[134]
Osteosarcoma	$CD133^+$	[149]
	ALDH	[157]
Ovarian adenocarcinoma	$CD44^+CD117^+$	[170]
	$CD44^+MyD88^+$	[4]
	$CD133^+$	[6]
Pancreatic carcinoma	$CD133^+$	[63]
	$CD133^+$	[95]
	$CD44^+CD24^+EpCAM^+$	[90]
Prostate carcinoma	$CD133^+CD44^+\alpha2\beta1^+$	[30]
	$CD44^+CD24^-$	[70]
	$CD44^+$	[114]
	$CD44^+$	[84]
Renal carcinoma	$CD105^+$	[18]
Head and neck SCC	$CD44^+$	[117]
Hepatocellular carcinoma	$CD133^+$	[136]
	$CD44^+CD90^+$	[165]
	$CD133^+$	[94]
	$CD24^+$	[87]

or Keratin 14 (K14)-expressing basal stem cells. Transgenic mice expressing *K5*-promoter-driven enhanced green fluorescent protein (EGFP) demonstrated EGFP expression restricted to these putative stem cell niches, confirming K5 expression in these self-renewing tracheal stem cells [130]. Lineage-tracing experiments further supported the stem cell properties of tracheal K5-expressing basal cells, and it was found that these cells could self-renew in addition to giving rise to ciliated cell progeny [15, 68]. In another mouse model, naphthalene-induced injury resulted in the rapid induction of K14-expressing basal cells. Transgenic mice harboring a *K14*-promoter linked to a lox-cre reporter were used to show that these rapidly proliferating K14-expressing basal cells gave rise to phenotypically heterogeneous progeny, including clonal species of the original K14-expressing basal cells [67, 68]. Therefore, the discovery that basal cells possess the capacity for multi-potent differentiation as well as self-renewal suggests these cells are the predominant stem cells of the proximal airways.

The bronchioles of middle airways are enriched with dome-shaped secretory cells termed as Clara cells. In rodent lung epithelia these cells are identified by the expression of secretoglobin, *scgb1a1,* also known as Clara cell secretory protein (CCSP) or CC10 [13, 105]. These cells are functionally equivalent to mature differentiated epithelial cells in the quiescent steady state. However, in response to nitrogen oxide/ozone inhalation injury, Clara cells become transit-amplifying cells (Clara type A cells) that can transiently de-differentiate and give rise to phenotypically different bronchiolar epithelial cells [42, 43, 81, 138] therefore serve as "facultative progenitor cells" [81, 139]. Because Clara cells are selectively damaged by naphthalene [17], the lung injury models induced by naphthalene identified other populations of bronchiolar cells that likely represent true bronchial airway stem cells [96, 139]. These models have shown that bifurcation zones of the bronchioles contain two separate populations of cells that proliferate in response to naphthalene-induced injury [139]. The first population consists of calcitonin gene-related peptide (CGRP)-positive pulmonary neuroendocrine cells (PNECs), the majority of which can be found in clusters and termed as neuroendocrine bodies (NEB). The second population consists of naphthalene resistant, CCSP-positive variant Clara cells that proliferate and repopulate bronchiole airways with phenotypically diverse progeny after naphthalene exposure [125, 137]. However, when airway epithelium regeneration was measured after selective ablation of variant Clara cells, PNECs expansion was not sufficient to reconstitute the airway epithelium [125]. These results imply that, while both PNECs and variant Clara cells are able to self-renew, only variant Clara cells have the capacity for multi-potent differentiation and thus represent true bronchial airway stem cells [66, 139].

The distal airways are composed of respiratory bronchioles and alveoli. The alveolar epithelium is composed of a thin layer of flattened type 1 pneumocytes (AT1) and cuboidal type 2 pneumocytes (AT2). Type 1 pneumocytes are terminally differentiated cells. In earlier rodent lung injury experiments, type 2 pneumocytes have been shown to have the properties of progenitor cells with the unipotent capacity to differentiate into type 1 pneumocytes [44]. Another class of lung stem cells was recently discovered in the putative stem cell niche of the bronchioalveolar duct junction (BADJ). In 2002, Giangreco and colleagues identified a naphthalene resistant CCSP-expressing population in the BADJ that was capable of self-renewal and regenerating the alveoli after naphthalene-induced injury [51]. Unlike variant Clara stem cells, these BADJ-associated Clara-like cells were not in proximity to the PNEC stem cells niche [51]. In 2005, Kim and colleagues reported that these $CCSP^+$ BADJ cells coexpressed surfactant protein C (SP-C) and termed them as "bronchioalveolar stem cells" (BASCs) [81]. These cells display a Sca-1^+/$CD45^-$/ platelet-endothelial cell adhesion molecule $(PECAM)^-$/$CD34^+$ cell-surface marker phenotype. The BASC population was capable of proliferation, self-renewal and multilineage differentiation in culture, suggesting them as true distal airway stem cells [81]. Recently, several groups have challenged these observations. One study suggested that $CD45^-$/$CD31^-$/$CD34^-$/Sca-$1^{low/AF\ low}$ is a more appropriate defining phenotype of BADJ stem cells [148]. Through lineage-tracing experiments, it was also demonstrated that Clara cells were not able to give rise to type II alveolar cells and suggested instead that $Id2^+$ cells are lung distal tip epithelial stem cells [120].

The identification of various lung epithelial stem cells largely relied on lung injuries caused by aromatic hydrocarbons and took advantage of mouse models. However, information regarding lineage decisions, self-renewal and clonogenic properties of resident stem cells is relatively less understood in human lung airways. Recently, stem cells were identified from human lung airway epithelium as $CD45^-$, side population (SP) cells [58] (More details on the Side Population cells are given in Sect. 24.4). $CD45^-$, SP cells represented approximately 0.1% of the total epithelial cell population. This subset of cells was able to maintain telomere length after several passages and sustain colony-forming capacity. Importantly, $CD45^-$, SP cells demonstrated multipotency by undergoing differentiation into basal, ciliated and mucus secreting cell types [58]. In addition, very recently, a pool of c-Kit-positive cells was demonstrated to have the fundamental properties of stem cells i.e. self-renewal, clonogenicity, and multipotentiality [78]. From 12 adults and 9 fetal human lung tissue specimens, the characteristic stem cells niches were observed in bronchioles and alveolar walls. *In vivo* transplantation of c-Kit^+ human lung stem cells adjacent to the cryo-injured lung tissue of immunosuppressed mice revealed the capacity of these cells to give rise to different populations of epithelial cells of endodermal origin as well as pulmonary vessels of mesodermal origin after coordinated differentiation. Therefore, these human lung stem cells are simultaneously activated to form both distal airways and distal pulmonary vasculature to form functionally competent gas exchange units [78].

24.3 Stem Cell Origin of Lung Cancers

The evidence of stem cell origin of lung cancers has been demonstrated using genetically modified mouse models of lung cancers. Classically, genetic modifications involved the transgenic expression or knockout of the specific transforming genes under the control of lung epithelial cell-specific promoters [40]. These genetic modifications generate identical mutations throughout the large proportion of the lung and therefore should be capable of generating cancers throughout the entire lung. However, surprisingly, in all models tested, the most originating sites for various lung cancers appear to coincide with identified airway stem cell niches, and therefore suggest its stem cell origin [121].

24.3.1 Cell of Origin for Small Cell Lung Carcinomas

SCLC is one of the most devastating forms of lung cancer with a very poor prognosis and a high rate of metastatic dissemination [73]. Human SCLCs predominately localize to mid-level bronchioles and typically express a range of neuroendocrine cell markers, including CGRP and other markers generally expressed within PNECs, while markers associated with other cell types are not expressed [48]. The loss of Retinoblastoma (Rb) and p53 functions are closely associated with human SCLC [98]. Despite Rb gene deletion throughout the airways, hyperplasia of PNECs was exclusive in a lung-specific conditional Rb inactivation model [99]. Similarly, deletion of both Rb and p53 in adult mice resulted in progressive epithelial hyperplasia that was restricted to the NEB microenvironment [99, 100]. Importantly, these lesions progressed to form metastatic tumors resembling human SCLC [99]. On the basis of these observations, SCLCs are proposed to originate from PNECs. In a recent report, a direct approach has been taken to identify the cell of origin for SCLCs using a cell-type specific Cre-loxP expression model in adult mouse lung [144]. Depletion of Rb and p53 specifically in tracheal and bronchial Clara cells, BASCs, AT-2 cells and PNECs suggested that PNECs are indeed the predominant cancer-initiating population in SCLC. In addition, data also supported the presence of an SPC-positive progenitor cell population that could give rise to SCLC following loss of p53 and Rb, although with a lower efficiency [144].

24.3.2 Cell of Origin for Squamous Cell Carcinomas of Lung

Compared to the mouse models available for other lung cancer types, genetically modified mouse models of human SCC do not exist. Mouse models for SCC are developed using chemical induced carcinogenesis. Although, with unknown mutations caused by the carcinogens, these mouse models yield hyperplastic lesions and SCCs that resemble the human disease [62, 158, 166, 167]. Carcinogen induced murine SCCs generally occur in the proximal airways down to the second or third bifurcation and are rarely observed distally, which coincide with the suggested niche for basal stem/progenitor cells [15, 45]. Histologically, carcinogen induced SCCs are found to be initiated with generalized basal cell hyperplasia [74]. Additionally, hyperproliferative and preneoplastic lesions in SCC were also found to be basal stem cells specific Keratin-14-positive [7]. Although direct approaches are needed to confirm the cell of origin for SCCs, current data support a direct relationship between proximal airway basal progenitors and SCC in murine models.

24.3.3 Cell of Origin for Adenocarcinoma/Bronchioalveolar Carcinoma of Lung

Bronchiolar adenocarcinomas and bronchioalveolar cell carcinomas are among the most common lung cancer types among smokers as well as non-smokers [73]. In murine models, adenocarcinomas arise from the junction between the terminal bronchiole and the alveolus termed the "bronchioalveolar duct junction" (BADJ) [81]. Murine as well as human adenocarcinoma cells express CCSP and/or SP-C markers suggesting that Clara or AT-2 cells function as the originating cells for adenocarcinomas [46, 81]. Additionally, in mouse models where CCSP or alveolar-SP-C promoters were utilized to express mutated proteins like mutated epidermal growth factor (EGF) receptor [75, 116], active K-Ras (G12D) [68, 69], dominant negative transforming growth factor-β [16], and large T antigen [38, 160], all resulted in tumors closely resembling human bronchioalveolar carcinoma. As described above, napthalene resistant, NEB-independent, CCSP and SP-C expressing, Sca-1-positive, BASCs were discovered in BADJ region of napthalene injured mouse lung [81]. In the same study, using the cell-type specific Cre-loxP expression model, Kim and colleagues have demonstrated selective, dose-dependent BASC expansion and bronchioalveolar carcinoma induction after introducing K-ras (G12D) mutation [81].

Overall, the above findings provide evidence that normal airway stem cells can directly act as originating cells for lung cancers. However, in an interesting study, specific expression of H-Ras to pulmonary PNECs using a CGRP promoter-driven transgenic construct resulted in the formation of bronchial adenocarcinomas but not SCLCs [142]. Similarly, in recent studies, introduction of an activated H-Ras (V12) could convert neuroendocrine type SCLC cells into a completely changed phenotype resembling more closely to NSCLC adenocarcinomas [19]. These results suggest that the physiological effect of different genetic lesions can also drive the same target cells into divergent differentiation paths. Moreover, it should also be emphasized that the cell of origin is described for those self-renewing stem cells which acquire first oncogenic mutation to destabilize its growth. However, CSC may also arise from restricted progenitors, which acquires the self-renewing properties through genetic or epigenetic mechanisms. For lung tumorigenesis, this proposal can be supported by a recent report where Sca-1-positive-BASCs were originally proposed as cell of origin for K-Ras (G12D) driven bronchioalveolar carcinoma. However, within the tumors, both Sca-1-positive as well as negative cells acquired cancer stem cells properties as demonstrated by their ability to initiate secondary tumors when implanted in recipient mice [32]. Therefore, cancer stem cells may show more distinct markers than their proposed cell of origin, which represents the major challenge in identification and isolation of these Cancer stem cells from tumors.

24.4 Identification of Lung Cancer Stem Cells

Based on a classical method described almost 40 years ago [60], the first experimental evidence for the existence of a stem-like clonogenic subpopulation in lung cancer was demonstrated in 1980s [20, 21]. In these pioneering studies in 1980s, only a small proportion of SCLC and lung adenocarcinoma cells from patient samples demonstrated their ability to generate colonies in soft agar [20, 21]. Transplantation of these colony forming cells displayed tumorigenic potential in athymic nude mice, suggesting them as lung cancer stem cells derived from the patient tumors [20, 21]. Since then, researchers have attempted to define lung cancer stem cells by various experimental approaches. Presently, these cells are isolated by flow cytometry using their expression of stem cell specific cell surface markers or by its functional properties as described below.

The side population phenotype is a specific functional property described to isolate normal human hematopoietic stem cells from bone marrow population [55]. It is based on the ability of drug transporters to efflux the Hoechst dye via the ATP-binding cassette (ABC) family of transporter proteins, mainly ABCB1 (P-glycoprotein, MDR1), ABCC1-5 (multidrug-resistant proteins, MRP1-5), and ABCG2 (breast cancer resistance protein, BRCP1). These proteins are specifically expressed within the cell membrane of stem cell populations [53, 171]. Hoechst 33342 dye excluding cells, termed 'Side Population' cells (SP cells), have been described in a variety of tumor types (Table 24.1), where they have been shown to display increased capacity of self-renewal and tumorigenicity when transplanted into immunocompromised mice [53, 164]. SP cells were detected in various human NSCLC cancer cell lines and had several *in vitro* properties typical of stem cells, including clonogenic proliferation, invasive phenotypes, multi-drug resistance, and increased telomerase (hTERT) as well as lower levels of DNA replication associated protein MCM7 as compared to rest of the non-SP (main population, MP) cells [64]. Similarly, SP cells were detected in primary tumors obtained from lung cancer patients. Like NSCLC cell lines, SCLC cell lines such as NCI-H82, H146 and H526 also demonstrated the presence of SP cells with tumorigenic potential. Our unpublished results also show a subset of side population cells in lung cancer cell lines and lung cancer specimens (Singh et al. in preparation, and Fig. 24.2). Gene expression profiling revealed the upregulation of several pluripotency associated genes such as KLF4, NANOG, NUMB, OCT4 and NOTCH1 in SP cells [128]. These lines of evidence support the notion that the side population assay selects for cancer stem cells in lung tumors; however, experimental variables such as incubation time, dye concentration, cell concentrations, and gating variability may produce different frequency of SP cells from the similar type of samples [53]. Therefore, a common experimental procedure needs to be proposed to avoid the variability from one laboratory to others.

In addition to the side population phenotype, another method for identifying and selecting stem cell population based on functional property is the specific high aldehyde dehydrogenase activity of stem cells. Aldehyde dehydrogenase (ALDH) is a family of intracellular enzymes that participates in cellular detoxification,

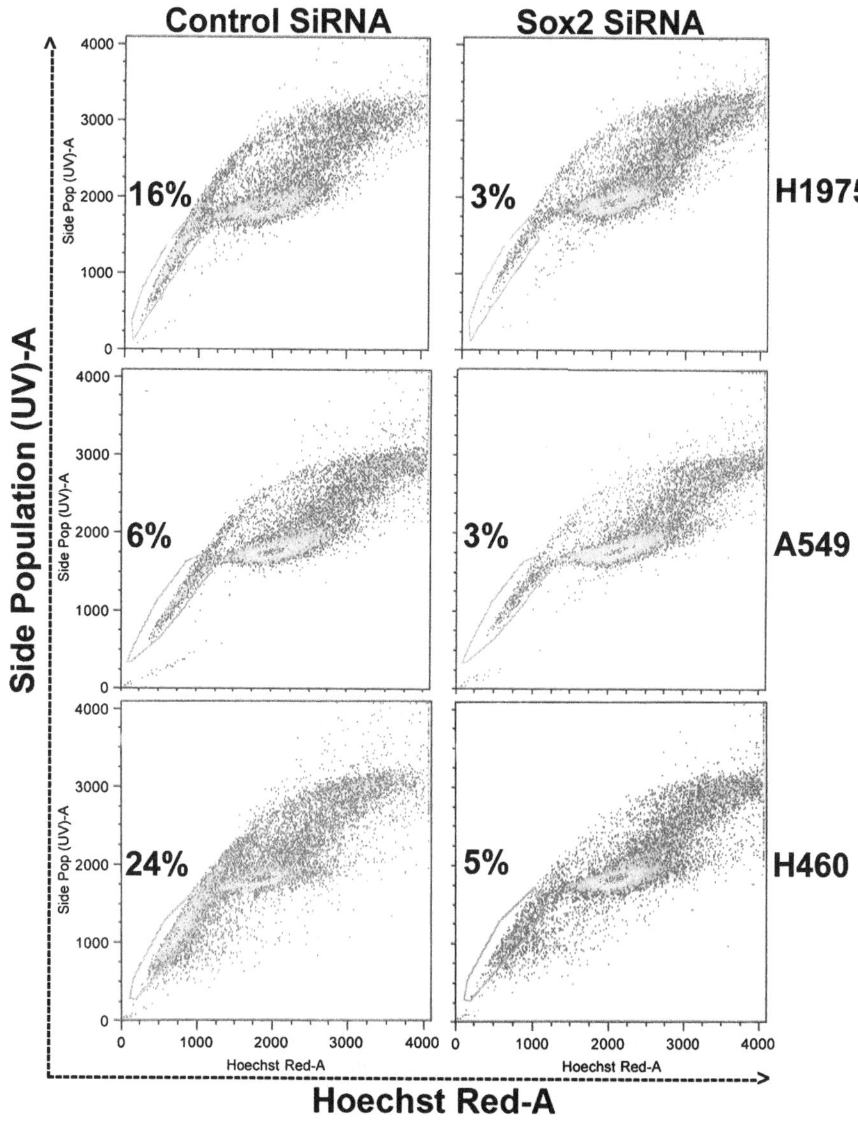

Fig. 24.2 Sox2 regulates side population phenotype. SiRNA mediated knockdown of Sox2 expression resulted in decreased frequency of side population cells in NSCLC cell lines H1975, A549 and H460

differentiation and drug resistance in stem cells [102]. ALDH activity is found to directly regulate the self-renewal of hematopoietic stem cells by inhibiting the endogenous retinoic acid biosynthesis [27]. Flow cytometry has been used to detect and isolate cells with elevated ALDH activity; this technique has led to the successful isolation of putative cancer stem cells from a variety of human cancers

(Table 24.1) [3, 103]. Evidence for ALDH as a relevant cancer stem cell marker for the lung came recently with the discovery of elevated levels of ALDH-isoforms, ALDH1A1 and ALDH3A1 protein expression in SCC and adenocarcinoma patients [76, 113]. Additionally, higher expression of these isoforms of ALDH was also detected in putative lung stem cell niche for adenocarcinoma [113]. In another study, more than 200 NSCLC tumor samples were analyzed for the expression of ALDH-isoforms and high expression of ALDH1A1 was strongly associated with reduced patient survival for Stage I NSCLC [141]. NSCLC cells with high ALDH activity showed more tumorigenic and clonogenic activity as compared to the cells with low its activity, supporting high ALDH activity as a putative lung cancer stem cell phenotype [141].

Identification and isolation of putative cancer stem cells is also commonly based on stem cell specific cell surface phenotypic markers. One of such markers successfully used to isolate lung cancer stem cell is CD133 (*Prom1*). It is a cell surface glycoprotein that consists of five transmembrane domains and two large glycosylated extracellular loops [101]. CD133 and its glycosylated epitope, AC133, have been useful in the selection of normal human hematopoietic and neural stem cells as well as for brain, colon and pancreatic cancer stem cells [80]. Highly tumorigenic, self-renewing $CD133^+$ cells in both NSCLC and SCLC specimens were isolated from single cell suspensions of whole tumor [41]. Alternatively, through retrospective approach, all the tumor forming cells from both SCLC and NSCLC were allowed to grow in serum free, stem cell selective media in a non-adherent condition. This strict culture condition allows the expansion of only self-renewing stem cells as spheres. All the sphere forming cells displayed CD133 expression, self-renewal and differentiation to specific lineage as well as recapitulated the hetrogeneous tumor in recipient mice [41]. The discovery of putative $CD133^+$ lung cancer stem cells in both SCLC and NSCLC indicate that CD133 may serve as a pan-lung cancer stem cell marker. CD133-expressing stem-like cells isolated from NSCLC patients were found to be resistant to Cisplatin treatment, suggesting the drug resistant phenotype of cancer stem cells [10]. However, the existence of variable CD133 isoforms and CD133 glycosylation states are also reported which complicate the detection of CD133 and AC133 in whole tumor as well at single cell level [11, 101]. Also, the clinical significance of CD133 expression in human various lung cancer types still needs to be validated [129].

24.5 Mechanisms for Self-renewal of Lung Cancer Stem Cells

In the stem cell model of cancer, the initiation and progression of cancer depends on the deregulated self-renewal of cancer stem cells. A number of genes like *notch, wnt and shh,* which are involved in maintenance and self-renewal of normal tissue stem cells, are found to be oncogenes in various cancers. These observations suggested that the pathways that govern normal stem cell self-renewal could also govern stem cell self-renewal in cancer as well. Therefore, identification of the developmental pathways involved in self-renewal of cancer stem cells for specific tumors has become an appealing strategy for finding suitable targets for treatment [33]. There is only a handful of information available for the mechanism of self-renewal of normal or cancer stem cells for lungs which is summarized below.

The Wnt protein-mediated activation of Frizzled receptors leads to β-catenin accumulation and nuclear translocation. This Wnt/β-catenin pathway regulates the self-renewal of hematopoietic stem cells [83, 123]; however, its role in lung epithelial stem cells is less understood [140]. Recently, activated Wnt/β-catenin signaling has been correlated with lung epithelium regeneration. Expression of constitutively active form of β-catenin, specifically in Clara cells revealed the expansion of BASCs upon naphthalene injury for lung epithelium regeneration [124]. However in another study, deletion of β-catenin specifically in Clara cells had no impact on repair of naphthalene-injured airways and BASC expansion [169]. These contrasting studies suggested a context dependent role of Wnt/β-catenin signaling in lung stem cell self-renewal and need further analysis. Higher expression of Wnt proteins and aberrant Wnt signaling has been reported in lung cancer progression [88, 153, 154]. Inhibition of Wnt signaling by a Wnt-2 monoclonal antibody demonstrated cell death in NSCLC cells [168]. However, mRNAs analysis for multiple Wnt in NSCLC cell lines and primary lung tumors revealed markedly decreased expression of Wnt-7a, compared to normal bronchial epithelial cell lines and normal lung tissue. Ectopic expression of Wnt-7a in NSCLC cell lines reversed cellular transformation, decreased anchorage-independent growth, and induced epithelial differentiation in a subset of the NSCLC cell lines, suggesting a tumor suppressor role of Wnt-7a [161, 162]. The possible association of Wnt signaling with stem cell self-renewal and lung tumorigenesis suggests its importance; however, further studies will be necessary to confirm the involvement of aberrant Wnt signaling in lung cancer stem cell self-renewal [36, 122].

The Hedgehog (Hh) signaling pathway is a key developmental pathway during embryogenesis [91]. The Hh signaling pathway is activated by sonic hedgehog (shh), a mammalian Hh ligand involved in pulmonary cell fate determination and branching morphogenesis [9, 115]. In response to naphthalene injury, activated Hh signaling was observed in airway repair and epithelial regeneration and mediated through increased numbers of neuroendocrine cells in PNECs niches [159]. Aberrations in expression and activation of this pathway also led to the development SCLCs [56, 106, 146]. Suppression of aberrant Hh signaling in some SCLCs resulted in a dramatic drop in cell viability and tumorigenicity [159], therefore representing a suitable therapeutic target against SCLC [155].

The Notch signaling pathway is one of the important cell fate determinants during organogenesis and tissue homeostasis. Upon the binding of Notch ligands to receptors on adjacent cells, the intracellular domain of the receptor is cleaved by a gamma-secretase, allowing for the activation of downstream targets, such as the inhibitory basic helix-loop-helix transcription factor Hes1 [5]. Notch signaling appears to be required for determining proximal and distal lung epithelial cell fates during lung development [31]. The indirect effect of Notch signaling has been demonstrated during lung development in a Hes1 knockout mouse model where inhibition of Notch signaling resulted in premature differentiation of pulmonary neuroendocrine stem cells [71]. In other studies, activation of Notch signaling, either through the ectopic expression of intracellular Notch domains or through gamma-secretase activation resulted in an increased number of distal airway stem cells through reduced differentiation of neuroendocrine and alveolar stem cells [35, 57, 152]. These results suggested that activated Notch signaling preserves the undifferentiated state of pulmonary stem cells. In lung cancer, while elevated Notch signaling transcripts have been described in NSCLC, the role of Notch in tumor maintenance remains poorly understood. Suppression of Notch signaling in some NSCLC cells by treatment with a gamma-secretase inhibitor induced cell death and decreased tumor growth in mice [61, 85]. Recently, elevated expression of Notch signaling responsive transcripts was reported in putative lung cancer stem cells with high ALDH activity. Using a γ-secretase inhibitor to suppress Notch signaling or ShRNA-mediated knockdown of NOTCH3 in lung cancer cells led to a significant reduction in $ALDH^+$ cells [141]. It suggests that Notch signaling may be activated in putative lung cancer stem cell populations and is required for tumor initiation capacity [10, 77, 89].

The transcription factors Oct4, Sox2 and Nanog have been identified as core transcription factors that maintain embryonic stem cell self-renewal [82]. Several lines of evidences suggest the involvement of Sox2 in normal lung development [54, 119]. Sox2 depletion in developing lung result in significant decrease in basal, ciliated and Clara cells as well as increased numbers of mucus-secreting cells, suggesting its role in normal differentiation during embryonic lung development [150]. In adult lung, Sox2 expression was found to be crucial for proper repair of airway epithelium upon SO_2 induced injury [119]. Sox2 depletion in basal stem cells resulted in suppressed undifferentiated proliferation *in vitro*, suggesting the role of Sox2 in self-renewal of basal stem cells in lung [119]. Several studies have demonstrated the amplification of Sox2 in SCCs of lung [8]. Further, high Sox2 expression was correlated with *in vitro* cell proliferation and anchorage independent growth of SCC cell lines, signifying its role as an oncogene [8]. The oncogenic role of Sox2 was demonstrated using Sox2 overexpressing *in vivo* mouse model of lung cancer [93]. Sox2 was specifically expressed in SP-C-positive developing lung or *scgb1a1*-positive adult lung airway stem cells which resulted in epithelial hyperplasia and adenocarcinoma development [93]. Further, immunohistochemical study on stage-I lung adenocarcinoma patient samples revealed a strong prognostic correlation [132]. Higher Sox2 expression was significantly associated with decreased overall survival for both male and female patients [132].These studies strongly suggest the role of Sox2 transcription factor for normal lung stem cells maintenance and lung development as well as correlated with the lung cancer progression. However, other experimental evidence to correlate Sox2 expression with the lung cancer stem cells maintenance and self-renewal is missing. Studies from our laboratory has found that SP cells from NSCLC cell lines express higher levels of *Sox2*. As shown in Fig. 24.2, depletion of Sox2 resulted in decreased side population frequency, indicating its direct role in maintaining the self-renewal of SP cells in NSCLCs (unpublished data). In addition to Sox2, few studies also suggest higher expression of Oct4 or Nanog in lung adenocarcinoma patients. Overexpression of Oct4 and Nanog in NSCLC cell lines is shown to induce stem cell properties like self-renewal, tumorigenesis, invasion and metastasis [25]. In another study depletion of Oct4 in CD133$^+$ cells resulted in decreased self-renewal of these cells [24]. Immunohistochemical studies support the role of Oct4 and Nanog in lung adenocarcinoma progression. The high levels of Oct4 and Nanog was positively associated with moderate and poorly differentiated grade of adenocarcinoma as well as poor overall survival of the patients [25]. However, lack of Oct4 and Nanog expression was reported in low grade as well as lower stage adenocarcinoma [25], whereas Sox2 was positively expressed irrespective of the stage or grade of lung cancer [132]. These observations strongly raise the possibility that Sox2 may regulate self-renewal of cancer stem cells independently of Oct4 and Nanog in lung cancer.

24.6 Conclusions

Our understanding of lung cancer stem cells and their association with cancer initiation, progression, metastasis and relapse after therapy is relatively very new and advancing. In depth studies are needed to better understand the causal role of this specific subset of cells. Accurate identification and molecular characterization of stem cells of diverse lung cancer subtypes has been challenging. In fact, the epithelial to mesenchymal transition is suggested as a mechanism for non-cancer stem cells to acquire stem cell like properties [97, 133]. It can be imagined that cytokines as well as various cell types present in the tumor microenvironment might accelerate these mechanisms and give rise to multiple sub-populations of self-renewing cancer stem cells within same tumor. Identification of cancer stem cells with innate ability to promote metastasis to specific secondary sites, if such cells do exist, would further advance our knowledge of lung cancer metastasis process. It is clear that the anti-apoptotic as well as drug resistance properties of cancer stem cells are associated with relapse of disease after therapy as well as the successful metastatic dissemination of cancers [133]. Therefore, it can be imagined that identifying the specific tumor-suppressor or oncogenic pathways that regulate the self-renewal of cancer stem cells would provide breakthroughs in lung cancer chemotherapy.

Acknowledgements The studies in the Chellappan lab are supported by the grants CA127725 and CA139612 from the NCI.

References

1. Al-Hajj M, Wicha MS, Benito-Hernandez A, Morrison SJ, Clarke MF (2003) Prospective identification of tumorigenic breast cancer cells. Proc Natl Acad Sci USA 100:3983–3988
2. Alison MR, Guppy NJ, Lim SM, Nicholson LJ (2010) Finding cancer stem cells: are aldehyde dehydrogenases fit for purpose? J Pathol 222:335–344

3. Alison MR, Islam S, Wright NA (2010) Stem cells in cancer: instigators and propagators? J Cell Sci 123:2357–2368
4. Alvero AB, Chen R, Fu HH, Montagna M, Schwartz PE, Rutherford T, Silasi DA, Steffensen KD, Waldstrom M, Visintin I, Mor G (2009) Molecular phenotyping of human ovarian cancer stem cells unravels the mechanisms for repair and chemoresistance. Cell Cycle 8:158–166
5. Artavanis-Tsakonas S, Rand MD, Lake RJ (1999) Notch signaling: cell fate control and signal integration in development. Science 284:770–776
6. Baba T, Convery PA, Matsumura N, Whitaker RS, Kondoh E, Perry T, Huang Z, Bentley RC, Mori S, Fujii S, Marks JR, Berchuck A, Murphy SK (2009) Epigenetic regulation of CD133 and tumorigenicity of CD133+ ovarian cancer cells. Oncogene 28:209–218
7. Barth PJ, Koch S, Muller B, Unterstab F, von Wichert P, Moll R (2000) Proliferation and number of Clara cell 10-kDa protein (CC10)-reactive epithelial cells and basal cells in normal, hyperplastic and metaplastic bronchial mucosa. Virchows Arch 437:648–655
8. Bass AJ, Watanabe H, Mermel CH, Yu S, Perner S, Verhaak RG, Kim SY, Wardwell L, Tamayo P, Gat-Viks I, Ramos AH, Woo MS, Weir BA, Getz G, Beroukhim R, O'Kelly M, Dutt A, Rozenblatt-Rosen O, Dziunycz P, Komisarof J, Chirieac LR, Lafargue CJ, Scheble V, Wilbertz T, Ma C, Rao S, Nakagawa H, Stairs DB, Lin L, Giordano TJ, Wagner P, Minna JD, Gazdar AF, Zhu CQ, Brose MS, Cecconello I, Ribeiro U Jr, Marie SK, Dahl O, Shivdasani RA, Tsao MS, Rubin MA, Wong KK, Regev A, Hahn WC, Beer DG, Rustgi AK, Meyerson M (2009) SOX2 is an amplified lineage-survival oncogene in lung and esophageal squamous cell carcinomas. Nat Genet 41:1238–1242
9. Bellusci S, Furuta Y, Rush MG, Henderson R, Winnier G, Hogan BL (1997) Involvement of Sonic hedgehog (Shh) in mouse embryonic lung growth and morphogenesis. Development 124:53–63
10. Bertolini G, Roz L, Perego P, Tortoreto M, Fontanella E, Gatti L, Pratesi G, Fabbri A, Andriani F, Tinelli S, Roz E, Caserini R, Lo Vullo S, Camerini T, Mariani L, Delia D, Calabro E, Pastorino U, Sozzi G (2009) Highly tumorigenic lung cancer CD133+ cells display stem-like features and are spared by cisplatin treatment. Proc Natl Acad Sci USA 106:16281–16286
11. Bidlingmaier S, Zhu X, Liu B (2008) The utility and limitations of glycosylated human CD133 epitopes in defining cancer stem cells. J Mol Med (Berl) 86:1025–1032
12. Boers JE, Ambergen AW, Thunnissen FB (1998) Number and proliferation of basal and parabasal cells in normal human airway epithelium. Am J Respir Crit Care Med 157:2000–2006
13. Boers JE, Ambergen AW, Thunnissen FB (1999) Number and proliferation of clara cells in normal human airway epithelium. Am J Respir Crit Care Med 159:1585–1591
14. Bonnet D, Dick JE (1997) Human acute myeloid leukemia is organized as a hierarchy that originates from a primitive hematopoietic cell. Nat Med 3:730–737
15. Borthwick DW, Shahbazian M, Krantz QT, Dorin JR, Randell SH (2001) Evidence for stem-cell niches in the tracheal epithelium. Am J Respir Cell Mol Biol 24:662–670
16. Bottinger EP, Jakubczak JL, Haines DC, Bagnall K, Wakefield LM (1997) Transgenic mice overexpressing a dominant-negative mutant type II transforming growth factor beta receptor show enhanced tumorigenesis in the mammary gland and lung in response to the carcinogen 7,12-dimethylbenz-[a]-anthracene. Cancer Res 57:5564–5570
17. Buckpitt A, Chang AM, Weir A, Van Winkle L, Duan X, Philpot R, Plopper C (1995) Relationship of cytochrome P450 activity to Clara cell cytotoxicity. IV. Metabolism of naphthalene and naphthalene oxide in microdissected airways from mice, rats, and hamsters. Mol Pharmacol 47:74–81
18. Bussolati B, Bruno S, Grange C, Ferrando U, Camussi G (2008) Identification of a tumor-initiating stem cell population in human renal carcinomas. FASEB J 22:3696–3705
19. Calbo J, van Montfort E, Proost N, van Drunen E, Beverloo HB, Meuwissen R, Berns A (2011) A functional role for tumor cell heterogeneity in a mouse model of small cell lung cancer. Cancer Cell 19:244–256
20. Carney DN, Gazdar AF, Bunn PA Jr, Guccion JG (1982) Demonstration of the stem cell nature of clonogenic tumor cells from lung cancer patients. Stem Cells 1:149–164

21. Carney DN, Gazdar AF, Minna JD (1980) Positive correlation between histological tumor involvement and generation of tumor cell colonies in agarose in specimens taken directly from patients with small-cell carcinoma of the lung. Cancer Res 40:1820–1823
22. Carpentino JE, Hynes MJ, Appelman HD, Zheng T, Steindler DA, Scott EW, Huang EH (2009) Aldehyde dehydrogenase-expressing colon stem cells contribute to tumorigenesis in the transition from colitis to cancer. Cancer Res 69:8208–8215
23. Charafe-Jauffret E, Ginestier C, Iovino F, Wicinski J, Cervera N, Finetti P, Hur MH, Diebel ME, Monville F, Dutcher J, Brown M, Viens P, Xerri L, Bertucci F, Stassi G, Dontu G, Birnbaum D, Wicha MS (2009) Breast cancer cell lines contain functional cancer stem cells with metastatic capacity and a distinct molecular signature. Cancer Res 69:1302–1313
24. Chen YC, Hsu HS, Chen YW, Tsai TH, How CK, Wang CY, Hung SC, Chang YL, Tsai ML, Lee YY, Ku HH, Chiou SH (2008) Oct-4 expression maintained cancer stem-like properties in lung cancer-derived CD133-positive cells. PLoS One 3:e2637
25. Chiou SH, Wang ML, Chou YT, Chen CJ, Hong CF, Hsieh WJ, Chang HT, Chen YS, Lin TW, Hsu HS, Wu CW (2010) Coexpression of Oct4 and Nanog enhances malignancy in lung adenocarcinoma by inducing cancer stem cell-like properties and epithelial-mesenchymal transdifferentiation. Cancer Res 70:10433–10444
26. Chu P, Clanton DJ, Snipas TS, Lee J, Mitchell E, Nguyen ML, Hare E, Peach RJ (2009) Characterization of a subpopulation of colon cancer cells with stem cell-like properties. Int J Cancer 124:1312–1321
27. Chute JP, Muramoto GG, Whitesides J, Colvin M, Safi R, Chao NJ, McDonnell DP (2006) Inhibition of aldehyde dehydrogenase and retinoid signaling induces the expansion of human hematopoietic stem cells. Proc Natl Acad Sci USA 103:11707–11712
28. Clarke MF, Dick JE, Dirks PB, Eaves CJ, Jamieson CH, Jones DL, Visvader J, Weissman IL, Wahl GM (2006) Cancer stem cells–perspectives on current status and future directions: AACR Workshop on cancer stem cells. Cancer Res 66:9339–9344
29. Clevers H (2005) Stem cells, asymmetric division and cancer. Nat Genet 37:1027–1028
30. Collins AT, Berry PA, Hyde C, Stower MJ, Maitland NJ (2005) Prospective identification of tumorigenic prostate cancer stem cells. Cancer Res 65:10946–10951
31. Collins BJ, Kleeberger W, Ball DW (2004) Notch in lung development and lung cancer. Semin Cancer Biol 14:357–364
32. Curtis SJ, Sinkevicius KW, Li D, Lau AN, Roach RR, Zamponi R, Woolfenden AE, Kirsch DG, Wong KK, Kim CF (2010) Primary tumor genotype is an important determinant in identification of lung cancer propagating cells. Cell Stem Cell 7:127–133
33. Dalerba P, Cho RW, Clarke MF (2007) Cancer stem cells: models and concepts. Annu Rev Med 58:267–284
34. Dalerba P, Clarke MF (2007) Cancer stem cells and tumor metastasis: first steps into uncharted territory. Cell Stem Cell 1:241–242
35. Dang TP, Eichenberger S, Gonzalez A, Olson S, Carbone DP (2003) Constitutive activation of Notch3 inhibits terminal epithelial differentiation in lungs of transgenic mice. Oncogene 22:1988–1997
36. Daniel VC, Peacock CD, Watkins DN (2006) Developmental signalling pathways in lung cancer. Respirology 11:234–240
37. Das B, Tsuchida R, Malkin D, Koren G, Baruchel S, Yeger H (2008) Hypoxia enhances tumor stemness by increasing the invasive and tumorigenic side population fraction. Stem Cells 26:1818–1830
38. DeMayo FJ, Finegold MJ, Hansen TN, Stanley LA, Smith B, Bullock DW (1991) Expression of SV40 T antigen under control of rabbit uteroglobin promoter in transgenic mice. Am J Physiol 261:L70–L76
39. Du L, Wang H, He L, Zhang J, Ni B, Wang X, Jin H, Cahuzac N, Mehrpour M, Lu Y, Chen Q (2008) CD44 is of functional importance for colorectal cancer stem cells. Clin Cancer Res 14:6751–6760
40. Dutt A, Wong KK (2006) Mouse models of lung cancer. Clin Cancer Res 12:4396s–4402s

41. Eramo A, Lotti F, Sette G, Pilozzi E, Biffoni M, Di Virgilio A, Conticello C, Ruco L, Peschle C, De Maria R (2008) Identification and expansion of the tumorigenic lung cancer stem cell population. Cell Death Differ 15:504–514
42. Evans MJ, Johnson LV, Stephens RJ, Freeman G (1976) Renewal of the terminal bronchiolar epithelium in the rat following exposure to NO2 or O3. Lab Invest 35:246–257
43. Evans MJ, Cabral-Anderson LJ, Freeman G (1978) Role of the Clara cell in renewal of the bronchiolar epithelium. Lab Invest 38:648–653
44. Evans MJ, Cabral LJ, Stephens RJ, Freeman G (1975) Transformation of alveolar type 2 cells to type 1 cells following exposure to NO2. Exp Mol Pathol 22:142–150
45. Evans MJ, Van Winkle LS, Fanucchi MV, Plopper CG (2001) Cellular and molecular characteristics of basal cells in airway epithelium. Exp Lung Res 27:401–415
46. Fisher GH, Wellen SL, Klimstra D, Lenczowski JM, Tichelaar JW, Lizak MJ, Whitsett JA, Koretsky A, Varmus HE (2001) Induction and apoptotic regression of lung adenocarcinomas by regulation of a K-Ras transgene in the presence and absence of tumor suppressor genes. Genes Dev 15:3249–3262
47. Friel AM, Sergent PA, Patnaude C, Szotek PP, Oliva E, Scadden DT, Seiden MV, Foster R, Rueda BR (2008) Functional analyses of the cancer stem cell-like properties of human endometrial tumor initiating cells. Cell Cycle 7:242–249
48. Gazdar AF, Carney DN, Nau MM, Minna JD (1985) Characterization of variant subclasses of cell lines derived from small cell lung cancer having distinctive biochemical, morphological, and growth properties. Cancer Res 45:2924–2930
49. Giangreco A, Arwert EN, Rosewell IR, Snyder J, Watt FM, Stripp BR (2009) Stem cells are dispensable for lung homeostasis but restore airways after injury. Proc Natl Acad Sci USA 106:9286–9291
50. Giangreco A, Groot KR, Janes SM (2007) Lung cancer and lung stem cells: strange bedfellows? Am J Respir Crit Care Med 175:547–553
51. Giangreco A, Reynolds SD, Stripp BR (2002) Terminal bronchioles harbor a unique airway stem cell population that localizes to the bronchoalveolar duct junction. Am J Pathol 161:173–182
52. Ginestier C, Hur MH, Charafe-Jauffret E, Monville F, Dutcher J, Brown M, Jacquemier J, Viens P, Kleer CG, Liu S, Schott A, Hayes D, Birnbaum D, Wicha MS, Dontu G (2007) ALDH1 is a marker of normal and malignant human mammary stem cells and a predictor of poor clinical outcome. Cell Stem Cell 1:555–567
53. Golebiewska A, Brons NH, Bjerkvig R, Niclou SP (2011) Critical appraisal of the side population assay in stem cell and cancer stem cell research. Cell Stem Cell 8:136–147
54. Gontan C, de Munck A, Vermeij M, Grosveld F, Tibboel D, Rottier R (2008) Sox2 is important for two crucial processes in lung development: branching morphogenesis and epithelial cell differentiation. Dev Biol 317:296–309
55. Goodell MA, Brose K, Paradis G, Conner AS, Mulligan RC (1996) Isolation and functional properties of murine hematopoietic stem cells that are replicating in vivo. J Exp Med 183:1797–1806
56. Goodrich LV, Scott MP (1998) Hedgehog and patched in neural development and disease. Neuron 21:1243–1257
57. Guseh JS, Bores SA, Stanger BZ, Zhou Q, Anderson WJ, Melton DA, Rajagopal J (2009) Notch signaling promotes airway mucous metaplasia and inhibits alveolar development. Development 136:1751–1759
58. Hackett TL, Shaheen F, Johnson A, Wadsworth S, Pechkovsky DV, Jacoby DB, Kicic A, Stick SM, Knight DA (2008) Characterization of side population cells from human airway epithelium. Stem Cells 26:2576–2585
59. Halpern MT, Warner KE (1993) Motivations for smoking cessation: a comparison of successful quitters and failures. J Subst Abuse 5:247–256
60. Hamburger AW, Salmon SE (1977) Primary bioassay of human tumor stem cells. Science 197:461–463

61. Haruki N, Kawaguchi KS, Eichenberger S, Massion PP, Olson S, Gonzalez A, Carbone DP, Dang TP (2005) Dominant-negative Notch3 receptor inhibits mitogen-activated protein kinase pathway and the growth of human lung cancers. Cancer Res 65:3555–3561
62. Henry CJ, Billups LH, Avery MD, Rude TH, Dansie DR, Lopez A, Sass B, Whitmire CE, Kouri RE (1981) Lung cancer model system using 3-methylcholanthrene in inbred strains of mice. Cancer Res 41:5027–5032
63. Hermann PC, Huber SL, Herrler T, Aicher A, Ellwart JW, Guba M, Bruns CJ, Heeschen C (2007) Distinct populations of cancer stem cells determine tumor growth and metastatic activity in human pancreatic cancer. Cell Stem Cell 1:313–323
64. Ho MM, Ng AV, Lam S, Hung JY (2007) Side population in human lung cancer cell lines and tumors is enriched with stem-like cancer cells. Cancer Res 67:4827–4833
65. Hochedlinger K, Yamada Y, Beard C, Jaenisch R (2005) Ectopic expression of Oct-4 blocks progenitor-cell differentiation and causes dysplasia in epithelial tissues. Cell 121:465–477
66. Hong KU, Reynolds SD, Giangreco A, Hurley CM, Stripp BR (2001) Clara cell secretory protein-expressing cells of the airway neuroepithelial body microenvironment include a label-retaining subset and are critical for epithelial renewal after progenitor cell depletion. Am J Respir Cell Mol Biol 24:671–681
67. Hong KU, Reynolds SD, Watkins S, Fuchs E, Stripp BR (2004) Basal cells are a multipotent progenitor capable of renewing the bronchial epithelium. Am J Pathol 164:577–588
68. Hong KU, Reynolds SD, Watkins S, Fuchs E, Stripp BR (2004) In vivo differentiation potential of tracheal basal cells: evidence for multipotent and unipotent subpopulations. Am J Physiol Lung Cell Mol Physiol 286:L643–L649
69. Huang EH, Hynes MJ, Zhang T, Ginestier C, Dontu G, Appelman H, Fields JZ, Wicha MS, Boman BM (2009) Aldehyde dehydrogenase 1 is a marker for normal and malignant human colonic stem cells (SC) and tracks SC overpopulation during colon tumorigenesis. Cancer Res 69:3382–3389
70. Hurt EM, Kawasaki BT, Klarmann GJ, Thomas SB, Farrar WL (2008) CD44+ CD24(−) prostate cells are early cancer progenitor/stem cells that provide a model for patients with poor prognosis. Br J Cancer 98:756–765
71. Ito T, Udaka N, Yazawa T, Okudela K, Hayashi H, Sudo T, Guillemot F, Kageyama R, Kitamura H (2000) Basic helix-loop-helix transcription factors regulate the neuroendocrine differentiation of fetal mouse pulmonary epithelium. Development 127:3913–3921
72. Jemal A, Siegel R, Ward E, Hao Y, Xu J, Murray T, Thun MJ (2008) Cancer statistics, 2008. CA Cancer J Clin 58:71–96
73. Jemal A, Thun MJ, Ries LA, Howe HL, Weir HK, Center MM, Ward E, Wu XC, Eheman C, Anderson R, Ajani UA, Kohler B, Edwards BK (2008) Annual report to the nation on the status of cancer, 1975–2005, featuring trends in lung cancer, tobacco use, and tobacco control. J Natl Cancer Inst 100:1672–1694
74. Jeremy George P, Banerjee AK, Read CA, O'Sullivan C, Falzon M, Pezzella F, Nicholson AG, Shaw P, Laurent G, Rabbitts PH (2007) Surveillance for the detection of early lung cancer in patients with bronchial dysplasia. Thorax 62:43–50
75. Ji H, Li D, Chen L, Shimamura T, Kobayashi S, McNamara K, Mahmood U, Mitchell A, Sun Y, Al-Hashem R, Chirieac LR, Padera R, Bronson RT, Kim W, Janne PA, Shapiro GI, Tenen D, Johnson BE, Weissleder R, Sharpless NE, Wong KK (2006) The impact of human EGFR kinase domain mutations on lung tumorigenesis and in vivo sensitivity to EGFR-targeted therapies. Cancer Cell 9:485–495
76. Jiang F, Qiu Q, Khanna A, Todd NW, Deepak J, Xing L, Wang H, Liu Z, Su Y, Stass SA, Katz RL (2009) Aldehyde dehydrogenase 1 is a tumor stem cell-associated marker in lung cancer. Mol Cancer Res 7:330–338
77. Jiang T, Collins BJ, Jin N, Watkins DN, Brock MV, Matsui W, Nelkin BD, Ball DW (2009) Achaete-scute complex homologue 1 regulates tumor-initiating capacity in human small cell lung cancer. Cancer Res 69:845–854

78. Kajstura J, Rota M, Hall SR, Hosoda T, D'Amario D, Sanada F, Zheng H, Ogorek B, Rondon-Clavo C, Ferreira-Martins J, Matsuda A, Arranto C, Goichberg P, Giordano G, Haley KJ, Bardelli S, Rayatzadeh H, Liu X, Quaini F, Liao R, Leri A, Perrella MA, Loscalzo J, Anversa P (2011) Evidence for human lung stem cells. N Engl J Med 364:1795–1806
79. Kauffman SL (1980) Cell proliferation in the mammalian lung. Int Rev Exp Pathol 22:131–191
80. Keysar SB, Jimeno A (2010) More than markers: biological significance of cancer stem cell-defining molecules. Mol Cancer Ther 9:2450–2457
81. Kim CF, Jackson EL, Woolfenden AE, Lawrence S, Babar I, Vogel S, Crowley D, Bronson RT, Jacks T (2005) Identification of bronchioalveolar stem cells in normal lung and lung cancer. Cell 121:823–835
82. Kim J, Chu J, Shen X, Wang J, Orkin SH (2008) An extended transcriptional network for pluripotency of embryonic stem cells. Cell 132:1049–1061
83. Kirstetter P, Anderson K, Porse BT, Jacobsen SE, Nerlov C (2006) Activation of the canonical Wnt pathway leads to loss of hematopoietic stem cell repopulation and multilineage differentiation block. Nat Immunol 7:1048–1056
84. Klarmann GJ, Hurt EM, Mathews LA, Zhang X, Duhagon MA, Mistree T, Thomas SB, Farrar WL (2009) Invasive prostate cancer cells are tumor initiating cells that have a stem cell-like genomic signature. Clin Exp Metastasis 26:433–446
85. Konishi J, Kawaguchi KS, Vo H, Haruki N, Gonzalez A, Carbone DP, Dang TP (2007) Gamma-secretase inhibitor prevents Notch3 activation and reduces proliferation in human lung cancers. Cancer Res 67:8051–8057
86. Lapidot T, Sirard C, Vormoor J, Murdoch B, Hoang T, Caceres-Cortes J, Minden M, Paterson B, Caligiuri MA, Dick JE (1994) A cell initiating human acute myeloid leukaemia after transplantation into SCID mice. Nature 367:645–648
87. Lee TK, Castilho A, Cheung VC, Tang KH, Ma S, Ng IO (2011) CD24(+) liver tumor-initiating cells drive self-renewal and tumor initiation through STAT3-mediated NANOG regulation. Cell Stem Cell 9:50–63
88. Lemjabbar-Alaoui H, Dasari V, Sidhu SS, Mengistab A, Finkbeiner W, Gallup M, Basbaum C (2006) Wnt and Hedgehog are critical mediators of cigarette smoke-induced lung cancer. PLoS One 1:e93
89. Levina V, Marrangoni AM, DeMarco R, Gorelik E, Lokshin AE (2008) Drug-selected human lung cancer stem cells: cytokine network, tumorigenic and metastatic properties. PLoS One 3:e3077
90. Li C, Heidt DG, Dalerba P, Burant CF, Zhang L, Adsay V, Wicha M, Clarke MF, Simeone DM (2007) Identification of pancreatic cancer stem cells. Cancer Res 67:1030–1037
91. Litingtung Y, Lei L, Westphal H, Chiang C (1998) Sonic hedgehog is essential to foregut development. Nat Genet 20:58–61
92. Lo Celso C, Prowse DM, Watt FM (2004) Transient activation of beta-catenin signalling in adult mouse epidermis is sufficient to induce new hair follicles but continuous activation is required to maintain hair follicle tumours. Development 131:1787–1799
93. Lu Y, Futtner C, Rock JR, Xu X, Whitworth W, Hogan BL, Onaitis MW (2010) Evidence that SOX2 overexpression is oncogenic in the lung. PLoS One 5:e11022
94. Ma S, Lee TK, Zheng BJ, Chan KW, Guan XY (2008) CD133+ HCC cancer stem cells confer chemoresistance by preferential expression of the Akt/PKB survival pathway. Oncogene 27:1749–1758
95. Maeda S, Shinchi H, Kurahara H, Mataki Y, Maemura K, Sato M, Natsugoe S, Aikou T, Takao S (2008) CD133 expression is correlated with lymph node metastasis and vascular endothelial growth factor-C expression in pancreatic cancer. Br J Cancer 98:1389–1397
96. Mahvi D, Bank H, Harley R (1977) Morphology of a naphthalene-induced bronchiolar lesion. Am J Pathol 86:558–572
97. Mani SA, Guo W, Liao MJ, Eaton EN, Ayyanan A, Zhou AY, Brooks M, Reinhard F, Zhang CC, Shipitsin M, Campbell LL, Polyak K, Brisken C, Yang J, Weinberg RA (2008) The epithelial-mesenchymal transition generates cells with properties of stem cells. Cell 133:704–715

98. Meuwissen R, Berns A (2005) Mouse models for human lung cancer. Genes Dev 19:643–664
99. Meuwissen R, Linn SC, Linnoila RI, Zevenhoven J, Mooi WJ, Berns A (2003) Induction of small cell lung cancer by somatic inactivation of both Trp53 and Rb1 in a conditional mouse model. Cancer Cell 4:181–189
100. Minna JD, Kurie JM, Jacks T (2003) A big step in the study of small cell lung cancer. Cancer Cell 4:163–166
101. Mizrak D, Brittan M, Alison MR (2008) CD133: molecule of the moment. J Pathol 214:3–9
102. Moreb J, Schweder M, Suresh A, Zucali JR (1996) Overexpression of the human aldehyde dehydrogenase class I results in increased resistance to 4-hydroperoxycyclophosphamide. Cancer Gene Ther 3:24–30
103. Moreb JS (2008) Aldehyde dehydrogenase as a marker for stem cells. Curr Stem Cell Res Ther 3:237–246
104. Morrison SJ, Kimble J (2006) Asymmetric and symmetric stem-cell divisions in development and cancer. Nature 441:1068–1074
105. Nakajima M, Kawanami O, Jin E, Ghazizadeh M, Honda M, Asano G, Horiba K, Ferrans VJ (1998) Immunohistochemical and ultrastructural studies of basal cells, Clara cells and bronchiolar cuboidal cells in normal human airways. Pathol Int 48:944–953
106. Nilsson M, Unden AB, Krause D, Malmqwist U, Raza K, Zaphiropoulos PG, Toftgard R (2000) Induction of basal cell carcinomas and trichoepitheliomas in mice overexpressing GLI-1. Proc Natl Acad Sci USA 97:3438–3443
107. Nowell PC (1976) The clonal evolution of tumor cell populations. Science 194:23–28
108. O'Brien CA, Pollett A, Gallinger S, Dick JE (2007) A human colon cancer cell capable of initiating tumour growth in immunodeficient mice. Nature 445:106–110
109. Oates JE, Grey BR, Addla SK, Samuel JD, Hart CA, Ramani VA, Brown MD, Clarke NW (2009) Hoechst 33342 side population identification is a conserved and unified mechanism in urological cancers. Stem Cells Dev 18:1515–1522
110. Pardal R, Clarke MF, Morrison SJ (2003) Applying the principles of stem-cell biology to cancer. Nat Rev Cancer 3:895–902
111. Parkin DM, Bray F, Ferlay J, Pisani P (2005) Global cancer statistics, 2002. CA Cancer J Clin 55:74–108
112. Passegue E, Jamieson CH, Ailles LE, Weissman IL (2003) Normal and leukemic hematopoiesis: are leukemias a stem cell disorder or a reacquisition of stem cell characteristics? Proc Natl Acad Sci USA 100(Suppl 1):11842–11849
113. Patel M, Lu L, Zander DS, Sreerama L, Coco D, Moreb JS (2008) ALDH1A1 and ALDH3A1 expression in lung cancers: correlation with histologic type and potential precursors. Lung Cancer 59:340–349
114. Patrawala L, Calhoun T, Schneider-Broussard R, Li H, Bhatia B, Tang S, Reilly JG, Chandra D, Zhou J, Claypool K, Coghlan L, Tang DG (2006) Highly purified CD44+ prostate cancer cells from xenograft human tumors are enriched in tumorigenic and metastatic progenitor cells. Oncogene 25:1696–1708
115. Pepicelli CV, Lewis PM, McMahon AP (1998) Sonic hedgehog regulates branching morphogenesis in the mammalian lung. Curr Biol 8:1083–1086
116. Politi K, Zakowski MF, Fan PD, Schonfeld EA, Pao W, Varmus HE (2006) Lung adenocarcinomas induced in mice by mutant EGF receptors found in human lung cancers respond to a tyrosine kinase inhibitor or to down-regulation of the receptors. Genes Dev 20:1496–1510
117. Prince ME, Sivanandan R, Kaczorowski A, Wolf GT, Kaplan MJ, Dalerba P, Weissman IL, Clarke MF, Ailles LE (2007) Identification of a subpopulation of cells with cancer stem cell properties in head and neck squamous cell carcinoma. Proc Natl Acad Sci USA 104:973–978
118. Proctor RN (2001) Tobacco and the global lung cancer epidemic. Nat Rev Cancer 1:82–86
119. Que J, Luo X, Schwartz RJ, Hogan BL (2009) Multiple roles for Sox2 in the developing and adult mouse trachea. Development 136:1899–1907

120. Rawlins EL, Clark CP, Xue Y, Hogan BL (2009) The Id2+ distal tip lung epithelium contains individual multipotent embryonic progenitor cells. Development 136:3741–3745
121. Rawlins EL, Hogan BL (2006) Epithelial stem cells of the lung: privileged few or opportunities for many? Development 133:2455–2465
122. Reya T, Clevers H (2005) Wnt signalling in stem cells and cancer. Nature 434:843–850
123. Reya T, Duncan AW, Ailles L, Domen J, Scherer DC, Willert K, Hintz L, Nusse R, Weissman IL (2003) A role for Wnt signalling in self-renewal of haematopoietic stem cells. Nature 423:409–414
124. Reynolds SD, Zemke AC, Giangreco A, Brockway BL, Teisanu RM, Drake JA, Mariani T, Di PY, Taketo MM, Stripp BR (2008) Conditional stabilization of beta-catenin expands the pool of lung stem cells. Stem Cells 26:1337–1346
125. Reynolds SD, Hong KU, Giangreco A, Mango GW, Guron C, Morimoto Y, Stripp BR (2000) Conditional clara cell ablation reveals a self-renewing progenitor function of pulmonary neuroendocrine cells. Am J Physiol Lung Cell Mol Physiol 278:L1256–L1263
126. Ricci-Vitiani L, Lombardi DG, Pilozzi E, Biffoni M, Todaro M, Peschle C, De Maria R (2007) Identification and expansion of human colon-cancer-initiating cells. Nature 445:111–115
127. Rutella S, Bonanno G, Procoli A, Mariotti A, Corallo M, Prisco MG, Eramo A, Napoletano C, Gallo D, Perillo A, Nuti M, Pierelli L, Testa U, Scambia G, Ferrandina G (2009) Cells with characteristics of cancer stem/progenitor cells express the CD133 antigen in human endometrial tumors. Clin Cancer Res 15:4299–4311
128. Salcido CD, Larochelle A, Taylor BJ, Dunbar CE, Varticovski L (2010) Molecular characterisation of side population cells with cancer stem cell-like characteristics in small-cell lung cancer. Br J Cancer 102:1636–1644
129. Salnikov AV, Gladkich J, Moldenhauer G, Volm M, Mattern J, Herr I (2010) CD133 is indicative for a resistance phenotype but does not represent a prognostic marker for survival of non-small cell lung cancer patients. Int J Cancer 126:950–958
130. Schoch KG, Lori A, Burns KA, Eldred T, Olsen JC, Randell SH (2004) A subset of mouse tracheal epithelial basal cells generates large colonies in vitro. Am J Physiol Lung Cell Mol Physiol 286:L631–L642
131. Shmelkov SV, Butler JM, Hooper AT, Hormigo A, Kushner J, Milde T, St Clair R, Baljevic M, White I, Jin DK, Chadburn A, Murphy AJ, Valenzuela DM, Gale NW, Thurston G, Yancopoulos GD, D'Angelica M, Kemeny N, Lyden D, Rafii S (2008) CD133 expression is not restricted to stem cells, and both CD133+ and CD133- metastatic colon cancer cells initiate tumors. J Clin Invest 118:2111–2120
132. Sholl LM, Barletta JA, Yeap BY, Chirieac LR, Hornick JL (2010) Sox2 protein expression is an independent poor prognostic indicator in stage I lung adenocarcinoma. Am J Surg Pathol 34:1193–1198
133. Singh A, Settleman J (2010) EMT, cancer stem cells and drug resistance: an emerging axis of evil in the war on cancer. Oncogene 29:4741–4751
134. Singh SK, Hawkins C, Clarke ID, Squire JA, Bayani J, Hide T, Henkelman RM, Cusimano MD, Dirks PB (2004) Identification of human brain tumour initiating cells. Nature 432:396–401
135. Smalley M, Ashworth A (2003) Stem cells and breast cancer: a field in transit. Nat Rev Cancer 3:832–844
136. Song W, Li H, Tao K, Li R, Song Z, Zhao Q, Zhang F, Dou K (2008) Expression and clinical significance of the stem cell marker CD133 in hepatocellular carcinoma. Int J Clin Pract 62:1212–1218
137. Stevens TP, McBride JT, Peake JL, Pinkerton KE, Stripp BR (1997) Cell proliferation contributes to PNEC hyperplasia after acute airway injury. Am J Physiol 272:L486–L493
138. Stripp BR (2008) Hierarchical organization of lung progenitor cells: is there an adult lung tissue stem cell? Proc Am Thorac Soc 5:695–698
139. Stripp BR, Maxson K, Mera R, Singh G (1995) Plasticity of airway cell proliferation and gene expression after acute naphthalene injury. Am J Physiol 269:L791–L799

140. Stripp BR, Reynolds SD (2008) Maintenance and repair of the bronchiolar epithelium. Proc Am Thorac Soc 5:328–333
141. Sullivan JP, Spinola M, Dodge M, Raso MG, Behrens C, Gao B, Schuster K, Shao C, Larsen JE, Sullivan LA, Honorio S, Xie Y, Scaglioni PP, DiMaio JM, Gazdar AF, Shay JW, Wistuba II, Minna JD (2010) Aldehyde dehydrogenase activity selects for lung adenocarcinoma stem cells dependent on notch signaling. Cancer Res 70:9937–9948
142. Sunday ME, Haley KJ, Sikorski K, Graham SA, Emanuel RL, Zhang F, Mu Q, Shahsafaei A, Hatzis D (1999) Calcitonin driven v-Ha-ras induces multilineage pulmonary epithelial hyperplasias and neoplasms. Oncogene 18:4336–4347
143. Sung JM, Cho HJ, Yi H, Lee CH, Kim HS, Kim DK, Abd El-Aty AM, Kim JS, Landowski CP, Hediger MA, Shin HC (2008) Characterization of a stem cell population in lung cancer A549 cells. Biochem Biophys Res Commun 371:163–167
144. Sutherland KD, Proost N, Brouns I, Adriaensen D, Song JY, Berns A (2011) Cell of origin of small cell lung cancer: inactivation of Trp53 and rb1 in distinct cell types of adult mouse lung. Cancer Cell 19:754–764
145. Suva ML, Riggi N, Stehle JC, Baumer K, Tercier S, Joseph JM, Suva D, Clement V, Provero P, Cironi L, Osterheld MC, Guillou L, Stamenkovic I (2009) Identification of cancer stem cells in Ewing's sarcoma. Cancer Res 69:1776–1781
146. Taipale J, Beachy PA (2001) The Hedgehog and Wnt signalling pathways in cancer. Nature 411:349–354
147. Takaishi S, Okumura T, Tu S, Wang SS, Shibata W, Vigneshwaran R, Gordon SA, Shimada Y, Wang TC (2009) Identification of gastric cancer stem cells using the cell surface marker CD44. Stem Cells 27:1006–1020
148. Teisanu RM, Lagasse E, Whitesides JF, Stripp BR (2009) Prospective isolation of bronchiolar stem cells based upon immunophenotypic and autofluorescence characteristics. Stem Cells 27:612–622
149. Tirino V, Desiderio V, d'Aquino R, De Francesco F, Pirozzi G, Graziano A, Galderisi U, Cavaliere C, De Rosa A, Papaccio G, Giordano A (2008) Detection and characterization of CD133+ cancer stem cells in human solid tumours. PLoS One e3469
150. Tompkins DH, Besnard V, Lange AW, Wert SE, Keiser AR, Smith AN, Lang R, Whitsett JA (2009) Sox2 is required for maintenance and differentiation of bronchiolar Clara, ciliated, and goblet cells. PLoS One 4:e8248
151. Tong L, Spitz MR, Fueger JJ, Amos CA (1996) Lung carcinoma in former smokers. Cancer 78:1004–1010
152. Tsao PN, Chen F, Izvolsky KI, Walker J, Kukuruzinska MA, Lu J, Cardoso WV (2008) Gamma-secretase activation of notch signaling regulates the balance of proximal and distal fates in progenitor cells of the developing lung. J Biol Chem 283:29532–29544
153. Uematsu K, He B, You L, Xu Z, McCormick F, Jablons DM (2003) Activation of the Wnt pathway in non small cell lung cancer: evidence of dishevelled overexpression. Oncogene 22:7218–7221
154. Uematsu K, Kanazawa S, You L, He B, Xu Z, Li K, Peterlin BM, McCormick F, Jablons DM (2003) Wnt pathway activation in mesothelioma: evidence of Dishevelled overexpression and transcriptional activity of beta-catenin. Cancer Res 63:4547–4551
155. Vestergaard J, Pedersen MW, Pedersen N, Ensinger C, Tumer Z, Tommerup N, Poulsen HS, Larsen LA (2006) Hedgehog signaling in small-cell lung cancer: frequent in vivo but a rare event in vitro. Lung Cancer 52:281–290
156. Visvader JE, Lindeman GJ (2008) Cancer stem cells in solid tumours: accumulating evidence and unresolved questions. Nat Rev Cancer 8:755–768
157. Wang L, Park P, Zhang H, La Marca F, Lin CY (2011) Prospective identification of tumorigenic osteosarcoma cancer stem cells in OS99-1 cells based on high aldehyde dehydrogenase activity. Int J Cancer 128:294–303
158. Wang Y, Zhang Z, Yan Y, Lemon WJ, LaRegina M, Morrison C, Lubet R, You M (2004) A chemically induced model for squamous cell carcinoma of the lung in mice: histopathology and strain susceptibility. Cancer Res 64:1647–1654

159. Watkins DN, Berman DM, Burkholder SG, Wang B, Beachy PA, Baylin SB (2003) Hedgehog signalling within airway epithelial progenitors and in small-cell lung cancer. Nature 422:313–317
160. Wikenheiser KA, Clark JC, Linnoila RI, Stahlman MT, Whitsett JA (1992) Simian virus 40 large T antigen directed by transcriptional elements of the human surfactant protein C gene produces pulmonary adenocarcinomas in transgenic mice. Cancer Res 52:5342–5352
161. Winn RA, Marek L, Han SY, Rodriguez K, Rodriguez N, Hammond M, Van Scoyk M, Acosta H, Mirus J, Barry N, Bren-Mattison Y, Van Raay TJ, Nemenoff RA, Heasley LE (2005) Restoration of Wnt-7a expression reverses non-small cell lung cancer cellular transformation through frizzled-9-mediated growth inhibition and promotion of cell differentiation. J Biol Chem 280:19625–19634
162. Winn RA, Van Scoyk M, Hammond M, Rodriguez K, Crossno JT Jr, Heasley LE, Nemenoff RA (2006) Antitumorigenic effect of Wnt 7a and Fzd 9 in non-small cell lung cancer cells is mediated through ERK-5-dependent activation of peroxisome proliferator-activated receptor gamma. J Biol Chem 281:26943–26950
163. Woodruff MF (1983) Cellular heterogeneity in tumours. Br J Cancer 47:589–594
164. Wu C, Alman BA (2008) Side population cells in human cancers. Cancer Lett 268:1–9
165. Yang ZF, Ho DW, Ng MN, Lau CK, Yu WC, Ngai P, Chu PW, Lam CT, Poon RT, Fan ST (2008) Significance of CD90+ cancer stem cells in human liver cancer. Cancer Cell 13:153–166
166. Yoshimoto T, Hirao F, Sakatani M, Nishikawa H, Ogura T (1977) Induction of squamous cell carcinoma in the lung of C57BL/6 mice by intratracheal instillation of benzo[a]pyrene with charcoal powder. Gann 68:343–352
167. Yoshimoto T, Inoue T, Iizuka H, Nishikawa H, Sakatani M, Ogura T, Hirao F, Yamamura Y (1980) Differential induction of squamous cell carcinomas and adenocarcinomas in mouse lung by intratracheal instillation of benzo(a)pyrene and charcoal powder. Cancer Res 40:4301–4307
168. You L, He B, Xu Z, Uematsu K, Mazieres J, Mikami I, Reguart N, Moody TW, Kitajewski J, McCormick F, Jablons DM (2004) Inhibition of Wnt-2-mediated signaling induces programmed cell death in non-small-cell lung cancer cells. Oncogene 23:6170–6174
169. Zemke AC, Teisanu RM, Giangreco A, Drake JA, Brockway BL, Reynolds SD, Stripp BR (2009) beta-Catenin is not necessary for maintenance or repair of the bronchiolar epithelium. Am J Respir Cell Mol Biol 41:535–543
170. Zhang S, Balch C, Chan MW, Lai HC, Matei D, Schilder JM, Yan PS, Huang TH, Nephew KP (2008) Identification and characterization of ovarian cancer-initiating cells from primary human tumors. Cancer Res 68:4311–4320
171. Zhou S, Schuetz JD, Bunting KD, Colapietro AM, Sampath J, Morris JJ, Lagutina I, Grosveld GC, Osawa M, Nakauchi H, Sorrentino BP (2001) The ABC transporter Bcrp1/ABCG2 is expressed in a wide variety of stem cells and is a molecular determinant of the side-population phenotype. Nat Med 7:1028–1034

Chapter 25
Stem Cell Characters in Primary and Metastatic Tumour Establishment

Maria Tsekrekou, Dimitris Mavroudis, Dimitris Kafetzopoulos, and Despoina Vassou

Contents

M. Tsekrekou • D. Kafetzopoulos (✉) • D. Vassou
Institute of Molecular Biology and Biotechnology, Foundation for Research and Technology-Hellas (IMBB-FORTH), Nikolaou Plastira 100, GR-70013 Heraklion, Crete, Greece
e-mail: kafetzo@imbb.forth.gr; dvassou@imbb.forth.gr

D. Mavroudis
Department of Medical Oncology, Laboratory of Tumour Cell Biology, University of Crete, Voutes GR-71110 Heraklion, Crete, Greece

R.K. Srivastava and S. Shankar (eds.), *Stem Cells and Human Diseases*,
DOI 10.1007/978-94-007-2801-1_25,

Abstract Continuously increasing evidence has shown that most primary tumours contain subpopulations of cells with stem-like characteristics and their existence is associated with resistance to therapy and formation of metastases at distant sites. This is further supported by the identification of several common key pathways and genes between cancer and normal stem cells involved mainly in proliferation, differentiation and de-differentiation steps. Cancer stem cells (CSCs) can self-renew as well as give rise to non-CSC progeny. CSCs may exist within the tumour since the early stages of cancer or develop during cancer progression through the developmental process of epithelial-to-mesenchymal transition (EMT). Cells with stem-like features have been recently identified within the circulating tumour cells (CTCs) that are responsible for the generation of metastases. Dissemination from the primary tumour can occur in the early stages of tumour development or later during tumour progression. Here, we review current knowledge concerning CSCs' development, metastatic potential, implications for therapy and mechanisms that govern these processes, including molecular pathways and the role of microenvironment.

Keywords Cancer stem cells • CSCs • Circulating tumour cells • CTCs • Tumour microenvironment • Metastasis • EMT

Abbreviations

ABC	ATP-binding cassette
ABCG5	ATP-binding cassette G5
ADCC	Antibody-dependent cellular cytotoxicity
ALDH	Aldehyde dehydrogenase
ALL	Acute lymphoblastic leukemia
AML	Acute myeloid leukaemia
ATP	Adenosine-5'-triphosphate
ATS	Adult tissue stem cells
BCR-ABL	Breakpoint cluster region-V-abl Abelson murine leukaemia viral oncogene
BMPs	Bone morphogenetic proteins
CGH	Comparative genomic hybridization
CK	Cytokeratins
CML	Chronic myeloid leukaemia
COX-2	Cyclooxygenase enzyme-2
CSCs	Cancer stem cells
CSF-1	Colony stimulating factor-1
CTCs	Circulating tumour cells
CTSCs	Circulating tumour stem cells
DGCR8	DiGeorge syndrome critical region gene 8
Dsh	Dishevelled protein
DTC	Disseminated tumour cell

ECM	Extracellular matrix
EGF	Epidermal growth factor
EMT	Epithelial-to-mesenchymal transition
EpCAM	Epithelial cell adhesion molecule
Erk	Extracellular signal-regulated kinase
ES	Embryonic stem cells
ESA	Epithelial specific antigen
FAK	Focal adhesion kinase
GDFs	Growth and differentiation factors
GSK-3	Glycogen-activated kinase-3
HER	Human epidermal growth factor receptor
hESC	Human embryonic stem cells
Hh	Hedgehog
HPC	Hematopoietic progenitor cells
IL-4	Interleukin-4
LOX	Lysyl oxidase
mAb	Monoclonal antibody
MAPK	Mitogen activated protein kinase
MaSC	Mammary stem cell
MEK	MAPK kinase
MET	Mesenchymal-to-epithelial transition
MRD	Minimal residual disease
miRNA	microRNA
MSCs	Mesenchymal stem cells
NF-κB	Nuclear factor-kappa-B
NICD	Notch intracellular domain
NOD–SCID	Non-obese diabetic–severe combined immunodeficient
Oct4	Octamer-binding transcription factor 4
PcG	Polycomb group
PI3K	Phosphoinositide 3-kinase
PROM1	Prominin 1
PTEN	Phosphatase and TENsin homologue
PTHC	Patched
ROS	Reactive oxygen species
R-SMADs	Receptor-regulated SMADs
RTKs	Receptor tyrosine kinases
SAPKs	Stress-activated protein kinases
SCID	Severe combined immunodeficient
SDF-1	Stromal cell-derived factor-1
SMOH	Smoothened
Sox2	SRY (sex determining region Y)-box 2
SSC	Somatic stem cells
STAT	Signal transducers and activators of transcription
TAF	Tumour-associated fibroblasts
T-ALL	T-lineage acute lymphoblastic leukemia
TAM	Tumour-associated macrophage

TDGF1	Teratocarcinoma-derived growth factor-1
TERT	Telomerase reverse transcriptase
TF	Transcription factors
TGF-β	Transforming growth factor beta
THY1	THYmocyte differentiation antigen 1
VEGF	Vascular endothelial growth factor
VEGFR1	Vascular endothelial growth factor receptor-1
VLA4	Very late antigen-4

25.1 Introduction

Although advances in the early detection of cancer and the development of new therapies are rapidly proceeding, mortality rates due to cancer are still high, mainly as a result of metastasis formation. Even patients with early stage disease have a significant risk to develop metastases despite state-of-the art "adjuvant" treatment. The metastatic cascade has been extensively studied for carcinomas (epithelial derived tumours), which account for almost the 80% of solid malignancies, but still remains ambiguous.

Tumours have long been considered as homogeneous cell populations. During the last years however it has become accepted that they consist of highly heterogeneous cell populations with respect to cellular morphology, proliferative potential, genetic lesions and treatment response. Continuously increasing evidence has shown that tumours are hierarchically organized with tumourigenic Cancer Stem Cells (CSCs) at the apex, generating the non-tumourigenic diverse tumour cell types. It is assumed that CSCs play a significant role in therapy (both chemotherapy and radiation) resistance, tumour recurrence and metastasis.

Metastasis is a complex process consisting of a series of steps, all of which must be successfully completed to give rise to a metastatic tumour [1, 2]. The very first step, for the dissemination of carcinoma cells, is the degradation of the underlying basal membrane and subsequent cancer cell invasion into the stroma. Then, cancer cells must intravasate in the lymph and blood vessels, and survive the journey through the circulation. Subsequently, they extravasate and home at distant sites. In the new environment, disseminated cancer cells initially form micrometastases, which are not macroscopically detectable, and later on, if the conditions are favourable, they grow further forming macrometastases.

Apart from micro- and macro-metastases formation, the late stages require cell motility, invasiveness and resistance to apoptosis, traits that are not exerted by epithelial cells. An attractive solution to explain these properties has been given by a developmental and wound healing program: the Epithelial-to-Mesenchymal Transition (EMT). During EMT epithelial cell layers lose polarity and cell-cell contacts and undergo a dramatic remodelling of the cytoskeleton [3, 4], which allows them to survive in an anchorage-independent environment and promotes migration and invasion, empowering cancer cells with stem-like characteristics.

Stem cells are able to differentiate into multiple cell types and maintain their pluri/multi-potency of differentiation through self-renewal. They are characterised by the potential of extensive proliferation although they remain relatively quiescent over time, thus preserving their self-renewal capacity. Self-renewed stem cells maintain their property of multi-potency and indefinitely supply progenitors required for replenishing relevant tissues. There are at least two types of stem cells: embryonic stem (ES) cells and adult tissue stem (ATS) or somatic stem (SSC) cells. ES cells are derived from the blastocyst, and generate, during development, the cells of all three primary germ layers: ectoderm, endoderm and mesoderm (pluri-potency). A number of transcription factors are required to maintain 'stemness' programs, such as *Oct4*, *Rex1*, *Sox2*, and *TDGF1* [5]. SSCs reside in various foetal and adult tissues and can differentiate into tissue-committed progenitors (multi-potency), which are transit amplifying cells, and will further differentiate into the diverse tissue cell types. In short, adult tissues are organised in a hierarchical manner with SSCs at the apex, generating multipotent progenitors with limited self-renewal capacity that will further differentiate to mature tissue cells. Although SSCs are multi-potent, there is increasing data supporting that they may be able to transdifferentiate [6].

SCs can divide either symmetrically, to produce two daughter cells sharing the same fate- stem or progenitor/apoptosis- or asymmetrically, where each stem cell divides to generate one daughter cell with stem-cell fate (self-renewal) and one daughter cell that differentiates [7]. Under physiological conditions, stem cells maintain a small but stable and highly efficient pool in tissues. The fate of stem cells is determined in a proper spatiotemporal manner by environmental signals, the stem cell niche, which consists of stromal or accessory cells, cytokines and developmental growth factors [8]. Niche cells provide a sheltering environment which protects stem cells from stimuli that would challenge stem cell reserves, such as differentiation or apoptotic stimuli. The niche also maintains a balance between quiescence and activity: SCs must be restricted from excessive SC production that could result in cancer, whereas they must periodically be activated to produce progenitor cells that will, in turn, produce mature cell lineages. Concluding, the SC niche regulates the SC fate and preserves the SC pool from diminishing or overexpansion.

25.2 Intratumoural Cell Heterogeneity

Tumours have long been considered as homogeneous cell populations, but it is now well established that they show remarkable heterogeneity in terms of cellular morphology, proliferative potential, genetic lesions and treatment response. A solid tumour may be regarded as an organ consisting of heterogeneous cancer cells, stromal cells and vasculature. As far as intratumoural heterogeneity of cancer cells and cancer development is concerned, several models have been proposed [9].

According to the stochastic model, during early tumour development, one or a few cells transform, grow and proliferate uncontrollably and eventually accumulate different mutations, resulting in the heterogeneity within the tumour. All cells are

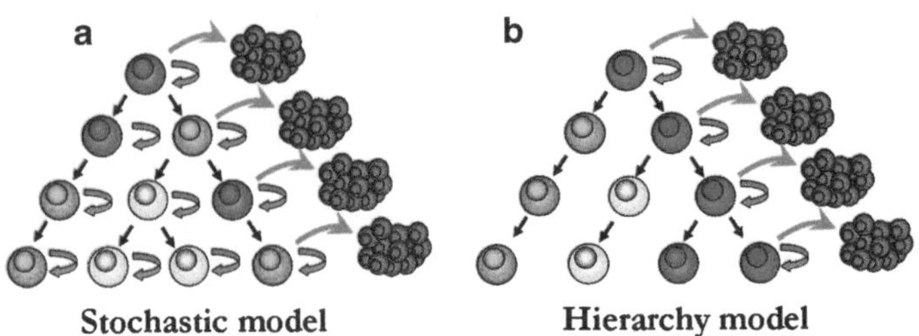

Fig. 25.1 The cancer stem cell hypothesis: the stochastic model (**a**) suggests that all cells can initiate tumour formation. Different colours represent cells in different stages of differentiation that have the ability to self-renew. The hierarchical model on the other hand (**b**), hypothesises that only a small portion of tumour cells, the cancer stem cells (CSCs, *blue*) have the ability to self renew and differentiate into tumour cells with limited proliferation capacity

considered to be tumourigenic, which means that any of the cancer cells can contribute to tumour growth or develop resistance to therapy and consequently give rise to metastasis (Fig. 25.1a).

According to the hierarchy model, only rare cells within the tumour have the capability to produce phenotypically heterogeneous cells. These cells sustain the tumour growth and exhibit stem cell properties, thus they are called Cancer Stem Cells (CSCs). Multi-potency of CSCs results in all the diverse cell types of a tumour, which only have limited proliferative capacity. Therefore, the tumour is consisted of two cell subpopulations: the tumourigenic one and the non-tumourigenic one (Fig. 25.1b). Thus, the tumour is organised in a manner similar to normal tissues, with cancer/tissue stem cells at the apex of the hierarchy, producing identical daughter stem cells, and committed progenitor cells that will, in turn, generate the different cell types of the tissue/tumour. It is assumed that the CSC subpopulation plays significant role in therapy resistance (chemotherapy and radiation) [10–12].

Interestingly, the stochastic model is in fact compatible with the existence of CSCs, as in this case self-renewal is an intrinsic property of all the cells within the tumour. It should be mentioned that neither of these models is necessarily the only correct model, since there are cancers that seem to be better described by the stochastic model (e.g. melanoma) or the hierarchical one (e.g. breast cancer).

Another source of heterogeneity that is applicable to both models described above is the aspect of clonal evolution, which claims that cancer cells, including CSCs, undergo evolution through natural selection [9, 13]. In the clonal evolution model, novel clones, with new genetic or epigenetic changes, develop in response to environmental pressure. Such clones may differ in their proliferation or metastatic potential, drug resistance or even self-renewal capability, leading to heterogeneity in phenotype, function, and response to therapy. The existence, however, of such heterogeneity does not imply that cancers must be hierarchically organised into tumourigenic and non-tumourigenic fractions, nor does the lack of hierarchical organization imply that cancers are homogeneous. The process of clonal evolution has been recently demonstrated in leukaemias [14].

25.3 The Cancer Stem Cell Hypothesis

The cancer stem cell hypothesis consists of two separate components (Fig. 25.1). The first one suggests that a (rare) population of cells with stem cell properties might drive and sustain cancer while the second one implicates a cell-of-origin responsible for the tumour formation [15].

The concept that tumours are driven by cellular components, which display stem cell properties, has been proposed almost 150 years ago [16–18]. According to these concepts, all cancers might arise from embryonic-like progenitor cells or germinal cells present in the wrong places in adult tissues [19]. Over the last 15 years, it has become evident that, at least in some cancers, a malignant stem cell compartment is responsible for tumour maintenance and initiation of relapse. The CSCs existence was first elegantly demonstrated for leukaemias [20] and a few years later CSCs were identified in solid tumours [9]: breast [21], brain [22, 23], colon [24], ovarian [25], prostate [26], lung [27, 28] and gastric cancers [29, 30].

The second constituent of the CSC theory refers to the cell-of-origin, which is a normal cell that acquires the first oncogenic hit. It is considered that the cell-of-origin is the source of heterogeneity, which occurs in tumours arising in the same organ, and results in the classification of distinct tumour subtypes. These subtypes are characterised by their morphology and molecular profile (expression of specific markers, such as hormone or growth factor receptors). The origins of CSCs have not yet been clarified. Cancers might arise from the oncogenic transformation of normal tissue stem cells to generate CSCs. This concept is quite appealing: SSCs, due to their longevity, may accumulate multiple mutations, undergo oncogenic transformation and yield CSCs. In this case, self-renewal is an inherent property of the cells and it is not required to be re-instigated. However, partially differentiated transit-amplifying cells could also be the targets of oncogenic transformation, providing they acquire the capacity of self-renewal. In fact, any cell in the tissue hierarchy that possesses proliferative capacity may serve as a cell-of-origin in cancer, providing it acquires mutations that restore self-renewal capability and impede its differentiation to a post-mitotic state [15, 31, 32].

25.4 Cancer Stem Cells Identification

In the last 5 years, it has become increasingly evident that CSCs may be responsible for tumour development, dissemination and metastasis [33]. CSCs may now be defined as the cells within a tumour that are able to self-renew in vitro and in vivo, generate the heterogeneous cell lineages of a tumour – both tumourigenic and non-tumourigenic – reveal aberrant differentiation, exhibit genetic alterations and can recapitulate the tumour in transplantation experiments. In fact, the latter parameter of the definition is the key means of identifying CSCs within a tumour [34].

The isolation of CSCs from several tumours has proved that they share several properties with the normal SCs, including the ability to self-renew and differentiate, active telomerase expression, resistance to apoptosis, increased membrane transporter activity and the capacity to migrate and metastasise. The latter, which is related to anchorage dependence, is a trait of transformed cells and it has been shown recently that it is also a property of normal tissue stem cells [15, 35–37].

CSCs have been identified in a variety of ways; however, the most convincing demonstration of their existence is the serial transplantation of tumour cellular subpopulations into immunodeficient animal models. The subpopulation containing CSCs must recapitulate the tumour, with all of the cellular heterogeneity found in the primary tumour; it should also maintain the self-renewing capability through serial passages [38]. Indeed, when inoculated into animal models at very low cell numbers, these CSCs frequently initiate tumours that are histologically similar to their parental tumours [32, 38, 39].

Cell-surface antigens have also been extensively used to isolate subpopulations, enriched in CSCs, from primary tumours. These include CD133 (also known as PROM1), CD44, CD24, epithelial cell adhesion molecule (EpCAM, also known as epithelial specific antigen, ESA), ALDH1 [40], THY1 and ATP-binding cassette G5 (ABCG5), as well as Hoechst 33342 exclusion by the side population cells [32], (described in Table 25.1). Most of these markers are specific for normal stem cells of the same organ. Unfortunately, a single 'stemness' signature based on cell-surface antigens is unlikely to be discovered, as spatiotemporal variation of marker expression has been shown among tissue types, cancer subtypes and even tumours of the same subtype. For example, [41] in gliomas CD133 expression is considered to characterise the CSC population but this is not always the case. In fact $CD133^-$ cells were equally tumourigenic as $CD133^+$ cells and could also initiate glioblastomas in nude mice. Stem cell markers that have been described in the literature for the identification of stem cells in different types of cancer are summarised in Table 25.2.

25.5 Epithelial-to-Mesenchymal Transition During Tumourigenesis: A Step Towards Metastasis

EMT is a cell-biological program that enables a polarised immotile epithelial cell to acquire a mesenchymal phenotype, which includes enhanced migratory capacity, invasiveness, heightened resistance to apoptosis and increased production of ECM components [63]. EMT is completed upon degradation of the underlying basement membrane, which allows the newly formed mesenchymal cell to migrate away from the epithelial layer in which it originated. A number of distinct molecular processes are required for the initiation and completion of an EMT, such as activation of transcription factors (EMT-TFs), expression of specific cell-surface proteins, extensive cytoskeleton remodelling, production of ECM-degrading enzymes and changes in the expression of miRNAs. Some of these factors are used as biomarkers indicating cell passaging through an EMT and include *CDH2* (N-cadherin), *VIM* (Vimentin), *FN1* (Fibronectin), *ZEB2* (SIP-1), *FOXC2*, *SNAIL1* (Snail), *SNAIL2* (Slug), *TWIST1* and *TWIST2* [64].

Table 25.1 Functions of solid tumour CSC markers

CSC marker from solid tumours	Function
CD133 (PROM1)	It is expressed in $CD34^+$ stem cells and progenitor cells in foetal liver, endothelial precursors, developing epithelium and foetal neural SCs. It is a five-transmembrane domain glyco-protein and it is thought to participate in the SC attachment to their niche
CD44 (PGP1)	It is an adhesion molecule with multiple roles in signalling, migration and homing. There are many isoforms: the standard one, CD44H, is highly affine for hyaluronate; CD44V administers metastatic properties
CD24 (HSA)	It is a highly glycosylated adhesion molecule, anchored on the membrane via glycosyl-phosphatidylinositol. Though it is not considered a CSC marker, low expression levels are related to Breast CSCs
CD90 (THY1)	It is a glycoprotein anchored on the membrane via glycosyl-phosphatidylinositol. It takes part in signal transduction and it is thought to play a role in stem cell differentiation and may mediate thymocyte adhesion to the stroma
EpCAM (ESA)	It is a homophillic Ca^{2+}-independent cell adhesion molecule expressed on the basolateral membrane of epithelial cells. It is thought to be overexpressed in cancer
ALDH1(Aldehyde dehydrogenase)	It belongs to the ALDH family which catalyzes the oxidation of aliphatic and aromatic aldehydes to carboxylic acids. It is involved in the oxidation of retinol to retinoic acid, important for proliferation, differentiation and survival, present in both somatic and cancer stem cells
ABC transporters	Members of the ABC (ATP binding cassette) transporters family are involved in multi-drug resistance via drug transport in the extracellular compartment. For example, ABCG5 is related with doxorubicin resistance
Hoechst 33342 exclusion	Exclusion of Hoechst 33342 dye is a common characteristic of cells with stem-like properties, both normal and malignant. It is thought that the dye is excluded due to quiescence that stem cells exert or the presence of ABC transporters

EMT normally occurs during embryological development and tissue regeneration. In the last years, EMT has been associated with tumour growth and progression and provides an appealing model that can adequately describe the dissemination process accomplished by malignant cells of epithelial origin during metastasis. Thus, EMTs take place in three distinct biological settings (development, wound healing and cancer) and, though they result in the formation of a mesenchymal cell, they are induced and progressed by diverse mechanisms.

Table 25.2 Representative cell surface markers for human cancer stem cells

Type of cancer stem cells	Cell surface markers	Reference
Acute myelogenous leukaemia (AML)	$CD34^+$, $CD38^-$	[20]
	$CD44^+$	[42]
	$CD123^+$	[43]
Chronic myeloid leukaemia (CML)	$CD34^+$, $CD38^-$, $CD123^+$	[44]
Glioblastoma	$CD133^+$	[39]
	$CD15^+$	[45, 46]
Medulloblastoma	$CD133^+$	[39]
Pilocytic astrocytoma	$CD133^+$	[23]
Anaplastic ependymoma	$CD133^+$	[23]
Breast	$CD44^+$, $CD24^{-/low}$, EpCAM+	[21]
Prostate	$CD133^+/\alpha2\beta1$integrin/$CD44^+$ $CD44^+/CD24^-$	[26] [47]
Ovarian cancer	$CD44^+$	[48]
	$MyD88^+$	[49, 50]
Colon cancer	$CD133^+$	[24, 51]
	$CD44^+$, $EpCAM^{high}$	[52]
	$CD166^+$	[53]
Pancreatic cancer	$CD133^+$	[54]
	$CD44^+$, $CD24^+$	[55]
Head and neck squamous cell carcinoma	$CD44^+$	[56]
Bone sarcomas	Stro-1^+, $CD105^+$, $CD44^+$	[57]
Melanoma	$CD20^+$	[58]
	$CD133^+$	[59]
Lung cancer	$CD133^+$	[28]
Liver	Bmi1, mutant β-catenin	[60]
	$CD133^+$	[61]
	$CD90^+$	[62]

The activation of EMT during tumourigenesis requires signalling between cancer cells and neighbouring stromal cells [65]. It is generally conceded that, in advanced primary carcinomas, cancer cells recruit a variety of cell types into the surrounding stroma, such as fibroblasts, myofibroblasts, granulocytes, macrophages, mesenchymal stem cells and lymphocytes; these recruited cells create an inflammatory microenvironment that seems to release EMT-inducing signals. The carcinoma cells, in respond to these environmental stimuli, activate the expression of EMT-TFs and consequently undergo EMT [66]. Accordingly, induction of an EMT program in carcinoma cells may be a feature of tumours able to recruit a reactive stroma.

Three signalling pathways are identified to be required simultaneously to induce an EMT. These pathways include TGF-β and canonical and non-canonical Wnt signalling and they are all initially stimulated in a paracrine manner from the reactive surrounding stroma. After the EMT completion, the mesenchymal/SC state is maintained by autocrine activation of these pathways [67]. This indicates that the mesenchymal/SC state is susceptible to destabilization by breaking the autocrine signalling loops required for its maintenance.

25.5.1 EMT Generates Stem-Like Cells

It is well accepted that cells, which have passed through EMT, acquire many of the traits of stem cells [67, 68]. Ectopic expression of EMT-inducing transcription factors in immortalised human mammary epithelial cells resulted in mesenchymal cells that have acquired the Mammary Stem Cell (MaSC) phenotype, i.e. $CD44^{+}/CD24^{low}$ and the ability to form mammospheres and self-renew [68]. Therefore, besides its role in metastasis, the physiological process of EMT may be a key step in CSC generation [69]. Recent studies have demonstrated that EMT can induce non-CSCs to enter into a CSC-like state [68, 70]. Induction of EMT in oncogenically transformed human mammary epithelial cells resulted in breast cancer stem-like cells with increased ability to form mammospheres and initiate tumours in immunocompromised mice [68].

Therefore, it is likely that two distinct sources of CSCs exist within a tumour: the intrinsic CSCs, which are formed and reside within tumours from their initiation, and the induced CSCs that are generated de novo by EMT, which in turn is often induced by contextual signals received from the tumour-associated reactive stroma. This raises the question of whether the two types of CSCs play identical roles in tumour progression and differ only in their origins. Regardless of their role, the presence of intrinsic and induced subpopulations of CSCs may address the issues related with heterogeneity of distinct tumour subtypes and early and late dissemination of carcinoma cells.

Concerning the intertumoural heterogeneity, carcinomas with a more differentiated phenotype (such as a tissue mature phenotype) seem to be less aggressive than those consisting of less differentiated tumour cells. For example, highly aggressive breast tumours (e.g. the claudin-low and basal-type breast cancers) have a normal mammary stem cell gene expression profile and comprise increased numbers of intrinsic CSCs. On the other hand, luminal-type breast cancers exhibit a mature mammary luminal cell phenotype and contain low numbers of intrinsic CSCs [71, 72].

Mesenchymal derivatives of carcinoma cells show a number of properties that favour metastasis such as separation from the collective as individual cells, increased migratory and invasive potential, increased survival in suspension and resistance to apoptosis in response to chemotherapy [73]. As such, EMT confers on epithelial cells precisely the set of traits that would empower them to disseminate from primary tumours and seed metastases [4]. Moreover, the heightened resistance to apoptosis, which is integral to cells generated by EMT, is critical to the ability of carcinoma cells to survive during the voyage from primary tumours to sites of metastasis [74]. In addition, the CSC-like state may be critical in the sites of dissemination for launching new colonies of cancer cells. Recent studies have shown that circulating tumour cells (CTCs) and micrometastasis show EMT and stem-like characteristics [75–80] and that residual breast tumour cell populations surviving after conventional treatment may be enriched for subpopulations of cells with both tumour-initiating and mesenchymal features [81]. In addition, EMT regulators alter the cell cycle machinery of cells, allowing the prolonged survival of residual cancer cells [82]. As an example, *Snail 1*, an EMT regulator was implicated in the emergence of local recurrence from residual disease after oncogene silencing [83].

25.5.2 *EMT Can Be Partial and Reversible*

It has been shown that cancer cells, in order to be able to complete the complex steps of metastasis, exhibit both mesenchymal- and epithelial-like properties at different times or even at the same time [84, 85]. Such carcinoma cells may have previously undergone a partial or complete EMT within primary tumours. Careful analysis of EMT-derived populations suggests that the 'hybrid' or metastable phenotype is more prevalent than pure mesenchymal derivatives. Indeed, coexpression of both epithelial and mesenchymal markers has been found in a breast cancer model [86] as well as other carcinomas (reviewed in [87]). This hybrid phenotype has become well recognised in cancer systems [88] and has been referred to as metastable phenotype [89] or activated epithelium [90].

Developmental EMT and cell migration is often followed by a later phase whereby the migrating cells cease migrating and aggregate through mesenchymal-to-epithelial transition (MET). Similarly, a MET-like reversion of EMT takes place after extravasation of cells to the site of metastasis [91]. The ability to convert between the epithelial and mesenchymal states is most effective in allowing cells to both leave the primary tumour and to establish a distant metastasis.

Plasticity between epithelial and mesenchymal states is therefore an important aspect of the initiation of metastatic colonies after extravasation of disseminated tumour cells. These cells may have previously undergone a partial or complete EMT within primary tumours, induced by the tumour-associated stroma. The acquired stem-like components may then enable their physical dissemination. Nevertheless, in the absence of an activated stroma after extravasation, these cells may well lapse back to a fully epithelial state. In order, however, to initiate a metastatic tumour, disseminated cells have to sustain some of their stem-like functions such as self-renewal, proliferation and resistance to apoptosis. This suggests that complex mechanisms operate to maintain the mesenchymal/CSC state, even in the induced rather than intrinsic CSCs.

25.6 Genetic Pathways Involved in 'Stemness' of Cancer Cells

It is becoming evident that the identification of CSC cell surface phenotypes is not adequate for understanding the functional nature of these cells. This could be achieved, at least in part, with a better understanding of the pathways promoting CSC maintenance and propagation. It is generally recognised that the stem-cell program operating in normal SCs appears to be also used by their neoplastic counterparts. This includes developmental pathways (Wnt, Hedgehog, Notch) [92] as well as pathways implicated in cellular growth and proliferation (TGF-β, MEK/Erk, PI3K/Akt) and other genes. Most of them are highly conserved among metazoan species. CSC maintenance, through activation of self-renewal pathways, has been demonstrated in numerous mouse transgenic models [93, 94], but only in a few human CSC xenograft models [95]. Recently a novel breast cancer mouse model has been described containing human breast and human bone [96]. Here, we will discuss what is currently known on the role of each pathway in cancer and stem cells, as well as recent data and their possible role in CSC maintenance.

25.6.1 Developmental Pathways

25.6.1.1 The Wnt Pathway

The Wnt signalling pathway is involved in multiple embryological, cellular and physiological activities from *C. elegans* to humans. It has three branches: the Ca++, the planar polarity and the canonical branch. The canonical branch of the pathway has been associated with the maintenance of pluri-potency in stem cells, enabling them to remain in an undifferentiated state, as well as tumourigenesis. The non-canonical pathways are implicated, together with other pathways, in inducing EMT, thus promoting motility and invasiveness of epithelial cancer cells. Wnt pathway can be interlinked with other pathways to activate similar targets.

An important component of the Wnt pathway is β-catenin, a protein that either binds to the cytoplasmic domain of cell-cell adhesion molecules, such as E-cadherin, implicated in loss of adhesion during EMT, or, upon Wnt signalling, acts as transcription co-regulator of proliferation stimulating genes [97]. The necessity of β-catenin for self-renewal of both normal HSCs and CSCs has been demonstrated in chronic myeloid leukaemia in a mouse model [98, 99]. Recently, it has also been shown that β-catenin activation is necessary for myeloid precursor transformation in a HoxA9/Meis1- transduced model of AML [100].

The canonical branch of the Wnt pathway has been shown to be necessary and sufficient to maintain the state of pluri-potency, in both ESCs and SSCs (demonstrated in HSCs of the bone marrow and epidermal stem cells) [95]. Alterations of this branch have been associated with a variety of tumours. The strongest evidence of the importance of the Wnt pathway to CSC biology has been reported in myeloid leukaemias. Activation of the Wnt signalling pathway has also been implicated in breast CSC maintenance [101].

25.6.1.2 The Notch Pathway

Notch signalling is involved in tissue homeostasis and cell-fate decisions during development, and in stem cells. Notch signalling mediates juxtacrine signalling and is activated by binding of Notch receptors (Notch 1-4) with their respective ligands on adjacent cells. It was shown that, in human breast stem cells, Notch signalling regulates asymmetric divisions of SCs [102, 103] and promotes SC self-renewal and differentiation of progenitor cells, while exhibiting negligible effects on mature differentiated epithelial cells [104]. Activating Notch1 mutations are found in more than 50% of T-ALL [105]. It is also reported that there might be specificity of different Notch receptors in the regulation of breast stem and progenitor cells. For example, Notch4 activity is increased in breast CSCs and inhibition of Notch4 signalling reduces breast CSCs and completely inhibits tumour-initiation. On the other hand, Notch1 activity is lower in breast CSCs compared to more differentiated progenitor cells [106].

25.6.1.3 The Hedgehog Pathway

The Hedgehog (Hh) pathway plays a critical role in a spatiotemporal-dependent manner during development by regulating patterns and maintenance of proliferative niches. Hh is known to signal through autocrine, juxtacrine, and paracrine mechanisms [107, 108]. A key molecule of the Hh pathway is Smoothened (SMOH), a G-coupled transmembrane protein. Hh regulates normal and malignant stem cells in both Drosophila and mammalian systems [109]. Abnormal activation of the Hh pathway is common in basal cell carcinoma, medulloblastoma and rhabdomyosarcoma. A role for Hh signalling in CSC was demonstrated by specific deletion of SMOH in BCR-ABL positive chronic myeloid leukaemia SCs, which prevented tumour-initiation. In addition, treatment with cyclopamine, a SMOH inhibitor, increased survival of mice transplanted with BCR-ABL leukaemia cells [110]. Hh signalling also plays a significant role in normal and malignant breast stem cells. Several Hh components are expressed in stem and progenitor cells when cultured as mammospheres, whereas their expression is substantially reduced when these cells undergo differentiation [101]. Moreover, they were expressed at higher levels in $CD44^{+}CD24^{-}$ breast CSCs compared to bulk cancer cells.

25.6.2 Proliferation and Survival Pathways

25.6.2.1 The TGF-β Pathway

The signalling pathway of the TGF-β superfamily is one of the most complicated pathways in metazoans and key for cancer development and progression. It was first discovered in tumours as an anti-proliferative signal, critical for maintaining tissue homeostasis by keeping cell proliferation in check. However, it was later demonstrated that it plays a critical and dual role in the progression of human cancer. During the early phase of tumour progression, TGF-β acts as a tumour suppressor. At a later stage, TGF-β also promotes processes that support tumour progression such as tumour cell invasion, dissemination, and immune evasion. Consequently, the TGF-β response is strongly context-dependent, including cell, tissue and cancer type [111].

The TGF-β signalling pathway consists of two main branches: (a) bone morphogenetic proteins (BMPs) and growth and differentiation factors (GDFs) activate SMAD1 and 5, (b) Activin and Nodal ligands activate SMAD2 and 3. Subsequently, activated SMADs form a higher-order complex with SMAD4, translocate to the nucleus and regulate transcription of a broad range of genes. It should be noted that apart from canonical signalling, which directly regulates the transcription of SMAD-dependent target genes, there is also extensive crosstalk between TGF-β and other pathways as well as alternative activities of the TGF-β pathway components. For example, it has been reported that R-SMADS (in a manner independent of SMAD4) are also involved in the regulation of miRNA maturation in the nucleus [112]. Furthermore, TGF-β signalling is implicated in RhoA–Rock1 signalling which is required for EMT [113, 114] and involved in increased cell motility [115]. Indeed, TGF-β activity has been associated with metastasis in breast, colon, and prostate cancer [116, 117]. The TGF-β receptor also activates MAP kinase signalling [118].

In ESCs, the Activin/Nodal branch is necessary to sustain pluri-potency, while the BMP branch appears to enhance differentiation [119–122]. A similar exertion is observed in glioblastoma, where TGF-β maintains tumourigenicity of glioma-initiating cells [123] and BMP suppresses it [124]. However, as already mentioned, the TGF-β functional response (tumour suppressing or promoting) is dependent on other environmental signals as well. Mutations or downregulation of TGF-β receptors/ligands or inactivation of *SMAD4* are detected in a variety of cancers. For example, *SMAD4* inactivation occurs in ~53% of human pancreatic ductal adenocarcinomas (PDAC) [125, 126] and BMP2 is dramatically overexpressed in 98% of lung carcinomas [127, 128].

25.6.2.2 The PI3K/Akt Pathway

The phosphatidylinositol 3-kinase/Akt (PI3K/Akt) pathway responds to a variety of extra- and intracellular signals, which are transduced via hormonal receptors, tyrosine kinase-linked receptors (RTKs) and intracellular factors. It regulates cellular proliferation, cell survival, energy metabolism and cytoskeletal rearrangements [129]. A major inhibitor of the PI3K/Akt pathway is PTEN (Phosphatase and TENsin homologue). Various PTEN mouse models have shown that PTEN/PI3K/Akt pathway controls stem cell homeostasis and malignancies in numerous tissues [93, 130–133].

Microarray analysis of hESCs indicates that components of the PI3K/Akt pathway are upregulated in undifferentiated ESC [134]. However, considering that the Wnt pathway is activated downstream of the PI3K/Akt pathway, pluri-potency could be attributed to Wnt activation. Indeed, genetic knockdown of PTEN in a xenograft model enriched for breast CSC markers and increased tumourigenicity; an effect that was mediated by Wnt/β-catenin signalling via Akt activation [40, 135].

25.6.2.3 The MEK/ERK Pathway

The MEK/ERK (Mitogen Activated Protein Kinase (MAPK) kinase/extracellular signal-regulated kinase) pathway transduces signals from cytokines and growth factors, through RTKs, to promote cell adhesion, proliferation, migration and survival [136]. Central components of the pathway are RAS and RAF that subsequently activate ERK. Given its central role in cell proliferation, it is consistent that the SOS-Ras-Raf-MAPK signalling cascade is deregulated in multiple human tumours. Most of these mutations occur in *RAS* and *RAF* and result in constitutive pathway activation and hyperproliferation. *RAS* mutations are detected in ~45% of colon cancers and ~90% of pancreatic cancers [136]. Furthermore, *RAF* mutations are found in almost two thirds of all melanoma [137]. MEK/ERK signalling is active in undifferentiated hESCs and downregulated upon differentiation [134]. Therefore it seems plausible that the ERK pathway is also active in the CSC compartment.

25.6.2.4 The ErbB (HER) Pathway

The ErbB (HER) receptors are well known mediators of cell proliferation, cell motility, migration, differentiation, adhesion, protection from apoptosis and transformation. The ErbB family consists of four cell surface receptors (ErbB1 or EGFR, ErbB2, 3 and 4), which are typical receptor tyrosine kinases and their ligands belong to the Epidermal Growth Factor (EGF) family. Downstream targets of this network include PLC- $\gamma1$, Ras-Raf-MEK-ERKs, PI3K-Akt-ribosomal S6 kinase, Src, the stress-activated protein kinases (SAPKs) and the signal transducers and activators of transcription (STAT). ErbB2 (HER2) modulates the transcription of cyclooxygenase enzyme (COX-2), a well known oncogene in breast cancer [138]. Even though ErbB family members are often over-expressed, amplified or mutated in many cancer types, there is no evidence concerning their role in CSC biology. ErbB3 amplification or overexpression is associated with prostate, bladder, and breast cancers. The ErbB2/3 dimer is the crucial mediator of ErbB2 signalling in tumours with ErbB2 amplification [139]. While ErbB4 is thought to be antiproliferative in some cancers, recurring activating mutations of ErbB4 have been identified in melanomas [139].

25.6.3 Other Genes Involved in Stem Cell Character

25.6.3.1 BMI-1

Another developmental gene implicated in stem cell biology is *BMI-1*, a member of the Polycomb group (PcG) proteins, which plays a significant role in self-renewal of normal SCs and CSCs [94, 140–144]. The PcG genes are transcriptional repressors and play an essential role in embryogenesis, regulation of the cell cycle and lymphopoiesis. BMI-1 represses genes that induce cellular senescence and cell death [145, 146] while being important in the regulation and maintenance of proliferative/self-renewal potential in both normal haematopoietic and leukemic stem cells [142]. Knockdown of *BMI-1* results in cells unable to engraft and reconstitute leukaemia in immunocompromised mice [94].

25.6.3.2 TP53

Apart from inducing and sustaining growth-stimulatory signals, cancer cells must also avoid programs that negatively regulate cell proliferation and depend on tumour suppressor genes. Multiple tumour suppressors have been discovered due to their inactivation observed in animal or human cancer. One of the most discussed is the *TP53* gene that encodes p53 protein. Since its discovery [147], there have been thousands of articles detailing numerous roles of p53 in cancer. The p53 is located in every cell nucleus, bound on DNA, where it integrates signals of cell stress and damages on the genome. If the degree of DNA damage is excessive, or if the levels of nucleotide pools, growth-stimulating signals, glucose, or oxygenation are not optimal, p53 can arrest the cell-cycle progression until these conditions return to normal. Alternatively, if the DNA damage cannot be repaired p53 induces apoptosis. Recently, it has been demonstrated that p53 regulates polarity of cell division in mammary SCs and that inactivation of p53 contributes to tumour growth by promoting symmetric divisions of CSCs [148].

25.6.3.3 miRNAs

MicroRNAs are endogenous non-coding ~20–23nt RNAs that lead to translational repression, mRNA destabilization or mRNA cleavage [149]. They are involved in the regulation of a variety of biological processes, including cell cycle [150], differentiation, development [151, 152], and metabolism [153] and can act as oncogenes or tumour suppressors [154]. In order to regulate their stemness programs, SCs make use of the flexibility that RNA interference provides. For example, the absence of DGCR8 or Dicer alters the G1-S transition and proliferation of ESCs [155, 156]. Moreover, the miRNA families that regulate self-renewal and pluri-potency are quickly down-regulated upon differentiation [156–158]. miRNA signatures that distinguish malignant from normal cells have been discovered and, at least some of them, may be associated with the prognosis and the progression of cancer [159–163].

25.7 The Metastatic Cascade

Metastasis is a complex multistep process [1, 2], where cancer cells disseminate from the primary tumour, invade the surrounding extracellular matrix (ECM), enter the blood circulation, and finally home in a distant organ where they proliferate.

Disseminated cancer cells circulate as small aggregates (Circulating tumour Cells, CTCs) and lodge in distal microvascular beds by passive mechanical or active mechanisms (extravasation). After successful homing (micrometastasis), cells proliferate giving rise to the metastatic tumour (macrometastasis). In order to be able to successfully complete all these steps, cancer cells may undergo EMT [65] that allows them to survive in an anchorage-independent manner and promotes migration and invasion through both ECM and vessel walls. Cells that have passed through EMT acquire stem-like characteristics [67, 68]. This transition facilitates not only invasion but also growth to the metastatic site. Metastatic cells may repeat the entire sequence of events to produce additional metastases and they can also colonise their tumours of origin, in a process referred to as 'tumour self-seeding' [164]. This process may explain local recurrence after tumour excision.

Metastasis is regarded as a highly inefficient process [165] since only a very small percentage of CTCs manage to successfully initiate cell growth in secondary organs [166]. Some tumour cells may remain dormant [167], while others are incapable of triggering the angiogenic switch necessary for tumour expansion [168]. It seems that the outcome of the metastatic process is finally determined by a complex series of interactions between metastatic cells and the microenvironment of the host organ.

25.8 Early vs. Late Cancer Cell Dissemination Models

Paget introduced the 'seed and soil' theory 100 years ago, according to which disseminating tumour cells – the seeds- are shed into the circulation from the primary tumour and home in a receptive microenvironment – the soil – where they form metastases [169]. His hypothesis has been confirmed by recent data that have shed new light on the generation of the 'seeds', suggesting that some of the features of the tumour cell phenotype are conveyed early in the process of oncogenesis, whereas others are selected for during cancer progression [170]. In early relapse, the presence of a small number of treatment-resistant tumour clones gives rise to a resistant recurrence. Late relapses are characterised by an early phase (tumour dormancy) in which little or no expansion of tumour cells takes place, followed by a late phase in which exponential tumour growth occurs [171].

It has been well documented that the risk of metastases increases with the larger size of a primary tumour. Therefore, tumour biology concepts have mainly postulated that cancer cell dissemination occurs late in tumour development and follows the classic multistep process of metastasis described above. Yet, numerous studies over the past two decades have revealed that single tumour cells might spread to distant sites much earlier than previously believed. Single disseminated tumour cells (DTCs) can be found in the lymph nodes or bone marrow even in the very early stages of different types of carcinomas such as breast, prostate, colorectal cancer and melanoma [172–179]. These single dormant tumour cells seem to be of prognostic relevance, as metastases may occur even more than a decade after primary tumour

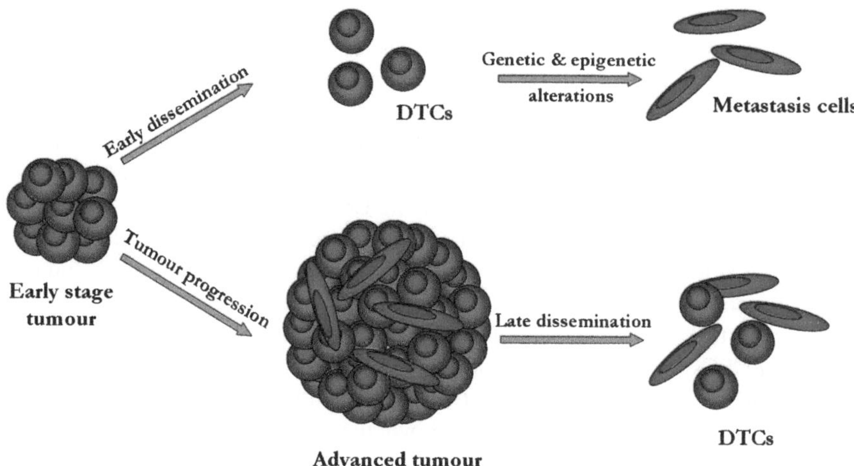

Fig. 25.2 Schematic representation of the early and late dissemination models of cancer cells. In the early dissemination model, cells disseminate from the primary tumour during the early stages of its development and reside to a distant site where they become adapted to the specific microenvironment that leads to site-specific selection of genetic and epigenetic alterations, in parallel with the progression of the primary tumour (parallel progression). The late dissemination model suggests that cells disseminate when the primary tumour is grown and the disseminated cells (DTCs) are fully malignant. Therefore, there is no need for additional genetic alterations and metastasis has the same characteristics as the primary tumour (linear progression)

surgery [180]. Thus, the concept of minimal residual disease (MRD), the microscopic tumour residues from which recurrence of the tumour originates and which is certainly well accepted for hematologic malignancies, has also been established for solid tumours [172]. Viewing DTCs as rare and late events during primary tumour progression, has been also challenged by expression profiling studies, in which a more ubiquitous 'metastatic phenotype' was proposed [181, 182].

To further disambiguate the issues concerning systemic cancer progression, two fundamental concepts have been proposed (Fig. 25.2) [183]. The first, linear progression, supports that the genetic development of metastasis founder cells occurs within the primary tumour, therefore, only fully malignant cells disseminate. These cells lead to metastasis after adaptation to the microenvironment of the host organ. In this case, cancer cells 'learn' to interpret signals from the microenvironment [184] and no further genetic development is required. In contrast, the second model, parallel progression, suggests that cells that are still evolving, disseminate from the primary tumour before they become fully malignant and reside to a distant site where they become adapted to the microenvironment that leads to site-specific selection of genetic and epigenetic alterations. Although both mechanisms may play a role in systemic cancer progression, they result in two opposing predictions about metastasis founder cells: The linear progression model predicts the genotype of the metastasis founder cells to be highly similar to the primary tumour, whereas the parallel progression model emphasises genetic divergence of primary tumours and disseminated tumour cells and metastases [185].

Regarding the early dissemination of tumour cells, it seems likely that intrinsic preneoplastic SCs already exist during the early stages of tumourigenesis. Such SCs may be able to disseminate long before malignant tumours have developed. Furthermore, it appears that induced CSCs occur in later stages of tumourigenesis when the reactive stroma, capable of inducing EMT, has developed.

Parallel progression also argues for a shift of research focus to the identification of metastasis founder cells. CSCs have usually been isolated from primary tumours and these populations are thought to consist of metastasis founder cells [186, 187]. The greater tumourigenic potential of CSCs over non-CSCs was established by the ability of CSCs to form tumours at low cell numbers within weeks in immunodeficient NOD–SCID (non-obese diabetic–severe combined immunodeficient) mice [188]. However, more recent data have shown that every melanoma cell could form tumours under different experimental conditions in some immunocompromised (NOD–SCID *Il2rg–/–*) mice [187, 189]. This highlights one artifactual drawback of xenotransplant models, as it is clear that neither every tumour cell [189] nor cells expressing given stem cell markers [187] proliferate in patients at rates similar to those observed in mouse models.

25.8.1 Genetic Disparity Between Primary Tumours and Metastasis

A more detailed genetic analysis of DTCs became feasible once single cell amplification methods were developed. Surprisingly, only few reports on genetic comparisons of primary tumours and DTCs have been published so far [190–193]. These reports have shown that in breast, prostate and oesophageal cancer, DTCs from bone marrow generally display fewer genetic abnormalities than matched primary tumours and that DTCs and primary tumours diverge to a large extent with respect to specific chromosomal gains and losses. While primary tumours often accumulate chromosomal gains and losses typical for the respective type of tumour, DTCs displaying these typical changes are mostly undetectable at the time of primary surgery, but later -when manifest metastasis is diagnosed- such changes will be found in almost all DTCs isolated at this stage. Comparative analysis with the primary tumour has revealed that these changes were not transmitted from the primary tumour to the manifest metastasis but must have been acquired independently [191, 193].

On the other hand, in other studies, genome and transcriptome analyses of single disseminated tumour cells demonstrated that the majority of DTCs are cells with genetic aberrations compatible with malignancy, and therefore most likely direct descendants of the primary tumour, although the genetic changes generally were incongruent with the dominant genotype of the corresponding primary tumour [191, 194, 195]. This indicates that shared aberrations do not prove clonality but may reflect selection of advantageous genomes that are acquired independently in primary tumours and metastases by convergent evolution [183].

An early study has shown that there is great clonal divergence between primary breast cancer tumours and their respective metastasis [196], while these results have been confirmed by more recent studies [197, 198]. For example, it has been shown that only 33% genomic alterations were shared between primary tumours and sentinel lymph node metastases [197]. In fact, 85% of the cases displayed isolated chromosomal changes that occurred exclusively either in the primary tumour or in the respective metastasis. In colorectal cancer, similarity, interpreted as clonality between primary tumours and matched metastases, seems to be higher than in breast cancer [199–201]. On the other hand, despite overlapping genetic changes, it was noted that in cases in which more than two lesions were analysed, the establishment of a simple series of theoretical precursors was difficult because of the extensive heterogeneity [199]. In another study that compared colorectal cancer lung metastases with their autologous primary tumours, 60% of the 10 pairs with a clonal relationship rated as high, shared only less than half of the observed aberrations, and none displayed an identical aberration pattern [201]. In addition, systematic comparative analyses of colorectal primary tumours and different metastatic sites [202–204] revealed a significant divergence between lymphatic and haematogenous metastases. Metastatic cells shared common chromosomal aberration patterns depending on the site of isolation (lymph node, liver or lung) that were distinct from the primary tumour. The genetic disparity between tumour cells isolated from different sites is also observed in patients with Barrett's oesophagus progressing to oesophageal adenocarcinoma [205, 206]. While comparative genomic hybridization (CGH) alterations accumulated in a stepwise mode from low-grade to high-grade dysplasia and then to carcinoma and metastasis, discordant patterns of aberrations were found in the different groups of morphological progression, with carcinomas and their lymph node metastases displaying the highest number of differences.

Finally, several studies revealed significant differences in gene expression profiles of primary tumours and their matched metastases in breast [207–210] and colorectal cancer [208, 211, 212]. Nevertheless, in a recent study, comparison between primary tumours and regional metastases showed statistically indistinguishable gene expression patterns. Comparison however between distant metastases versus primary tumours or regional metastases showed that the distant metastases were distinct and distinguished by the lack of expression of fibroblast/mesenchymal genes, and by the high expression of a 13-gene profile, the 'vascular endothelial growth factor (VEGF) profile', that showed prognostic significance in patients with breast and lung cancer and glioblastomas [213].

In a comparison between primary colon cancer vs. lung metastasis, it was shown that while primary tumours were similar, the metastatic tumours were highly diverse. A pair-wise comparison of the matching primary-metastatic tumours showed that different groups of genes were activated in the lung metastases and a number of these genes showed similar differential expression patterns in all the patients. These were the cancer cell-, the microenvironment- and the stem cell-specific gene groups [212].

To be taken into consideration, however, is the fact that the origin of the discrepancies reported may reside in the different bioinformatic approaches, because, when gene expression analysis is performed for several thousand genes, it is clear that transcriptional quantitative trait loci (genetic loci that have a quantitative impact on the expression of a phenotype), that are measured by microarray experiments, depend on the genetic background. The genes that are altered in cancer are masked by quantitative differences of thousands of genes between individuals [214]. Another aspect to be taken into account is the fact that primary tumours and metastases share many biological functions such as proliferation, induction of angiogenesis and differentiation related to the histogenetic origin, which may be regulated by divergent genomes. Moreover, stromal and infiltrating cells contribute to a significant extent to the gene expression profile of tumour tissue. These cells are influenced by the tumour and provide signals to the tumour, effects that impact on the gene expression profile but not immediately on the genome [185].

25.9 Metastatic Potential of CSCs

Experimental and clinical data associate the presence of CTCs with poor prognosis, due to the development of metastases in cancer patients [215]. However, neither all patients with detectable CTCs will develop metastases nor all patients without detectable CTCs will escape relapse, suggesting that stochastic events (mutations or activations) and specific conditions and/or particular subpopulations of CTCs with high metastatic potential may be responsible for the establishment of metastasis. The biological characteristics of these micrometastatic cells are poorly understood, as they often have distinct characteristics compared to the primary tumour and they form a heterogeneous population even in the same patient [216]. So far, scarce reports exist on profiling and genotyping of CTCs isolated from blood [195, 217, 218] and no final conclusion can be drawn regarding their molecular profiling.

Cells with stem cell properties may be present within the CTC population leading to metastases. Therefore, the presence of circulating tumour stem cells (CTSCs) among CTCs might explain the heterogenic metastatic potential of patients with early stage cancer who present CTCs in their blood. Furthermore, since stem cells are refractory to chemotherapy, their presence could explain the incurable nature of metastatic cancer. Liu et al. [219] was the first to identify a 'metastatic cancer stem cell' gene signature that was correlated with poor prognosis.

As a consequence, CTCs are heterogeneous and can possess epithelial phenotype, go through EMT or acquire a stem cell character. The existing methods for CTCs detection are usually biased towards epithelial phenotype and molecular characterization of CTCs is limited (reviewed in [220]) because of the heterogeneity and the small numbers of CTCs. CSCs have been demonstrated to exist as a subpopulation in both DTCs [221] and CTCs [78] of breast cancer patients. Furthermore, patients with triple-negative breast cancer, more commonly overexpressed EMT genes in peripheral blood compared to non-triple-negative patients, while EMT of CTCs

Table 25.3 Stem cell markers in CTCs detection

	Reference	Method of analysis
Mesenchymal markers		
Vimentin	[222]	IF
	[79]	IF
	[75]	RT-PCR
N-cadherin	[222]	IF, WB
O-cadherin	[222]	IF, WB
Twist1	[79]	IF
	[223]	Multiplex RT-PCR
	[76]	Real time qPCR
Fibronectin	[75]	RT-PCR
Akt2	[223]	Multiplex RT-PCR
PI3Kalpha	[223]	Multiplex RT-PCR
SNAIL1	[76]	Real time qPCR
SLUG		
ZEB1		
FOXC2		
Stem cell markers		
CD133	[222]	IF
ALDH1	[75]	RT-PCR
	[223]	Multiplex RT-PCR
	[78]	IF
CD44 CD24 (−)	[78]	IF

IF immunofluoresence, *WB* Western blot, *RT-PCR* reverse transcription PCR, *qPCR* quantitative PCR

may result in their underdetection by conventional methods [77]. In addition, it has been found that in metastatic prostate and breast cancer patients the majority (>80%) of the detected CTCs co-express epithelial proteins such as EpCAM, cytokeratins (CK) and E-cadherin, mesenchymal proteins, including vimentin, N-cadherin, and O-cadherin, and the stem cell marker CD133 [222]. Patients with metastatic breast cancer have significantly higher numbers of CTCs expressing Twist and Vimentin, suggestive of EMT, compared to patients with early stage disease [79]. The variable expression of these molecules in CTCs at different stages of disease, imply the predominance of EMT phenotype during disease evolution. This hypothesis is further supported by the observation that cytokeratin positive CTCs in patients with early breast cancer are more heterogeneous concerning the expression of EMT markers [79]. These data strongly support the notion that EMT is involved in the metastatic potential of CTCs. A number of markers have already been described for the identification of tumour stem cells within the CTCs population of different tumour types with different detection methods [75, 76, 78, 79, 222, 223] summarised in Table 25.3, but the 'optimal' stem cell marker remains elusive.

So far, it has been shown that both tumours and CTCs are heterogeneous and CSCs form a subpopulation within them with tumour-initiating and metastasis-forming capacity. The question however whether all CSCs are equally competent for generating metastasis remains. Indeed, 'metastasis-founding cells' may simply be defined as a subpopulation of cells from a broader tumour-propagating CSC population, or a different cell lineage altogether. This idea is supported by studies from different solid cancer models. Human cancer stem cells can be isolated in various tumour types according to their expression of the surface antigen CD133 [24, 39, 51, 224], although it is noted that the methodology associated with CD133 detection remains somewhat controversial [225]. In human pancreatic cancer cell lines, most $CD133^+$ cells can uniquely form tumours when transplanted into mice but only a small fraction of these colonise the liver [54]. This subpopulation of metastasis-founding cells are located at the invasive edge of pancreatic tumours and are characterised by increased expression of CXCR4, a chemokine receptor known to mediate metastatic lung and bone colonization [188, 226]. Consequently, in this model, metastatic $CXCR4^+$ cells are migratory subsets of $CD133^+$ CSCs. $CD133^+$ CSCs are also present in colorectal tumours isolated directly from patients. In addition, several other unique surface antigens could be used to isolate partially overlapping CSC populations from matched colorectal tumours and liver metastasis [227]. However, only CSCs marked by the CD26 antigen possessed both tumourigenic and metastatic competence, and this potential was independent of any other surface antigen examined (including $CD133^+$) [227].

Concluding, there is considerable heterogeneity within a tumour's CSC content. Tumourigenic stem cells may have varying degrees of metastatic competence, depending on the context. The discrimination of metastasis-founding cells through therapeutic resistance or predilection for anatomical sites is of great importance. The fact that different molecular subtypes of breast cancer display unique organ preferences [228], may be related to their original cellular lineages. Understanding the mechanisms that regulate the emergence of metastasis-founding cells in general is therefore fundamental. These metastasis-initiating subpopulations may appear after additional genetic or epigenetic alterations within an initial tumour-propagating population, or alternatively, they can pre-exist separately as untransformed precursors that are intrinsically more 'invasive'. The existence of disseminating premalignant cells is implied by the observation that some normal mammary epithelial cells can extravasate into the lungs of mice and persist there well before oncogenic transformation [229].

25.10 Stem Cell Niche in Cancer and the Establishment of Metastasis

Stem cells reside in niches that regulate SC stemness, proliferation and apoptosis resistance. The SC niche is composed of different stromal cells, such as fibroblasts and mesenchymal cells, immune cells, vasculature, soluble factors and extracellular matrix components [230]. Similarly, there is accumulating evidence that there is a CSC niche that plays a crucial role in virtually every step of the tumourigenic cascade and helps CSCs to preserve their unique abilities to self-renew and generate more differentiated progenitor cells, while maintaining an undifferentiated state themselves [231]. Moreover, the CSC niche protects CSCs from diverse genotoxic insults and therefore contributes to their resistance to treatment [232, 233].

In primary tumours, the CSC niche is an important regulator of stemness: it has been demonstrated in a variety of different tumours that the loss of a niche environment leads to the loss of CSCs [42, 234]. Apart from maintaining the CSC pool and supporting the growth of primary tumours, the niche is also involved in tumour invasion and dissemination via reverse of non-tumourigenic cells into CSCs by processes related to EMT. Finally, it is also thought that a premetastatic niche is formed to enable successful homing of cancer cells to distant sites and the development of metastasis. The supporting role of the microenvironment in tumour growth, progression and metastasis formation, renders the CSC niche as an upcoming therapeutic target. Detailed descriptions of SC niches exist for many tissues such as the mammary gland, epidermis, brain, colon and bone marrow [8, 38, 230].

It is known that metastases selectively occur in certain organs such as lungs, liver, brain, and bones, and each cancer type displays different predilections [235]. For example, breast cancer metastasises to bone, lung, brain and liver [228, 236], while prostate carcinomas preferentially metastasise to bone and colorectal carcinomas to the liver [237]. This observation led to the 'seed and soil' hypothesis, first introduced by Paget [169], as discussed earlier. This hypothesis in its modern context [238, 239] states that malignant cancer cells (seed) acquire mutations in oncogenes or tumour suppressor genes that confer the ability to egress from the tissue of origin, survive in the blood or lymphatic circulation, and grow in certain organs (soil). The local microenvironment of these organs seems to be more receptive to DTCs from particular malignancies than other organs [240]. Thus, occurrence of metastasis does not happen randomly, but DTCs need to meet a hospitable microenvironment in order to initiate a secondary tumour (reviewed in [241]). To address the mechanisms underlying metastatic cell homing associated with spread through the blood circulation, several investigators took parental cancer cells and selected tissue-specific homing variants through serial passage in mice [238, 242, 243]. More recent studies have used the transcriptome profiling to identify cohorts of genes for which expression confers both enhanced metastatic potential and altered microenvironment [244, 245]. These data (reviewed in [235]) show that the 'seed' can have a preferred site for growth that is encoded by genetic alterations in the tumour cell itself. Furthermore, in recent years, evidence suggested that the primary tumour itself is actively involved in adapting these so-called premetastatic niches for tumour cells to come, by secreting systemic factors and directing bone marrow–derived cells and macrophages to certain tissues, thereby priming certain tissues for tumour cell engraftment [246, 247].

Accordingly, VEGFR1-positive bone marrow–derived hematopoietic progenitor cells (HPC) were shown to localise to premetastatic sites and form clusters before the arrival of tumour cells, initiating the pre-metastatic niche [246] and to be involved in angiogenesis [248]. Eradication of these cells from the bone marrow prevents the formation of premetastatic clusters and, subsequently, tumour metastasis. In addition to homing of HPCs, pre-existing fibroblasts are noted to increase fibronectin deposition on these sites, which most likely binds to VLA4, a fibronectin receptor expressed on HPCs, and facilitates accumulation of these cells [246]. Furthermore, activated fibroblasts were shown to induce remodelling of stroma required for liver metastasis in a murine melanoma model [249]. Fibroblast-secreted TGF-β seems to increase tumour cell motility [250] and SDF-1 enhances tumour growth [251]. Thus, in addition to their contribution to the CSC niche at the primary tumour site, fibroblasts are suggested to have a critical role in premetastasis niche formation as well.

Mature monocyte and macrophage cells and neutrophils are also important in primary tumour and metastatic microenvironments [252, 253]. Tumours release multiple chemoattractants, such as CSF-1, which are thought to recruit Tumour-Associated Macrophages (TAMs) at the tumour microenvironment [254]. There, TAMs secrete pro-invasive and pro-angiogenic cytokines that promote tumour aggressiveness [255]. Moreover, all these cell populations secrete factors, chemokines and matrix-degrading enzymes that modulate the local microenvironment and mediate the chemoattraction of other inflammatory cells to the pre-metastatic niche. In addition, endothelial progenitor cells are mobilised from the bone marrow during angiogenesis [248]. It has been suggested that recruitment of endothelial progenitor cells instigates the micrometastatic to macrometastatic switch [246, 256].

Finally, mesenchymal stem cells (MSCs) give rise to tumour-associated fibroblasts (TAFs), and vascular pericytes [257–261]. They may also directly interact with tumour cells to enhance their metastatic phenotype [262]. In an in vivo model, in which a human breast cancer cell line was mixed with MSCs and subsequently implanted into a SCID mouse, both primary tumour size and lung metastases increased significantly. This was the effect of the tumour-induced MSC secretion of the chemokine CCL5 [263].

The preference of metastatic cells for specific organs may be also mediated by chemokines. The local expression of these chemoattractants might guide cognate chemokine receptor-expressing tumour cells to specific destinations, as a result of locally induced chemotaxis and invasion of tumour cells [226, 264]. For example, signalling through the chemokine receptors CXCR4 and CCR7, expressed by breast cancer cells, mediates actin polymerization and pseudopod formation that contributes to a chemotactic and invasive response [226]. The specificity of homing was achieved through expression of the cognate ligands, SDF-1 and CCl-21, respectively, in the sites of metastasis, but not other organs. These chemokine responses are important, as neutralizing antibodies to either SDF-1 or CXCR4 impaired breast cancer metastasis in an experimental metastasis model [226]. Of all the chemokine receptors, the most prevalently overexpressed in human tumours is CXCR4, which correlates with poor prognosis [265]. CXCR4 signalling is also essential for ErbB2-induced breast cancer metastasis [266].

Another aspect on metastatic cell homing is that tissues colonised by cancer cells may themselves become modified [267]. Initially the cells may react similar to damage or infection, giving rise to an inflammatory response. Indeed, malignancy has been described as a wound that does not heal [268] and many of the molecular processes used are similar. Sites of metastasis may become hypoxic, enhancing local invasion and angiogenesis via the upregulation of key genes such as VEGF, lysyl oxidase (LOX) cell membrane receptors, and particularly proteases which release local growth factors from sequestration in the matrix [269]. Interaction with existing stromal fibroblasts can enhance cancer cell survival.

These data demonstrate the significance of the microenvironment in homing of metastatic cells. There is an alternative view however, based upon intravital imaging, that proposes a passive role of the microenvironment in homing, at least for metastases through blood circulation [1]. This suggests that the predominant sites of metastases simply reflect the first pass of cells in the circulation and their entrapment in local capillaries. Thus, breast cancer cells spread predominantly to lung, lymph nodes and bone, whereas colon cancer cells travel to the liver through the hepatic–portal circulation. Deposition of cells in specific sites in this view is not due to active homing, but instead to the ability of a small portion of the many cells that become lodged in various tissues to survive, invade and grow in a particular environment (soil) either by chance or because they have appropriate mutations [241].

Concluding, successful metastatic outgrowth depends on the cumulative ability of cancer cells to adapt to distinct microenvironments at each step of the metastatic cascade, from the primary tumour to the final metastatic site. In combination with the role of the niche in the preservation of CSCs unique abilities, targeting the microenvironment seems to be a promising strategy for novel therapies.

25.11 Stem Cells and Implications in Therapy

Conventional anti-cancer therapy, including chemotherapy, radiotherapy and tumour-targeting agents, displays low efficacy and either fails to eliminate the tumour or is followed by relapse quickly after initial remission. Possible reasons for this failure include the inefficiency of the treatment, the inherent drug resistance of some tumour cells and/or the genetic instability of cancer cells. Treatment failure can also be related to the inefficiency of the treatment itself. A representative example is the case of adjuvant therapy, which is usually approved after testing the efficacy on patients with overt metastatic disease. It is now clear that not all therapies successful in the overt metastatic context are successful in the adjuvant (micrometastatic) context and vice-versa. For example, the anti-VEGF monoclonal antibody Bevacizumab, demonstrated improved overall survival for patients with metastatic colorectal cancer but failed to improve the disease-free survival period in the adjuvant colorectal cancer setting [270–272].

Current anti-cancer therapies treat tumours as if they were homogenous and all cancer cells respond in the same way [273]. Typically, the therapy initially shrinks the tumour and appears to be successful, but later on the tumour grows back. It has been suggested that a small population of residual cells that are intrinsically resistant to radiation and drugs persist after radiotherapy and chemotherapy respectively, ultimately resulting in disease relapse and the formation of metastasis. Continuously increasing evidence has shown that the therapy-resistant population consists of CSCs [10]. For example, CSC chemoresistance has been reported in human leukaemias [274–280], in malignant melanoma [281, 282] and in brain [283], breast [163, 284], pancreatic [54], and colorectal [285] cancers. Moreover, CSC radioresistance has been detected in brain [11] and breast [12, 286] cancers. Mechanisms of CSC resistance to therapy (Fig. 25.3) include their quiescent or slowly proliferating

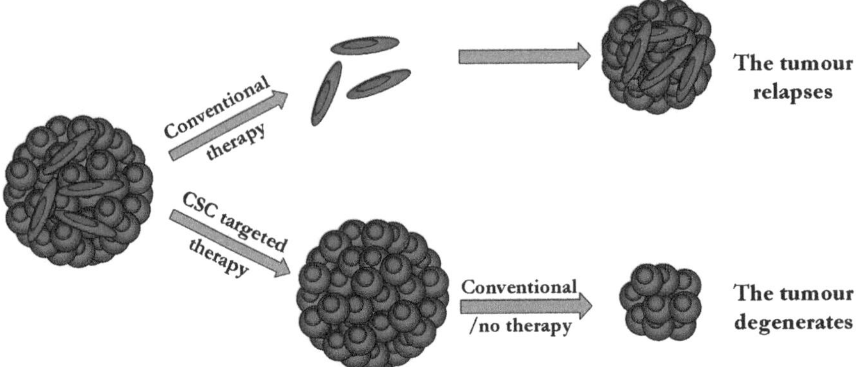

Fig. 25.3 Conventional anticancer therapies target predominantly bulk tumour cells (*red*), leading to tumour shrinkage. However, chemo- and radio- resistant CSCs (*blue*) survive and re-establish the tumour leading to relapse. On the contrary, a CSC-targeted approach will eliminate CSCs, without greatly reducing the size of the tumour. Subsequent administration of a conventional therapy will eradicate the tumour resulting in cure. Therefore, combination therapies that target both CSCs and bulk tumour populations seem to provide the means to increase disease-free survival of cancer patients

nature [287], expression of ATP-Binding Cassette (ABC) drug efflux pumps [275, 287], impaired function of apoptotic pathways [288], resistance to oxidative damage [277, 286] and increased recognition and repair of DNA damaged by the drug or ionizing radiation [11].

Prospective analysis of primary colorectal and breast cancers suggests that the frequency of stem cell populations that are resistant to chemotherapy is proportional to tumour grade and metastasis incidence [227, 289, 290]. This is consistent with the fact that gene signatures specific of CSCs can be detected in the bulk of primary breast tumours and correlate with poor prognosis [219, 291]. Moreover, even though the genetic aberrations that drive metastasis may be acquired by independent cell clones, the downstream pathways associated with these cells might still be conserved. Indeed, common gene signatures can be derived from normal embryonic and somatic stem cells and these correlate with poor differentiation and recurrence across different cancer subtypes [292, 293]. Although these studies are consistent with the notion that relapse is related to the quantity of cells with a particular molecular phenotype within a primary tumour, they cannot exclude the possibility that metastasis-founding cells with similar antigenic or transcriptional profiles can arise independently in different organs.

It is worth mentioning that targeting CSCs is not a panacea in treating cancer. Problems that have been encountered during treatment of bulk tumour cell populations, such as the development of drug resistance and the subsequent selection for cells acquiring resistance, remain unresolved. This seems plausible, given the plasticity with respect to genetic, epigenetic, or cellular properties of CSCs, as well as the effect of the microenvironment on the stem-cell state. Nevertheless, understanding the CSC biology may enable the design of new strategies, including combination therapies to counter drug resistance [278, 294].

Table 25.4 Agents, targeting self-renewal signalling or CSC surface markers, currently under clinical evaluation

Target	Agent (company)	Function?	Reference
Hedgehog pathway	GDC-0449 (Genetech/Curis)	Small-molecule SMO antagonist	[295–297]
	IPI-926 (Infinity Pharmaceuticals)	Cyclopamine- derived SMO antagonist	[295, 298, 299], Infinity Ph website (www.infi.com)
	PF-04449913 (Pfizer)	Small-molecule hedgehog inhibitor	[295]
	XL-139/ BMS-833923 (Exelixis/Bristol-Myers Squibb)	Small-molecule SMO antagonist	[295, 300]
	LDE225 (Novartis)	SMO-antagonist	[295]
Notch pathway	MK0752 (Merck)	γ-secretase inhibitor	[295]
	RO4929097 (NCI)	γ-secretase inhibitor	[295]
	MDX-1106 (Bristol-Myers Squibb)	Small-molecule Notch inhibitor	[295]
	PF-03084014 (Pfizer)	Small-molecule Notch inhibitor	[295]
EpCAM	Removab/Catumaxomab (Fresenius Biotech GmbH)	mAb against EpCAM and CD3	[295], Fresenius Biotech website (www.fresenius-biotech.com)
	MT110 (Micromet, Inc.)	mAb	[295], Micromet website (www.micromet.de)
TGF-β 1	Trabedersen/AP 12009 (Antisense Pharma)	Antisense oligonucleotide	[295], Antisense Pharma website (www.antisense-pharma.com)

A few therapeutic strategies targeting CSCs are currently validated and are expected to increase responsiveness to current anticancer treatment and hopefully reduce the risk of relapse and metastasis. These approaches include inhibition of CSC pathways, eradication by using antitumour agents targeting CSC markers, reversal of chemo- or radio-resistance mechanisms active in CSCs, differentiation therapy, disruption of pro-tumourigenic CSC-microenvironment interactions, anti-angiogenic or anti-vasculogenic therapy and disruption of immunoevasion pathways. Currently evaluated agents are summarised in Table 25.4.

25.11.1 Therapeutic Targets

25.11.1.1 Developmental Pathways

As already mentioned a number of developmental pathways are implicated in CSC self-renewal ability and therefore provide targets for therapy development. Therapeutic targeting via inhibiting agents of Wnt, Notch and Hedgehog pathways is currently being tested in the clinical setting. Both design and use of such agents require consideration regarding their possible side effects, especially on long term administration, on normal SSCs, where self-renewal signalling is crucial for tissue renewal. The challenge is therefore to define the therapeutic window in which these inhibitors exhibit efficiency as anti-cancer agents while minimizing the impact on normal tissues. Furthermore, Wnt, Hedgehog and Notch signalling interact with one another and with additional stimuli, such as growth factors, produced by the CSCs themselves or the surrounding bulk tumour or stroma cells. Integration of these signals produces the CSC traits (self-renewal, proliferation, apoptotic resistance, multi-potency). The various interactions among these pathways are not yet completely determined, therefore it is unclear whether inhibition of an individual pathway would differentiate CSCs, impair their proliferation or eradicate them.

A number of Wnt inhibitors have been discovered, but their clinical application still remains elusive [301]. Their active derivatives, however, can be designed to have desirable pharmacokinetic properties, in order to be developed into antitumour agents. Inhibition of the Hedgehog signalling pathway is also a viable therapeutic strategy and several Hedgehog inhibitors are currently evaluated under clinical trials [295]. Notch signalling is another potential target in order to eliminate CSCs. Notch inhibitors are under evaluation for various cancer types, such as breast and pancreatic cancer and ALL. For example, pharmacological blockade of the protease γ-secretase, which cleaves Notch, demonstrates remarkable antineoplastic effects in Notch-expressing transformed cells in vitro and in xenograft models [105, 302, 303] and depletes CSCs in brain tumours [304]. Nevertheless, γ-secretase inhibition affects multiple Notch pathways. As such, the therapeutic window is narrow due to possible effect on normal stem cells. In order to avoid toxicity related to γ-secretase, antagonist antibodies against specific Notch receptors are also being explored [305].

25.11.1.2 Differentiation Pathways

Another approach to eliminate CSCs would be to promote their differentiation and consequently attenuate their tumourigenicity and increase their susceptibility to conventional cytotoxic anticancer therapies [124, 306]. To this direction, BMPs and miRNAs may prove useful. For example, it has been demonstrated that BMPs induce the differentiation of $CD133^+$ glioblastoma SCs to astrocyte-like cells, which markedly reduced their tumourigenicity in a preclinical model [124]. In addition, in breast cancer, enforced expression of let-7 miRNA promoted the differentiation of $CD44^+CD24^{-/low}$ and reduced their ability to generate tumours when implanted in mice [163].

25.11.1.3 Proliferation Pathways

Other potential therapeutic targets include the gene products that are implicated in sustained growth, enhanced survival and invasion, such as telomerase reverse transcriptase (TERT), Cripto-1, tenacin C, NF-κB, PI3K/Akt/mTOR, IL-4/IL-4Rα, Bcl-2, survivin, snail, slug, twist and ALDH [307–314]. It has been recently reported that targeting these signalling elements could prevent tumour growth [282, 311, 315–317].

25.11.1.4 Reversal of Resistance Mechanisms Active in CSCs

So far, there is experimental evidence that inhibition of the mechanisms responsible for CSC resistance to chemo- and radio-therapy provides a potential CSC-directed therapeutic approach. Chemoresistance can be inversed by impeding multidrug resistance ABC transporters, as shown in human melanoma [281, 318]. Radioresistance, critical for curing solid tumours, could be also inversed by targeting DNA damage checkpoint response, as demonstrated for glioma SCs [11, 319], or by depleting ROS (reactive oxygen species) scavengers, as shown in breast CSCs [12, 286].

25.11.1.5 Cell Surface Markers

Cell surface markers may also provide a therapeutic target, through monoclonal antibody (mAb) binding and subsequent antibody-dependent cellular cytotoxicity (ADCC), mediated by immune cells or complement-dependent cytotoxicity [253, 282]. For instance, when an anti-ABCB5 mAb was administered, prior to xenotransplantation, to nude mouse recipients of human melanoma xenografts, it inhibited tumour formation and growth. Moreover, the same mAb reduced the growth of established tumours through CSC-specific ADCC [282]. In addition, agents directed on CSC surface markers could possibly result in CSC ablation through impairment of their interactions with the niche. For example, it has been reported that a mAb against CD44 eliminated AML-SCs by impeding their communication with the surrounding supportive microenvironment, which ultimately led to their differentiation [42].

While treating cancer remains a challenging field, the stem cell hypothesis provides an important framework for drug discovery. Combination treatment strategies have emerged from preclinical studies and hopefully even more will follow. Targeting all tumour cell subpopulations through combinations of drugs that target both bulk and metastasis-initiating cancer cells and their microenvironment may provide a more efficient way to increase relapse-free survival and yield long-term benefits for cancer patients.

References

1. Chambers AF, Groom AC, MacDonald IC (2002) Dissemination and growth of cancer cells in metastatic sites. Nat Rev Cancer 2(8):563–572
2. Bacac M, Stamenkovic I (2008) Metastatic cancer cell. Annu Rev Pathol 3:221–247

3. Acloque H, Adams MS, Fishwick K, Bronner-Fraser M, Nieto MA (2009) Epithelial-mesenchymal transitions: the importance of changing cell state in development and disease. J Clin Invest 119(6):1438–1449. doi:10.1172/JCI38019
4. Thiery JP, Acloque H, Huang RY, Nieto MA (2009) Epithelial-mesenchymal transitions in development and disease. Cell 139(5):871–890. doi:10.1016/j.cell.2009.11.007
5. Koestenbauer S, Zech NH, Juch H, Vanderzwalmen P, Schoonjans L, Dohr G (2006) Embryonic stem cells: similarities and differences between human and murine embryonic stem cells. Am J Reprod Immunol 55(3):169–180
6. Masip M, Veiga A, Izpisua Belmonte JC, Simon C (2010) Reprogramming with defined factors: from induced pluripotency to induced transdifferentiation. Mol Hum Reprod 16(11):856–868
7. Morrison SJ, Kimble J (2006) Asymmetric and symmetric stem-cell divisions in development and cancer. Nature 441(7097):1068–1074
8. Moore KA, Lemischka IR (2006) Stem cells and their niches. Science 311(5769):1880–1885
9. Bomken S, Fiser K, Heidenreich O, Vormoor J (2010) Understanding the cancer stem cell. Br J Cancer 103(4):439–445
10. Dean M, Fojo T, Bates S (2005) Tumour stem cells and drug resistance. Nat Rev Cancer 5(4):275–284
11. Bao S, Wu Q, McLendon RE, Hao Y, Shi Q, Hjelmeland AB, Dewhirst MW, Bigner DD, Rich JN (2006) Glioma stem cells promote radioresistance by preferential activation of the DNA damage response. Nature 444(7120):756–760
12. Phillips TM, McBride WH, Pajonk F (2006) The response of CD24(-/low)/CD44+ breast cancer-initiating cells to radiation. J Natl Cancer Inst 98(24):1777–1785
13. Greaves M (2010) Cancer stem cells: back to Darwin? Semin Cancer Biol 20(2):65–70
14. Anderson K, Lutz C, van Delft FW, Bateman CM, Guo Y, Colman SM, Kempski H, Moorman AV, Titley I, Swansbury J, Kearney L, Enver T, Greaves M (2011) Genetic variegation of clonal architecture and propagating cells in leukaemia. Nature 469(7330):356–361
15. Wicha MS, Liu S, Dontu G (2006) Cancer stem cells: an old idea–a paradigm shift. Cancer Res 66(4):1883–1890, discussion 1895-1886
16. Durante F (1874) Nesso fisio-pathologico tra la struttura dei nei materni e la genesi di alcuni tumori maligni. Arch Memor Observ Chir Pract 11:217–226
17. Cohnheim J (1867) Ueber entzundung und eiterung. Path Anat Physiol Klin Med 40:1–79
18. Cohnheim J (1875) Congenitales, quergestreiftes Muskelsarkon der Nireren. Virchows Arch 65:64
19. Sell S (2004) Stem cell origin of cancer and differentiation therapy. Crit Rev Oncol Hematol 51(1):1–28
20. Bonnet D, Dick JE (1997) Human acute myeloid leukemia is organized as a hierarchy that originates from a primitive hematopoietic cell. Nat Med 3(7):730–737
21. Al-Hajj M, Wicha MS, Benito-Hernandez A, Morrison SJ, Clarke MF (2003) Prospective identification of tumorigenic breast cancer cells. Proc Natl Acad Sci USA 100(7):3983–3988
22. Hemmati HD, Nakano I, Lazareff JA, Masterman-Smith M, Geschwind DH, Bronner-Fraser M, Kornblum HI (2003) Cancerous stem cells can arise from pediatric brain tumors. Proc Natl Acad Sci USA 100(25):15178–15183
23. Singh SK, Clarke ID, Terasaki M, Bonn VE, Hawkins C, Squire J, Dirks PB (2003) Identification of a cancer stem cell in human brain tumors. Cancer Res 63(18):5821–5828
24. O'Brien CA, Pollett A, Gallinger S, Dick JE (2007) A human colon cancer cell capable of initiating tumour growth in immunodeficient mice. Nature 445(7123):106–110
25. Bapat SA, Mali AM, Koppikar CB, Kurrey NK (2005) Stem and progenitor-like cells contribute to the aggressive behavior of human epithelial ovarian cancer. Cancer Res 65(8): 3025–3029
26. Collins AT, Berry PA, Hyde C, Stower MJ, Maitland NJ (2005) Prospective identification of tumorigenic prostate cancer stem cells. Cancer Res 65(23):10946–10951
27. Ho MM, Ng AV, Lam S, Hung JY (2007) Side population in human lung cancer cell lines and tumors is enriched with stem-like cancer cells. Cancer Res 67(10):4827–4833

28. Eramo A, Lotti F, Sette G, Pilozzi E, Biffoni M, Di Virgilio A, Conticello C, Ruco L, Peschle C, De Maria R (2008) Identification and expansion of the tumorigenic lung cancer stem cell population. Cell Death Differ 15(3):504–514
29. Fukuda K, Saikawa Y, Ohashi M, Kumagai K, Kitajima M, Okano H, Matsuzaki Y, Kitagawa Y (2009) Tumor initiating potential of side population cells in human gastric cancer. Int J Oncol 34(5):1201–1207
30. Takaishi S, Okumura T, Tu S, Wang SS, Shibata W, Vigneshwaran R, Gordon SA, Shimada Y, Wang TC (2009) Identification of gastric cancer stem cells using the cell surface marker CD44. Stem Cells 27(5):1006–1020
31. McDermott SP, Wicha MS (2010) Targeting breast cancer stem cells. Mol Oncol 4(5):404–419
32. Visvader JE, Lindeman GJ (2008) Cancer stem cells in solid tumours: accumulating evidence and unresolved questions. Nat Rev Cancer 8(10):755–768
33. Charafe-Jauffret E, Ginestier C, Birnbaum D (2009) Breast cancer stem cells: tools and models to rely on. BMC Cancer 9:202
34. Rahman M, Deleyrolle L, Vedam-Mai V, Azari H, Abd-El-Barr M, Reynolds BA (2011) The cancer stem cell hypothesis: failures and pitfalls. Neurosurgery 68(2):531–545, discussion 545
35. Reynolds BA, Weiss S (1996) Clonal and population analyses demonstrate that an EGF-responsive mammalian embryonic CNS precursor is a stem cell. Dev Biol 175(1):1–13
36. Weiss S, Reynolds BA, Vescovi AL, Morshead C, Craig CG, van der Kooy D (1996) Is there a neural stem cell in the mammalian forebrain? Trends Neurosci 19(9):387–393
37. Dontu G, Abdallah WM, Foley JM, Jackson KW, Clarke MF, Kawamura MJ, Wicha MS (2003) In vitro propagation and transcriptional profiling of human mammary stem/progenitor cells. Genes Dev 17(10):1253–1270
38. Burness ML, Sipkins DA (2010) The stem cell niche in health and malignancy. Semin Cancer Biol 20(2):107–115
39. Singh SK, Hawkins C, Clarke ID, Squire JA, Bayani J, Hide T, Henkelman RM, Cusimano MD, Dirks PB (2004) Identification of human brain tumour initiating cells. Nature 432(7015):396–401
40. Ginestier C, Hur MH, Charafe-Jauffret E, Monville F, Dutcher J, Brown M, Jacquemier J, Viens P, Kleer CG, Liu S, Schott A, Hayes D, Birnbaum D, Wicha MS, Dontu G (2007) ALDH1 is a marker of normal and malignant human mammary stem cells and a predictor of poor clinical outcome. Cell Stem Cell 1(5):555–567
41. Beier D, Hau P, Proescholdt M, Lohmeier A, Wischhusen J, Oefner PJ, Aigner L, Brawanski A, Bogdahn U, Beier CP (2007) CD133(+) and CD133(-) glioblastoma-derived cancer stem cells show differential growth characteristics and molecular profiles. Cancer Res 67(9):4010–4015
42. Jin L, Hope KJ, Zhai Q, Smadja-Joffe F, Dick JE (2006) Targeting of CD44 eradicates human acute myeloid leukemic stem cells. Nat Med 12(10):1167–1174
43. Du X, Ho M, Pastan I (2007) New immunotoxins targeting CD123, a stem cell antigen on acute myeloid leukemia cells. J Immunother 30(6):607–613
44. Florian S, Sonneck K, Hauswirth AW, Krauth MT, Schernthaner GH, Sperr WR, Valent P (2006) Detection of molecular targets on the surface of CD34+/CD38– stem cells in various myeloid malignancies. Leuk Lymphoma 47(2):207–222
45. Son MJ, Woolard K, Nam DH, Lee J, Fine HA (2009) SSEA-1 is an enrichment marker for tumor-initiating cells in human glioblastoma. Cell Stem Cell 4(5):440–452
46. Read TA, Fogarty MP, Markant SL, McLendon RE, Wei Z, Ellison DW, Febbo PG, Wechsler-Reya RJ (2009) Identification of CD15 as a marker for tumor-propagating cells in a mouse model of medulloblastoma. Cancer Cell 15(2):135–147
47. Hurt EM, Kawasaki BT, Klarmann GJ, Thomas SB, Farrar WL (2008) CD44+ CD24(-) prostate cells are early cancer progenitor/stem cells that provide a model for patients with poor prognosis. Br J Cancer 98(4):756–765
48. Baba T, Convery PA, Matsumura N, Whitaker RS, Kondoh E, Perry T, Huang Z, Bentley RC, Mori S, Fujii S, Marks JR, Berchuck A, Murphy SK (2009) Epigenetic regulation of CD133 and tumorigenicity of CD133+ ovarian cancer cells. Oncogene 28(2):209–218

49. Silasi DA, Alvero AB, Illuzzi J, Kelly M, Chen R, Fu HH, Schwartz P, Rutherford T, Azodi M, Mor G (2006) MyD88 predicts chemoresistance to paclitaxel in epithelial ovarian cancer. Yale J Biol Med 79(3–4):153–163
50. Alvero AB, Chen R, Fu HH, Montagna M, Schwartz PE, Rutherford T, Silasi DA, Steffensen KD, Waldstrom M, Visintin I, Mor G (2009) Molecular phenotyping of human ovarian cancer stem cells unravels the mechanisms for repair and chemoresistance. Cell Cycle 8(1):158–166
51. Ricci-Vitiani L, Lombardi DG, Pilozzi E, Biffoni M, Todaro M, Peschle C, De Maria R (2007) Identification and expansion of human colon-cancer-initiating cells. Nature 445(7123):111–115
52. Du L, Wang H, He L, Zhang J, Ni B, Wang X, Jin H, Cahuzac N, Mehrpour M, Lu Y, Chen Q (2008) CD44 is of functional importance for colorectal cancer stem cells. Clin Cancer Res 14(21):6751–6760
53. Dalerba P, Dylla SJ, Park IK, Liu R, Wang X, Cho RW, Hoey T, Gurney A, Huang EH, Simeone DM, Shelton AA, Parmiani G, Castelli C, Clarke MF (2007) Phenotypic characterization of human colorectal cancer stem cells. Proc Natl Acad Sci USA 104(24): 10158–10163
54. Hermann PC, Huber SL, Herrler T, Aicher A, Ellwart JW, Guba M, Bruns CJ, Heeschen C (2007) Distinct populations of cancer stem cells determine tumor growth and metastatic activity in human pancreatic cancer. Cell Stem Cell 1(3):313–323
55. Huang P, Wang CY, Gou SM, Wu HS, Liu T, Xiong JX (2008) Isolation and biological analysis of tumor stem cells from pancreatic adenocarcinoma. World J Gastroenterol 14(24):3903–3907
56. Prince ME, Sivanandan R, Kaczorowski A, Wolf GT, Kaplan MJ, Dalerba P, Weissman IL, Clarke MF, Ailles LE (2007) Identification of a subpopulation of cells with cancer stem cell properties in head and neck squamous cell carcinoma. Proc Natl Acad Sci USA 104(3):973–978
57. Gibbs CP, Kukekov VG, Reith JD, Tchigrinova O, Suslov ON, Scott EW, Ghivizzani SC, Ignatova TN, Steindler DA (2005) Stem-like cells in bone sarcomas: implications for tumorigenesis. Neoplasia 7(11):967–976
58. Fang D, Nguyen TK, Leishear K, Finko R, Kulp AN, Hotz S, Van Belle PA, Xu X, Elder DE, Herlyn M (2005) A tumorigenic subpopulation with stem cell properties in melanomas. Cancer Res 65(20):9328–9337
59. Monzani E, Facchetti F, Galmozzi E, Corsini E, Benetti A, Cavazzin C, Gritti A, Piccinini A, Porro D, Santinami M, Invernici G, Parati E, Alessandri G, La Porta CA (2007) Melanoma contains CD133 and ABCG2 positive cells with enhanced tumourigenic potential. Eur J Cancer 43(5):935–946
60. Chiba T, Zheng YW, Kita K, Yokosuka O, Saisho H, Onodera M, Miyoshi H, Nakano M, Zen Y, Nakanuma Y, Nakauchi H, Iwama A, Taniguchi H (2007) Enhanced self-renewal capability in hepatic stem/progenitor cells drives cancer initiation. Gastroenterology 133(3):937–950
61. Suetsugu A, Nagaki M, Aoki H, Motohashi T, Kunisada T, Moriwaki H (2006) Characterization of CD133+ hepatocellular carcinoma cells as cancer stem/progenitor cells. Biochem Biophys Res Commun 351(4):820–824
62. Yang ZF, Ho DW, Ng MN, Lau CK, Yu WC, Ngai P, Chu PW, Lam CT, Poon RT, Fan ST (2008) Significance of CD90+ cancer stem cells in human liver cancer. Cancer Cell 13(2):153–166
63. Kalluri R, Neilson EG (2003) Epithelial-mesenchymal transition and its implications for fibrosis. J Clin Invest 112(12):1776–1784
64. Kalluri R, Weinberg RA (2009) The basics of epithelial-mesenchymal transition. J Clin Invest 119(6):1420–1428
65. Yang J, Weinberg RA (2008) Epithelial-mesenchymal transition: at the crossroads of development and tumor metastasis. Dev Cell 14(6):818–829. doi:10.1016/j.devcel.2008.05.009
66. Chaffer CL, Weinberg RA (2011) A perspective on cancer cell metastasis. Science 331(6024): 1559–1564

67. Scheel C, Eaton EN, Li SH, Chaffer CL, Reinhardt F, Kah KJ, Bell G, Guo W, Rubin J, Richardson AL, Weinberg RA (2011) Paracrine and autocrine signals induce and maintain mesenchymal and stem cell States in the breast. Cell 145(6):926–940. doi:10.1016/j.cell.2011.04.029
68. Mani SA, Guo W, Liao MJ, Eaton EN, Ayyanan A, Zhou AY, Brooks M, Reinhard F, Zhang CC, Shipitsin M, Campbell LL, Polyak K, Brisken C, Yang J, Weinberg RA (2008) The epithelial-mesenchymal transition generates cells with properties of stem cells. Cell 133(4):704–715. doi:10.1016/j.cell.2008.03.027
69. Kang Y, Massague J (2004) Epithelial-mesenchymal transitions: twist in development and metastasis. Cell 118(3):277–279
70. Morel AP, Lievre M, Thomas C, Hinkal G, Ansieau S, Puisieux A (2008) Generation of breast cancer stem cells through epithelial-mesenchymal transition. PLoS One 3(8):e2888. doi:10.1371/journal.pone.0002888
71. Lim E, Vaillant F, Wu D, Forrest NC, Pal B, Hart AH, Asselin-Labat ML, Gyorki DE, Ward T, Partanen A, Feleppa F, Huschtscha LI, Thorne HJ, Fox SB, Yan M, French JD, Brown MA, Smyth GK, Visvader JE, Lindeman GJ (2009) Aberrant luminal progenitors as the candidate target population for basal tumor development in BRCA1 mutation carriers. Nat Med 15(8):907–913
72. Prat A, Parker JS, Karginova O, Fan C, Livasy C, Herschkowitz JI, He X, Perou CM (2010) Phenotypic and molecular characterization of the claudin-low intrinsic subtype of breast cancer. Breast Cancer Res 12(5):R68
73. Thiery JP (2002) Epithelial-mesenchymal transitions in tumour progression. Nat Rev Cancer 2(6):442–454
74. Gal A, Sjoblom T, Fedorova L, Imreh S, Beug H, Moustakas A (2008) Sustained TGF beta exposure suppresses Smad and non-Smad signalling in mammary epithelial cells, leading to EMT and inhibition of growth arrest and apoptosis. Oncogene 27(9):1218–1230. doi:10.1038/sj.onc.1210741
75. Raimondi C, Gradilone A, Naso G, Vincenzi B, Petracca A, Nicolazzo C, Palazzo A, Saltarelli R, Spremberg F, Cortesi E, Gazzaniga P (2011) Epithelial-mesenchymal transition and stemness features in circulating tumor cells from breast cancer patients. Breast Cancer Res Treat 130(2):449–455
76. Mego M, Mani SA, Lee BN, Li C, Evans KW, Cohen EN, Gao H, Jackson SA, Giordano A, Hortobagyi GN, Cristofanilli M, Lucci A, Reuben JM (2011) Expression of epithelial-mesenchymal transition-inducing transcription factors in primary breast cancer: the effect of neoadjuvant therapy. Int J Cancer doi: 10.1002/ijc.26037
77. Mego M, Mani SA, Li C, Andreoupolou E, Tin S, Jackson S, Cohen EN, Gao H, Cristofanilli M, JMR Circulating tumor cells (CTCs) and epithelial mesenchymal transition (EMT) in breast cancer: describing the heterogeneity of microscopic disease. In: CTRC-AACR San Antonio Breast Cancer Symposium, San Antonio, Texas, USA, 2009. Cancer Res 69(24 Suppl) Abstract nr 3011
78. Theodoropoulos PA, Polioudaki H, Agelaki S, Kallergi G, Saridaki Z, Mavroudis D, Georgoulias V (2010) Circulating tumor cells with a putative stem cell phenotype in peripheral blood of patients with breast cancer. Cancer Lett 288(1):99–106
79. Kallergi G, Papadaki MA, Politaki E, Mavroudis D, Georgoulias V, Agelaki S (2011) Epithelial-mesenchymal transition markers expressed in circulating tumor cells of early and metastatic breast cancer patients. Breast Cancer Res 13(3):R59. doi:10.1186/bcr2896
80. Willipinski-Stapelfeldt B, Riethdorf S, Assmann V, Woelfle U, Rau T, Sauter G, Heukeshoven J, Pantel K (2005) Changes in cytoskeletal protein composition indicative of an epithelial-mesenchymal transition in human micrometastatic and primary breast carcinoma cells. Clin Cancer Res 11(22):8006–8014
81. Creighton CJ, Li X, Landis M, Dixon JM, Neumeister VM, Sjolund A, Rimm DL, Wong H, Rodriguez A, Herschkowitz JI, Fan C, Zhang X, He X, Pavlick A, Gutierrez MC, Renshaw L, Larionov AA, Faratian D, Hilsenbeck SG, Perou CM, Lewis MT, Rosen JM, Chang JC (2009) Residual breast cancers after conventional therapy display mesenchymal as well as tumor-initiating features. Proc Natl Acad Sci USA 106(33):13820–13825

82. Barrallo-Gimeno A, Nieto MA (2005) The Snail genes as inducers of cell movement and survival: implications in development and cancer. Development 132(14):3151–3161
83. Moody SE, Perez D, Pan TC, Sarkisian CJ, Portocarrero CP, Sterner CJ, Notorfrancesco KL, Cardiff RD, Chodosh LA (2005) The transcriptional repressor Snail promotes mammary tumor recurrence. Cancer Cell 8(3):197–209
84. Hugo H, Ackland ML, Blick T, Lawrence MG, Clements JA, Williams ED, Thompson EW (2007) Epithelial–mesenchymal and mesenchymal–epithelial transitions in carcinoma progression. J Cell Physiol 213(2):374–383
85. Thiery JP, Sleeman JP (2006) Complex networks orchestrate epithelial-mesenchymal transitions. Nat Rev Mol Cell Biol 7(2):131–142
86. Ackland ML, Newgreen DF, Fridman M, Waltham MC, Arvanitis A, Minichiello J, Price JT, Thompson EW (2003) Epidermal growth factor-induced epithelio-mesenchymal transition in human breast carcinoma cells. Lab Invest 83(3):435–448
87. Christiansen JJ, Rajasekaran AK (2006) Reassessing epithelial to mesenchymal transition as a prerequisite for carcinoma invasion and metastasis. Cancer Res 66(17):8319–8326
88. Zavadil J, Haley J, Kalluri R, Muthuswamy SK, Thompson E (2008) Epithelial-mesenchymal transition. Cancer Res 68(23):9574–9577
89. Lee JM, Dedhar S, Kalluri R, Thompson EW (2006) The epithelial-mesenchymal transition: new insights in signaling, development, and disease. J Cell Biol 172(7):973–981
90. Klymkowsky MW, Savagner P (2009) Epithelial-mesenchymal transition: a cancer researcher's conceptual friend and foe. Am J Pathol 174(5):1588–1593
91. Soon L, Tachtsidis A, Fok S, Williams ED, Newgreen DF, Thomson EW (2011) The continuum of epithelial mesenchymal transition – implication of hybrid states for migration and survival in development and cancer. In: Lyden D, Welch DR, Psaila B (eds) Cancer metastasis: biologic basis and therapeutics. Cambridge University Press, New York
92. O'Brien CA, Kreso A, Jamieson CH (2010) Cancer stem cells and self-renewal. Clin Cancer Res 16(12):3113–3120
93. Yilmaz OH, Valdez R, Theisen BK, Guo W, Ferguson DO, Wu H, Morrison SJ (2006) Pten dependence distinguishes haematopoietic stem cells from leukaemia-initiating cells. Nature 441(7092):475–482
94. Lessard J, Sauvageau G (2003) Bmi-1 determines the proliferative capacity of normal and leukaemic stem cells. Nature 423(6937):255–260
95. Dreesen O, Brivanlou AH (2007) Signaling pathways in cancer and embryonic stem cells. Stem Cell Rev 3(1):7–17
96. Xia TS, Wang GZ, Ding Q, Liu XA, Zhou WB, Zhang YF, Zha XM, Du Q, Ni XJ, Wang J, Miao SY, Wang S (2011) Bone metastasis in a novel breast cancer mouse model containing human breast and human bone. Breast Cancer Res Treat doi: 10.1007/s10549-011-1496-0
97. Tetsu O, McCormick F (1999) Beta-catenin regulates expression of cyclin D1 in colon carcinoma cells. Nature 398(6726):422–426
98. Zhao C, Blum J, Chen A, Kwon HY, Jung SH, Cook JM, Lagoo A, Reya T (2007) Loss of beta-catenin impairs the renewal of normal and CML stem cells in vivo. Cancer Cell 12(6):528–541
99. Malanchi I, Peinado H, Kassen D, Hussenet T, Metzger D, Chambon P, Huber M, Hohl D, Cano A, Birchmeier W, Huelsken J (2008) Cutaneous cancer stem cell maintenance is dependent on beta-catenin signalling. Nature 452(7187):650–653
100. Wang Y, Krivtsov AV, Sinha AU, North TE, Goessling W, Feng Z, Zon LI, Armstrong SA (2010) The Wnt/beta-catenin pathway is required for the development of leukemia stem cells in AML. Science 327(5973):1650–1653
101. Liu S, Dontu G, Mantle ID, Patel S, Ahn NS, Jackson KW, Suri P, Wicha MS (2006) Hedgehog signaling and Bmi-1 regulate self-renewal of normal and malignant human mammary stem cells. Cancer Res 66(12):6063–6071
102. Clarke RB, Anderson E, Howell A, Potten CS (2003) Regulation of human breast epithelial stem cells. Cell Prolif 36(Suppl 1):45–58

103. Clarke RB, Spence K, Anderson E, Howell A, Okano H, Potten CS (2005) A putative human breast stem cell population is enriched for steroid receptor-positive cells. Dev Biol 277(2):443–456
104. Dontu G, Jackson KW, McNicholas E, Kawamura MJ, Abdallah WM, Wicha MS (2004) Role of Notch signaling in cell-fate determination of human mammary stem/progenitor cells. Breast Cancer Res 6(6):R605–R615
105. Weng AP, Ferrando AA, Lee W, Morris JP, Silverman LB, Sanchez-Irizarry C, Blacklow SC, Look AT, Aster JC (2004) Activating mutations of NOTCH1 in human T cell acute lymphoblastic leukemia. Science 306(5694):269–271
106. Harrison H, Farnie G, Howell SJ, Rock RE, Stylianou S, Brennan KR, Bundred NJ, Clarke RB (2010) Regulation of breast cancer stem cell activity by signaling through the Notch4 receptor. Cancer Res 70(2):709–718
107. Rubin LL, de Sauvage FJ (2006) Targeting the Hedgehog pathway in cancer. Nat Rev Drug Discov 5(12):1026–1033
108. Theunissen JW, de Sauvage FJ (2009) Paracrine Hedgehog signaling in cancer. Cancer Res 69(15):6007–6010
109. Jiang J, Hui CC (2008) Hedgehog signaling in development and cancer. Dev Cell 15(6):801–812
110. Dierks C, Beigi R, Guo GR, Zirlik K, Stegert MR, Manley P, Trussell C, Schmitt-Graeff A, Landwerlin K, Veelken H, Warmuth M (2008) Expansion of Bcr-Abl-positive leukemic stem cells is dependent on Hedgehog pathway activation. Cancer Cell 14(3):238–249
111. Meulmeester E, Ten Dijke P (2011) The dynamic roles of TGF-beta in cancer. J Pathol 223(2):205–218
112. Hata A, Davis BN (2009) Control of microRNA biogenesis by TGFbeta signaling pathway-A novel role of Smads in the nucleus. Cytokine Growth Factor Rev 20(5–6):517–521
113. Hutchison N, Hendry BM, Sharpe CC (2009) Rho isoforms have distinct and specific functions in the process of epithelial to mesenchymal transition in renal proximal tubular cells. Cell Signal 21(10):1522–1531
114. Bhowmick NA, Ghiassi M, Bakin A, Aakre M, Lundquist CA, Engel ME, Arteaga CL, Moses HL (2001) Transforming growth factor-beta1 mediates epithelial to mesenchymal transdifferentiation through a RhoA-dependent mechanism. Mol Biol Cell 12(1):27–36
115. Parri M, Chiarugi P (2010) Rac and Rho GTPases in cancer cell motility control. Cell Commun Signal 8:23
116. Wikstrom P, Stattin P, Franck-Lissbrant I, Damber JE, Bergh A (1998) Transforming growth factor beta1 is associated with angiogenesis, metastasis, and poor clinical outcome in prostate cancer. Prostate 37(1):19–29
117. Picon A, Gold LI, Wang J, Cohen A, Friedman E (1998) A subset of metastatic human colon cancers expresses elevated levels of transforming growth factor beta1. Cancer Epidemiol Biomarkers Prev 7(6):497–504
118. Lee MK, Pardoux C, Hall MC, Lee PS, Warburton D, Qing J, Smith SM, Derynck R (2007) TGF-beta activates Erk MAP kinase signalling through direct phosphorylation of ShcA. EMBO J 26(17):3957–3967
119. Sato N, Sanjuan IM, Heke M, Uchida M, Naef F, Brivanlou AH (2003) Molecular signature of human embryonic stem cells and its comparison with the mouse. Dev Biol 260(2): 404–413
120. James D, Levine AJ, Besser D, Hemmati-Brivanlou A (2005) TGFbeta/activin/nodal signaling is necessary for the maintenance of pluripotency in human embryonic stem cells. Development 132(6):1273–1282
121. Xu RH, Chen X, Li DS, Li R, Addicks GC, Glennon C, Zwaka TP, Thomson JA (2002) BMP4 initiates human embryonic stem cell differentiation to trophoblast. Nat Biotechnol 20(12):1261–1264
122. Besser D (2004) Expression of nodal, lefty-a, and lefty-B in undifferentiated human embryonic stem cells requires activation of Smad2/3. J Biol Chem 279(43):45076–45084

123. Ikushima H, Todo T, Ino Y, Takahashi M, Miyazawa K, Miyazono K (2009) Autocrine TGF-beta signaling maintains tumorigenicity of glioma-initiating cells through Sry-related HMG-box factors. Cell Stem Cell 5(5):504–514
124. Piccirillo SG, Reynolds BA, Zanetti N, Lamorte G, Binda E, Broggi G, Brem H, Olivi A, Dimeco F, Vescovi AL (2006) Bone morphogenetic proteins inhibit the tumorigenic potential of human brain tumour-initiating cells. Nature 444(7120):761–765
125. Hansel DE, Kern SE, Hruban RH (2003) Molecular pathogenesis of pancreatic cancer. Annu Rev Genomics Hum Genet 4:237–256
126. Bardeesy N, Cheng KH, Berger JH, Chu GC, Pahler J, Olson P, Hezel AF, Horner J, Lauwers GY, Hanahan D, DePinho RA (2006) Smad4 is dispensable for normal pancreas development yet critical in progression and tumor biology of pancreas cancer. Genes Dev 20(22):3130–3146
127. Langenfeld EM, Calvano SE, Abou-Nukta F, Lowry SF, Amenta P, Langenfeld J (2003) The mature bone morphogenetic protein-2 is aberrantly expressed in non-small cell lung carcinomas and stimulates tumor growth of A549 cells. Carcinogenesis 24(9):1445–1454
128. Langenfeld EM, Kong Y, Langenfeld J (2006) Bone morphogenetic protein 2 stimulation of tumor growth involves the activation of Smad-1/5. Oncogene 25(5):685–692
129. Hennessy BT, Smith DL, Ram PT, Lu Y, Mills GB (2005) Exploiting the PI3K/AKT pathway for cancer drug discovery. Nat Rev Drug Discov 4(12):988–1004
130. Guo W, Lasky JL, Chang CJ, Mosessian S, Lewis X, Xiao Y, Yeh JE, Chen JY, Iruela-Arispe ML, Varella-Garcia M, Wu H (2008) Multi-genetic events collaboratively contribute to Pten-null leukaemia stem-cell formation. Nature 453(7194):529–533
131. Hill R, Wu H (2009) PTEN, stem cells, and cancer stem cells. J Biol Chem 284(18): 11755–11759
132. Wang S, Garcia AJ, Wu M, Lawson DA, Witte ON, Wu H (2006) Pten deletion leads to the expansion of a prostatic stem/progenitor cell subpopulation and tumor initiation. Proc Natl Acad Sci USA 103(5):1480–1485
133. Zhang J, Grindley JC, Yin T, Jayasinghe S, He XC, Ross JT, Haug JS, Rupp D, Porter-Westpfahl KS, Wiedemann LM, Wu H, Li L (2006) PTEN maintains haematopoietic stem cells and acts in lineage choice and leukaemia prevention. Nature 441(7092):518–522
134. Armstrong L, Hughes O, Yung S, Hyslop L, Stewart R, Wappler I, Peters H, Walter T, Stojkovic P, Evans J, Stojkovic M, Lako M (2006) The role of PI3K/AKT, MAPK/ERK and NFkappabeta signalling in the maintenance of human embryonic stem cell pluripotency and viability highlighted by transcriptional profiling and functional analysis. Hum Mol Genet 15(11):1894–1913
135. Korkaya H, Paulson A, Charafe-Jauffret E, Ginestier C, Brown M, Dutcher J, Clouthier SG, Wicha MS (2009) Regulation of mammary stem/progenitor cells by PTEN/Akt/beta-catenin signaling. PLoS Biol 7(6):e1000121
136. Katz M, Amit I, Yarden Y (2007) Regulation of MAPKs by growth factors and receptor tyrosine kinases. Biochim Biophys Acta 1773(8):1161–1176
137. Sebolt-Leopold JS, Herrera R (2004) Targeting the mitogen-activated protein kinase cascade to treat cancer. Nat Rev Cancer 4(12):937–947
138. Howe LR, Howe LR (2007) Inflammation and breast cancer. Cyclooxygenase/prostaglandin signaling and breast cancer. Breast Cancer Res 9(4):210
139. Moasser MM (2007) The oncogene HER2: its signaling and transforming functions and its role in human cancer pathogenesis. Oncogene 26(45):6469–6487
140. Iwama A, Oguro H, Negishi M, Kato Y, Morita Y, Tsukui H, Ema H, Kamijo T, Katoh-Fukui Y, Koseki H, van Lohuizen M, Nakauchi H (2004) Enhanced self-renewal of hematopoietic stem cells mediated by the polycomb gene product Bmi-1. Immunity 21(6):843–851
141. Molofsky AV, Pardal R, Iwashita T, Park IK, Clarke MF, Morrison SJ (2003) Bmi-1 dependence distinguishes neural stem cell self-renewal from progenitor proliferation. Nature 425(6961):962–967

142. Park IK, Qian D, Kiel M, Becker MW, Pihalja M, Weissman IL, Morrison SJ, Clarke MF (2003) Bmi-1 is required for maintenance of adult self-renewing haematopoietic stem cells. Nature 423(6937):302–305
143. Raaphorst FM (2003) Self-renewal of hematopoietic and leukemic stem cells: a central role for the Polycomb-group gene Bmi-1. Trends Immunol 24(10):522–524
144. Dick JE (2003) Stem cells: self-renewal writ in blood. Nature 423(6937):231–233
145. Cao R, Tsukada Y, Zhang Y (2005) Role of Bmi-1 and Ring1A in H2A ubiquitylation and Hox gene silencing. Mol Cell 20(6):845–854
146. Park IK, Morrison SJ, Clarke MF (2004) Bmi1, stem cells, and senescence regulation. J Clin Invest 113(2):175–179
147. Hollstein M, Sidransky D, Vogelstein B, Harris CC (1991) p53 mutations in human cancers. Science 253(5015):49–53
148. Cicalese A, Bonizzi G, Pasi CE, Faretta M, Ronzoni S, Giulini B, Brisken C, Minucci S, Di Fiore PP, Pelicci PG (2009) The tumor suppressor p53 regulates polarity of self-renewing divisions in mammary stem cells. Cell 138(6):1083–1095
149. Bartel DP (2009) MicroRNAs: target recognition and regulatory functions. Cell 136(2): 215–233
150. Carleton M, Cleary MA, Linsley PS (2007) MicroRNAs and cell cycle regulation. Cell Cycle 6(17):2127–2132
151. Harfe BD (2005) MicroRNAs in vertebrate development. Curr Opin Genet Dev 15(4):410–415
152. Boehm M, Slack F (2005) A developmental timing microRNA and its target regulate life span in C. elegans. Science 310(5756):1954–1957
153. Boehm M, Slack FJ (2006) MicroRNA control of lifespan and metabolism. Cell Cycle 5(8):837–840
154. Garzon R, Calin GA, Croce CM (2009) MicroRNAs in cancer. Annu Rev Med 60:167–179
155. Murchison EP, Partridge JF, Tam OH, Cheloufi S, Hannon GJ (2005) Characterization of Dicer-deficient murine embryonic stem cells. Proc Natl Acad Sci USA 102(34):12135–12140
156. Wang Y, Baskerville S, Shenoy A, Babiarz JE, Baehner L, Blelloch R (2008) Embryonic stem cell-specific microRNAs regulate the G1-S transition and promote rapid proliferation. Nat Genet 40(12):1478–1483
157. Judson RL, Babiarz JE, Venere M, Blelloch R (2009) Embryonic stem cell-specific microRNAs promote induced pluripotency. Nat Biotechnol 27(5):459–461
158. Xu N, Papagiannakopoulos T, Pan G, Thomson JA, Kosik KS (2009) MicroRNA-145 regulates OCT4, SOX2, and KLF4 and represses pluripotency in human embryonic stem cells. Cell 137(4):647–658
159. Lu J, Getz G, Miska EA, Alvarez-Saavedra E, Lamb J, Peck D, Sweet-Cordero A, Ebert BL, Mak RH, Ferrando AA, Downing JR, Jacks T, Horvitz HR, Golub TR (2005) MicroRNA expression profiles classify human cancers. Nature 435(7043):834–838
160. Volinia S, Calin GA, Liu CG, Ambs S, Cimmino A, Petrocca F, Visone R, Iorio M, Roldo C, Ferracin M, Prueitt RL, Yanaihara N, Lanza G, Scarpa A, Vecchione A, Negrini M, Harris CC, Croce CM (2006) A microRNA expression signature of human solid tumors defines cancer gene targets. Proc Natl Acad Sci USA 103(7):2257–2261
161. Shimono Y, Zabala M, Cho RW, Lobo N, Dalerba P, Qian D, Diehn M, Liu H, Panula SP, Chiao E, Dirbas FM, Somlo G, Pera RA, Lao K, Clarke MF (2009) Downregulation of miRNA-200c links breast cancer stem cells with normal stem cells. Cell 138(3):592–603
162. Wellner U, Schubert J, Burk UC, Schmalhofer O, Zhu F, Sonntag A, Waldvogel B, Vannier C, Darling D, Zur Hausen A, Brunton VG, Morton J, Sansom O, Schuler J, Stemmler MP, Herzberger C, Hopt U, Keck T, Brabletz S, Brabletz T (2009) The EMT-activator ZEB1 promotes tumorigenicity by repressing stemness-inhibiting microRNAs. Nat Cell Biol 11(12):1487–1495
163. Yu F, Yao H, Zhu P, Zhang X, Pan Q, Gong C, Huang Y, Hu X, Su F, Lieberman J, Song E (2007) let-7 regulates self renewal and tumorigenicity of breast cancer cells. Cell 131(6):1109–1123

164. Kim MY, Oskarsson T, Acharyya S, Nguyen DX, Zhang XH, Norton L, Massague J (2009) Tumor self-seeding by circulating cancer cells. Cell 139(7):1315–1326. doi:10.1016/j.cell.2009.11.025
165. Fidler IJ (1970) Metastasis: guantitative analysis of distribution and fate of tumor embolilabeled with 125 I-5-iodo-2'-deoxyuridine. J Natl Cancer Inst 45(4):773–782
166. Chambers AF, MacDonald IC, Schmidt EE, Koop S, Morris VL, Khokha R, Groom AC (1995) Steps in tumor metastasis: new concepts from intravital videomicroscopy. Cancer Metastasis Rev 14(4):279–301
167. Barkan D, Green JE, Chambers AF (2010) Extracellular matrix: a gatekeeper in the transition from dormancy to metastatic growth. Eur J Cancer 46(7):1181–1188
168. Hedley BD, Chambers AF (2009) Tumor dormancy and metastasis. Adv Cancer Res 102:67–101
169. Paget S (1889) The distribution of secondary growths in cancer of the breast. Lancet 1:571–573
170. Pantel K, Brakenhoff RH (2004) Dissecting the metastatic cascade. Nat Rev Cancer 4(6):448–456
171. Alix-Panabieres C, Muller V, Pantel K (2007) Current status in human breast cancer micrometastasis. Curr Opin Oncol 19(6):558–563
172. Pantel K, Alix-Panabieres C, Riethdorf S (2009) Cancer micrometastases. Nat Rev Clin Oncol 6(6):339–351
173. Riethmuller G, Klein CA (2001) Early cancer cell dissemination and late metastatic relapse: clinical reflections and biological approaches to the dormancy problem in patients. Semin Cancer Biol 11(4):307–311. doi:10.1006/scbi.2001.0386
174. Davis JW, Nakanishi H, Kumar VS, Bhadkamkar VA, McCormack R, Fritsche HA, Handy B, Gornet T, Babaian RJ (2008) Circulating tumor cells in peripheral blood samples from patients with increased serum prostate specific antigen: initial results in early prostate cancer. J Urol 179(6):2187–2191, discussion 2191
175. Lang JE, Hall CS, Singh B, Lucci A (2007) Significance of micrometastasis in bone marrow and blood of operable breast cancer patients: research tool or clinical application? Expert Rev Anticancer Ther 7(10):1463–1472
176. Gerber B, Krause A, Muller H, Richter D, Reimer T, Makovitzky J, Herrnring C, Jeschke U, Kundt G, Friese K (2001) Simultaneous immunohistochemical detection of tumor cells in lymph nodes and bone marrow aspirates in breast cancer and its correlation with other prognostic factors. J Clin Oncol 19(4):960–971
177. Weitz J, Kienle P, Magener A, Koch M, Schrodel A, Willeke F, Autschbach F, Lacroix J, Lehnert T, Herfarth C, von Knebel DM (1999) Detection of disseminated colorectal cancer cells in lymph nodes, blood and bone marrow. Clin Cancer Res 5(7):1830–1836
178. Izbicki JR, Hosch SB, Pichlmeier U, Rehders A, Busch C, Niendorf A, Passlick B, Broelsch CE, Pantel K (1997) Prognostic value of immunohistochemically identifiable tumor cells in lymph nodes of patients with completely resected esophageal cancer. N Engl J Med 337(17):1188–1194
179. Ulmer A, Fischer JR, Schanz S, Sotlar K, Breuninger H, Dietz K, Fierlbeck G, Klein CA (2005) Detection of melanoma cells displaying multiple genomic changes in histopathologically negative sentinel lymph nodes. Clin Cancer Res 11(15):5425–5432. doi:10.1158/1078-0432.CCR-04-1995
180. Rocken M (2010) Early tumor dissemination, but late metastasis: insights into tumor dormancy. J Clin Invest 120(6):1800–1803. doi:10.1172/JCI43424
181. Bernards R, Weinberg RA (2002) A progression puzzle. Nature 418(6900):823
182. van de Vijver MJ, He YD, van't Veer LJ, Dai H, Hart AA, Voskuil DW, Schreiber GJ, Peterse JL, Roberts C, Marton MJ, Parrish M, Atsma D, Witteveen A, Glas A, Delahaye L, van der Velde T, Bartelink H, Rodenhuis S, Rutgers ET, Friend SH, Bernards R (2002) A gene-expression signature as a predictor of survival in breast cancer. N Engl J Med 347(25):1999–2009

183. Klein CA (2009) Parallel progression of primary tumours and metastases. Nat Rev Cancer 9(4):302–312
184. Scheel C, Onder T, Karnoub A, Weinberg RA (2007) Adaptation versus selection: the origins of metastatic behavior. Cancer Res 67(24):11476–11479, discussion 11479–11480
185. Stoecklein NH, Klein CA (2010) Genetic disparity between primary tumours, disseminated tumour cells, and manifest metastasis. Int J Cancer 126(3):589–598
186. Brabletz T, Jung A, Spaderna S, Hlubek F, Kirchner T (2005) Migrating cancer stem cells [mdash] an integrated concept of malignant tumour progression. Nature Rev Cancer 5:744–749
187. Visvader JE, Lindeman GJ (2008) Cancer stem cells in solid tumours: accumulating evidence and unresolved questions. Nature Rev Cancer 8:755–768
188. Kang Y (2003) A multigenic program mediating breast cancer metastasis to bone. Cancer Cell 3:537–549
189. Quintana E (2008) Efficient tumour formation by single human melanoma cells. Nature 456:593–598
190. Schardt JA (2005) Genomic analysis of single cytokeratin-positive cells from bone marrow reveals early mutational events in breast cancer. Cancer Cell 8:227–239
191. Schmidt-Kittler O, Ragg T, Daskalakis A, Granzow M, Ahr A, Blankenstein TJ, Kaufmann M, Diebold J, Arnholdt H, Muller P, Bischoff J, Harich D, Schlimok G, Riethmuller G, Eils R, Klein CA (2003) From latent disseminated cells to overt metastasis: genetic analysis of systemic breast cancer progression. Proc Natl Acad Sci USA 100(13):7737–7742
192. Stoecklein NH (2008) Direct genetic analysis of single disseminated cancer cells for prediction of outcome and therapy selection in esophageal cancer. Cancer Cell 13:441–453
193. Weckermann D (2009) Perioperative activation of disseminated tumor cells in bone marrow of patients with prostate cancer. J Clin Oncol 27(10):1549–1556
194. Klein CA, Blankenstein TJ, Schmidt-Kittler O, Petronio M, Polzer B, Stoecklein NH, Riethmuller G (2002) Genetic heterogeneity of single disseminated tumour cells in minimal residual cancer. Lancet 360(9334):683–689
195. Klein CA, Seidl S, Petat-Dutter K, Offner S, Geigl JB, Schmidt-Kittler O, Wendler N, Passlick B, Huber RM, Schlimok G, Baeuerle PA, Riethmuller G (2002) Combined transcriptome and genome analysis of single micrometastatic cells. Nat Biotechnol 20(4):387–392
196. Kuukasjarvi T, Karhu R, Tanner M, Kahkonen M, Schaffer A, Nupponen N, Pennanen S, Kallioniemi A, Kallioniemi OP, Isola J (1997) Genetic heterogeneity and clonal evolution underlying development of asynchronous metastasis in human breast cancer. Cancer Res 57(8):1597–1604
197. Santos SC, Cavalli IJ, Ribeiro EM, Urban CA, Lima RS, Bleggi-Torres LF, Rone JD, Haddad BR, Cavalli LR (2008) Patterns of DNA copy number changes in sentinel lymph node breast cancer metastases. Cytogenet Genome Res 122(1):16–21
198. Torres L, Ribeiro FR, Pandis N, Andersen JA, Heim S, Teixeira MR (2007) Intratumor genomic heterogeneity in breast cancer with clonal divergence between primary carcinomas and lymph node metastases. Breast Cancer Res Treat 102(2):143–155
199. Alcock HE, Stephenson TJ, Royds JA, Hammond DW (2003) Analysis of colorectal tumor progression by microdissection and comparative genomic hybridization. Genes Chromosomes Cancer 37(4):369–380
200. Al-Mulla F, Keith WN, Pickford IR, Going JJ, Birnie GD (1999) Comparative genomic hybridization analysis of primary colorectal carcinomas and their synchronous metastases. Genes Chromosomes Cancer 24(4):306–314
201. Jiang JK, Chen YJ, Lin CH, Yu IT, Lin JK (2005) Genetic changes and clonality relationship between primary colorectal cancers and their pulmonary metastases–an analysis by comparative genomic hybridization. Genes Chromosomes Cancer 43(1):25–36
202. Knosel T, Petersen S, Schwabe H, Schluns K, Stein U, Schlag PM, Dietel M, Petersen I (2002) Incidence of chromosomal imbalances in advanced colorectal carcinomas and their metastases. Virchows Arch 440(2):187–194

203. Knosel T, Schluns K, Dietel M, Petersen I (2005) Chromosomal alterations in lung metastases of colorectal carcinomas: associations with tissue specific tumor dissemination. Clin Exp Metastasis 22(7):533–538
204. Knosel T, Schluns K, Stein U, Schwabe H, Schlag PM, Dietel M, Petersen I (2004) Chromosomal alterations during lymphatic and liver metastasis formation of colorectal cancer. Neoplasia 6(1):23–28
205. Walch AK, Zitzelsberger HF, Bink K, Hutzler P, Bruch J, Braselmann H, Aubele MM, Mueller J, Stein H, Siewert JR, Hofler H, Werner M (2000) Molecular genetic changes in metastatic primary Barrett's adenocarcinoma and related lymph node metastases: comparison with nonmetastatic Barrett's adenocarcinoma. Mod Pathol 13(7):814–824
206. Walch AK, Zitzelsberger HF, Bruch J, Keller G, Angermeier D, Aubele MM, Mueller J, Stein H, Braselmann H, Siewert JR, Hofler H, Werner M (2000) Chromosomal imbalances in Barrett's adenocarcinoma and the metaplasia-dysplasia-carcinoma sequence. Am J Pathol 156(2):555–566
207. Feng Y, Sun B, Li X, Zhang L, Niu Y, Xiao C, Ning L, Fang Z, Wang Y, Zhang L, Cheng J, Zhang W, Hao X (2007) Differentially expressed genes between primary cancer and paired lymph node metastases predict clinical outcome of node-positive breast cancer patients. Breast Cancer Res Treat 103(3):319–329
208. Suzuki M, Tarin D (2007) Gene expression profiling of human lymph node metastases and matched primary breast carcinomas: clinical implications. Mol Oncol 1(2):172–180
209. Vecchi M, Confalonieri S, Nuciforo P, Vigano MA, Capra M, Bianchi M, Nicosia D, Bianchi F, Galimberti V, Viale G, Palermo G, Riccardi A, Campanini R, Daidone MG, Pierotti MA, Pece S, Di Fiore PP (2008) Breast cancer metastases are molecularly distinct from their primary tumors. Oncogene 27(15):2148–2158
210. Hao X, Sun B, Hu L, Lahdesmaki H, Dunmire V, Feng Y, Zhang SW, Wang H, Wu C, Wang H, Fuller GN, Symmans WF, Shmulevich I, Zhang W (2004) Differential gene and protein expression in primary breast malignancies and their lymph node metastases as revealed by combined cDNA microarray and tissue microarray analysis. Cancer 100(6):1110–1122
211. Ki DH, Jeung HC, Park CH, Kang SH, Lee GY, Lee WS, Kim NK, Chung HC, Rha SY (2007) Whole genome analysis for liver metastasis gene signatures in colorectal cancer. Int J Cancer 121(9):2005–2012
212. Kim SH, Choi SJ, Cho YB, Kang MW, Lee J, Lee WY, Chun HK, Choi YS, Kim HK, Han J, Kim J (2011) Differential gene expression during colon-to-lung metastasis. Oncol Rep 25(3):629–636
213. Hu Z, Fan C, Livasy C, He X, Oh DS, Ewend MG, Carey LA, Subramanian S, West R, Ikpatt F, Olopade OI, van de Rijn M, Perou CM (2009) A compact VEGF signature associated with distant metastases and poor outcomes. BMC Med 7:9
214. Hunter K (2006) Host genetics influence tumour metastasis. Nat Rev Cancer 6(2):141–146
215. Mavroudis D (2010) Circulating cancer cells. Ann Oncol 21(Suppl 7):vii95–vii100
216. Lin H, Balic M, Zheng S, Datar R, Cote RJ (2010) Disseminated and circulating tumor cells: role in effective cancer management. Crit Rev Oncol Hematol 77(1):1–11
217. Smirnov DA, Zweitzig DR, Foulk BW, Miller MC, Doyle GV, Pienta KJ, Meropol NJ, Weiner LM, Cohen SJ, Moreno JG, Connelly MC, Terstappen LW, O'Hara SM (2005) Global gene expression profiling of circulating tumor cells. Cancer Res 65(12):4993–4997. doi:10.1158/0008-5472.CAN-04-4330
218. Paris PL, Kobayashi Y, Zhao Q, Zeng W, Sridharan S, Fan T, Adler HL, Yera ER, Zarrabi MH, Zucker S, Simko J, Chen WT, Rosenberg J (2009) Functional phenotyping and genotyping of circulating tumor cells from patients with castration resistant prostate cancer. Cancer Lett 277(2):164–173. doi:10.1016/j.canlet.2008.12.007
219. Liu R, Wang X, Chen GY, Dalerba P, Gurney A, Hoey T, Sherlock G, Lewicki J, Shedden K, Clarke MF (2007) The prognostic role of a gene signature from tumorigenic breast-cancer cells. N Engl J Med 356(3):217–226. doi:10.1056/NEJMoa063994
220. Pantel K, Alix-Panabieres C (2010) Circulating tumour cells in cancer patients: challenges and perspectives. Trends Mol Med 16(9):398–406

221. Balic M, Lin H, Young L, Hawes D, Giuliano A, McNamara G, Datar RH, Cote RJ (2006) Most early disseminated cancer cells detected in bone marrow of breast cancer patients have a putative breast cancer stem cell phenotype. Clin Cancer Res 12(19):5615–5621
222. Armstrong AJ, Marengo MS, Oltean S, Kemeny G, Bitting R, Turnbull J, Herold CI, Marcom PK, George D, Garcia-Blanco M (2011) Circulating tumor cells from patients with advanced prostate and breast cancer display both epithelial and mesenchymal markers. Mol Cancer Res. doi:10.1158/1541-7786.MCR-10-0490
223. Aktas B, Tewes M, Fehm T, Hauch S, Kimmig R, Kasimir-Bauer S (2009) Stem cell and epithelial-mesenchymal transition markers are frequently overexpressed in circulating tumor cells of metastatic breast cancer patients. Breast Cancer Res 11(4):R46
224. Bertolini G, Roz L, Perego P, Tortoreto M, Fontanella E, Gatti L, Pratesi G, Fabbri A, Andriani F, Tinelli S, Roz E, Caserini R, Lo Vullo S, Camerini T, Mariani L, Delia D, Calabro E, Pastorino U, Sozzi G (2009) Highly tumorigenic lung cancer CD133+ cells display stem-like features and are spared by cisplatin treatment. Proc Natl Acad Sci USA 106(38):16281–16286
225. Shmelkov SV, Butler JM, Hooper AT, Hormigo A, Kushner J, Milde T, St Clair R, Baljevic M, White I, Jin DK, Chadburn A, Murphy AJ, Valenzuela DM, Gale NW, Thurston G, Yancopoulos GD, D'Angelica M, Kemeny N, Lyden D, Rafii S (2008) CD133 expression is not restricted to stem cells, and both CD133+ and CD133- metastatic colon cancer cells initiate tumors. J Clin Invest 118(6):2111–2120
226. Muller A, Homey B, Soto H, Ge N, Catron D, Buchanan ME, McClanahan T, Murphy E, Yuan W, Wagner SN, Barrera JL, Mohar A, Verastegui E, Zlotnik A (2001) Involvement of chemokine receptors in breast cancer metastasis. Nature 410(6824):50–56
227. Pang R, Law WL, Chu AC, Poon JT, Lam CS, Chow AK, Ng L, Cheung LW, Lan XR, Lan HY, Tan VP, Yau TC, Poon RT, Wong BC (2010) A subpopulation of CD26+ cancer stem cells with metastatic capacity in human colorectal cancer. Cell Stem Cell 6(6):603–615
228. Smid M, Wang Y, Zhang Y, Sieuwerts AM, Yu J, Klijn JG, Foekens JA, Martens JW (2008) Subtypes of breast cancer show preferential site of relapse. Cancer Res 68(9):3108–3114
229. Podsypanina K, Du YC, Jechlinger M, Beverly LJ, Hambardzumyan D, Varmus H (2008) Seeding and propagation of untransformed mouse mammary cells in the lung. Science 321(5897):1841–1844
230. Borovski T, De Sousa EMF, Vermeulen L, Medema JP (2011) Cancer stem cell niche: the place to be. Cancer Res 71(3):634–639
231. Calabrese C, Poppleton H, Kocak M, Hogg TL, Fuller C, Hamner B, Oh EY, Gaber MW, Finklestein D, Allen M, Frank A, Bayazitov IT, Zakharenko SS, Gajjar A, Davidoff A, Gilbertson RJ (2007) A perivascular niche for brain tumor stem cells. Cancer Cell 11(1):69–82
232. Hovinga KE, Shimizu F, Wang R, Panagiotakos G, Van Der Heijden M, Moayedpardazi H, Correia AS, Soulet D, Major T, Menon J, Tabar V (2010) Inhibition of notch signaling in glioblastoma targets cancer stem cells via an endothelial cell intermediate. Stem Cells 28(6):1019–1029
233. Folkins C, Man S, Xu P, Shaked Y, Hicklin DJ, Kerbel RS (2007) Anticancer therapies combining antiangiogenic and tumor cell cytotoxic effects reduce the tumor stem-like cell fraction in glioma xenograft tumors. Cancer Res 67(8):3560–3564
234. LaBarge MA (2010) The difficulty of targeting cancer stem cell niches. Clin Cancer Res 16(12):3121–3129
235. Nguyen DX, Bos PD, Massague J (2009) Metastasis: from dissemination to organ-specific colonization. Nat Rev Cancer 9(4):274–284
236. Patanaphan V, Salazar OM, Risco R (1988) Breast cancer: metastatic patterns and their prognosis. South Med J 81(9):1109–1112
237. Hess KR, Varadhachary GR, Taylor SH, Wei W, Raber MN, Lenzi R, Abbruzzese JL (2006) Metastatic patterns in adenocarcinoma. Cancer 106(7):1624–1633
238. Fidler IJ (2003) The pathogenesis of cancer metastasis: the 'seed and soil' hypothesis revisited. Nat Rev Cancer 3(6):453–458

239. Fokas E, Engenhart-Cabillic R, Daniilidis K, Rose F, An HX (2007) Metastasis: the seed and soil theory gains identity. Cancer Metastasis Rev 26(3–4):705–715
240. Suzuki M, Mose ES, Montel V, Tarin D (2006) Dormant cancer cells retrieved from metastasis-free organs regain tumorigenic and metastatic potency. Am J Pathol 169(2):673–681
241. Joyce JA, Pollard JW (2009) Microenvironmental regulation of metastasis. Nat Rev Cancer 9(4):239–252
242. Fidler IJ, Kripke ML (1977) Metastasis results from preexisting variant cells within a malignant tumor. Science 197:893–895
243. Langley RR, Fidler IJ (2007) Tumor cell-organ microenvironment interactions in the pathogenesis of cancer metastasis. Endocr Rev 28(3):297–321
244. Clark EA, Golub TR, Lander ES, Hynes RO (2000) Genomic analysis of metastasis reveals an essential role for RhoC. Nature 406(6795):532–535
245. Nguyen DX, Massague J (2007) Genetic determinants of cancer metastasis. Nat Rev Genet 8(5):341–352
246. Kaplan RN, Riba RD, Zacharoulis S, Bramley AH, Vincent L, Costa C, MacDonald DD, Jin DK, Shido K, Kerns SA, Zhu Z, Hicklin D, Wu Y, Port JL, Altorki N, Port ER, Ruggero D, Shmelkov SV, Jensen KK, Rafii S, Lyden D (2005) VEGFR1-positive haematopoietic bone marrow progenitors initiate the pre-metastatic niche. Nature 438(7069):820–827
247. Hiratsuka S, Watanabe A, Aburatani H, Maru Y (2006) Tumour-mediated upregulation of chemoattractants and recruitment of myeloid cells predetermines lung metastasis. Nat Cell Biol 8(12):1369–1375
248. Lyden D (2001) Impaired recruitment of bone-marrow-derived endothelial and hematopoietic precursor cells blocks tumor angiogenesis and growth. Nature Med 7:1194–1201
249. Olaso E, Santisteban A, Bidaurrazaga J, Gressner AM, Rosenbaum J, Vidal-Vanaclocha F (1997) Tumor-dependent activation of rodent hepatic stellate cells during experimental melanoma metastasis. Hepatology 26(3):634–642
250. Stuelten CH, Busch JI, Tang B, Flanders KC, Oshima A, Sutton E, Karpova TS, Roberts AB, Wakefield LM, Niederhuber JE (2010) Transient tumor-fibroblast interactions increase tumor cell malignancy by a TGF-Beta mediated mechanism in a mouse xenograft model of breast cancer. PLoS One 5(3):e9832
251. Orimo A, Gupta PB, Sgroi DC, Arenzana-Seisdedos F, Delaunay T, Naeem R, Carey VJ, Richardson AL, Weinberg RA (2005) Stromal fibroblasts present in invasive human breast carcinomas promote tumor growth and angiogenesis through elevated SDF-1/CXCL12 secretion. Cell 121(3):335–348
252. Pollard JW (2004) Tumour-educated macrophages promote tumour progression and metastasis. Nature Rev Cancer 4:71–78
253. Yang L, Moses HL (2008) Transforming growth factor b: tumor suppressor or promoter? Are host immune cells the answer? Cancer Res 68:9107–9111
254. Lin EY, Nguyen AV, Russell RG, Pollard JW (2001) Colony-stimulating factor 1 promotes progression of mammary tumors to malignancy. J Exp Med 193(6):727–740
255. Adams GB, Chabner KT, Alley IR, Olson DP, Szczepiorkowski ZM, Poznansky MC, Kos CH, Pollak MR, Brown EM, Scadden DT (2006) Stem cell engraftment at the endosteal niche is specified by the calcium-sensing receptor. Nature 439(7076):599–603
256. Gao D (2008) Endothelial progenitor cells control the angiogenic switch in mouse lung metastasis. Science 319:195–198
257. Cheng JD, Weiner LM (2003) Tumors and their microenvironments: tilling the soilA Phase I dose-escalation study of sibrotuzumab in patients with advanced or metastatic fibroblast activation protein-positive cancer. Clin Cancer Res 9:1590–1595
258. Kalluri R, Zeisberg M (2006) Fibroblasts in cancer. Nature Rev Cancer 6:392–401
259. Hata N, Shinojima N, Gumin J, Yong R, Marini F, Andreeff M, Lang FF (2010) Platelet-derived growth factor BB mediates the tropism of human mesenchymal stem cells for malignant gliomas. Neurosurgery 66(1):144–156; discussion 156–147. doi:10.1227/01.NEU.0000363149.58885.2E

260. Kidd S, Spaeth E, Dembinski JL, Dietrich M, Watson K, Klopp A, Battula VL, Weil M, Andreeff M, Marini FC (2009) Direct evidence of mesenchymal stem cell tropism for tumor and wounding microenvironments using in vivo bioluminescent imaging. Stem Cells 27(10):2614–2623
261. Spaeth EL, Dembinski JL, Sasser AK, Watson K, Klopp A, Hall B, Andreeff M, Marini F (2009) Mesenchymal stem cell transition to tumor-associated fibroblasts contributes to fibrovascular network expansion and tumor progression. PLoS One 4(4):e4992
262. Karnoub AE (2007) Mesenchymal stem cells within tumour stroma promote breast cancer metastasis. Nature 449:557–563
263. Karnoub AE, Dash AB, Vo AP, Sullivan A, Brooks MW, Bell GW, Richardson AL, Polyak K, Tubo R, Weinberg RA (2007) Mesenchymal stem cells within tumour stroma promote breast cancer metastasis. Nature 449(7162):557–563
264. Zlotnik A (2004) Chemokines in neoplastic progression. Semin Cancer Biol 14(3):181–185
265. Zlotnik A (2008) New insights on the role of CXCR4 in cancer metastasis. J Pathol 215(3):211–213
266. Li YM, Pan Y, Wei Y, Cheng X, Zhou BP, Tan M, Zhou X, Xia W, Hortobagyi GN, Yu D, Hung MC (2004) Upregulation of CXCR4 is essential for HER2-mediated tumor metastasis. Cancer Cell 6(5):459–469
267. Eccles SA (2011) Growth regulatory pathways contributing to organ selectivity of metastasis. In: Lyden D, Welch DR, Psaila B (eds) Cancer metastasis: biologic basis and therapeutics. Cambridge University Press, New York
268. Schafer M, Werner S (2008) Cancer as an overhealing wound: an old hypothesis revisited. Nat Rev Mol Cell Biol 9(8):628–638
269. Erler JT, Weaver VM (2009) Three-dimensional context regulation of metastasis. Clin Exp Metastasis 26(1):35–49
270. Giantonio BJ, Catalano PJ, Meropol NJ, O'Dwyer PJ, Mitchell EP, Alberts SR, Schwartz MA, Benson AB 3rd (2007) Bevacizumab in combination with oxaliplatin, fluorouracil, and leucovorin (FOLFOX4) for previously treated metastatic colorectal cancer: results from the Eastern Cooperative Oncology Group Study E3200. J Clin Oncol 25(12):1539–1544
271. Allegra CJ, Yothers G, O'Connell MJ, Sharif S, Petrelli NJ, Colangelo LH, Atkins JN, Seay TE, Fehrenbacher L, Goldberg RM, O'Reilly S, Chu L, Azar CA, Lopa S, Wolmark N (2011) Phase III trial assessing bevacizumab in stages II and III carcinoma of the colon: results of NSABP protocol C-08. J Clin Oncol 29(1):11–16
272. Van Cutsem E, Lambrechts D, Prenen H, Jain RK, Carmeliet P (2011) Lessons from the adjuvant bevacizumab trial on colon cancer: what next? J Clin Oncol 29(1):1–4
273. Reya T, Morrison SJ, Clarke MF, Weissman IL (2001) Stem cells, cancer, and cancer stem cells. Nature 414(6859):105–111
274. Costello RT, Mallet F, Gaugler B, Sainty D, Arnoulet C, Gastaut JA, Olive D (2000) Human acute myeloid leukemia CD34+/CD38- progenitor cells have decreased sensitivity to chemotherapy and Fas-induced apoptosis, reduced immunogenicity, and impaired dendritic cell transformation capacities. Cancer Res 60(16):4403–4411
275. de Grouw EP, Raaijmakers MH, Boezeman JB, van der Reijden BA, van de Locht LT, de Witte TJ, Jansen JH, Raymakers RA (2006) Preferential expression of a high number of ATP binding cassette transporters in both normal and leukemic CD34+CD38- cells. Leukemia 20(4):750–754
276. Ishikawa F, Yoshida S, Saito Y, Hijikata A, Kitamura H, Tanaka S, Nakamura R, Tanaka T, Tomiyama H, Saito N, Fukata M, Miyamoto T, Lyons B, Ohshima K, Uchida N, Taniguchi S, Ohara O, Akashi K, Harada M, Shultz LD (2007) Chemotherapy-resistant human AML stem cells home to and engraft within the bone-marrow endosteal region. Nat Biotechnol 25(11):1315–1321
277. Ito K, Hirao A, Arai F, Matsuoka S, Takubo K, Hamaguchi I, Nomiyama K, Hosokawa K, Sakurada K, Nakagata N, Ikeda Y, Mak TW, Suda T (2004) Regulation of oxidative stress by ATM is required for self-renewal of haematopoietic stem cells. Nature 431(7011):997–1002

278. Jordan CT (2009) Cancer stem cells: controversial or just misunderstood? Cell Stem Cell 4(3):203–205
279. Michor F, Hughes TP, Iwasa Y, Branford S, Shah NP, Sawyers CL, Nowak MA (2005) Dynamics of chronic myeloid leukaemia. Nature 435(7046):1267–1270
280. Viale A, De Franco F, Orleth A, Cambiaghi V, Giuliani V, Bossi D, Ronchini C, Ronzoni S, Muradore I, Monestiroli S, Gobbi A, Alcalay M, Minucci S, Pelicci PG (2009) Cell-cycle restriction limits DNA damage and maintains self-renewal of leukaemia stem cells. Nature 457(7225):51–56
281. Frank NY, Margaryan A, Huang Y, Schatton T, Waaga-Gasser AM, Gasser M, Sayegh MH, Sadee W, Frank MH (2005) ABCB5-mediated doxorubicin transport and chemoresistance in human malignant melanoma. Cancer Res 65(10):4320–4333
282. Schatton T, Murphy GF, Frank NY, Yamaura K, Waaga-Gasser AM, Gasser M, Zhan Q, Jordan S, Duncan LM, Weishaupt C, Fuhlbrigge RC, Kupper TS, Sayegh MH, Frank MH (2008) Identification of cells initiating human melanomas. Nature 451(7176):345–349
283. Eramo A, Ricci-Vitiani L, Zeuner A, Pallini R, Lotti F, Sette G, Pilozzi E, Larocca LM, Peschle C, De Maria R (2006) Chemotherapy resistance of glioblastoma stem cells. Cell Death Differ 13(7):1238–1241
284. Li X, Lewis MT, Huang J, Gutierrez C, Osborne CK, Wu MF, Hilsenbeck SG, Pavlick A, Zhang X, Chamness GC, Wong H, Rosen J, Chang JC (2008) Intrinsic resistance of tumorigenic breast cancer cells to chemotherapy. J Natl Cancer Inst 100(9):672–679
285. Dylla SJ, Beviglia L, Park IK, Chartier C, Raval J, Ngan L, Pickell K, Aguilar J, Lazetic S, Smith-Berdan S, Clarke MF, Hoey T, Lewicki J, Gurney AL (2008) Colorectal cancer stem cells are enriched in xenogeneic tumors following chemotherapy. PLoS One 3(6):e2428
286. Diehn M, Cho RW, Lobo NA, Kalisky T, Dorie MJ, Kulp AN, Qian D, Lam JS, Ailles LE, Wong M, Joshua B, Kaplan MJ, Wapnir I, Dirbas FM, Somlo G, Garberoglio C, Paz B, Shen J, Lau SK, Quake SR, Brown JM, Weissman IL, Clarke MF (2009) Association of reactive oxygen species levels and radioresistance in cancer stem cells. Nature 458(7239): 780–783
287. Gottesman MM (2002) Mechanisms of cancer drug resistance. Annu Rev Med 53:615–627
288. Gimenez-Bonafe P, Tortosa A, Perez-Tomas R (2009) Overcoming drug resistance by enhancing apoptosis of tumor cells. Curr Cancer Drug Targets 9(3):320–340
289. Charafe-Jauffret E, Ginestier C, Iovino F, Wicinski J, Cervera N, Finetti P, Hur MH, Diebel ME, Monville F, Dutcher J, Brown M, Viens P, Xerri L, Bertucci F, Stassi G, Dontu G, Birnbaum D, Wicha MS (2009) Breast cancer cell lines contain functional cancer stem cells with metastatic capacity and a distinct molecular signature. Cancer Res 69(4):1302–1313
290. Pece S, Tosoni D, Confalonieri S, Mazzarol G, Vecchi M, Ronzoni S, Bernard L, Viale G, Pelicci PG, Di Fiore PP (2010) Biological and molecular heterogeneity of breast cancers correlates with their cancer stem cell content. Cell 140(1):62–73
291. Shipitsin M, Campbell LL, Argani P, Weremowicz S, Bloushtain-Qimron N, Yao J, Nikolskaya T, Serebryiskaya T, Beroukhim R, Hu M, Halushka MK, Sukumar S, Parker LM, Anderson KS, Harris LN, Garber JE, Richardson AL, Schnitt SJ, Nikolsky Y, Gelman RS, Polyak K (2007) Molecular definition of breast tumor heterogeneity. Cancer Cell 11(3):259–273
292. Ben-Porath I, Thomson MW, Carey VJ, Ge R, Bell GW, Regev A, Weinberg RA (2008) An embryonic stem cell-like gene expression signature in poorly differentiated aggressive human tumors. Nat Genet 40(5):499–507
293. Wong DJ, Liu H, Ridky TW, Cassarino D, Segal E, Chang HY (2008) Module map of stem cell genes guides creation of epithelial cancer stem cells. Cell Stem Cell 2(4):333–344
294. Zhou BB, Zhang H, Damelin M, Geles KG, Grindley JC, Dirks PB (2009) Tumour-initiating cells: challenges and opportunities for anticancer drug discovery. Nat Rev Drug Discov 8(10):806–823
295. Health. CgBMNIo. Available from: http://clinicaltrialsgov/

296. Robarge KD, Brunton SA, Castanedo GM, Cui Y, Dina MS, Goldsmith R, Gould SE, Guichert O, Gunzner JL, Halladay J, Jia W, Khojasteh C, Koehler MF, Kotkow K, La H, Lalonde RL, Lau K, Lee L, Marshall D, Marsters JC Jr, Murray LJ, Qian C, Rubin LL, Salphati L, Stanley MS, Stibbard JH, Sutherlin DP, Ubhayaker S, Wang S, Wong S, Xie M (2009) GDC-0449-a potent inhibitor of the hedgehog pathway. Bioorg Med Chem Lett 19(19):5576–5581
297. Von Hoff DD, LoRusso PM, Rudin CM, Reddy JC, Yauch RL, Tibes R, Weiss GJ, Borad MJ, Hann CL, Brahmer JR, Mackey HM, Lum BL, Darbonne WC, Marsters JC Jr, de Sauvage FJ, Low JA (2009) Inhibition of the hedgehog pathway in advanced basal-cell carcinoma. N Engl J Med 361(12):1164–1172
298. Tremblay MR, Nevalainen M, Nair SJ, Porter JR, Castro AC, Behnke ML, Yu LC, Hagel M, White K, Faia K, Grenier L, Campbell MJ, Cushing J, Woodward CN, Hoyt J, Foley MA, Read MA, Sydor JR, Tong JK, Palombella VJ, McGovern K, Adams J (2008) Semisynthetic cyclopamine analogues as potent and orally bioavailable hedgehog pathway antagonists. J Med Chem 51(21):6646–6649
299. Olive KP, Jacobetz MA, Davidson CJ, Gopinathan A, McIntyre D, Honess D, Madhu B, Goldgraben MA, Caldwell ME, Allard D, Frese KK, Denicola G, Feig C, Combs C, Winter SP, Ireland-Zecchini H, Reichelt S, Howat WJ, Chang A, Dhara M, Wang L, Ruckert F, Grutzmann R, Pilarsky C, Izeradjene K, Hingorani SR, Huang P, Davies SE, Plunkett W, Egorin M, Hruban RH, Whitebread N, McGovern K, Adams J, Iacobuzio-Donahue C, Griffiths J, Tuveson DA (2009) Inhibition of Hedgehog signaling enhances delivery of chemotherapy in a mouse model of pancreatic cancer. Science 324(5933):1457–1461
300. Siu LL PK, Alberts SR et al (2009) A first-in-human, phase I study of an oral Hedgehog pathway antagonist, BMS-833923 (XL139), in subjects with advanced or metastatic solid tumors. Mol Cancer Ther 8(A):55
301. Watanabe K, Dai X (2011) Winning WNT: race to Wnt signaling inhibitors. Proc Natl Acad Sci USA 108(15):5929–5930
302. Weijzen S, Rizzo P, Braid M, Vaishnav R, Jonkheer SM, Zlobin A, Osborne BA, Gottipati S, Aster JC, Hahn WC, Rudolf M, Siziopikou K, Kast WM, Miele L (2002) Activation of Notch-1 signaling maintains the neoplastic phenotype in human Ras-transformed cells. Nat Med 8(9):979–986
303. Bocchetta M, Miele L, Pass HI, Carbone M (2003) Notch-1 induction, a novel activity of SV40 required for growth of SV40-transformed human mesothelial cells. Oncogene 22(1):81–89
304. Fan X, Matsui W, Khaki L, Stearns D, Chun J, Li YM, Eberhart CG (2006) Notch pathway inhibition depletes stem-like cells and blocks engraftment in embryonal brain tumors. Cancer Res 66(15):7445–7452
305. Li K, Li Y, Wu W, Gordon WR, Chang DW, Lu M, Scoggin S, Fu T, Vien L, Histen G, Zheng J, Martin-Hollister R, Duensing T, Singh S, Blacklow SC, Yao Z, Aster JC, Zhou BB (2008) Modulation of Notch signaling by antibodies specific for the extracellular negative regulatory region of NOTCH3. J Biol Chem 283(12):8046–8054
306. van Es JH, van Gijn ME, Riccio O, van den Born M, Vooijs M, Begthel H, Cozijnsen M, Robine S, Winton DJ, Radtke F, Clevers H (2005) Notch/gamma-secretase inhibition turns proliferative cells in intestinal crypts and adenomas into goblet cells. Nature 435(7044):959–963
307. Mimeault M, Batra SK (2007) Interplay of distinct growth factors during epithelial mesenchymal transition of cancer progenitor cells and molecular targeting as novel cancer therapies. Ann Oncol 18(10):1605–1619
308. Mimeault M, Batra SK (2008) Recent progress on normal and malignant pancreatic stem/progenitor cell research: therapeutic implications for the treatment of type 1 or 2 diabetes mellitus and aggressive pancreatic cancer. Gut 57(10):1456–1468
309. Mimeault M, Batra SK (2010) New promising drug targets in cancer- and metastasis-initiating cells. Drug Discov Today 15(9–10):354–364
310. Mimeault M, Mehta PP, Hauke R, Batra SK (2008) Functions of normal and malignant prostatic stem/progenitor cells in tissue regeneration and cancer progression and novel targeting therapies. Endocr Rev 29(2):234–252

311. Todaro M, Alea MP, Di Stefano AB, Cammareri P, Vermeulen L, Iovino F, Tripodo C, Russo A, Gulotta G, Medema JP, Stassi G (2007) Colon cancer stem cells dictate tumor growth and resist cell death by production of interleukin-4. Cell Stem Cell 1(4):389–402
312. Wang X, Belguise K, Kersual N, Kirsch KH, Mineva ND, Galtier F, Chalbos D, Sonenshein GE (2007) Oestrogen signalling inhibits invasive phenotype by repressing RelB and its target BCL2. Nat Cell Biol 9(4):470–478
313. Strizzi L, Bianco C, Normanno N, Salomon D (2005) Cripto-1: a multifunctional modulator during embryogenesis and oncogenesis. Oncogene 24(37):5731–5741
314. Hu XF, Xing PX (2005) Cripto as a target for cancer immunotherapy. Expert Opin Ther Targets 9(2):383–394
315. Sarkar FH, Li Y, Wang Z, Kong D (2008) NF-kappaB signaling pathway and its therapeutic implications in human diseases. Int Rev Immunol 27(5):293–319
316. Chen JS, Pardo FS, Wang-Rodriguez J, Chu TS, Lopez JP, Aguilera J, Altuna X, Weisman RA, Ongkeko WM (2006) EGFR regulates the side population in head and neck squamous cell carcinoma. Laryngoscope 116(3):401–406
317. Zhou J, Zhang H, Gu P, Bai J, Margolick JB, Zhang Y (2008) NF-kappaB pathway inhibitors preferentially inhibit breast cancer stem-like cells. Breast Cancer Res Treat 111(3):419–427
318. Elliott AM, Al-Hajj MA (2009) ABCB8 mediates doxorubicin resistance in melanoma cells by protecting the mitochondrial genome. Mol Cancer Res 7(1):79–87
319. Chang CJ, Hsu CC, Yung MC, Chen KY, Tzao C, Wu WF, Chou HY, Lee YY, Lu KH, Chiou SH, Ma HI (2009) Enhanced radiosensitivity and radiation-induced apoptosis in glioma CD133-positive cells by knockdown of SirT1 expression. Biochem Biophys Res Commun 380(2):236–242

Chapter 26
Plasticity of Cancer Stem Cells

Zhizhong Li

Contents

Abstract Cancer stem cell (CSC), also named cancer initiating cells and cancer propagating cells, are cancer cells that are highly undifferentiated and drive cancer growth *in vivo*. CSCs represent important targets for developing novel anti-cancer therapies, because they are highly tumorigenic and frequently resistant to traditional treatments. CSCs share many characteristics with normal stem cells such as self-renewal and multi-lineage differentiation capacity. CSCs were once thought to be a defined subpopulation in a given cancer samples. However, recent research has identified unexpected plasticity of CSCs. Through differentiation and dedifferentiation processes, CSC and non-stem cancer cell status are interchangeable. Also, CSCs may transdifferentiate into other cell lineages to help tumor growth. The plasticity of CSC is primarily controlled by tumor microenvironment signals. Those discoveries underline the importance of studying the interaction between CSC and its microenvironment, which may lead to identification of novel drug candidates in treating relapsing malignancies.

Keywords Cancer stem cell • Cell plasticity • Differentiation and dedifferentiation • Transdifferentiation • Tumor microenvironment

---

Z. Li Ph.D. (✉)
Novartis Institutes for Biomedical Researches, 100 Technology Square, Cambridge, MA 02139, USA
e-mail: zhizhong.li@novartis.com

R.K. Srivastava and S. Shankar (eds.), *Stem Cells and Human Diseases*,
DOI 10.1007/978-94-007-2801-1_26, © Springer Science+Business Media B.V. 2012

26.1 Cancer Stem Cell: Definition and Characterization

An emerging and exciting field in cancer research is the identification, characterization and validation of cancer stem cell (CSC) in various tumor types [1, 2]. Previously, cancer research was largely focusing on identifying genetic mutations associated with cancer development [3]. However, this strategy fails to explain the observation that subpopulation of cells have an extraordinary ability to initiate tumor growth in animal model, comparing to other cells bearing same genetic lesions. This indicates that like in development, there must be important epigenetic regulations in cancer as well. CSC theory is aiming at deciphering cancer epigenetics. It hypothesizes that cancer cells are heterogeneous and those that display stem-like epigenetic signatures maintain cancer growth *in vivo* [1].

CSC is also known as tumor initiating cells or tumor prorogating cells, as defined by their strong and unique capability to initiate the growth of secondary tumor, which recapitulate the features of original tumors (Fig. 26.1). Original CSC hypothesis can be tracked back to more than 150 years ago when physicians suggest that cancer may arise from rare cell population with stem cell property [4]. However, prospective isolation of CSC was not achieved thanks to limited knowledge and available technique tools to isolate different cell population. The breakthrough came in late 1990s when Bonnet and Dick successfully isolated a CD34+CD38- population from leukemic cells that displayed extraordinary to initiate tumor formation in NOD/SCID mice [5]. Since that, cell populations that demonstrate CSC properties have been isolated and characterized from multiple cancer types including tumors from breast, brain, colon, etc. [2, 6–8].

CSCs frequently share many characteristics with normal stem cells, such as self-renewal, multiple lineage differentiation and existence/absence of unique surface markers. For instance, CSCs in malignant glioma frequently express neural stem cell markers CD133 and Nestin [8, 9]. However, the similarity between CSC and normal stem cells are not sufficient to define a CSC. The most definitive feature of CSC should be its tumorigenic ability. By definition, a cancer cell is a CSC only if this single cell could drive the growth of secondary tumors. This is a strictly functional definition. Much effort has been put into attempting to identify molecular signatures of a CSC. This led to numerous publications claiming a long list of CSC signature genes, including surfacing proteins, transcriptional factors, and microRNAs [10–14]. It seems that CSC specific genes are highly tumor type-dependent and none of the markers identified so far could be widely used across tumor types. CD133 is a widely used CSC marker for malignant gliomas and some of recent researches applied this marker in other malignancies such as colon cancer, liver cancer and stomach cancer [15, 16]. However, there has been strong controversy about whether CD133 is a reliable CSC marker and questions are raised from both sides constantly [17]. There are also data from mice tumor model that suggest CSC markers may highly dependent on gene mutation profile, i.e. CSC markers are depending on what oncogenes and tumor suppressors are altered in tumors [18]. Unfortunately, due to the limitation of robust molecular analysis at single cell level, we could expect this kind of argument continues in the near future.

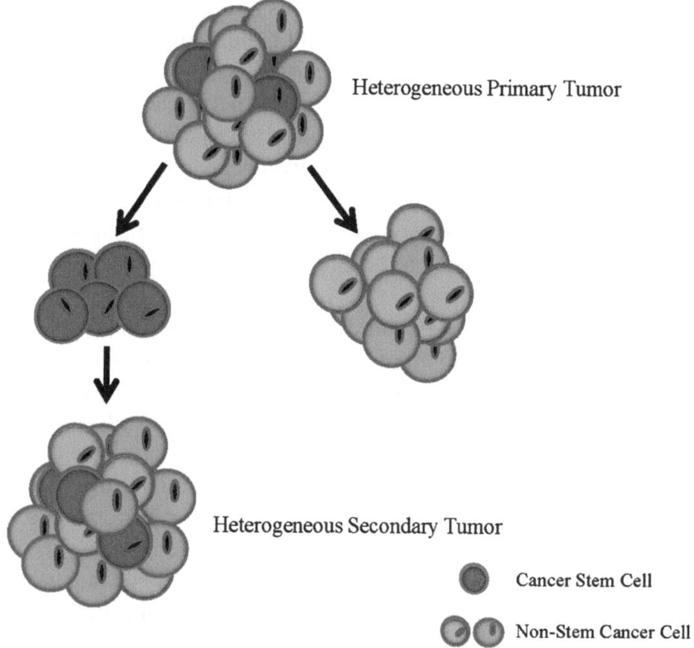

Fig. 26.1 The cancer stem cell model. Cancer cells are heterogeneous and consist of cells with various differentiation statuses. Cancer stem cells (*CSCs*) are defined by their undifferentiated status and extraordinary capacity to initiate secondary tumor growth. In contrast, non-stem cancer cells usually fail to drive secondary tumor growth *in vivo*

Researchers used to think CSCs as a rare population in a given tumor. However, increasing evidence shows that frequency of CSCs are highly dependently on many factors, including tumor type, disease stage, genomic alterations, etc. [19]. In metastatic melanoma, cells with CSC characteristics may account for more than 75% of tumor cells [20]. Importantly, CSCs have been shown to be more resistant to chemotherapy and radiotherapy [9, 21]. The molecular mechanism underlying such resistance and how it is linked to the stem cell property are largely unclear. Nevertheless, the combination of treatment resistance and strong tumor initiation capacity of CSCs offers a plausible explanation why recurrence frequently happens. Not surprisingly, a strong CSC signature in cancer is usually associated with unfavorable outcome [4, 22, 23].

26.2 Plasticity of Normal Stem Cells

The stem cell differentiation process was once thought one-direction and rigid, meaning that stem cells can give rise to differentiated daughter cells but not vice versa. However, researchers later found that lower organisms, such as worms and amphibians, constantly reverse this process during regeneration [24]. This phenomenon, termed dedifferentiation, could happen in various tissue types. In addition, in Drosophila, if stem cell population is somehow damaged or even depleted, cells that have committed to become sperm can repopulate the stem cell [25]. In mammals, dedifferentiation is much less common and the best evidence is probably from liver tissue. During liver regeneration, mature liver cells can once again acquire stem cell property and give rise to more hepatocytes [26]. In summary, stem cell differentiation is a much more plastic process than we thought and many cell types have the ability to reverse differentiation process under certain circumstance.

In addition to the plasticity of the differentiation process, normal stem cell could also exhibit another layer of plasticity: differentiate into cell type of another lineage or transdifferentiation. Most adult stem cells are thought to be tissue specific and certain stem cell can only differentiate into lineage-specific terminal cell types. For instance, neural stem cells only give rise to neuron and glia cells, while hematopoietic stem cells also reconstitute blood system. However, there have been observations that hematopoietic stem cells may actually be able to contribute to other lineages such as muscle and liver [27, 28]. In mouse model as well as human patients that receive bone marrow stem cell transplantation, researchers found that those stem cells not only contribute to hematopoietic system as they suppose to, but they also integrate into liver and muscle [27, 28]. This phenotype is called “trans-differentiation”. People later found that neural stem cells may also be able to transdifferentiate into muscle lineage [29]. If this is real, it would have great impact in clinic. For example, bone marrow transplantation would not only helpful for patients with hematopoietic disease or after treatment-induced hematopoietic depletion (e.g. chemotherapy), but those with liver degeneration or muscle atrophy. However, this transdifferentiation hypothesis drew lots of critics because it is a very rare event in natural condition and it is possible the so called “transdifferentiation” is just a result of contamination from other stem cells (stem cells for other lineage or multipotent stem cells) or cell fusion [30].

While the existing of natural transdifferentiation debate is still ongoing, the recent explosion of researches about inducible pluripotent stem (iPS) cells have clearly demonstrate that it is possible to change the lineage commitment of cells [31]. Fibroblast cells have been successfully converted into not only pluripotent stem cells, but also multipotent hematopoietic progenitors, cardiocyte, hepatocyte and neuron [32–36]. This strongly suggests that cells, including stem cells, hold the potential to change their lineage identify and switch into another cell type. The precise control of cell plasticity is still largely unclear, but numerous reports have shown that multiple layers of epigenetic regulations such a histone acetylation and methylation play crucial roles and are key determinants of successful cell-fate conversion [37–39]. Nevertheless, the unexpected but robust plasticity of stem cells and other cell types have been changing our view of differentiation process and will have great impact on regenerative medicine research.

26.3 Cancer Stem Cell Plasticity: Differentiation and Dedifferentiation

Cancer stem cells share characteristics with normal stem cells, including their self-renewal and differentiation capacity [13]. Even though in general, gene mutations in CSC make them hard to fully differentiate comparing to normal stem cells, most CSC do have retain the capability of differentiation. To promote CSC differentiation has been widely proposed and tried as an anti-cancer treatment strategy. For instance, upon imatinib (Gleevec) treatment, some Chronic Myeloid Leukemia (CML) stem cell can differentiate into mature granulocyte-like cells, expressing surface markers of terminal differentiated cells [40]. Similarly, glioma stem cells have been shown to be able to partially differentiate and express multiple mature cell markers of glial and neuron [9]. However, in most cases, the differentiation of CSCs is often incomplete and those "differentiated cells" still retain some stem cell markers, even when they seem to lose their tumorigenic property.

Traditional pathology suggests that a given cancer sample is usually a mixture of malignant cells with different differentiation status. Some cells are highly undifferentiated and express various stem cell markers; some look like completely differentiated, while most lie in between. This conclusion is further supported by recent development of single cell expression profile [41]. The percentage of each group greatly varies among samples. The original CSC hypothesis suggest that those "undifferentiated cells" are CSCs and eliminate them may greatly improve therapy efficiency [42]. While lots of evidence verified that CSCs are more tumorigenic, it is unclear whether kill CSC population only is sufficient to stop tumor growth in clinic. Recent data suggest that there may be a balance of highly undifferentiated CSCs and "differentiated cancer cells", controlled by microenvironment [13, 43, 44]. It has been well-known that tumor microenvironment, such as growth factors, nutrition, PH and oxygen have great impact on tumor biology. We recently find in gliomas, all those factors can contribute to CSC biology. For instance, hypoxia and low PH environment favor the reprogramming of "differentiated cancer cells" into CSCs [43, 44]. The researches in other cancer types also indicate that CSC and differentiated cancer cells constantly switch identify, both *in vitro* and *in vivo*. For instance, Roesch and colleagues report that Jarid1b+ cells represent cancer stem cells in some melanoma samples [45]. More importantly, the expression of Jarid1b is highly dynamic and doesn't follow the classic hierarchical model. While only Jarid1b+ cells are highly tumorigenic, Jarid1b- cells can become Jarid1b+ at certain circumstance [45]. Therefore, not only CSC gives rise to differentiated cancer cells, but those cancer cells with differentiated markers can rapidly gain CSC property upon proper microenvironment induction. This phenotype is similar to what observed in some normal tissue, where accidental loss of stem cell pool could induce the dedifferentiated of post-mitotic cells [25]. Observations like this strongly challenged the original CSC hypothesis and promote researchers to rethink and modify the CSC model.

Currently, it is widely agreed that CSCs don't represent a defined "population" like we original thought. Rather, CSC is a "status" that is highly dynamic and governed by microenvironment. The ease of dedifferentiation by cancer cells represent a great challenge in developing anti-CSC treatment and suggest that in order to inhibit cancer growth, besides killing the existing CSCs, blocking other cancer cells from dedifferentiation is also crucial. This also suggests that a deep understanding of molecular mechanisms that control the "CSC status" is critical in order to develop efficient anti-cancer drugs. Many pathways have been reported to be differentially regulated in CSCs and non-stem cancer cells, including Hedgehog, Wnt, Notch, HIFα, etc. [13, 46–48]. However, it is unclear how many of them are essential components of a "core CSC network". Besides, many of those pathways may be redundant in CSCs and therefore, inhibit one of them wouldn't be sufficient to eliminate CSCs. It is therefore important to determine how those pathways contribute to the CSC phenotype and whether they converge into a shared downstream program, which may represent a more attractive candidate for anti-cancer drug development.

26.4 Cancer Stem Cell Plasticity: Transdifferentiation

Dedifferentiation of differentiate cancer cells indicates that cancer cells are highly plastic in term of their differentiation status. However, a couple of recent report suggested another unexpected plasticity of cancer stem cell: transdifferentiation. In malignant glioma, researchers show that CSC is able to give rise to endothelial cells and help form new blood vessels that feed the tumor growth [49]. To my knowledge, this is the first report indicating that CSCs can transdifferentiate into other cell lineages. The question now is whether this is biologically relevant. Normal adult stem cells such as neural stem cell, hematopoietic stem cells have shown some transdifferentiation ability but it is usually very inefficient [27]. Researchers propose that this may due to the fact that epigenetic hurdles are very hard to pass. Given that cancers have unstable genome in comparison to normal cells and therefore usually show much faster evolution, it would be reasonable to think they are better at changing epigenetic landmarks and their identify. However, this has not been approved so far by experiments. It remains to be seen that whether transdifferentiation is a common and frequent events during cancer progression. This finding, if further proved in clinic, could have important impact in multiple areas.

First, it changed the traditional idea that cancer-associated stroma cells are genetically stable. Much effect has been made to study tumor-associated stroma and develop drugs against special pathways that control the crosstalk of tumor-associated stroma cells and tumor cells [15, 50]. Many scientists are enthusiastic about the idea of targeting stroma cells, because they do not have unstable genome as cancer cells and therefore unlikely to develop rapid drug resistance. Anti-angiogenesis therapy is one of the leading areas in developing drugs against tumor stroma cells instead of tumor cells themselves. Indeed, one of the first clinical approved anti-angiogenesis drugs, Avastin, has demonstrated good efficacy against various tumors in clinic including colon cancer, malignant gliomas, breast cancer [51–53]. This has served as a successful proof-of-concept for the strategy of targeting tumor-associated stromas. However, if indeed, CSCs can transdifferentiate into blood vessels and maybe other stroma cells, then the strategy certainly requires careful re-evaluation. In addition, our unpublished data show that malignant cancer stem cells actually express VEGF receptors and therefore could be a direct target of Avastin. This complicates the potential mechanism through which Avastin works and raises the question whether the efficacy of Avastin is solely due to its role of inhibiting normal blood vessel growth. For instance, does it also inhibit CSC survival and transdifferentiation?

Secondly, this may also shed light on some new mechanism regarding tumor metastasis, the major reason for cancer mortality [54]. Metastasis is a highly inefficient process and the tumor cells have to be exposed to various microenvironment that are totally different from each other along the process [54]. For instance, for a lung metastasis from primary breast cancer, they have to sequentially adapt to microenvironment in breast, blood vessel and lung. Given their strong tumor initiating capacity in animal models, CSCs have been proposed to be the seeds for metastasis [55]. If CSCs can transdifferentiate, then it is possible that they could sequentially change their identities alone to metastasis process to mimic cells that normally locate in that microenvironment. It may significantly help their survival at metastatic sites. The diagnosis of primary non-small cell lung cancer versus a lung metastasis from breast cancer can be challenging some times because they look similar at morphology level and both are CK7+ and CK20– [56]. It is possible that breast cancer cells have transdifferentiated and acquired some lung epithelial cell property. It could only be proven (or disproven) if we can track the epigenetic features of single cancer cell along the way. Novel reporter and imaging systems are required in order to do this. Besides, it is well-known that certain types of cancer prefer to metastasize to some tissues [57]. For instance, breast cancers predominantly metastasize to lung while colon cancer prefers liver site. How this happens is still a mystery even though many theories have been proposed. A bald but exciting hypothesis would be that CSCs from a given cancer type favor the "transdifferentiation" into certain lineage and thus survive much better at the distant site. Further experiments are required to access this possibility.

How does "transdifferentiation" happen and whether it is a common phenotype in different cancers are still unknown. Recent studies on inducible pluripotent stem (iPS) cells prove that the barriers between two cell types are much smaller than we thought [31]. Moreover, not only can fully differentiated cells be reverted to become stem cells, but terminal differentiated fibroblasts can easily transdifferentiated into other lineages like neurons, muscle cells, hepatocytes, using different combination of factors [32–36]. Therefore, there are two major ways for a cell to change lineage identify. First, cells can be first fully reprogrammed into pluripotent cells and then differentiate into another lineage. Or, it can directly be converted into another lineage without going through the iPS stage. One way or the other, transdifferentiation happens. Neither of these two procedures commonly happens in normal cells under normal development, because it will lead to devastating consequences to tissues.

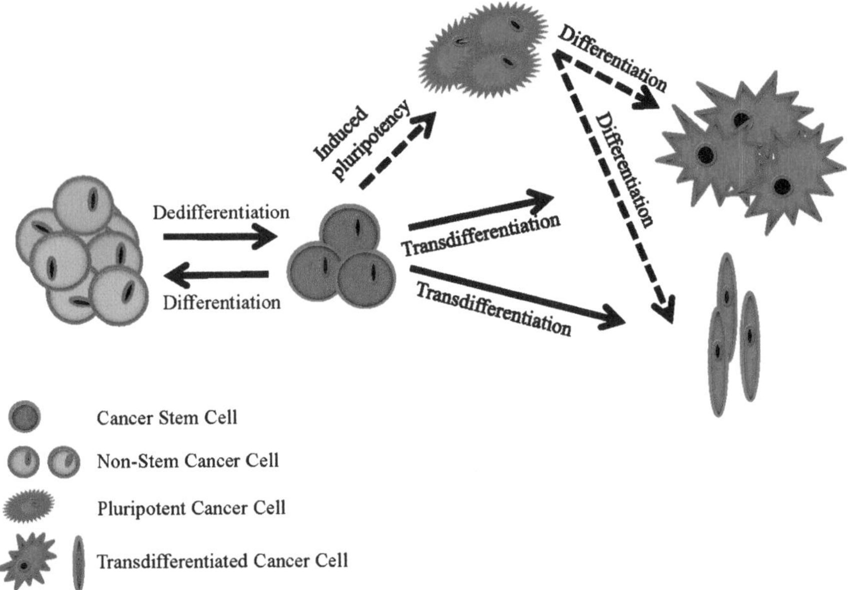

Fig. 26.2 Plasticity of cancer stem cell. CSCs could give rise to non-stem cancer cells through differentiation, while the reverse process called dedifferentiation also happens upon certain microenvironment stimulus, such as hypoxia and low PH. CSCs could also produce daughter cells with features of other cell lineages through two potential mechanisms. First, CSCs could direct transdifferentiate into other cell lineages. In addition, CSCs may first convert into a more pluripotent stem cell stage and then differentiate into other lineages

However, given that cancer cells' genomes are usually much more unstable than normal cells, it is plausible to think that CSCs are easier to get "reprogrammed" into embryonic stem cell-like or another cell lineage (Fig. 26.2).

If transdifferentiation does happen to CSCs, it is of great interest to understand which pathway it takes and what molecular mechanisms they use. On one hand, it helps us understand cancer biology. On the other hand, it may serve as a study model and give us hints on how we can manipulate the fate of normal cells. By studying how cancer cells turn themselves into another cell type, we may be able to use similar and more controlled way to change identity of normal cells.

Some preliminary experiment data support the idea that CSCs could take the iPS-and-differentiation path. Traditionally, researchers assume that CSCs resemble their normal tissue counterpart: tissue specific adult stem cells. However, through comprehensive gene-expressing microarray analysis, couple of reports suggesting that CSCs are most similar to embryonic stem cells, instead of hematopoietic stem cells [58, 59]. Whether this has a broader implication in other tumor types are still unclear. Nevertheless, the identification and study of CSC transdifferentiation phenotype deserves our attentions because it is not only important for basic research but has a direct impact in translational medicine.

26.5 Cancer Stem Cell Plasticity: Final Remark

The list of cancer types that contain CSCs has dramatically grown in the past decade. However, many fundamental questions remain. For instance, can we cure cancer by simply eliminate CSCs? Are there universal markers for CSCs in a given tumor type? What are the "minimal molecular program" that control CSC phenotype? The first step of CSC research is to isolate and enrich CSCs from tumor samples. Unfortunately, up to now, there have been no consensus markers for CSCs in any solid tumors, including breast cancer and malignant gliomas, the best studied models for CSC research. This becomes one of the biggest hurdles in CSC research field. However, there is no question that CSC represents a great opportunity in oncology research and many researchers have dedicated themselves to this exciting area.

The recognition of CSC plasticity is critical because it modified the original CSC hypothesis. It suggests that we have to getting deeper into the fundamental mechanism that controls cell self-renewal and differentiation. CSC plasticity raises bigger challenge to the already extremely hard CSC research. We now not only need to study CSC at a "static" stage, but have to investigate how CSCs and other cancer cells dynamically respond upon microenvironment changes. For instance, does anti-angiogenesis drug treatment increases the incidence of CSC transdifferentiation into blood vessels? Do breast CSCs change epigenetic landscapes to mimic lung tissue when they form lung metastatic? Do hypoxia and/or low PH increase the percentage of cells in CSC status in vivo? Many of those are open questions and require extensive investigation and collaboration from researchers with various background and expertise. With improved understanding of CSC dynamic and plasticity, we could expect another evolution of cancer research, which hopefully lead to better anti-cancer therapies.

References

1. Jordan CT, Guzman ML, Noble M (2006) Cancer stem cells. N Engl J Med 355(12):1253–1261
2. Visvader JE, Lindeman GJ (2008) Cancer stem cells in solid tumours: accumulating evidence and unresolved questions. Nat Rev Cancer 8(10):755–768
3. Hanahan D, Weinberg RA (2000) The hallmarks of cancer. Cell 100(1):57–70
4. Wicha MS, Liu S, Dontu G (2006) Cancer stem cells: an old idea – a paradigm shift. Cancer Res 66(4):1883–1890, discussion 1895–1896
5. Bonnet D, Dick JE (1997) Human acute myeloid leukemia is organized as a hierarchy that originates from a primitive hematopoietic cell. Nat Med 3(7):730–737
6. Al-Hajj M et al (2003) Prospective identification of tumorigenic breast cancer cells. Proc Natl Acad Sci USA 100(7):3983–3988
7. O'Brien CA et al (2007) A human colon cancer cell capable of initiating tumour growth in immunodeficient mice. Nature 445(7123):106–110
8. Singh SK et al (2003) Identification of a cancer stem cell in human brain tumors. Cancer Res 63(18):5821–5828
9. Bao S et al (2006) Glioma stem cells promote radioresistance by preferential activation of the DNA damage response. Nature 444(7120):756–760

10. Cairo S et al (2010) Stem cell-like micro-RNA signature driven by Myc in aggressive liver cancer. Proc Natl Acad Sci USA 107(47):20471–20476
11. Shimono Y et al (2009) Downregulation of miRNA-200c links breast cancer stem cells with normal stem cells. Cell 138(3):592–603
12. Ji Q et al (2009) MicroRNA miR-34 inhibits human pancreatic cancer tumor-initiating cells. PLoS One 4(8):e6816
13. Li Z et al (2009) Hypoxia-inducible factors regulate tumorigenic capacity of glioma stem cells. Cancer Cell 15(6):501–513
14. Zimmerman AL, Wu S (2011) MicroRNAs, cancer and cancer stem cells. Cancer Lett 300(1):10–19
15. Gilbertson RJ, Rich JN (2007) Making a tumour's bed: glioblastoma stem cells and the vascular niche. Nat Rev Cancer 7(10):733–736
16. Wu Y, Wu PY (2009) CD133 as a marker for cancer stem cells: progresses and concerns. Stem Cells Dev 18(8):1127–1134
17. Cheng JX, Liu BL, Zhang X (2009) How powerful is CD133 as a cancer stem cell marker in brain tumors? Cancer Treat Rev 35(5):403–408
18. Curtis SJ et al (2010) Primary tumor genotype is an important determinant in identification of lung cancer propagating cells. Cell Stem Cell 7(1):127–133
19. Dick JE (2009) Looking ahead in cancer stem cell research. Nat Biotechnol 27(1):44–46
20. Quintana E et al (2008) Efficient tumour formation by single human melanoma cells. Nature 456(7222):593–598
21. Cheng L, Bao S, Rich JN (2010) Potential therapeutic implications of cancer stem cells in glioblastoma. Biochem Pharmacol 80(5):654–665
22. Gupta PB et al (2009) Identification of selective inhibitors of cancer stem cells by high-throughput screening. Cell 138(4):645–659
23. Merlos-Suarez A et al (2011) The intestinal stem cell signature identifies colorectal cancer stem cells and predicts disease relapse. Cell Stem Cell 8(5):511–524
24. Stocum DL (2004) Amphibian regeneration and stem cells. Curr Top Microbiol Immunol 280:1–70
25. Sheng XR, Brawley CM, Matunis EL (2009) Dedifferentiating spermatogonia outcompete somatic stem cells for niche occupancy in the Drosophila testis. Cell Stem Cell 5(2):191–203
26. Elaut G et al (2006) Molecular mechanisms underlying the dedifferentiation process of isolated hepatocytes and their cultures. Curr Drug Metab 7(6):629–660
27. Spyridonidis A et al (2005) Stem cell plasticity: the debate begins to clarify. Stem Cell Rev 1(1):37–43
28. Jahagirdar BN, Verfaillie CM (2005) Multipotent adult progenitor cell and stem cell plasticity. Stem Cell Rev 1(1):53–59
29. Gritti A, Vescovi AL, Galli R (2002) Adult neural stem cells: plasticity and developmental potential. J Physiol Paris 96(1–2):81–90
30. Wurmser AE, Gage FH (2002) Stem cells: cell fusion causes confusion. Nature 416(6880):485–487
31. Takahashi K, Yamanaka S (2006) Induction of pluripotent stem cells from mouse embryonic and adult fibroblast cultures by defined factors. Cell 126(4):663–676
32. Szabo E et al (2010) Direct conversion of human fibroblasts to multilineage blood progenitors. Nature 468(7323):521–526
33. Pfisterer U et al (2011) Direct conversion of human fibroblasts to dopaminergic neurons. Proc Natl Acad Sci USA 108(25):10343–10348
34. Kim J et al (2011) Direct reprogramming of mouse fibroblasts to neural progenitors. Proc Natl Acad Sci USA 108(19):7838–7843
35. Ieda M et al (2010) Direct reprogramming of fibroblasts into functional cardiomyocytes by defined factors. Cell 142(3):375–386
36. Huang P et al (2011) Induction of functional hepatocyte-like cells from mouse fibroblasts by defined factors. Nature 475(7356):386–389

37. Kim K et al (2010) Epigenetic memory in induced pluripotent stem cells. Nature 467(7313):285–290
38. Lister R et al (2011) Hotspots of aberrant epigenomic reprogramming in human induced pluripotent stem cells. Nature 471(7336):68–73
39. Djuric U, Ellis J (2010) Epigenetics of induced pluripotency, the seven-headed dragon. Stem Cell Res Ther 1(1):3
40. Sell S (2005) Leukemia: stem cells, maturation arrest, and differentiation therapy. Stem Cell Rev 1(3):197–205
41. Novak R et al (2011) Single-cell multiplex gene detection and sequencing with microfluidically generated agarose emulsions. Angew Chem Int Ed Engl 50(2):390–395
42. Reya T et al (2001) Stem cells, cancer, and cancer stem cells. Nature 414(6859):105–111
43. Heddleston JM et al (2009) The hypoxic microenvironment maintains glioblastoma stem cells and promotes reprogramming towards a cancer stem cell phenotype. Cell Cycle 8(20):3274–3284
44. Hjelmeland AB et al (2011) Acidic stress promotes a glioma stem cell phenotype. Cell Death Differ 18(5):829–840
45. Roesch A et al (2010) A temporarily distinct subpopulation of slow-cycling melanoma cells is required for continuous tumor growth. Cell 141(4):583–594
46. Espada J et al (2009) Wnt signalling and cancer stem cells. Clin Transl Oncol 11(7):411–427
47. Pannuti A et al (2010) Targeting Notch to target cancer stem cells. Clin Cancer Res 16(12):3141–3152
48. Merchant AA, Matsui W (2010) Targeting Hedgehog–a cancer stem cell pathway. Clin Cancer Res 16(12):3130–3140
49. Wang R et al (2010) Glioblastoma stem-like cells give rise to tumour endothelium. Nature 468(7325):829–833
50. Hofmeister V, Schrama D, Becker JC (2008) Anti-cancer therapies targeting the tumor stroma. Cancer Immunol Immunother 57(1):1–17
51. [PZ Innovation Prize for Avastin: Bevacizumab-new therapy option in colon carcinoma]. Krankenpfl J, 2005. 43(7–10): p. 234
52. Miles DW et al (2010) Phase III study of bevacizumab plus docetaxel compared with placebo plus docetaxel for the first-line treatment of human epidermal growth factor receptor 2-negative metastatic breast cancer. J Clin Oncol 28(20):3239–3247
53. Sathornsumetee S et al (2010) Phase II trial of bevacizumab and erlotinib in patients with recurrent malignant glioma. Neuro Oncol 12(12):1300–1310
54. Klein CA (2008) Cancer. The metastasis cascade. Science 321(5897):1785–1787
55. Hermann PC, Huber SL, Heeschen C (2008) Metastatic cancer stem cells: a new target for anti-cancer therapy? Cell Cycle 7(2):188–193
56. Hirano T et al (2003) Usefulness of TA02 (napsin A) to distinguish primary lung adenocarcinoma from metastatic lung adenocarcinoma. Lung Cancer 41(2):155–162
57. Hart IR (1982) 'Seed and soil' revisited: mechanisms of site-specific metastasis. Cancer Metastasis Rev 1(1):5–16
58. Ben-Porath I et al (2008) An embryonic stem cell-like gene expression signature in poorly differentiated aggressive human tumors. Nat Genet 40(5):499–507
59. Wong DJ et al (2008) Module map of stem cell genes guides creation of epithelial cancer stem cells. Cell Stem Cell 2(4):333–344

Chapter 27
Cross Talks Among Notch, Wnt, and Hedgehog Signaling Pathways Regulate Stem Cell Characteristics

Su-Ni Tang, Sharmila Shankar, and Rakesh K. Srivastava

Contents

Abstract The use of stem cells as medicines is a promising area of research as they may help the body to replace damaged or lost tissue in a host of diseases including cancer. The integration of intrinsic and extrinsic signals is required to preserve the self-renewal and tissue regenerative capacity of adult stem cells, while protecting them from malignant conversion or loss of proliferative potential by death, differentiation or senescence. It is now clear that malignant tumors are heterogeneous and

---

S.-N. Tang
Department of Pharmacology, Toxicology and Therapeutics, and Medicine,
The University of Kansas Cancer Center, The University of Kansas Medical Center,
3901 Rainbow Boulevard, Kansas City, KS 66160, USA

S. Shankar
Department of Pathology and Laboratory Medicine, The University of Kansas Cancer Center,
The University of Kansas Medical Center, Kansas City, KS, USA

R.K. Srivastava (✉)
Department of Pharmacology, Toxicology and Therapeutics, and Medicine,
The University of Kansas Cancer Center, The University of Kansas Medical Center,
3901 Rainbow Boulevard, Kansas City, KS 66160, USA

Department of Pathology and Laboratory Medicine, The University of Kansas Cancer Center,
The University of Kansas Medical Center, Kansas City, KS, USA
e-mail: rsrivastava@kumc.edu

R.K. Srivastava and S. Shankar (eds.), *Stem Cells and Human Diseases*,
DOI 10.1007/978-94-007-2801-1_27,

contain diverse subpopulations of cells with unique characteristics including the ability to initiate a tumor and metastasize. This phenomenon might be explained by the so-called cancer stem cell (CSC) theory. Recent technological developments have allowed a deeper understanding and characterization of CSCs. The CSCs share some of the common signaling pathways of self-renewal with that of normal stem cells or progenitor cells. Signaling pathways such as Notch, Sonic hedgehog and Wnt play major roles in stem cell self-renewal and metastasis. These pathways cross-talk and allow stem cells to balance their regenerative potential and the initiation of terminal differentiation programs, ensuring appropriate tissue homeostasis. Understanding the signaling circuitries regulating stem cell fate decisions might provide insights into cancer initiation and progression that involve the progressive loss of tissue-specific adult stem cells. Efficacious therapeutic approaches targeting the CSC population should be explored to overcome therapeutic failure and improve patient outcomes. This review will focus on the signaling pathways required for regulation of CSCs, and development of therapeutic approaches to target specifically CSCs.

Keywords Stem cells • Cancer stem cells • Notch • Hedgehog • Wnt

27.1 Stem Cells and Cancer Stem Cells

Multicellular organisms consist of three types of cells: somatic cells, germ cells, and stem cells. They are different in several fundamental ways. Somatic cells include more than 200 different types of cells which compose the major part of body of organism. Their proliferative capacity is limited. The somatic cells undergo 50–70 population doubling before entering the senescence state, in which they maintains metabolic activity but stops the further divisions. Germ cells are cells which give rise to gametes through spermatogenesis and oogenesis. They are pluripotent as stem cells but divide through meiosis. Stem cells are the origin of cells and exist in embryonic and adult tissues. They have the ability to self-renew, differentiate into mature cells of a particular tissue, repair damaged organs, and even regenerate in some cases. They divide through mitosis and give rise to germ cells and somatic cells. In turn, germ cells can fuse to create genetically different offspring, while somatic cell can support, protect and nurture the populations of the other two types of cells.

Based on their ability to proliferate and differentiate into various mature functional cells, the stem cells have been categorized into three types, embryonic stem cells (ESC), cord blood/placental stem cells (CBPSC), and somatic stem cells (SSC). In the human body, all of the cells come from the fertilized ovum, termed as the primordial stem cell. ESCs are the immediate progeny of the primordial stem cells and have been shown to be totipotent [61]. They have the ability to give rise to all kinds of somatic as well as stem cells. In contrast to the symmetric proliferation of ESCs, the progenitor cells of ESCs in early embryo will endure the asymmetric division, begin losing lost potency and gain differentiated characteristics during the process of determination. The half of daughter cells retains the potential to produce different cell lineages as their parental stem cells. The other half will start the process of differentiation and known as SSC, which is multipotent and responsible for normal tissue renew. The stem cells from cord blood, umbilical cord, placenta and amniotic fluid have been reported and known as CBPSC. However their potency is still under scrutiny, it is believed that CBPSCs are pluripotent and represent a highly potential source of stem cells [36]. SSCs are the major contributor to most of cellular renewal and found at bone marrow, skin, brain, heart, skeletal muscle, gut epithelium and other tissues [5, 27, 46]. They are most limited in the potency among three types of stem cells but still play a crucial in regulating tissue homeostasis, neural plasticity and maintenance and repair of the tissue in which they are present.

Since the hypothesis that cancer arise from the maturation arrest of stem cell differentiation [56], the role of stem cells in carcinogenesis has been widely studied in every major theory of cellular origin of cancer. In field theory, both the germinal stem cells and the ESC have the potential to express the malignant phenotype if they are placed in abnormal environment favor the tumorigenesis [57]. In the theory of chemical carcinogenesis, viral infections, and mutations, the stem cells are more susceptible to molecular lesion and infection than somatic cells [8, 23, 69].

The development of cancer is a multi-step procedure. The major differences between stem cells and somatic cells are the abilities to self-renew and increased longevity. The tissues repaired and replenished by stem cell proliferation is a crucial step to cancer development. The increased lifespan of stem cells provides the opportunity to the sequential accumulation of the "essential" mutations and prevent the loss of alteration during normal tissue turnover [64].

The discovery of cancer stem cells (CSC) shed a new light on our understanding of tumorigenesis in stem cells. Many studies indicated that acute myeloid leukemia (AML) cells have limited proliferative ability, which suggested a possible presence of a population of stems cells within tumor cells [19, 53]. In 1994, a population of AML-initiation cells were revealed from bone marrow of AML patients [31]. They could successfully produce the colony-forming progenitors after transplanting into severe combined immune-deficient mice. Further research indicates that there is a hierarchical organization present within AML cells and originating from a subset of primitive cells [7]. Since then, the similar pluripotent "tumor initiating cells" have been reported in breast, brain, colon, head and neck, lung, liver, pancreatic, prostate, breast, and other solid tumors [1, 11, 13, 15, 25, 35, 41, 45, 47, 49, 54].

CSCs are generally defined as a subpopulation of stem cells within a tumor that have the capacity to self-renewal, can develop into all cell types in the overall tumor population, the ability to initiate, regenerate, and drive the proliferation of cancer cells. The CSC theory might explain why tumors often reappear even after the initial successful treatment. Because of the pivotal role of CSC in tumorigenesis, the cancer therapies targeting CSCs offer the potential to eradicate the tumor population. So far at least 14 CSC-targeting drugs have entered clinical trials [68]. However, Tumor cells often acquire resistance to chemo and radiation therapies after the initial non-lethal treatment, which might present the natural selection over the cancer population.

Consistent with this notion, it has been suggested that the heterogeneity shown in CSC populations might result from the evolution of resistance against the initial treatments [20]. Many reports have indicated CSCs have exhibited numerous genetic and cellular adaptations, including, the over-amplification of apoptosis inhibitors s, DNA repair system, and the overexpression of drug efflux membrane pumps, against conventional anti-tumor therapeutics [4, 14, 32]. Therefore it is necessary to fully elucidate the mechanisms underlying CSC-induced tumorigenesis because CSCs have already exhibit resistant to several chemo and radiation therapies. The poor understandings of CSCs might limit the actual clinical benefits of those novel anticancer therapies.

Since the molecular pathways are important to the self-renewal and differentiation of both stem cells and CSCs, the comparisons of those signaling pathway will benefit the development of therapeutic approach targeting. So far, eight major signaling pathways, which include the NFκB pathway, the Wnt pathway, the Notch pathway, the Jak/Stat pathway, the mitogen-activated protein (MAP) kinase pathway, the TGF-β pathway, the Hedgehog (Hh) pathway, and the PI3K/AKT pathway, have been implicated in both stem cells and CSCs [9]. Among those pathways, the Wnt, Notch and HH pathways have been well studied because they are commonly activated in many types of cancer. The knowledge learned from them might provide valuable insight in how to design the therapeutic agents targeting these pathways.

27.2 Notch Signaling Pathway

Notch signaling is an evolutionary conserved pathway mediating signaling between adjacent cells, which can regulate many cellular processes, such as cell proliferation, apoptosis, migration, invasion, and angiogenesis during the development of tissue. In 1917, Notch was firstly genetically identified from a mutant fly with 'notches' in its wings [37]. Molecular cloning studies in the mid-1980s revealed that the Notch gene codes for a single-pass transmembrane receptor which undergoes proteolytic cleavage upon the bind of ligands present on the surface and neighboring cells and releases a transcription factor to induce the expression of target genes [66]. Since then, the homologs of Notch genes have been identified in numerous organisms. So far, four Notch receptors (Notch 1–4) and five structure-related Notch ligands (Delta 1, 2 and 3, Jagged 1 and 2) have been identified in mammals [16].

Activation of Notch signaling is initiated by the binding of Notch ligands to Notch receptors, which leads to a series of proteolytic cleavages catalyzed by a disintegrin and metalloproteinase (ADAM) family metalloproteases and γ-secretase. Then the active notch intracellular domain (NICD) is liberated and subsequently shuttles into the nucleus, where it forms a transcription initiation complex by binding CBF1/Suppressor of Hairless/Lag-1 (CSL) and Mastermind-like 1 (MAM) [2].

Notch signaling plays an integral role in coordinating the communication between neighboring cells. The Notch pathway emerges as an important regulator of cell differentiation and fate decision. It is reported that it has been involved in breast development, colorectal epithelial maturation, and immune regulation. Depending on the different cell types and developmental stages, the Notch signaling may exhibit opposite effects. For example, Notch can promote the adjacent cell to adopt the same cell fate in Drosophila wing growth or it can inhibit the spread of neurogenesis in the Drosophila embryo. Moreover, the Notch pathway is also responsible for the stem cell maintenance. In hematopoietic stem cells (HSC), the Notch signaling is dispensable for primitive and secondary hematopoiesis [22, 30]. In neural stem cells, the Notch signaling controls the differentiation of neural stem cells in two steps. It initially inhibits the neuronal fate while induce the glial cell fate. Sequentially the Notch signaling promotes the differentiation of astrocytes while inhibit the differentiation to both neurons and oligodendrocytes [18].

Deregulated Notch signaling has been observed in a number of malignancies. Down-regulated Notch activity has been reported in T-cell acute lymphoblastic leukemia, which might result from impaired differentiation of hematolymphopoietic cells from HSCs in the deregulation of Notch pathway [65]. Moreover, recent studies indicate that the aberrant overexpression of Notch signaling can promote the transform SCC into CSC in gliomas [58] (Fig. 27.1).

27.3 Wnt Signaling Pathway

The Wnt pathway is also among the most evolutionary conserved pathways and regulates a variety of cellular activities which include morphogenesis, embryogenesis, cell polarization, cell proliferation and differentiation. The first Wnt gene, Int-1, was initially recognized as the murine oncogene [39]. The consequent studies revealed that *Drosophila* segment polarity gene, the wingless (wnt-1) gene, is the homologue of Int-1 [50]. So far 19 mammalian Wnts have been identified and have distinct biological properties even they are structure related [28].

Wnts are a group of secreted glycoproteins of 350–400 amino acids in length which bind to cell surface receptors to initiate signaling downstream pathways [34]. The Wnt pathway possesses three downstream branches: the canonical pathways, the planar polarity pathway, and the Ca^{2+} pathway [38]. Among these three pathways, the canonical pathway is responsible for the tumorigenesis of CSCs as well as the maintenance of stem cells. The activity of canonical pathway depends on the amount of β-catenin in the cytoplasm (Fig. 27.2). When Wnt is absent, β-catenin is generally maintained at a low level and tagged by adenomatous polyposis coli (APC) for ubiquitination and degradation. When Wnt is secreted, the Wnt ligands will bind the Frizzed (Fz) receptor, a seven trans-membrane repeat protein, and lead to phosphorylation of Dishevelled (DVL) through its PDZ domain [6]. Then, the activated DVL will bind to Axin and prevent the Glycogen Synthase Kinase-3 (GSK-3) from phosphorylating the APC. Consequently, β-catenin is stabilized and free to accumulate in the cytoplasm. Then, the free β-catenin is translocated into nucleus, in which it interacts with T-cell factor (TCF) family and Lymphocyte enhancer binding factor (LEF) of transcription factors and recruits various transcriptional co-activators,

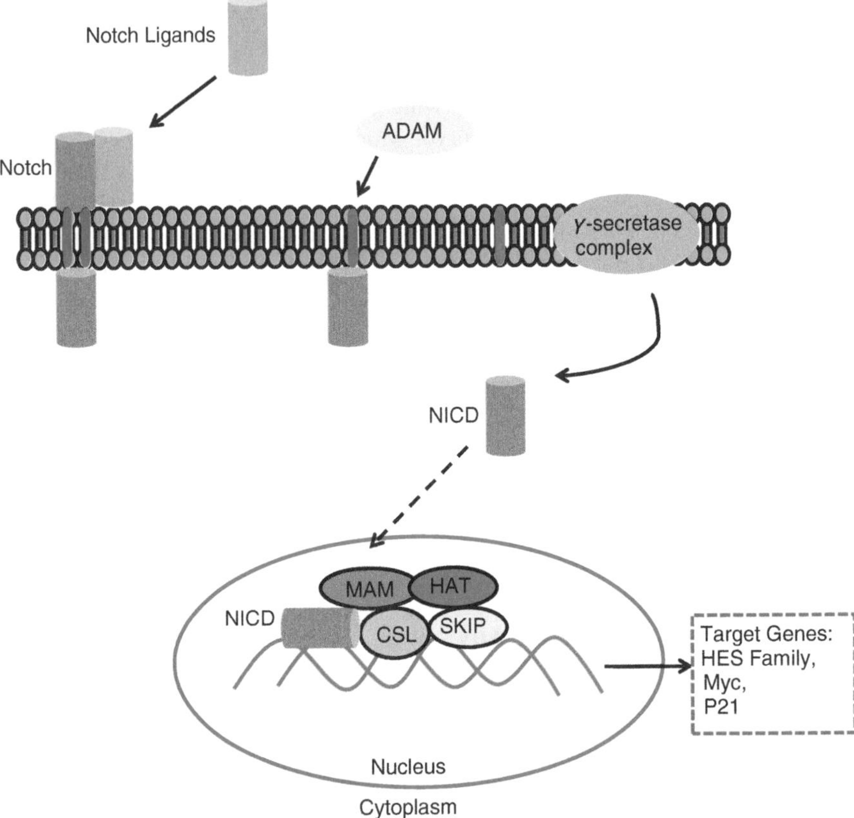

Fig. 27.1 Notch signaling pathway. Notch signaling is initiated by the binding of Notch ligands to Notch receptor, which leads to a conformational change of Notch receptor. Following the cleavages by a disintegrin and metalloproteinase (ADAM) and γ-secretase complex, the active notch intracellular domain (NICD) is released from Notch receptor and translocated into the nucleus. NICD binds to the transcription initiation complex-CBF1/Su(H)/Lag-1 (CSL), Mastermind-like 1 (MAM), histone acetyltransferase (HAT) and ski-interacting protein (SKIP), which promotes the expression of target genes (HES family, Myc, P21, etc.)

e.g. BCL9/Legless, cAMP response element-binding protein (CREB)-binding protein (CBP) and its homolog p300, through its N- and C-terminal transactivation domains. In turn, they will lead to the expression of a series of cell proliferation genes, e.g. cyclin D1, surviving, and c-myc [24, 60].

In general, the Wnt signaling activates proliferation and inhibits apoptosis. In both ESC and SSC, the Wnt pathway is critical to sustain the self-renewal of many tissues. It plays an important role in the development and renewal of the intestinal epithelium cells through the signaling gradient from the crypt base to the villus [55]. Similar to the intestinal epithelium, the Wnt signaling is also important in the skin homeostasis and hair follicle regeneration [26]. Recent studies have begun to reveal the role of

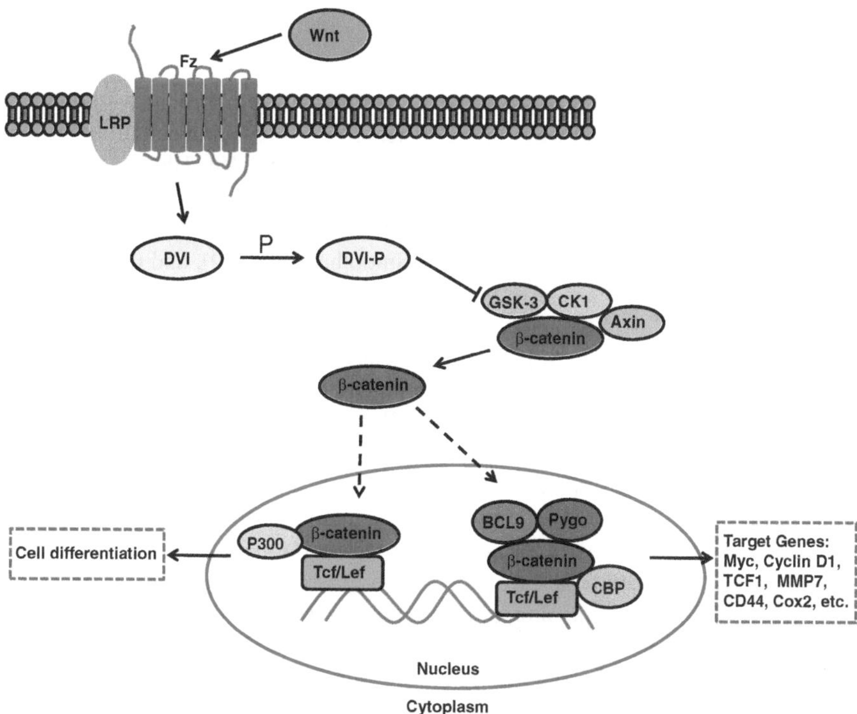

Fig. 27.2 Wnt signaling pathway. Wnt proteins act as the ligands for the Frizzled (Fz) transmembrane receptor. In the absent of Wnt ligands, β-catenin binds to the tumour suppressors-adenomatous polyposis coli (APC) and axin, and its amino terminus is phosphorylated by casein kinase Ia (CKI) and glycogen synthase kinase 3h (GSK-3). Canonical Wnt signaling is initiated by the binding of Wnt to the cognate receptor complex (Fz and LRP). Dishevelled (DVL) is phosphorylated by CKI resulting in the release of β-catenin from the protein complex. The free β-catenin can be translocated to the nucleus, form complex with P300 or cyclic AMP response element-binding protein (CBP), and promote the expression of target genes

Wnt signaling in hematopoiesis. The early impaired α/β T cell differentiation was observed in LEF1$^{-/-}$TCF1$^{-/-}$ mice [42]. Wnt3a protein induces self-renewal of HSC [67]. β-catenin overexpression promoting the mobilization of HSC [29]. Additionally, the Wnt singling pathway is also implicated in the maintenance of neural, prostate gland and stem cells [62].

In CSC, the accumulated DNA damages might be enhanced by genomic instability induced by the Wnt pathway. Increased Wnt signaling has been observed in colon CSCs. It is consistent with the incident of colorectal cancer in APC-loss mice since APC gene is a tumor suppressor in the Wnt signaling pathway by degrading β-catenin [59]. The important role of Wnt signaling in cell fate decisions including proliferation, differentiation, and apoptosis, coupling with the fact that the alteration of the Wnt pathway is associated with numerous solid tumors, makes it a key target for anti-tumor agents.

27.4 Hedgehog Signaling Pathway

The Hedgehog (Hh) genes were initially identified for embryonic patterning in *Drosophila*, and the Hh pathway plays a critical role in regulating cell proliferation, migration and differentiation, and stem cell maintenance [40]. Hh family ligands include Indian (Ihh), Desert (Dhh) and Sonic (Shh) in mammals. In the absent of Hh the transmembrane receptor patched (PTCH1 and PTCH2) can inhibit the activity of another transmembrane protein Smoothened (SMO), and the regulation of Glioblastoma (Gli) family of transcription factors (Gli1, Gli2 and Gli3) is suppressed (Fig. 27.3). Gli1 and Gli2 act as transcriptional activators, while Gli3 is a transcriptional repressor [52]. In the present of Hh ligands the binding of Hh ligands to patched protein can release the inhibition of SMO and initiate Hh pathway activation, which results in that Gli1 and Gli2 are active and translocated to the nucleus to induce the transcription of target genes such as Cyclin D, Cyclin E, Myc, Gli1, Ptch and HIP [63] (Fig. 27.3).

Previous studies have shown a role of Hh signaling in self-renewal and maintenance of stem cells in human skin, nervous system and hematopoietic system. In skin hair follicles there is a population of bulge stem cells, which regenerate hair follicles during hair cycles and repair skin injury [10]. Transcription of hair follicle stem cell markers KRT15, KRT19, and Gas1 can be induced by lentivirus-mediated overexpression of transcription factor Gli1 in keratinocytes, demonstrating that Hh signaling is required for the maintenance of bulge stem cells in young and aged human skin [51]. The investigation of Hh signaling in the developing mouse neocortex has revealed a crucial mechanism in regulating the number of embryonic and postnatal mouse neocortical cells with stem cell properties and their proliferation [43]. Moreover, deletion of *Smoothened* gene function within the subventricular zone of the adult neural stem inhibited their proliferation and neurogenesis, which indicates that Hh signaling is important to maintain adult neural stem cells [3]. Interestingly, some studies have found that Hh signaling is an important regulator in hematopoietic stem cells, however, removal of *Smoothened* gene has no distinct effect on adult hematopoietic stem cells, and Hh signaling is not required in adult hematopoietic stem cells [17].

According to direct evidence from a lot of recent studies in different human tumors including breast cancer, colon cancer and chronic myeloid leukemia (CML), Hh signaling pathway has been implicated in cancer stem cells. *In vivo* and *in vitro* experiments supported that Hh signaling controls self-renewal of both normal and malignant human mammary stem cells through *BMI-1* gene [33]. Hh signaling regulated the expression of CD133 in colon cancer stem cells and increase the tumor growth *in vitro* and *in vivo*, indicating that it is required for the survival of colon cancer stem cells [21]. The increase of the frequency of CML stem cells by constitutively active Smo and the decrease of CML growth by loss of Smo have showed that Hh signaling is essential for maintenance of cancer stem cells in CML [70]. Mutations in Hh signaling can cause human medulloblastoma and rhabdomyosarcoma, for example, mutations in *PTCH1* and *SMOH* were identified in 10% of basal

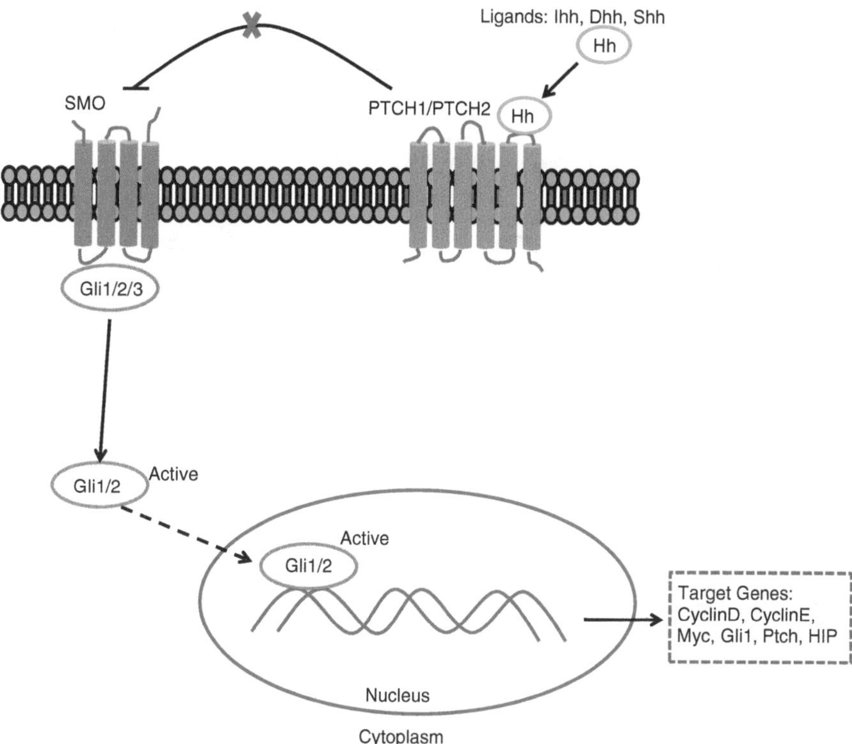

Fig. 27.3 Hedgehog signaling pathway. Hedgehog (Hh) family ligands include Indian (Ihh), Desert (Dhh) and Sonic (Shh) in mammals. In the absence of Hh ligands, the transmembrane receptor patched (PTCH1 and PTCH2) can inhibit the activity of another transmembrane protein Smoothened (SMO), and the regulation of Glioblastoma (Gli) family of transcription factors (Gli1, Gli2, Gli3) are suppressed. Gli3 is a transcriptional repressor. When Hh ligands bind to patched, the inhibition of SMO is released. Gli1 and Gli2 are active and translocated to the nucleus to induce the transcription of target genes such as Cyclin D, Cyclin E, Myc, Gli1, Ptch and HIP

cell carcinomas [48]. Hh signaling pathway may represent an important regulator of cancer stem cell carcinogenesis. The increased understanding of the role of Hh pathway in stem cells and cancer stem cells provide a good opportunity for cancer therapeutic intervention.

27.5 Conclusions

Cancer stem cells and normal stem cells share some of the biological properties including indefinite self-renewal, symmetric cell division and an innate resistance to cytotoxic therapeutics. According to cancer stem cells and tumor formation, three models have been proposed [12] (Fig. 27.4). A progenitor cell has the capacity to

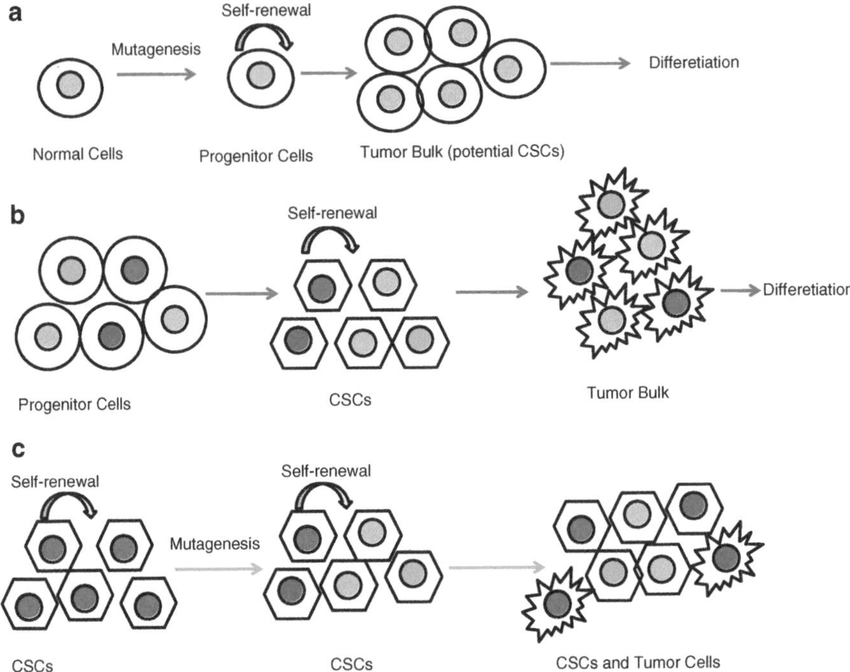

Fig. 27.4 Cancer stem cells and tumor formation. A progenitor cell has the capacity to differentiate into a specific type of cell, but is already more specific than a stem cell, which can only divide a limited number of times. (**a**) The stochastic model: Tumors are initiated from rare cancer stem cells. DNA mutations in any normal cell may initiate tumor formation. Tumors are composed of a group of homogeneous cells, and they develop their functional heterogeneity through stochastic events. Cancer stem cells can arise from all tumor cells under the certain conditions. (**b**) The cancer stem cell (CSC) model: Tumors arise from cancer stem cells and consist of a group of heterogeneous cells. After differentiation these cells have different characteristics and form a cellular hierarchy. (**c**) The third model: Originally, CSCs are a group of homogeneous cells. Mutagenesis within the CSC population and extrinsic environmental factors result in CSC heterogeneity. Tumor cells are a minor group within CSC population

differentiate into a specific type of cell, but is already more specific than a stem cell, which can only divide a limited number of times. In the stochastic model, tumors are initiated from rare cancer stem cells, and DNA mutations in any normal cell may initiate tumor formation. Tumors are composed of a group of homogeneous cells, and they develop their functional heterogeneity through stochastic events. Cancer stem cells can arise from all tumor cells under the certain conditions. The CSC model states that tumors arise from cancer stem cells and consist of a group of heterogeneous cells. After differentiation these cells have different characteristics and form a cellular hierarchy. The third model is based on the stochastic model and the CSC model. Originally, CSCs are a group of homogeneous cells. Mutagenesis within the stem cell population and extrinsic environmental factors result in CSC heterogeneity. Tumor cells are a minor group within CSC population. While a specialized progeny arising from the process of differentiation initiated by a normal stem cell has no proliferative potential, a progeny origin of a CSC exhibits uncontrolled proliferation. Because of the similarities to normal stem cells, CSCs are supposed to lie on the pathways that govern development, self-renewal and cell fate.

In many human tumors including the breast, lung, colon, pancreas, prostate, skin, head/neck and brain, CSCs have been identified. Tissue stem cells markers, including CD133 (promimin-1), nestin, c-kit, sox2, Oct4, and musashi-1also express in CSCs from various primary tumors [44]. Many studies indicate that CSCs also display the resistance to chemotherapy and radiation therapy. With new discoveries related to CSCs, we will have another effective anti-tumor approach to be integrated into the current therapeutic strategies. Besides, the understanding of Notch, Wnt and Hh signaling network, which orchestrate the potency of stem cells, and their molecular mechanisms will be an necessary step towards the development of rational therapeutic approaches.

Acknowledgements This work was supported in part by the grants from the National Institutes of Health (R01CA125262, RO1CA114469 and RO1CA125262-02S1), Susan G. Komen Breast Cancer Foundation, and Kansas Bioscience Authority.

References

1. Al-Hajj M, Wicha MS, Benito-Hernandez A, Morrison SJ, Clarke MF (2003) Prospective identification of tumorigenic breast cancer cells. Proc Natl Acad Sci USA 100:3983–3988
2. Artavanis-Tsakonas S, Rand MD, Lake RJ (1999) Notch signaling: cell fate control and signal integration in development. Science 284:770–776
3. Balordi F, Fishell G (2007) Mosaic removal of Hedgehog signaling in the adult SVZ reveals that the residual wild-type stem cells have a limited capacity for self-renewal. J Neurosci 27:14248–14259
4. Bao S, Wu Q, McLendon RE, Hao Y, Shi Q, Hjelmeland AB, Dewhirst MW, Bigner DD, Rich JN (2006) Glioma stem cells promote radioresistance by preferential activation of the DNA damage response. Nature 444:756–760
5. Beltrami AP, Barlucchi L, Torella D, Baker M, Limana F, Chimenti S, Kasahara H, Rota M, Musso E, Urbanek K, Leri A, Kajstura J, Nadal-Ginard B, Anversa P (2003) Adult cardiac stem cells are multipotent and support myocardial regeneration. Cell 114:763–776
6. Bhanot P, Brink M, Samos CH, Hsieh J-C, Wang Y, Macke JP, Andrew D, Nathans J, Nusse R (1996) A new member of the frizzled family from Drosophila functions as a Wingless receptor. Nature 382:225–230
7. Bonnet D, Dick JE (1997) Human acute myeloid leukemia is organized as a hierarchy that originates from a primitive hematopoietic cell. Nat Med 3:730–737
8. Braakhuis BJM, Tabor MP, Kummer JA, Leemans CR, Brakenhoff RH (2003) A genetic explanation of slaughter's concept of field cancerization. Cancer Res 63:1727–1730
9. Brivanlou AH, Darnell JE (2002) Signal transduction and the control of gene expression. Science 295:813–818
10. Claudinot S, Nicolas M, Oshima H, Rochat A, Barrandon Y (2005) Long-term renewal of hair follicles from clonogenic multipotent stem cells. Proc Natl Acad Sci USA 102:14677–14682
11. Collins AT, Berry PA, Hyde C, Stower MJ, Maitland NJ (2005) Prospective identification of tumorigenic prostate cancer stem cells. Cancer Res 65:10946–10951

12. Dalerba P, Cho RW, Clarke MF (2007) Cancer stem cells: models and concepts. Annu Rev Med 58:267–284
13. Dalerba P, Dylla SJ, Park I-K, Liu R, Wang X, Cho RW, Hoey T, Gurney A, Huang EH, Simeone DM, Shelton AA, Parmiani G, Castelli C, Clarke MF (2007) Phenotypic characterization of human colorectal cancer stem cells. Proc Natl Acad Sci USA 104:10158–10163
14. Dean M (2009) ABC transporters, drug resistance, and cancer stem cells. J Mammary Gland Biol Neoplasia 14:3–9
15. Eramo A, Lotti F, Sette G, Pilozzi E, Biffoni M, Di Virgilio A, Conticello C, Ruco L, Peschle C, De Maria R (2007) Identification and expansion of the tumorigenic lung cancer stem cell population. Cell Death Differ 15:504–514
16. Fortini ME (2009) Notch signaling: the core pathway and its posttranslational regulation. Dev Cell 16:633–647
17. Gao J, Graves S, Koch U, Liu S, Jankovic V, Buonamici S, El Andaloussi A, Nimer SD, Kee BL, Taichman R, Radtke F, Aifantis I (2009) Hedgehog signaling is dispensable for adult hematopoietic stem cell function. Cell Stem Cell 4:548–558
18. Grandbarbe L, Bouissac J, Rand M, Hrabé de Angelis M, Artavanis-Tsakonas S, Mohier E (2003) Delta-Notch signaling controls the generation of neurons/glia from neural stem cells in a stepwise process. Development 130:1391–1402
19. Griffin J, Lowenberg B (1986) Clonogenic cells in acute myeloblastic leukemia. Blood 68:1185–1195
20. Guenechea G, Gan OI, Dorrell C, Dick JE (2001) Distinct classes of human stem cells that differ in proliferative and self-renewal potential. Nat Immunol 2:75–82
21. Gulino A, Ferretti E, De Smaele E (2009) Hedgehog signalling in colon cancer and stem cells. EMBO Mol Med 1:300–302
22. Hadland BK, Huppert SS, Kanungo J, Xue Y, Jiang R, Gridley T, Conlon RA, Cheng AM, Kopan R, Longmore GD (2004) A requirement for Notch1 distinguishes 2 phases of definitive hematopoiesis during development. Blood 104:3097–3105
23. Hahn WC, Counter CM, Lundberg AS, Beijersbergen RL, Brooks MW, Weinberg RA (1999) Creation of human tumour cells with defined genetic elements. Nature 400:464–468
24. He T-C, Sparks AB, Rago C, Hermeking H, Zawel L, da Costa LT, Morin PJ, Vogelstein B, Kinzler KW (1998) Identification of c-MYC as a target of the APC pathway. Science 281:1509–1512
25. Hemmati HD, Nakano I, Lazareff JA, Masterman-Smith M, Geschwind DH, Bronner-Fraser M, Kornblum HI (2003) Cancerous stem cells can arise from pediatric brain tumors. Proc Natl Acad Sci USA 100:15178–15183
26. Ito M, Yang Z, Andl T, Cui C, Kim N, Millar SE, Cotsarelis G (2007) Wnt-dependent de novo hair follicle regeneration in adult mouse skin after wounding. Nature 447:316–320
27. Jiang Y, Jahagirdar BN, Reinhardt RL, Schwartz RE, Keene CD, Ortiz-Gonzalez XR, Reyes M, Lenvik T, Lund T, Blackstad M, Du J, Aldrich S, Lisberg A, Low WC, Lergaespada DA, Verfaillie CM (2002) Pluripotency of mesenchymal stem cells derived from adult marrow. Nature 418:41–49
28. Katoh Y, Katoh M (2005) Identification and characterization of rat Wnt6 and Wnt10a genes in silico. Int J Mol Med 15:527–531
29. Kim K-i, Cho H-J, Hahn J-Y, Kim T-Y, Park K-W, Koo B-K, Soo Shin C, Kim C-H, Oh B-H, Lee M-M, Park Y-B, Kim H-S (2006) β-catenin overexpression augments angiogenesis and skeletal muscle regeneration through dual mechanism of vascular endothelial growth factor–mediated endothelial cell proliferation and progenitor cell mobilization. Arterioscler Thromb Vasc Biol 26:91–98
30. Kumano K, Chiba S, Kunisato A, Sata M, Saito T, Nakagami-Yamaguchi E, Yamaguchi T, Masuda S, Shimizu K, Takahashi T, Ogawa S, Hamada Y, Hirai H (2003) Notch1 but not Notch2 is essential for generating hematopoietic stem cells from endothelial cells. Immunity 18:699–711
31. Lapidot T, Sirard C, Vormoor J, Murdoch B, Hoang T, Caceres-Cortes J, Minden M, Paterson B, Caligiuri MA, Dick JE (1994) A cell initiating human acute myeloid leukaemia after transplantation into SCID mice. Nature 367:645–648

32. Levis M, Murphy KM, Pham R, Kim K-T, Stine A, Li L, McNiece I, Smith BD, Small D (2005) Internal tandem duplications of the FLT3 gene are present in leukemia stem cells. Blood 106:673–680
33. Liu S, Dontu G, Mantle ID, Patel S, N-s A, Jackson KW, Suri P, Wicha MS (2006) Hedgehog signaling and Bmi-1 regulate self-renewal of normal and malignant human mammary stem cells. Cancer Res 66:6063–6071
34. Logan CY, Nusse R (2004) The Wnt signaling pathway in development and disease. Annu Rev Cell Dev Biol 20:781–810
35. Ma S, Chan K, Guan X-Y (2008) In search of liver cancer stem cells. Stem Cell Rev Rep 4:179–192
36. Matikainen T, Laine J (2005) Placenta–an alternative source of stem cells. Toxicol Appl Pharmacol 207:544–549
37. Morgan TH (1917) The theory of the gene. Am Nat 51:513–544
38. Nelson WJ, Nusse R (2004) Convergence of Wnt, ß-catenin, and cadherin pathways. Science 303:1483–1487
39. Nusse R, Varmus HE (1982) Many tumors induced by the mouse mammary tumor virus contain a provirus integrated in the same region of the host genome. Cell 31:99–109
40. Nusslein-Volhard C, Wieschaus E (1980) Mutations affecting segment number and polarity in Drosophila. Nature 287:795–801
41. O'Brien CA, Pollett A, Gallinger S, Dick JE (2007) A human colon cancer cell capable of initiating tumour growth in immunodeficient mice. Nature 445:106–110
42. Okamura RM, Sigvardsson M, Galceran J, Verbeek S, Clevers H, Grosschedl R (1998) Redundant regulation of T cell differentiation and TCR[alpha] gene expression by the transcription factors LEF-1 and TCF-1. Immunity 8:11–20
43. Palma V, Altaba ARi (2004) Hedgehog-GLI signaling regulates the behavior of cells with stem cell properties in the developing neocortex. Development 131:337–345
44. Pannuti A, Foreman K, Rizzo P, Osipo C, Golde T, Osborne B, Miele L (2010) Targeting Notch to target cancer stem cells. Clin Cancer Res 16:3141–3152
45. Peters R, Leyvraz S, Perey L (1998) Apoptotic regulation in primitive hematopoietic precursors. Blood 92:2041–2052
46. Pittenger MF, Mackay AM, Beck SC, Jaiswal RK, Douglas R, Mosca JD, Moorman MA, Simonetti DW, Craig S, Marshak DR (1999) Multilineage potential of adult human mesenchymal stem cells. Science 284:143–147
47. Prince ME, Sivanandan R, Kaczorowski A, Wolf GT, Kaplan MJ, Dalerba P, Weissman IL, Clarke MF, Ailles LE (2007) Identification of a subpopulation of cells with cancer stem cell properties in head and neck squamous cell carcinoma. Proc Natl Acad Sci USA 104:973–978
48. Reifenberger J, Wolter M, Weber RG, Megahed M, Ruzicka T, Lichter P, Reifenberger G (1998) Missense mutations in SMOH in sporadic basal cell carcinomas of the skin and primitive neuroectodermal tumors of the central nervous system. Cancer Res 58:1798–1803
49. Ricci-Vitiani L, Lombardi DG, Pilozzi E, Biffoni M, Todaro M, Peschle C, De Maria R (2007) Identification and expansion of human colon-cancer-initiating cells. Nature 445:111–115
50. Rijsewijk F, Schuermann M, Wagenaar E, Parren P, Weigel D, Nusse R (1987) The Drosophila homology of the mouse mammary oncogene int-1 is identical to the segment polarity gene wingless. Cell 50:649–657
51. Rittié L, Stoll SW, Kang S, Voorhees JJ, Fisher GJ (2009) Hedgehog signaling maintains hair follicle stem cell phenotype in young and aged human skin. Aging Cell 8:738–751
52. Sasaki H, Nishizaki Y, Hui C, Nakafuku M, Kondoh H (1999) Regulation of Gli2 and Gli3 activities by an amino-terminal repression domain: implication of Gli2 and Gli3 as primary mediators of Shh signaling. Development 126:3915–3924
53. Sawyers CL, Denny CT, Witte ON (1991) Leukemia and the disruption of normal hematopoiesis. Cell 64:337–350
54. Schatton T, Murphy GF, Frank NY, Yamaura K, Waaga-Gasser AM, Gasser M, Zhan Q, Jordan S, Duncan LM, Weishaupt C, Fuhlbrigge RC, Kupper TS, Sayegh MH, Frank MH (2008) Identification of cells initiating human melanomas. Nature 451:345–349

55. Scoville DH, Sato T, He XC, Li L (2008) Current view: intestinal stem cells and signaling. Gastroenterology 134:849–864
56. Sell S (1993) Cellular origin of cancer: dedifferentiation or stem cell maturation arrest? Environ Health Perspect 101:15–26
57. Sell S, Pierce GB (1994) Maturation arrest of stem cell differentiation is a common pathway for the cellular origin of teratocarcinomas and epithelial cancers. Lab Invest 70:6–22
58. Shiras A, Chettiar ST, Shepal V, Rajendran G, Prasad GR, Shastry P (2007) Spontaneous transformation of human adult nontumorigenic stem cells to cancer stem cells is driven by genomic instability in a human model of glioblastoma. Stem Cells 25:1478–1489
59. Sparks AB, Morin PJ, Vogelstein B, Kinzler KW (1998) Mutational analysis of the APC/β-catenin/Tcf pathway in colorectal cancer. Cancer Res 58:1130–1134
60. Tetsu O, McCormick F (1999) [beta]-catenin regulates expression of cyclin D1 in colon carcinoma cells. Nature 398:422–426
61. Thomson JA (1998) Embryonic stem cell lines derived from human blastocysts. Science 282:1145–1147
62. Valkenburg KC, Graveel CR, Zylstra-Diegel CR, Zhong Z, Williams BO (2011) Wnt/β-catenin signaling in normal and cancer stem cells. Cancers 3:2050–2079
63. Varnat F, Duquet A, Malerba M, Zbinden M, Mas C, Gervaz P, Altaba RiA (2009) Human colon cancer epithelial cells harbour active HEDGEHOG-GLI signalling that is essential for tumour growth, recurrence, metastasis and stem cell survival and expansion. EMBO Mol Med 1:338–351
64. Wang T-L, Rago C, Silliman N, Ptak J, Markowitz S, Willson JKV, Parmigiani G, Kinzler KW, Vogelstein B, Velculescu VE (2002) Prevalence of somatic alterations in the colorectal cancer cell genome. Proc Natl Acad Sci USA 99:3076–3080
65. Weng AP, Ferrando AA, Lee W, Morris JP, Silverman LB, Sanchez-Irizarry C, Blacklow SC, Look AT, Aster JC (2004) Activating mutations of NOTCH1 in human T cell acute lymphoblastic leukemia. Science 306:269–271
66. Wharton KA, Johansen KM, Xu T, Artavanis-Tsakonas S (1985) Nucleotide sequence from the neurogenic locus Notch implies a gene product that shares homology with proteins containing EGF-like repeats. Cell 43:567–581
67. Willert K, Brown JD, Danenberg E, Duncan AW, Weissman IL, Reya T, Yates JR, Nusse R (2003) Wnt proteins are lipid-modified and can act as stem cell growth factors. Nature 423:448–452
68. Winquist RJ, Boucher DM, Wood M, Furey BF (2009) Targeting cancer stem cells for more effective therapies: taking out cancer's locomotive engine. Biochem Pharmacol 78:326–334
69. Wu X, Ding S, Ding Q, Gray NS, Schultz PG (2004) Small molecules that induce cardiomyogenesis in embryonic stem cells. J Am Chem Soc 126:1590–1591
70. Zhao C, Chen A, Jamieson CH, Fereshteh M, Abrahamsson A, Blum J, Kwon HY, Kim J, Chute JP, Rizzieri D, Munchhof M, VanArsdale T, Beachy PA, Reya T (2009) Hedgehog signalling is essential for maintenance of cancer stem cells in myeloid leukaemia. Nature 458:776–779

Chapter 28
Global OMICs Profiling and Functional Analysis of CD44+/CD24– Stem-Like Cells in Normal Human Breast Tissue and Breast Cancer

Marina Bessarabova, Kornelia Polyak, and Yuri Nikolsky

Contents

Abstract It is generally believed that spontaneous tumors originate from a single cell and evolve into complex tissues composed of multiple cell types and characterized by very high morphological, physiological and genetic heterogeneity (Marusyk A, Polyak K, Biochim Biophys Acta 1805(1):105–117, 2010). In the process, cancer cells acquire six core biological capabilities (hallmarks) via evolutionary fueled by high rate of somatic mutations and genomic instability and rigorous Darwinian selection (Hanahan D, Weinberg RA, Cell 144(5):646–674, 2011). Two leading concepts explaining tumor heterogeneity are cancer stem cells hypothesis and clonal evolution theory, which are considered rather complementary than mutually exclusive (Campbell LL, Polyak K, Cell Cycle 6(19):2332–2338, 2007). Substantial evidence has been accumulated for the existence of cancer of cells with characteristics of stem cells such as self-renewal and ability to efficiently seed new tumors in recipient mice.

M. Bessarabova • Y. Nikolsky (✉)
Thomson Reuters, 169 Saxony Road, #104, Encinitas, CA 92024, USA
e-mail: marina.bessarabova@thomsonreuters.com; yuri.nikolsky@thomsonreuters.com

K. Polyak
Department of Medical Oncology, Dana-Farber Cancer Institute, Harvard Medical School, 450 Brookline Ave. D740C, Boston, MA 02115, USA
e-mail: kornelia_polyak@dfci.harvard.edu

R.K. Srivastava and S. Shankar (eds.), *Stem Cells and Human Diseases*,
DOI 10.1007/978-94-007-2801-1_28,

The study of cancer stem cells is relatively novel and fast-growing area in basic research and anti-cancer drug discovery. In this book chapter we will focus on global OMICs profiling and functional analysis of CD44+/CD24– stem-like cells in normal human breast tissue and breast cancer.

Keywords Cancer stem cells • Breast cancer • CD44+/CD24– stem-like cells • OMICs profiling • Functional analysis

28.1 Cancer Stem Cells (CSC)

The origin of CSCs is not yet resolved and may differ between tumor types. In some tissues, normal adult stem cells may undergo oncogenic transformation; in other types partially differentiated progenitor cells may yield CSCs [2]. Adult stem cells share many characteristics with cancer cells, including a capacity to self-renew, give rise to heterogeneous progeny, migrate and invade into surrounding tissues. "Stem-like" properties are thought to be responsible for the growth, progression, and recurrence of a tumor [4–8]. It has also been proposed that cancer stem cells, just as normal adult stem cells from a tumor tissue of origin, produce daughter cells with limited proliferative capacity and more differentiated state [9]. All these properties and findings imply stem or progenitor cells as candidate tumor-initiating cells.

The concept of tumor stem cells explains well many aspects of tumorigenesis, such as intratumor heterogeneity, which drives the tumor's evolution towards developing traits responsible for therapeutic resistance, recurrence, and tumor progression [1, 10]. The existence of tumor stem-like cells explains therapeutic failures and recurrences, as almost all commonly used chemotherapeutic drugs do not target this cell population [11–13].

The study of cancer stem cells is a relatively novel and fast-growing area in basic research and anti-cancer drug discovery. Cancer stem cells were first identified in hematopoietic malignancies, where it was found that only a phenotypically distinct subset of human acute myeloid leukemia cells (CD34+/CD38–) could form a tumor upon transplantation into immunodeficient mice in limiting dilution assays [14, 15]. Subsequently, cancer stem cells were identified in multiple tumor types and believed to be a common feature of oncogenesis [7]. In breast cancer, CD44+/CD24– tumor cells were identified as CSCs based on their ability to initiate tumors upon transplantation into the mammary fat pads of immunocompromised mice [16]. In other studies breast cancer cells with different phenotypes (e.g., CD133+) could also initiate tumors. The overall procedure for the isolation of cancer stem cells has been rather similar across the studies and types of cancer and consisted of affinity purification of individual cells from primary tumors using cell surface markers specific for the normal stem cells of the same organ [4–8]. Follow-up *in vivo* tumorigenicity and *in vitro* clonogenicity studies demonstrated both tumorigenicity and the "stemness" of these isolated "stem-like" cancer cells.

However, several recent studies have questioned the validity of the cancer stem cell model, at least in its strictest interpretation. First, the xenotransplant assay used for the identification of CSCs is highly subject to experimental conditions. The strain of mice used and the way cells are injected (e.g., mixed with matrix or not), highly influences the outcome [17, 18]. Second, tumor cells in their physiologic environment exist as part of a complex structure, not as individual cells. Thus, their behavior when isolated from their environment may not reflect their properties in patients. Third, CSCs and non-CSCs in the same tumor display high degree of genetic heterogeneity both within and between populations. This was shown in breast cancer [19] and more recently in leukemias as well [20, 21]. Therefore, CSCs are more likely a reflection of epigenetic heterogeneity within tumors.

28.2 CD44+/CD24− and CD44−/CD24+ Cells in Normal Breast Tissue and Breast Cancer

According to *in vitro* clonogenicity studies, in breast tissue there are bipotential stem cells that can give rise to both luminal epithelial and myoepithelial cells [5, 22–27]. A series of unique surface protein markers were identified for differentiated and progenitor cells. Thus, MUC1 and CD10 (CALLA/MME) are thought to be surface markers of luminal epithelial and myoepithelial cells, respectively [23, 26, 27]. CD44 and CD24 are also markers for progenitor and luminal epithelial cells, respectively. Al-Hajj et al. demonstrated that in breast cancer lin^{-}/CD44−/CD24$^{-/low}$ (subsequently referred to as CD44+/CD24−) cells from malignant pleural effusions of breast cancer patients were more tumorigenic in NOD/SCID mice than CD44−/CD24+ (subsequently referred to as CD44−/CD24+) cells, and the resulting xenografts reproduced the heterogeneity of the original tumors. This observation led to the hypothesis that CD44+/CD24− cells may represent breast cancer stem cells [16]. Follow-up studies in breast and prostate cancers confirmed that CD44+/CD24− cells have progenitor-like properties and are tumorigenic when injected into immunodeficient mice [28, 29].

Several studies implicated certain signaling pathways in self-renewal and survival of normal tissue-specific stem cells and embryonic stem cells, such as Hedgehog, Notch and Wnt/β-catenin, and suggested that these development pathways may be involved in maintaining the “stemness” of CD44+/CD24− cells. However, these pathways have to be considered in a specific biological context, as their functions are pleiotropic, and they can be initiated by and trigger complex cellular networks.

The issue of biological context can be to some extend assessed in high-throughput experiments such as “genome-wide” profiling of gene expression and epigenetic alterations, SNP analysis and exon resequencing. Biological interpretation of large “high-throughput” datasets is facilitated by specialized “knowledge-based” platforms, which combine comprehensive databases of protein interactions and pathways with software tools of functional analysis such as ontology enrichment, network-generation modules and interactome topology algorithms [30]. Several comprehensive whole-genome profiling studies comparing CD44+/CD24− and CD44−/CD24+ cells were conducted recently, addressing clonogenicity, molecular

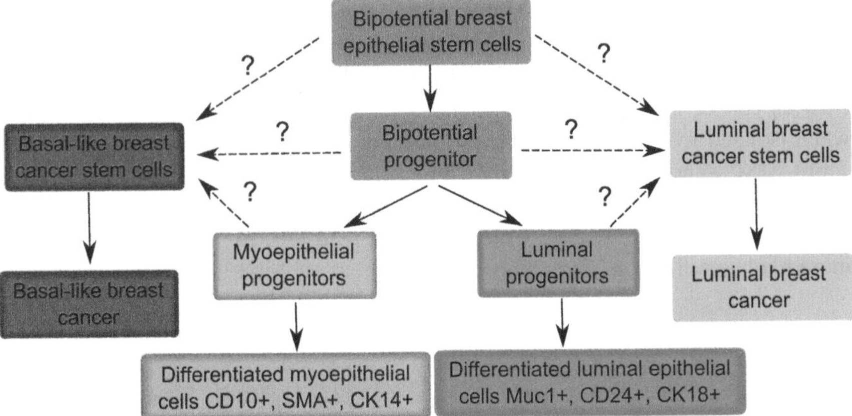

Fig. 28.1 Hypothetical model of solid tumor cancer stem cells using breast cancer as an example [37]. The central part of the scheme represents a hypothetical differentiation pathway for normal mammary epithelial cells. Bipotential stem cells give rise to bipotential progenitors. Bipotential progenitors can produce myoepithelial or luminal progenitors differentiating into mature myoepithelial and luminal epithelial cells, respectively. Cancer stem cells could potentially be derived from bipotential stem cells or from more differentiated cells that acquired self-renewal capabilities. Cancer stem cells produce tumors with features of a particular lineage: luminal or basal like

characteristics, and clinical relevance of cells in normal and tumor breast [31–36]. In general, functional properties derived from global gene expression and epigenetic patterns of CD44+/CD24– and CD44–/CD24+ cells were consistent with the hypothesis that these cells represent progenitor-enriched and luminal-committed cells, respectively. For example, the gene expression profiles of CD44+/CD24– and CD44–/CD24+ cells in breast tumors were consistently different at both gene and pathway levels [32, 35]. In CD44+/CD24– cells, highly expressed genes were involved in invasion and angiogenesis and displayed activated TGF-β–, Hh-, and PLAU-signaling pathways. In CD44–/CD24+ cells, expression of markers and pathways of luminal epithelial differentiation were more pronounced. A hypothetical model of breast cancer stem cells suggests that bipotential stem cells give rise to bipotential progenitors that can become committed myoepithelial or luminal progenitors differentiating into mature myoepithelial and luminal epithelial cells, respectively. During the differentiation process, the self-renewal capacity of the cells gradually decreases. Cancer stem cells could potentially be derived from bipotential stem cells or from more differentiated cells that acquired self-renewal capabilities [37] (Fig. 28.1).

The results of multiple studies suggest that CD44+/CD24– cells are more tumorigenic than CD44–/CD24+ cells. Thus, the CD44+/CD24– breast cancer cell population was enriched in tumor-initiating [16] and chemotherapy-resistant [38, 39] cells. It was also shown that breast cancer cell lines of high lymphatic metastatic ability have a higher fraction of cells with a cancer stem cell-like

CD44+/CD24− phenotype than breast cancer cells with low lymphatic metastasis ability [40]. Based on immunohistochemical analyses of a large cohort of invasive and in situ breast carcinomas, distribution of CD44+/CD24− and CD44−/CD24+ markers in breast cancer cells varies with significant diversity both among and within tumors. Overall, CD44+/CD24− cells are more frequent in basal-like breast cancer, whereas luminal tumors are enriched in CD44−/CD24+ cells [19, 41–43]. It was also shown that "gene signatures" for CD44+/CD24− and CD44−/CD24+ cells' gene expression correlate with clinical outcomes; tumors composed of mostly CD44+/CD24− cells may have worse clinical behavior than tumors enriched in CD44−/CD24+ cells [31, 32, 44]. As basal-like breast cancer currently is the only major breast tumor subtype without effective targeted treatment strategies and with poor prognosis [45], therapies specifically eliminating CD44+/CD24− cells may represent a new approach for their more effective treatment.

The high therapeutic potential of CD44+/CD24− cells emphasizes the importance of understanding the pathways specific for CD44+/CD24− cells and the underlying mechanisms of contribution of these cells to tumorigenesis. Here we describe the results of several "whole-genome" studies carried out on global gene expression and epigenetic profiling of CD44+/CD24− and CD44+CD24+ cells and the follow-up functional analysis, which revealed possible applications in breast cancer treatment. Based on functional analysis of gene expression data, several genes and pathways were identified as specifically required for CD44+/CD24− breast cancer cells in primary tumors. Among them are the TGF-beta and JAK2/STAT3 pathways, and the IL6, PTGIS, HAS1, CXCL3, ISG15, PFKFB3 and IGFBP7 genes [32, 36]. It was shown that therapeutic inhibition of these pathways/genes can be used to obtain effective therapeutic response accompanied by the elimination of CD44+/CD24− cells.

28.3 Gene Expression Profiling of CD44+/CD24− and CD44−/CD24+ Cells

In our first comprehensive gene expression profiling study [32], we aimed to understand the differences in gene expression between CD44+/CD24− and CD44−/CD24+ cells and the differential mechanisms required for the functionality of these two cell types. SAGE (serial analysis of gene expression) libraries were generated from CD44+/CD24− and CD44−/CD24+ cells purified from normal human breast epithelium as well as from breast carcinomas. Unsupervised hierarchical clustering of gene expression profiles demonstrated that normal and cancer CD44+/CD24− stem cells are more similar to each other than to CD44−/CD24+ cells from the same tissue and that CD44+/CD24− and CD44−/CD24+ cells represent different cell populations with distinct gene expression patterns.

Moreover, in breast cancer, CD44+/CD24− and CD44−/CD24+ gene expression patterns correlated with clinical outcome. We identified two gene expression signatures differentially expressed in CD44+/CD24− cells in comparison with

CD44–/CD24+ cells: signature “A” consisted of genes upregulated in CD44+/CD24– in comparison with CD44–/CD24+ (characteristic of breast cancer CD44+/CD24– cells) and signature “B” consisted of genes downregulated in CD44+/CD24– cells in comparison with CD44–/CD24+ cells (characteristic of breast cancer CD44–/CD24+ cells). As expected for statistically derived cancer gene signatures [46], functional enrichment of the gene content of signature “A” and signature “B” had little correlation with functionality of the complete list of differentially expressed genes (DEGs). High expression of signature “A” genes was associated with shorter-distance metastasis-free survival times, while high expression of signature “B” genes with longer-distance metastasis-free survival times. Signatures “A” and “B” were also associated with a shorter and longer relapse-free and overall survival times, respectively. The correlation of signatures “A” and “B” with outcome was independent of ER (estrogen receptor) expression and tumor grade.

These results are in agreement with another gene expression study [31], where a gene-expression profile of CD44+/CD24– tumorigenic breast-cancer cells was compared with that of normal breast epithelium. The resulting differentially expressed genes (DEGs) were used to generate a 186-gene “invasiveness” gene signature, which was strongly associated with both overall and metastasis-free survival ($P<0.001$, for both) in breast cancer patients. The “invasiveness” gene signature divided patients with high-risk early breast cancer into “good” and “poor” prognostic categories; good prognosis meant 81% of 10-years of metastasis-free survival, compared with 57% for poor prognosis. This signature of tumorigenic breast-cancer cells was associated even more strongly with clinical outcomes when combined with the wound-response signature in breast cancer [47].

Functional analysis of differentially expressed genes (DEGs) in CD44+/CD24– and CD44–/CD24+ cells [32] showed activation of distinct signaling pathways and supported the hypothesis that CD44–/CD24+ and CD44+/CD24– cells represent more differentiated luminal epithelial and progenitor-like cells, respectively, as known markers of these lineages were nearly mutually exclusive in the respective SAGE libraries. Functional classification shown that genes expressed in the various SAGE libraries from cancer and normal CD44+/CD24– cells are enriched in cell motility, chemotaxis, hemostasis, and angiogenesis processes, while genes preferentially expressed in CD44–/CD24+ cells implicated in basic “housekeeping” processes such as carbohydrate metabolism and RNA splicing.

In order to identify signaling pathways based on the expression data that are specifically activated in more tumorigenic CD44+/CD24– cells, we applied the MetaCore platform (Thomson Reuters) [30]. Enrichment analysis was performed in several functional ontologies such as “canonical pathway maps”, GO and GeneGo functional processes for normal and cancer CD44+/CD24– SAGE libraries and compared to the corresponding CD44–/CD24+ SAGE libraries. Cell motility, cell adhesion, protein biosynthesis, protein folding, and cell proliferation were identified as strongly upregulated processes both in normal CD44+/CD24– and cancer CD44+/CD24– cells. Among “canonical” pathways, TGF-β and WNT signaling, cytoskeleton remodeling, integrin-mediated processes, reverse signaling by ephrin B, and chemokines and cell adhesion were upregulated in both normal CD44+/CD24– and

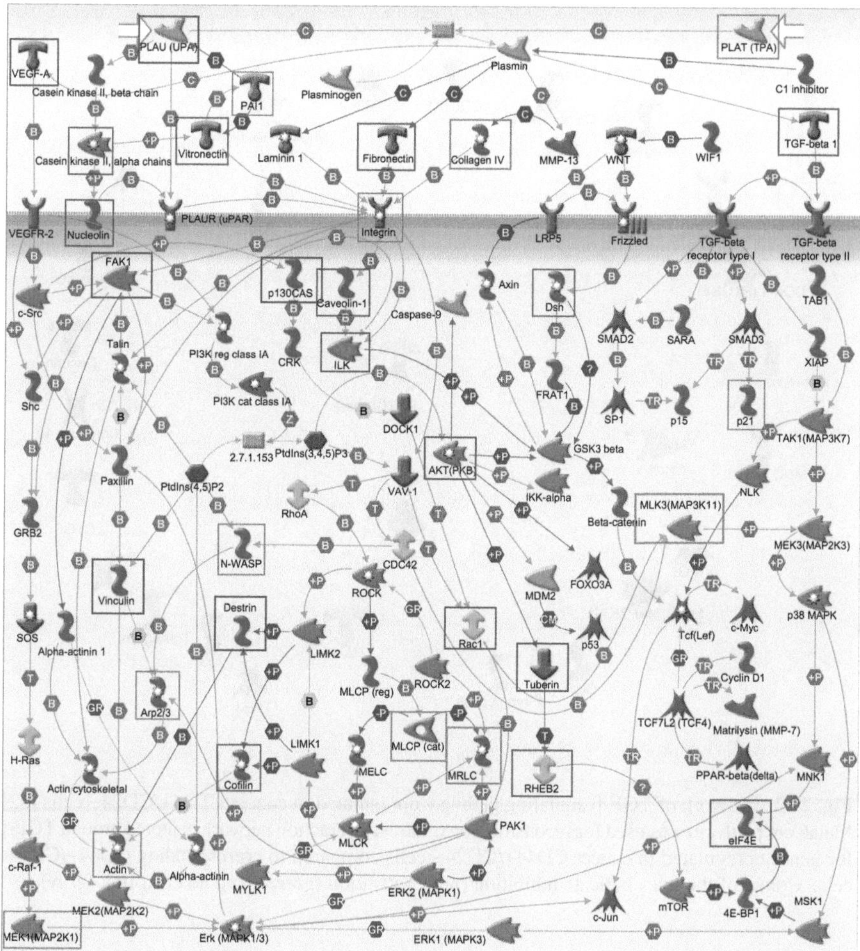

Fig. 28.2 TGF-β and WNT signaling and cytoskeleton remodeling pathway is upregulated in CD44+/CD24– cells [32]. TGF-β and WNT signaling and cytoskeleton remodeling is strongly upregulated processes in more tumorigenic CD44+/CD24– cells in comparison with CD44–/CD24+ cells in both normal and cancer cells. *Rectangles* indicate upregulation of the indicated genes in breast cancer (*red*) or normal (*blue*) CD44+/CD24– cells or both (*yellow*) compared to corresponding CD44–/CD24+ cells

cancer CD44+/CD24– cells (Fig. 28.2). This functional profile is consistent with CD44+/CD24– cells demonstrating a more mesenchymal, motile, and less proliferative stem-cell-like profile.

A more detailed analysis of different functionality in CD44+/CD24– and CD44–/CD24+ cells was performed by building direct interaction (DI) networks for the lists of genes highly overrepresented in normal and cancer CD44+/CD24– cell SAGE libraries compared to the corresponding CD44–/CD24+ SAGE libraries. A DI

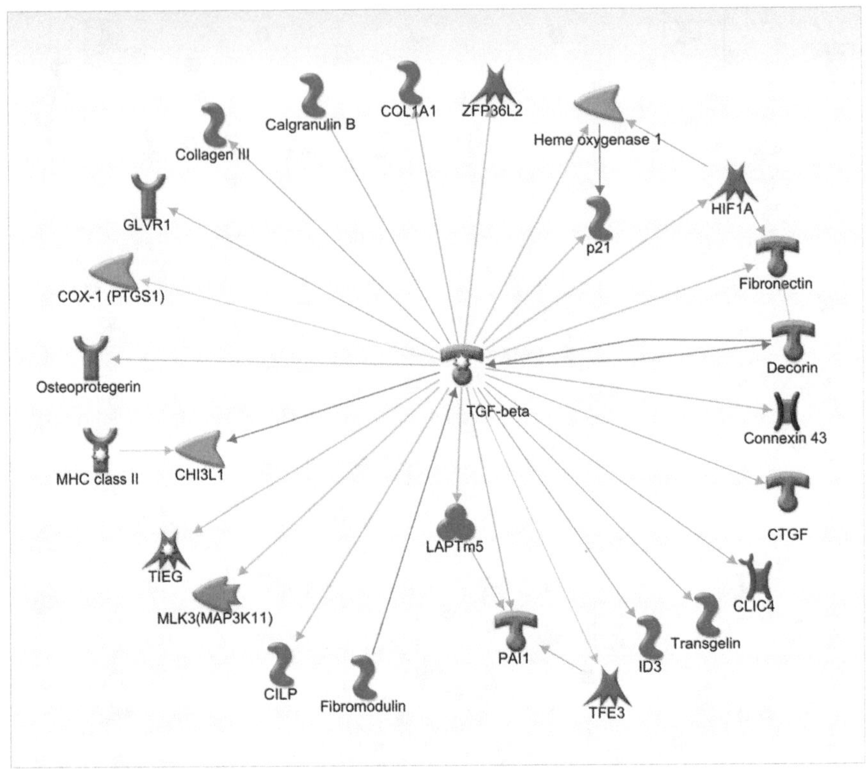

Fig. 28.3 Network of TGF-β signaling pathway upregulated in cancer CD44+/CD24– cells [32]. MetaCore platform was used for reconstruction of direct interaction network centered around TGF-β for genes upregulated in cancer CD44+/CD24– cells compared to corresponding CD44–/CD24+ cells. Colors of the lines indicate inhibition (*red*), activation (*green*), and no clear link (*gray*)

network built for cancer CD44+/CD24– cell-specific genes was centered around TGF-β1 (Fig. 28.3). The key genes on the CD44+/CD24– network were dynamin, fibronectin, caveolin, casein kinase II, collagen 1, transcription factors HIF1A and ETS1, VEGF-A, IL-1, NF-kb, AP-1, Rac1, SMAD3, Notch pathway and TGF-β3.

We found that TGF-β signaling is a key pathway specifically activated in CD44+/CD24– breast cancer cells. This pathway plays an important role in human embryonic stem cells as well as in tumorigenesis [48–51]. However, TGF-β plays a dual role in tumor progression depending on biological context: on the one hand, it is one of the most potent inhibitors of cell proliferation; on the other hand, it promotes epithelial-mesenchymal transition (EMT), invasion, angiogenesis, and metastasis [52]. The TGF-β pathway was shown to inhibit tumorigenesis when it is the only pathway activated in cells; in the malignant state the TGF-β pathway cooperates with other pathways, which leads to tumorigenesis [53].

In order to clarify the role of the TGF-β pathway in the case of breast cancer CD44+/CD24– cells, we analyzed cellular response to treatment of CD44+/CD24– and

CD44–/CD24+ cells with a dual TGFBR1/TGFBR2 kinase inhibitor, which effectively shuts down the TGF-β pathway. One possible reason for the specific activation of TGF-β signaling in CD44+/CD24– breast cancer cells is its epigenetic silencing in CD44–/CD24+ cells. In agreement with this hypothesis, the TGFBR kinase inhibitor specifically affected CD44+/CD24– tumor cells, whereas there was no response in CD44–/CD24+ cells. In CD44+/CD24– cells, TGFBR inhibitor treatment lead to dramatic morphological changes associated with mesenchymal-to-epithelial transition. While untreated CD44+/CD24– cells are round-shaped and dispersed, CD44+/CD24– cells treated with the TGFBR inhibitor are more epithelial in appearance. At a subcellular level, TGFBR inhibitor treatment caused transfer of β-catenin, E-cadherin, and ZO-1 to the cell membrane, which is consistent with the induction of a more epithelial cell phenotype. These results suggest that the TGF-β pathway is specifically activated in CD44+/CD24– breast cancer cells and that it regulates, at least in part, their more mesenchymal appearance. TGFBR inhibitor treatment of CD44–/CD24+ cells did not cause any significant morphological effects, which is in agreement with lack of expression of TGFBR2 in CD44–/CD24+ cells.

Therefore, both functional analysis (ontology enrichment and networks) and inhibitor treatment experiments suggested an important role of TGF-β pathway in establishing the invasive phenotype of CD44+/CD24– cells. In good accordance with these data, high expression of the TGF-β pathway was associated with a shorter-distance metastasis-free survival in a set of breast cancer patients. Particularly high expression of the "TGF-β cassette" (15 genes more highly expressed in cancer CD44+/CD24– cells than in cancer CD44–/CD24+ cells) was statistically significantly associated with a shorter-distance metastasis-free survival time in one data set. It suggests that activation of the TGF-β pathway in CD44+/CD24– cancer cells may be relevant for disease progression in a subset of breast cancer patients and that therapy targeting this pathway may lead to more effective treatment of these patients, probably in a combination with other drugs.

Additional details on different functionality in CD44+/CD24– and CD44–/CD24+ cells were revealed in a follow-up study [33]. GSEA (gene set enrichment analysis) of CD44+/CD24– and CD44–/CD24+ expression profiles confirmed differential expression of the TGF-β pathway in CD44+/CD24– cell populations and identified several other CD44+/CD24– specific tumorigenic pathways, including oncogenic Ras signaling, TNF and IFN response pathways. Gene-ontology process classification revealed that genes involved in 'stemness', cell proliferation/maintenance, cell adhesion, cell motility, invasion, angiogenesis, growth factor/cytokine, immune response/suppression, and metabolism were highly expressed in CD44+/CD24– cells. All these genes may contribute to oncogenesis, for example: Notch2, a 'stemness'-related gene, is in the TGF-β pathway; LAMA3, a cell invasion- or adhesion-related gene, KLF5, EPAS1, and VEGF, angiogenesis-related genes, are in the oncogenic Ras pathway. It was also shown that activity of NF-kB is higher in CD44+/CD24– cells than in CD44–/CD24+ cells. Moreover, DHMEQ, a highly specific inhibitor for NF-kB, suppressed tumorigenesis in the CD44+/CD24– cells in a mouse model. Thus, NF-kB as well as the TGF-β pathway could be a promising target for treatment of breast cancer patients.

In another expression microarray study of CD44+/CD24− and CD44−/CD24+ cells, a subset of 32 genes was shown to be associated with EMT in CD44+/CD24− cells [34]. This gene set consists of AKT3, BDNF, CDH2, CTGF, DAB2, FGFR1, FYN, HMGA2, IL8, ILK, ITF2/TCF4, JAG1, JAK2, MAP4K4, MMP-2, NR2F1/COUP-TF1, Periostin, PIK3R1, PRKC, S100A4, SMAD3, SMAD7, SMURF2, SNAI2/SLUG, SPARC, TGFb1, TGFb2, TWIST2, Wnt5A, Wnt5B, ZEB-1/TCF8, ZEB-2/ZFHX1B. This study also revealed the role of tumor necrosis factor (TNF) in altering the phenotype of breast cancer stem cells. Treatment with TNF, which induces NF-kB and represses E-cadherin, or overexpression of SLUG in CD44−/CD24+ cells gave rise to a subpopulation of CD44+/CD24− cells.

In order to identify new ways to promote more effective treatment of breast cancers with a predominant CD44+/CD24− phenotype, we recently performed a large-scale shRNA screen, which represents an unbiased screening strategy [36]. Based on the assumption that signaling pathways on which CD44+/CD24− cells depend upon are enriched with genes differentially expressed between CD44+/CD24− and CD44−/CD24+ cells, we subjected 1,576 genes differentially expressed between CD44+/CD24− and CD44−CD24+ breast cancer cells to shRNA screening in basal-like and luminal breast cell lines. The screening identified 15 genes as basal-like specific hits: PTGIS, CXCL3, MED27, IL6, C2, SERPING1, ISG15, PFKFB3, THBS2, COL6A3, HAS1, KRT8, VIM, CPB1, IGFB7. These genes represent potential therapeutic targets for CD44+/CD24− cells which are enriched in basal-like breast tumors, the only major breast tumor subtype without effective targeted treatment strategies and with poor prognosis.

All 15 basal-like-specific hits are relevant both to tumorigenesis and "stem cell-likeness" features, as several of the signaling pathways populated with these genes are required for the survival of stem cells in breast or other organ types and associated with poor breast cancer prognosis. In particular, high IL-6 levels have been associated with poor clinical outcomes in breast cancer patients [54]. IL-6 is involved in the maintenance of stem cell–like cancer cells [55] and it is required for mammosphere formation [56]. Stat3, a downstream transcription factor in the IL-6 signaling pathways, enforces the undifferentiated state in murine embryonic stem cells [57]. IL-6 and Stat3 play critical roles in the survival of intestinal epithelial cells in colitis-associated cancer [55, 58, 59]. A similar link was observed between IL-6–mediated inflammation and cellular transformation in mammary epithelial cells [60].

PTGIS, HAS1, CXCL3, ISG15, PFKFB3, and IGFBP7 are the other genes on the list of basal-like–specific hits that are important in CD44+/CD24– breast cancer cells. PTGIS encodes prostacyclin synthase and is related to PTGES, which is required for hematopoietic stem cell maintenance [61]. It was also reported that the use of NSAIDs, which inhibit prostaglandin production catalyzed by PTGIS and PTGES, is associated with a lower risk of breast cancer [62] and an improved clinical outcome in breast cancer patients, largely due to decreased risk of distant metastasis [63]. Thus, inhibition of the prostaglandin pathway may be effective in breast cancer treatment by reducing the number of CD44+/CD24– stem cell–like cells.

HAS1 encodes hyaluronan synthase 1, which catalyzes production of hyaluronic acid, a ligand of CD44 [64]. This is in agreement with the hypothesis that CD44 not only marks stem cell–like breast cancer cells but also promotes their viability [65]. Supporting the latter, it was shown that CD44 is required for the survival of leukemia stem cells [66]. Moreover, hyaluronic acid synthases have been linked to invasiveness and metastatic behavior in multiple cancer types [67].

CXCR1, a homolog of the CXCL3 receptor gene CXCR2, is implicated in breast cancer stem cell survival [68, 69]. Similarly, ISG15 was proposed to be associated with poor prognosis in breast cancer patients [70]. The higher expression and specificity for basal-like breast cancer cells of PFKFB3, encoding a glycolytic enzyme, correlate with higher glycolytic activity in basal mammary epithelial cells than in luminal ones [71]. Furthermore, PFKFB3 is one of the genes in the CD44+/CD24– cell gene signature ("signature A") discussed above and linked to increased risk of distant metastasis and poor clinical outcomes in breast cancer patients [32]. IGFBP7 is a target of the TGF-β pathway [72], which, as was already discussed here, is specifically activated in CD44+/CD24– breast cancer cells [32].

We applied the MetaCore network analysis tool and protein interaction database to connect 15 basal-like–specific proteins in a compact "causal" network with Stat3 as a key downstream transcriptional mediator (Fig. 28.4). Based on this network model, we predicted that inhibition of upstream objects that regulate Stat3 in this network will lead to downregulation of Stat3 activity. For example, one of its upstream regulators, HAS1, is responsible for production of hyaluronic acid. Hyaluronic acid links binds to the hyaluronic acid receptor CD44 and activates downstream signaling pathways resulting in activation of Stat3. We followed up with an experimental validation of the network by inhibiting key nodes using small molecule compounds. Inhibition of PTGIS, CXCR2, HAS1, and PFKFB3 decreases both pStat3 protein levels and transcriptional activity. The link between the enzymatic activities of PTGIS and HAS1 and Stat3 signaling was also reported in several studies [73, 74]. The importance of Stat3 in CD44+/CD24–cells and the clinical relevance of this cell type was also proposed based on the results of comprehensive gene expression profiling of basal-like breast cancer cells treated with STAT3 siRNAs and various inhibitors [36]. In the study, it was shown that the Stat3 gene signature is commonly affected by inhibitors and demonstrated that this is associated with increased risk of distant metastasis in breast cancer patients.

The JAK2/Stat3 pathway has been intensely investigated in breast and other cancer types [75]. In various malignancies, constitutive activation of Stat3 signaling was demonstrated due to mutations in JAK1 and JAK2 [76]. A recent whole-genome sequencing study of a basal-like breast tumor identified a JAK2 mutation, although the functional importance of JAKs mutations in breast cancer has not been studied yet [77]. Thus, mutational activation of the JAK2/Stat3 pathway is unlikely to be responsible for its frequent activation in breast cancer. We hypothesized that CD44+/CD24– and even CD44+/CD24+ breast cancer cells have high pStat3 levels due to their expression of genes that increase it, such as IL6, PTGIS, and HAS1,

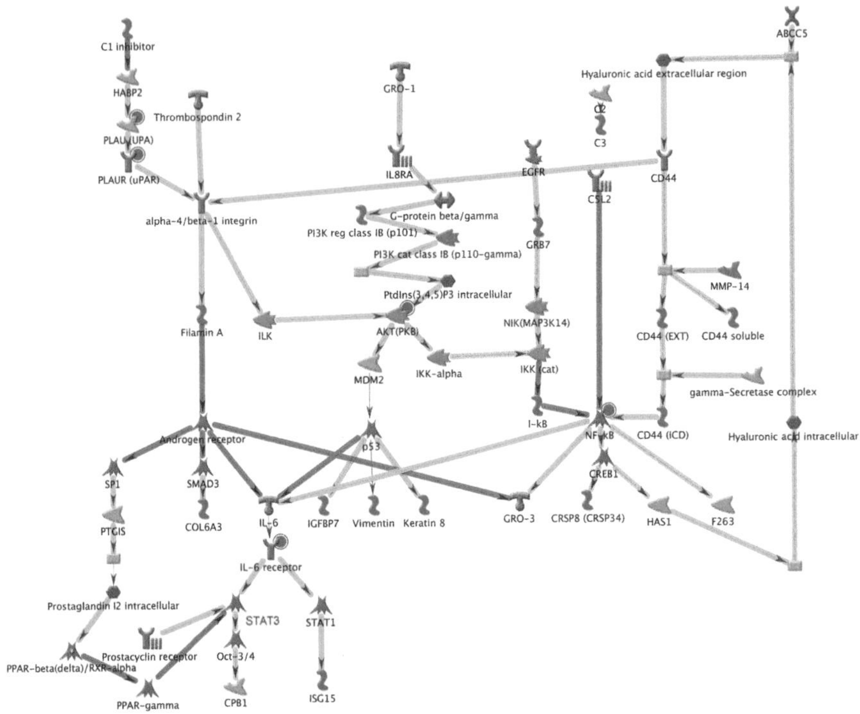

Fig. 28.4 "Causal" network for 15 basal-like–specific screening hits with Stat3 as a key downstream transcriptional mediator. MetaCore platform was used for reconstruction of "causal" network connecting 15 basal-like–specific genes in one functional module. Genes targeted by screening hits are marked with *red concentric circles*. *Red arrows* indicate inhibition, *green* – activation. *Gray arrows* indicate interactions between substrate and reaction or reaction and product or indicate unspecified effects

activating an autocrine loop. In turn, CD44–/CD24+ and CD44–/CD24– breast cancer cells are pStat3+ due to their response to IL-6 (or other cytokines) secreted by neighboring CD44+/CD24– cells and stromal inflammatory cells and fibroblasts. Systemic inhibition of the JAK pathway appears to be nontoxic, as several JAK inhibitors are already in different phases of clinical trials for the treatment of cancer diseases and have been well tolerated with minimal side effects [78].

In summary, the shRNA screen in combination with network modeling elucidated multiple signaling pathways that are specifically required for the viability of CD44+/CD24– breast cancer cells and regulation of Stat3 activation in those cells. Inhibition of these pathways is a promising strategy for targeting these stem cell–like breast cancer cells in all tumors that contain them. This type of therapy may be effective in conjunction with other treatments designed to eliminate other breast cancer cell types, and such a combined treatment strategy may also help prevent therapeutic resistance and limit side effects of cancer treatment.

28.4 Differential Epigenetic Regulation in CD44+/CD24– and CD44–/CD24+ Cells

Epigenetic regulation of gene expression is becoming increasingly important for understanding mechanisms of tumorigenesis and stem cell biology. Epigenetic alterations, such as DNA methylation and chromatin modification, play a key role in embryonic stem cell functionality and differentiation [79, 80]. Several genes important in pluripotency and self-renewal are hypomethylated and expressed in stem cells and silenced by methylation in differentiated cells [81]. However, many aspects of epigenetic regulation in normal breast tissues and breast cancer are poorly understood, including the identity of mammary epithelial progenitor cell-specific epigenetic programs and their relatedness to those in embryonic stem cells and breast carcinomas.

In order to investigate these issues, we conducted the first comprehensive study by performing both DNA methylation and gene expression profiling of four distinct cell populations from normal human breast tissue: CD10+, MUC1+, CD44+/CD24– and CD44–/CD24+ cells [35]. This study showed that four cell types correspond to different differentiation stages and have distinct DNA methylation profiles. In general, CD44+/CD24– cells, characterized by stem cell properties, were more hypomethylated compared with each of the other three cell types. We have also demonstrated that the methylation profile in tumor CD44+/CD24– and CD44–/CD24+ cells was similar to the corresponding CD44+/CD24– and CD44–/CD24+ profiles from normal breast cells. Moreover, epigenetic patterns defining progenitor-like cells were distinct in different breast cancer subtypes. Methylation profiles (specifically, sets of genes uniquely methylated in different cell types) demonstrated good correlation with clinical data, i.e. patients with CD44+/CD24– cell-like tumors had statistically significantly shorter distant metastasis-free survival than patients with CD44–/CD24+ cell-like tumors.

Enrichment analysis of differentially methylated genes between CD44+/CD24– and each of the other three cell types in gene ontology (GO) processes revealed transcription-related functions and regulation of cell proliferation and differentiation as the top processes regulated by DNA methylation. In particular, several genes like HOXA10 and TCF7L1(TCF3) were hypomethylated in CD44+/CD24– cells compared to CD44–/CD24+ cells. These genes encode a homeobox and a polycomb transcription factor known to regulate stem cell function, respectively. These data are consistent with the hypotheses that CD44+/CD24– cells include mammary epithelial progenitors and that their phenotype and differentiation, at least in part, are epigenetically regulated.

A more detailed pathway analysis revealed that differentially methylated and expressed genes in both CD44+/CD24– and CD44–/CD24+ take part in the same pathways and processes, demonstrating functional synergy. The pathways and processes enriched in expression and methylation data are cell type-specific and this specificity is in agreement with progenitor-like and luminal epithelial phenotypes of CD44+/CD24– and CD44–/CD24+ cells, respectively. The genes overexpressed in

CD44+/CD24– were enriched for cytoskeleton and extracellular matrix remodeling, integrin-mediated cell adhesion, immune response processes and IL-4-mediated signaling, whereas CD44–/CD24+ overexpressed genes were enriched for insulin-regulated pathways, mitochondrial metabolism, and apoptosis. Similarly, networks reconstructed by MetaCore algorithms [30] as the most relevant for expression data, reflected phenotypes of CD44+/CD24– and CD44–/CD24+ cell lines: MYC, AR, TGF-β/SMAD pathways predominated in the CD44+/CD24– cell-related network, whereas the CD44–/CD24+ cell-related network contained several DNA damage check-point genes.

We hypothesized that transcription factors hypomethylated and highly expressed in CD44+/CD24– compared with CD44–/CD24+ could play a role in progenitor cell function. Transcription factor FOXC1 was both the most hypomethylated and highly expressed in CD44+/CD24– in comparison with CD44–/CD24+. To confirm this hypothesis, a FOXC1 interaction network was reconstructed. The network was enriched with genes differentially hypomethylated in CD44+/CD24– in comparison with CD44–/CD24+. This network also included several key pathways regulating progenitor cell function: FGF, TGF-β, Notch and WNT signaling pathways. All this data suggests FOXC1 as an important candidate progenitor cell phenotype regulator. Experimental validation of this conclusion revealed that stable expression of FOXC1 in MCF-12A cells results in conversion of the differentiated epithelial phenotype to a CD44+/CD24– cell-like mesenchymal phenotype.

In a follow-up study [36], we have shown that in both CD44+ and CD24+ cells, gene hypermethylation immediately upstream and near promoters is negatively associated with gene expression, indicating a repressive effect. On the contrary, hypermethylation in the gene body is positively associated with gene expression levels, indicating an activating effect. Functional ontology analysis demonstrated a significant difference at the pathway level between genes hypermethylated in their promoter region and genes hypermethylated in the gene body region for both CD44+/CD24– and CD44–/CD24+ cells. In agreement with previous observations suggesting DNA methylation of transcription factors as one of the key mechanisms responsible for differentiation, it was shown that the genes with hypermethylated promoters in CD44+/CD24– and genes with gene body hypermethylation in CD44–/CD24+ cells were enriched in transcription factors. The latter suggests that the expression of transcription factors relevant in CD44–/CD24+ cells (e.g., GATA3) is suppressed by promoter methylation in CD44+/CD24– cells and is positively regulated by gene body hypermethylation in CD44–/CD24+ cells.

Another mechanism of epigenetic regulation of gene expression - histone protein modifications - is also important for regulation of mammary epithelial and luminal linage commitment in CD44+/CD24– and CD44–/CD24+ cells [36]. Histone H3K27me3 (K27) profiling identified 716 genes K27-enriched in both CD44+/CD24– and CD44–/CD24+ cells, whereas about 2,000 genes were cell type-specifically K27-enriched during luminal lineage commitment: 466 genes had a K27 mark in CD44+/CD24– cells and lost it in CD44–/CD24+ cells and 1,502 genes gained a K27 mark in CD44–/CD24+ cells. This result suggests that K27 regulation of gene expression is involved in mammary epithelial linage commitment and differentiation. Indeed, genes enriched for K27 marks in both or in each of the two cells types were functionally distinct based on functional ontology analysis. The highest ranked pathways and processes unique for genes enriched for K27 only in CD44−/CD24+ cells were related to stem cell function such as cyclic AMP, WNT, PIP3K and TGF-β signaling. This indicates that K27 modifications regulate key signaling pathways in the two cell types relevant to progenitor and luminal epithelial cell functions.

Histone H3K27 modification profiles and expression profiles for both CD44+/CD24− and CD44−/CD24+ demonstrated a high degree of correlation, i.e. K27-enriched genes showed low expression levels. Moreover, genes overexpressed in CD44+/CD24− cells were enriched for the K27 mark in CD44−/CD24+ cells, whereas genes overexpressed in CD44−/CD24+ cells were enriched for the K27 mark in CD44+/CD24− cells. This indicates that 10–20% of genes, differentially expressed between CD44+/CD24− and CD44−/CD24+ cells, may be regulated by K27 modification. Furthermore, based on functional analysis, genes overexpressed in CD44+/CD24− cells with a K27 mark in CD44−/CD24+ cells are highly enriched with transcription factors, indicating that K27 modification may preferentially regulate transcription factors important for mammary epithelial lineage commitment and differentiation. Among these transcription factors are HOX genes, GLI1, HES3, HES7, HEYL, and TCF4, all known regulators of stem cells, as well as several epithelial to mesenchymal transition inducing transcription factors such as GSC, SNAI2, TWIST1, and ZEB2.

K27 enriched regions tend to form blocks with profiles consistent within cell line origin and distinct between CD44+/CD24− and CD44−/CD24+ cells. K27 enrichment and gene density in these blocks, as well as gene expression were mutually exclusive. Many genes important for development and stem cells were located within such blocks only in CD44−/CD24+ cells, whereas the opposite pattern was observed for several genes important for luminal linage differentiation. Localization of genes in K27 blocks may ensure the coordinated cell type-specific regulation of these genes and, as a result, coordinated differentiation process.

28.5 Summary

We summarize several recent studies on CD44+/CD24− "stem-like" cells in normal and neoplastic breast tissues. These cells may play a role in the initiation and progression of breast cancer, and, therefore, they are highly important as the source of protein and pathway targets for novel therapies, especially in the basal sub-type of breast cancer.

Experimentation with "stem-like" cancer cell populations is complicated, as stem cell biology and cancerogenesis are among the most complex biological phenomena, with thousands of genes and hundreds of pathways involved at different stages in a well orchestrated, regulated way. Deconvolution of such complexity requires two technologies: "whole genome" OMICs experimentation to measure the response of all the genes and proteins involved, and a computational "systems biology" toolkit capable of functional analysis of thousands of data points in the context of known biology, i.e. pathways, cellular processes, networks and interactions. We have started with global gene expression studies of differential gene expression between CD44+/CD24– and CD44–/CD24+ cell lineages. They show a significant difference between gene expression patterns in progenitor and luminal cell types, with CD44+/CD24– expression enriched with development and tumorigenic pathways. Activation of a TGF-β signaling network was the key feature in CD44+/CD24– differential expression.

It is important to emphasize the significance of applying multiple "OMICs" assays on the same samples, as biological systems function at multiple levels simultaneously, and no single level (for instance, mRNA gene expression) is sufficient to reveal the whole complexity of phenotype. In the next two studies, we analyzed and compared both global gene expression and two types of epigenetic regulation of transcription in CD44+/CD24– cells in comparison with other cell types – specifically, more differentiated luminal CD44–/CD24+ cells. The analysis allowed us to identify key genes, pathways and transcription regulation mechanisms responsible for the CD44+/CD24– stem-like phenotype. In the first genome-wide study of epigenetic programs in progenitor enriched CD44+/CD24–, and more differentiated luminal epithelial CD44–/CD24+ cell populations, we revealed the differential impact of gene methylation and histone modification in regulation of gene expression relevant to mammary epithelial and luminal lineage commitment. Network analysis of the resulting profiles identified key regulators of human mammary epithelial cell type specification.

As is the case with any "high-throughput" experimentation, the studies described above resulted in a series of hypotheses on the most relevant genes, pathways and networks responsible for a certain phenotype (in this case, "stem-like" properties of CD44+/CD24– cells). These findings are a rich resource for the further analysis of human breast development in norm and cancer and for new therapy development. However, any findings deduced from "whole genome" assays have to be thoroughly validated in small-scale experiments to be of any value. Importantly, some of key results were, indeed, validated experimentally, for example the "causal network" connecting 15 genes found in shRNA screening for basal breast cancer [36]. The main validation technique consisted of small molecule inhibition of the key protein hubs on a Stat3 network, which validates the concept of "causal networks" as an important source of novel therapeutic targets.

In conclusion, we believe that multi-data-type OMICs experimentation coupled with a "knowledge-based" systems level analysis approach is a powerful technology combination for studying complex biological phenomena such as tumorigenesis and stemness. We hope that the arsenal of high-throughput methods will be expanded to "effector" levels of cellular processes, such as proteomics and metabolomics.

References

1. Marusyk A, Polyak K (2010) Tumor heterogeneity: causes and consequences. Biochim Biophys Acta 1805(1):105–117
2. Hanahan D, Weinberg RA (2011) Hallmarks of cancer: the next generation. Cell 144(5): 646–674
3. Campbell LL, Polyak K (2007) Breast tumor heterogeneity: cancer stem cells or clonal evolution? Cell Cycle 6(19):2332–2338
4. Clarke MF, Fuller M (2006) Stem cells and cancer: two faces of eve. Cell 124(6):1111–1115
5. Lynch MD, Cariati M, Purushotham AD (2006) Breast cancer, stem cells and prospects for therapy. Breast Cancer Res 8(3):211
6. Polyak K, Hahn WC (2006) Roots and stems: stem cells in cancer. Nat Med 12(3):296–300
7. Weissman IL (2005) Normal and neoplastic stem cells. Novartis Found Symp 265:35–50; discussion 50–54, 92–97
8. Wicha MS (2006) Identification of murine mammary stem cells: implications for studies of mammary development and carcinogenesis. Breast Cancer Res 8(5):109
9. Ailles LE, Weissman IL (2007) Cancer stem cells in solid tumors. Curr Opin Biotechnol 18(5):460–466
10. Merlo LM et al (2006) Cancer as an evolutionary and ecological process. Nat Rev Cancer 6(12):924–935
11. Dean M, Fojo T, Bates S (2005) Tumour stem cells and drug resistance. Nat Rev Cancer 5(4):275–284
12. Eckfeldt CE, Mendenhall EM, Verfaillie CM (2005) The molecular repertoire of the 'almighty' stem cell. Nat Rev Mol Cell Biol 6(9):726–737
13. Singh A, Settleman J (2010) EMT, cancer stem cells and drug resistance: an emerging axis of evil in the war on cancer. Oncogene 29(34):4741–4751
14. Lapidot T et al (1994) A cell initiating human acute myeloid leukaemia after transplantation into SCID mice. Nature 367(6464):645–648
15. Bonnet D, Dick JE (1997) Human acute myeloid leukemia is organized as a hierarchy that originates from a primitive hematopoietic cell. Nat Med 3(7):730–737
16. Al-Hajj M et al (2003) Prospective identification of tumorigenic breast cancer cells. Proc Natl Acad Sci USA 100(7):3983–3988
17. Quintana E et al (2008) Efficient tumour formation by single human melanoma cells. Nature 456(7222):593–598
18. Quintana E et al (2010) Phenotypic heterogeneity among tumorigenic melanoma cells from patients that is reversible and not hierarchically organized. Cancer Cell 18(5):510–523
19. Park SY et al (2010) Heterogeneity for stem cell-related markers according to tumor subtype and histologic stage in breast cancer. Clin Cancer Res 16(3):876–887
20. Notta F et al (2011) Evolution of human BCR-ABL1 lymphoblastic leukaemia-initiating cells. Nature 469(7330):362–367
21. Anderson K et al (2011) Genetic variegation of clonal architecture and propagating cells in leukaemia. Nature 469(7330):356–361
22. Bocker W et al (2002) Common adult stem cells in the human breast give rise to glandular and myoepithelial cell lineages: a new cell biological concept. Lab Invest 82(6):737–746
23. Clayton H, Titley I, Vivanco M (2004) Growth and differentiation of progenitor/stem cells derived from the human mammary gland. Exp Cell Res 297(2):444–460
24. Dontu G et al (2003) Stem cells in normal breast development and breast cancer. Cell Prolif 36(Suppl 1):59–72
25. Dontu G et al (2003) In vitro propagation and transcriptional profiling of human mammary stem/progenitor cells. Genes Dev 17(10):1253–1270
26. Stingl J et al (2005) Epithelial progenitors in the normal human mammary gland. J Mammary Gland Biol Neoplasia 10(1):49–59

27. Stingl J et al (1998) Phenotypic and functional characterization in vitro of a multipotent epithelial cell present in the normal adult human breast. Differentiation 63(4):201–213
28. Patrawala L et al (2005) Side population is enriched in tumorigenic, stem-like cancer cells, whereas ABCG2+ and ABCG2- cancer cells are similarly tumorigenic. Cancer Res 65(14):6207–6219
29. Ponti D et al (2005) Isolation and in vitro propagation of tumorigenic breast cancer cells with stem/progenitor cell properties. Cancer Res 65(13):5506–5511
30. Nikolsky Y et al (2009) Functional analysis of OMICs data and small molecule compounds in an integrated "knowledge-based" platform. Methods Mol Biol 563:177–196
31. Liu R et al (2007) The prognostic role of a gene signature from tumorigenic breast-cancer cells. N Engl J Med 356(3):217–226
32. Shipitsin M et al (2007) Molecular definition of breast tumor heterogeneity. Cancer Cell 11(3):259–273
33. Murohashi M et al (2010) Gene set enrichment analysis provides insight into novel signalling pathways in breast cancer stem cells. Br J Cancer 102(1):206–212
34. Bhat-Nakshatri P et al (2010) SLUG/SNAI2 and tumor necrosis factor generate breast cells with CD44+/CD24– phenotype. BMC Cancer 10:411
35. Bloushtain-Qimron N et al (2008) Cell type-specific DNA methylation patterns in the human breast. Proc Natl Acad Sci USA 105(37):14076–14081
36. Marotta LL et al (2011) The JAK2/STAT3 signaling pathway is required for growth of CD44CD24 stem cell-like breast cancer cells in human tumors. J Clin Invest 121(7): 2723–2735
37. Shipitsin M, Polyak K (2008) The cancer stem cell hypothesis: in search of definitions, markers, and relevance. Lab Invest 88(5):459–463
38. Creighton CJ et al (2009) Residual breast cancers after conventional therapy display mesenchymal as well as tumor-initiating features. Proc Natl Acad Sci USA 106(33):13820–13825
39. Li X et al (2008) Intrinsic resistance of tumorigenic breast cancer cells to chemotherapy. J Natl Cancer Inst 100(9):672–679
40. Pandit TS et al (2009) Lymphatic metastasis of breast cancer cells is associated with differential gene expression profiles that predict cancer stem cell-like properties and the ability to survive, establish and grow in a foreign environment. Int J Oncol 35(2):297–308
41. Park SY et al (2010) Cellular and genetic diversity in the progression of in situ human breast carcinomas to an invasive phenotype. J Clin Invest 120(2):636–644
42. Honeth G et al (2008) The CD44+/CD24– phenotype is enriched in basal-like breast tumors. Breast Cancer Res 10(3):R53
43. Nakshatri H, Srour EF, Badve S (2009) Breast cancer stem cells and intrinsic subtypes: controversies rage on. Curr Stem Cell Res Ther 4(1):50–60
44. Neumeister V et al (2010) In situ identification of putative cancer stem cells by multiplexing ALDH1, CD44, and cytokeratin identifies breast cancer patients with poor prognosis. Am J Pathol 176(5):2131–2138
45. Schneider BP et al (2008) Triple-negative breast cancer: risk factors to potential targets. Clin Cancer Res 14(24):8010–8018
46. Shi W et al (2010) Functional analysis of multiple genomic signatures demonstrates that classification algorithms choose phenotype-related genes. Pharmacogenomics J 10(4):310–323
47. Chang HY et al (2004) Gene expression signature of fibroblast serum response predicts human cancer progression: similarities between tumors and wounds. PLoS Biol 2(2):E7
48. James D et al (2005) TGFbeta/activin/nodal signaling is necessary for the maintenance of pluripotency in human embryonic stem cells. Development 132(6):1273–1282
49. Moses HL, Serra R (1996) Regulation of differentiation by TGF-beta. Curr Opin Genet Dev 6(5):581–586
50. Serra R, Moses HL (1996) Tumor suppressor genes in the TGF-beta signaling pathway? Nat Med 2(4):390–391
51. Siegel PM, Massague J (2003) Cytostatic and apoptotic actions of TGF-beta in homeostasis and cancer. Nat Rev Cancer 3(11):807–821

52. Bates RC, Mercurio AM (2005) The epithelial-mesenchymal transition (EMT) and colorectal cancer progression. Cancer Biol Ther 4(4):365–370
53. Massague J (2008) TGFbeta in cancer. Cell 134(2):215–230
54. Knupfer H, Preiss R (2007) Significance of interleukin-6 (IL-6) in breast cancer (review). Breast Cancer Res Treat 102(2):129–135
55. Bromberg J, Wang TC (2009) Inflammation and cancer: IL-6 and STAT3 complete the link. Cancer Cell 15(2):79–80
56. Sansone P et al (2007) IL-6 triggers malignant features in mammospheres from human ductal breast carcinoma and normal mammary gland. J Clin Invest 117(12):3988–4002
57. Matsuda T et al (1999) STAT3 activation is sufficient to maintain an undifferentiated state of mouse embryonic stem cells. EMBO J 18(15):4261–4269
58. Grivennikov S et al (2009) IL-6 and Stat3 are required for survival of intestinal epithelial cells and development of colitis-associated cancer. Cancer Cell 15(2):103–113
59. Bollrath J et al (2009) gp130-mediated Stat3 activation in enterocytes regulates cell survival and cell-cycle progression during colitis-associated tumorigenesis. Cancer Cell 15(2):91–102
60. Iliopoulos D, Hirsch HA, Struhl K (2009) An epigenetic switch involving NF-kappaB, Lin28, Let-7 MicroRNA, and IL6 links inflammation to cell transformation. Cell 139(4):693–706
61. North TE et al (2007) Prostaglandin E2 regulates vertebrate haematopoietic stem cell homeostasis. Nature 447(7147):1007–1011
62. Ulrich CM, Bigler J, Potter JD (2006) Non-steroidal anti-inflammatory drugs for cancer prevention: promise, perils and pharmacogenetics. Nat Rev Cancer 6(2):130–140
63. Holmes MD et al (2010) Aspirin intake and survival after breast cancer. J Clin Oncol 28(9):1467–1472
64. Wai PY, Kuo PC (2008) Osteopontin: regulation in tumor metastasis. Cancer Metastasis Rev 27(1):103–118
65. Gotte M, Yip GW (2006) Heparanase, hyaluronan, and CD44 in cancers: a breast carcinoma perspective. Cancer Res 66(21):10233–10237
66. Jin L et al (2006) Targeting of CD44 eradicates human acute myeloid leukemic stem cells. Nat Med 12(10):1167–1174
67. Adamia S, Maxwell CA, Pilarski LM (2005) Hyaluronan and hyaluronan synthases: potential therapeutic targets in cancer. Curr Drug Targets Cardiovasc Haematol Disord 5(1):3–14
68. Ginestier C et al (2010) CXCR1 blockade selectively targets human breast cancer stem cells in vitro and in xenografts. J Clin Invest 120(2):485–497
69. Wicha MS, Hayes DF (2011) Circulating tumor cells: not all detected cells are bad and not all bad cells are detected. J Clin Oncol 29(12):1508–1511
70. Bektas N et al (2008) The ubiquitin-like molecule interferon-stimulated gene 15 (ISG15) is a potential prognostic marker in human breast cancer. Breast Cancer Res 10(4):R58
71. Hu M et al (2008) Regulation of in situ to invasive breast carcinoma transition. Cancer Cell 13(5):394–406
72. Pen A et al (2008) Glioblastoma-secreted factors induce IGFBP7 and angiogenesis by modulating Smad-2-dependent TGF-beta signaling. Oncogene 27(54):6834–6844
73. Lo RK, Wise H, Wong YH (2006) Prostacyclin receptor induces STAT1 and STAT3 phosphorylations in human erythroleukemia cells: a mechanism requiring PTX-insensitive G proteins, ERK and JNK. Cell Signal 18(3):307–317
74. Bourguignon LY et al (2008) Hyaluronan-CD44 interaction activates stem cell marker Nanog, Stat-3-mediated MDR1 gene expression, and ankyrin-regulated multidrug efflux in breast and ovarian tumor cells. J Biol Chem 283(25):17635–17651
75. Bollrath J, Greten FR (2009) IKK/NF-kappaB and STAT3 pathways: central signalling hubs in inflammation-mediated tumour promotion and metastasis. EMBO Rep 10(12):1314–1319
76. Vainchenker W, Dusa A, Constantinescu SN (2008) JAKs in pathology: role of Janus kinases in hematopoietic malignancies and immunodeficiencies. Semin Cell Dev Biol 19(4):385–393
77. Ding L et al (2010) Genome remodelling in a basal-like breast cancer metastasis and xenograft. Nature 464(7291):999–1005

78. Pardanani A (2008) JAK2 inhibitor therapy in myeloproliferative disorders: rationale, preclinical studies and ongoing clinical trials. Leukemia 22(1):23–30
79. Jackson M et al (2004) Severe global DNA hypomethylation blocks differentiation and induces histone hyperacetylation in embryonic stem cells. Mol Cell Biol 24(20):8862–8871
80. Pasini D et al (2007) The polycomb group protein Suz12 is required for embryonic stem cell differentiation. Mol Cell Biol 27(10):3769–3779
81. Wernig M et al (2007) In vitro reprogramming of fibroblasts into a pluripotent ES-cell-like state. Nature 448(7151):318–324

Index

R.K. Srivastava and S. Shankar (eds.), *Stem Cells and Human Diseases*, DOI 10.1007/978-94-007-2801-1, © Springer Science+Business Media B.V. 2012

S

T

W

Lightning Source UK Ltd.
Milton Keynes UK
UKOW06f0607110816

280441UK00002B/4/P